THE ROUTLEDGE HANDBOOK OF CONSCIOUSNESS

There has been an explosion of work on consciousness in the last 30–40 years from philosophers, psychologists, and neurologists. Thus, there is a need for an interdisciplinary, comprehensive volume in the field that brings together contributions from a wide range of experts on fundamental and cutting-edge topics. *The Routledge Handbook of Consciousness* fills this need and makes each chapter's importance understandable to students and researchers from a variety of backgrounds. Designed to complement and better explain primary sources, this volume is a valuable "first-stop" publication for undergraduate or graduate students enrolled in any course on "Consciousness," "Philosophy of Mind," or "Philosophy of Psychology," as well as a valuable handbook for researchers in these fields who want a useful reference to have close at hand. The 34 chapters, all published here for the first time, are divided into three parts:

- Part I covers the "History and Background Metaphysics" of consciousness, such as dualism, materialism, free will, and personal identity, and includes a chapter on Indian philosophy.
- Part II is on specific "Contemporary Theories of Consciousness," with chapters on representational, information integration, global workspace, attention-based, and quantum theories.
- Part III is entitled "Major Topics in Consciousness Research," with chapters on psychopathologies, dreaming, meditation, time, action, emotion, multisensory experience, animal and robot consciousness, and the unity of consciousness.

Each chapter begins with a brief introduction and concludes with a list of "Related Topics," as well as a list of "References," making the volume indispensable for the newcomer and experienced researcher alike.

Rocco J. Gennaro is Professor of Philosophy and Chairperson of the Philosophy Department at the University of Southern Indiana. Two of his more recent books are *The Consciousness Paradox: Consciousness, Concepts, and Higher-Order Thoughts* (2012) and *Consciousness* (Routledge, 2017). He is also editor of *Disturbed Consciousness: New Essays on Psychopathology and Theories of Consciousness* (2015).

ROUTLEDGE HANDBOOKS IN PHILOSOPHY

Routledge Handbooks in Philosophy are state-of-the-art surveys of emerging, newly refreshed, and important fields in philosophy, providing accessible yet thorough assessments of key problems, themes, thinkers, and recent developments in research.

All chapters for each volume are specially commissioned, and written by leading scholars in the field. Carefully edited and organized, *Routledge Handbooks in Philosophy* provide indispensable reference tools for students and researchers seeking a comprehensive overview of new and exciting topics in philosophy. They are also valuable teaching resources as accompaniments to textbooks, anthologies, and research-orientated publications.

For a full list of published *Routledge Handbooks in Philosophy*, please visit https://www.routledge.com/Routledge-Handbooks-in-Philosophy/book-series/RHP

Recently published:

The Routledge Handbook of Metaethics
Edited by Tristram McPherson and David Plunkett

The Routledge Handbook of Evolution and Philosophy
Edited by Richard Joyce

The Routledge Handbook of Libertarianism
Edited by Jason Brennan, Bas van der Vossen, and David Schmidtz

The Routledge Handbook of Collective Intentionality
Edited by Marija Jankovic and Kirk Ludwig

The Routledge Handbook of Pacifism and Nonviolence
Edited by Andrew Fiala

The Routledge Handbook of Consciousness
Edited by Rocco J. Gennaro

THE ROUTLEDGE HANDBOOK OF CONSCIOUSNESS

Edited by Rocco J. Gennaro

Routledge
Taylor & Francis Group
New York London

First published 2018
by Routledge
52 Vanderbilt Avenue, New York, NY 10017

and by Routledge
2 Park Square, Milton Park, Abingdon, Oxon, OX14 4RN

First issued in paperback 2020

Routledge is an imprint of the Taylor & Francis Group, an informa business

Library of Congress Cataloging-in-Publication Data
A catalog record for this book has been requested

ISBN 13: 978-0-367-57190-0 (pbk)
ISBN 13: 978-1-138-93621-8 (hbk)

Typeset in Bembo
by Deanta Global Publishing Services, Chennai, India

CONTENTS

FIGURES

CONTRIBUTORS

Sean Allen-Hermanson is Associate Professor of Philosophy at Florida International University in Miami, Florida. Animal consciousness lies within his general interest in topics at the intersection of philosophy of mind and cognitive science. He is also Director of the No Stone Age Unturned project.

Adam Alonzi is an independent researcher and interdisciplinary analyst who has worked in the fields of biotechnology, publishing, film production, roboethics, financial modeling, artificial intelligence, futures research, and consciousness studies.

Bernard J. Baars is a former senior fellow in theoretical neurobiology at The Neurosciences Institute in La Jolla, California. He is best known as the originator of the global workspace theory, a theory of human cognitive architecture and consciousness. Baars co-founded the Association for the Scientific Study of Consciousness, and the Academic Press journal *Consciousness and Cognition* with the late William P. Banks.

David Barrett is an instructor of philosophy at the University of Arkansas. He is the author of articles on the philosophy of mind, philosophy of psychology, and consciousness.

Tim Bayne is an Australian Research Council Future Fellow at Monash University in Melbourne, Australia. He is the author of *The Unity of Consciousness* (Oxford University Press, 2010) and *Thought: A Very Short Introduction* (Oxford University Press, 2013).

Berit "Brit" Brogaard is Professor of Philosophy at University of Miami, Florida and Professor II at University of Oslo. Her areas of research include philosophy of perception, philosophy of emotions, and philosophy of language. She is the author of *Transient Truths* (Oxford University Press, 2012), *On Romantic Love* (Oxford University Press, 2015) and *The Superhuman Mind* (Penguin, 2015).

Andrew Brook is Chancellor's Professor of Philosophy and Cognitive Science Emeritus at Carleton University, Ottawa, Canada. He is former President of the Canadian Philosophical Association and the current Treasurer of the International Psychoanalytic Association.

He founded the Institute of Cognitive Science at Carleton, which houses Canada's only free-standing PhD in Cognitive Science, and was Director for more than ten years. He has about 130 publications, including seven authored or edited books.

Gregg D. Caruso is Associate Professor of Philosophy at SUNY Corning, New York and Co-Director of the Justice without Retribution Network housed at the University of Aberdeen School of Law, Scotland. He is the author of *Free Will and Consciousness: A Determinist Account of the Illusion of Free Will* (Lexington Books, 2012), co-editor of *Neuroexistentialism: Meaning, Morals, and Purpose in the Age of Neuroscience* (Oxford University Press, 2017), and editor of *Exploring the Illusion of Free Will and Moral Responsibility* (Rowman & Littlefield, 2013).

Philippe Chuard is Associate Professor of Philosophy at SMU in Dallas, Texas. He has published several articles on the dispute between conceptualism and nonconceptualism and is currently writing a book on temporal experiences.

Elijah Chudnoff is Associate Professor of Philosophy at the University of Miami, Florida. He works primarily on epistemology and the philosophy of mind. He has published papers on intuition, perception, phenomenal intentionality, theories of knowledge, and cognitive phenomenology. His books include *Intuition* (Oxford University Press, 2013) and *Cognitive Phenomenology* (Routledge, 2015).

Christian Coseru is Associate Professor of Philosophy at the College of Charleston, South Carolina, working in the fields of philosophy of mind, phenomenology, and cross-cultural philosophy, especially Indian and Buddhist philosophy in dialogue with Western philosophy and cognitive science. He is the author of *Perceiving Reality: Consciousness, Intentionality, and Cognition in Buddhist Philosophy* (Oxford University Press, 2012), and is currently working on a book manuscript on the intersections between perceptual and affective consciousness, tentatively entitled *Sense, Self-Awareness, and Sensibility*, and on an introduction to Buddhist Philosophy of Mind, entitled *Moments of Consciousness*.

Jake H. Davis is Postdoctoral Associate with the Virtues of Attention project at New York University. He is editor of the volume *A Mirror Is for Reflection: Understanding Buddhist Ethics* (Oxford University Press, 2017), and has authored and co-authored articles at the intersection of Buddhist philosophy, moral philosophy, and cognitive science.

Francis Fallon is Assistant Professor of Philosophy at St. John's University, New York City. His publications on consciousness include "Dennett on Consciousness: Realism without the Hysterics" (*Topoi*, forthcoming), and "Integrated Information Theory (IIT) and Artificial Consciousness" (in *Advanced Research on Cognitively Inspired Architecture,* IGI Global, 2017).

Shaun Gallagher is the Lillian and Morrie Moss Professor of Excellence in Philosophy at the University of Memphis, Tennessee. His has a secondary research appointment at the University of Wollongong, Australia. Professor Gallagher holds the Humboldt Foundation's Anneliese Maier Research Award [Anneliese Maier-Forschungspreis] (2012–18). He is a founding editor and a co-editor-in-chief of the journal *Phenomenology and the Cognitive Sciences*. His publications include *How the Body Shapes the Mind* (Clarendon Press, 2005), *The Phenomenological Mind* (with Dan Zahavi, Routledge, 2nd ed., 2012), and *Enactivist Interventions: Rethinking the Mind* (Oxford University Press, 2017).

Rocco J. Gennaro is Professor of Philosophy and Chairperson of the Philosophy Department at the University of Southern Indiana. Two of his more recent books are *The Consciousness Paradox: Consciousness, Concepts, and Higher-Order Thoughts* (MIT Press, 2012) and *Consciousness* (Routledge, 2017). He is also the editor of *Disturbed Consciousness: New Essays on Psychopathology and Theories of Consciousness* (MIT Press, 2015).

Chad Gonnerman is Assistant Professor of Philosophy at the University of Southern Indiana. He has written, among other things, on the nature of concepts, methodology of philosophical intuitions, egocentric biases in mindreading, and philosophy's ability to enhance cross-disciplinary research.

Michael S. Graziano is Professor of Psychology and Neuroscience at Princeton University, New Jersey. He has made contributions to three main areas of neuroscience: the neural representation of the space around the body, the control of complex movements in the motor cortex, and the brain basis of consciousness. His most recent books include *Consciousness and the Social Brain* (Oxford University Press, 2013) and *The Spaces Between Us* (Oxford University Press, 2017). He has also published several award-winning novels and children's books including, with Leapfrog Press, *The Last Notebook of Leonardo* (Leapfrog Press, 2010).

Valerie Gray Hardcastle is Professor of Philosophy, Psychology, and Psychiatry and Behavioral Neuroscience at the University of Cincinnati, Ohio. She is currently Scholar-in-Residence at the Weaver Institute for Law and Psychiatry and is the founding director of the Medicine, Health, and Society Program.

Larry M. Jorgensen is Associate Professor of Philosophy at Skidmore College in Saratoga Springs, New York. His main research is on Leibniz's philosophy of mind and on the development of the uniquely modern conception of consciousness that emerged during the seventeenth century.

Amy Kind is Russell K. Pitzer Professor of Philosophy at Claremont McKenna College in Claremont, California. She is the author of *Persons and Personal Identity* (Wiley, 2015) and the editor of the *Routledge Handbook of Philosophy of Imagination* (Routledge, 2016). With Peter Kung, she also edited the collection *Knowledge through Imagination* (Oxford University Press, 2016). She is also currently editing *Philosophy of Mind in the Twentieth and Twenty-First Centuries*, a collection that is forthcoming with Routledge.

Janet Levin is Professor of Philosophy at the University of Southern California. She works primarily in the philosophy of mind and the theory of knowledge, and has published articles on the nature of conscious experience, the norms of assertion, and the role of thought experiments in philosophical inquiry.

Victor Loughlin is a Postdoctoral Research Fellow with the Research Foundation Flanders (FWO). His research interests include philosophy of mind, cognitive science, and Wittgenstein. He currently works at the University of Antwerp, Belgium.

Corey J. Maley is Assistant Professor in the Philosophy Department at the University of Kansas. His work focuses on the philosophy of mind, psychology, and cognitive science.

Erik Myin is Professor of Philosophy at the University of Antwerp and Director of the Centre for Philosophical Psychology. He has recently published two books, *Radicalizing Enactivism: Basic Minds without Content* (MIT Press, 2013) and *Evolving Enactivism: Basic Minds Meet Content* (MIT Press, 2017), both written with Daniel Hutto.

Andrew Peterson is a Research Assistant Professor at George Mason University and Research Fellow in the Institute for Philosophy and Public Policy in the Washington, D.C., metro area.

Ian Phillips is Professor in Philosophy of Psychology in the Department of Philosophy at the University of Birmingham. He is also currently a Visiting Research Scholar in the Program in Cognitive Science at Princeton University, New Jersey. His work primarily focuses on topics at the intersection of philosophy of mind and cognitive science, most notably issues concerning temporal experience, the nature and limits of perceptual consciousness, and the metaphysics of perception. He has just published *The Routledge Handbook of Philosophy of Temporal Experience* (Routledge, 2017).

Gualtiero Piccinini is Professor of Philosophy and Associate Director of the Center for Neurodynamics at the University of Missouri–St. Louis. He has published over 50 articles in the philosophy of mind and related sciences. His book, *Physical Computation: A Mechanistic Account*, was published in 2015 by Oxford University Press.

David Pitt is Professor of Philosophy at California State University, Los Angeles. He has published papers on topics in the philosophy of mind, the philosophy of language, and metaphysics. He is currently at work on a manuscript, *The Quality of Thought*, to be published by Oxford University Press.

Paavo Pylkkänen is Senior Lecturer in Theoretical Philosophy, Vice Dean of Faculty of Arts (Research) and Head of Department of Philosophy, History and Art Studies at the University of Helsinki, Finland. He is also an Associate Professor in Theoretical Philosophy at the University of Skövde, Sweden. His research areas are philosophy of mind and the foundations of quantum theory. In particular, he has studied the relevance of David Bohm's interpretation of quantum theory to problems in scientific metaphysics. He is the author of *Mind, Matter and the Implicate Order* (Springer, 2007).

Vicente Raja is a PhD Candidate in the Philosophy Department at the University of Cincinnati, Ohio. His main field of research is philosophy of cognitive science, paying special attention to embodied approaches to perception, action, and cognition.

Antti Revonsuo is Professor of Cognitive Neuroscience at the University of Skövde, Sweden, and Professor of Psychology at the University of Turku, Finland. He has been conducting both philosophical and empirical research on consciousness since the early 1990s. His empirical work focuses on dreaming as a conscious state and on the neural correlates of visual consciousness. His philosophical views are presented in two books, *Inner Presence: Consciousness as a Biological Phenomenon* (The MIT Press, 2006) and *Consciousness: The Science of Subjectivity* (Routledge/Psychology Press, 2010).

William S. Robinson is Emeritus Professor of Philosophy at Iowa State University. He writes on issues in philosophy of mind, with special attention to consciousness, mental causation, and artificial intelligence. His books include *Understanding Phenomenal Consciousness* (Cambridge

University Press, 2004) and an introduction for non-specialists, *Your Brain and You: What Neuroscience Means for Us* (Goshawk Books, 2010).

Adina L. Roskies is Helman Family Distinguished Professor of Philosophy and Chair of Cognitive Science at Dartmouth College. She has PhDs in both neuroscience and philosophy, and a law degree. She is co-editor, with Stephen Morse, of *A Primer on Criminal Law and Neuroscience* (Oxford University Press, 2013).

Elizabeth Schechter is Assistant Professor in the Department of Philosophy and with the Philosophy-Neuroscience-Psychology program at Washington University in St. Louis. She has recently published a book entitled *Self-Consciousness and "Split" Brains: The Minds' I* (Oxford University Press, 2018) on minds, selves, and self-consciousness in split-brain subjects.

William Seager is Professor of Philosophy at the University of Toronto Scarborough in Toronto, Canada. He works mainly in the philosophy of mind, with a special interest in consciousness studies. His most recent books are *Theories of Consciousness*, 2nd ed. (Routledge, 2016) and *Natural Fabrications: Science, Emergence and Consciousness* (Springer, 2012).

Jonathan Waskan is formerly Associate Professor of Philosophy at the University of Illinois, Urbana-Champaign. His work in the philosophy of science, cognitive science, and experimental philosophy largely concerns the nature and role of models in human thought processes and in science.

Demian Whiting is a Senior Lecturer based in the Department of Philosophy and Hull York Medical School at the University of Hull, United Kingdom. His research interests include philosophy of emotion, phenomenal consciousness, moral psychology, and various issues in medical ethics.

Jennifer M. Windt is a Lecturer in Philosophy at Monash University in Melbourne (Australia). Her research centers on philosophy of mind and philosophy of cognitive science, especially on the topics of dreaming, sleep, and self-consciousness. She is the author of *Dreaming* (The MIT Press, 2015) and edited, with Thomas Metzinger, *Open MIND* (MIT Press, 2016; an open access version is available at open-mind.net). She is the author of the forthcoming *Consciousness: A Contemporary Introduction* (Routledge).

Wayne Wu is Associate Professor in and Associate Director of The Center for the Neural Basis of Cognition at Carnegie Mellon University. He has published *Attention* with Routledge and has written articles on the philosophy of mind and of cognitive science on agency, attention, consciousness, perception, and schizophrenia.

ACKNOWLEDGMENTS

I would like to thank Andy Beck and Vera Lochtefeld at Routledge Press for their guidance and support throughout this project. I would also like to thank all of the contributors to this volume for their work.

INTRODUCTION

Rocco J. Gennaro

1 The Rationale

There has been an explosion of work on consciousness in the last few decades from philosophers, psychologists, and neurologists. Because of the large volume and interdisciplinary nature of this research, there is a need for a wide-ranging collection of essays that brings together fundamental and cutting-edge topics on consciousness, making their philosophical import understandable to researchers with various backgrounds. Such an approach can also appeal to upper-level undergraduates, who may have had only one or two courses in philosophy. *The Routledge Handbook of Consciousness* will work as a valuable reference for such students enrolled in courses on "Consciousness," "Philosophy of Mind," or "Philosophy of Psychology," designed to complement and better explain primary sources. Even seasoned philosophers of mind and philosophers of psychology will likely find this book useful, since it is very difficult to claim expertise in all of the areas covered. Still, the overall emphasis is to introduce the uninitiated to cutting-edge interdisciplinary work, which is at least one way that this collection will stand out among its competitors.[1] Of course, due to the very nature of some topics, some chapters are understandably more advanced or technical than others.

Consciousness is arguably the most important area within contemporary philosophy of mind. It is also perhaps the most puzzling aspect of the world, despite the fact that it is so very familiar to each of us. Although some features of mental states can perhaps be explained without reference to consciousness, it is consciousness which seems most resistant to a straightforward explanation. Can conscious experience be explained in terms of brain activity? Is the conscious mind physical or non-physical? What is the relationship between consciousness and attention or between consciousness and free will? What do psychopathologies and disorders of consciousness tell us about the normal conscious mind? Are animals conscious? Could a robot be conscious? These and many other questions are explored in the chapters that follow. Although there is much of contemporary interest on consciousness in Eastern thought, especially Indian philosophy (e.g. Siderits et al. 2011; Coseru 2012), virtually all chapters in this volume are restricted to Western philosophy and fairly recent work in philosophy of mind.[2]

2 Terminology

Part of the problem can be that the concept of consciousness is notoriously ambiguous. This adds to the complexity of the debate and can result in unnecessary confusion. Thus, it is important to make several distinctions and to define key terms. The noun 'consciousness,' especially in some abstract sense, is not used very often in the contemporary literature, though it originally derives from the Latin *con* (with) and *scire* (to know). One can have knowledge of the external world or one's own mental states through introspection. The primary contemporary interest lies more in the use of the expressions 'x is conscious' or 'x is conscious of y.' Under the former category, perhaps most important is the distinction between *state* and *creature* consciousness (Rosenthal 1993). We sometimes speak of an individual mental state, such as a desire or perception, as being conscious. On the other hand, we also often talk about organisms or creatures as conscious, such as when we say that "human beings are conscious" or "dogs are conscious." Creature consciousness is simply meant to refer to the fact that an organism is awake, as opposed to sleeping or in a coma. However, some kind of state of consciousness is normally implied by creature consciousness; that is, if a creature is conscious, then it must have conscious mental states. There are of course some possible exceptions, such as one who is sleepwalking. Perhaps there can also be state consciousness without creature consciousness, such as in the case of vivid dreams. Due to the lack of a direct object in the expression 'x is conscious,' this is usually referred to as *intransitive* consciousness, in contrast to *transitive* consciousness where the phrase 'x is conscious of y' is used (Rosenthal 1993). We might say that a person is conscious or aware of a dog in front of her. Most contemporary theories of consciousness are aimed at explaining state consciousness, that is, what makes a mental state conscious.

One might think that the term 'conscious' is synonymous with, say, 'awareness,' or 'experience,' or 'attention.' However, it is important to recognize that this is not generally accepted in some circles. For example, one might hold that there are unconscious experiences depending on how the term 'experience' is defined (Carruthers 2000). More common is the belief that we can be aware of external objects in some unconscious sense, such as during instances of subliminal perception. The expression 'conscious awareness' does not seem to be redundant. Finally, it is not clear that consciousness ought to be restricted to attention. It seems plausible to suppose, for example, that one is conscious of objects to some extent in one's peripheral visual field, even though one is attending to a narrower (or focal) set of objects within that visual field. Some of the disagreement can be purely terminological but some is also more substantial. Needless to say, contemporary philosophers and psychologists are nearly unanimous in allowing for unconscious mental states or representations, though they sometimes differ as to whether this applies to all kinds of mental states including, say, pains and emotions.

Probably the most commonly used notion of "conscious" is captured by Thomas Nagel's famous "what it is like" sense (Nagel 1974). When I am in a conscious mental state, there is "something it is like" for me to be in that state from the subjective or first-person point of view. When I smell a flower or have a conscious auditory sensation, there is something it "seems" or "feels like" from my perspective. An organism such as a bat is conscious if it is able to experience the world through its echolocation senses. There is also something it is like to be a conscious creature, whereas there is nothing it is like to be a table or tree. This is primarily the sense of "conscious" used by the authors in this book.

There are still a cluster of other expressions and technical terms associated with Nagel's sense. For example, philosophers often refer to conscious states as *phenomenal* or *qualitative* states. More technically, philosophers describe such states as having qualitative properties called "qualia" (singular, quale). Chalmers explains that a "mental state is conscious if there is something it is like to be in that mental state... We can say that a mental state is conscious if it has a qualitative feel... These qualitative feels are also known as phenomenal qualities, or qualia for short" (1996: 4). There is significant disagreement over the nature, and even the existence, of qualia, but they are often understood as the felt qualities of conscious states (Kind 2008). Others might, more neutrally, say that qualia are qualitative features present in experience. What it feels like, experientially, to see a red rose is different from what it feels like to see a yellow rose. Likewise, for hearing a musical note played by a piano and hearing the same musical note played by a tuba. The qualia of these experiences are what give each of them its characteristic "feel" and also what distinguishes them from one another. In any case, qualia are most often treated as properties of some mental states, though some do use the term "qualia" in the more external sense of "the qualities of what is represented."

One also finds closely allied expressions like "phenomenal character" and "subjective character" in the literature. Tye (2009), for example, explains that the phenomenal character of an experience is what it is like subjectively to undergo the experience. Kriegel (2009) distinguishes what he calls "qualitative character" from "subjective character" under the larger umbrella of "phenomenal character." He explains that "a phenomenally conscious state's qualitative character is what makes it the phenomenally conscious state it is, while its subjective character is what makes it a phenomenally conscious state at all" (Kriegel 2009: 1). In his view, then, the phenomenally conscious experience of the blue sky should be divided into two components: (1) its *qualitative* character, which is the "bluish" component of the experience (or the what of the experience), and (2) its *subjective* character, which is what he sometimes calls the "for-me" or "mine-ness" component (or what determines that it is conscious at all).

Ned Block (1995) makes a well-known distinction between *phenomenal* consciousness (or "phenomenality") and *access* consciousness. Phenomenal consciousness is very much in line with Nagel's notion described earlier. However, Block defines the quite different notion of access consciousness in terms of a mental state's relationship with other mental states, for example, a mental state's "availability for use in reasoning and rationality guiding speech and action" (Block 1995: 227). This view would, for example, count a visual perception as (access) conscious not because it has the "what it's likeness" of phenomenal states, but because it carries visual information that is generally available for use by the organism, regardless of whether or not it has any qualitative properties. Access consciousness is therefore a functional notion concerned with what such states do. Although something like this idea is certainly important in cognitive science and philosophy of mind generally, not everyone agrees that access consciousness deserves to be called "consciousness" in any important sense. Block himself argues that neither sense of consciousness implies the other, while others urge that a more intimate connection holds between the two.

Finally, it is helpful to distinguish between consciousness and *self-consciousness*, which plausibly involves some kind of awareness or consciousness of one's own mental states (instead of something out in the world). Self-consciousness itself arguably comes in degrees of sophistication, ranging from minimal bodily self-awareness to the ability to reason and reflect on one's own mental states, such as one's beliefs and desires. The term 'introspection' is often used for this latter, more reflective, notion. Some important historical figures have even held that consciousness entails some form of self-consciousness (Kant 1781/1965, Sartre 1956), a view shared by some contemporary philosophers (Gennaro 1996, Kriegel 2004).

3 The Major Themes and Topics

This handbook contains three parts, the first of which covers the "History and Background Metaphysics" of consciousness. Part II covers "Contemporary Theories of Consciousness" and

Part III is entitled "Major Topics in Consciousness Research." The main criterion for selecting most of the topics (especially in Parts II and III) was whether they are cutting-edge and "live," that is, whether innovative and provocative debate on the topic is underway in the research community. Part III has by far the most chapters.

In general, it is always worth keeping in mind the two most common and opposing metaphysical positions on the nature of mind and consciousness: dualism and materialism. While there are many versions of each, dualism generally holds that the conscious mind or a conscious mental state is non-physical in some sense. On the other hand, materialists believe that the mind is the brain, or, as "identity theorists" would put it, that conscious mental activity is identical with neural activity. These views are critically discussed at length in Part I (especially in Chapters 3 and 4 by Janet Levin and William S. Robinson). They include discussion of many different flavors of materialism and dualism, including identity theory, eliminative materialism, functionalism, substance dualism, property dualism, and epiphenomenalism. Some form of materialism is probably more widely held today than in centuries past. Perhaps part of the reason has to do with an increase in scientific knowledge about the brain and its intimate connection with consciousness, including the clear correlations between brain damage and various states of consciousness. Stimulation to very specific areas of the brain results in very specific conscious experiences. Nonetheless, some major difficulties remain such as the so-called "hard problem of consciousness" (Chalmers 1995), which basically refers to the difficulty of explaining just how or why physical processes in the brain give rise to subjective conscious experiences.

There are also a number of other anti-materialist metaphysical views discussed by William Seager in Chapter 5, including panpsychism, idealism, and emergentism. The bigger picture and a more historical overview is presented by Larry M. Jorgensen in Chapter 2. Part I also contains essays by Amy Kind (Chapter 1) and Gregg D. Caruso (Chapter 6), which address such questions as: How is consciousness related to one's personal identity and the possibility of immortality? Is consciousness necessary for free will and moral responsibility? In Chapter 7, Christian Coseru examines a range of Indian philosophical conceptions of consciousness, including the naturalist theories of Nyāya, the (largely phenomenalist) accounts of mental activity and consciousness of Abhidharma and Yogācāra Buddhism, and the subjective transcendental theories of consciousness of Advaita Vedānta.

Part II ("Contemporary Theories of Consciousness") contains chapters on many of the leading and currently active theories of consciousness. They address questions such as: What makes a mental state a *conscious* mental state? Can conscious mental states be understood solely in terms of representational states? Can consciousness be reduced to neurophysiology? Can consciousness be understood as some kind of information integration? How closely related are consciousness and attention? Are conscious states intimately connected with having sensorimotor abilities? Can results in quantum physics shed light on the nature of consciousness?

To be more specific, in Chapter 8, Rocco J. Gennaro focuses his discussion on widely discussed "representational theories of consciousness," such as the "higher-order thought (HOT) theory of consciousness," which attempt to reduce consciousness to "mental representations" rather than directly to neural or other physical states. Various representational theories are critically discussed. In Chapter 9, Bernard J. Baars and Adam Alonzi explain and elaborate on Baars's very influential "Global Workspace Theory" (GWT) of consciousness (beginning with Baars 1988). According to Baars, we should think of the entire cognitive system as built on a "blackboard architecture," which is a kind of global workspace (i.e. a functional hub of signal integration and propagation). Unconscious cognitions compete for the spotlight of attention from which information is "broadcast globally" throughout the system. Consciousness consists in such global broadcasting and functions as a dynamic and adaptable global workspace. Francis Fallon (in Chapter 10) critically discusses the Integrated Information Theory (IIT) developed by the neuroscientist Giulio Tononi. On this view, consciousness depends upon a kind of information – integrated information – which is understood via a quantifiable metric. In Chapter 11, Francis Fallon and Andrew Brook examine Daniel Dennett's Multiple Drafts Model (MDM), which denies that consciousness involves an inner observer of a single linear stream of consciousness (the "Cartesian Theater"). Instead, the brain composes multiple drafts of a narrative. David Barrett (Chapter 12) critically discusses Jesse Prinz's account of consciousness, which is called the "Intermediate Level Theory of Consciousness." It holds that consciousness arises when representations at the intermediate level of processing are attended to. In Chapter 13, Michael S. Graziano presents an overview of his own Attention Schema Theory of Consciousness, which describes how an information-processing machine can be understood as being conscious of something. In the theory, the brain is an information processor that is captive to the information constructed within it.

Antti Revonsuo (Chapter 14) explains both John Searle's Biological Naturalism (BN) and his own Biological Realism (BR). They have in common the view that consciousness is the inner presence of unified qualitative subjectivity, which constitutes a real biological phenomenon happening in our brain at a higher level of neurophysiological organization. The strengths and weaknesses of BN and BR are described and weighted. In Chapter 15, Erik Myin and Victor Loughlin explain and defend the "sensorimotor approach" to consciousness, which holds that perceptual experience is something we do, not something that happens in us. That is, having perceptual experience is fundamentally a matter of engaging with our environments in particular ways. Paavo Pylkkänen (Chapter 16) expounds on the idea that the holistic and non-mechanical notion of physical reality implied by quantum theory could help us to find a place for mind and consciousness in nature. He provides an introduction to some of the main theories that have arisen from these explorations.

Each of the theories discussed in Part II is currently the subject of vigorous debate and continued development. Most of them are in competition with others, but some could instead serve to complement others.

Part III ("Major Topics in Consciousness Research") contains chapters on many cutting-edge, and even sometimes provocative, topics frequently encountered in contemporary work on consciousness. Authors explore answers to such questions as: What are the candidates for the neural correlates of consciousness? What is the precise relationship between consciousness and attention? What can various disorders of consciousness tell us about normal consciousness? Are animals, or at least most animals, conscious? Could a robot ever be conscious? Are dreams conscious? What is the "unity of consciousness"? Are sensory experiences essentially conceptual in nature? What is the relationship between consciousness and intentionality? How does time or temporal experience manifest itself in conscious experience? What is special about multisensory consciousness and the fascinating phenomenon of "synesthesia"? What is the role of consciousness in action? Are emotions always conscious? What does meditation tell us about consciousness? How can we know when a post-comatose patient is conscious and what ethical problems arise in such cases?

In Chapter 17, Valerie Gray Hardcastle and Vicente Raja critically examine the quest for the so-called "neural correlates of consciousness" (NCC) and explain why there is still no agreement among scientists or philosophers regarding what the NCC might be. Wayne Wu (in Chapter 18) explores the many different relations between attention and phenomenal consciousness, such as if attention is necessary for consciousness, if attention is sufficient for it, whether attention changes consciousness, and how attention might give us access to consciousness. In Chapter 19, David Pitt explains that although mainstream analytic philosophy of mind has long held that consciousness is not required for the intentionality of mental states, there has been increasing recent support for the idea that intentionality is essentially experiential. This chapter summarizes the views and arguments for and against such a claim. Philippe Chuard (Chapter 20) asks: Does sensory consciousness require conceptualization so that what one is sensorily aware of in conscious perception is partly a function of what one conceptually identifies? This chapter critically reviews some of the central considerations advanced against this conceptualist doctrine. Ian Phillips (Chapter 21) explores the notion that our capacity for conscious awareness of temporal aspects of reality depends essentially on memory. He ultimately argues that the idea that memory is involved in all temporal experience can be sustained across all plausible accounts of temporal experience. In doing so, he critically engages with Dainton's influential carving of the landscape into three models of temporal experience, i.e. cinematic, retentional, and extensional. In Chapter 22, Shaun Gallagher adduces evidence for the view that consciousness plays a significant role before, during, and after action. This can also be seen as an argument against epiphenomenalism, which holds that consciousness does not have a causal impact on our behavior. Demian Whiting (Chapter 23) focuses on the questions: "What exactly is meant by saying that emotions are conscious and why does it matter?" and "Are emotions *always* conscious?"

In Chapter 24, Berit Brogaard and Elijah Chudnoff carefully distinguish between two kinds of ordinary multisensory experience, explain the virtues of this distinction, and then examine synesthesia, which is a more atypical multisensory experience. Rocco J. Gennaro (Chapter 25) reviews the growing interdisciplinary field sometimes called "philosophical psychopathology," which is also related to so-called "philosophy of psychiatry" (covering the overlapping topics of psychopathy and mental illness). The focus is on various psychopathologies with special attention to how they negatively impact conscious experience, such as amnesia, somatoparaphrenia, schizophrenia, visual agnosia, autism, and dissociative identity disorder (DID). In Chapter 26, Andrew Peterson and Tim Bayne review the use of neuroimaging and electroencephalographic methods to assess covert consciousness in patients who are diagnosed as being in a vegetative or minimally conscious state. They conclude with a discussion of the moral relevance of consciousness in this patient group. Elizabeth Schechter (Chapter 27) points out how at any moment in time, an experiencing subject's perspective encompasses a multitude of elements: sights and sounds, on the face of it, as well as thoughts, feelings, and so on. Thus, questions about the "unity of consciousness" concern the relations between these elements and how to account for them. Questions about conscious unity also concern the identity of experiencing subjects or "selves." In Chapter 28, Corey J. Maley and Gualtiero Piccinini critically examine the notion that phenomenal consciousness might have evolved in one of three ways. If phenomenal consciousness performs a function—that is, if it has physical effects that confer an adaptive advantage—it was probably selected for. If phenomenal consciousness has no function, it is either a byproduct of some other trait or a frozen evolutionary accident.

Sean Allen-Hermanson (Chapter 29) focuses on the questions: "Are animals conscious?" and "If so, to what extent?" This chapter surveys some historical views on animal consciousness but then includes significant discussion of recent inferential and non-inferential approaches as well as neuro-reductive and representational theories. In Chapter 30, Jonathan Waskan considers various answers to the following questions: "Can robots be made to have conscious experiences?" "Will they ever see red or feel pain?" "How would we know?" "What moral obligations would we have towards them?" "Could we create beings vastly more sophisticated than ourselves, such as hyper-intelligent robots?" Jennifer M. Windt (Chapter 31) introduces a version of the "simulation view" that defines dreaming through its immersive, here-and-now structure. She focuses on minimal forms of dreaming, arguing that they coincide with minimal forms of self-experience including bodily experiences. In Chapter 32, Jake H. Davis examines the philosophical value of a proposal arising out of a specific Buddhist meditative practice: the claim that we can and ought to experience "passing away." He aims to demonstrate by example how engaging with a line of thought from a specific meditative tradition can help to advance debates in the analytic philosophy of consciousness. Adina L. Roskies (Chapter 33) discusses scientific and ethical questions about the diagnosis, treatment, and end of life issues of patients with disorders of consciousness. In Chapter 34, Chad Gonnerman reviews recent research in experimental philosophy of consciousness. He first addresses recent debates about just how to characterize "experimental philosophy," and then examines two strands of subsequent research: the folk psychology of group phenomenal minds and the cognitive systems responsible for ordinary attributions of phenomenal states to others.

I hope you enjoy the journey through these fascinating topics. Debate and discussion is of course ongoing.

Notes

1 Other anthologies on consciousness are Block, Flanagan, and Güzeldere (1997), Baars, Banks, and Newman (2003), Zelazo, Moscovitch, and Thompson (2007), Velmans and Schneider (2007), Bayne, Cleeremans, and Wilken (2009), and Alter and Howell (2012). For a sample of single author introductions, see Revonsuo (2010), Blackmore (2012), P.M. Churchland (2013), Weisberg (2014), Seager (2016), and Gennaro (2017). There are also many useful overview articles with expansive references in the online *Stanford Encyclopedia of Philosophy* (http://plato.stanford.edu/) and the *Internet Encyclopedia of Philosophy* (http://www.iep.utm.edu/). Annual interdisciplinary conferences such as "The Science of Consciousness," and the "Association for the Scientific Study of Consciousness," as well as the journals *Philosophical Psychology*, *Journal of Consciousness Studies*, and *Consciousness and Cognition* have offered quality places for disseminating work in the field. The same is true for the wonderful database and bibliography PhilPapers (http://philpapers.org/).

2 The main exceptions in this volume being C. Coseru's "Consciousness and the Mind-Body Problem in Indian Philosophy" and J.H. Davis's "Meditation and Consciousness."

References

Alter, T., and Howell, R. (eds.) (2012) *Consciousness and the Mind-Body Problem*, New York: Oxford University Press.

Baars, B. (1988) *A Cognitive Theory of Consciousness*, Cambridge, MA: Cambridge University Press.

Baars, B., Banks, W., and Newman, J. (eds.) (2003) *Essential Sources in the Scientific Study of Consciousness*, Cambridge, MA: MIT Press.

Bayne, T., Cleeremans, A., and Wilken, P. (eds.) (2009) *Oxford Companion to Consciousness*, New York: Oxford University Press.

Blackmore, S. (2012) *Consciousness: An Introduction*, 2nd edition, Oxford: Oxford University Press.

Block, N. (1995) "On a Confusion about the Function of Consciousness," *Behavioral and Brain Sciences* 18: 227–247.

Block, N., Flanagan, O., and Güzeledere, G. (eds.) (1997) *The Nature of Consciousness*, Cambridge, MA: MIT Press.

Carruthers, P. (2000) *Phenomenal Consciousness*, Cambridge: Cambridge University Press.

Chalmers, D. (1995) "Facing Up to the Problem of Consciousness," *Journal of Consciousness Studies* 2: 200–219.

Chalmers, D. (1996) *The Conscious Mind*, Oxford: Oxford University Press.

Churchland, P.M. (2013) *Matter and Consciousness*, 3rd edition, Cambridge, MA: MIT Press.

Coseru, C. (2012) *Perceiving Reality: Consciousness, Intentionality, and Cognition in Buddhist Philosophy*, New York: Oxford University Press.

Gennaro, R. (1996) *Consciousness and Self-Consciousness: A Defense of the Higher-Order Thought Theory of Consciousness*, Amsterdam and Philadelphia: John Benjamins.

Gennaro, R. (2017) *Consciousness*, New York: Routledge.

Kant, I. (1781/1965) *Critique of Pure Reason*. Translated by N. Kemp Smith. New York: MacMillan.

Kind, A. (2008) Qualia. *Internet Encyclopedia of Philosophy*, http://www.iep.utm.edu/qualia/

Kriegel, U. (2004) "Consciousness and Self-Consciousness," *Monist* 87: 182–205.
Kriegel, U. (2009) *Subjective Consciousness*, New York: Oxford University Press.
Nagel, T. (1974) "What Is It Like to Be a Bat?" *Philosophical Review* 83: 435–456.
Revonsuo, A. (2010) *Consciousness: The Science of Subjectivity*, New York: Psychology Press.
Rosenthal, D.M. (1993) "State Consciousness and Transitive Consciousness," *Consciousness and Cognition* 2: 355–363.
Sartre, J. (1956) *Being and Nothingness*, New York: Philosophical Library.
Seager, W. (2016) *Theories of Consciousness*, 2nd edition, New York and London: Routledge.
Siderits, M., Thompson, E., and Zahavi, D. (eds.) (2011) *Self, No Self?: Perspectives from Analytical, Phenomenological, and Indian Traditions*, New York: Oxford University Press.
Tye, Michael, "Qualia," *The Stanford Encyclopedia of Philosophy (Summer 2009 Edition)*, Edward N. Zalta (ed.), http://plato.stanford.edu/archives/sum2009/entries/qualia/
Velmans, M., and Schneider, S. (eds.) (2007) *The Blackwell Companion to Consciousness*, Malden, MA: Blackwell.
Weisberg, J. (2014) *Consciousness*, Malden, MA: Polity Press.
Zelazo, P., Moscovitch, M., and Thompson, E. (eds.) (2007) *The Cambridge Handbook of Consciousness*, Cambridge, MA: Cambridge University Press.

PART I

Consciousness

History and Background Metaphysics

1

CONSCIOUSNESS, PERSONAL IDENTITY, AND IMMORTALITY

Amy Kind

Introduction

Several different intersecting questions are in play in philosophical discussions of personal identity. One such question concerns the nature of persons: What makes someone a person? Another question concerns the nature of self-identification: What makes someone the particular person that she is? And yet a third question concerns the nature of a person's existence through time: What makes a person the same person over time?[1]

In this chapter we focus primarily on the third question and, in particular, the role that consciousness has played in philosophical attempts to answer it. We begin in Section 1 with the memory-based view of personal identity offered by John Locke. Though this view faces various objections, we turn in Section 2 to various adjustments that can be made to the view to make it considerably more plausible. In Section 3 we turn away from these psychologically-based approaches to physical alternatives. Finally, in Section 4, we turn to a consideration of how issues related to immortality help shed light on the debate about personal identity.

1 The Lockean View

John Locke (1632–1704) is often considered the father of philosophical discussion of personal identity. In his *Essay Concerning Human Understanding* he offered an account of personal identity over time that has proved particularly influential:

> since consciousness always accompanies thinking, and it is that which makes every one to be what he calls *self*, and thereby distinguishes himself from all other thinking things, in this alone consists personal Identity, i.e. the sameness of a rational being: And as far as this consciousness can be extended backwards to any past Action or Thought, so far reaches the Identity of that Person; it is the same *self* now as it was then; and 'tis by the same *self* with this present one that now reflects on it, that that Action was done.
>
> *(Locke 1689/1975: 335)*

In talking about extending one's consciousness backward, Locke seems to have memory in mind. In particular, it seems that he is here focusing on what is often called *episodic* or *experience* memory. Episodic memory involves memories of events that were personally experienced, and it thereby contrasts with purely factual memory. When taking a geography test you're likely to be relying primarily on your factual memories – all those state capitals and river names that you've previously memorized. In contrast, when writing your autobiography you're likely to be relying primarily on your episodic memories – all those life experiences that you've previously undergone.

With this distinction in place, we can see why episodic memory might naturally be described as a backward extension of consciousness. We can also see why episodic memory might naturally be invoked to explain personal identity over time. It seems plausible that we're each connected to our past selves by our episodic memories of those selves' experiences. Consider an example. On November 2, 2004, someone was elected as the United States senator representing Illinois. On January 20, 1999, someone was sworn in as the 44th President of the United States. And on January 13, 2016, someone gave a State of the Union address. What makes it the case that each of these three "someones" is the very same person, the person known as Barack Obama? According to Locke's theory, it's due to the fact that they share the very same consciousness. The person giving the State of the Union address can remember being sworn in as President, and can remember being elected as the senator from Illinois, and it's in these connections of episodic memory – in the sharing of consciousness – that personal identity consists.

Of course, in an ordinary case like this one, not only do the three someones share the same consciousness but they also share the same body. So we might wonder why a theorist like Locke would privilege sameness of consciousness over sameness of body in accounting for personal identity. Here it's perhaps most helpful to consider hypothetical cases in which sameness of consciousness and sameness of body come apart. Many such cases are presented to us in fiction and film, and they tend to involve the transfer of consciousness from one human body to another (or from one human body to some other kind of body altogether). In thinking about these cases, it may be helpful to keep in mind an analogy used by Locke: Just as we wouldn't think that someone becomes a different person by changing their clothes, we also shouldn't think that someone becomes a different person just by changing their bodies.

Take *Freaky Friday*, for example – be it the 1972 book by Mary Rodgers or any of the various film versions. Though the details vary somewhat from book to film to remake, the basic body swap plotline remains the same across all the different versions. Let's consider the 2003 film starring Lindsay Lohan as a teenage girl named Anna and Jamie Lee Curtis as Anna's mother Tess. One morning, after having received cryptic fortunes while out for dinner the night before, Tess and Anna awake to discover that Tess's consciousness is now in Anna's body, and Anna's consciousness is now in Tess's body. As the plot unfolds, it's clear that viewers are meant to think that each of the characters goes where her consciousness goes – that the person with Tess's body and Anna's consciousness is really Anna, while the person with Anna's body and Tess's consciousness is really Tess – and indeed, this seems to most people to be the most natural description of what happens.

Or consider James Cameron's film *Avatar*, released in 2009. Jake Sully, a disabled former Marine, ends up having his consciousness transferred into a different body, while on a mission in outer space to Pandora. Here there's an added wrinkle: the body to which his consciousness is transferred is not even a human one. Rather, it's Na'vi, a species native to Pandora. As in *Freaky Friday*, we're meant to believe – and it seems natural to believe – that Sully goes where his consciousness goes. Though his human body dies, Sully – the very same person – survives in the Na'vi body that now houses his consciousness.

Though some philosophers have disputed that we should trust our intuitions in these kinds of cases (see especially Williams 1970), many philosophers take them to make a strong case that personal identity does not consist in sameness of body. But that's not yet to say that Locke's view has been established, for we might think that there is another possible view consistent with the body swap and consciousness transfer scenarios. Perhaps what's important for personal identity is not sameness of consciousness, but sameness of immaterial substance, i.e., sameness of soul (see, e.g., Swinburne 1973–4 and Madell 1981). This kind of view about the nature of personal identity is often associated with dualist views about the nature of mind (see Robinson, Chapter 4, this volume). In the contemporary literature about personal identity, it is often referred to as *the simple view*.

In defending against the simple view, Locke asks us to consider a different kind of case. Consider Thersites, a figure from Greek mythology who was supposedly present at the siege of Troy. Now suppose that souls exist, and that someone existing today – call him Sunil – happens to have Thersites' soul. Is that alone enough to make Sunil the same person as Thersites? Locke suggests that such a supposition would be absurd. For example, as depicted by Homer in the *Iliad*, Thersites was struck across the back and shoulders by Odysseus in response to his having sharply criticized Agamemnon; after being hit, he sat cowering, crying, and in pain. But presumably Sunil can't extend his consciousness backward to that experience no matter how he tries. On Locke's view, merely having the same soul as Thersites is as incidental to Sunil's personal identity as if Sunil's body happened to be made up of some of the same particles of matter that once constituted Thersites' body. As he argues, "the same immaterial Substance, without the same consciousness, no more [makes] the same Person, by being united to any Body, than the same particle of matter, without consciousness united to any Body, makes the same Person" (Locke 1689/1975: 339–340).

As these considerations suggest, there is something eminently plausible about a theory that explains personal identity in terms of episodic memory. But that said, Locke's own specification of the view is threatened by several counterexamples. Recall that Locke requires that for a presently existing individual to be the same person as some past existing individual, the present individual must be able to extend her consciousness backward to the experiences of that past individual. Since persons are prone to forgetting all sorts of experiences they've once had, this requirement runs into serious trouble. Writing in the late 18th century, Thomas Reid forcefully spelled out the problem with what's become known as the *Brave Officer case* (Reid 1785). Consider a brave officer, who while on a military campaign, engages in a heroic act. As a young boy, this same man had stolen some apples from a neighbor's orchard. And now, as a retired old man, he has become senile. Though he still remembers his military career, including his act of heroism, he no longer remembers stealing the apples. But assuming that while he was in the military he still could remember this childhood theft, we're presented with a paradox. Since the retired old man can extend his consciousness backward to recall the experiences of the brave officer, they are the same person. Since the brave officer can extend his consciousness backward to the young thieving boy, they too are the same person. According to the principle of the transitivity of identity, if a is identical to b, and b is identical to c, then a is identical to c. So it seems to follow that the retired old man is identical to the young thieving boy. But since the retired old man cannot extend his consciousness backward to recall the experiences of the young thieving boy, Locke's theory would deny that the retired old man is identical to the young thief. The theory thus seems to lead to a contradiction.

2 The Continuity of Consciousness View

In response to this kind of worry, philosophers sympathetic with the spirit of Locke's view tend to suggest a modification of it. Rather than requiring that there be direct connections of memories between two individuals for them to count as the same person we can instead require simply that there be a continuity of memory between them. It doesn't matter, then, that the retired old man can't directly extend his consciousness backward to the experiences of the young thief. Since he can extend his consciousness backward to the experiences of the brave officer, who in turn can extend his consciousness backward to the experiences of the young boy, the experiences of all three stages are part of the same shared continuity of consciousness. Let's call this the *continuity of consciousness* view.

While the continuity of consciousness view avoids the problem posed by the Brave Officer case, views in the Lockean spirit face another objection that cannot be dealt with so easily. The Brave Officer problem arises essentially due to cases of forgetting.[2] But in addition to the fact that some of our memories can be forgotten, there is also the fact that some of our memories can be false. When I extend my consciousness backward to some event, I take that to be an event that I myself experienced. But what if I'm wrong?

Cases of mistaken memories are not at all uncommon. Consider this scenario: Jordan starts recounting a story about a time he beat up a bully who was taunting a group of younger kids. In fact, his brother Zach was the one who pummeled the bully. ("Hey, that wasn't you, that was me!" he might say.) Of course, this might just be a case of boastfulness on Jordan's part. But it also might be a case where he sincerely believes that he was the one to deliver those punches. From the inside, his apparent memory of doing so seems the same as all his genuine memories. But we don't want to take this apparent memory to imply that Jordan is identical to the person who beat up the bully. False memories cannot make you into someone that you're not.

It may seem that there is an easy fix here. Why not simply require that the continuity of consciousness be *real* rather than apparent? Unfortunately, however, things are not quite so simple. As pointed out by the 18th-century philosopher and theologian Bishop Joseph Butler (1736), views in the Lockean tradition run into a problem of circularity. We are trying to use the notion of memory to explicate personal identity. But, as we have seen, doing so will only be plausible if the memory in question is *genuine* and not merely apparent. So how are we going to distinguish those cases in which memories are merely apparent from those in which they are genuine? Going back to the case of Jordan and Zach, it seems natural to say something like this: Since Zach was the person who pummeled the bully, his memory is real, and since Jordan wasn't the person who pummeled the bully, his memory is merely apparent. Now look what's happened: To explain personal identity, we've invoked continuity of memory. But to explain continuity of memory, we've invoked personal identity. As Butler put it, we can't define personal identity in terms of memory if it turns out that a proper understanding of memory presupposes the notion of personal identity.

Philosophers in the Lockean tradition have made various attempts to solve this problem. One particularly promising line of defense invokes a causal theory of memory. On this kind of theory, there must be an appropriate causal connection between a mental state and an experience in order for the mental state to count as a genuine memory of that experience (see Perry 1975 for discussion). Perhaps this defense is successful, perhaps not. But even if it is, there is yet one more problem facing the continuity of consciousness view that we need now to consider, namely, what's often known as the *problem of reduplication*.

To motivate the problem, it will be helpful to return to the example we saw above of consciousness transfer. Recall that in the movie *Avatar*, Sully's consciousness is transferred from his human body to a Na'vi body. But now we might wonder: once his consciousness is transferred out of his original body, why can't it be transferred back into two or more bodies? As Reid noted in the 18th century, "if the same consciousness can be transferred from one intelligent being to another ... *then two or twenty intelligent beings may be the same person*" (Reid 1785: 114).[3]

In recent philosophical discussion, this problem is often motivated by way of an example introduced by Derek Parfit (1984: 199–200). Suppose in the future we are offered the opportunity to travel off-world to the moon or to Mars by way of teleportation. Rather than spending years cooped up on a space shuttle, one could simply step onto a teleporter pad, press a button, and wake up on Mars. How does the teleporter work? It scans the traveler, records her mental and physical "pattern," destroys the traveler's body on Earth and then imposes that mental and physical pattern on new matter on Mars. The person now on Mars has a new body – the old one was destroyed – but has a continuity of consciousness with the person who stepped onto the teleporter pad on Earth.

Given just what we've said so far, the teleporter example may well seem to *support* the continuity of consciousness view. Insofar as we're inclined to view such a procedure as a form of transportation rather than a means of suicide, the case provides us with another case where our intuitions favor continuity of consciousness over sameness of body. (Though the body on Mars may be *qualitatively identical* to the body on Earth, i.e., while it might look exactly the same as the Earth body, it is not *numerically identical* to the body on Earth.) But Parfit adds an additional wrinkle that calls those intuitions into doubt: What if the teleporter neglected to destroy the body on Earth? Though it still scans the mental and physical pattern and imposes it on new matter on Mars, suppose that it leaves completely intact the body that steps onto the transporter. With this added wrinkle, it no longer seems clear that the person on Mars is the same person as the one who stepped onto the transporter – despite the continuity of consciousness between them.

So how should we assess this case? There seem to be four options: (1) The traveler exists on Earth and not on Mars; (2) she exists on Mars and not on Earth; (3) she exists in both places; or (4) she exists in neither place. For someone who subscribes to the continuity of consciousness view, there seem to be problems with all four of these options. To make the case easier to talk about, let's refer to the person who stepped on the transporter as *Initial-Traveler* and to the two resulting persons as *Earth-Traveler* and *Mars-Traveler*. Earth-Traveler and Mars-Traveler do not share consciousness with one another, but at the moment Mars-Traveler comes into existence, her consciousness is continuous with Initial-Traveler in exactly the same way that Earth-Traveler's consciousness is continuous with Initial-Traveler. Because of this, there is no reason to privilege one of the first two options over the other, and they thus must both be discarded.

Might we instead choose option (3) and claim that the traveler exists in both places? Though this may initially appear to be an appealing option, recall that identity is a transitive relation: if a is identical to b and b is identical to c, then a is identical to c. So if we say that Initial-Traveler is identical to Earth-Traveler and we say that Initial-Traveler is identical to Mars-Traveler, then it follows that Earth-Traveler is identical to Mars-Traveler. But this seems wrong – after all, as we've just noted, Earth-Traveler and Mars-Traveler do not share consciousness with one another. Thus, our third option seems just as problematic as the first two.

What about the last remaining option, the claim that the traveler exists in neither place, i.e., that the traveler ceases to exist. From the perspective of the individual on Earth, this conclusion would no doubt seem absurd. How could stepping onto a teleporter pad and being (non-destructively) scanned make you go out of existence? The conclusion seems equally absurd from the perspective of the individual on Mars. The relationship between Mars-Traveler and Initial-Traveler in this case seems just like the relationship between Mars-Traveler and Initial-Traveler in the case as first introduced, before the added wrinkle of reduplication. In both cases there is continuity of consciousness between them. If this continuity is good enough for Mars-Traveler to be the same person as Initial-Traveler in the case as first introduced, then why shouldn't this continuity be good enough for Mars-Traveler to be the same person as Initial-Traveler in the case with the added wrinkle of reduplication? As Parfit has sometimes put the point about cases such as this: How can a double success be a failure? (See Parfit 1984: 256.)

In response to this problem, some proponents of the continuity of consciousness view adopt what's often referred to as a *non-branching requirement*: Rather than claiming that personal identity consists in continuity of consciousness, we should instead claim that personal identity consists in continuity of consciousness *only when it takes a non-branching form*. Adoption of this requirement is often conjoined with further claims about the unimportance of identity. Consider Parfit's own assessment of this kind of reduplication case. Though neither Earth-Traveler nor Mars-Traveler is the same person as Initial-Traveler, it will still be true that Initial-Traveler *survives* as both of them. Moreover, this survival is just about as good as ordinary cases of survival where there is no reduplication. The surviving travelers will navigate the world just as Initial-Traveler would. They will approach new situations just as she would, carry out her projects and plans just as she would, and so on. On Parfit's view, identity is not what matters in survival (Parfit 1984: Ch. 12).

While there is considerably more to be said about the problem of reduplication and potential solutions to it, the preceding discussion should give at least a general sense of the kind of challenge it poses for the continuity of consciousness view.[4] At this point, then, it seems worth stepping outside the Lockean tradition and considering alternatives to consciousness-based theories of personal identity. In the next section, we turn our attention to the options that are available for someone who thinks that we should understand personal identity in physical rather than psychological terms.

3 Physical Approaches

As we saw above, when we think about *Freaky Friday*-like cases of body swaps, consciousness looks to be central to personal identity. But we might wonder whether we should put much stock in our intuitions about these hypothetical cases.[5] And in fact, once we turn back from science fiction to real-life, we might be struck by one particularly salient fact: At the very beginning of our lives – when we are fetuses in the womb – we are not yet conscious. None of us can extend our consciousness backward to our experiences as a fetus. Indeed, we can't even extend our consciousness backward to our experiences as a very young infant! So how could any kind of consciousness-based approach have any hope of explaining our identity over time?

A similar conclusion might be reached by considering various scenarios that occur at the end of life. Consider someone – call her Beatrice – who, as a result of serious cardiac arrest and the subsequent loss of oxygen to the brain, is in what's often called a *persistent vegetative state*. While the subcortical parts of Beatrice's brain controlling respiration and blood circulation continue to function, her cerebral cortex has been destroyed. She has thus irretrievably lost all higher mental functions. She is no longer conscious, nor is there any hope that she ever will be again. Setting aside all the thorny legal issues that such cases raise, we might nonetheless ask: Does Beatrice still exist? And here there seems to be good reason to answer yes. Just as we're inclined to say that Beatrice herself was once a fetus, we're also inclined to say that Beatrice herself is now in a persistent vegetative state. So here again we have a case that seems inexplicable from the perspective of a consciousness-based approach.

In recent years, Eric Olson has used these sorts of considerations about both the beginning and end of life to support a biologically-based approach to identity over time (see, e.g., Olson 1997). On this approach, a view that he calls *animalism*, what matters for our persistence through time is biological continuity, i.e., the continuation of one's purely animal functions such as metabolism, respiration, circulation, etc. For the animalist, beings such as you and I should be seen first and foremost not as conscious subjects but as living organisms, as human animals. In this view, consciousness is neither necessary nor sufficient for our identity through time.

An individual in a persistent vegetative state, though no longer conscious, is still the same living organism. Likewise, a fetus is the same living organism as the child and adult that it will eventually become. But while the animalist view nicely explains our intuitions in the sorts of beginning and end of life cases just considered, it gives us a counterintuitive result in the sorts of body swap cases we considered in the previous section. Consider the individual with Tess's consciousness and Anna's body. Though we are inclined to classify that individual as Tess, the animalist cannot do so. Confronted with the question, "which living organism is it?" the animalist's answer is clearly Anna.

Olson addresses this sort of challenge to the animalist view in the context of a transplant case involving Prince, a rich and tyrannical ruler, and Cobbler, a poor but healthy working-class man. When Prince's body is severely damaged in a yachting accident, the royal servants kidnap Cobbler, and the royal medical staff then proceeds with a complicated transplant procedure. After destroying Cobbler's cerebrum, they remove Prince's cerebrum and transplant it into Cobbler's body, having attached it to Cobbler's brainstem. As Olson describes the resulting scenario:

> Two human beings resulted from this. One of them, called "Brainy," had Cobbler's arms, legs, trunk, and other parts, but Prince's cerebrum. Brainy looked just like Cobbler, but he had Prince's personality and character, and was able to remember as much of Prince's past as Prince could; and he knew nothing about Cobbler's past. The other offshoot, "Brainless," had all of Prince's parts except for his missing cerebrum. Although Brainless could wake and sleep, cough and sneeze, and even make reflex movements with his arms and legs, his eyes could only stare vacantly at the ceiling. He was in roughly the sort of persistent vegetative state that sometimes results from massive cerebral damage.
>
> (Olson 1997: 43)

Confronted with this sort of case, most people have what Olson calls *the transplant intuition*, namely, that Brainy is Prince.[6] This intuition seems to cause trouble for the animalist, however, since the animalist view identifies Brainy with Cobbler. On this view, a cerebrum transplant is no different from a kidney transplant. Just as getting Prince's kidney doesn't affect Cobbler's identity, neither does getting Prince's cerebrum. Since the brainstem was never removed from Cobbler's body, the same living organism continues to exist.

In responding to this sort of case, Olson does not deny the force of the transplant intuition; indeed, he notes that he himself feels its pull: "It seems to me too, at first glance, that Prince survives the operation as Brainy" (Olson 1997: 44). As we know, however, first appearances are often deceiving, and Olson argues that such is the case here. To his mind, we have good theoretical reasons to think that both Prince and Cobbler are living organisms, and that no single living organism can be identified first as Prince and then as Brainy. He thus offers the following diagnosis of the situation: The transplant intuition derives most of its force from certain underlying principles about practical matters. But while the transplant intuition is incompatible with animalism, the underlying practical principles are not. In particular, it may be rational for people to treat Brainy as Prince and for Brainy to be held responsible for Prince's actions. The defense of animalism thus depends on showing that practical matters involving moral responsibility, personal concern, and so on can come apart from facts about numerical identity.[7]

To many philosophers, however, giving up the transplant intuition seems like too hard a pill to swallow. Is there any other way to accommodate the intuition that we were once fetuses – or at least, that we were once very young infants? Here one might look to the kind of view that in recent years has been put forth by Jeff McMahan. Like continuity of consciousness theorists, McMahan claims that our continued existence depends on our having the same consciousness – or as he typically puts it, *the same mind*. Unlike such theorists, however, McMahan assigns critical importance to the embodiment of minds in brains. On this view, often referred to as *the embodied mind view*, a person's continued existence through time consists in the continued existence and functioning of the brain. In particular, the brain must continue to function in such a way that it supports the capacity for consciousness.

This emphasis on the *capacity* for consciousness is another way in which the embodied mind view departs from the continuity of consciousness view. Theorists in the Lockean tradition tend to require continuity of at least some psychological contents. In contrast, McMahan does not. On his view, even if a brain has been completely "deprogrammed," with its contents systematically wiped and destroyed, there is no threat to personal identity as long as the brain retains its capacity for conscious functioning (McMahan 2002: 68). This emphasis on capacity also allows the embodied mind view to defuse the threat of the fetus problem discussed above. Current scientific understanding suggests that a fetus's brain will develop the capacity for consciousness somewhere between 24 and 28 weeks of pregnancy. Thus, the embodied mind view can accommodate the claim that we were once fetuses – or at least, that we were once third-trimester fetuses. In this way, the embodied mind view occupies a nice middle ground between the continuity of consciousness view and the animalist view. Though it assigns consciousness central importance in thinking about our personal identity over time, it also accommodates the intuition that there is more to our continued identity than just the contents of that consciousness.

But now recall the teleporter cases that we encountered above. When functioning normally, a teleporter dematerializes the original body that steps onto the transporter pad. It doesn't simply wipe out the contents of the brain; it destroys it entirely. The embodied mind view thus entails that teleportation is not simply a method of high-speed transport. Rather, it is more akin to a suicide machine.

Relatedly, cases of uploading also pose trouble for the embodied mind view. In these early years of the 21st century, it seems that it is only a matter of time before technology has sufficiently developed to enable us to leave behind our physical bodies and upload our consciousness to machines or to the cloud. While some cognitive scientists question the technological feasibility of this vision, the embodied mind view rules it out in principle; without the continued functioning of your brain, you cease to exist. This position runs counter to that of many futurists writing on the possibility of uploading. In their view, the prospect of uploading shouldn't be viewed as death but rather as a way of achieving potential immortality.

Of course, the mere fact that immortality might be desirable does not in itself make it possible.[8] Anyone who argued in favor of the continuity of consciousness view over a physical approach, like animalism or the embodied view, solely on the grounds that they desired immortality would be guilty of fallacious reasoning. It's also worth noting that the simple view of personal identity that we encountered briefly above – a view that sees personal identity as simply a matter of sameness of soul – is also compatible with immortality. In fact, traditionally, it's this kind of soul-based view that has been most closely associated with claims about immortality.[9]

The relationship between the continuity of consciousness view and immortality is thus slightly more nuanced than it might have initially appeared. But that said, insofar as there are reasons to believe that your uploaded consciousness would still really be you, or insofar as there are other reasons to believe in the possibility of immortality, the physical approach to personal identity does seem to be threatened. So are there any such reasons? In the final section of this essay, we examine in more detail the case for the conceptual possibility of achieving life after bodily death, by way of uploading or other means.

4 Immortality

As detailed in his book *The Singularity Is Near,* inventor and futurist Ray Kurzweil has been taking aggressive steps to survive in good health for long enough to experience "the full blossoming of the biotechnology revolution" (Kurzweil 2005: 212). Not only does he take a daily regimen of 250 pill supplements, but he also receives approximately six intravenous therapies per week. In Kurzweil's view, we will someday be able to overcome the limitations of our frail and cumbersome current bodies – what he calls our "version 1.0 biological bodies" (Kurzweil 2005: 9). In particular, once we achieve the Singularity – a point of such rapidly accelerating technological innovation that our whole way of life will be drastically rewritten – we will be able to transcend our biological limitations and take our mortality into our own hands.

Kurzweil describes his own philosophical position on personal identity as *patternism*: one's identity as a person lies principally in a pattern that persists through time. The pattern is independent of the susbstrate in which it is realized. As Kurzweil notes, "the particles containing my brain and body change within weeks, but there is a continuity to the pattern that these particles make" (Kurzweil 2005: 5). Moreover, such continuity could exist even were the pattern to be realized in a different physical substrate – a robotic body made up principally of a network of nanobots, say. For the patternist, it's the continuity that fundamentally matters. As long as a particular personal pattern continues to exist that person also continues to exist.

While the proponent of the continuity of consciousness approach need not adopt this Kurzweilian patternism, the two approaches seem broadly consonant with one another. Moreover, as our discussion has already suggested, there is at least *prima facie* reason to believe that both of these approaches seem to support the possibility of survival through uploading. But what kind of uploading? Here it is perhaps worth considering different varieties of uploading scenarios, to see which would be most conducive to the preservation of personal identity after bodily death (and thus to possible immortality).

In a recent discussion of the issue, David Chalmers distinguishes three different kinds of uploading that might one day be possible: destructive uploading, gradual uploading, and nondestructive uploading. In destructive uploading, the brain is frozen and then its precise state is analyzed and recorded – perhaps by way of serial sectioning where scientists analyze its structure one layer at a time. After all of the information about the distribution of neurons and their interconnections is retrieved, it is then loaded onto a computer model, where a simulation is produced. Gradual uploading, in contrast, occurs by way of nanotransfer:

> One or more nanotechnological devices (perhaps tiny robots) are inserted into the brain and each attaches itself to a single neuron, learning to simulate the behavior of the associated neuron and also learning about its connectivity. Once it simulates the neuron's behavior well enough, it takes the place of the original neuron, perhaps leaving receptors and transmitters in place and uploading the relevant processing to a computer via radio transmitters. It then moves to other neurons and repeats the procedure, until eventually every neuron has been replaced by an emulation, and perhaps all processing has been uploaded to a computer.
>
> *(Chalmers 2014: 103)*

Finally, nondestructive uploading works in a similar fashion to gradual uploading, only without the destruction of the original neurons.

As Chalmers notes, two different kinds of questions arise as we assess these three methods of uploading. Firstly: Will the resulting uploaded entity be conscious? And secondly: Will the resulting uploaded entity be *me*? Given that we are conscious beings, an affirmative answer to the second question depends on an affirmative answer to the first question – and this will be true whatever theory of personal identity one adopts. But as it would here take us too far afield to do the requisite survey of theories of consciousness needed to answer the first question, we will here focus solely on the second question.[10] Do we have grounds to believe that any of these methods of uploading would successfully preserve personal identity?

In Chalmers's view, there are good grounds for pessimism with respect to both destructive and nondestructive uploading. Consider first nondestructive uploading, and call the system that results from a nanotransfer *DigiDave*. In such a case, since the original Dave still exists, Dave and DigiDave cannot be numerically identical to one another. But if Dave does not survive as DigiDave in the case where the original system is preserved, we might wonder why Dave would survive as DigiDave in the case where the original system is destroyed.

Here we see a similar dialectic as in the case of teleportation, and similar moves could be redeployed to make a case for the preservation of personal identity in the destructive uploading case. But rather than rehearse those moves, let's instead turn to gradual uploading – the uploading scenario about which Chalmers suggests we have the most reason to be optimistic. The case for optimism rests on a simple argument. It seems pretty plausible that when just one percent of someone's brain is replaced by functionally isomorphic robotic technology, personal identity is preserved. To deny this would seem to commit one to the result that the destruction of even a single neuron of someone's brain would lead to their death. But now suppose that, having replaced one percent of someone's brain, we slowly repeat the process once a month, so that by the end of 100 months the original system is completely destroyed and we are left with a system that has been wholly uploaded to nanotechnology. It seems plausible that the original person still exists after month two, and after month three, and so on. So where would we draw the line? Letting $Dave_n$ stand for the system after n months, we get the following argument (see Chalmers 2014: 111):

1 For all $n < 100$, $Dave_{n+1}$ is identical to $Dave_n$.
2 If for all $n < 100$, $Dave_{n+1}$ is identical to $Dave^n$, then $Dave_{100}$ is identical to Dave.
3 Therefore, $Dave_{100}$ is identical to Dave.

Chalmers himself finds this argument reasonably convincing, though not everyone agrees (see, e.g., Pigliucci 2014: 126-128; Corrabi and Schneider 2014: 138–140). Ultimately, then, it may be that we've reached something of a stalemate. At the end of the previous section, I noted that the conceptual possibility of uploading would count against the physical approach to personal identity and in favor of the continuity of consciousness approach. Insofar as our intuitions about the possibility of uploading are inextricably intertwined with our intuitions about personal identity, we might not be able to use the former to help us sort out the latter.

That said, if there were other reasons to believe in the possibility of immortality – reasons arising from considerations other than the kinds of uploading scenarios we've just discussed – such reasons could indeed be relevant to the personal identity debate. In fact, we wouldn't even need full-on immortality for such reasons to be effective. Evidence that one could exist without one's physical body, even if only for a brief stretch and not for an immortal afterlife, would still count against the physical approach.

Some philosophers point to Near Death Experiences (NDEs) as a possible source of such evidence. While NDEs are often described as having an ineffable quality, in typical cases the individual experiences emotional detachment – and, in particular, an absence of fear – and has the impression of having left his or her body. As described by David Lund (2009), NDEs often involve the sensation that one is traveling in some kind of tunnel or dark passageway, towards a bright light or ephemeral presence. It's when one turns back, or is pulled back, from the bright light or presence that one has the sensation of re-entering one's body.[11]

When such experiences occur merely *near* death, their value in supporting the possibility of life after death may seem questionable. It seems just as (if not more) plausible that such experiences are hallucinatory in nature than that they provide a glimpse into an afterlife. But in many such cases the experiences are reported after a patient has been (at least briefly) clinically dead. For example, in a much-cited Dutch study, 62 of 344 patients (18%) who were successfully resuscitated after cardiac arrest reported having had NDEs (Van Lommel 2001). Given the absence of neural activity at the time that such experiences were reportedly occurring, these kinds of cases may seem to point to the conclusion that we can survive the death of our bodies.[12]

In response, various alternative explanations of NDE are open to the proponent of the physical approach. Most notably, they might deny that the NDE really occurred in the absence of neural activity. Firstly, one might question how one can accurately pinpoint the timing of an NDE. Perhaps it occurs slightly earlier than reported, before neural activity has ceased, or perhaps it occurs slightly later than reported, once neural activity has resumed. Secondly, even if the timing of the NDE was accurately pinpointed, one might still question whether all neural activity had really ceased at that moment. Perhaps some brain activity continues, undetectable by current instrumentation. Given these possibilities, it seems questionable that NDEs can be taken as decisive evidence for post-bodily survival.[13]

At this point in time, then, it seems that considerations about immortality cannot be used effectively to help settle the debate about personal identity. That said, if futurists like Kurzweil are correct, it may not be too long before the technologies arrive that will force the issue upon us. Indeed, Kurzweil has predicted that the Singularity will be upon us by 2045. In the meantime, however – and perhaps even after – the debate about what exactly constitutes personal identity will undoubtedly continue.

Notes

1 Elsewhere I call these the *identification question*, the *characterization question*, and the *reidentification question* (Kind 2015).

2 To deal with other more dramatic cases of forgetting, such as amnesia, contemporary proponents of this view tend to broaden their conception of the continuity of consciousness, so that it requires not just continuity of memories, but also of other psychological states more broadly (see, e.g., Parfit 1984: 205).

3 The problem of reduplication was raised earlier in the 18th century by British philosopher Samuel Clarke in his correspondence with Anthony Collins. See Uzgalis (2008) for some of the relevant portions of this correspondence.

4 Another common approach to the problem of reduplication is to adopt four-dimensionalism, a metaphysical view about the general nature of an object's survival through time. For a development and defense of this view, see Sider (2001). For further discussion of the problem of reduplication, see Kind (2015: Ch. 3) and Noonan (2003: Ch. 7).

5 For an extended discussion that is motivated by this worry, see Wilkes (1988).

6 Schechtman (2014: 152) suggests that the transplant intuition is almost universally held. For at least one dissenting voice, see Thomson (1997). In John Perry's *Dialogue on Personal Identity and Immortality* (1978), the fictional character Gretchen Weirob also makes a case against this intuition.

7 Schechtman (1996) offers a different set of arguments to show that these sorts of practical matters can come apart from the facts of numerical identity.

8 Not everyone thinks that an immortal life would be desirable (see, e.g., Williams 1973).

9 But here recall Locke's objections (mentioned in Section 1 above) that mere continuation of immaterial substance would really preserve personal identity. For detailed further discussion of this issue, see Perry (1978).

10 For issues relevant to the first question, see Janet Levin's discussion in Chapter 3 of this volume. See also Chalmers (2014: 103–07).

11 For a recent fictionalized depiction of NDEs that raises interesting philosophical questions, see the Netflix series *The OA*.

12 For another source arguing in favor of NDEs, see van Lommel (2010).

13 Arguments against NDEs are presented in Mitchell-Yellin and Fischer (2014) and Augustine (2015).

References

Augustine, K. (2015) "Near-Death Experiences Are Hallucinations," in M. Martin and K. Augustine (eds.) *The Myth of an Afterlife*, New York: Rowman and Littlefield.

Butler, J. (1736) "Of Personal Identity," in J. Perry (ed.) *Personal Identity* (revised edition), Berkeley, CA: University of California Press.

Chalmers, D. J. (2014) "Uploading: A Philosophical Analysis," in *Intelligence Unbound: The Future of Uploaded and Machine Minds*, Chichester, UK: John Wiley and Son.

Corabi, J. and Schneider, S. (2014) "If You Upload, Will You Survive?" in *Intelligence Unbound: The Future of Uploaded and Machine Minds*, Chichester, UK: John Wiley and Sons.

Kind, A. (2015) *Persons and Personal Identity*, Cambridge: Polity Press.

Locke, J. (1689/1975) *An Essay Concerning Human Understanding*, edited with an introduction by Peter H. Nidditch, Oxford: Oxford University Press.

Lund, D. (2009) *Persons, Souls and Death*, Jefferson, MO: McFarland and Company, Inc.

McMahan, J. (2002) *The Ethics of Killing: Problems at the Margins of Life*, Oxford: Oxford University Press.

Madell, G. (1981) *The Identity of the Self*, Edinburgh: Edinburgh University Press.

Mitchell-Yellin, B. and Fischer, J. (2014) "The Near-Death Experience Argument Against Physicalism," *Journal of Consciousness Studies* 21: 158–183.

Olson, E. T. (1997) *The Human Animal: Personal Identity Without Psychology*, New York: Oxford University Press.

Parfit, D. (1984) *Reasons and Persons*, Oxford: Oxford University Press.

Perry, J. (ed.) (2008) *Personal Identity* (revised edition), Berkeley, CA: University of California Press.

Perry, J. (ed.) (1975) "Personal Identity, Memory, and the Problem of Circularity," in J. Perry (ed.) *Personal Identity* (revised edition), Berkeley, CA: University of California Press.

Perry, J. (ed.) (1978) *A Dialogue on Personal Identity and Immortality*, Indianapolis, IN: Hackett Publishing Company.

Pigliucci, M. (2014) "Mind Uploading: A Philosophical Counter-Analysis," in *Intelligence Unbound: The Future of Uploaded and Machine Minds*, Chichester, UK: John Wiley and Sons.

Reid, T. (1785) "Of Mr. Locke's Account of Our Personal Identity," in J. Perry (ed.) *Personal Identity* (revised edition), Berkeley, CA: University of California Press.

Schechtman, M. (1996) *The Constitution of Selves*, New York: Cornell University Press.

Schechtman, M. (2014) *Staying Alive: Personal Identity, Practical Concerns, and the Unity of a Life*, Oxford: Oxford University Press.

Sider, T. (2001) *Four-Dimensionalism: An Ontology of Persistence and Time.* New York: Oxford University Press.

Swinburne, R. (1973–4) "Personal Identity," *Proceedings of the Aristotelian Society, New Series* 74: 231–247.

Thomson, J. J. (1997) "People and Their Bodies," in J. Dancy (ed.) *Reading Parfit*, Oxford: Blackwell Publishers.

Uzgalis, W. (2008) "Selections from the Clarke-Collins Correspondence," in J. Perry (ed.) *Personal Identity* (revised edition), Berkeley, CA: University of California Press.

van Lommel, P. (2001) "Near-Death Experience in Survivors of Cardiac Arrest: A Prospective Study in the Netherlands," *The Lancet* 358: 2039–45.

van Lommel, P. (2010) *Consciousness Beyond Life: The Science of the Near-Death Experience*, New York: Harper Collins.

Wilkes, K. (1988) *Real People: Personal Identity Without Thought Experiments*, Oxford: Oxford University Press.

Williams, B. (1970) "The Self and the Future," *Philosophical Review* 79: 161–80.

Williams, B. (1973) "The Makropoulos Case: Reflections on the Tedium of Immortality," in B. Williams, *Problems of the Self*, Cambridge: Cambridge University Press.

Related Topics

Dualism
Materialism
Consciousness in Western Philosophy
The Unity of Consciousness
Consciousness and Psychopathology

2

CONSCIOUSNESS IN WESTERN PHILOSOPHY

Larry M. Jorgensen

A fully naturalized philosophy of mind is often held up as a gold standard. As one person has noted, "a casual observer of recent philosophy of mind would likely come to the conclusion that, amidst all of the disagreements between specialists in this field, there is at least one thing that stands as more or less a consensus view: the commitment to a *naturalistic* philosophy of mind" (Horst 2009: 219). In this pursuit of a naturalized philosophy of mind, consciousness often receives concentrated attention, in part because the phenomena of consciousness seem particularly recalcitrant, difficult to explain in the terms of the physical and biological sciences. There is an expectation that consciousness will turn out to be compatible with the natural sciences, but for now just how remains a mystery. One version of this expectation is that consciousness is compatible with a fully physicalist metaphysics. If consciousness is explicable in terms of purely physical interactions, then it seems easily explicable in terms of the natural sciences.

However, a quick historical survey will show that naturalism has not always been combined with physicalism. Insofar as we can identify a common project under the heading of "naturalism," it is a project that can unfold in quite a few ways. Attempts at naturalizing consciousness turn out to be compatible with versions of dualism and idealism, and there is reason to expect that even today a fully naturalized theory of consciousness might be incompatible with physicalism.

This survey of consciousness in Western philosophy will focus on one particular thread: the search for a naturalized theory of consciousness. Of course, there are many *non*-naturalists in the history of Western philosophy, philosophers who argue for some degree of divine influence in nature or who argue that humans are exceptional and can act in ways that should not be conceived of in terms of natural causation. And many of these philosophers have interests in understanding and theorizing about consciousness. So, I do not intend to argue that the history of consciousness is exhausted by a survey of the efforts to naturalize consciousness. But I think that the efforts to make consciousness intelligible in natural terms encompasses a broader swath of philosophers in the West than has previously been allowed. For example, the mere fact that a philosopher is a theist (as many, going back to Ancient Greece, were) is not an indication that they are not interested in a naturalized philosophy of mind.

In what follows, I will begin by characterizing what I take the goal of naturalism to be, characterizing it in a way that will identify a common project from Ancient Greece through to today.[1] Second, I will look at Aristotle as a prime mover in articulating a naturalized theory of consciousness. Third, I will argue that as the Aristotelian physics and metaphysics were overturned in the early modern period, consciousness came to the fore as a philosophical issue, and the uniquely modern conception of consciousness became a focus of concentrated attention. Fourth, I will consider how Kantian views redirected the discussion of consciousness. I will close with some brief considerations of how the historical development of a naturalized theory of consciousness might inform today's efforts.

1 Naturalism and Consciousness

A *naturalized theory* is a theory that has no irresolvable "mysteries"—mysteries like those presented by phenomenal consciousness or subjective experience. As Fred Dretske has put it, a naturalized theory may not "remove *all* the mysteries [but] it removes enough of them… to justify putting one's money on the nose of this philosophical horse" (Dretske 1997: xliii). While many of today's defenders of naturalism will define naturalism in terms of the natural sciences, there is reason in a survey to articulate a broader definition. What is it *about* continuity with the natural sciences that would make this a desirable goal? I will identify two principal constraints that I believe to be at the core of what makes naturalism desirable.

One way to recognize a naturalized theory is that it provides plausible or satisfactory *explanations* of all mental states and events. This is evident from the claim that we want to remove *mysteries* from our theories. Naturalism is about discharging explanatory demands. Any explanation of consciousness should make it *intelligible*. Call this the *intelligibility constraint.*

Of course, there may be *non-natural* ways to make something intelligible. When natural events are conceived of in terms of the behavior of the gods, this is a way of making those events intelligible. However, if appeal to the gods makes the explanation *more* mysterious (because the gods are fickle and unpredictable), then it would not satisfy the intelligibility constraint. What we want is a way of making the events intelligible without introducing new mysteries: making them intelligible in ways that would allow us to (at least in principle if not in practice) make predictions and govern our behavior accordingly.[2]

However, some conceptions of divine activity are fully consistent and predictable.[3] Would such a theory that incorporated divine activity into natural explanations be a naturalized theory? Surely not. Rather, a second requirement on naturalism seems to require intelligibility in terms of the *natures* of the things themselves. That is, an explanation should be *immanent* to the things being explained. This is not to say that natural events must be intelligible in terms of *intrinsic* properties. Rather, any properties invoked should be properties (intrinsic, dispositional, relational) of the kinds of things being explained. For example, rain would be more *naturally* explained by appealing to the properties of the atmosphere and water cycle rather than by appealing to divine activity. Call this the *immanence constraint.*[4]

The naturalizing project, then, will be to satisfy these two constraints, and so a naturalized theory of consciousness would be one that makes consciousness *intelligible* in terms of features of the mind and body. Or, to put it differently, it will provide an explanatory *framework* that ensures intelligibility, consistency, and immanence, and in which consciousness plays its unique role. Consciousness then becomes an intelligible aspect of nature. Without such a framework, consciousness remains somewhat mysterious.[5]

With this understanding of naturalism, we can now turn to the topic of consciousness.[6] Tracing the development of consciousness in Western philosophy is complicated by the fact that the *term* consciousness was not coined until the seventeenth century.[7] Even more problematic is that once the modern term "consciousness" is in use, the term becomes an umbrella for several different phenomena. My approach, then, will be to identify passages in which it is clear that the philosopher is grappling with what we today identify under the heading of "phenomenal consciousness."

Phenomenal consciousness is typically described as the "what it's like" aspect of experience, the first-personal aspect of experience.[8] In this survey, I will focus on passages where it is somewhat clear that the philosopher is grappling with the subjective mental *seeming* of world or imagination.

2 Ancient Greek Conceptions of Consciousness

While ancient philosophers had much to say about the soul (*psyche*), consciousness as such was not a primary focus of theoretical work. Some argue that these issues are wholly absent from Ancient Greek concerns. As one scholar wrote of Aristotle:

> The general account of sense-perception remains for the most part basically physiological… There is an almost total neglect of any problem arising from psycho-physical dualism and the facts of consciousness. The reason appears to be that concepts like that of consciousness do not figure in his conceptual scheme at all; they play no part in his analysis of perception, thought, etc. (Nor do they play any significant role in Greek thought in general.)
>
> *(Hamlyn 1993: xii–xiii)*[9]

The search for consciousness in Ancient Greek philosophical texts may well be a fool's errand. However, other scholars have noted some overlapping concepts or concerns in Ancient Greek texts, which—with the necessary translation—can be seen as in the family of issues related to consciousness.

For example, although Plato never provides an analysis of consciousness, his theories have implications for a theory of consciousness. Plato makes use of a conscious-unconscious divide, most frequently in reference to knowledge. In reply to the oracle at Delphi, Socrates replies, "I am very conscious that I am not wise at all; what then does he mean by saying I am the wisest?" (Plato 1997: *Apology* 21b),[10] and in *Charmides* Socrates expresses a "fear of unconsciously thinking I know something when I do not" (166d). Similarly, in *Philebus*, Plato presents the intellectual faculties as necessary to having some sort of unified experience, since, with respect to pleasurable experiences, you would need a kind of judgment to "realize that you are enjoying yourself even while you do" and you would need memory in order to unify it in a common experience, "for any pleasure to survive from one moment to the next" (21c, cf. 60d-e). This role for the intellect in awareness connects with Plato's theory of recollection, which holds that we have in our minds ideas of which we are unaware, needing only the right triggers to bring them to the surface as if remembering them (*Meno* 81b and following).[11]

While someone might be able to work with these threads to develop a Platonic conception of consciousness, Plato himself left the theory rather sketchy. Even with respect to sensation, Plato gives more attention to bodily motions rather than the states of the soul that result from these motions (see *Timaeus* 42a, 43c, and 61d-68d). What *is* clear is that Plato would not be inclined to *reduce* sensation to the motions in the body.[12]

Aristotle, by contrast, gives extended treatment of the nature of the soul, perception, and the intellect in *De Anima* and other works (*Sense and Sensibilia* and *On Sleep*). Some scholars have seen the resources here to construct a theory of consciousness that maps somewhat faithfully onto what we would call phenomenal consciousness. One particularly key passage is *De Anima* 3.2, where Aristotle says:

> Since we perceive that we see and hear, it must either be by sight that one perceives that one sees or by another [sense]. But in that case there will be the same [sense]

> for sight and the color which is the subject for sight. So that either there will be two [senses] for the same thing or [the sense] itself will be the one for itself. Again, if the sense concerned with sight were indeed different from sight, either there will be an infinite regress or there will be some [sense] which is concerned with itself; so that we had best admit this of the first in the series.
>
> *(Hamelyn 1993: 425b11–15)*

This cryptic passage has yielded a variety of interpretations of Aristotle's conception of consciousness. Since the language here suggests we are aware of our sensations by means of a perception of a perception, Aristotle must have in mind some sort of "inner sense." The inner sense reading is a version of what is called a *higher-order theory of consciousness*: a higher-order perception takes a lower-order perception as its object, rendering it conscious.[13]

However, Victor Caston interprets the *De Anima* passage differently. Here is a reconstruction of the argument Caston thinks Aristotle is presenting here (Caston 2002):[14]

1 We perceive that we see the color red, which means that there is a dual content: we perceive that we are seeing red, and we perceive the color red.
2 We perceive that we are seeing red either (i) by means of a distinct perception or (ii) by means of the initial perception (the act of seeing).
3 *Therefore*, either (a) there will be two perceptions of the same thing (namely, the color red, since perceiving that we are seeing red is a perception of red just as the primary perception of red is), or (b) the one perception will also be of itself.
4 But there are not two perceptions of one and the same thing (namely, the color red). (No Double Vision thesis)
5 *Therefore*, the one perception will also be of itself.

Caston concludes that Aristotle has something *similar to* a higher-order theory, where consciousness is grounded in the content of the perception (grounded in intentionality). However, Aristotle would deny that the higher-order state is distinct from the original perception. Rather, the original perception is reflexive: it is a perception of red *and* a perception that I am seeing red. The regress argument in *De Anima* suggests that if the higher-order perception were distinct from the lower-order perception, then the theory would be incoherent. Aristotle rejects this view and says instead that "we had best admit this of the first of the series." That is, we had best admit that the first perception is reflexive and includes itself as an object of its perception. Thus, Aristotle's theory would be incoherent on the internal sense theory.

What the internal sense interpretation and Caston's interpretation have in common is that they both see Aristotle as grounding phenomenal consciousness in intentionality. The difference is in whether the grounding of phenomenal consciousness in intentionality requires a distinct perception or not. But the basic move is one that will be common among those that look for a naturalized theory of consciousness: it satisfies the intelligibility constraint, since it provides a way of explaining consciousness in terms of something more fundamental, and it satisfies the immanence constraint, since the explanation of consciousness is fully in terms of other aspects of the mind. We might be inclined here to press further for an account of intentionality, which Aristotle would answer in terms of his hylomorphism and causal relations between perceivers and intelligible forms, again satisfying the intelligibility and immanence constraints.[15] Minds, perception, and consciousness are explained in an integrated way with the whole of nature in a hylomorphic framework.[16] Thus, we have one example of a naturalized theory that is not straightforwardly a physicalist theory of mind.

3 The Seventeenth-Century Awakening

With the advent of revolutions in astronomy and physics, the early modern philosophers were not satisfied that the Aristotelian framework offered an intelligible account of the world, and in general they regarded explanations in terms of forms particularly *un*illuminating. Seventeenth-century philosopher Nicolas Malebranche argued that such "ways of speaking are not false: it is just that in effect they mean nothing" (Malebranche 1997: 444, cf. 242). One general trajectory of early modern natural philosophy was to dispense with substantial forms and to provide explanations in terms of merely material interactions. Whether this materialist mode of explanation could also explain human mentality then became a controversial matter, and it forms the backdrop of our conversations today.

It is during the seventeenth century that we first find explicit introduction of concepts and terms related to consciousness. Prior to the seventeenth century, the language of consciousness was bound up with the language of conscience (a moral sensibility, and "internal witness" to one's own integrity). But in the seventeenth century, and beginning particularly with Descartes, these two concepts began to diverge, resulting in the more purely psychological concept of consciousness, separated from its moral sense. This shift required the introduction of a new vocabulary. The English language acquired the word "consciousness" in the seventeenth century.[17]

This conceptual and linguistic shift more closely aligns with the way consciousness is framed as a philosophical problem today. As such, it is worth noting just what led to the introduction of this more distinctively modern conception of consciousness.

The most concise story that I can tell of this seventeenth-century innovation focuses on Descartes and Leibniz. We will see a similar relation between Descartes and Leibniz as there was between Plato and Aristotle. While Plato's philosophy had *implications* for a theory of consciousness, he left it largely unanalyzed, but Aristotle developed the idea and presented a wholly integrated and naturalized philosophy of mind. Similarly, Descartes's philosophy made use of the concept of consciousness in its modern sense, but he did not go very far in presenting an analysis of the concept. Leibniz was the first major philosopher to give focused attention to this task, and his account of consciousness goes much farther than Descartes's in integrating perception and consciousness into the natural order.[18]

Starting with Descartes, we see that he defines thought *in terms of* consciousness:

> *Thought.* I use this term to include everything that is within us in such a way that we are immediately aware [*conscii*] of it.
>
> (*Descartes 1985: 2.113, cf. 1.195*)

Descartes uses the term "thought" broadly, to include all mental states such as doubt, understanding, affirmation, denial, willingness [*volo*], refusal [*nolo*], imagination, and sense-perception (2.19). And so, Descartes's definition of "thought" entails that all mental states are conscious. Consciousness, for Descartes, *is* the mark of the mental.

While this passage is not intended as an analysis of consciousness itself, Descartes makes the important shift from a moral notion of *conscience* to a purely psychological notion of *consciousness*. In the famous *cogito* argument, Descartes's "internal witness" (the older sense of conscience) testifies to the existence and nature of an active mind (the modern psychological sense of consciousness). In his *Meditations on First Philosophy*, Descartes explicitly sets aside moral concerns[19] and turns inward to discover a psychological criterion for truth, giving rise to an emphasis on the more purely psychological sense of consciousness.

This much is clear. It is less clear whether Descartes provided a *naturalized* theory of consciousness. But he does suggest that consciousness has a structure:

> *Idea*. I understand the term to mean the form of any given thought, immediate perception of which makes me aware [*conscius*] of the thought.
>
> *(2.113)*

The proposition that the "immediate perception" of a thought "makes me aware of the thought" might suggest a higher-order theory of consciousness. However, Descartes actually has a model closer to what Caston was arguing for on behalf of Aristotle. Each thought involves self-reference. As Descartes says in reply to a Jesuit, Pierre Bourdin, who raised objections to Descartes's views:

> My critic says that to enable a substance to be superior to matter and wholly spiritual…, it is not sufficient for it to think: it is further required that it should think that it is thinking, by means of a reflexive act, or that it should have awareness [*conscientia*] of its own thought. This is…deluded…. [T]he initial thought by means of which we become aware of something does not differ from the second thought by means of which we become aware that we were aware of it, any more than this second thought differs from the third thought by means of which we become aware that we were aware that we were aware.
>
> *(2.382)*

Notice, first of all, that this exchange is couched as a worry about physicalism—Bourdin thinks that Descartes has not provided enough of a distinction between the material and the mental. What more is needed? The mental substance "should think that it is thinking, by means of a reflexive act," that is to say, "it should have awareness of its own thought." In his response, Descartes argues that the awareness of a thought comes from the thought itself by means of what Alison Simmons describes as "a form of immediate acquaintance" that a thought has of itself (Simmons 2012: 8). Each thought is reflexive in this minimal sense, and so for Descartes all thought is conscious thought.

This structure may not provide for a naturalized theory of consciousness in the sense described above. What exactly is this "immediate acquaintance," and how is it to be understood in terms of other features of the natural world? One way we might understand what is going on is to say that the thought *represents* both an external object and itself, in which case consciousness would be explicable in terms of representation. However, Alison Simmons argues that "Cartesian representation [is] tied to the notion of objective being, so that a thought represents whatever has objective being in it…, and there is no indication that Descartes thinks that thoughts exist objectively within themselves" (8). Thus, while thoughts seem to be self-intimating for Descartes, this is not by means of any representational content. And so consciousness is not explained in terms of representation for Descartes. Simmons concludes that "consciousness does not seem to be analyzable into any other features of thought" (8).[20]

If this is right, then consciousness does not have any further explanation. Consciousness in its most basic sense, for Descartes, is a kind of immediate acquaintance a thought has of itself. While acquaintance requires a structure—the thought is *about* itself in some way—this structure is not representational. But what else could it be? Descartes does not give us much more to go on. There are similar limits in Descartes's account of mind-body interaction that relate to current discussions of qualia, the qualitative aspects of experience. Descartes says that it is *possible* that the same motions of the body could have been represented in the mind differently (for example, the feeling of pain in the foot could have been represented in the mind as "the actual motion occurring in the brain, or in the foot," or "it might have indicated something else entirely" [Descartes 1985: 2.61]). That is, there is no way of explaining why certain motions of the brain give rise to certain qualitative experiences, other than appealing to divine teleology when devising the mind-body union. This arbitrariness and the limits of explanation for consciousness and qualia entail a *non*-naturalized theory of mind and consciousness.

John Locke's account of consciousness also includes a self-referential aspect. As Shelley Weinberg has recently argued, each mental act for Locke is a complex state involving, "at the very least, an act of perception, an idea perceived, and consciousness (that *I am perceiving*)" (Weinberg 2016: xi). Locke makes innovative *use* of this reflexive account of consciousness in his accounts of sensation, memory, and personal identity, but he provides no deep *analysis* of the concept. Although Locke does define consciousness as "the perception of what passes in a man's own mind" (Locke 1975: 2.1.19), this definition does not yield a full theory. As such, although Locke parts ways with Descartes on important matters, they are alike in that neither has given a full analysis or a fully naturalized theory of consciousness.[21]

Gottfried Wilhelm Leibniz turned the Cartesian mind upside down, and he argues that neither Locke nor Descartes has provided a fully naturalized theory of the mind. Contrary to Descartes's view that consciousness is the mark of the mental, Leibniz argues that *representation* is the mark of the mental and that consciousness is grounded in representation. For Descartes, all mental states are conscious and some are representational; for Leibniz, all mental states are representational and some are conscious.

Leibniz is the first major philosopher to introduce a systematic argument for *non*-conscious mental states, and he argued that the failure to recognize non-conscious mental states is a significant mistake. Leibniz says,

> It is good to distinguish between *perception*, which is the internal state of the monad [that is, a simple substance] representing external things, and *apperception*, which is *consciousness*, or the reflective knowledge of this internal state, something not given to all souls, nor at all times to a given soul. Moreover, it is because they lack this distinction that the Cartesians have failed, disregarding the perceptions that we do not apperceive, in the same way that people disregard imperceptible bodies.
>
> *(Leibniz 1989: 208)*

Leibniz coins a new term, *apperception*, which is the nominalization of the French verb for "to be aware of," in order to point out what the Cartesians missed. While the Cartesians properly speak of perception, which, *by definition* for Leibniz, is a representational state of a simple substance, they fail to recognize that some perceptions are not *apperceived*. That is, there are some perceptions of which we are not aware.

Leibniz makes his desire to *naturalize* the mind explicit—it is an animating principle in his philosophy of mind. Leibniz saw himself as providing a more consistently natural account of physics and of mind than the Cartesians. For example, in a letter to Arnauld, Leibniz says,

> The ordinary Cartesians confess that they cannot account for [the union of mind and body]; the authors of the hypothesis of occasional causes think that it is a "difficulty worthy of a liberator, for which the intervention of a *Deus ex machina* is necessary;" for myself, *I explain it in a natural manner.*
>
> *(Leibniz 1967: 145, emphasis mine)*

And Leibniz posits a general rule:

> This vulgar opinion—that we ought in philosophy to avoid, as much as possible, what surpasses the natures of creatures—is a very reasonable opinion. Otherwise, nothing

> will be easier than to account for anything by bringing in the deity, *Deum ex machina*, without minding the natures of *things*.
>
> *(Leibniz and Clarke 1956: Letter 5, §107, translation altered)*

Leibniz regarded the philosophies of Descartes and others who followed him as general failures in providing a *naturalized* philosophy of mind, and he aimed to do better.

Leibniz sought to provide a naturalized theory by arguing that all changes result from the immanent natures of the things themselves. That is, although Leibniz was a theist, he did not countenance divine meddling in natural occurrences.[22] And so Leibniz developed some heuristic principles that would enable him to test for the intelligibility of a system, to see whether the system had rid itself of mysteries. One such principle is the principle of continuity, which says that any natural change proceeds by degrees and not "by a leap" (see Leibniz 1969: 351–354). Leibniz applied this principle to Cartesian physics to show that Descartes's laws of impact yielded gaps in the explanation. That is, there were unexplained mysteries remaining in the system, and so the theory ought to be rejected in favor of one that makes all of the changes intelligible in terms of the natures of things. Leibniz also explicitly applies this to his theory of mind, which, for him, is a simple substance: "Since all natural change is produced by degrees, something changes and something remains. As a result, there must be a plurality of properties and relations in the simple substance, although it has no parts" (Leibniz 1989: 214, see also Leibniz 1996: 51–59).

What this means for Leibniz's theory of consciousness is that conscious states must arise by degrees from states that are not conscious. Some have interpreted Leibniz's theory of consciousness as requiring a higher-order perception, as is suggested by the quotation above, where Leibniz describes consciousness as "the reflective knowledge of [the] internal state." However, if the higher-order theory requires a distinct higher-order perception (as most interpretations have it), then it is difficult to see how such a perception could arise by degrees.[23]

Recent interpreters instead read Leibniz as articulating an account of consciousness that arises from variations in what he calls "perceptual distinctness." The concept of "perceptual distinctness" plays several roles in Leibniz's philosophy, but the central aspect of the concept for his theory of consciousness is that a perception becomes distinct when it is distinct*ive*, that is, it *stands out* from the background of other perceptions. This happens when there is enough similarity in what smaller perceptions represent that, when aggregated, they present their contents together more forcefully. (A process Leibniz describes as the "confusion" of their representational contents.) Here is one frequently repeated example from Leibniz:

> [T]he roaring noise of the sea… impresses itself on us when we are standing on the shore. To hear this noise as we do, we must hear the parts which make up this whole, that is the noise of each wave, although each of these little noises makes itself known only when combined confusedly with all the others, and would not be noticed if the wave which made it were by itself.
>
> *(Leibniz 1996: 54)*

In this example, Leibniz says that the *petites perceptions*—each little wave noise—aggregates into the full experience of the sound of the wave. And he describes sensation identically:

> Also evident is the nature of the perception…, namely the expression of many things in one, which differs widely from expression in a mirror or in a corporeal organ, which is not truly one. If the perception is more distinct, it makes a sensation.
>
> *(Leibniz 1973: 85; see also Leibniz 1996: 134)*

Sensation and other forms of phenomenal consciousness are functions of the combination of representational contents of perceptions. Once a perception has passed a sufficient threshold of distinctness against background perceptions such that it stands out, then that perception will be a conscious perception. Of course, the threshold will vary by context, since it will take more to exceed a very noisy background versus a tranquil background. Passing the threshold "makes a sensation." Call this the *threshold interpretation*.

The threshold interpretation as presented here may oversimplify matters a bit, since it doesn't spell out how perceptual distinctness also works across time and involves memory. There is some interpretive controversy around this point, but some of the basics seem to be agreed on by scholars today.[24] Scholars tend to agree that what accounts for consciousness, for Leibniz, is representational features of the underlying unconscious perceptions. This account of consciousness will allow for a number of interesting claims: (a) consciousness comes in degrees; (b) at a particular threshold consciousness arises; (c) the threshold and degrees of distinctness are sensitive to context; and (d) the theory of consciousness bears a strong analogy to what is going on in Leibniz's dynamics: the same underlying smaller forces may or may not have their effect depending on other variables. And this view is a naturalized theory in that consciousness is explained by the underlying intentionality of perception and so satisfies the intelligibility and immanence constraints of naturalism.[25]

What we have from Leibniz is the first concerted attempt at an analysis of consciousness in terms of more fundamental features. Leibniz presents a representational theory of mind and consciousness, which bears interesting relations to contemporary discussions of representational theories. But what is additionally remarkable is that Leibniz presents a naturalized theory of mind that is broadly idealist. The most fundamental elements of reality, for Leibniz, are "monads," which are minds or mind-like substances that are fully representational. Other features of nature, such as inter-substantial causal relations, are explained in terms of representational relations among these mind-like substances. And so, we have another example of a naturalized, non-physicalist theory of consciousness.

4 Kantian Consciousness

Kant famously introduced a systematic division in philosophy, a result of what he calls a new Copernican Revolution. In astronomy, Copernicus's great insight was that we should factor into our astronomical calculations how the movement of the earth affects our observations. Kant had a similar insight. Metaphysics had sought to describe the world *as it really is,* and the project consistently hit dead ends. And so, Kant proposed a new Copernican Revolution: in order to make sense of our observations of the world, we have to factor in what *we* contribute to our knowledge of things.

At its most basic, Kant's system is a philosophy of mind: what are the features of our own minds that enable us to experience the world? Kant argued that our minds actively structure our experiences so that things can become objects of experience *for us*. That is, when we are affected by something, our minds structure the experience. But Kant went beyond what we might ordinarily think—for example, we might think that certain subjective perspectives or points of view might distort our experience in some ways, but in general we are able to experience things as they are. Kant argues for a more radical conclusion: space and time are themselves the basic ways our minds structure and organize our experiences, which allows us to experience things coherently, connected, and causally related. And so, objects of our experience are in space and time because our minds *must* structure things according to the forms of space and time. But the things *as they are in themselves* are not spatiotemporally structured. This creates a division in Kant's philosophy between phenomena, objects of our experience, and noumena, things as they are in themselves. Kant then claims that we can *know* phenomena, but we cannot know noumena, things as they are in themselves, since they can never be objects of our experience. There is more to the story, but this is enough background for us to see what is at issue in Kant's philosophy of mind.

One consequence of Kant's division is that now the meaning of the word "nature" is put into question. Kant argues that "if nature meant the existence of things *in themselves*, we would never be able to cognize it" (Kant 2004: 46). That is, since we don't have cognitive access to things as they are *in themselves,* we would never be able to know anything about nature in this sense. But, he says, nature has "yet another meaning, namely one that determines the *object*," and so nature in this sense is "the *sum total of all objects of experience*" (Kant 2004: 47–48). That is, when we seek knowledge of an object, it will always be knowledge *as* a possible object of experience. Judgments based on this condition are objectively valid, since we are identifying the necessary conditions by which the things become *objects of experience*. But we should not be confused and regard these objectively valid judgments to describe things *as they really are*, independent of the conditions of experience.

And so, to *naturalize* a theory for Kant requires paying attention to the conditions of experience, in order to determine the necessary conditions and relations that objects must have. And this will be true of our own minds as well. When we do empirical psychology, we will be attending to the conditions under which we become objects of our own experience. Introspection becomes the basis of empirical psychology.[26]

One aspect of consciousness that precedes empirical psychology is, what Kant calls, the *transcendental unity of apperception* (borrowing Leibniz's word). What he argues is that there is a condition of *unity* that must be applied to consciousness in order for it to provide a single experience:

> This original and transcendental condition is nothing other than the *transcendental apperception*. The consciousness of oneself in accordance with the determinations of our state in internal perception is merely empirical, forever variable; it can provide no standing or abiding self in this stream of inner appearances, and is customarily called *inner sense* or *empirical apperception*... There must be a condition that precedes all experience and makes the latter itself possible.
>
> *(Kant 1997: A106–107)*

Thus, Kant gives us an argument for the unity of consciousness, a formal unity that is a condition for experience. But this is different from the "merely empirical" consciousness that yields inner sense. We should not confuse the formal condition of *unity* with a claim about what we *are* as minds, however. Kant says that "apart from this logical significance of the *I*, we have no acquaintance with the subject in itself..." (Kant 1997: A350). That is, the logical unity necessary for us to have experience at all does not give us cognition of our own mind *as it is in itself.* We are always an appearance, even to ourselves.

But, given this, we can still differentiate levels of consciousness *within* empirical experience for Kant. The transcendental unity of apperception is conceptually prior to nature (in Kant's sense), since it performs the synthesis that allows us to have experience of objects in the first place.[27] But the main function of empirical consciousness is to provide differentiation of objects, and he appeals to the relative clarity and distinctness of a perception to explain this differentiation, and so Kant follows a broadly Leibnizian analysis of consciousness in terms of the distinctness of a mental state.[28] With respect to empirical consciousness, since all objects of experience have been synthesized according to the forms of space and time, there will always be *some* differentiating factor among them, even if it is relatively obscure. But the relative distinctness of the mental state will allow Kant to differentiate low-level consciousness (obscure and indistinct) from higher degrees or levels of consciousness (which are more distinct).[29]

Thus, Kant gives us a robust *naturalized* theory, provided that nature is understood as within the domain of experience itself, but we must always acknowledge that such a theory is limited, remaining at the level of phenomena. Within phenomena, events are intelligible in terms of immanent laws and structures. But Kant doesn't provide a much more robust account of empirical consciousness, making use of some of the Cartesian and Leibnizian theories bequeathed to him. In general, the systematic attention to an analysis of consciousness will have to wait for another century or so.[30]

5 Naturalized Theories of Consciousness Today

None of the philosophers I have looked at provide physicalist theories of the mind, and yet, arguably, some of them do make attempts to *naturalize* the mind. One main project of much recent philosophy of mind has been to discover how the mind fits into a physical world, and so naturalism has been regarded as coextensive with physicalism. But I think this is a mistake. And this brief historical tour provides some examples of ways we can aim for the goals of naturalism without prejudging the debate between physicalism and its detractors.

It might turn out that the best naturalized theory of consciousness will also be a reductive or physicalist theory of consciousness. But it may not. Recently some have argued for naturalized versions of dualism (Gertler 2012) and panpsychism (Brogaard 2015). By distinguishing the aims of naturalism from those of physicalism, we may be able better to articulate what we want from a naturalized theory without presupposing the outcome.[31]

Notes

1 This is not to imply that the understanding of the naturalizing project from within each historical context was common. Rather, I mean to say that the proper contextual understanding of each philosopher yields a common thread that we can recognize as overlapping and forming historical precedents for later ways of thinking about the mind.

2 I say, "in principle if not in practice," since many of the natural causes are so complex that it is *practically* impossible to make a prediction if not *theoretically* impossible. Quantum mechanics is often mentioned in this context, raising the question of just how strong the intelligibility constraint ought to be. I don't have a fully formed answer to this, but if the natural sciences become as unpredictable as the fickle gods, then I am not sure any more what the project of naturalism will be. There does seem to be *some* condition of intelligibility required even in these cases, and interpretations of quantum theory seem to support this claim.

3 One historical model of this is the theory of Occasionalism, which explains all causal interactions in terms of fully consistent and unchanging divine activity. We can expect law-like regularity in causal interactions because God's activity is regular. See Adams (2013).

4 There are perhaps other ways to formulate the constraints of a naturalized theory, and indeed someone trying to articulate how naturalism is understood today would likely identify different constraints. However, I intend to identify constraints in a way that are sufficiently neutral to the theory that results from them. For example, I would not want to identify a constraint of intelligibility *in terms of common or universal natural laws*, since that is a modern concept and Aristotle would be rejected as providing naturalized theory from the outset.

5 These two constraints might create problems for a theory in which consciousness turns out to be a *basic* property of the mind. But intuitively this seems right. If consciousness turns out to be basic, then consciousness will be able to play a role in explaining *other* features of mentality, but it will not *itself* be explained. Someone might try to save naturalism here by positing it as a basic fact that consciousness is a property of the mind, in which case it satisfies the immanence constraint—there is no appeal to other things to explain the presence of consciousness besides the fundamental nature of the minds themselves. But this leaves open the question of intelligibility.

One prominent response of this sort is found in Descartes's reply to Princess Elisabeth of Bohemia. When Elisabeth asks how mind and body interact, Descartes appeals to a primitive notion of mind-body union, which is not explicable in terms of any more fundamental notions. Many historians of philosophy have found this rather *un*intelligible. And, as we will see below, Descartes hits other obstacles of this kind when he discusses consciousness. For the exchange between Descartes and Elisabeth, see Princess Elisabeth and Descartes (2007: 61–73), Garber (1983), and Yandell (1997).

6 For more discussion of how naturalism is used in today's context, see Carruthers (2000), De Caro (2010), Dretske (1997), and Horst (2009). As I mentioned, my own characterization here differs in important respects from the positions defended in these texts since my goal is to find the *core* of naturalism that would allow us to make an informed survey of historical theories.

7 For an exposition of the Greek and Latin lexical history, see Lewis (1960).

8 The *locus classicus* for this description of phenomenal consciousness is Nagel (1974).

9 See also Wilkes (1984: 242):

> I would point out that the Greeks, who by the fifth century BC had a rich, flexible and sophisticated psychological vocabulary, managed quite splendidly without anything approximating to our notion of 'consciousness'...

10 On this passage and its connection with the theory of recollection in the *Meno*, see Brancacci (2011).

11 However, even here it is difficult to say just how *un*conscious these ideas are since the *Charmides*, a dialogue about the nature of temperance, claims that if temperance really resides in you then it "provides a sense of its presence" (159a). And so, while the theory of recollection might imply unconscious ideas, it might also merely imply *obscured* but conscious ideas.

12 See Plato's discussion of the Protagorean claim that "all things are in motion" in *Theaetetus* 152c–d, 156a and 181d–183c; see also his discussion of material vs. psychological causes in *Phaedo* 97c–99b.

13 Thomas Johansen has provided a careful argument for this view (Johansen 2005).

14 For another close reading of *De Anima* 3.2 that does not entail a kind of "post-Cartesian, post-Kantian" self-consciousness, see Kosman (1975).

15 For more on Aristotle's philosophy of mind, see Irwin (1991) and Shields (2007: ch. 7).

16 Peter A. Morton also makes this claim, describing Aristotle's theory as a *naturalized* theory, meaning that Aristotle "constructs a theory wherein the soul is an integral part of the natural order of material objects, plants, and animals" (Morton 2010: 37).

17 The Oxford English Dictionary lists some early uses of the word, "consciousness," the first being in 1605, although these earlier uses still retained the sense of "conscience." But later, in 1678 and 1690, Cudworth and Locke use the term "consciousness" to refer to a more purely psychological capacity (OED 2017).

18 For a fuller story of what was going on in the seventeenth century, see Jorgensen (2014).

19 "It should be noted in passing that I do not deal at all with sin, i.e., the error which is committed in pursuing good and evil, but only with the error that occurs in distinguishing truth from falsehood" (Descartes 1985: 2.11).

20 Scholars have argued that there are other forms of consciousness in Descartes, but if this most basic form of consciousness cannot be made intelligible, then other forms will have similar problems. For more on consciousness in Descartes, see Lähteenmäki (2007), Radner (1988), and Simmons (2012).

21 For more on Locke's innovative use of the concept of consciousness but also some of its limitations, see Weinberg (2016) and Jorgensen (2016).

22 The extent to which Leibniz allowed for *any* miracles is controversial, although he does claim that a non-natural theory requires "perpetual miracles" to fill in the gaps in explanation. This is a charge Leibniz leveled at Descartes, Malebranche, and Isaac Newton. And so, even if Leibniz would grant an isolated miracle, it would not be a part of the natural theory to allow for this given that it would be an event that has its source *outside* of the natures of finite things.

23 For arguments in favor of the higher-order reading, see Gennaro (1999), Kulstad (1991), and Simmons (2001). For discussion of the criticism from the principle of continuity and some possible ways around this for the higher-order interpreters, see Jorgensen (2009).

24 For recent work on this controversy, see Bolton (2011), Jorgensen (2011a), Jorgensen (2011b), and Simmons (2011).

25 One might ask what explains the intentionality of thought. For Leibniz this was explained in causal terms, by the internal causes of each individual mind. That is, each mind's present perceptual state causes subsequent perceptual states, which have a complex structure that present to the mind similar structures external to the mind.

26 For more discussion of this, see Brook (2016).

27 See Kant (1997: B 414–415n), where Kant says that there are infinitely many degrees of consciousness down to its vanishing. But the way he characterizes consciousness suggests that the vanishing is simply a limit case of obscurity. Qtd. in Dyck (2011: 47).

28 Mediated by Christian Wolff and others. See Dyck (2011).

29 For illuminating discussions of Kant's theory of consciousness, see Dyck (2011) and Sturm and Wunderlich (2010).

30 For a sampling of the discussion of consciousness and unconscious thinking, see (Taylor and Shuttleworth 1998: Introduction and contents of Section II).

31 I would like to thank Rocco Gennaro and three of my students—Rachel Greene, Landon Miller, and Jonathan Stricker—for helpful comments on drafts of this essay.

References

Adams, R.M. (2013) "Malebranche's Causal Concepts," In E. Watkins (ed.) *The Divine Order, the Human Order, and the Order of Nature*, Oxford: Oxford University Press.

Bolton, M.B. (2011) "Leibniz's Theory of Cognition," In B. Look (ed.) *The Continuum Companion to Leibniz*, London: Continuum International Publishing Group.

Brancacci, A. (2011) "Consciousness and Recollection: From the Apology to Meno," In *Inner Life and Soul: Psyche in Plato.* Lecturae Platonis. Sankt Augustin: Academia.

Brogaard, B. (2015) "The Status of Consciousness in Nature," In S. Miller (ed.) *The Constitution of Phenomenal Consciousness: Toward a Science and Theory*, Amsterdam: John Benjamins.

Brook, A. (2016) "Kant's View of the Mind and Consciousness of Self," *The Stanford Encyclopedia of Philosophy*. https://plato.stanford.edu/archives/win2016/entries/kant-mind/.

Carruthers, P. (2000) *Phenomenal Consciousness: A Naturalistic Theory*, Cambridge: Cambridge University Press.

Caston, V. (2002) "Aristotle on Consciousness," *Mind: A Quarterly Review of Philosophy* 111: 751–815.

De Caro, M. (2010) "Varieties of Naturalism," In R. Koons and G. Bealer (eds.) *The Waning of Materialism*, Oxford: Oxford University Press.

Descartes, R. (1985) *The Philosophical Writings of Descartes*, translated by John Cottingham, Robert Stoothoff, Dugald Murdoch and Anthony Kenny (vol. 3). 3 vols., Cambridge: Cambridge University Press.

Dretske, F. (1997) *Naturalizing the Mind*, Cambridge, MA: MIT Press.

Dyck, C. (2011) "A Wolff in Kant's Clothing: Christian Wolff's Influence on Kant's Accounts of Consciousness, Self-Consciousness, and Psychology," *Philosophy Compass* 6: 44–53.

Garber, D. (1983) "Understanding Interaction: What Descartes Should Have Told Elisabeth," *Southern Journal of Philosophy* 21, no. SUPP: 15–32.

Gennaro, R.J. (1999) "Leibniz on Consciousness and Self-Consciousness," In R. Gennaro and C. Huenemann (eds.) *New Essays on the Rationalists*, Oxford: Oxford University Press.

Gertler, B. (2012) "In Defense of Mind-Body Dualism," In T. Alter and R. Howell (eds.) *Consciousness and the Mind-Body Problem*, Oxford: Oxford University Press.

Hamlyn, D.W. (1993) *Aristotle: De Anima Books II and III.*

Horst, S. (2009) "Naturalisms in Philosophy of Mind," *Philosophy Compass* 4: 219–254.

Irwin, T.H. (1991) "Aristotle's Philosophy of Mind," In S. Everson (ed.) *Companions to Ancient Thought 2: Psychology*, Cambridge: Cambridge University Press.

Johansen, T. (2005) "In Defense of Inner Sense: Aristotle on Perceiving That One Sees," *Proceedings of the Boston Area Colloquium in Ancient Philosophy* 21: 235–276.

Jorgensen, L.M. (2009) "The Principle of Continuity and Leibniz's Theory of Consciousness," *Journal of the History of Philosophy* 47: 223–248.

Jorgensen, L.M. (2011a) "Leibniz on Memory and Consciousness," *British Journal for the History of Philosophy* 19: 887–916.

Jorgensen, L.M. (2011b) "Mind the Gap: Reflection and Consciousness in Leibniz," *Studia Leibnitiana* 43: 179–195.

Jorgensen, L.M. (2014) "Seventeenth-Century Theories of Consciousness," *Stanford Encyclopedia of Philosophy*. https://plato.stanford.edu/archives/win2014/entries/consciousness-17th/.

Jorgensen, L.M. (2016) "Review of Shelley Weinberg's *Consciousness in Locke*." *Notre Dame Philosophical Reviews*. http://ndpr.nd.edu/news/68666-consciousness-in-locke/.

Kant, I. (1997) *Critique of Pure Reason*, transl. and ed. by Paul Guyer and Allen W. Wood. Cambridge: Cambridge University Press.

Kant, I. (2004) *Prolegomena to Any Future Metaphysics*, Edited by Gary Hatfield. Revised ed. Cambridge: Cambridge University Press.
Kosman, L.A. (1975) "Perceiving That We Perceive 'on the Soul' III,2," *Philosophical Review* 84: 499–519.
Kulstad, M. (1991) *Leibniz on Apperception, Consciousness, and Reflection*, Munich: Philosophia Verlag.
Lähteenmäki, V. (2007) "Orders of Consciousness and Forms of Reflexivity in Descartes," In S. Heinämaa, V. Lähteenmäki and P. Remes (eds.) *Consciousness: From Perception to Reflection in the History of Philosophy*, Dordrecht: Springer.
Leibniz, G.W. (1969) *Philosophical Papers and Letters*, Translated by Leroy E. Loemker. 2nd ed. Boston: D. Reidel.
Leibniz, G.W. (1973) *Philosophical Writings*, Translated by Mary Morris and G.H.R. Parkinson. 2nd ed. London: J.M. Dent & Sons.
Leibniz, G.W. (1989) *Philosophical Essays*, Translated by Roger Ariew and Daniel Garber. Indianapolis, IN: Hackett Publ.
Leibniz, G.W. (1996) *New Essays on Human Understanding*, Translated by Peter Remnant and Jonathan Bennett. Cambridge: Cambridge University Press.
Leibniz, G.W., and Arnauld, A. (1967) *The Leibniz-Arnauld Correspondence*, Translated by H.T. Mason. Manchester: Manchester University Press.
Leibniz, G.W., and Clarke, S. (1956) *The Leibniz-Clarke Correspondence*, Edited by H.G. Alexander Manchester: Manchester University Press.
Lewis, C.S. (1960) "Conscience and Conscious," In *Studies in Words*, Cambridge: Cambridge University Press.
Locke, J. (1975) *An Essay Concerning Human Understanding*, Edited by Peter H. Nidditch. Oxford: Clarendon Press.
Malebranche, N. (1997) *The Search after Truth*, Edited by Thomas M. Lennon and Paul J. Olscamp. Cambridge: Cambridge University Press.
Morton, P.A. (2010) *A Historical Introduction to the Philosophy of Mind*, 2nd ed. Ontario, Canada: Broadview Press.
Nagel, T. (1974) "What Is It Like to Be a Bat?," *Philosophical Review* 83: 435-450.
Plato (1997) *Complete Works*, John M. Cooper (ed.), Indianapolis, IN: Hackett Publishing Co.
Princess Elisabeth and R. Descartes. (2007) *The Correspondence between Princess Elisabeth of Bohemia and René Descartes*, translated by Lisa Shapiro. Chicago: University of Chicago Press.
Radner, D. (1988) "Thought and Consciousness in Descartes," *Journal of the History of Philosophy* 36: 439–452.
Shields, C. (2007) *Aristotle*, New York: Routledge.
Simmons, A. (2001) "Changing the Cartesian Mind: Leibniz on Sensation, Representation, and Consciousness," *Philosophical Review* 101: 31–75.
Simmons, A. (2011) "Leibnizian Consciousness Reconsidered," *Studia Leibnitiana* 43: 196–215.
Simmons, A. (2012) "Cartesian Consciousness Reconsidered," *Philosophers' Imprint* 12: 1–21.
Sturm, T., and F. Wunderlich. (2010) "Kant and the Scientific Study of Consciousness," *History of the Human Sciences* 23: 48–71.
Taylor, J.B., and Shuttleworth, S. (eds.) (1998) *Embodied Selves: An Anthology of Psychological Texts 1830–1890*, Oxford: Clarendon Press.
Weinberg, S. (2016) *Consciousness in Locke*, Oxford: Oxford University Press.
Wilkes, K.V. (1984) "Is Consciousness Important?" *British Journal for the Philosophy of Science* 35: 223–243.
Yandell, D. (1997) "What Descartes Really Told Elisabeth: Mind-Body Union as a Primitive Notion," *British Journal for the History of Philosophy* 5: 249–273.

Related Topics

Materialism
Dualism
Idealism, Panpsychism, and Emergentism
Representational Theories of Consciousness
Consciousness and Intentionality
The Unity of Consciousness

3
MATERIALISM

Janet Levin

1 Introduction

We humans have a great variety of conscious experiences: seeing the colors of the sunset, hearing thunder, feeling pain, tasting vegemite, hallucinating a dagger, or being in altered states of consciousness that are far from routine. It's hard to doubt, moreover, that many non-human animals have a variety of conscious experiences—some familiar, and some (e.g. the perceptual experiences of bats and octopuses) radically unlike any of our own. Nevertheless, there is a common feature, shared by all these states, that is essential to their being conscious experiences: they have a certain feel, or qualitative character; *there is something that it's like* to have them. Moreover, the distinctive *what* it's like to be in pain or hallucinate a dagger seems essential to their being conscious experiences of that type: one cannot be in pain, or hallucinate a dagger, unless one has an experience with a particular type of qualitative character, or feel.

Given this characterization of conscious experiences, the question naturally arises: what kinds of things could conscious experiences be, and what is their relation to the *physical* states and processes that occur in bodies and brains? One answer to this question, most closely associated with Descartes (1641), is that the locus of one's conscious experiences (and conscious thoughts) is an immaterial substance—a mind or (equivalently) a soul—that is distinct from, but able to interact with, bodies. A related view, held primarily by more contemporary theorists, is that while conscious mental states are states of the brain and body, their "feels" or qualitative features are special, non-physical, properties of those states. Both views are species of Dualism, the thesis that, in one way or another, the mental is distinct from the physical.

Dualism effectively captures the intuition that the qualitative features of conscious states and processes are radically unlike, and impossible to be explained by, any properties that occur elsewhere in the physical world, including neural processes such as the release of neurotransmitters, or the synchronized firing of certain neurons in the brain. As T.H. Huxley, a 19th-century Dualist, dramatically puts it (1881): "How it is that anything as remarkable as a state of consciousness comes about as a result of irritating nerve tissue, is just as unaccountable as the appearance of the Djin, where Aladdin rubbed his lamp in the story."

Almost as dramatically, G.W. Leibniz (1714) expresses a similar worry about any materialistic explanation of perception:

> If we imagine that there is a machine whose structure makes it think, sense, and have perceptions, we could conceive it enlarged, keeping the same proportions, so that we could enter into it, as one enters a mill. Assuming that, when inspecting its interior, we will find only parts that push one another, and we will never find anything to explain a perception.

Nevertheless (and this is acknowledged even by its sympathizers), if Dualism were true, it would be hard to explain the occurrence of mental-physical *causation*. For example, I put my hand on the hot stove, I feel pain, I say "ouch". This seems to involve a familiar causal sequence from physical to mental and then again to physical events, but it is hard to explain how a physical event could have effects on something non-physical—and even more seriously, how a non-physical state or event could have any sort of effect in the physical realm, given that we accept that every physical change that occurs in the world has a sufficient physical cause. There has been concern about mental-physical causation ever since Princess Elisabeth of Bohemia posed the question to Descartes in their (1643/1985) correspondence, and it has never been given a fully satisfactory answer.

Another serious question for Dualism concerns where, and how, consciousness arises on the phylogenetic spectrum in an otherwise physical world. Finally, Dualism raises epistemological worries: if conscious mental states or their qualitative properties are not physical, then they do not exist in space and cannot be perceived by anyone other than the subject who has them. But this means that we have no direct evidence that anyone other than ourselves ever sees the colors of the sunset, or feels pain, or for that matter has any conscious mental states at all—and, in addition, that scientists investigating the role of conscious mental states in the production of behavior have no way to determine which states are occurring in their subjects (if any) other than the introspective reports of those subjects themselves.

In contrast, the thesis of Materialism (often called "Physicalism") maintains that there is nothing required for having conscious mental states and processes besides the occurrence of various types of physical states in the conscious creature's brain and body (and perhaps in the world around it). It is easy to see why Materialism, in general, is an attractive view. If conscious mental states and processes can be fully characterized as various sorts of physical states and processes, then there is no need to explain how (or why) non-physical features arise in the natural world, and how they could be genuine causes of behavior. Materialism therefore seems to be a simpler and more economical theory than Dualism. In addition, if conscious mental states and their qualitative features are physical, then it is possible in principle for them to be observed by others.

On the other hand, there are well-known arguments, both classical and contemporary, that purport to show that *no* materialistic theory could provide an adequate account of the qualitative character of conscious experience, of *what it's like* to see red or feel pain. Thus, although Materialism may seem to have promise for integrating mental states into the physical world, and connecting the study of mental states to the sciences of chemistry, biology, and neurophysiology, many contend that this cannot be done.

The primary goal of this chapter is to explore the prospects for a materialistic theory of conscious mental states and processes—or, more precisely, the prospects for a number of different materialistic theories that began to be proposed during the beginning of the 20th century—in particular, Behaviorism, the Type-Identity Theory, Functionalism, and (in passing) other versions of what has come to be known as Non-Reductive Physicalism.

This chapter will focus on the strengths and weaknesses of each of these theories—while considering whether or not any of them could explain how, as Huxley puts it, "anything as remarkable as a state of consciousness comes about as a result of irritating nerve tissue." It will also explore the viability of Eliminativism, the thesis that despite popular belief and the deliverances of introspection, our bodies and brains have no real and robust qualitative features at all.

Contemporary Materialism has antecedents in both the Classical and Modern periods. Leucippus (5th century BCE) and his student Democritus—and later Epicurus (341–270 BCE) and Lucretius (d.c. 50 BCE) all contend that everything that exists in the world can be explained as configurations of, and interactions among, atoms in the void. In the Modern period, Descartes's contemporary, Hobbes (1668/1994), and later La Mettrie (1747/1994), articulate what can be identified as materialistic theories of mental states. However, because the current debates about the pros and cons of Materialism focus primarily on the more contemporary versions of the doctrine, they will be the topics of discussion here.

2 Behaviorism

Behaviorism achieved prominence in the early to mid-20th century, both as a scientific theory of behavior (associated primarily with Watson, 1930, and Skinner, 1953) and as a philosophical theory of the meanings of our mental state terms or concepts. According to scientific behaviorism, the best explanation of human (and animal) behavior appeals not to a subject's internal mental states, but rather to its *behavioral dispositions*—that is, its tendencies to behave in certain specified ways given certain environmental stimulations, which are shaped by the contingencies of its past interactions with the environment. A major attraction of scientific behaviorism is its promise to explain behavior by appeal to states and processes that are indisputably physical, and also intersubjectively observable, rather than accessible (by introspection) only to the subjects of those mental states themselves.

In contrast, philosophical (or logical) behaviorism, associated primarily with Malcolm (1968), Ryle (1949), and more contentiously, Wittgenstein (1953), is not a scientific thesis subject to empirical disconfirmation, but rather the product of conceptual analysis. According to logical behaviorism, reflection on our mental state terms or concepts suggests that our ordinary claims about mental states and processes can be translated, preserving meaning, into statements about behavioral dispositions. For (an overly simplified) example, "S believes that it is raining" would be equivalent to "If S were to leave the house, S would take an umbrella, and if S had been heading to the car wash, S would turn around," and "R is thirsty" would be equivalent to "If R were offered some water, then R would drink it quickly."

However, as many philosophers have argued (see Chisholm 1957, Putnam 1968), statements about behavioral dispositions are unlikely to provide adequate translations of our claims about mental states, since, intuitively, a subject can have the mental states in question without the relevant behavioral dispositions—and vice versa—if they have other mental states of various sorts. For example, S could believe that it's going to rain, and avoid taking an umbrella when leaving the house if S *enjoys getting wet*—and S may take an umbrella, even if she does *not* believe it will rain, if she superstitiously believes that carrying an umbrella will prevent it from raining (or wants to assure her mother that she is planning for all contingencies). In short, the arguments continue, it is impossible to specify a subject's mental states as pure behavioral dispositions; they can only be specified as dispositions to behave in certain ways, given the presence or absence of *other* mental states and processes.

Similar worries have been raised (perhaps most influentially by Chomsky 1959) about the explanatory prospects of scientific behaviorism. Although scientific behaviorism had (and continues to have) some success in explaining certain types of learning, these successes, arguably, depend on the implicit control of certain variables: experimenters implicitly assume, usually correctly, that (human) subjects *want* to cooperate with them, and *understand* and *know how* to follow the instructions; in the absence of these controls, it is unclear that the subjects would be disposed to behave in the ways that they do. It seemed to the critics of behaviorism, therefore, that theories that explicitly take account of an organism's beliefs, desires, and other mental states, as well as stimulations and behavior, would provide a fuller and more accurate account of why organisms behave as they do.

In addition, it seems that both experimental practice and conceptual analysis suggest that mental states are genuine *causes* of behavior: when I put my hand on a hot stove, feel pain, and say "ouch", my saying "ouch" is not a manifestation of a behavioral disposition, but rather an event produced by my feeling pain. Therefore, despite its attractions, most philosophers and psychologists have abandoned behaviorism and attempt to provide other materialistic theories of conscious mental states and processes. One such theory is the Type-Identity Theory, another is Functionalism; these will be the topics of the next two sections.

3 The Type-Identity Theory

The Type-Identity Theory, first articulated by U.T. Place (1956), H. Feigl (1958), and J.J.C. Smart (1959; also see his 2007), contends that for each type of mental state or process M, there is a type of brain state or process B, such that M is identical with B. For example, *pain* is identical with *C-fiber stimulation*. These claims are to be understood as property identities: *being a state of Type M* is just *being a state of Type B*—which entails that every instance of an M is a B, and vice versa. Thus, for the Type-Identity Theory to be true, there must be (at minimum) a *correlation* between instances of mental Type M (determined by the introspective reports of the individuals who are in them)—and physical Type B (determined by instruments such as brain scans).

Place, Feigl, Smart, and other early Type-Identity theorists recognized that the science of the time was nowhere near discovering any such universal correlations, but they were most concerned to establish, against intuitions (and arguments) to the contrary, that mental state–brain state identities are *possible*; that there are no logical or conceptual reasons to think that they could not be true. If these identities are possible, they argued, and if there are in fact correlations between instances of mental and physical states, then identity theorists could argue that the simplest and most economical explanation of these correlations—and the one that avoids the other difficulties of Dualism—is that the correlated mental and physical properties are identical.

Early identity theorists suggested that many objections to the possibility of mental-physical identities arise from the mistaken assumption that if mental-physical identity statements are true, then they should be knowable *a priori*; that is, solely by reason and reflection, without need for empirical investigation. They went on, however, to challenge this assumption, and to liken statements such as "Pain is C-fiber stimulation" to scientific identity statements such as "Lightning is electrical discharge" or "Water is H_2O"—statements that we believe to be true, but that can be known only *a posteriori*; only by appeal to observations of the world as it is.

However, early identity theories also faced another important objection, the "Distinct Property Objection", articulated by Smart (1959), namely, that the only way that an *a posteriori* identity statement $A = B$ can be true is for both A and B to pick out their common referent by means of logically distinct (that is, conceptually unconnected) properties, or "modes of presentation," of that object that entail, respectively, its being an A and its being a B. For example, "water" picks out its referent as *the colorless odorless liquid that comes out of our faucets*; "H_2O" picks out its referent as *the compound of two hydrogen atoms and one oxygen atom*—and if, in fact, it turns out that the colorless odorless stuff that comes out of our faucets is composed of that compound of hydrogen and oxygen atoms, then we have an explanation of how "water is H_2O," though *a posteriori*, can be true.

However, the objection continues, in the case of mental-physical identities, the only sorts of properties that could entail being a *conscious mental* state of the relevant type (e.g. a pain, or an experience of a sunset) are *qualitative* properties (e.g. feeling a certain distinctive way, or being qualitatively reddish-orange). But then one can establish the identity of mental and physical states or processes only by attributing an irreducibly qualitative property to that state or process—and so one has not established a purely materialistic theory of conscious mental states.

Smart's solution is to argue that mental state terms can be translated, preserving meaning, into "topic-neutral" terms, that is, terms that describe certain properties or relations that can be satisfied by *either* mental *or* physical states, processes, or events. He suggests, for example, that "I see a yellowish-orange after-image" can be translated into "There is something going on [in me] which is like what goes on when I have my eyes open, am awake, and there is an [unripe] orange illuminated in good light in front of me." This term picks out a relational property that is "logically distinct" from any physical (or mental) property, and—if there really is a meaning equivalence between mental and topic-neutral terms—a state's having that topic-neutral property will indeed entail its being a mental state of the relevant sort.

This particular suggestion for a topic-neutral translation, however, is generally regarded as unsatisfactory, since such topic-neutral terms are not sufficiently specific to serve as translations of our ordinary mental state terms. After all, many different mental states can be *like*, in some way or another, what goes on in me when I'm looking at an unripe orange; I could be having an after-image of a banana, or a perception of a faded basketball—or the thought that the orange juice I'm about to make for breakfast will be sour. One needs to say more about the *way* in which my having an experience is like what goes on when I'm seeing an unripe orange, and—as many have argued—it's unclear that the relevant sort of resemblance can be specified in topic-neutral terms.

However, other Type-Identity theorists have attempted, with greater success, to provide topic-neutral equivalents of our ordinary mental state vocabulary; for example, David Armstrong (1981) attempts to characterize mental states in terms of their "aptness" to cause certain sorts of behavior. The most developed account of this sort is David Lewis's (1966) suggestion that topic-neutral translations of our mental state terms can be extracted from our "common sense theory" of the mind, which can be understood to define mental states "all at once" by specifying (what we commonly believe to be) their causal interactions with environmental stimulations, behavior, and one another. For (an overly simplified) example:

> *Pain* is the state that tends to be caused by bodily injury, to produce the belief that something is wrong with the body and the desire to be out of that state, to produce anxiety, and, in the absence of any stronger, conflicting desires, to cause wincing or moaning.

This way of characterizing mental states and processes is often called a *functional specification*, since it specifies the way these states, together, function to produce behavior. If this specification indeed provides a translation (or close enough) of "pain," and if it is uniquely satisfied by C-fiber stimulation, then "pain = C-fiber stimulation" is true—and so on for other mental-physical identity statements. Moreover, Lewis explicitly argues, it would thereby be unnecessary to invoke simplicity or economy to establish the Type-Identity Theory: if these causal-relational descriptions indeed capture the meanings of our mental state terms, then any brain states that (uniquely) satisfy those descriptions will automatically be instances of those mental states.

Not surprisingly, there is skepticism about whether these sorts of "common sense" functional specifications can provide logically necessary and sufficient conditions for the occurrence of conscious mental states. Isn't it possible, many ask, for a creature to satisfy such a specification, but not feel pain—or indeed not have any conscious mental states at all—or, conversely, for a creature to be in pain without satisfying the common sense specification? These questions are similar to the classic objections to logical behaviorism, and will be discussed further in Section 5. However, there is another worry about the Type-Identity Theory put forward by materialists themselves that needs to be addressed, namely, that it is too restrictive, or "chauvinistic," in that it restricts the range of those who can possess mental states to humans, or at least mammals with neural structures similar to our own.

After all, it seems that there could be creatures that respond to the environment much like humans—who cry out when they're injured, and report feeling pain or hearing thunder in the same circumstances as we do, and whose other mental states interact with one another and with environmental stimulations like our own—but whose internal states are physically quite different from ours. Presumably, some argue, certain non-human animals (perhaps dolphins or octopuses) are like this, and certainly we can imagine silicon-based life forms with different types of physical states that satisfy the same functional specification as ours (think of androids, familiar from science fiction). But if some sort of experiential-neural identity thesis is true, then we could not consider these creatures to share our conscious mental states.

This worry has motivated some materialists to propose a related theory of what it is for someone to be in a particular type of mental state: Role Functionalism, or the Functional State Identity Theory. This theory will be addressed in the next section.

4 Role Functionalism

Consider (a fragment of) the functional specification presented earlier as a topic-neutral characterization of pain, namely, "*Pain* is the state that tends to be caused by bodily injury… and, in the absence of any stronger, conflicting desires, to cause wincing or moaning." This specification depicts the causal role of pain in our so-called "common sense theory" of the mind, and may be satisfied, in humans, by C-fiber stimulation, and by different types of physical states in other, non-human, creatures. However, an alternative to maintaining that these other creatures are not in the same type of state as we are—or that pain is the *disjunctive* property that comprises whichever states satisfy the functional specification in different creatures—is to contend that pain is not to be identified with any particular type (or disjunction of types) of physical states that satisfy that description (or occupy that causal role), but rather with that causal role property itself.

Role Functionalism, that is, maintains that S is in pain just in case S is in the (higher-order) state of being in *one or another first-order state* that plays the causal role specified by the relevant functional description. Pain itself is to be identified with that higher-order state; those first-order states that occupy that role in some creature (e.g. C-fiber stimulation) are said to *realize* that state, and if different types of states can occupy the "pain role" in different creatures, pain is said to be *multiply realized*.

A major attraction of Role Functionalism, in contrast to the Type-Identity Theory, is that it permits humans, octopuses, silicon-based creatures—and even the non-biological but human-like androids familiar from science fiction—to count, literally, as being in the same mental state, as long as their first-order internal states occupy the same causal roles. Role Functionalism would thereby avoid the (alleged) human chauvinism of the Type-Identity Theory, although it would be compatible with a "token" identity theory, in which each *instance* (or token) of a mental state of some type (e.g. pain) is identical with an instance (token) of some type of physical state or other.

Role Functionalism, it should be noted, comes in two versions: one that derives from our "common sense" theory of the causal roles of mental states, and another (often called *Psychofunctionalism*; see Block 1980) that derives from empirical theories, developed by experimental psychologists and cognitive scientists, that include generalizations that may depart from the deliverances of common sense. Psychofunctionalist theories can provide more precise and detailed characterizations of mental states than our commonsense theories, which makes them less likely to be satisfied by systems (such as the economy of Bolivia; see Block 1980) that do not seem to have mental states at all. On the other hand, while psychofunctional characterizations can be topic-neutral, if specified solely in causal and relational language, they may not provide *translations*, however loose, of our mental state terms. Therefore the resulting identity statements linking mental and functional states will have no claim to being *a priori*, and thus may be subject to the "Distinct Property Objection." Whether or not these identity statements—or any mental-physical identity statements—need to be *a priori* to avoid Dualism will be discussed later (in Section 5), but there is a further worry about Role Functionalism that threatens both versions of the view.

The worry is that Role Functionalism (like property Dualism) cannot account for the causal efficacy of mental states. Once again, it seems that if I put my hand on a hot stove, feel pain, and then say "ouch," my feeling pain causes my saying "ouch." However, if every physical event has a complete, sufficient physical cause, then my saying "ouch" will be caused by the physical, presumably neural, state that satisfies the functional specification of (or "realizes") pain. But then my being in pain, if this is identified with a higher-order functional state, seems causally irrelevant. This is regarded as a problem not only for Role Functionalism (and property Dualism), but also for any materialistic view that treats the relation between mental and physical states as anything other than identity—for example, the view (Pereboom 2011) that mental states are *constituted by* physical states (in just the way that, as some suggest, a statue is constituted by, but not identical with, the material from which it is made).

Many Role Functionalists, in response, argue that this worry arises from the assumption that a genuine cause must "generate" or "produce" its effect, where this involves some sort of transfer of energy. However, they continue, this is not the only way to think about causation. Instead, causation should be regarded as a special sort of counterfactual dependence between effects and their causes (Loewer 2002), or as a special sort of regularity that holds between them (Melnyk 2003). If this is correct, then functional role properties and the physical events or states that realize them could *both* count as causally efficacious.

To be sure, property dualists could avail themselves of this defense as well. However, there is a further worry about causation (articulated by Kim 1989, 1998) that may differentiate the views, namely, that if mental and physical events (or properties) are both causally sufficient for producing behavior, then any behavior that has a mental cause would be causally *overdetermined*; that is, there would be more than one event that could have caused it by itself. But overdetermination occurs elsewhere in the world only rarely—for example, when two individuals simultaneously hit a window with a hardball, each with enough force to break it (or when more than one member of a firing squad hits the victim with lethal force)—and so it is counterintuitive to suggest that this is a routine and widespread occurrence in the causation of behavior.

One response to this worry (developed in different ways by Yablo 1992 and Bennett 2008) is to argue that the causation of behavior by a lower-level neural state and a functional role state does not fit the profile of classic overdetermination because lower-level neural states *necessitate* the functional states they realize; that is, if N is a realization of R, then necessarily, if some individual were to be in state N, then that individual would be in state R. If this is so, there is an explanation for the ubiquity of the production of behavior by both a mental and physical cause.

This response is available to Role Functionalists and other non-reductive physicalists such as those who maintain that mental states are constituted by physical states of various types. But this response would not be available to property dualists, who (usually) maintain that there is no necessary connection between mental and physical properties. Nevertheless, this response remains controversial—and thus the question of whether mental causation provides an insurmountable problem for Role Functionalism (or any materialistic theory other than the Type-Identity Theory) remains a matter of debate.

There are other recent theorists (Bechtel and Mundale 1999; Polger and Shapiro 2016) who attempt to "split the difference" between Type-Identity and Functionalism by arguing that Type-Identity Theory can achieve nearly as much universality as Role Functionalism, at least in its characterization of the mental states of actual existing creatures. These theorists argue, first, that a closer look at the functional organization of humans and other species such as dolphins and octopuses reveals that there is less functional similarity between these species and ourselves as philosophers once assumed. In addition, they continue, a closer look at the way neural states and processes are individuated in practice by neuroscientists shows that the neural states of different species that initially may seem to be quite different have certain properties in common that are more abstract or general—but are still decidedly physical, rather than functional. If this is so, then the Type-Identity Theory would allow for a greater range of creatures that could share the same mental states—but it still would not (presumably) include silicon-based life forms, or non-biological androids, as creatures capable of having mental states like our own. It remains a controversial issue among materialists whether an adequate theory needs to account for such creatures—and thus there is no consensus about which theory is most promising.

Moreover, as noted in Section 1, there are some well-known arguments directed against *all* materialistic theories of conscious mental states that must be considered in evaluating the pros and cons of Materialism. These arguments purport to show that no materialistic theory, no matter how detailed and comprehensive in specifying the internal structure of our physical states *and* their causal and other topic-neutral relations, can provide an adequate account of the qualitative character of conscious experience, of *what it's like* to see red, feel pain, or be in any other kind of conscious mental state. The best-known contemporary arguments against all forms of Materialism are the so-called Zombie Argument, presented by David Chalmers (1996, 2010), and the Knowledge Argument, presented by Frank Jackson (1982). (See Kripke 1980, Block 1980, and Searle 1980 for arguments similar to the Zombie Argument, and Nagel 1974 for an argument similar to the Knowledge Argument.) These arguments will be addressed in the next section.

5 General Arguments against Materialism

In the Zombie Argument, Chalmers defines a zombie as a molecule for molecule duplicate of a conscious human being—that is, a creature that is exactly like us both physically and functionally—but which has no conscious mental states whatever: there is *nothing that it's like* to be a zombie. He then argues as follows:

1 Zombies are conceivable.
2 If zombies are conceivable, then zombies are genuinely possible.
3 If zombies are genuinely possible, then Materialism is false.
(C) *Therefore*, Materialism is false.

The general idea behind Premise (1) is that we can think of a body in all its physical (and functional) detail—and think about what it's like to be in a conscious state in all its *qualitative* detail—and see no connection whatsoever between the two. The general idea behind Premise (2) is that such a radical disconnect between our conceptions of the physical and the qualitative is evidence that physical (including functional) and qualitative states and properties must be radically different types of things—and this is because what we can (carefully) conceive to be possible or impossible is our only source of knowledge about possibility and necessity; about what can, or cannot, be.

The Knowledge Argument, although superficially different, relies on similar ideas. Jackson describes a brilliant neuroscientist, Mary, who has been born and raised in a black-and-white room, but has nevertheless managed to learn all the physical and functional facts about human color experience via achromatic textbooks and videos. However, Jackson continues, it seems clear that if she were released from her room and presented with a ripe strawberry, she would be surprised by her experience and consider herself to have learned something new, namely, *what it's like* to see red. Jackson then argues as follows:

1 Mary knows all the physical and functional facts about human color experience while still in her black-and-white room, but does not know what it's like to see red (since she learns this only when she actually experiences red).
2 If Mary knows all the physical and functional facts about human color experience before leaving the black-and-white room, but does not know what it's like to see red, then there is a fact about human color experience that is not a physical or functional fact.
3 If there is a fact about human color experience that is not a physical or functional fact, then Materialism is false.
(C) *Therefore*, Materialism is false.

Here too (Premise 1) the contention is that no amount of knowledge of the physical (and functional) features of the brains of those who are seeing colors could provide knowledge about the qualitative features of color experiences (and by analogy any type of state that there is *something it is like* to be in) and (Premise 2) that this lack of connection entails that there is something about these qualitative features that is different from anything physical (or functional).

To challenge these arguments, some materialists (e.g. Dennett 1988; Van Gulick 1993) and later Jackson himself, who (2004) eventually rejects the Knowledge Argument and its relatives, challenge Premise (1) of these arguments. They argue that although it may initially seem plausible that we can conceive of a zombie, on second thought this should seem implausible, since doing so would require that we have in mind, and be able to attend to, all the details of the physical structure and functional organization of our molecular duplicates, which is exceedingly hard to do. If we could do this, however, then we would recognize that such creatures were indeed having conscious mental states with qualitative properties just like our own. Similarly, they suggest, if Mary could internalize and concentrate sufficiently on all her physical knowledge about color experiences while still in her black-and-white room, then she would be able to know what it's like to have those experiences before she actually sees colors. These views maintain that there is an *a priori* link between our concepts of the qualitative and the physical (or functional), even though it may be difficult to discern. Chalmers (2002b) calls this *Type A Materialism*. He also discusses a related view—called *Type C Materialism*—which maintains that there are *a priori* connections between the qualitative and physical-functional features of our experiences, but that we haven't yet, or (McGinn, 1989) because of certain inescapable conceptual limitations *cannot*, form the concepts that are required to see that this is so.

However, many theorists—both dualist and materialist—(e.g. Chalmers 2002b; Stoljar 2001; Alter 2016) remain skeptical, and contend that learning, internalizing, and attending to more physical and functional information about our brains and bodies could not possibly provide knowledge of what it's like to feel pain, see red or have any other sort of conscious mental state. The reason, they argue, is that physical and functional descriptions provide information solely about the "structure and dynamics" of what goes on in our brains and bodies, and these are all *relational* properties, whereas the distinctive qualitative features of our conscious mental states—as we can tell from introspection—are *intrinsic* properties.

Some Type A materialists question whether introspection reveals that the distinctive qualitative properties of conscious mental states are exclusively intrinsic—after all, they ask, would we really count an experience as *pain* if we didn't experience it as something we want to get rid of? And would we really count an experience as a *yellow-orange after-image* if we didn't experience its qualitative features as fading in certain ways over time, and being similar to and different from other color experiences? In short, they argue that the claim that the qualitative properties of experience are intrinsic is itself a product of inattentive (or biased) introspection.

However, there are other materialists—in Chalmers's locution, *Type B Materialists* (e.g. Loar 1997; Hill and McLaughlin 1999; Papineau 2002; Levin 2007; Balog 2012)—who accept Premise (1) of both the Zombie and the Knowledge Arguments, and challenge Premise (2) instead. They argue that our ability to *conceive* of a zombie does not show that zombies are genuinely possible, but only that our qualitative or phenomenal concepts of experience, derived by "pointing" introspectively at some feature of an experience one is currently having, are radically different from any physical-functional characterizations of what is going on in the brain. Similarly, they argue that when Mary first sees colors, she does not gain access to any new, non-physical, facts about human color experience, but only (via introspection) to new qualitative or phenomenal *concepts* of the neurophysiological processes that she learned about in her black-and-white room.

These views, in short, concede that there is no *a priori* link between our introspection-derived and physical-functional concepts of our conscious experiences—but deny that this shows that they cannot be concepts of the very same things. In addition, these materialists respond to the "Distinct Property Objection" to the Type-Identity Theory discussed by Smart (see Section 3) by contending that the concepts of our conscious mental states acquired by introspection can pick out those states *directly*, by demonstration, without need for any modes of presentation that entail that what has been picked out is a mental state of a particular qualitative type.

There are a number of different versions of Type B Materialism, but all face a common objection, namely, that while scientific identity statements such as "Water = H_2O" or "Heat = Mean molecular kinetic energy" seem perfectly intelligible after we learn more about the composition of the items in the world around us, it remains mysterious how, in Huxley's terms, "anything as remarkable as a state of consciousness comes about as a result of irritating nerve tissue"—even as we come to know more and more about the brain and nervous system. Type B materialists, in response, argue that the fact that our qualitative or phenomenal concepts derive from introspection, and are therefore radically disconnected from our physical and functional concepts, provides a compelling explanation of why there remains a hint of mystery in these cases, and these cases alone. In addition, some theorists (e.g. Nagel 1965; Brown 2010) argue that if there were a developed theory of *immaterial* substances and properties, then dualists would face a similar problem.

Type B Materialism nevertheless remains a controversial view. However, there is yet another way for materialists to avoid any unsatisfying consequences of the materialistic alternatives presented so far—namely, to embrace *Eliminativism* about conscious mental states. This view (which also, to be sure, has counterintuitive consequences) will be discussed in the next section.

6 Eliminativism

To embrace Eliminativism about some category of things is to deny that those things exist. One of the best-known eliminativists about mental states is Paul Churchland (1981) who argues that our common sense views about the role played by beliefs and desires in explaining behavior and other psychological phenomena are radically false, and, moreover, that they do not mirror, even approximately, the empirically established generalizations of a truly explanatory psychological theory. Thus, he concludes, it is reasonable to deny the existence of beliefs and desires, and take our routine attributions of such states no more literally than our talk of the sun's rising and setting.

Churchland's contention is highly controversial, but—regardless of its plausibility—he does not extend it to conscious mental states such as after-images, perceptual experiences, and sensations. There are a few radical eliminativists about such states; for example, Georges Rey (1983) denies outright that there are any properties that have the features that we ascribe to our conscious experiences. But most materialists who consider themselves eliminativists endorse what we may call Partial Eliminativism. Dennett (2002) argues that our common sense conception of conscious experience includes elements that further reflection will reveal to be incompatible—and argues that those theses that conflict with a broadly functionalist account of conscious experiences should be rejected. More recently, Par Sundstrom (2008) argues that we may be more willing than we think to be eliminativists: we start by being willing to deny that our color experiences possess qualities like (what seems to be) the yellow-orangeness of a yellow-orange after-image—and go on to recognize that it's far from clear, even by means of introspection, what the qualitative properties of our sensations and perceptual experiences are supposed to be. (See also Schwitzgebel 2008, for more general skepticism about the deliverances of introspection.)

In the end, both materialists and dualists may have to concede that there are, and always will be, some unsatisfying consequences of the views they endorse, and leave things at that. Indeed, Eric Schwitzgebel (2014) argues that all (well-developed) metaphysical theories of the nature of mental states, be they dualist or materialist, are "crazy," in the sense that they include at least some important ("core") theses that conflict with common sense—which we are given no compelling evidence to believe. Whether or not further reflection (or acculturation) will alleviate the bizarreness of some of these theses—or, alternatively, provide a compelling explanation of why they may always seem bizarre)—this view needs to be taken seriously.

7 Conclusion

However, even if all extant theories of the nature of conscious experience are crazy, in Schwitzgebel's sense, materialists can argue that adopting Dualism has, overall, too high a price: one has to accept two types of fundamental entities in the world, with little explanation of how non-physical properties arise in humans and certain non-human animals, and how they can have causal efficacy. Surely, materialists (or at least Type B materialists) argue, it is reasonable to accept that qualitative-physical identity statements may retain a hint of "mystery"—as long as there is an explanation for why such mystery may arise in these, and only these, cases.

But even if the pros of Materialism outweigh the cons, the materialists' work is far from done, since it is far from settled *which* materialist view is most promising. Does the greater universality of Functionalism (or Psychofunctionalism) outweigh its potential problems with mental causation, or are Type-Identity theories superior, even if they may not seem sufficiently universal? If Functionalism is superior, just what are the relations among mental states, stimulations, and behavior that make them *conscious* states: must these states be somehow "scanned" by the individual who is in them, or be the objects of that individual's thoughts (see Lycan 1996;

Rosenthal 1986; Gennaro 2004)? And which relations make mental states conscious states of particular types, e.g. experiences of red versus experiences of green? Moreover, perceptual experiences seem to *represent* items in the world: is this to be taken at face value, and if so, can there be an adequate materialistic account of what it is for a mental state to represent some object or property that allows for illusion and hallucination? These are just some of the questions that need to be answered to provide an adequate theory of conscious mental states, and therefore, even for those who believe that there are good grounds for embracing Materialism, there is still a lot of work to be done.

References

Alter, T. (2016) "The Structure and Dynamics Argument," *Noûs* 50: 794–815.
Balog, K. (2012) "In Defense of the Phenomenal Concept Strategy," *Philosophy and Phenomenological Research* 84: 1–23.
Bechtel, W. and Mundale, J. (1999) "Multiple Realizability Revisited: Linking Neural and Cognitive States," *Philosophy of Science* 66: 175–207.
Block, N. (1980) "Troubles with Functionalism," in N. Block (ed.) *Readings in Philosophy of Psychology, Volume One*, Cambridge, MA: Harvard University Press.
Chalmers. D. (1996) *The Conscious Mind: In Search of a Fundamental Theory*, New York: Oxford University Press.
Chalmers, D. (ed.) (2002a) *Philosophy of Mind*, New York: Oxford University Press.
Chalmers, D. (2002b) "Consciousness and Its Place in Nature," in Chalmers 2002a.
Chisholm, R. (1957) *Perceiving*, Ithaca, NY: Cornell University Press.
Chomsky, N. (1959) "Review of Skinner's *Verbal Behavior*," *Language* 35: 26–58.
Dennett, D. (1988) "Quining Qualia," Reprinted in Chalmers (2002a).
Descartes, R. (1641) *Meditations on First Philosophy*. Reprinted in Cottingham, Stoothoff, Murdoch (tr.), (1985). *The Philosophical Writings of Descartes*, Vol. 2, Cambridge: Cambridge University Press.
Descartes, R., Elisabeth, Princess of Bohemia (1643/1985) Correspondence with Descartes. In Cottingham, Stoothoff, Murdoch, Kenny (tr.), *The Philosophical Writings of Descartes*, Vol. 3. Cambridge: Cambridge University Press.
Feigl, H. (1958) "The 'Mental' and the 'Physical,'" in H. Feigl, M. Scriven and G. Maxwell (eds.) *Concepts, Theories and the Mind-Body Problem* (Minnesota Studies in the Philosophy of Science, Volume 2), Minneapolis, MN: University of Minnesota Press.
Gennaro, R. (2004) "Higher-Order Theories of Consciousness: An Overview," in R. Gennaro (ed.) *Higher-Order Theories of Consciousness*, Amsterdam and Philadelphia: John Benjamins Publishers.
Hill, C. and McLaughlin, B. (1999) "There Are Fewer Things in Reality than Are Dreamt of in Chalmers' Philosophy," *Philosophy and Phenomenological Research* 59: 445–454.
Hobbes, T. (1668/1994) *Leviathan*, Indianapolis, IN: Hackett Publishing Company; Underlined, Notations edition.
Huxley, T.H. (1881) *Lessons in Elementary Physiology*, Macmillan and Co.
Jackson, F. (1982) "Epiphenomenal Qualia," *Philosophical Quarterly* 32: 127–136.
Jackson, F. (2004) "Mind and Illusion," In P. Ludlow, Y. Nagasawa, and D. Stoljar (eds.) *There's Something About Mary*, Cambridge, MA: MIT Press.
Kim, J. (1998) *Mind in a Physical World*, Cambridge, MA: MIT Press.
Kripke, S. (1980) *Naming and Necessity*, Cambridge, MA: Harvard University Press.
La Mettrie, J. (1747/1994) *Man, a Machine*, Indianapolis, IN: Hackett Publishing.
Levin, J. (2007) "What Is a Phenomenal Concept?," in T. Alter and S. Walter (eds.) *Phenomenal Concepts and Phenomenal Knowledge*, Oxford: Oxford University Press.
Leibniz, G.W. (1714) Reprinted in D. Garber (tr.) and R. Ariew (ed.) (1991). *Discourse on Metaphysics and Other Essays*, Indianapolis, IN: Hackett Publishing.
Loar, B. (1997) "Phenomenal States (revised)," in Chalmers 2002a.
Lycan, W.G. (1996) *Consciousness and Experience*, Cambridge, MA: Bradford Books, MIT Press.
Malcolm, N. (1968) "The Conceivability of Mechanism," *Philosophical Review* 77: 45–72.
McGinn, C. (1989) "Can We Solve the Mind-Body Problem?" *Mind* 98: 349–366.
Nagel, T. (1965) "Physicalism," *The Philosophical Review* 74: 339–356.

Nagel, T. (1974) "What Is It Like to Be a Bat?" *The Philosophical Review* 83: 435–450.
Papineau, D. (2002) *Thinking About Consciousness*, Oxford: Clarendon Press.
Pereboom, D. (2011) *Consciousness and the Prospects of Physicalism*, New York: Oxford University Press.
Polger, T. and Shapiro, L. (2016) *The Multiple Realization Book*, Oxford: Oxford University Press.
Place, U.T. (1956) "Is Consciousness a Brain Process?" *British Journal of Psychology* 47: 44–50.
Putnam, H. (1968) "Brains and Behavior," in R.J. Butler (ed.) *Analytical Philosophy, Second Series*. Blackwell: 1–19.
Rey, G. (1983) "A Reason for Doubting the Existence of Consciousness," in R. Davidson, G. Schwartz, and D. Shapiro (eds.) *Consciousness and Self-Regulation*, Vol. 3, New York: Plenum.
Rosenthal, D. (1986) "Two Concepts of Consciousness," *Philosophical Studies* 49: 329–359.
Ryle, G. (1949) *The Concept of Mind*, London: Hutcheson.
Schwitzgebel, E. (2008) "The Unreliability of Naïve Introspection," *Philosophical Review* 117: 245–273.
Schwitzgebel, E. (2014) "The Crazyist Metaphysics of Mind," *Australasian Journal of Philosophy* 92: 665–682.
Searle, J. (1980) "Minds, Brains, and Programs," *The Behavioral and Brain Sciences* 3: 417–457.
Skinner, B.F. (1953) *Science and Human Behavior*, New York: Macmillan.
Smart, J.J.C. (1959) "Sensations and Brain Processes," *Philosophical Review* 68: 141–156
Smart, J.J.C. (2007) "The Mind/Brain Identity Theory," *The Stanford Encyclopedia of Philosophy* (Winter 2014 Edition), Edward N. Zalta (ed.), URL = https://plato.stanford.edu/archives/win2014/entries/mind-identity/.
Stoljar, D. (2001) "Two Conceptions of the Physical," *Philosophical and Phenomenological Research* LXII: 253–281.
Sundstrom, P. (2008) "A Somewhat Eliminativist Proposal about Phenomenal Consciousness," in A. Hieke and H. Leitgeb (eds.) *Reduction and Elimination in Philosophy and the Sciences: Papers of the 31st International Wittgenstein Symposium*, Kirchberg am Wechsel: The Austrian Ludwig Wittgenstein Society.
Van Gulick, R. (1993) "Understanding the Phenomenal Mind: Are We All Just Armadillos?" In M. Davies and G. Humphreys (eds.) *Consciousness: Psychological and Philosophical Essays*. Oxford: Blackwell.
Watson, J. (1930) *Behaviorism*, Norton: New York.
Wittgenstein, L. (1953) *Philosophical Investigations*, New York: Macmillan.
Yablo, S. (1992) "Mental Causation," *Philosophical Review* 101: 245–280.

Related Topics

Consciousness in Western Philosophy
Dualism
Biological Naturalism and Biological Realism
The Neural Correlates of Consciousness

4
DUALISM

William S. Robinson

Dualism is the view that our world contains two irreducible kinds of entities, the physical and the non-physical. Its main contemporary rival is physicalism (also known as "materialism"). According to this view, everything there is, notably including conscious minds, is physical. To understand the distinction between the physical and what is not physical, let us begin with something that is uncontroversially physical – say, a rock. It is uncertain what the very smallest parts of a rock are, but in this article, I'll assume that the Standard Model of physics gives us the *fundamental* physical things. These are particles such as photons, electrons, and quarks. Physical objects, then, are the fundamental physical objects together with everything that is composed exclusively of those. A representative list that can be generated from this definition includes electrons, protons, atoms, molecules, crystals, cells, rocks, corals, bricks, buildings, planets, and stars.

The physical includes more than physical objects. It includes events that happen in physical objects, such as lightning flashes, muscle contractions, and landslides. It includes properties of fundamental physical objects, such as charge, mass, and spin. It includes properties of composites, such as liquidity of water and the temperature of the air. Spatial properties (e.g., distance, shape), temporal properties (e.g., age) and spatio-temporal properties (e.g., velocity) are also physical properties.

The key dualist claim is that when it comes to minds – in particular, our consciousness – we cannot give a full accounting that uses only physical objects and laws among physical events and properties. Dualists hold that something needs to be added to what physical science provides, if we are to have a satisfactory account of everything there is.

There are several versions of dualism, and several kinds of arguments for supposing that dualism is true. The first two sections below introduce the main divisions among dualistic views. Later sections will examine some important arguments.

1 Types of Dualism (A)

"Consciousness," "minds," and "mental" are often applied to a large and somewhat diverse set of items. These include bodily sensations, such as pains and itches; sensations we have during perceptual experiences, such as the ways things look, sound, taste, and so on; beliefs, desires, hopes, fears, and similar states; and selves, conceived as what *has* sensations, experiences, and mental states. Dualistic claims and arguments sometimes concern all of these aspects of the mental, but sometimes concern only one or another aspect.

Substance Dualism claims that our minds are substances that are distinct from any physical substance. The use of the term "substance" in philosophy follows this rule: if A and B are distinct substances, then neither one is required in order for the other to exist. So, substance dualism says that our minds are a kind of thing that could exist without anything physical existing – in particular, without our bodies existing. Of course, while we are alive, we are composites of two substances, mind and body.

Substance dualism leaves open the possibility of survival of our conscious mind after the death of our body. This implication provides a motivation that may lead some to *hope* that substance dualism is true. It does not, however, provide an argument for that view, since survivability after bodily death is itself controversial. Later in this article, we shall look at some arguments for substance dualism that do not rest on a prior assumption of survivability.

A more popular view among contemporary dualists is *property dualism*, the view that there are non-physical properties of events that take place in our bodies. Property dualists hold that instantiation of non-physical properties cannot happen without bodies, but nonetheless, the properties themselves are not physical properties.

To understand this view, we need to understand two ways in which a property can count as "physical." First, the properties of fundamental physical objects are physical. We accept that there are these properties because the physical theories that propose them provide the best explanations of events that we can observe, either in laboratories or in everyday life.

Second, non-fundamental properties are counted as physical if they can be explained by the laws of interaction of fundamental physical properties plus facts about how things are composed. The liquidity of water or alcohol, for example, is explained by their being composed of parts that are able to pass by each other without much resistance. When we can give explanations like this, we can say that a property (liquidity, in this case) has been "reduced to," or "constructed from" physical properties of its parts (in this case, properties of atoms or molecules that are held to compose the liquid). The essential claim of property dualism is that there are some properties that are *not* reducible to (or constructible from) physical properties.

Property dualism does not require a non-physical substance. A property dualist can consistently say that some physical objects or events have both some physical and some non-physical properties. So long as the properties themselves are not reducible, there will be something that actually exists but cannot be fully accounted for solely by physical objects, events, and properties.

Event Dualism understands non-physical properties in the same way as property dualism. Its distinctive claim is that non-physical properties do not need to be instantiated in – need not be properties *of* – objects, physical or otherwise. The smell you experience when entering a bakery, for example, is an instance of a particular odor property. That property is in your stream of consciousness during a certain interval of time, but it is not a property of your brain, nor of an event in your brain, nor of molecules in the air.

When we refer to facts involving a property, we usually attribute the property to a thing – a thing that, we say, "has" the property. So, event dualism may seem puzzling at first sight. Physicists, however, often talk of fields, for example the magnetic field that surrounds the Earth. The strength and direction of that field are properties that differ at different points. We can say that a point in space has a magnetic field of a certain strength and direction. A point, however, is just a location, and neither a point nor a location is a *thing*. Analogously, event dualism proposes that an occurrence of a non-physical property does not require a *thing* to "have" it.

The pull to find a thing to have non-physical properties is a powerful one. It sometimes leads critics of property dualism or event dualism to invent a bearer for non-physical properties, and a popular name for this alleged bearer is "ectoplasm." Readers, Beware! Ectoplasm is a caricature drawn by physicalists. It is supposed to be a special kind of stuff. But property dualists attribute non-physical properties to physical bearers, i.e. things that also have physical properties, and thus do not need a special kind of "stuff." Event dualists *deny* that there is a "stuff" that has non-physical properties.

2 Types of Dualism (B)

Many arguments concerning dualism depend upon assumptions about causal relations between mental and physical items. This section divides dualisms according to their claims about causation. In making this division, I will adopt the common view that *if* there are causal relations between something physical and something that is mental and non-physical, then the "something physical" is a brain, or a part of a brain, or some events in a brain.

Interactionism says that some brain events cause mental events, and some mental events cause physical events.

To illustrate the first clause, stimulations of our sense organs (when they and the rest of our neural systems are in normal conditions) are held to cause pains, color sensations, sounds, tastes, smells, feelings of pressure, and so on. To illustrate the second clause, in normal conditions our decisions are held to cause our actions, and having a sensation of a particular kind is held to causally contribute to our reporting having that kind of sensation.

Epiphenomenalism says that some brain events cause mental events, but no mental event causes a physical event.

Parallelism says that there are no causal connections either way between physical events and mental events.

An obvious problem for parallelism is to account for why there should be correlations between physical events and mental events. For example, whenever one is cut (in normal conditions, e.g., in absence of an anesthetic), one feels a pain. Why would there be such a regularity, if the cuts are not causing the pains?

Historically, advocates of parallelism have had theological motives, and have explained correlations between the mental and the physical by appealing to agency on the part of a deity. There are very few today who hold the required theological views. The options for dualism that remain standing in current debates are thus interactionism and epiphenomenalism.

On first hearing, epiphenomenalism strikes most people as highly counterintuitive. There are several more formal objections to it, the most important of which are based on evolution, and on self-stultification. The key point about evolution is that a trait can be shaped by natural selection only if it causally contributes to behavior that increases or decreases an organism's fitness. If sensations have no physical effects, it seems that natural selection cannot explain why our sensations are appropriate to our circumstances, or even why we evolved to have any sensations at all.

Self-stultification is held to follow from two assumptions. The first is agreeable to epiphenomenalists, namely: (a) Epiphenomenalists claim to know something about their sensations. The second is a generalization of a principle that uncontroversially holds for perception, namely: (b) A person can know about a thing only if that thing causes the person to form a belief about it. (For example, if an object is not causing your belief that you see it, you don't know it is there, even if you make a lucky guess that it is.) If epiphenomenalists could be forced to accept (b), they would be committed to making knowledge claims that their own view would imply they cannot know.

Epiphenomenalists believe they have adequate responses to these objections.[1] But readers may wonder why anyone would bother to defend epiphenomenalism, when there is a rival view that seems obviously true, namely interactionism. The answer is that there is also a strong objection to interactionism.

This objection arises from the very wide acceptance of the principle of Physical Causal Closure:

(PCC) Every physical event that has a cause has a sufficient physical cause.

Support for this principle comes from the success of physical science and, in particular, success in discovering the details of the mechanisms by which brain parts change their states and influence other brain parts.

Most actions require movement of our bodies. Our bodies are physical objects, and their movements are physical events. So, accepting PCC entitles us to infer that when we act, the movements of our bodies have a sufficient physical cause, if they have any cause at all; and they do seem to have a cause. For example, if I raise my arm to vote for someone, various muscles contract. Those contractions are physical events, and have physical causes, such as release of neurotransmitter molecules into junctions with muscle fibers. That release is caused by events in the neurons coming into muscle tissue from the spinal cord. Those neurons are activated by other neurons that descend into the spinal cord from the brain. And so on. To accept PCC is to accept that there is a continuation of this story that can, in principle, completely explain why my arm goes up, and that consists entirely of a series of physical events in sense organs and in various parts of the brain.

If this account is correct, then no non-physical events are needed to give a causal explanation of our behavior. Moreover, if physical events alone are sufficient to cause our behavior, then non-physical events, even if they are present, do not *make a difference* to our behavior, where making a difference requires that without those mental events, our behavior would not have been what it was. If one accepts that non-physical events do not make a difference to our behavior in the required sense, then one has become an epiphenomenalist in all but name.

To summarize the current debate, consider the following four statements, each of which is currently found plausible by a substantial number of thinkers (and 4 is accepted by all parties).

1 Any complete account of our mentality requires us to include non-physical events. (Dualism)
2 All events that are required for a complete account of our mentality make causal contributions to our behavior. (Mental Efficacy)
3 The only kind of thing that can causally affect a physical event is a physical event. (PCC plus requirement to make a difference)
4 Our behavior consists of changes in our bodies, which are physical events.

This quartet is mutually inconsistent. For example, dualism plus mental efficacy implies that some non-physical property has an effect on our behavior. Since our behavior consists of physical events, this implies that some non-physical property has an effect on some physical event; which contradicts 3.

Since we cannot consistently accept all four of these statements, we must give up at least one of them (and giving up any one is enough to remove inconsistency). Physicalism rejects 1. Interactionism rejects 3. Epiphenomenalism rejects 2.

3 Arguments for Dualism

Perhaps the most famous argument for (substance) dualism was given by Descartes (1596–1650). This argument rests on some claims about *certainty*. Descartes worried that there might be an Evil Genius who gets his jollies from deceiving him. He knew that he could be deceived about many things, including even the existence of his own body. He was aware that some people suffer from "phantom limb," a condition in which patients feel that they still possess a limb that has in fact been amputated. So, maybe a powerful deceiver could make him feel as if he had a whole body, when in fact he had none at all. But when Descartes asked himself whether he could be deceived when he thinks to himself *I exist*, his answer was that he certainly could not be so deceived. Indeed, he would have to exist in order for an Evil Genius to be deceiving *him*.

The argument itself has been stated in many ways. A simple formulation is this:

D1. I am certain that I exist.
D2. I am not certain that anything physical exists (including what I'm in the habit of thinking of as my own body).
D3. I cannot be certain and uncertain of the same thing at the same time.

Therefore,

D4. I am not the same thing as my body.

Contemporary dualists do not offer this kind of argument. They recognize that D3 is false. So long as we have two names or descriptions, we can indeed be certain and uncertain of the same thing at the same time. For example, one can be sure one has read something written by Mark Twain, but uncertain, or even doubtful, whether one has read anything written by Samuel Clemens. Before the Babylonians discovered that the Morning Star and the Evening Star are the same body (namely, Venus), it would have been entirely reasonable to be certain one was observing the Morning Star, while doubting that one was observing the Evening Star.

A second kind of argument is based on *intentionality*. "Intentionality" is a Latinate word that means *aboutness*. We have beliefs about where the economy is going, about who will get elected, about where various cities are located. We have desires about foods, about potential mates, about social justice, and so on. So, beliefs and desires are about things, and they can be said to have aboutness. Philosophers usually express this point by saying that beliefs and desires have intentionality.

When we intend to act in a certain way, our intention is about our action (or about the result we want to produce). So, our intentions have intentionality. But the term is somewhat confusing, because many things have intentionality that are not intentions; for example, beliefs, desires, doubts, wonderings, and fears.

The intentionality of some of our mental states and events has been taken by some thinkers as providing a reason for dualism. The reason turns on two peculiar properties of aboutness. One is this: A thought (belief, desire, and so on) can be about things that do not exist – for example, fictional entities such as Sherlock Holmes or unicorns, posits of failed theories, such as humors or the luminiferous aether, and even impossible things such as perpetual motion machines or round squares.

The other peculiarity of intentionality comes out in the following argument.

1 Jones believes that Mark Twain wrote *The War Prayer*.
2 Mark Twain is Samuel Clemens.

Therefore,

3 Jones believes that Samuel Clemens wrote *The War Prayer*.

This inference is plainly invalid. If Jones is not aware that 2 is true, the premises will still be true, but 3 may very well be false. The same kind of invalidity occurs whether we talk about what

Jones believes, or what Jones desires, hopes, fears, or knows. To generalize: When dealing with mental states, we cannot count on having a valid argument, even when all we do is replace a term in the first premise by another term that refers to the exact same thing.

Why did I call these properties "peculiar"? Let us first notice that aboutness seems to be a relation. Relations characteristically relate two (or more) items – for example, "X is a brother of Y," or "X is taller than Y." We use the same grammatical form with aboutness: X (e.g., a belief) is about Y (the state of the economy, Sherlock Holmes, etc.).

But wait! Relations are supposed to *relate*. How can a mental state be in a *relation* to something that doesn't even exist?

Regarding the second property, the peculiarity is this. All the relations that we find in our natural sciences allow inferences of the kind that are not allowed when states that have intentionality are involved. For example, if (i) Jones is taller than Mark Twain, and (ii) Mark Twain is Samuel Clemens, it does follow that (iii) Jones is taller than Samuel Clemens. It doesn't matter who believes or doesn't believe what: if (i) and (ii) are true, then (iii) must be true. Similarly, if spoilage caused a cheese to turn green, and green is in fact Aunt Tillie's favorite color, then it follows that spoilage caused the cheese to turn the color that is Aunt Tillie's favorite.

The argument for dualism that is based on intentionality should now be obvious. Relations among physical objects require existence, and allow inferences when we substitute terms that refer to the same thing. Relations between mental states and what they are about do not require existence of what they are about and do not allow inferences, even when all we do is to substitute a term that refers to the same thing. The conclusion is that intentionality is not a physical relation. There must be something very special, and non-physical, about the mind if it can stand in this special sort of "relation" to other things, and even to non-existent things.

This argument would fail if intentionality could be "naturalized," that is, constructed from physical relations. Although proposals for such construction involve very complex networks of relations, and although there are disagreements about details, a majority of contemporary philosophers think that intentionality can be naturalized, and thus they do not accept this argument for dualism.

A third kind of argument rests on claims about *conceivability*, and its relation to possibility. To understand arguments of this kind, we may begin by trying to conceive of a unicorn. What would it be like for there to be one? Well, there would be something that is mostly like a horse, except that it would have a single horn emerging from its forehead. Moreover, its horn would not be held on by glue, or even by a bone graft. A unicorn would have to have its horn naturally – it would have to be a member of a species that regularly produced offspring that would develop horns at roughly the same age.

Are unicorns possible? Well, didn't we just conceive such a possibility? Bulls have horns, narwhals have a single horn. Why couldn't there be unicorns? If we can conceive something in clear detail, as we just did with unicorns, and it is obvious that there is no contradiction in what we are conceiving, isn't that the same as showing that it is really possible?

Unicorns are generally regarded as possible (even though known to be non-actual). But it is controversial how we should answer the general question – whether conceivability, or conceivability with some restriction regarding the clarity and detail of the conception, is enough to establish genuine possibility.

A conceivability principle is a principle that says that conceivability (suitably restricted) is sufficient to establish genuine possibility. A conceivability argument is an argument that has such a principle as a premise. "Suitably restricted" is needed to indicate that care is needed in defining "conceivability." We can make grammatical sentences using the phrases "round square" or "perpetual motion machine," but we cannot provide a clear and detailed account of how to construct them. A suitably restricted definition of "conceivability" must count these as not genuinely conceivable, despite the fact that we can understand what they are well enough to know they cannot exist.

There are two kinds of conceivability argument that have been proposed in recent decades, one for substance dualism, and one for property dualism. A Conceivability Argument for Substance Dualism (CSD) goes as follows:

CSD1. I can clearly conceive of my stream of consciousness continuing after the destruction of my body.
CSD2. Conceivability implies possibility.

So,

CSD3. It is possible for my stream of consciousness to continue after the destruction of my body.
CSD4. It cannot be that my stream of consciousness continues to exist without *me* existing.

So,

CSD5. It is possible for me to continue to exist after the destruction of my body.
CSD6. It is not possible for the same thing to be both destroyed and to continue to exist at the same time.

So,

CSD7. I am not the same thing as my body.[2]

The same argument would show that I am not the same thing as any of my bodily organs, including my brain. (Just specify that destruction of my body is thoroughgoing, i.e. involves the destruction of all my bodily parts down to their atoms.) It is not remotely plausible that I am the same thing as some physical object outside my body. So, the force of the conclusion can be easily extended to the claim that I am not a physical object of any kind whatsoever.

There are many things to be said about this argument, but I will limit my discussion to likely responses from physicalists. They will have doubts about the first two premises. Regarding the first, they may argue as follows. Unless we beg the question against physicalism (in which case the argument fails) we cannot suppose that we *know* that our stream of consciousness is not dependent upon, or even identical with, events in our brains. If they are identical, then we cannot *really* conceive of our stream of consciousness outlasting the destruction of our brains. So, we do not know that the first premise is true; and so, we do not know, by this argument, that its conclusion is true.

A slightly more accommodating response concedes that this case is not like the round square case. I can not only grammatically *say* "stream of consciousness that survives bodily destruction"; it seems that I can form a robust "picture" of thinking my thoughts, enjoying my memories, and wondering what will happen next, even though I am no longer associated with a body. But then, it can be doubted that the second premise is true. Why ever should we think that forming such a picture shows real possibility? If my thoughts are identical with events in my brain, CSD3 is false. If CSD3 is false, then in whatever sense of "conceivability" it may be in which CSD1 is true, CSD2 (using the same sense of "conceivability") would have to be false.

Proponents of the above argument may respond that the only reason to doubt the first two premises is the question-begging assumption that physicalism is true. Such exchanges of charges of mutual question-begging are never easily resolved.

Another kind of conceivability argument aims to establish property dualism, and is often called the Zombie Argument. To understand this argument, we must distinguish between Hollywood zombies, and zombies as philosophers understand them. Hollywood zombies walk stiffly, stare vacantly, and aim to harm you. In contrast, Zombies in philosophy behave exactly – *exactly* – like a normal person, and they are anatomical duplicates of ordinary human beings. What makes them zombies is that they live in a world with different laws of nature. In their world, unlike ours, brain events do not cause sensations. So, although zombies wince when they're stuck with a needle, they have no pains. They complain of hunger, and eat with all the behavioral signs of pleasure, but they have no hunger pangs, and their foods have no actual tastes for them.

The Zombie Argument goes like this.

Z1. Zombies are conceivable.
Z2. Conceivability implies possibility.

So,

Z3. Zombies are possible.
Z4. If zombies are possible, then some properties in our sensations (painfulness, tastes, colors, and other properties like these) cannot be the same properties as any physical properties.

Remember, zombies are physical duplicates of humans. If our sensations were nothing but physical constructions, zombies would have the same physical constructions, and thus the same sensations that we do. But that would contradict the assumption that we are describing zombies. So, if zombies are so much as possible, our sensations must involve a property that is not reducible to (or constructible from) physical properties.

From Z3 and Z4, it follows that:

Z5. Some properties in our sensations are not the same properties as any physical properties.

So,

Z6. Physicalism is false.[3]

This argument does not say that sensations could exist without brain events – it says only that the latter could exist (in some possible world) without sensations. So, it is not an argument for minds (or, entities that have sensations) that could exist without bodies. It is an argument that our sensations involve properties that, unlike liquidity, cannot be explained through constitution by physical parts plus laws of nature that apply to the relations among such parts.

As in the previous argument, the first two premises of the Zombie Argument are controversial. Physicalists often concede that we do not presently have a theory that explains how sensations of red, or of chocolate taste, or of pain can be constructed from the assumption that they are composed of events in brain parts (events in neurons, for example) plus laws governing the relations among such events. They can offer this lack of theory as a reason that makes Z1 seem plausible, while consistently denying that zombies are *really* conceivable. And with or without this concession, they can either deny that Z2 is true, or deny that we know that Z2 is true.

For dualists, this stance seems question-beggingly ideological. If we have no ghost of an inkling of how sensations of red or chocolate could be constructed out of brain events, it is downright *un*scientific to declare that nonetheless they must somehow be thus constructible.

A fifth argument for (property) dualism is the *Knowledge Argument*. This argument was advanced by Frank Jackson in 1982, and it begins by introducing us to Mary, a brilliant scientist. Her specialty was color vision, and she knew everything that our natural sciences can tell us about that subject.

What was distinctive about Mary, aside from her brilliance and dedication, was that during her whole life she had been confined to a room in which everything was black, white, or some shade of gray. Her TV and educational materials were all black and white. As a result of her confinement, she had never had a color experience. She knew everything there is to know about what happens in people's brains when they look at, say red roses, and everything about what would happen in her own brain if she were to see one. But she had never actually had an experience of red, or of any other chromatic color.

Jackson imagined a day on which Mary is finally to be let out of her room, and allowed to see something red for the first time. The Knowledge Argument concerns this moment, and goes as follows:

KA1. Mary already knows all the physical facts about what will happen in her visual systems when the door is opened.

KA2. Mary will learn a new fact when the door is opened – namely what red is.

So,

KA3. The new fact is not a physical fact.

So,

KA4. Not all facts about the world are physical facts.

The literature in response to this argument is far too large to be summarized here.[4] I will mention just one source of doubt about it that is related to several of the more formal replies that have been made.

KA2 gives "what red is" as the fact that Mary is about to learn. "What it is like to see red" is also a common phrase that is used to identify this fact. Both formulations have this peculiarity: they are not sentences. But facts are usually stated as sentences. For example, it is a fact that *Brazil is in South America*, it is a fact that *water boils at 100° C*, and so on. It is natural to expect a new fact to be stated in the form of a sentence; but it is not clear what sentence properly expresses the fact that Mary is supposed to learn.

This peculiarity leads to a worry. Maybe what happens to Mary is not correctly described as her *learning* (coming to know) a new fact. There is certainly something new that happens to her. What must be allowed by everyone is that, for the first time, *she experiences red*. That is compatible with holding that a red experience is identical with a brain state – for, as again all will agree, her brain has never before been in the state it enters when she first sees something red. Physicalists can hold without contradiction that what happens to Mary is not that she comes to know a new fact, but instead that she comes to stand in a new relation to a fact she already knows. That is, instead of just knowing what state she would be in if she saw something red she is now actually in that state.

These remarks will be as controversial as the more formal replies in the literature. I will close the discussion of the Knowledge Argument by noting that Jackson has subsequently rejected its conclusion. In 1982 (and 1986), he followed the presentation of the Knowledge Argument with a recommendation to adopt epiphenomenalism, as the best view to take, given the conclusion of the Knowledge Argument. Epiphenomenalism, as noted earlier, is counterintuitive, and Jackson is no longer content to accept it. In a 1996 book (with David Braddon-Mitchell) he defended the Knowledge Argument against several replies in criticism of it. He did not claim to see exactly why it failed, but offered the "There must be a reply" reply to it. That is, he thought that there must be something wrong with the argument, even if we cannot explain what the error is. Naturally, advocates of the Knowledge Argument find this stance unsatisfying. "There must be a solution to a problem for my account (even though I can't think of one)" is not generally accepted as an adequate defense of views in philosophy or in science.

A sixth kind of argument turns on the *Relative Simplicity* of properties in our sensations. To understand this argument, we may begin with a less dramatic version of Jackson's starting point. Consider that congenitally blind people usually know many things about colors, and some know a great deal about light waves, stimulation of retinal cells by light, optic nerves, and visual processing in the brain. Yet it is extremely plausible that something is missing from their experience. They may know that a red light means one should stop, but they have never had the experience that gives "red" its meaning, in normally sighted people. Advising them to repair this lack by studying harder would be an exercise in grim humor.

These remarks can be generalized to apply to congenitally deaf people, who may know about compression waves in the air; and to people who know about molecular structures of molecules they are unable to smell, even though those molecules cause distinctive odor experiences in most people. A few people are born without the ability to experience pain, but that does not affect their intelligence, or their ability to understand anatomy.

The properties to which these considerations apply – colors, sound qualities, scents, flavors, pains and others – are collectively known as *phenomenal qualities* or *qualia*. (The latter is pronounced 'kwah´-lee-uh' and its singular form is "quale," pronounced 'kwah´-lay'.) Qualia are properties, and they are the most intuitive candidates for non-physical properties.

Many qualia have some degree of complexity. For example, some sounds are chords, most colors are mixtures (orange, for example, is a mixture of red and yellow), and cooks are often complimented for the complexity of the tastes of their food. Qualia do not, however, have the same degree of complexity as the physical properties with which they are correlated. For example, they are not as complex as properties of compression waves, or patterns of light energies at various wavelengths, or arrangements of bonding among atoms. Neither are qualia as complex as the multitude of neural events that are required for us to have experiences.

This difference of complexity gives rise to the Relative Simplicity argument for a dualism of properties (i.e., either property dualism or event dualism).

RS1. The physical properties with which qualia are correlated are complex.
RS2. Qualia are relatively simple properties (i.e., they are simple relative to their physical correlated properties).
RS3. No property can be both complex and relatively simple (i.e., no property can be simpler than itself).

So,

RS4. Qualia are not identical with their physical correlated properties.
RS5. Qualia are not identical with physical properties with which they are not at least correlated.

So,

RS6. Qualia are not identical with any physical properties.

Some physicalists resist this conclusion by pointing to water, which is in fact composed of H_2O molecules even though the way it appears to us gives no hint of that. Analogously, they say, RS2 may be false; maybe qualia are not relatively simple properties, but merely appear to us as being so. Dualists, however, think that physicalists who take this line are missing the point of their own analogy. Water has a shiny, clear appearance. Alcohol looks the same; so shiny clarity cannot be the same property as being composed of H_2O. Thus, the pattern in the water case is that when a thing does not appear as what it is, a distinct property is involved in the way it does appear. Applying this pattern to qualia should lead physicalists to say that qualia are complex properties that have a distinct property involved in the way that they appear. But this result concedes the need for properties that are distinct from the complex properties with which they are correlated.

Other physicalists reject the argument from Relative Simplicity of qualia by proposing that experiences have no qualia, but only *represent* properties; and the properties that are represented are all physical properties such as patterns of compression waves, patterns of energies at various wavelengths of light, molecular structures, and so on. Dualists can respond that experience does not represent such properties *as* having the complexity that they actually have, and that relatively simple qualia will have to be introduced in order to explain how a complex property can be represented *as* relatively simple by an experience.

4 Motivations for Dualism

Arguments for dualism aim to support dualism by relying on premises that are at least claimed to be less controversial than dualism. By "motivations" for dualism, I mean reasons for hoping that dualism is true, where those reasons rest on assumptions that are at least as controversial as dualism. We have already seen one such motivation – the fit between dualism and our hope for survival after bodily death.[5] This section introduces three other kinds of motivations.

The first of these concerns the issue of *free will*. If everything is physical, and the physical world is deterministic (i.e., every event has a sufficient cause), then all my actions are determined by a series of causes that stretch back to times as early as you like to consider, up to the big bang. This view of our world seems to leave no room for free will.

Our most powerful physical theory is quantum mechanics, and leading interpretations of that theory hold that some events have no cause. It is widely held, however, that mere quantum mechanical indeterminacy also leaves no room for free will in any meaningful sense. Free will is often connected with the notion of moral responsibility. It is not evident how people could be responsible for their actions if it turned out that whether they did them or not depended on whether some uncaused event in their brains occurred or did not occur.

Some thinkers have concluded that there must be a non-physical self that is capable of making uncaused, but morally responsible decisions. However, it is not evident how this proposal escapes the dilemma that decisions are either caused (which some thinkers take to be incompatible with being morally responsible) or uncaused (and again, not something for which one is responsible).

Many philosophers have held that the traditional notion of free will is confused beyond repair. Others have tried to clarify, and thus rescue, free will. Since the status of free will is highly controversial, one cannot expect reflections upon it to provide a non-controversial argument for or against physicalism.[6]

Another motivation concerns the *unity of consciousness*. This motivation starts with the observation that we generally have more than one quale at a time. For example, when watching a conductor lead an orchestra, we have both visual and auditory experiences. We often have complex non-sensory mental states. For example, we may find a stranger attractive and entertain strategies of approach, all the while doubting that any approach would be successful and chiding ourselves for our lack of confidence.

Elements of complex mental states of this kind do not seem to us to be mere items on a list. They seem to have a unity with each other, something about them that makes them all obviously *my* perceptions, desires, thoughts and doubts. This unity of our consciousness has seemed to some thinkers to provide a reason for a non-physical self – a self that would explain the sense of unity by being the common possessor of the several mental states. Such a view can allow that different states depend on events in different parts of the brain, while denying that occurrence in the same brain at the same time is sufficient by itself to explain the unity of consciousness.

This view is, however, controversial. An alternative view notes that mental states have many relations among themselves. For example, we may desire what we also see, our thoughts may be about means to satisfy our desires, our lack of confidence may be based on unpleasant memories. This alternative view holds that relations of these kinds among our several mental states are sufficient to bundle them into a unified consciousness.

Similar controversy concerns *personal identity*, the continuity of the same person over a period of time. There is a host of respects in which I am different from what I was when I was 10 years old, but it seems compelling to say that I am the same person. Perhaps there is not a single atom in my brain that was there when I was 10, and the distribution of synaptic connection strengths between my neurons is undoubtedly quite different now from what it was then. If there is something the same about me – something that grounds the fact that I am the same person – then, it seems, it must be a non-physical self whose possession of all my mental states is what makes them all mine.

Once again, an alternative view holds that sameness of me-now and me-at-10 is sufficiently explained by both the existence of a few memories of episodes that happened when I was 10, and the gradualness of the changes as I have aged. To explain this last point a little: If one compares the mental organization of a person at times differing by, say, one month, one can expect a massive – but of course not perfectly complete – overlap of opinions, desires, abilities and memories.

As with unity of consciousness, the issue of what is the best theory of personal identity is controversial. To some thinkers, these features of our mental life suggest a non-physical self. But if we state this suggestion as an argument, the premises will be as controversial as the dualistic conclusion that may be based on them.

5 Conclusion

There is a large literature on the debate between dualism and physicalism. There are replies to everything I have said in the section on arguments, counter-replies to those replies, and so on. The foregoing discussion, however, provides an understanding of what dualism claims, and of the issues that figure most prominently in current discussions of dualism.

Notes

1 For these responses, see the article "Epiphenomenalism" in the online *Stanford Encyclopedia of Philosophy*, and several of the papers referred to therein.

2 For a developed version and defense of this kind of argument, see Swinburne (1997).

3 For a fully developed version and discussion of this kind of argument (including a complication concerning Russellian Monism), see Chalmers (2010, Chs. 5 and 6).

4 Several important papers about the KA are collected in Ludlow *et al.* (2004).

5 This hope may be tempered by reflection on what kind of mind would survive in those who have suffered brain damage due to Alzheimer's disease, strokes, etc. See Gennaro and Fishman (2015) for explanation and discussion of this issue.
6 A good source for issues concerning free will is Kane (2005).

References

Chalmers, D. J. (2010) *The Character of Consciousness*, Oxford: Oxford University Press.

Gennaro, R. J., and Fishman, Y. I. (2015) "The Argument from Brain Damage Vindicated," in M. Martin and K. Augustine (eds.) *The Myth of Afterlife: The Case against Life after Death*, Lanham, MD: Rowman & Littlefield.

Jackson, F. C. (1982) "Epiphenomenal Qualia," *Philosophical Quarterly* 32: 127–136.

Jackson, F. C. (1986) "What Mary Didn't Know," *Journal of Philosophy* 83: 291–295.

Jackson, F. C., and Braddon-Mitchell, D. (2007) *The Philosophy of Mind and Cognition* (2nd edition), Oxford: Blackwell.

Kane, R. (ed.) (2005) *The Oxford Handbook of Free Will* (2nd edition), Oxford: Oxford University Press.

Ludlow, P., Nagasawa, Y. and Stoljar, D. (eds.) (2004) *There's Something About Mary: Essays on Phenomenal Consciousness and Frank Jackson's Knowledge Argument*, Cambridge, MA: MIT Press.

Robinson, W. S., "Epiphenomenalism," *The Stanford Encyclopedia of Philosophy* (Fall 2015 Edition), Edward N. Zalta (ed.), URL = https://plato.stanford.edu/archives/fall2015/entries/epiphenomenalism/.

Swinburne, R. (1997) *The Evolution of the Soul* (revised edition), Oxford: Clarendon Press.

Related Topics

Consciousness, Personal Identity, and Immortality
Consciousness in Western Philosophy
Materialism
Consciousness, Free Will, and Moral Responsibility
Idealism, Panpsychism, and Emergentism
The Unity of Consciousness

Further Reading

Alter, T., and Howell, R. (eds.) (2012) *Consciousness and the Mind-Body Problem*, New York: Oxford University Press.

Chalmers, D. J. (1996) *The Conscious Mind*, Oxford: Oxford University Press. (Foundational source for classification of physicalist and dualist views, and extensive discussion of arguments in this field.)

Kirk, R. (2005) *Zombies and Consciousness*, Oxford: Oxford University Press.

Papineau, D., and Selina, H. (2005) *Introducing Consciousness*, Cambridge: Icon Books. (Papineau's text and Selina's cartoons give a highly accessible introduction to issues about consciousness.)

Robinson, W. S. (2004) *Understanding Phenomenal Consciousness*, Cambridge: Cambridge University Press. (Clarification of many views about consciousness, culminating in an argument for epiphenomenalistic event dualism.)

5

IDEALISM, PANPSYCHISM, AND EMERGENTISM

The Radical Wing of Consciousness Studies

William Seager

1 Why Consider Radical Approaches?

There is always a legitimate philosophical interest in the history of significant doctrines and there is no doubt that all of idealism, panpsychism and emergentism have illustrious pasts. But, unlike topics that have purely historical interest (e.g. Aristotle on spontaneous generation), the problem of consciousness remains the subject of intense investigation. Despite staggering advances in the scientific study of the brain, it remains fundamentally unsolved. Why is that? The answer lies in a certain understanding of the physical and the roadblock this throws up when we try to integrate subjective experience into a world whose nature is restricted to that conception of the physical. The modern locus of this concern is Thomas Nagel's (1974) famous reflection on our inability to get a grip on the subjective nature of non-human consciousness, despite the openness to investigation of the objective world specified in our physical theories.

Thus problematizing consciousness shows that it can be understood in quite simple terms: not 'self-consciousness' or 'transcendental subjectivity,' or awareness of the self as a subject, or awareness of one's own mental states, or the ability to conceptualize one's own mental states as such. Consciousness is simply sentience, or the way things are present to the mind (abstracting from the question of whether anything exists that matches what is present). So, there should be no difficulty about wondering whether bees, for example, are conscious (which I'm pretty sure they are).

The worry is hardly new. The mismatch between the nature of the physical as revealed by science and the subjective nature of consciousness was frequently pointed out in the 19th century. Thomas Huxley wrote that, "...how it is that anything so remarkable as a state of consciousness comes about as the result of irritating nervous tissue, is just as unaccountable as the appearance of the Djin when Aladdin rubbed his lamp" (1866: 210). John Tyndall was more blunt: "the passage from the physics of the brain to ... consciousness is inconceivable" (1879, v. 2: 86–87).

If we think that advances in physics and the brain sciences have erased this worry, we will be disappointed. Nothing that modern physicalist philosophers have to say about how consciousness arises through 'nervous irritation' could not equally have been adduced to defend a hypothetical mechanistic theory of consciousness advanced in 1875. Of course, there are novel quantum and 'information' based theories of consciousness. We have uncovered a host of brain mechanisms undreamt of before the 20th, sometimes even the 21st, century. But the philosophical arguments linking these to the nature of consciousness do not essentially depend on any scientific advances. Instead, new accounts of consciousness either lead towards one of our radical options, as in Hameroff and Penrose (1996) or Integrated Information Theory (Tononi 2012), which tend towards panpsychism or, more commonly, endorse the hope for a standard emergentist account. For example, in recent work on a 'Semantic Pointer' theory of consciousness (Thagard and Stewart 2014), the qualitative aspect of consciousness is regarded as an emergent property, but it is claimed that "there is nothing mysterious about emergent properties," which "result from the interactions of the parts" (78). These authors offer no account of how consciousness could result from the interactions of, ultimately, mass, spin and charge. One might satirize the physicalist attitude as: "I don't know how matter generates consciousness, but I am a physicalist for other reasons. It somehow works. You can't prove I am wrong."

That last point is true. But what someone not already committed to physicalism needs is an intelligible account of how consciousness is a purely physical phenomenon, just as we have an intelligible outline of how, for example, the liquidity of water is purely physical, even though liquidity is not a property found within fundamental physics. Such an identity might be regarded as inexplicable, but harmlessly so. Even though it was a surprising astronomical discovery, there is no question of how it could be that Hesperus is identical to Phosphorus (Block and Stalnaker 1999). This is wrong for at least two reasons. First, suppose that, to all appearances, Hesperus had a property which Phosphorus should, by its scientifically given nature, lack. This is the situation with consciousness and the physicalist thus owes an account of how subjectivity attaches to a physical nature which is fundamentally entirely bereft of it. Second, the brain is a complicated organ with a multitude of parts. If consciousness is not a fundamental physical feature, we need a story of how it emerges from the interactivity of the brain's purely physical constituents, whether or not the final complex state is *identical* to a conscious state, just as we need (and to a great extent have) an account of how it is that water is liquid, given the entirely non-liquid nature of its constituents.

The famous anti-physicalist arguments all stem from considerations that highlight the disconnect between the received understanding of 'the physical' and our direct acquaintance with the subjective aspect of the world revealed in consciousness. These arguments are so well known that they need not be repeated here.[1] Granted the intuitive difficulty of understanding consciousness as a purely physical phenomenon, could we audaciously deny the very *existence* of consciousness? Obviously, we could be wrong about many things connected to our states of consciousness, but not about the existence of an immediately available source of information present to the mind. Consider your belief that something is happening right now. As Descartes famously noted, this proposition is in a different category from most quotidian knowledge. It is in the category of things that you *could not* be wrong about. So, there must be some source of information that vouchsafes your unassailable claim that something is happening. This source is the 'present to mind' we call consciousness. It is real, but how it could be or arise from an entirely un-present physical reality is a complete mystery.

The problem of consciousness can thus be summed up in a simple inconsistent triad:

1 Fundamental reality is entirely un-present.
2 There is presence.
3 There is no way to generate presence from the un-present.

Proposition 2 is not negotiable. The radical approaches to the problem of consciousness which this chapter addresses stem from denying either Proposition 1 or Proposition 3.

2 Idealism

Idealism is the view that consciousness is a fundamental feature of reality (denying Proposition 1). Idealism goes further by asserting that consciousness is all there is to reality. Historical idealism is a famous doctrine, championed in one form or another by Leibniz, Berkeley, Kant, Hegel (and a host of associated German philosophers), Mill, Bradley (and a host of associated British philosophers), not to mention serious proponents beyond the Western philosophical tradition. The history of idealism is necessarily complex (see Guyer and Horstmann 2015), it still retains some defenders and may be due for a resurgence of philosophical interest (see e.g. Sprigge 1983; Foster 2008; Pelczar 2015; Chalmers forthcoming). I have not the space nor the expertise to survey this history, but will situate idealism in the modern debates about consciousness.

Leaving aside suspect epistemological motivations,[2] what would lead one to endorse idealism? It is natural to consider that if the physical world has no place for consciousness, then perhaps the realm of consciousness can assimilate the physical. Budding philosophers delight to think of ways that identical experiences can be produced by many different possible 'underlying' situations (the world, dreams, the matrix, the evil genius). This may suggest that what we call the physical world, the world we experience in everyday life, has its core being in the realm of experience itself rather than some remote background, which can vary independent of experience. Following John Foster (2008), let us define 'physical realism' as the view that the physical world is (1) independent of consciousness and (2) not reducible to anything non-physical. This is evidently a way of stating some of the core theses of physicalism, which would typically add that the basic nature of the physical is exhaustively revealed by the science of physics and, crucially, that there is nothing 'over and above' the physical. That is to say: once the fundamental physical features of the world are put into place, everything else in the world is logically necessitated.[3]

Foster argued that physical realism could not support what he called the "empirical immanence" of the world we experience. This means that physical realism does not support a view of the world "which allows it to be the world which our ordinary physical beliefs are about" (Foster 2008: 164). To support this claim, consider two worlds: one of whose physical underpinning is in accord with perception; and another in which two regions of physical space are exchanged with instantaneous, video-game like, transfer from the boundaries of the exchanged regions. There is no perceptible difference between the worlds (Foster 2008: 125ff.), but in the underlying space Oxford is in a region east of Cambridge. Such a world would, of course, violate physical laws but that is irrelevant to Foster's point. His claim is that in that world reality would correspond not to the bizarre underlying state but rather to standard conceptions of locations and paths of travel. Oxford would really be west of Cambridge. In general, reality would be correctly aligned with experience, not the putative underlying reality. As Foster says:

> The physical world, to qualify as *the* physical world ... has to be *our* world, and it can only be our world in the relevant sense, if it is ours *empirically* – if it is a world that is, as we might put it, *empirically immanent*.
>
> *(138)*

There is something right about this thought. The world which science uncovers has got to match up with the world we experience, not the other way around. Even if the world as physics reveals it is mighty strange, in the end the scientific conception answers to our experience. But surely this only shows that there must be an intelligible route from what physical science reveals to the world as we experience it. This does not seem to require that the world be *constituted* by experience. But Foster takes his thought experiment (and considerable argumentation) to show that experience, and its organization, is metaphysically fundamental; experience itself is what "ultimately determine[s] what physically obtains" (191).

Idealism does not then deny that the physical world exists. It lays out the metaphysical ground for this world which turns out to be ultimately experiential. This means there will always be two ways of thinking about the physical world and its inhabitants. One is from the point of view of the metaphysical ground, which sustains the physical world: experience. The other is the 'internal' viewpoint from within the physical world itself (cf. Foster 2008: 183ff.). A number of traditional objections can be tackled in this framework. For example, one must distinguish metaphysical from physical time. The metaphysical basis for physical time is the world-suggestive system of experience. But within physical time itself, consciousness comes after the Big Bang. Connections between neural states and states of consciousness are similarly a feature of the physical world's causal structure, even as that entire world constitutively depends on experience. The unity of the physical world is also explicable within this framework, roughly along Leibnizian lines. The experiential metaphysical foundation comprises many minds, whose totality of different viewpoints underpins a *single* physical world by joint concordance and consilience. Sometimes idealists are supposed to have particular difficulty with the problem of other minds. But since mind is constitutive of the world for idealism, the only problem is about the plurality of minds and the mere refractoriness of the world we all experience would seem to offer a ground for believing in many minds. These minds are then assigned to appropriate physical bodies in standard ways from within the physical worldview.

All these objections, however, point to a central issue. For Foster it is the world-suggestiveness of the system of experience that metaphysically underpins the existence of the physical world. But, as he recognized, this leaves open the question of what controls or generates the world-suggestive system of experience. The physicalist can here almost agree with Foster, and grant that in a way the system of experience provides a mandatory outline of a world which must be accepted as metaphysically primary in the sense that any full conception of the world must be in accord with it. However, the physicalist account of the generator of world-suggestiveness will be the familiar one: the arrangement of the basic physical entities along with the laws which govern them (quantum field theory for the 'small,' general relativity for the 'large'). This we might call the Proud Kantian position, which asserts that physics has revealed to us the nature of the thing-in-itself "beneath" and generating the empirically accessible and rightfully called "real world."

Unfortunately, Proud Kantianism carries a terrible load of perpetual failure, leading to the pessimistic induction (Laudan 1981). The history of science shows us that our current understanding of physical reality is always eventually falsified. Maxwell wrote that "there can be no doubt" about the existence of the "luminiferous aether," whose properties "have been found to be precisely those required to explain electromagnetic phenomena" (1878). The equally famous chemist Antoine Lavoisier wrote that the phenomena of heat "are the result of a real, material substance, of a very subtile fluid, that insinuates itself throughout the molecules of all bodies and pushes them apart" (Lavoisier 1790: 5). These apparently solid results of physical science turned out to be not only false but deeply false, at least according to *our* lights. There is no reason to think that finally, *now*, we have got to the 'real truth.' Science is manifestly still incomplete and our grandest and deepest theories are not merely disconnected, they are jointly *inconsistent*.

This history of epistemic woe is compounded by a more general and philosophically significant feature of science, which is that it reveals only the structural or relational properties of the world. The structuralist insight goes back a good way, at least to Poincaré (1901/1905), Russell (1927b) and Eddington (1928).[4] Bertrand Russell lamented that "physics is mathematical not because we know so much about the physical world, but because we know so little: it is only its mathematical properties that we can discover. For the rest, our knowledge is negative" (1927a: 125).

Arthur Eddington concurred: "physical science consists of purely structural knowledge, so that we know only the structure of the universe which it describes" (1939: 142). We can think of structural features in terms of dispositional properties. Science maps out a network of dispositions, ultimately of the kind that tell us that in such-and-such a configuration so-and-so will happen.

What, for example, is an electron? Leaving aside its "true" nature as mere probability excitation of a certain matter-field, the electron is an entity of mass 9.1×10^{-31} kg, charge -1.6×10^{-19} C and intrinsic angular momentum of $\pm\hbar/2$. But mass is defined as the 'resistance' a body has to acceleration when a force is applied; electric charge is that property in virtue of which a body is disposed to move in a certain way in an electromagnetic field; angular momentum is defined directly in terms of position, motion and mass. All the properties dealt with by physics are dispositional in this way, and the dispositions are all ultimately encountered and measured in Foster's immanent empirical world.

This is nicely in line with what is often called "Kantian Humility" (see Lewis 2009; Langton 1998, 2004), which says that although we have vast knowledge of the mathematical structure of the system of dispositions which define the fundamental physical properties science deals with, we know nothing about their intrinsic natures. Don't let the everyday familiarity of garden variety physical objects mislead you. They resolve into mystery. The odyssey of physics from the mechanical world view of discrete objects interacting by contact to the system of quantum fields possessed of non-local holistic features is the external image of this mystery. The world is not made of miniature Lego pieces or tiny bouncing billiard balls. It is evidently more akin to David Bohm's characterization in which the "entire universe must, on a very accurate level, be regarded as a single indivisible unit in which separate parts appear as idealizations" (Bohm 1951: 167). The bottom line is that we have absolutely no positive conception of the basic nature of the physical world.

The retreat to a humble structuralism is hard to avoid. The question of the background which generates the world-suggestiveness of our experiences remains open. Foster's own answer was to make a giant leap to a theistically grounded idealism. The *minimal* answer would be that the background, as intrinsically characterized, is restricted to generating the dispositions which are revealed in fundamental physics, and no more. Once these dispositions in the empirical realm are set up then, hopefully, all other phenomena we could ever encounter would then be metaphysically determined. This entails that all properties other than those referred to in fundamental physics are purely relational or structural properties. In the philosophy of mind, for example, this would amount to an endorsement of a broadly understood functionalism for all mental properties. Whatever the details, on this view all mental properties can be completely characterized in relational or structural terms with no residual appeal to intrinsic properties beyond those grounding the dispositions of physics.

Of course, the difficulty with this approach is that it leaves the problem of consciousness in exactly the same place we started. The primary challenge that consciousness intuitively presents is precisely that there seems to be an intrinsic residue left over after we have tried to characterize it in purely structural or relational terms. The venerable inverted color-spectrum thought experiment is clearly supposed to illustrate this unavoidable lacuna. Experiential qualities do not reduce without remainder to their place in some abstractly definable structure.

In fact, we can prove this. Let us suppose a species, not so different from our own but with a perfectly symmetrical experiential color space.[5] For reductio, suppose that the abstract structure of these creatures' color quality space is an exhaustive representation of the phenomenology associated with their color vision. Then we can immediately adapt an argument of Hilbert and Kalderon (2000). If the quality space is perfectly symmetrical then any wholesale transformation, such as inversion (or even small shifts), will make no difference to the overall relational structure. Then by our assumption there can be no difference in experiential quality due to the shift, which is absurd since one region of the space maps to, say, the reddish quality and another to the green. The situation would be akin to having a sphere with one red hemisphere and the other green, but where it is claimed that the features of every point on the sphere are exhaustively represented by the relational properties of that point with respect to all other points on the sphere. Since every point stands in exactly the same such relation to its fellows, rotating the sphere should not change anything, yet one such sphere set beside a rotated one would obviously be different.

Opponents of the idea that experiential qualities outstrip relational structure, such as Hilbert and Kalderon, will read the argument the other way: if the relational structure is an exhaustive representation of phenomenology, then a perfectly symmetrical quality space will be qualitatively uniform, and inversion will be impossible. Each side will accuse the other of begging the question.

But without a preexisting commitment to physicalism, the view that in consciousness there are intrinsic features present to the mind is the natural option. However, while this may cast doubt on the minimal answer it does not force acceptance of idealism. Two alternative responses that respect the problem of consciousness are panpsychism and some form of emergentism.

3 Panpsychism

A picture of the world grounded on physics may not fund a satisfactory answer to the problem of consciousness. But it is a vastly intricate and staggeringly comprehensive view of the natural world, in which an awful lot of what it suggests is going on has little or nothing to do with consciousness. One way to acknowledge the gravity of the problem of consciousness, while respecting the advances of physical science, is to adopt panpsychism.

Panpsychism is the view that some form of consciousness is a fundamental and ubiquitous feature of nature. But, unlike idealism, panpsychism denies that consciousness exhausts fundamental reality. To the modern sensibility, steeped in materialism and sometimes an unfortunately scientistic cultural background, panpsychism is, as we used to say, hard to get your head around. Like idealism, panpsychism is a venerable doctrine with philosophically important defenders down through the 20th century (Skrbina 2005) which fell out of favor with the general rise of materialism. It has enjoyed a remarkable renaissance over the last 20 years or so, especially after David Chalmers tentatively explored panpsychism as a possible response to his famous "hard problem" of consciousness (Chalmers 1996, ch. 8; Seager 1995).[6]

There is a straightforward argument in favor of panpsychism which was nicely codified by Thomas Nagel (1979) and which in basic form closely resembles the inconsistent triad above:

1 Consciousness is either a fundamental feature or it emerges from the fundamental.
2 Consciousness is not an emergent feature.
3 *Therefore*, consciousness is a fundamental feature.

Of course, this does not get us quite all the way to panpsychism since fundamentality does not entail ubiquity. However, if we maintain our respect for physical science we would expect that the fundamental psychic feature will be coupled to some fundamental physical feature and will thus be more or less spread out across the entire universe. For example, if – as current theory has it – the world is made of a small number of interacting quantum fields, which pervade all of spacetime then the panpsychist should hold that some or all of these fields partake in some measure of consciousness.

Panpsychism is hard to believe, or worse. John Searle (2013) calls it "absurd" and claims that the view "does not get up to the level of being false"; Colin McGinn (1999: 97) labels panpsychism as "ludicrous." Neither critic seems to have really given much sympathetic thought to the doctrine however. But they illustrate some common misconceptions. McGinn (1999: 95ff.) presents one as a dilemma for panpsychism: either it is wildly implausible or trivial.

Panpsychism is absurd, says the critic, because it claims that rocks are conscious beings. This is somewhat like the claim that since electric charge is a fundamental feature of the world, everything must be charged and have more or less the same charge. That would indeed be absurd. The panpsychist should hold that the relation between the "elementary units" of consciousness and more complex forms is not identity.

Now the charge will be vacuity. According to this complaint, the panpsychist is only saying that matter possesses an indefinable *something*, which "grounds" consciousness, a claim shared with orthodox physicalism. This complaint misses the mark if we are able to point to some common feature of consciousness: what I called "presence" or the "what it is likeness" of experience that constitutes the subjective aspect of nature.[7] Bare subjectivity in this sense does not call for complexity or an introspecting sophisticated subject, but it is far from a mere empty name for what explains consciousness without consciousness.

It is also objected that the simple physical entities of the world exhibit no sign of consciousness. There is just no empirical evidence in favor of panpsychism. Now, there is question of what counts as evidence here. Exactly what kind of behavior shows that something has a subjective aspect? Notoriously, it is possible for something to act conscious without being conscious and for something to be conscious without being able to act conscious. Consider another analogy with the physical case. What empirical evidence is there that individual electrons gravitate? They give, one by one, absolutely no detectable trace of a gravitational field. Why expect the elementary units of consciousness to give signs of consciousness discernible to us? We believe that electrons gravitate because of their place in our overall theoretical scheme. Similarly, the panpsychist assigns to fundamental entities a 'weak' consciousness, presumably of a form of unimaginable simplicity and self-opacity.

There is a kind of reverse of this negative argument in favor of panpsychism. Complex consciousness exists, and it is hard to see how it would leap into existence by some small change in material organization. In the words of William Kingdon Clifford, since "we cannot suppose that so enormous a jump from one creature to another should have occurred at any point in the process of evolution as the introduction of a fact entirely different and absolutely separate from the physical fact" (Clifford 1886: 266), consciousness must be presumed to exist at the fundamental level of reality.[8]

Of course, the fundamental features of physics are discovered via a system of experimentation and theorizing in which mental features play no part.[9] Does that mean that consciousness – or any other physically non-fundamental aspect of the world – must be epiphenomenal? That is a large philosophical question. If all the motion which matter undergoes is fully explained, or at least determined, by the fundamental interactions then there is *never* any need to appeal to consciousness to explain any behavior, or its determination at least, no less of human beings than of electrons.

But this line of thought ignores a critical incident in the history of physics. At its inception, consciousness was self-consciously excluded: the experiential side of nature was quarantined from scientific investigation as a recalcitrant realm resistant to mathematization (because not purely structural). In the words of Galileo, at the birth of mathematical physics:

> tastes, odors, colors, and so on are no more than mere names so far as the object in which we place them is concerned, and ... they reside only in the consciousness.

> Hence if the living creature were removed all these qualities would be wiped away and annihilated.
>
> *(Galilei 1623/1957: 274)*

Physics henceforth concerned itself with material motion and its causes. Physics is built, so to speak, to describe and explain a world without consciousness. Physics provides the recipe for building a world of philosophical zombies, creatures whose bodies, and the particles which make up their bodies, move exactly as we do but who entirely lack any subjective aspect. Within such a picture of the world, subjectivity has got to appear as something which has no effect on the motion of matter and, essentially, the motion of matter is all there is.

One intriguing reply to the charge of epiphenomenalism begins by recalling that science is restricted to revealing the structure of the world but not its intrinsic nature. Since structure requires something non-structural in order to make the transition from mere abstraction to concrete existence, presence, the core of subjectivity common to all consciousness, can be postulated as the intrinsic ground of the structural features outlined by physical science.[10] One of the main historical advocates of such a view was Bertrand Russell, and in its various forms the view has become known as Russellian Monism. It too has seen a remarkable renaissance of interest as the problem of consciousness refuses to release its bite (Alter and Nagasawa 2015).

Panpsychist Russellian Monism holds that consciousness, in its most basic form of pure presence or bare subjectivity, is the intrinsic nature which 'grounds' or makes concrete the system of relationally defined structure discerned by physics. We have no access to this level of reality, except for a limited acquaintance in our own experience, which is why Russell wrote that we really only ever perceive our own brains (1927b: 383).[11] Michael Lockwood explains the point as "consciousness...provides a kind of 'window' on to our brains" thereby revealing "some at least of the intrinsic qualities of the states and processes which go to make up the material world" (1989: 159). This view undercuts the charge of epiphenomenalism by giving consciousness a role in the metaphysical grounding of causal powers, while leaving the relational structure of causation entirely within the realm of physical science.

A natural question to ask within the context of panpsychist Russellian Monism is just how much humility is advisable. Granting that in consciousness we catch a glimpse of the intrinsic bedrock of the world, are there further, unknown and unknowable intrinsic natures lurking behind our structural understanding of the physical world? Such there may be, but it's a good policy not to add unnecessary hypotheses to one's theories. An intrinsic nature is needed to concretize otherwise abstract structure. We have one already to hand: presence or basic subjectivity. In the absence of positive reasons to posit additional and distinct intrinsic natures, we should refrain from such excesses of theoretical zeal.

In the face of this general scheme, what is perhaps the most serious objection to panpsychism unavoidably looms and it leads to our final subject.

4 Emergence

Panpsychism does not ascribe consciousness as we know it to everything. In fact, it is compatible with panpsychism that very few physical entities are in any way conscious at all. This is because most entities are not fundamental and are composite. Consider that although the fundamental entities (electrons, quarks) which physics posits as the constituents of familiar composites are electrically charged, the composites themselves generally lack charge. Mass is another feature possessed of these constituents, but in this case, it steadily, though not purely additively, increases as larger bodies are formed. Evidently, there is some system of relatedness that governs how the fundamental features combine in composite entities. Throughout nature there are intricate systems of relatedness leading to ever more complex properties increasingly remote from, though based upon, the properties deployed in fundamental physics.

Since panpsychism introduces an elementary form of consciousness (presence or bare subjectivity) which is associated with elementary physical entities, and since it wants to allow for a distinction between conscious and non-conscious composites, panpsychism too faces the challenge of explicating how 'mental chemistry' works, or is even possible. This is the "combination problem" (Seager 1995).[12]

The general problem which both the deceptively familiar physical and contentious mentalistic cases point to is that of emergence. In very broad terms, a property of X is emergent if none of X's constituents possess it. Liquidity is an emergent feature of water; neither oxygen nor hydrogen atoms have the property of being liquid. Our world is awash in emergence since almost no macroscopic properties of interest are shared by the fundamental entities of physics.

It is impossible here to give a comprehensive survey of the vast literature on emergence, which remains controversial in both science and philosophy (see O'Connor and Wong 2015; Gillett 2016). I will focus on a distinction between two forms of emergence and apply it to the problem of consciousness. The distinction is necessary to understand why emergence belongs within the 'radical wing' of consciousness studies.

The idea of 'mental chemistry' as an explicit system describing the emergence of complex states of consciousness goes back to John Stuart Mill (1843/1963: ch. 4). His views on emergence prefigure the more sophisticated and worked out accounts of the so-called British Emergentists (see McLaughlin 1992). The essence of this form of emergence is that it denies that the emergent properties of X are determined solely by the properties of X's constituents and the laws that govern their interactions. That is, in order for the emergent property to appear, there must be 'extra' laws of nature which specifically govern ontological emergence.

A useful way to think about this is in terms of computer simulations. We can imagine a fundamental physics simulation of parts of the world. Emergence of the kind we are considering predicts that the simulation will fail to duplicate real world behavior because it neglects the extra, cross level, laws. We can call this 'radical emergence' to contrast it with the uncontroversial and very widespread 'conservative emergence,' by which emergents are fully determined by their submergent domain.

The linchpin and supposedly obvious example which these emergentists used was that of chemistry. They regarded it as evident that chemical properties were not determined by, and *a fortiori* could not be explained by, the physical properties of the elementary constituents of a chemical substance. Taking the case of chemistry as given, they advanced the view that a host of properties "above" the chemical were also radically emergent, especially including the case of consciousness.

After 1925, the success of quantum mechanics in explaining chemical properties largely undercut any claim that radical emergence was commonplace and made it unlikely that it existed at all. Although the exact relation between physics and chemistry remains controversial, it seems that Dirac expressed the basic situation correctly, if somewhat hyperbolically, when he wrote that the "underlying physical laws necessary for … the whole of chemistry are thus completely known" (Dirac 1929: 714). Note that there is no claim here that chemistry is reducible in the sense that there is a translation and hence eliminability of chemistry in favor of physics, nor that there is no need for distinctive chemical concepts and theories to aid explanation and prediction. Rather the claim is that the entities of physics and the laws that govern them at the fundamental physical level suffice to strictly determine the chemical features of the world.

Perhaps it is not deeply surprising to find chemistry depending upon physics insofar as they both reside within the structural domain. There is no metaphysical barrier blocking determination of the complex structural patterns that chemistry picks out by the more basic structural patterns fixed on by fundamental physics. At the same time, the conservation laws militate against radical emergence. For example, if a radically emergent property is to be causally efficacious it will have to in some way alter the motion of physical matter. This requires some flux of energy, which would appear to come from nowhere and thus prima facie violate the conservation of energy. At a high level of generality, this is why we should expect that all the structure in the world should be determined by the fundamental physical structures discovered by physics.[13]

However, if consciousness cannot be exhaustively characterized in purely structural terms, then this does put up a kind of metaphysical barrier between it and what physics can describe. The panpsychist thus sees basic consciousness or bare subjectivity as ontologically fundamental in its own right. It is also evident that there is complex consciousness, which has its own relational structure, in terms of which it can be largely but not exhaustively described (as in color experience). The combination problem for panpsychism is to explain, or even make plausible, how complex consciousness can conservatively emerge from the postulated simpler forms.

Here we can note another misplaced complaint against panpsychism which is often presented as a dilemma. Since there are complex states of consciousness, panpsychism must either declare them to be a fundamental form, and hence ubiquitous in nature, or develop some account of how the complex states emerge from some much simpler forms of consciousness. If the former, panpsychism becomes even more implausible, supposing that electrons, say, enjoy a rich interior life. If the latter, then panpsychism, embracing emergence, should be replaced with the orthodox view that consciousness emerges from the physical. We can see that panpsychism should embrace emergence. It should embrace conservative emergence. The emergence of consciousness from the purely structural features outlined in physics would, however, be a very strange form of radical emergence, of doubtful coherence insofar as it holds that intrinsics emerge from the relational. On the other hand, if consciousness is, so to speak, already in play then we can hope for an account of mental chemistry, which appeals to a more plausible conservative emergence, the general existence of which everyone should accept. But this approach only works if the combination problem can be solved.

It is impossible here to canvass all the efforts to solve the combination problem, and the criticisms of them, which have been advanced (see work referred to in note 6). Let me conclude here with some basic approaches to the problem. One sort of solution is "constitutive" in the sense that the elements of basic consciousness are synchronically present in the resultant state of complex consciousness, perhaps in some way blended or "added" (Coleman 2012, Roelofs 2014). Our own experience of the unity of consciousness already hints that diverse simpler conscious states can unite into a more complex form in an intelligible way.

The second approach sees mental chemistry as a kind of "fusion" of the elementary states into a new resultant, in which the original states are eliminated (Mørch 2014; Seager 2016). This is not a retreat to radical emergence if the fusion operation is a feature of the laws that govern these elementary states. One analogy is that of the classical black hole, in which the properties of the constituents are 'erased' and all that remains are the total mass, charge and angular momentum. This obliteration is the consequence of underlying laws of nature. Another is that of quantum entanglement, in which new systems irreducible to their parts are formed under certain conditions, again, as a consequence of the basic laws governing the basic entities of quantum physics.[14]

Another approach takes the combination problem to be looking at things backwards. On this view, sometimes called "cosmopsychism," the fundamental entity is the entire world regarded as metaphysically primary, and the problem is then one of de-combining cosmic consciousness into individual minds of the sort we are introspectively familiar with (Goff forthcoming; Miller 2017).

Radical emergentist options remain open as well. In light of the distinction between structural and intrinsic features, an emergentist could hold that there are non-mentalistic intrinsic features, which ground the relational structures that science investigates. Then, upon attaining certain configurations, these intrinsic features have the power to generate wholly novel properties – those of consciousness. Although a logical possibility, both parsimony and theoretical elegance would suggest that a conservatively emergentist panpsychism is preferable.

Of course, those of a standard physicalist persuasion will hold out hope for a conservative emergentist account of consciousness based solely upon the structural features of the world as revealed by fundamental physics. One should 'never say never,' but our growing knowledge of the brain and its intimate connections to states of consciousness gives no indication of a theoretical apparatus which makes subjective consciousness an intelligible product of basic physical processes. The investigation of radical approaches remains both interesting and essential to progress in our search to understand consciousness and its place in nature.

Notes

1 The three major strands of argumentation are conveniently associated with Nagel (1974), Jackson (1982) and the triumvirate of Descartes (1641/1985, Meditation 6), Kripke (1980, Lecture 3) and Chalmers (1996, especially ch. 4).

2 Without doubt, one motivation for idealism has been epistemological: fear of skepticism. I don't think that this motivation is especially compelling however. Why not go all the way to a solipsism of the present moment if one wishes to secure an indubitable system of beliefs? Or, at least, what stops the slide towards this lonely and stultifying endpoint?

3 Perhaps we should also add that everything is *constitutively* physical, to avoid the (faint) chance that there are some rogue brute absolute necessities which link the physical to some non-physical aspect of nature (see Wilson 2005; 'correlative' vs. 'constitutive' supervenience is discussed in Seager 1991).

4 For the history see French (2014, ch. 4). A forceful presentation of this viewpoint in the context of the problem of consciousness can be found in Galen Strawson (2003, 2006).

5 The human color space of hue, saturation and brightness is asymmetrical. For example, there are more discriminable colors between blue and red than between yellow and green, even though inversion should take blue into yellow and red into green (see Byrne 2016). The issue here is clearest in the case of a symmetrical quality space, but it does not really matter since there are (rather trivial) mathematical ways to generate correspondence between asymmetrical spaces that preserve reactive dispositions by widening the scope of allowable transformations (Hoffman 2006).

6 Evidence of renewed interest can be found in dedicated publications: Rosenberg (2004); Freeman (2006); Skrbina (2009); Blamauer 2011; Brüntrup and Jaskolla (2016); Seager (forthcoming).

7 Of course, the more 'watered down' one's idea of the pan-X ground of consciousness the more on-target the charge of vacuity appears (see Chalmers 2015).

8 An interesting contrast here is with the emergence of life. As we now know, life is fully and intelligibly explicated in terms of purely chemical processes. Unlike the case of consciousness, these exhibit no 'enormous jump' as they increase in structural complexity from the non-living to the living.

9 This is actually controversial. Some interpretations of quantum mechanics hold that consciousness is a fundamental feature of reality required to make measurements of quantum systems determinate (see Wigner 1962, London and Bauer 1939/1983).

10 It is possible to question this 'argument from concreteness' (Ladyman *et al.* 2007), but then some account of 'concrete structure' is required which makes mathematics, some of it but not all of it, 'real.' One must do this carefully to avoid making all possible structures trivially instantiated because of what is known as Newman's Problem (1928): structure is abstractly definable in terms of ordered sets which exist as soon as their members do. Structure unconstrained by some intrinsic reality is too easy to come by.

11 While Russellian Monism is nicely adaptable to panpsychism, Russell himself was not a panpsychist. Following William James, he endorsed Neutral Monism, in which the most fundamental features of

reality are neither mental nor physical. These latter are constructs from the neutral material (see Tully 2003). James's relation to panpsychism is somewhat murky but it seems that he ends up accepting it (see Cooper 1990).

12 The problem was first noted by William James (1890/1950, ch. 6). For discussions see Brüntrup and Jaskolla (2016), Seager (forthcoming). For a sustained investigation of the general problem of whether conscious subjects could 'combine' see Roelofs (2015).

13 This is not to say that radical emergence lacks contemporary defenders; see O'Connor (1994); O'Connor and Wong (2005); Silberstein and McGeever (1999).

14 Although developed in a different context, something like the idea of fusion is presented in work of Paul Humphreys (1997b, 1997a).

References

Alter, T. and Nagasawa, Y. (eds.) (2015) *Consciousness in the Physical World: Perspectives on Russellian Monism*, Oxford: Oxford University Press.

Blamauer, M. (ed.) (2011) *The Mental as Fundamental: New Perspectives on Panpsychism*, Frankfurt: Ontos Verlag.

Block, Ned and Stalnaker, R. (1999) "Conceptual Analysis, Dualism, and the Explanatory Gap," *Philosophical Review* 108: 1–46.

Bohm, D. (1951) *Quantum Theory*, Englewood Cliffs, NJ: Prentice-Hall.

Brüntrup, G. and Jaskolla, L. (eds.) (2016) *Panpsychism*, Oxford: Oxford University Press.

Byrne, Alex (2016) "Inverted Qualia," In Edward N. Zalta (ed.), *The Stanford Encyclopedia of Philosophy*. Stanford, CA: Metaphysics Research Lab, Stanford University, Winter ed. URL: https://plato.stanford.edu/archives/win2016/entries/qualia-inverted/.

Chalmers, D. (1996) *The Conscious Mind: In Search of a Fundamental Theory*, Oxford: Oxford University Press.

Chalmers, D. (2015) "Panpsychism and Panprotopsychism," In T. Alter and Y. Nagasawa (eds.) *Consciousness in the Physical World: Essays on Russellian Monism*, Oxford: Oxford University Press.

Chalmers, D. (forthcoming) "Idealism," In W. Seager (ed.) *The Routledge Handbook of Panpsychism*, London: Routledge.

Clifford, W. (1886) "Body and Mind," In L. Stephen and F. Pollock (eds.) *Lectures and Essays*, London: Macmillan, 2nd ed. (Originally published in the December 1874 issue of *Fortnightly Review*).

Coleman, S. (2012) "Mental Chemistry: Combination for Panpsychists," *Dialectica* 66: 137–166.

Cooper, W. (1990) "William James's Theory of Mind," *Journal of the History of Philosophy* 28: 571–593.

Descartes, R. (1641/1985) "Meditations of First Philosophy," In J. Cottingham, R. Stoothoff and D. Murdoch (eds.) *The Philosophical Writings of Descartes, Vol. 2*. Cambridge: Cambridge University Press. (J. Cottingham, R. Stoothoff, D. Murdoch, trans.).

Dirac, P. A. M. (1929) "Quantum Mechanics of Many-Electron Systems," *Proceedings of the Royal Society of London. Series A, Containing Papers of a Mathematical and Physical Character* 123 (792): 714–733.

Eddington, A. (1928) *The Nature of the Physical World*, New York: Macmillan & Co.

Eddington, A. (1939) *The Philosophy of Physical Science*, New York: Macmillan & Co.

Foster, J. (2008) *A World for Us: The Case for Phenomenalistic Idealism*, Oxford: Oxford University Press.

Freeman, A. (ed.) (2006) *Consciousness and Its Place in Nature*, Exeter: Imprint Academic.

French, S. (2014) *The Structure of the World: Metaphysics and Representation*, Oxford: Oxford University Press.

Galilei, G. (1623/1957) "The Assayer," In D. Stillman (ed.) *Discoveries and Opinions of Galileo*, New York: Anchor Books (D. Stillman, trans.).

Gillett, C. (2016) *Reduction and Emergence in Science and Philosophy*, Cambridge: Cambridge University Press.

Goff, P. (forthcoming) "Cosmopsychism, Micropsychism and the Grounding Relation," In W. Seager (ed.) *The Routledge Handbook of Panpsychism*, London: Routledge.

Guyer, P. and Horstmann, R. (2015) "Idealism," In Edward N. Zalta (ed.), *The Stanford Encyclopedia of Philosophy*. Stanford, CA: Metaphysics Research Lab, Stanford University, Fall ed. URL: https://plato.stanford.edu/archives/fall2015/entries/idealism/.

Hameroff, S. and Penrose, R. (1996) "Conscious Events as Orchestrated Space-Time Selections," *Journal of Consciousness Studies* 3: 36–53.

Hilbert, D. and Kalderon, M. E. (2000) "Color and the Inverted Spectrum," In S. Davis (ed.) *Color Perception: Philosophical, Psychological, Artistic and Computational Perspectives*, Oxford: Oxford University Press.

Hoffman, D. (2006) "The Scrambling Theorem: A Simple Proof of the Logical Possibility of Spectrum Inversion," *Consciousness and Cognition* 15: 31–45.

Humphreys, P. (1997a) "Emergence, Not Supervenience," *Philosophy of Science* 64: S337–345.
Humphreys, P. (1997b) "How Properties Emerge," *Philosophy of Science* 64: 1–17.
Huxley, T. (1866) *Lessons in Elementary Physiology*, London: Macmillan.
Jackson, F. (1982) "Epiphenomenal Qualia," *Philosophical Quarterly* 32: 127–136.
James, W. (1890/1950) *The Principles of Psychology*, vol. 1, New York: Henry Holt and Co. Reprinted in 1950, New York: Dover. (Page references to the Dover edition.)
Kripke, S. (1980) *Naming and Necessity*, Cambridge: Cambridge University Press.
Ladyman, J., Ross, D., Spurrett, D., and Collier, J. (2007) *Everything Must Go: Metaphysics Naturalized*, Oxford: Oxford University Press.
Langton, R. (1998) *Kantian Humility: Our Ignorance of Things in Themselves*, Oxford: Oxford University Press.
Langton, R. (2004) "Elusive Knowledge of Things in Themselves," *Australasian Journal of Philosophy* 82: 129–136.
Laudan, L. (1981) "A Confutation of Convergent Realism," *Philosophy of Science* 48: 19–49.
Lavoisier, A. (1790) *Elements of Chemistry*, Edinburgh: William Creech. R. Kerr (trans.).
Lewis, D. (2009) "Ramseyan Humility," In D. Braddon-Michell and R. Nola (eds.) *Conceptual Analysis and Philosophical Naturalism*, Cambridge, MA: MIT Press (Bradford Books).
Lockwood, M. (1989) *Mind, Brain and the Quantum*, Oxford: Blackwell.
London, F. and Bauer, E. (1939/1983) "The Theory of Observation in Quantum Mechanics," In J. Wheeler and W. Zurek (eds.) *Quantum Theory and Measurement*, Princeton: Princeton University Press. Originally published as 'La théorie de l'observation en mécanique quantique' in *Actualités scientifiques et industrielles*, no. 775, Paris: Heinemann, 1939.
Maxell, J. C. (1878) "Ether," In T. Baynes (ed.) *Encyclopedia Britannica, Ninth Edition*, Vol. 8, Edinburgh: A. & C. Black.
McGinn, C. (1999) *The Mysterious Flame: Conscious Minds in a Material World*, New York: Basic Books.
McLaughlin, B. (1992) "The Rise and Fall of British Emergentism," In A. Beckermann, H. Flohr and J. Kim (eds.) *Emergence or Reduction*, Berlin: De Gruyter.
Mill, J. S. (1843/1963) *A System of Logic*, vols. 7-8 of *The Collected Works of John Stuart Mill*, Toronto: University of Toronto Press.
Miller, G. (2017) "Can Subjects Be Proper Parts of Subjects? The De-Combination Problem," *Ratio*. URL: http://dx.doi.org/10.1111/rati.12166. DOI = 10.1111/rati.12166.
Mørch, H. (2014) *Panpsychism and Causation: A New Argument and a Solution to the Combination Problem*, Ph.D. thesis, University of Oslo. URL: https://philpapers.org/rec/HASPAC-2.
Nagel, T. (1974) "What Is It Like to Be a Bat?" *Philosophical Review* 83: 435–450. (This article is reprinted in many places, notably in Nagel's *Mortal Questions*, Cambridge: Cambridge University Press, 1979.)
Nagel, T. (1979) "Panpsychism," In *Mortal Questions*, Cambridge: Cambridge University Press. (Reprinted in D. Clarke *Panpsychism: Past and Recent Selected Readings*, Albany, NY: SUNY Press, 2004.)
Newman, M. (1928) "Mr. Russell's Causal Theory of Perception," *Mind* 37: 137–148.
O'Connor, T. (1994) "Emergent Properties," *American Philosophical Quarterly* 31: 91–104.
O'Connor, T. and Wong, H.Y. (2005) "The Metaphysics of Emergence," *Noûs* 39: 658–678.
O'Connor, Timothy and Hong Yu Wong (2015). 'Emergent Properties'. In E. Zalta (ed.), *The Stanford Encyclopedia of Philosophy*. Stanford University: The Metaphysics Research Lab, Summer ed. URL: https://plato.stanford.edu/archives/sum2015/entries/properties-emergent.
Pelczar, M. (2015) *Sensorama: A Phenomenalist Analysis of Spacetime and Its Contents*, Oxford: Oxford University Press.
Poincaré, H. (1901/1905) *Science and Hypothesis*, New York: Walter Scott Publishing Co. Ltd. (W. J. Greenstreet, trans.).
Roelofs, L. (2014) "Phenomenal Blending and the Palette Problem," *Thought: A Journal of Philosophy* 3: 59–70.
Roelofs, L. (2015) *Combining Minds: A Defence of the Possibility of Experiential Combination*, Ph.D. thesis, Univ. of Toronto, Toronto. URL: https://tspace.library.utoronto.ca/handle/1807/69449.
Rosenberg, G. (2004) *A Place for Consciousness: Probing the Deep Structure of the Natural World*, Oxford: Oxford University Press.
Russell, B. (1927a) *An Outline of Philosophy*, London: George Allen & Unwin.
Russell, B. (1927b) *The Analysis of Matter*, London: K. Paul, Trench, Trubner.
Seager, W. (1991) *Metaphysics of Consciousness*, London: Routledge.
Seager, W. (1995) "Consciousness, Information and Panpsychism," *Journal of Consciousness Studies* 2 (3): 272–288. (Reprinted in J. Shear (ed.) *Explaining Consciousness*, Cambridge, MA: MIT Press, 1997.)

Seager, W. (2016) "Panpsychist Infusion," In G. Brüntrup and L. Jaskolla (eds.) *Panpsychism*, Oxford: Oxford University Press.
Seager, W. (ed.) (forthcoming) *The Routledge Handbook on Panpsychism*, London: Routledge.
Searle, J. (2013) "Can Information Theory Explain Consciousness," *New York Review of Books* (January 10).
Silberstein, M. and McGeever, J. (1999) "The Search for Ontological Emergence," *Philosophical Quarterly* 49: 182–200.
Skrbina, D. (2005) *Panpsychism in the West*, Cambridge, MA: MIT Press.
Skrbina, D. (ed.) (2009) *Mind That Abides: Panpsychism in the New Millennium*, Amsterdam: John Benjamins.
Sprigge, T. (1983) *The Vindication of Absolute Idealism*, Edinburgh: Edinburgh University Press.
Sprigge, T. (2010) "Absolute Idealism," In L. McHenry (ed.) *The Importance of Subjectivity: Selected Essays in Metaphysics and Ethics*, Oxford: Oxford University Press.
Strawson, G. (2003) 'Real Materialism," In L. Anthony and N. Hornstein (eds.) *Chomsky and His Critics*, Oxford: Blackwell. (Reprinted, with new postscript, in T. Alter and Y. Nagasawa (eds.) *Consciousness in the Physical World: Perspectives on Russellian Monism*, Oxford: Oxford University Press, 2015, 161–208.)
Strawson, G. (2006) "Realistic Monism: Why Physicalism Entails Panpsychism," *Journal of Consciousness Studies* 13 (10-11): 3–31. (Reprinted in A. Freeman (ed.) *Consciousness and Its Place in Nature*, Exeter: Imprint Academic, 2006.)
Thagard, P. and Stewart, T.C. (2014) "Two Theories of Consciousness: Semantic Pointer Competition vs. Information Integration," *Consciousness and Cognition* 30: 73–90.
Tononi, G. (2012) *Phi: A Voyage from the Brain to the Soul*, New York: Pantheon Books.
Tully, R. (2003) "Russell's Neutral Monism," In N. Griffin (ed.) *Cambridge Companion to Bertrand Russell*, 332–370. Cambridge: Cambridge University Press.
Tyndall, J. (1879) *Fragments of Science: A Series of Detached Essays, Addresses and Reviews*, London: Longmans, Green & Co., 6th edition.
Wigner, E. (1962) "Remarks on the Mind-Body Problem," In I. Good (ed.) *The Scientist Speculates*, London: Heinemann. (Reprinted in J. Wheeler and W. Zurek (eds.) *Quantum Theory and Measurement*, Princeton: Princeton University Press, 1983, 168–181.)
Wilson, J. (2005) "Supervenience-Based Formulations of Physicalism," *Nous* 39: 426–459.

Related Topics

Dualism
Materialism
Consciousness in Western Philosophy
Quantum Theories of Consciousness

6

CONSCIOUSNESS, FREE WILL, AND MORAL RESPONSIBILITY

Gregg D. Caruso

In recent decades, with advances in the behavioral, cognitive, and neurosciences, the idea that patterns of human behavior may ultimately be due to factors beyond our conscious control has increasingly gained traction and renewed interest in the age-old problem of free will. To properly assess what, if anything, these empirical advances can tell us about free will and moral responsibility, we first need to get clear on the following questions: Is consciousness necessary for free will? If so, what role or function must it play? For example, are agents morally responsible for actions and behaviors that are carried out automatically or without conscious control or guidance? Are they morally responsible for actions, judgments, and attitudes that are the result of implicit biases or situational features of their surroundings of which they are unaware? Clarifying the relationship between consciousness and free will is imperative if we want to evaluate the various arguments for and against free will.

In this chapter, I will outline and assess several distinct views on the relationship between consciousness and free will, focusing in particular on the following three broad categories:

1 The first maintains that consciousness is a necessary condition for free will and that the condition can be satisfied. Such views affirm the existence of free will and claim conscious control, guidance, initiation, broadcasting, and/or awareness are essential for free will. Different accounts will demand and impart different functions to consciousness, so this category includes a number of distinct views.

2 The second category also maintains that consciousness is a necessary condition for free will, but believes that recent developments in the behavioral, cognitive, and neurosciences either shrinks the realm of free and morally responsible action or completely eliminates it. I include here two distinct types of positions: (2a) The first denies the causal efficacy of *conscious will* and receives its contemporary impetus from pioneering work in neuroscience by Benjamin Libet, Daniel Wegner, and John-Dylan Haynes; the second (2b) views the real challenge to free will as coming not from neuroscience, but from recent work in psychology and social psychology on *automaticity, situationism, implicit bias*, and the *adaptive unconscious*. This second class of views does not demand that *conscious will* or *conscious initiation of action* is required for free will, but rather conscious awareness, broadcasting, or integration of certain relevant features of our actions, such as their morally salient features.

It further maintains that developments in psychology and social psychology pose a threat to this consciousness condition (see Caruso 2012, 2015b; Levy 2014).

3 A third class of views simply thinks consciousness is irrelevant to the free will debate. I include here traditional conditional analyses approaches as well as many *deep self* and *reasons-responsive* accounts that either ignore or explicitly reject a role for consciousness. Classical compatibilism, for example, typically focused on the correct semantic analysis of the expression "could have done otherwise," without any reference to consciousness or experience. More recently, a growing number of contemporary philosophers have explicitly rejected a consciousness condition for free will, focusing instead on features of the agent that are presumably independent of consciousness. Prominent examples include: Nomy Arplay (2002), Angela Smith (2005), and George Sher (2009). These philosophers typically rely on everyday examples of agents who appear free and morally responsible in the relevant sense but who act for reasons of which they are apparently unconscious.

1 Free Will and Moral Responsibility

Before discussing each of the categories in detail, let me begin by defining what I mean by free will and moral responsibility. The concept of *free will*, as it is typically understood in the contemporary debate, is a term of art referring to the control in action required for a core sense of moral responsibility. This sense of moral responsibility is traditionally set apart by the notion of *basic desert* and is purely backward-looking and non-consequentialist (see Feinberg 1970; Pereboom 2001, 2014; G. Strawson 1994; Caruso and Morris 2017). Understood this way, free will is a kind of power or ability an agent must possess in order to justify certain kinds of desert-based judgments, attitudes, or treatments in response to decisions or actions that the agent performed or failed to perform. These reactions would be justified on purely backward-looking grounds, and would not appeal to consequentialist or forward-looking considerations—such as future protection, future reconciliation, or future moral formation.

Historically, the problem of free will has centered on *determinism*—the thesis that every event or action, including human action, is the inevitable result of preceding events and actions and the laws of nature. *Hard determinists* and *libertarians* argue that causal determinism is incompatible with free will—either because it precluded the *ability to do otherwise* (leeway incompatibilism), or because it is inconsistent with one's being the "ultimate source" of action (source incompatibilism). The two views differ, however, on whether or not they accept determinism. Hard determinists claim that determinism is true and hence there is no free will, while libertarians reject determinism and defend an indeterminist conception of free will. *Compatibilists*, on the other hand, attempt to reconcile determinism and free will. They hold that what is of utmost importance is not the falsity of determinism, nor that our actions are uncaused, but that our actions are voluntary, free from constraint and compulsion, and caused in the appropriate way.

More recently a new crop of *free will skeptics*—i.e., those who doubt or deny the existence of free will—has emerged, who are agnostic about the truth of determinism. Most argue that while determinism is incompatible with free will and moral responsibility, so too is *indeterminism*, especially the variety posited by quantum mechanics (Pereboom 2001, 2014; Caruso 2012). Others argue that regardless of the causal structure of the universe, we lack free will and moral responsibility because free will is incompatible with the pervasiveness of *luck* (Levy 2011). Others (still) argue that free will and ultimate moral responsibility are incoherent concepts, since to be free in the sense required for ultimate moral responsibly we would have to be *causa sui* (or "cause of oneself") and this is impossible (Strawson 1994, 1986). What all these arguments for free will skepticism have in common is the claim that what we do, and the way we are, is ultimately the result of factors beyond our control, and because of this we are never morally responsible for our actions in the *basic desert* sense.

In addition to these philosophical arguments, there have also been recent developments in the behavioral, cognitive, and neurosciences that have caused many to take free will skepticism seriously. Chief among them have been findings in neuroscience that appear to indicate that unconscious brain activity causally initiates action prior to the conscious awareness of the intention to act (Libet et al. 1993; Soon et al. 2008), and recent findings in psychology and social psychology on automaticity, situationism, and the adaptive unconscious (Nisbett and Wilson 1997; Bargh 1997; Bargh and Chartrand 1999; Bargh and Ferguson 2000; Doris 2002; Wilson 2002). Viewed collectively, these developments suggest that much of what we do takes place at an automatic and unaware level and that our commonsense belief that we consciously initiate and control action may be mistaken. They also indicate that the causes that move us are often less transparent to ourselves than we might assume—diverging in many cases from the conscious reasons we provide to explain and/or justify our actions. No longer is it believed that only "lower level" or "dumb" processes can be carried out non-consciously. We now know that the higher mental processes that have traditionally served as quintessential examples of "free will"—such as evaluation and judgment, reasoning and problem solving, and interpersonal behavior—can and often do occur in the absence of conscious choice or guidance.

For some, these findings represent a serious threat to our everyday folk understanding of ourselves as conscious, rational, responsible agents—since they indicate that the conscious mind exercises less control over our behavior than we have traditionally assumed. In fact, even some compatibilists now admit that because of these behavioral, cognitive, and neuroscientific findings, "free will is at best an occasional phenomenon" (Baumeister 2008: 17). This is an important concession because it acknowledges that the *threat of shrinking agency*—as Thomas Nadelhoffer (2011) calls it—remains a serious one, independent of any traditional concerns over determinism. That is, even if one believes free will can be reconciled with determinism, chance, or luck, the deflationary view of consciousness that emerges from these empirical findings must still be confronted, including the fact that we often lack transparent awareness of our true motivational states. Such a deflationary view of consciousness is potentially agency undermining and must be dealt with independent of, and in addition to, the traditional compatibilist/incompatibilist debate (see e.g. Sie and Wouters 2010; Nadelhoffer 2011; King and Carruthers 2012; Caruso 2012, 2015b; Levy 2014).

2 Is Consciousness Necessary for Free Will?

Turning now to the relationship between consciousness and free will, the three categories outlined above are largely defined by how they answer the following two questions: (1) Is consciousness necessary for free will? And if so, (2) can the consciousness requirement be satisfied given the threat of shrinking agency and recent developments in the behavioral, cognitive, and neurosciences? Beginning with the first question, we can identify two general sets of views—those that reject and those that accept a *consciousness condition* on free will. The first group includes philosophers like Nomy Arpaly (2002), Angela Smith (2005), and George Sher (2009), who explicitly deny that consciousness is needed for agents to be free and morally responsible. The second group, which includes Neil Levy (2014), Gregg Caruso (2012, 2015b), and Joshua Shepherd (2012, 2015), argue instead that consciousness *is* required and that accounts that downplay, ignore, or explicitly deny a role for consciousness are significantly flawed and missing something important.

Among those who deny that consciousness is necessary for free will are many proponents of the two leading theories of free will and moral responsibility: *deep self* and *reasons-responsive* accounts. Contemporary proponents of deep self-accounts, for instance, advocate for an updated version of what Susan Wolf (1990) influentially called the *real self*-view, in that they ground an agent's moral responsibility for her actions "in the fact...that they express who she is as an agent" (Smith 2008: 368). According to deep self-accounts, an agent's free and responsible actions should bear some kind of relation to the features of the psychological structure constitutive of the agent's *real* or *deep* self (Arpaly and Schroeder 1999; Arpaly 2002; Wolf 1990). Deep self theorists typically disagree on which psychological elements are most relevant, but importantly none of them emphasize consciousness. In fact, some explicitly deny that expression of who we are as agents requires that we be conscious either of the attitudes we express in our actions or the moral significance of our actions (see e.g. Arpaly 2002; Smith 2005). Deep self accounts, therefore, generally fall into the third category identified in the introduction.

Reasons-responsive accounts also tend to dismiss the importance of consciousness. According to John Martin Fischer and Mark Ravizza's (1998) influential account, responsibility requires not *regulative* control—actual access to alternative possibilities—but only *guidance* control. And, roughly speaking, an agent exercises guidance control over her actions if she recognizes reasons, including moral reasons, as motivators to do otherwise, and she would actually do otherwise in response to some such reasons in a counterfactual scenario. But, as Shepherd (2015) and Levy (2014) have noted, such accounts typically impart no significant role to consciousness. Indeed, Gideon Yaffe claims that "there is no reason to suppose that consciousness is required for reasons-responsiveness" (2012: 182). Given this, reasons-responsive accounts can also be placed in the third category.

Let me take a moment to briefly discuss Sher and Smith's accounts, since they are representative of the kinds of views that reject a consciousness requirement on free will. Most accounts of moral responsibility maintain an *epistemic condition* along with a *control condition*—with perhaps some additional conditions added. The former demands that an agent know what they are doing in some important sense, while the latter specifies the kind of control in action needed for moral responsibility. In *Who Knew? Responsibility Without Awareness* (2009), Sher focuses on the epistemic condition and criticizes a popular but, in his view, inadequate understanding of it. His target is the "searchlight view," which assumes that agents are responsible only for what they are aware of doing or bringing about—i.e., that their responsibility extends only as far as the searchlight of their consciousness. Sher argues that the searchlight view is (a) inconsistent with our attributions of responsibility to a broad range of agents who *should but do not realize* that they are acting wrongly or foolishly, and (b) not independently defensible. Sher defends these criticisms by providing everyday examples of agents who intuitively appear morally responsible, but who act for reasons of which they are ignorant or unaware. The basic idea behind Sher's positive view is that the relation between an agent and her failure to recognize the wrongness of what she is doing should be understood in causal terms—i.e., the agent is responsible when, and because, her failure to respond to her reasons for believing that she is acting wrongly has its origins in the same constitutive psychology that generally does render her reasons-responsive.

Angela Smith (2005) likewise argues that we are justified in holding ourselves and others responsible for actions that do not appear to reflect a conscious choice or decision. Her argument, however, is different than Sher's, since she attacks the notion that voluntariness (or active control) is a precondition of moral responsibility rather than the epistemic condition. She writes, "our commonsense intuitions do not, in fact, favor a volitionalist criterion of responsibility, but a rationalist one." That is to say, "the kind of activity implied by our moral practices is not the activity of [conscious] choice, but the activity of evaluative judgment." She argues that this distinction is important, "because it allows us to say that what makes an attitude 'ours' in the sense relevant to questions of responsibility and moral assessment is not that we have voluntarily chosen it or that we have voluntary control over it, but that it reflects our own evaluative judgments or appraisals" (2005: 237). Smith then proceeds by considering various examples designed to bring out the intuitive plausibility of the *rational relations view*, while at the same time casting doubt upon the claim that we ordinarily take conscious choice or voluntary control to be a precondition of legitimate moral assessment.

Contrary to these views, Neil Levy (2014), Joshua Shepherd (2012, 2015), and Gregg Caruso (2012, 2015b) have argued that consciousness *is* in fact required for free will and moral responsibility—and accounts like those described above that deny or reject a consciousness condition are untenable, flawed, and perhaps even incoherent. Neil Levy, for example, has argued for something he calls the *consciousness thesis*, which maintains that "consciousness of some of the facts that give our actions their moral significance is a necessary condition for moral responsibility" (2014: 1). He contends that since consciousness plays the role of integrating representations, behavior driven by non-conscious representations is inflexible and stereotyped, and only when a representation is conscious "can it interact with the full range of the agent's personal-level propositional attitudes" (2014: vii). This fact entails that consciousness of key features of our actions is a necessary (though not sufficient) condition for moral responsibility, since consciousness of the morally significant facts to which we respond is required for these facts to be assessed by and expressive of the agent him/herself.

Levy further argues that the two leading accounts of moral responsibility outlined above—*deep self* (or what he calls *evaluative accounts*) and *reasons-responsive* (or *control-based*) accounts—are committed to the truth of the consciousness thesis, despite what proponents of these accounts maintain. And this is because: (a) only actions performed consciously express our evaluative agency, and that expression of moral attitudes requires consciousness of that attitude; and (b) we possess reasons-responsive control only over actions that we perform consciously, and that control over their moral significance requires consciousness of that moral significance.

In assessing Levy's consciousness thesis, a couple of things are important to keep in mind. First, the kind of consciousness Levy has in mind is *not* phenomenal consciousness but rather states with *informational* content. That is, he limits himself to philosophically arguing for the claim that "contents that might plausibly ground moral responsibility are *personally* available for report (under report-conducive conditions) and for driving further behavior, but also occurrent [in the sense of] shaping behavior or cognition" (2014: 31).

Second, on Levy's account, information of the right kind must be personally available to ground moral responsibility. But what kind of information is the right kind? Rather than demanding consciousness of all relevant mental states, Levy argues that when agents are morally blameworthy or praiseworthy for acting in a certain manner, they must be conscious of certain facts which play an especially important role in explaining the *valence* of responsibility. Valence, in turn, is defined in terms of moral significance: "facts that make the action bad play this privileged role in explaining why responsibility is valenced negatively, whereas facts that make the action good play this role in explaining why the responsibility is valenced positively" (2014: 36). Additionally, the morally significant facts that determine the valence need not track the actual state of affairs that pertain, but the facts that the agent *takes* to pertain. According to the consciousness thesis, then, if an action is morally bad the agent must be conscious of (some of) the aspects that make it bad, and conscious of those aspects under appropriate descriptions in order to be blameworthy for the action.

I should note that in *Free Will and Consciousness* (Caruso 2012), I also argued for a consciousness thesis—though there I argued for the claim that conscious control and guidance were of utmost importance. That is, I argued that, "for an action to be free, consciousness must be involved in *intention* and *goal formation*" (2012: 100). My reasoning was motived by cases of somnambulism and concerns over automaticity and the adaptive unconscious (2012: 100–130) where conscious executive control and guidance are largely absent. More recently, however, I have come to think that Levy's consciousness thesis, or something close to it, is more accurate (see Caruso 2015a, b). This is because, first, I no longer think that the empirical challenges to *conscious will* from neuroscience are all that relevant to the problem of free will (see Pereboom and Caruso 2018). Second, many of the arguments I presented in the book are captured just as well, perhaps better, by Levy's version of the consciousness thesis—including my internal challenge to compatibilism based on recent developments in the psychology, social psychology, and cognitive science. Finally, Levy's consciousness thesis has the virtue of capturing what I believe is an intuitive component of the epistemic condition on moral responsibility (contra Sher)—i.e., that agents must be aware of important moral features of their choices and actions to be responsible for them. The one remaining difference between us is that I still prefer to understand and explain consciousness in terms of the *Higher-Order Thought* (HOT) theory of consciousness (Caruso 2012, 2005; Rosenthal 2005), while Levy favors the Global Workspace Theory (Levy 2014; see also Baars 1988, 1997; Dehaene and Naccache 2001; Dehaene, Changeux, and Naccache 2011).

Joshua Shepherd (2012, 2015) has also argued that consciousness is a necessary condition for free will, but his argument is based on taking our folk psychological commitments seriously. In a series of studies, he provides compelling evidence that ordinary folk accord a central place to consciousness when it comes to free will and moral responsibility—furthermore, "the way in which it is central is not captured by extant [Real or] Deep Self Views" (2015: 938).

3 If Consciousness Is Necessary for Free Will, Can We Ever Be Free and Morally Responsible?

Assuming for the moment that consciousness is required for free will, the next question would be: Can the consciousness requirement be satisfied given the threat of shrinking agency and empirical findings in the behavioral, cognitive, and neurosciences? In the literature, two leading empirical threats to the consciousness condition are identifiable. The first maintains that recent findings in neuroscience reveal that unconscious brain activity causally initiates action prior to the conscious awareness of the intention to act and that this indicates *conscious will* is an illusion. The pioneering work in this area was done by Benjamin Libet and his colleagues. In their groundbreaking study on the neuroscience of movement, Libet et al. (1983) investigated the timing of brain processes and compared them to the timing of conscious intention in relation to self-initiated voluntary acts and found that the conscious intention to move (which they labeled *W*) came 200 milliseconds before the motor act, but 350–400 milliseconds after *readiness potential*—a ramp-like buildup of electrical activity that occurs in the brain and precedes actual movement. Libet and others have interpreted this as showing that the conscious intention or decision to move cannot be the cause of action because it comes too late in the neuropsychological sequence (see Libet 1985, 1999). According to Libet, since we become aware of an intention to act only after the onset of preparatory brain activity, the conscious intention cannot be the true cause of the action.

Libet's findings, in conjunction with additional findings by John-Dylan Haynes (Soon et al. 2008) and Daniel Wegner (2002), have led some theorists to conclude that conscious will is an illusion and plays no important causal role in how we act. Haynes and his colleagues, for example, were able to build on Libet's work by using functional magnetic resonance imaging (fMRI) to predict with 60% accuracy whether subjects would press a button with either their right or left hand up to 10 seconds before the subject became aware of having made that choice (Soon et al. 2008). For some, the findings of Libet and Haynes are enough to threaten our conception of ourselves as free and responsible agents since they appear to undermine the causal efficacy of the types of willing required for free will.

Critics, however, maintain that there are several reasons for thinking that these neuroscientific arguments for free will skepticism are unsuccessful. First, critics contend that there is no direct way to tell which conscious phenomena, if any, correspond to which neural events. In particular, in the Libet studies, it is difficult to determine what the readiness potential corresponds to—for example, is it an *intention formation* or *decision*, or is it merely an *urge* of some sort? Al Mele (2009) has argued that the readiness potential (RP) that precedes action by a half-second or more need not be construed as the *cause* of the action. Instead, it may simply mark the beginning of forming an *intention* to act. On this interpretation, the RP is more accurately characterized as an "urge" to act or a preparation to act. That is, it is more accurately characterized as the advent of items in what Mele calls the *preproximal-intention group* (or PPG). If Mele is correct, this would leave open the possibility that conscious intentions can still be causes.

A second criticism is that almost everyone on the contemporary scene who believes we have free will, whether compatibilist or libertarian, also maintains that freely willed actions are caused by a chain of events that stretch backwards in time indefinitely. At some point in time these events will be such that the agent is not conscious of them. Thus, all free actions are caused, at some point in time, by unconscious events. However, as Eddy Nahmias (2011) points out, the concern for free will raised by Libet's work is that *all* of the relevant causing of action is (typically) non-conscious, and consciousness is not causally efficacious in producing action. Given determinist compatibilism, however, it's not possible to establish this conclusion by showing that non-conscious events that precede conscious choice causally determine action, since such compatibilists hold that every case of action will feature such events, and that this is compatible with free will. And given most incompatibilist libertarianisms, it's also impossible to establish this conclusion by showing that there are non-conscious events that render actions more probable than not by a factor of 10% above chance (Soon et al. 2008), since almost all such libertarians hold that free will is compatible with such indeterminist causation by unconscious events at some point in the causal chain (De Caro 2011).

Other critics have noted the unusual nature of the Libet-style experimental situation—i.e., one in which a conscious intention to flex at some time in the near future is already in place, and what is tested for is the specific implementation of this general decision. Nahmias (2011), for example, points out that it's often the case—when, for instance, we drive or play sports or cook meals—that we form a conscious intention to perform an action of a general sort, and subsequent specific implementations are not preceded by more specific conscious intentions. But in such cases, the general conscious intention is very plausibly playing a key causal role. In Libet-style situations, when the instructions are given, subjects form conscious intentions to flex at some time or other, and if it turns out that the specific implementations of these general intentions are not in fact preceded by specific conscious intentions, this would be just like the kinds of driving and cooking cases Nahmias cites. It seems that these objections cast serious doubts on the potential for neuroscientific studies to undermine the claim that we have the sort of free will at issue.

But even if neuroscience is not able to refute free will, there are other empirical threats to free will and moral responsibility that remain. And these threats challenge a different sort of consciousness thesis—the one proposed by Neil Levy. In fact, Levy argues that those who think the work of Libet and Wegner undermine free will and moral responsibility are "wrong in claiming that it is a conceptual truth that free will (understood as the power to act such that we are morally responsible for our actions) requires the ability consciously to initiate action" (2014: 16). Instead, for Levy, what is of true importance is the causal efficacy of deliberation. Levy's consciousness thesis therefore demands not the conscious initiation of action, but rather consciousness of the facts that give our actions their moral significance.

In defending the consciousness thesis, Levy argues that the integration of information that consciousness provides allows for the flexible, reasons-responsive, online adjustment of behavior. Without such integration, "behaviors are stimulus driven rather than intelligent responses to situations, and their repertoire of responsiveness to further information is extremely limited" (2014: 39). Consider, for example, cases of *global automatism*. Global automatisms may arise as a consequence of frontal and temporal lobe seizures and epileptic fugue, but perhaps the most familiar example is somnambulism. Take, for instance, the case of Kenneth Parks, the Canadian citizen who on May 24, 1987 rose from the couch where he was watching TV, put on his shoes and jacket, walked to his car, and drove 14 miles to the home of his parents-in-law where he proceeded to strangle his father-in-law into unconsciousness and stab his mother-in-law to death. He was charged with first-degree murder but pleaded not guilty, claiming he was sleep-walking and suffering from "non-insane automatism." He had a history of sleepwalking, as did many other members of his family, and the duration of the episode and Parks' fragmented memory were consistent with somnambulism. Additionally, two separate polysomnograms indicated abnormal sleep. At his trial, Parks was found not guilty and the Canadian Supreme Court upheld the acquittal.

While cases like this are rare, they are common enough for the defense of non-insane automatism to have become well established (Fenwick 1990; Schopp 1991; McSherry 1998). Less dramatic, though no less intriguing, are cases involving agents performing other complex actions while apparently asleep. Siddiqui et al. (2009), for example, recently described a case of sleep emailing. These cases illustrate the complexity of the behaviors in which agents may engage in the apparent absence of awareness. Levy argues that such behaviors tend to be inflexible and insensitive to vital environmental information. The behaviors of somnambulists, for instance, exhibit some degree of responsiveness to the external environment, but they also lack genuine flexibility of response. To have genuine flexibility of response, or sensitivity to the content of a broad range of cues at most or all times, consciousness is required. With regard to free will and moral responsibility, Levy argues that the functional role of awareness "entails that agents satisfy conditions that are widely plausibly thought to be candidates for necessary conditions of moral responsibility only when they are conscious of facts that give to their actions their moral character" (2014: 87). More specifically, Levy argues that deep self and reasons-responsive accounts are committed to the truth of the consciousness thesis, despite what proponents of these accounts maintain.

Assuming that Kenneth Parks was in a state of global automatism on the night of May 24, 1987, he acted without consciousness of a range of facts, each of which gives to his actions moral significance: "he is not conscious *that he is stabbing an innocent person*; he is not conscious *that she is begging him to stop*, and so on" (2014: 89). These facts, argues Levy, "entail that his actions do not express his evaluative agency or indeed any morally condemnable attitude" (2014: 89). Because Parks is not conscious of the facts that give to his actions their moral significance, these facts are not globally broadcast—and because these facts are not globally broadcast, "they do not interact with the broad range of the attitudes constitutive of his evaluative agency" (2014: 89). This means that they do not interact with his *personal-level* concerns, beliefs, commitments, or goals. Because of this, Levy maintains that Parks' behavior is "not plausibly regarded as an expression of his evaluative agency"—agency caused or constituted by his personal-level attitudes (2014: 90).

Now, it's perhaps easy to see why agents who lack creature consciousness, or are in a very degraded global state of consciousness, are typically excused moral responsibility for their behaviors, but what about more common everyday examples where agents *are* creature conscious, but are not conscious of a fact that gives an action its moral significance? Consider, for instance, an example drawn from the experimental literature on implicit bias. Uhlmann and Cohen (2005) asked subjects to rate the suitability of two candidates for police chief, one male and one female. One candidate was presented as "streetwise" but lacking in formal education, while the other one had the opposite profile. Uhlmann and Cohen varied the sex of the candidates across conditions, so that some subjects got a male, streetwise candidate and a female, well-educated candidate, while other subjects got the reverse. What they found was that in both conditions subjects considered the male candidate significantly better qualified than the female, with subjects shifting their justification for their choice. That is, they rated being "streetwise" or being highly educated as a significantly more important qualification for the job when the male applicant possessed these qualifications than when the female possessed them. These results indicate a preference for a male police chief was driving subjects' views about which characteristics are needed for the job, and not the other way around (Levy 2014: 94).

Is this kind of implicit sexism reflective of an agent's *deep self*, such that he should be held morally responsible for behaviors stemming from it? Levy contend that, "though we might want to say that the decision was a sexist one, its sexism was neither an expression of evaluative agency, nor does the attitude that causes it have the right kind of content to serve as grounds on the basis of which the agent can be held (directly) morally responsible" (2014: 94). Let us suppose for the moment that the agent does not consciously endorse sexism in hiring decisions—i.e., that had the agent been conscious that the choice had a sexist content he would have revised or abandoned it. Under this scenario, the agent was not conscious of the facts that give his choice its moral significance. Rather, "they were conscious of a confabulated criterion, which was itself plausible (it is easy to think of plausible reasons why being streetwise is essential for being police chief; equally, it is easy to think of plausible reasons why being highly educated might be a more relevant qualification)" (Levy 2014: 95). Since it was this confabulated criterion that was globally broadcast (in the parlance of Levy's preferred Global Workspace Theory of consciousness), and which was therefore assessed in the light of the subjects' beliefs, values, and other attitudes, the agent was unable to evaluate and assess the implicit sexism against his personal-level attitudes. It is for this reason that Levy concludes that the implicit bias is "not plausibly taken to be an expression of [the agent's] evaluative agency, their deliberative and evaluative perspective on the world" (2014: 95).

Levy makes similar arguments against reasons-responsive accounts of moral responsibility. He argues that in both the case of global automatism and implicit bias, reasons-responsive control requires consciousness. This is because (a) reasons-responsiveness requires creature consciousness, and (b) the agent must be conscious of the moral significance of their actions in order to exercise responsibility-level control over them.

Levy's defense of the consciousness condition and his assessment of the two leading accounts of moral responsibility entail that people are less responsible than we might think. But how much less? In the final section of his book, he addresses the concerns of theorists like Caruso (2012) who worry that the ubiquity and power of non-conscious processes either rule out moral responsibility completely, or severely limit the instances where agents are justifiably blameworthy and praiseworthy for their actions. There he maintains that adopting the consciousness thesis need not entail skepticism of free will and basic desert moral responsibility, since the consciousness condition can be (and presumably often is) met. His argument draws on an important distinction between cases of global automatism and implicit bias, on the one hand, and cases drawn from the *situationist* literature on the other. Levy maintains that in the former cases (global automatism and implicit bias), agents are excused moral responsibility since they either lack creature consciousness or they are creature conscious but fail to be conscious *of* some fact or reason, which nevertheless plays an important role in shaping their behavior. In situational cases, however, Levy maintains that agents *are* morally responsible, despite the fact that their actions are driven by non-conscious situational factors, since the moral significance of their actions remains consciously available to them and globally broadcast (Levy 2014: 132; for a reply, see Caruso 2015b).

4 Volitional Consciousness

Let me end by noting one last category of views—i.e., those that maintain that consciousness is a necessary condition for free will and that the condition can be satisfied. In order to be concise, I will limit my discussion to two leading libertarian accounts of volitional consciousness, those of John Searle and David Hodgson.

Both Searle (2000, 2001) and Hodgson (2005, 2012) maintain that consciousness is physically realized at the neurobiological level and advocate naturalist accounts of the mind. Yet they also maintain that there is true (not just psychological) indeterminism involved in cases of rational, conscious decision-making. John Searle's *indeterminist* defense of free will is predicated on an account of what he calls *volitional consciousness*. According to Searle, consciousness is essential to rational, voluntary action. He boldly proclaims: "We are talking about conscious processes. The problem of freedom of the will is essentially a problem about a certain aspect of consciousness" (2000: 9). Searle argues that to make sense of our standard explanations of human behavior, explanations that appeal to reasons, we have to postulate "an entity which is conscious, capable of rational reflection on reasons, capable of forming decisions, and capable of agency, that is, capable of initiating actions" (2000: 10). Searle maintains that the problem of free will stems from volitional consciousness—our consciousness of the apparent gap between determining reasons and choices. We experience the gap when we consider the following: (1) our reasons and the decision we make, (2) our decision and action that ensues, (3) our action and its continuation to completion (2007: 42). Searle believes that, if we are to act freely, then our experience of the gap cannot be illusory: it must be the case that the causation at play is non-deterministic.

Searle attempts to make sense of these requirements by arguing that consciousness is a system-feature and that the whole system moves at once, but not on the basis of causally sufficient conditions. He writes:

> What we have to suppose, if the whole system moves forward toward the decision-making, and toward the implementation of the decision in actual actions; that the conscious rationality at the top level is realized all the way down, and that means that the whole system moves in a way that is causal, but not based on causally sufficient conditions.
>
> *(2000: 16)*

According to Searle, this account is only intelligible "if we postulate a conscious rational agent, capable of reflecting on its own reasons and then acting on the basis of those reasons" (2000: 16). That is, this "postulation amounts to a postulation of a self. So, we can make sense of rational, free conscious actions, only if we postulate a conscious self" (2000: 16). For Searle, the *self* is a primitive feature of the system that cannot be reduced to independent components of the system or explained in different terms.

David Hodgson (2005, 2012) presents a similar defense of free will, as the title of his book makes clear: *Rationality + Consciousness = Free Will* (2012). On Hodgson's account, a free action is determined by the conscious subject him/herself and not by external or unconscious factors. He puts forth the following *consciousness requirement*, which he maintains is a requirement for any intelligible account of indeterministic free will: "[T]he transition from a pre-choice state (where there are *open alternatives* to choose from) to a single post-choice state is a conscious process, involving the interdependent existence of a subject and contents of consciousness." For Hodgson, this associates the exercise of free will with consciousness and "adopts a view of consciousness as involving the interdependent existence of a self or subject and contents of consciousness" (2005: 4). In the conscious transition process from pre- to post-choice, Hodgson maintains, the subject grasps the availability of alternatives and knows how to select one of them. This, essentially, is where free will gets exercised. For Hodgson, it is essential to an account of free will that subjects be considered as capable of being *active*, and that this activity be reflected in the contents of consciousness.

There are, however, several important challenges confronting libertarian accounts of volitional consciousness. First, Searle and Hodgson's understanding of the *self* is hard to reconcile with our current understanding of the mind, particularly with what we have learned from cognitive neuroscience about reason and decision-making. While it is perhaps true that we *experience* the self as they describe, our sense of a unified self, capable of acting on conscious reasons, may simply be an illusion (see e.g. Dennett 1991; Klein et al. 2002). Second, work by Daniel Kahneman (2011), Jonathan Haidt (2001, 2012), and others (e.g. Wilson 2002) has shown that much of what we take to be "unbiased conscious deliberation" is at best rationalization. Third, Searle's claim that the *system itself* is indeterminist makes sense only if you think a quantum mechanical account of consciousness (or the system as a whole) can be given. This appeal to quantum mechanics to account for conscious rational behavior, however, is problematic for three reasons.

First, it is an empirically open question whether quantum indeterminacies can play the role needed on this account. Max Tegmark (1999), for instance, has argued that in systems as massive, hot, and wet as neurons of the brain, any quantum entanglements and indeterminacies would be eliminated within times far shorter than those necessary for conscious experience. Furthermore, *even if* quantum indeterminacies could occur at the level needed to affect consciousness and rationality, they would also need to exist at precisely the right *temporal* moment—for Searle and Hodgson this corresponds to the gap between determining reasons and choice. These are not inconsequential empirical claims. In fact, Searle acknowledges that there is currently no proof for them.

Second, Searle and Hodgson's appeal to quantum mechanics and the way it is motivated comes off as desperate. When Searle, for instance, asks himself, "How could the behavior of the conscious brain be indeterminist? How exactly would the neurobiology work on such an hypothesis?" He candidly answers, "I do not know the answer to that question" (2000: 17). Well, positing one mystery to account for another will likely be unconvincing to many.

Lastly, it's unclear that appealing to quantum indeterminacy in this way is capable of preserving free will in any meaningful way. There is a long-standing and very powerful objection to such theories. The *luck objection* (or *disappearing agent objection*) maintains that if our actions are the result of indeterminate events, then they become matters of luck or chance in a way that undermines our free will (see e.g. Mele 1999; Haji 1999; Pereboom 2001, 2014; Levy 2011; Caruso 2015c). The core objection is that because libertarian agents will not have the power to *settle* whether or not the decision will occur, they cannot have the role in action basic desert moral responsibility demands. Without smuggling back in mysterious agent-causal powers that go beyond the naturalistic commitments of Searle and Hodgson, what does it mean to say that the agent "selects" one set of reasons (as her motivation for action) over another? Presumably this "selection" is not within the active control of the agent, since it is the result of indeterminate events that the agent has *no ultimate control over.*

5 Conclusion

In this survey I have provided a rough taxonomy of views regarding the relationship between consciousness, free will, and moral responsibility. We have seen that there are three broad categories of views, which divide on how they answer the following two questions: (1) Is consciousness necessary for free will? And if so, (2) can the consciousness requirement be satisfied, given the threat of shrinking agency and recent developments in the behavioral, cognitive, and neurosciences? With regard to the first question, we find two general sets of views—those that reject and those that accept a consciousness condition on free will. The first group explicitly denies that consciousness is needed for agents to be free and morally responsible, but disagree on the reasons why. The second group argues that consciousness *is* required, but then divides further over whether and to what extent the consciousness requirement can be satisfied. I leave it to the reader to decide the merits of each of these accounts. In the end I leave off where I began, with questions: Is consciousness necessary for free will and moral responsibility? If so, what role or function must it play? Are agents morally responsible for actions and behaviors that are carried out automatically or without conscious control or guidance? And are they morally responsible for actions, judgments, and attitudes that are the result of implicit biases or situational features of their surroundings of which they are unaware? These questions need more attention in the literature, since clarifying the relationship between consciousness and free will is imperative if one wants to evaluate the various arguments for and against free will.

References

Arplay, N. (2002) *Unprincipled Virtues: An Inquiry into Moral Agency*, New York: Oxford University Press.

Arplay, N., and T. Schroeder. (1999) "Praise, Blame and the Whole Self," *Philosophical Studies* 93: 161–199.

Baars, B. (1988) *A Cognitive Theory of Consciousness*, Cambridge: Cambridge University Press.

Baars, B. (1997) *In the Theater of Consciousness*, New York: Oxford University Press.

Bargh, J.A. (1997) "The Automaticity of Everyday Life," in R. S. Wyer, Jr. (ed.) *The Automaticity of Everyday Life: Advances in Social Cognition*, Vol. 10, Mahwah, NJ: Erlbaum.

Bargh, J.A., and Chartrand, T.L. (1999) "The Unbearable Automaticity of Being," *American Psychologist* 54: 462–479.

Bargh, J.A., and Ferguson, M.J. (2000) "Beyond Behaviorism: On the Automaticity of Higher Mental Processes," *Psychological Bulletin* 126: 925–945.

Baumeister, R.F. (2008) "Free Will in Scientific Psychology," *Perspectives of Psychological Science* 3: 14–19.

Caruso, G.D. (2005) "Sensory States, Consciousness, and the Cartesian Assumption," in N. Smith and J. Taylor (eds.) *Descartes and Cartesianism*, UK: Cambridge Scholars Press.

Caruso, G.D. (2012) *Free Will and Consciousness: A Determinist Account of the Illusion of Free Will*, Lanham, MD: Lexington Books.

Caruso, G.D. (2015a) "Précis of Neil Levy's *Consciousness and Moral Responsibility*," *Journal of Consciousness Studies* 22 (7–8): 7–15.

Caruso, G.D. (2015b) "If Consciousness Is Necessary for Moral Responsibility, then People Are Less Responsible than We Think," *Journal of Consciousness Studies* 22 (7–8): 49–60.

Caruso, G.D. (2015c) "Kane Is Not Able: A Reply to Vicens' 'Self-Forming Actions and Conflicts of Intention'," *Southwest Philosophy Review* 31: 21–26.

Caruso, G.D., and Morris, S.G. (2017) "Compatibilism and Retributive Desert Moral Responsibility: On What Is of Central Philosophical and Practical Importance," *Erkenntnis* 82: 837–855.

Dehaene, S., and Naccache, L. (2001) "Toward a Cognitive Neuroscience of Consciousness: Basic Evidence and a Workspace Framework," *Cognition* 79: 1–37.

Dehaene, S., Changeux, J.P. and Naccache, L. (2011) "The Global Neuronal Workspace Model of Conscious Access: From Neuronal Architecture to Clinical Applications," in S. Dehaene and Y. Christen (eds.) *Characterizing Consciousness: From Cognition to the Clinic?* Berlin: Springer-Verlag.

Dennett, D.C. (1991) *Consciousness Explained*, London: Penguin Books.

Doris, J.M. (2002) *Lack of Character: Personality and Moral Behavior*, Cambridge: Cambridge University Press.

Feinberg, J. (1970) "Justice and Personal Desert," in his *Doing and Deserving*, Princeton: Princeton University Press.

Fenwick, P. (1990) "Automatism, Medicine and the Law," *Psychological Medicine Monograph* 17: 1-–27.

Fischer, J.M., and Ravizza, M. (1998) *Responsibility and Control: A Theory of Moral Responsibility*, Cambridge: Cambridge University Press.

Haidt, J. (2001) "The Emotional Dog and Its Rational Tail: A Social Intuitionist Approach to Moral Judgment," *Psychological Review* 108: 814–834.

Haidt, J. (2012) *The Righteous Mind: Why Good People Are Divided by Politics and Religion*, New York: Pantheon.

Haji, I. (1999) "Indeterminism and Frankfurt-Type Examples," *Philosophical Explorations* 1: 42–58.

Hodgson, D. (2005) "A Plain Person's Free Will," *Journal of Consciousness Studies* 12 (1): 1–19.

Hodgson, D. (2012) *Rationality + Consciousness = Free Will*, New York: Oxford University Press.

Kahneman, D. (2011) *Thinking Fast and Slow*, New York: Farrar, Straus, and Giroux.

King, M., and Carruthers, P. (2012) "Moral Responsibility and Consciousness," *Journal of Moral Philosophy* 9: 200–228.

Klein, S., Rozendal, K., and Cosmides, L. (2002) "A Social-Cognitive Neuroscience Analysis of the Self," *Social Cognition* 20: 105–135.

Levy, N. (2011) *Hard Luck: How Luck Undermines Free Will and Moral Responsibility*, New York: Oxford University Press.

Levy, N. (2014) *Consciousness and Moral Responsibility*, New York: Oxford University Press.

Libet, B. (1985) "Unconscious Cerebral Initiative and the Role of Conscious Will in Voluntary Action," *Behavioral and Brain Science* 8: 529–566.

Libet, B. (1999) "Do We Have Free Will?" *Journal of Consciousness Studies* 6 (8–9): 47–57, reprinted in R. Kane (ed.) *The Oxford Handbook of Free Will*, New York: Oxford University Press, 2002.

Libet, B., Gleason, C.A., Wright, E. W., and Pearl, D. K. (1983) "Time of Conscious Intention to Act in Relation to Onset of Cerebral Activity (Readiness-Potential): The Unconscious Initiation of a Freely Voluntary Act," *Brain* 106: 623–642.

McSherry, B. (1998) "Getting Away with Murder: Dissociative States and Criminal Responsibility," *International Journal of Law and Psychiatry* 21: 163–176.

Mele, A. (1999) "Ultimate Responsibility and Dumb Luck," *Social Philosophy and Policy* 16: 274–293.

Mele, A. (2009) *Effective Intentions*, New York: Oxford University Press.

Nadelhoffer, T. (2011) "The Threat of Shrinking Agency and Free Will Disillusionism," in L. Nadel and W. Sinnott-Armstrong (eds.) *Conscious Will and Responsibility: A Tribute to Benjamin Libet*, New York: Oxford University Press.

Nahmias, E. (2011) "Intuitions about Free Will, Determinism, and Bypassing," in R. Kane (ed.) *The Oxford Handbook of Free Will, 2nd ed.*, New York: Oxford University Press.

Nisbett, R., and Wilson, T. (1997) "Telling More Than We Can Know: Verbal Reports on Mental Processes," *Psychological Review* 84: 231–258.

Pereboom, D. (2001) *Living Without Free Will*, Oxford: Cambridge University Press.

Pereboom, D. (2014) *Free Will, Agency, and Meaning in Life*, Oxford: Oxford University Press.

Pereboom, D., and Caruso, G.D. (2018) "Hard-Incompatibilism Existentialism: Neuroscience, Punishment, and Meaning in Life," in G.D. Caruso and O. Flanagan (eds.) *Neuroexistentialism: Meaning, Morals, and Purpose in the Age of Neuroscience*, New York: Oxford University Press.

Rosenthal, D. (2005) *Consciousness and Mind*, New York: Oxford University Press.

Schopp, R.F. (1991) *Automatism, Insanity, and the Psychology of Criminal Responsibility: A Philosophical Inquiry*, Cambridge: Cambridge University Press.

Searle, J. (2000) "Consciousness, Free Action and the Brain," *Journal of Consciousness Studies* 7 (10): 3–22.

Searle, J. (2001) *Rationality in Action*, Cambridge, MA: MIT Press.

Searle, J. (2007) *Freedom and Neurobiology: Reflections on Free Will, Language and Political Power*, New York: Columbia University Press.

Shepherd, J. (2012) "Free Will and Consciousness: Experimental Studies," *Consciousness and Cognition* 21: 915–927.

Shepherd, J. (2015) "Consciousness, Free Will, and Moral Responsibility: Taking the Folk Seriously," *Philosophical Psychology* 28: 929–946.

Sher, G. (2009) *Who Knew? Responsibility Without Awareness*, New York: Oxford University Press.

Siddiqui, F., Osuna, E., and Chokroverty, S. (2009) "Writing Emails as Part of Sleepwalking After Increase in Zolpidem," *Sleep Medicine* 10: 262–264.

Sie, M., and Wouters, A. (2010) "The BCN Challenge to Compatibilist Free Will and Personal Responsibility," *Neuroethics* 3: 121–133.

Smith, A. (2005) "Responsibility for Attitudes: Activity and Passivity in Mental Life," *Ethics* 115: 236–271.

Smith, A. (2008) "Control, Responsibility, and Moral Assessment," *Philosophical Studies* 138: 367–392.

Soon, C.S., Brass, M., Heinze, H.J., and Haynes, J.D. (2008) "Unconscious Determinants of Free Decisions in the Human Brain," *Nature Neuroscience* 11: 543–545.

Strawson, G. (1986) *Freedom and Belief*, Oxford: Oxford University Press [revised edition 2010].

Strawson, G. (1994) "The Impossibility of Moral Responsibility," *Philosophical Studies* 75: 5–24.

Tegmark, M. (1999) "The Importance of Quantum Decoherence in Brain Processes," *Physics Review E* 61: 4194–4206.

Uhlmann, E.L., and Cohen, G.L. (2005) "Constructed Criteria: Redefining Merit to Justify Discrimination," *Psychological Science* 16: 474–480.

Wegner, D. (2002) *The Illusion of Conscious Will*, Cambridge, MA: MIT Press.

Wilson, T. (2002) *Strangers to Ourselves: Discovering the Adaptive Unconscious*, Cambridge, MA: The Belknap Press of Harvard University Press.

Wolf, S. (1990) *Freedom and Reason*, Oxford: Oxford University Press.

Yaffe, G. (2012) "The Voluntary Act Requirement," in M. Andrei (ed.) *Routledge Companion to Philosophy of Law*, New York: Routledge.

Related Topics

Consciousness and Action
Global Workspace Theory
Representational Theories of Consciousness
Quantum Theories of Consciousness

7

CONSCIOUSNESS AND THE MIND-BODY PROBLEM IN INDIAN PHILOSOPHY

Christian Coseru

1 Introduction

The thriving contemporary enterprise of Consciousness Studies owes its success in large measure to two late 20th-century intellectual developments in cognitive science and its allied philosophy of mind: a growing interest in the study of the neurobiological processes that underlie consciousness and cognition, and the rehabilitation of first-person approaches to the study of consciousness associated with the 20th-century European tradition of phenomenological philosophy. The first development marks a shift away from preoccupations with the status of mental representation to understanding the function of perception, attention, action, and cognition in embodied and enactive, rather than purely representational, terms. The second acknowledges the importance of fine-grained accounts of experience for the purpose of mapping out the neural correlates of consciousness. Both developments recognize that empirical research is essential to advancing any robust philosophical and scientific theory of consciousness. At the same time, these developments also open up the possibility that there may be aspects of consciousness that are not empirically tractable, aspects whose understanding require that we revise the way we conceptualize both the easy and hard problems of consciousness. It is this revisionary approach that has opened the door to systematic contributions to the study of consciousness that take its phenomenological and transcendental dimensions seriously.

Indian philosophy is host to a rich tradition of such systematic examinations of consciousness that focus primarily, though not exclusively, on its phenomenological and transcendental dimensions. Indeed, one could go as far as to argue that the nature and function of consciousness is perhaps the single most contentious issue among the different schools of Indian philosophy, a development without parallel in the West, prior to Descartes, Kant, and the British empiricists. From its earliest association in the Upaniṣads with the principle of individuation or the self (*ātman*), to its indispensability to any theory of knowledge, the concept of consciousness (variously rendered in Sanskrit as *cit*, *citta*, *vijñāna*) has been at the center of debates about personal identity, agency, and the grounds of epistemic reliability. Not only are analyses of the different aspects of consciousness essential to the problem of self-knowledge, they are also fundamental in settling metaphysical claims about the nature of reality (Siderits 2015). Much of the debate follows the familiar terrain of inquiries into such pressing matters as the reach of perception, the nature of mental content, and the character of veridical states of cognitive awareness. But the tradition is also host to a vast repertoire of first-person methods and to a rich vocabulary of phenomenal concepts meant to capture dimensions of consciousness that are not ordinarily available to empirical scrutiny.

Considering the sheer amount of literature associated with the exploration of consciousness in Indian philosophy, coming anywhere near a comprehensive survey within the limits of this chapter would be impossible. I have therefore chosen to focus on a range of methodological and conceptual issues, drawing on three main sources: (i) the naturalist theories of mind of Nyāya and Vaiśeṣika, (ii) the mainly phenomenological accounts of mental activity and consciousness of Abhidharma and Yogācāra Buddhism, and (iii) the subjective transcendental theory of consciousness of Advaita Vedānta. The contributions of Indian philosophers to the study of consciousness are examined here not simply as a contribution to intellectual history, but rather with a view to evaluating their relevance to contemporary issues, specifically to the mind-body problem.

It is worth noting from the outset that there are no explicit articulations of the Cartesian mind-body problem in Indian philosophy. In India, defenders of metaphysical dualism operate with conceptions of substance that do not admit of a strict dichotomy between *res extensa* and *res cogitans*. Dualist schools of thought such as Sāṃkhya, for instance, take substance (*dravya*) to be reducible neither to the category of quality (*guṇa*) nor to that of action (*karman*). On this view, matter has emergent properties but lacks internal dynamism, which is provided by the activity of consciousness. And while pure consciousness itself lacks extension, in the process of being and becoming, it reaches out (or 'extends') into the world through reason, experience, and the ability to entertain first-person thoughts. Similarly, for Nyāya and Vaiśeṣika thinkers following in the footsteps of Jayanta Bhatta (fl. 850 C.E.), Bhasarvajña (fl. 950 C.E.) and Udayana (fl. 1050 C.E.), selves can be said to have extension (*vibhu*), by virtue of possessing a rather unique property known as pervasion (*vyāpti*). Furthermore, the conception of mind (*manas*) at work in Indian philosophy differs in significant ways from the prevailing Cartesian notion of an immaterial thinking substance (Ganeri 2012: 218–221). Mind is largely conceived as a faculty that occupies an intermediary place between the senses and the intellect, and is defined primarily in terms of its capacity to organize and integrate the raw experiential data available to conscious cognition.

Given a general preoccupation with overcoming the limitations of the human condition, conceived largely in terms of constraints imposed by our embodied condition on our psychology, the absence of the mind-body problem in Indian philosophy might seem like an inexplicable lacuna. How could Indian thinkers, prior to their encounter with European philosophy, have overlooked such an essential problem? One possible answer would be to make the case that, as stated, the problem can only arise in the context of scientific discoveries about human physiology and the brain, coupled with a commitment to the sort of mechanistic conception of reality prevalent in Europe at the dawn of modern science. Another possibility, which is in keeping with critics of the Cartesian legacy in contemporary philosophy of mind, is to say that the mind-body problem is really a pseudo-problem, the outcome of metaphysical commitments to some version of mechanistic dualism. But the presence of dualist positions with strong naturalist undercurrents in Indian philosophy, especially in the Nyāya and Sāṃkhya traditions, rules out the possibility of treating the mind-body problem as an idiosyncratic feature of Cartesian metaphysics. As current debates in the metaphysics of mind have demonstrated, even assuming different varieties of dualism (predicate, property, and substance), there are ways of conceiving of the relation between mind and matter that avoid the Cartesian interactionist model, with new forms of hylomorphism (Jaworski 2016), psychophysical parallelism (Heidelberger 2003), and non-Cartesian substance dualism (Lowe 2006) as the main alternatives.

2 Epistemology and the Metaphysics of Consciousness

Is there some persistent aspect of human experience, something that originates at birth or even at conception and continues through the various stages of one's life, and perhaps beyond?

Metaphysical speculations about the existence and nature of such an entity, known in classical Indian sources as the *ātman* or the self, are the principal concern of the Upaniṣads, a group of texts in the style of Platonic dialogues composed around the middle of the first millennium B.C.E. In one of the earliest such accounts, from the *Bṛhadaranyaka Upaniṣad* (3.4.2), we come across a systematic refutation of epistemological reflexivity. The formula, which appears in several other locations in the same text, reads as follows: "You cannot see the seer who does the seeing; you cannot hear the hearer who does the hearing; you cannot think of the thinker who does the thinking; and you can't perceive the perceiver who does the perceiving" (Olivelle 1998: 83).

The view articulated here, which will eventually come to inform the subjective transcendental theory of consciousness of Advaita Vedānta (see below, Section 5), is that the principle of cognitive awareness, that which makes possible knowledge in all its modalities (perceptual, inferential, introspective, etc.), cannot itself be known or cognized by those very faculties whose cognizing it makes possible. Whether this principle is taken to be a self or consciousness itself with its intentional and subjective aspects, it is not something that can be made known or manifest. What serves as the basis for something cannot itself be made manifest or present by the very thing that it makes possible.[1] Indian metaphysics of mind has it that ultimately, the nature of reality is such that it must be constituted as an immutable dimension of consciousness. To the extent that cognition is intimately connected to consciousness, then, consciousness is what ultimately makes cognition possible. If consciousness itself is what makes cognition possible, the conditions for cognition being reliable are internal to cognition itself, which suggests that the earliest Indian philosophical speculations about consciousness point to epistemic internalism.

What is it about consciousness that determines how a subject comes to have veridical experiences? Classical Indian discussions of consciousness take cognitive events to be individual states of consciousness whose epistemic status depends on the reliability of access consciousness. A cognition of blue is simply a case of consciousness taking the form of the object cognized or of having that form superimposed on it. Given the close connection between consciousness and cognition, and considering that knowledge is a matter of consciousness undergoing the sort of transformation that results in the occurrence of reliable cognitive events, epistemological concerns are never altogether absent from considerations about the nature of consciousness. If that which we call the self cannot itself be seen or thought, even though it is present whenever we see or think, then it is not something that can become an object of consciousness. As Yājñavalkya explains to his wife and philosophical interlocutor, Maitreyī, in a seminal passage of the *Bṛhadaranyaka Upaniṣad* (2.4.14): "When there is duality of some kind...then the one can see the other...then the one can think the other, and the one can perceive the other. When, however, the whole has become one's very self...then who is there for one to see and by what means?" (Olivelle 1998: 69). What we have here is a clear example of transcendental subjectivity: the thinker itself cannot be thought. Rather, thought, much like sensation and perception, is an irreflexive or anti-reflexive relation, at least with regard to the consciousness whose thinking episode it is.

One of the problems with the anti-reflexivity principle is that it cannot bridge the explanatory gap between the physical and mental domains. If cognition is but a transformation of consciousness, on the assumption that consciousness cannot be understood in non-phenomenal terms, it would follow that all cognitions have a distinct phenomenal character (a rather controversial position). While it is obvious that perceptual awareness has its attendant phenomenology, it is not at all clear that propositional attitudes have their own proprietary phenomenology, if at all. The fragrance of a lotus flower, the taste of freshly brewed coffee, and the bathing hues of a summer sunset are distinct phenomenal types: there is something it is like to experience them. It is not at all clear, however, that thoughts of the sort, 'Paris is the capital of France' or 'Sanskrit is a fusional language' have any phenomenological character: rather, they are discerned on the basis of their propositional content. Of course, one may abstract from experience the concept that coffee is an aromatic substance, but this is primarily a phenomenal concept grounded in a specific phenomenal experience (coffee drinking), not an abstract concept whose mastery depends on knowing the chemical composition of the *coffea* genus.

That consciousness is central to cognition, and to veridical cognition, in particular, is a commonly shared view among Indian philosophers. Disagreements arise, however, when considering whether cognition (*jñāna*) is just an aspect of consciousness (*cit*), and thus not different from it, or a distinct event in the mental stream occasioned by the availability of a particular object. One way to frame this problem is to consider the different ways in which the problem of consciousness may be conceptualized. In general, Indian philosophers operate with three distinct concepts of consciousness: (i) as a quality of the self; (ii) as an act of the self; and (iii) as identical with the self or as the self itself. Taking consciousness to be a quality of the self raises additional questions: is it an essential or merely an accidental quality, and if the latter, what are the specific conditions under which consciousness becomes manifest? (This is an issue with implications for the mind-body problem.) Likewise, the view that consciousness is an act of the self or the self itself confronts a different set of issues, mainly concerning the nature of agency, and the problem of composition and metaphysical grounding.

Unlike consciousness, whose function of illuminating or making present is unmistakable, cognition may be either true or false. Since only valid cognitions count as knowledge, the Sanskrit term for a cognition that is epistemically warranted is *pramā*. The indubitability of conscious experience suggests that Indian philosophers by and large endorse the immunity to error through misidentification thesis: there is no mistaking the fact that one is conscious, irrespective of whether the contents of one's consciousness are reliably apprehended or not. But the immunity to error through misidentification thesis assumes that phenomenality is the unmistakable character of consciousness: to be conscious is for there to be something it is like. But this locution, at least as initially employed by Nagel (1974), assumes the presence of a subjective point of view, which is incompatible with some Indian philosophical perspectives, specifically those of Sāṃkhya and Vedānta, which take consciousness to be ultimately lacking any structure.

The analysis that follows considers three different approaches to the problem of how consciousness and cognition are related, and its implications for the mind-body problem and the problem of personal identity.

3 Consciousness as an Attribute of the Self: Nyāya Naturalistic Dualism

If knowledge is an epistemic relation, the question naturally arises: how can it be ascertained that the state in question is a conscious rather than an unconscious state? The absence of any testimony while such states endure makes it more plausible to consider that their occurrence is inferred rather than directly experienced. In seeking to articulate various intuitions about the nature of consciousness, one of the most common strategies in Indian philosophy is to examine the difference between waking, dreaming, and dreamless states of consciousness. While waking states provide the norm for consciousness in all its aspects, and dreaming suggests that consciousness persists beyond wakefulness, it is an open question whether consciousness persists in some latent form in dreamless sleep. Assuming the presence of an indeterminate consciousness in dreamless sleep, and perhaps of an even deeper state of consciousness beyond dreamless sleep, raises the question: how is the presence of consciousness in such states to be ascertained? The Upaniṣads, the principal source for this idea, fail to provide a positive account. While such states are assumed, their mode of ascertainment is not at all clear. Texts like the *Māṇḍūkya Upaniṣad* (vs. 7) tell us that these indeterminate states of awareness are "ungraspable...without distinguishing marks...unthinkable...indescribable" and something "whose essence is the perception of itself alone" (Olivelle 1998: 475).

Philosophers associated with the so-called "Method of Reasoning" School or Nyāya take a different view about the relation between consciousness and the self. Beginning with Gautama in the 2nd century C.E., continuing with the seminal works of Vātsyāyana and Uddyotakara in the 4th and 5th centuries, and concluding with the mature contributions of Jayanta and Udayana in the 9th and 10th centuries, Nyāya philosophers insist on setting more stringent requirements for ascertaining the relation between consciousness and cognition. Instead of assuming an experiential level of nonconceptual or even non-cognitive awareness, they reason that it is more apt to say that we infer the absence or presence of consciousness in states of deep sleep or swoon. We do not recollect it. In taking consciousness to be a property of the self, Naiyāyikas argue that certain necessary causal conditions must be satisfied for ascertaining the phenomenal character and content of a mental state: first, there must be contact between the sense and a given object, then, the mind must attend to the sense experience, and finally, the self must be in contact with mind (*Nyāyasūtrabhāṣya* 2.1.14; *Nyāyavārttika* 2.12; Jha 1939: 124). Since Naiyāyikas reject the reflexivity thesis, cognitions can grasp an object, but they cannot grasp themselves. What makes an object-directed cognition (*vyavasāya*) known to the cognizing subject is not some intrinsic aspect or property of that cognition, such as its luminosity or self-reflexivity, but a second order cognition (*anuviyavasāya*), which takes the first one for its object (Chakrabarti 1999: 34).

But this account of the relation between consciousness and cognition is regressive: if it takes a secondary or second order cognition to make the first cognition known, then, this second cognition would require a third cognition to be known and so on. How does Nyāya answer the charge of infinite regress? Assuming cognition C_1 requires a second cognition C_2 does not entail that C_2 itself must be made manifest by a subsequent cognition C_3. Rather, C_2 may do its work of making C_1 known without itself becoming known unless there is a subsequent desire to manifest C_2 as an instance of metacognitive awareness. On the general Nyāya rule that a cognition operates by fixing the intentionality of a token mental state, only C_1 needs to be made known, for in disclosing to the individual that a cognition of a certain object has occurred, the infinite regress is blocked. In perceiving (C_1) the tree outside the window, all that a subject requires is that contact between the visual system and the object be made manifest (C_2). There is no requirement that C_2 must itself be introspectively available. If Nyāya philosophers have an explanation for why their account of intentional mental states is not regressive, their understanding of the relation between consciousness and cognition is problematic. The occurrence of a primary C_1 type cognition does not necessitate the occurrence of a secondary C_2 type cognition. In other words, unless one is conscious and desires to know by directing one's attention to whatever is perceptually or introspectively available, cognitions that merely make their object known will never become available to the subject. But to want to know C_1 by attending to what is perceptually available requires that one is already acquainted in some direct capacity with what one desires to know. For we cannot desire to know something we have no acquaintance of. For this account of cognition to work, Nyāya philosophers would have to assume the existence of pre-reflective modes of acquaintance. But such assumptions run counter to the theory (cf. Mohanty 1999: 12).

What blocks this seemingly intuitive move to ground cognition into more basic pre-predicative and pre-reflective modes of awareness is a commitment to direct realism. Indeed, one of the key features of the Nyāya theory of consciousness is that for cognition to be conscious or available to consciousness is for it to have objective content. Hence, the phenomenal character of cognition is provided by its intentional content. In cognizing a pot, both the phenomenal character and the phenomenal content of the cognition is provided by the object's specific features. In Lockian terms, the object furnishes cognition with both its primary and secondary qualities: that a pot is apprehended as having a particular shape, color, and weight is a function of cognition's directness toward the object and of its specific mode of apprehension. Since cognitions cannot be self-revealing or about themselves, their content is fixed by the object. At the same time, they become known only in so far as a relation between the self and the mind obtains, for although cognitions are about their object, they are made manifest only as qualities or properties (*guṇas*) of the self. For Nyāya, then, cognition makes its object known only in so far as it presents itself as a quality of the conscious self.

What implication does the Nyāya theory of consciousness have for the mind-body problem? First, we must specify that philosophers pursuing this line of inquiry share the ontological stance of their partner school, Vaiśeṣika, which admits nine types of substances and several kinds of properties in its ontology. Just as physical objects have real properties like shape, color, and mass, so also consciousness and cognition are real properties of the self, one of the nine substances of Vaiśeṣika ontology. How do these different substances and properties relate or correlate? Specifically, how does Nyāya account for the properties of physical objects becoming the qualities of conscious experience? The general picture is something like this: the senses reach out and apprehend the specific properties of objects. But although these properties are disclosed by cognition, they are still the intrinsic properties of the things themselves. For example, a cognition in which the color and shape of a jar are apprehended is due to the inherence (*samavāya*) of the color property in the jar and to contact between the eye and the jar (*Nyāyavārttika ad Nyāyasūtra* 1.2.4). In other words, perception apprehends not only unique particulars, but also their properties and relations. But this epistemological solution to the question of how mental and physical properties relate or correlate is too stringent to allow for cognitive error. By itself the relation of inherence (*samavāya*) cannot tell whether the properties in question belong to the object or to the cognition of the object. It cannot tell us whether cognition gets its phenomenal content from the object or from itself.

4 Consciousness without a Self: Buddhist Phenomenalism

An altogether different line of inquiry about whether cognitive events can become instances of knowledge in the absence of a subject of knowledge is the hallmark of the Buddhist tradition. Of course, Buddhist metaphysics is well known for its rejection of a permanent self as the agent of sensory activity (Collins 1982; Harvey 1995). It is worth emphasizing that while Buddhists reject the notion of an enduring or permanent self, they do not deny the reality of the elements of existence (*dharmas*) (Bodhi 1993: chapter 2). But this is a metaphysics of experience (rather than of causally efficacious particulars) that takes the body to be an instrument (*karaṇa*) of sensory activity, and not simply a causally determined physical aggregate. As such, the body is both the medium of contact with the world *and* the world with which it comes in contact (a view that finds an interesting parallel in Husserl's account of the paradoxical nature of the body as revealed through phenomenological reduction). This intuition about the dual nature of embodied awareness (as locus of lived experience) discloses a world of lived experience, whose boundaries are not fixed but constantly shifting in relation to the desires, actions, and attitudes of an agent (Husserl 1970: III, A). The question that both Buddhist philosophers and phenomenologists must address is whether intentional experiences—of the sort that disclose a world as pre-reflectively but meaningfully given—presuppose that consciousness itself, as the disclosing medium, is a knowable object.

Unlike the Naiyāyikas, Buddhists typically argue that conscious cognitive events are not apprehended diachronically (or inferentially) in a subsequent instance of cognitive awareness.

Rather, by virtue of being conscious episodes, they are inherently self-aware, even if only minimally so. Although we may intend a previous moment of conscious awareness in introspection, this retrospective apprehension of consciousness as an object cannot be its essential feature.

Let us briefly consider one of the key problems that the reductionist account of experience must necessarily confront: the project of reductive analysis, which aims to identify those elements (sensations, volitions, dispositions, patterns of habituation) that are constitutive of what we ordinarily designate as 'persons,' has an important, and perhaps unintended, consequence. It assumes that an awareness which arises in conjunction with the activity of a given sensory system is itself impermanent and momentary: visual awareness and visual object, for instance, are both events within a mental stream of continuing relations. What, then, accounts for the sense of recollection that accompanies these cognitive series? In other words, if discrete, episodic cognitive events are all that constitutes the mental domain, how does appropriation, for instance, occur? I refer here specifically to the basic mode of givenness, or for-me-ness, of our experience (Zahavi and Kriegel 2015), which presents its objects to reflective awareness.

The causal account, it seems, gives only an incomplete picture of the mental. The Buddhist Sanskrit term for cognitive awareness, *vijñāna*, conveys the sense of differentiation and discernment. But it is not exactly clear how such discernment also sorts between an inner and an outer domain of experience. Indeed, consciousness is not merely a faculty for discerning and sorting through the constitutive elements of experience, but is itself an event in a series of interdependent causal and conditional factors. Other than positing a continuity of awareness or a stream of mental events, early Buddhist solutions to this conundrum do not offer a satisfactory answer to how accounts of causal generation in the material domain can explain the phenomenal features of cognitive awareness.

It is largely in response to this need to provide an account for the continuity of awareness that the self-reflexivity thesis finds its first articulation in the work of the influential Buddhist philosopher Dignāga (480–540). As he claims, we must assume that cognitions are inherently self-reflexive if we are to account adequately for the phenomenal character of conscious experience. By singling out self-reflexivity as a constitutive aspect of perception, Dignāga seeks to account for the specific mode of presentation of all mental states insofar as they arise bearing a distinct mode of givenness: to perceive is to be implicitly or non-thematically present to the perceptual occasion. For Dignāga, the intentional structure of consciousness is a relational feature of its mode of presentation. Indeed, by stating that each cognitive event arises in the form of a dual-aspect relation between apprehending subject and apprehended object, Dignāga posits the aspectual nature of intentional reference (Williams 1998; Garfield 2006; Chadha 2011; MacKenzie 2011; Arnold 2012; Coseru 2012).

Unlike the Nyāya thinkers we discussed above, Dignāga and all those who follow the tradition of epistemic inquiry that he helped to initiate take the opposite view: a reliable source of cognition is to be taken, not as an instrument that makes knowledge (or the acquisition thereof) possible, but rather as the result, that is, as knowledge itself. As he notes, "a source of knowledge is effective only as a result, because of being comprehended along with its action" (*Pramāṇasamuccaya* I, 8; Hattori 1968: 97). In containing the image or aspect of its object, cognition may well appear to have a representational structure, but while appearing to comprise the act of cognizing or to enact it in some way, it is in effect nothing but the result of cognitive activity. For instance, in apprehending an object, say a lotus flower, all that we are aware of on this model is the internal aspect (*ākāra*) of that cognitive event, or, in phenomenological terms, the intended object just as it is intended (Dhammajoti 2007). It is obvious that cognitions are contentful, but what makes them epistemically reliable is the fact that comprehension or the result of cognitive activity are nothing but cognition in its dual-aspect form. Dignāga's understanding of what counts as a reliable cognition comes very close to something like Husserl's notion of noematic content, or the *perceived as such*, which is what we get after performing the epoché or phenomenological reduction. For Dignāga, just as for Husserl, perception is ultimately constituted by intentional content: perceiving is an intentional (that is, object-directed) and self-revealing (*svaprakaśa*) cognition.

Dignāga appears to be making two important claims here. First, all cognitions are self-intimating: regardless of whether an object is present or not, and of whether the present object is real or imagined, cognition arises having this dual appearance. Second, Dignāga tells us that the determination of the object, that is, how the object appears in cognition, conforms in effect to how it is intended: for example, as something desirable or undesirable. It should be possible therefore to interpret Dignāga's descriptive account of cognition as providing support for the dual-aspect nature of intentional acts. On the one hand, intentional experiences span a whole range of cognitive modalities: perceiving, remembering, judging, etc. On the other, each intentional experience is also about a specific object, whether it be something concrete, like a pot, or something imagined, like a unicorn.

What does it mean for cognitive awareness to be self-revealing? One perfectly acceptable way to answer this question is to say that self-reflexivity is a feature of each cognitive event by virtue of arising together with it. It is precisely this aspect of the Buddhist epistemologist's theory of cognition that is the main target of criticism by philosophers like Candrakīrti (600–650), the champion of a particular interpretation of the scope of Middle Way or Madhyamaka philosophy. One of the axiomatic principles of Madhyamaka, as conceived by its founder, Nāgārjuna (fl. c. 150 C.E.), is that all things, including all cognitive episodes, by virtue of being the product of cause and conditions, lack inherent existence (*svabhāva*) and are thus empty (*Mūlamadhyamaka-kārikā* 3. 6–9; 4, 1–8; Siderits and Katsura 2013). In setting out to defend this principle, Candrakīrti reiterates the view that no mental state could be such as to be inherently self-presenting or self-disclosing (Candrakīrti 1960: 62). Thus, Candrakīrti's critique targets the knowledge intimation thesis, specifically the notion that there is a class of cognitive events that are essentially self-reflexive: they reveal their own character and sense of ownership without recourse to an additional instance of cognitive awareness, an object, or the positing of a subject of experience. More to the point, Candrakīrti rejects the notion that reflexive awareness has this unique property of giving access to the pure datum of experience (Duerlinger 2012; Tillemans 1996: 49). Self-knowledge, on this view, is a matter of achieving a conceptually mediated understanding of what is introspectively available: instead of depending on the elusive and seemingly irreducible capacity of consciousness to make known, cognition becomes an instance of self-knowledge only metacognitively, that is, only when cognition takes a previous instance of cognition as its object.

In setting out to reject the thesis that consciousness consists in conscious mental states being implicitly self-aware, Candrakīrti and his Buddhist followers share a common ground with Nyāya realists: that cognition occurs for someone is not something that is immediately available. Rather, cognition's subjective aspect is inferred from the effects of that cognition. Whereas the reflexivist thinks that I can know something only to the extent that each instance of cognition is inherently self-revealing or self-illuminating, the opponent counters that such cognitive acts as "seeing something" are transparent with regard to their own operations. If knowing is an act, we are only aware of it indirectly, when reflection turns within and toward its own operations. We see the tree outside the window, not the seeing of that tree. But we can infer that seeing has occurred for someone from the tree that is now seen.

Readers familiar with contemporary debates in phenomenology and philosophy of mind would immediately recognize these positions as versions of conceptualism versus non-conceptualism regarding perceptual content, and of the Higher-Order versus First-Order theories of consciousness (Janzen 2008; Gennaro 2012; Bayne 2012). In their effort to respond to the challenge posed by the Higher-Order theorists (both within and outside the Buddhist tradition), champions of the reflexivity thesis, such as Śāntarakṣita (725–788), turn to two main arguments: one concerning the character of consciousness and the other pertaining to the character of cognition. While sympathetic to the project of Middle Way or Madhyamaka metaphysics, and its critique of the very notion of an inherently existing entity (*svabhāva*), Śāntarakṣita does concede that consciousness has perforce a distinctive character that sets it apart from unconscious phenomena: it is something contrary to insentient objects. As he notes, "Consciousness arises as something that is excluded from all insentient objects. The self-reflexive awareness of that cognition is none other than its non-insentience" (*Tattvasaṃgraha* 2000; Coseru 2012: 239). This view that consciousness is contrary to insentience is meant to do double duty: on the one hand, it captures the notion that the conditions for the possibility of self-knowledge must be part of the structure of self-awareness. If self-awareness is a conceptually mediated process, then individuals who have not yet mastered a natural language or the requisite concepts of mind would lack the capacity for self-awareness. But infants and non-human animals, who lack such conceptual capacities, do behave in ways that suggest they have immediate access to their own mental states.

In taking consciousness to be something radically opposed to insentient objects, Buddhist philosophers following in the footsteps of Śāntarakṣita offer an ingenious way of conceptualizing the mind-body problem. In response to a largely emergentist picture championed by the Indian 'physicalists' or the Cārvākas (Bhattacharya 2009; Coseru 2017), they propose a conception of the mind-body relation as part of a complex causal chain of dependently arisen phenomena. Simply put, the causal principle at work states that a causal relation cannot be established between two things, if changes in one do not result in changes in the other. For something to count as the effect of a cause, it must be brought about by changes in the immediately preceding instance in the causal chain. For phenomenal consciousness to be the effect of a body and its sensory organs, its presence must be causally dependent on the latter. But, as the argument goes, experience suggests otherwise. For instance, loss of cognitive function in specific domains (hearing, sight, etc.) and other kinds of sensory and motor impairment do not impact the self-reflexive character of phenomenal consciousness. Thus, phenomenal consciousness is dependent neither on the body and the senses working together, nor on each of them taken individually.

5 Transcendental Subjectivity and the Problem of Witness Consciousness

Tasked with providing an account of the structure of consciousness barring any metaphysical commitment to enduring or persistent selves, Buddhist philosophers, specifically those associated with the Yogācāra tradition, developed the first phenomenology of consciousness and subjectivity in Indian philosophy (Kellner 2010; Dreyfus 2011; Coseru 2012). Two key ideas in particular define this phenomenological enterprise: (i) the notion that reflexivity must be a constitutive feature of both First-Order and Second-Order cognitive events; and (ii) a dual-aspect theory of mind, which takes intentionality and subjectivity or first-personal givenness to be constitutive features of the structure of cognitive awareness. It is worth noting that the reflexivity thesis only holds for a narrow class of cognitive events, specifically those that guarantee that consciousness is unified, that despite its specialized operations and multiplicity of content consciousness presents us with a unified phenomenal field.

But while these Buddhists did not think it necessary to postulate an ontological basis for the self-reflexive dimension of consciousness, philosophers associated with Advaita—the non-dualist school of thought pioneered by Śaṅkara (c. 700–750)—do. Drawing their inspiration from the Upaniṣads, Advaitins take the principle of self-luminosity to its logical conclusion.

Consciousness is no longer just an attribute of the self or a property that certain mental states have, but rather its own ultimate metaphysical ground. The Advaita theory of consciousness rests on the claim that, ultimately, mind and world are an irreducibly singular reality, in which the ultimate principle of things (*brahman*) and the principle of individuation (*ātman*) are one and the same (Bhattacharya 1973; Hulin 1978). There is nothing else besides this consciousness and its world-projecting capacities. Not only is there no ontology of mind-independent particulars, there is no ontology of subjects either. To the extent that Advaita recognizes and seeks to give an account of objects, these must ultimately belong in consciousness.

How does Advaita reconcile this conception of pure consciousness as the ultimate ground of being and what there is with our ordinary account of experience, which is irreducibly first-personal and embodied? Despite its seemingly radical metaphysics, the Advaita position on the phenomenology of subjectivity is quite straightforward: it is the result of an account of the sort of relations that obtain among intentional mental states when seen through the lens of consciousness's own constitutive features. The postulation of a pure consciousness lacking in any content and character would seem to preclude any attempt to offer a coherent account of intentionality, of how mental states come to be about things other than their own operations (as Avramides 2001 has convincingly argued, similarly, the Cartesian legacy of postulating privileged access to our own minds confronts us with the problem of other minds). The workaround solution is to claim that consciousness can be transitively self-reflexive about its occurrence but not about its operations. In short, for the cognition of an object to become an instance of knowledge, all that is required is for cognition to be aware that it is about an object of some kind. Its subjectivity, or subjective character, is not a matter of consciousness taking itself or its operations as an object in reflection or introspection. Rather, self-consciousness is a constitutive cognition (*svarūpajñāna*) of the sort that manifests as a capacity (*yogyatva*) whose association with mental content results in epistemically warranted cognitive events (*vṛttijñāna*). That consciousness has this constitutive capacity to apprehend its content first-personally or through a process of 'I-making' (*ahaṃkāra*) is just what it means for consciousness to be self-luminous or self-intimating. Advaitins thus share with the Yogācāra Buddhists the view that we have immunity to error through misidentification: what the notion that consciousness is constitutively self-luminous (*svataḥ prakāśa*) proves is simply that we have infallible access to the *occurrence* of our own mental states. It does not prove that our grasp of the *content* of those mental states is epistemically warranted (Gupta 2003; Ram-Prasad 2007; Timalsina 2009).

Let us consider some of the key features of this conception of consciousness, specifically as articulated by its most influential proponents—Śaṅkara (788–820), Śrīharṣa (fl. c. 12 C.E.), and Citsukha (fl. c. 1220). To begin with, the idea that consciousness becomes manifest by its own light goes back to the Upaniṣads, where one comes across statements to the effect that "the self itself is its light" (*Bṛhadaranyaka* 4.3.6; Olivelle 1998). For Śaṅkara, what this light manifests are the contents of the mind, which cannot be known on the basis of their own operations. There is no other source of illumination besides this self, which is itself pure cognition (*viśuddhavijñāna*) or cognition only (*vijñānamātra*) (*Kāṭhaupanisadbhāṣya* 12–14). This conception of 'consciousness only,' then, stands for the non-dual, self-reflexive awareness that is none other than the self.

In order to buttress their conception of a non-dual reflexive consciousness, Advaita philosophers use the analogy of a witness. Consider being a witness at a trial or racing event: while the experience of witnessing is immersive, it is non-participatory. The witness does not engage with the relevant actors, but simply observes from the sideline. Nor is the witness in any way affected by the outcome of the events that are witnessed. Advaitins use this analogy to make the case that cognition is an event to which consciousness simply bears witness. It is something that is made manifest by the witnessing consciousness (*sākṣin*), not something that consciousness itself does (*Brahmasūtra-bhāṣya* 2.2.28; Timalsina 2009: 21).

This parallelism between the Advaita conception of the luminosity of consciousness (*svaprakāśatā*) and the Yogācāra notion of self-reflexive consciousness (*svasaṃvedana*) should be obvious. Indeed, while acknowledging its deep root in the Buddhist tradition, Śrīharṣa thinks the notion that consciousness has this unique character of illuminating or revealing the operations of cognition is a self-established fact (*svatahḥsiddha*) (*Khaṇḍana Khaṇḍakhādya*; Dvivedi 1990: 69). Following a line of reasoning that owes a great deal to Dharmakīrti's account of reflexivity (Dharmakīrti 1989: III, 485–503), Śrīharṣa makes the seemingly obvious point that we cannot meaningfully talk about cognitive episodes that are unknown before they are thematized, any more than we can talk about unconscious pleasure and pain. Concepts such as pleasure and pain cannot be grasped outside the phenomenal experiences that instantiate them (Chalmers 2003 makes the case that, while corrigible, phenomenal beliefs of the sort 'I am in pain' depend on phenomenal concepts that are not themselves corrigible). Similarly, the cognition of an object does not and cannot occur so to speak in the dark, without being known, as Nyāya realists and Mādhyamika Buddhists have claimed.

Debates about how best to understand the luminosity thesis are the hallmark of late Indian philosophical accounts of the relation between consciousness and cognition. Concerned with the need to provide an adequate account of the nature of reflexivity, and aware that the reflexivity thesis could be taken to entail such obviously incoherent positions as that cognition serves as its own object, Advaitins came up with different solutions. One of the most representative of such solutions comes from Citsukha, who offers a three-pronged definition of the luminosity thesis: (i) self-luminosity itself is not something that is known, on account of not being an object: (ii) self-luminosity serves as an enabling condition for consciousness's own manifestation as witnessing; (iii) self-luminosity gives consciousness its own immediacy (*Tattvapradīpikā* 5–6; Ram-Prasad 2007: 78). What we have here is a clear attempt to argue that, while a cognition, say of a pot, can become the object of another cognition in introspection or thought recollection, it does so by virtue of the presence of witness consciousness. The enabling condition and the immediacy clause, likewise, are meant to show that, although consciousness itself cannot form an object of cognition, it does not mean that cognition is not intentionally constituted as being about an object of some kind. Citsukha is thus concerned to preserve for the Advaitin a conception of cognition as pertaining to objects, regardless of whether these objects are taken to be ontologically discrete particulars, or simply the intentional contents of awareness. For the Advaitin, thus, the reflexivity or self-luminosity thesis is simply a statement about the unity of consciousness: whatever its nature, and however it may come to illuminate the non-cognitive (*jaḍa*) processes of mental activity, consciousness itself is such that it cannot admit any duality (of 'knower' and 'known' or of 'subject' and 'object') within itself.

Advaita's non-dual metaphysics of mind would seem to preclude the sort of Cartesian phenomenology that assigns consciousness to the internal domain of thought, while postulating an external world of objects (Descartes 1996: 75ff.). Rather than arguing that the mind-body problem is ill conceived, because our experience of objects is not a phenomenon external to the mind, the Advaitin might be seen as arguing for a different conception of the hard problem. The really hard problem, on this account, is not to explain how consciousness could arise from something insentient such as the body. Rather, the problem is why consciousness, as the "light up" or illuminating aspect of mind, cannot itself become an object.

6 Conclusion

Indian philosophy is host to a rich tradition of reflection about the nature of consciousness, that incorporates both causal theories of mental content and detailed phenomenological analyses of the structure and operations of consciousness and cognition. While there is no clear indication that Indian philosophers conceived of something analogous to Descartes' mind-body problem, their solutions to the problem of agency, the problem of self-consciousness, and the problem of personal identity offer new ways to conceive the experiential features of our surface and deep phenomenology, a naturalistic epistemology grounded in pragmatic rather than normative concerns that echoes recent developments in embodied and enactive cognitive science, and a sophisticated conceptual vocabulary for thinking about the mind and mental phenomena in both egological and non-egological terms.

Note

1 For an interesting contrastive analysis with how classical Western metaphysics, specifically in the Neoplatonic tradition, conceives of the function of consciousness, see Hacker (1977).

References

Arnold, D. (2012) *Brains, Buddhas, and Believing*, New York: Columbia University Press.
Avramides, A. (2001) *Other Minds*, London: Routledge.
Bhattacharya, K. (1973) *L'ātman-Brahman dans le bouddhisme ancien*, Paris: École Française d'Extrême Orient.
Bhattacharya, R. (2009) *Studies on the Cārvāka/Lokāyata*, Florence: Società Editrice Fiorentina.
Bodhi, B. (1993) *A Comprehensive Manual of Abhidharma*, Kandy: Buddhist Publication Society.
Candrakānta (1991 [1869]) *Vaiśeṣika-sūtra-bhāṣya*, Varanasi: Vyasa Publishers.
Candrakīrti (1960) *Prasannapadā*, in P. L. Vaidya (ed.) *Madhyamakaśāstra of Nāgārjuna with the Commentary Prasannapadā by Candrakīrti*, Dharbanga: The Mithila Institute.
Chadha, M. (2011) "Self-Awareness: Eliminating the Myth of the 'Invisible Subject'," *Philosophy East and West* 61: 453–467.
Chakrabarti, K. K. (1999) *Classical Indian Philosophy of Mind*, Albany, NY: SUNY Press.
Chalmers, D. (2003) "The Content and Epistemology of Phenomenal Belief," in *Consciousness: New Philosophical Perspectives*, Oxford: Oxford University Press.
Collins, S. (1982) *Selfless Persons: Imagery and Thought in Theravāda Buddhism*, Cambridge: Cambridge University Press.
Coseru, C. (2012) *Perceiving Reality: Consciousness, Intentionality, and Cognition in Buddhist Philosophy*, New York: Oxford University Press.
Coseru, C. (2017) "Consciousness and Causal Emergence: Śāntarakṣita Against Physicalism," in J. Ganeri (ed.) *The Oxford Handbook to Indian Philosophy*, Oxford: Oxford University Press.
Descartes, R. (1996) *Meditations on First Philosophy: With Selections from the Objections and Replies*, edited by J. Cottingham (2nd ed.), Cambridge: Cambridge University Press.
Dhammajoti, K. L. (2007) "*Ākāra* and Direct Perception," *Pacific World Journal* 3: 245–272.
Dharmakīrti. (1989) *Pramāṇavārttika*, ed. P. C. Pandeya, Delhi: Motilal Banarsidass.
Dreyfus, G. (2011) "Self and Subjectivity: A Middle Way Approach," in M. Siderits, E. Thompson, and D. Zahavi (eds.) *Self, No-Self? Perspectives from Analytical, Phenomenological, and Indian Traditions*, Oxford: Oxford University Press.
Duerlinger, J. (2012) *The Refutation of the Self in Indian Buddhism: Candrakīrti on the Selflessness of Persons*, London: Routledge Press.
Ganeri, J. (2012) *The Self: Consciousness, Intentionality, and the First-Person Stance*, Oxford: Oxford University Press.
Garfield, J. (2006) "The Conventional Status of Reflexive Awareness: What's at Stake in a Tibetan Debate?" *Philosophy East and West* 56: 201–228.
Gennaro, R. (2012) *The Consciousness Paradox: Consciousness, Concepts, and Higher Order Thoughts*, Cambridge, MA: MIT Press.
Gupta, B. (2003) *Cit: Consciousness*, New Delhi: Oxford University Press.
Hacker, P. (1977) "Cit and Nous," in *Kleine Schriften*, edited by L. Schmithausen, Wiesbaden: Franz Steiner Verlag.
Harvey, P. (1995) *The Selfless Mind: Personality, Consciousness and Nirvāṇa in Early Buddhism*, Richmond, Surrey: Curzon Press.

Hattori, M. (1968) *Dignāga, on Perception*, Cambridge, MA: Harvard University Press.
Heidelberger, M. (2003) "The Mind-Body Problem in the Origin of Logical Positivism," in P. Parrini, W. C. Salmon, and M. H. Salmon (eds.) *Logical Empiricism: Historical and Contemporary Perspectives*, pp. 233–262. Pittsburgh, PA: University of Pittsburgh Press.
Hulin, M. (1978) *Le Principe de l'ego dans la pensée indienne classique: La Notion d'ahaṃkāra*, Publication de l'Institut de Civilisation Indienne, Paris: Diffusion
Husserl, E. (1970) *The Crisis of European Sciences and Transcendental Phenomenology*, Translated by David Carr. Evanston, IL: Northwestern University Press.
Jaworski, W. (2016) *Structure and the Metaphysics of Mind: How Homomorphism Solves the Mind-Body Problem*, Oxford: Oxford University Press.
Jha, Ganganatha, trans. (1939) *The Nyāya-sūtras of Gautama*. Vols 1–4. Delhi: Motilal Banarsidass.
Kellner, B. (2010) "Self-Awareness (*Svasaṃvedana*) in Dignāga's *Pramāṇsamuccaya* and *–vṛtti*: A Close Reading," *Journal of Indian Philosophy* 38: 203–231.
Lowe, E. J. (2006) "Non-Cartesian Substance Dualism and the Problem of Mental Causation," *Erkenntnis* 65 (1): 5–23.
MacKenzie, M. (2011) "Enacting the Self: Buddhist and Enactivist Approaches to the Emergence of the Self," in M. Siderits, E. Thompson, and D. Zahavi (eds.) *Self, No-Self? Perspectives from Analytical, Phenomenological, and Indian Traditions*, Oxford: Oxford University Press.
Nagel, T. (1974) "What Is It Like to Be a Bat?" *Philosophical Review* 83: 435–450.
Siderits, M. (2015) *Personal Identity and Buddhist Philosophy: Empty Persons*, Burlington, VT: Ashgate.
Tillemans, T. (1996) "What Would It Be Like to Be Selfless? Hīnayānist Versions, Mahāyānist Versions and Derek Parfit," *Études Asiatiques / Asiatische Studien* 50: 835–852.
Timalsina, S. (2009) *Consciousness in Indian Philosophy: The Advaita Doctrine of 'Awareness Only,'* London: Routledge.
Williams, P. (1998) *The Reflexive Nature of Awareness*, London: Curzon Press.
Zahavi, D. and Kriegel, U. (2015) "For-me-ness: What It Is and What It Is Not," in D. Dahlstrom, A. Elpidorou, and W. Hopp (eds.) *Philosophy of Mind and Phenomenology: Conceptual and Empirical Approaches*, London: Routledge.

Related Topics

Consciousness in Western Philosophy
Consciousness and Conceptualism
Consciousness and Intentionality
Meditation and Consciousness
The Unity of Consciousness

PART II

Contemporary Theories of Consciousness

8

REPRESENTATIONAL THEORIES OF CONSCIOUSNESS

Rocco J. Gennaro

A question that should be answered by any theory of consciousness is: What makes a mental state a conscious mental state? The focus of this chapter is on "representational theories of consciousness," which attempt to reduce consciousness to "mental representations" instead of directly to neural states. Examples of representational theories include *first-order* representationalism (FOR), which attempts to explain conscious experience primarily in terms of world-directed (or first-order) intentional states, and *higher-order* representationalism (HOR), which holds that what makes a mental state M conscious is that a HOR is directed at M. A related view, often called "self-representationalism," is also critically discussed in this chapter.

1 Representational Theories of Consciousness

Some theories attempt to reduce consciousness *in mentalistic terms* such as "thoughts" and "awareness." One such approach is to reduce consciousness to mental representations. The notion of a "representation" is of course very general and can be applied to photographs and various natural objects, such as the rings inside a tree. Indeed, this is part of the appeal of representational theories, since much of what goes on in the brain might also be understood in a representational way. Further, mental events are thought to represent outer objects partly because they are caused by such objects in, say, cases of veridical visual perception. Philosophers often call these mental states "intentional states" which have representational content, that is, mental states which are "directed at something," such as when one has a thought about a tree or a perception of a boat. Although intentional states, such as beliefs and thoughts, are sometimes contrasted with phenomenal states, such as pains and color experiences, many conscious states have both phenomenal and intentional properties, such as in visual perceptions.

The view that we can explain conscious mental states in terms of representational states is called "representationalism." Although not automatically reductionistic, most representationalists believe that there is room for a second-step reduction to be filled in later by neuroscience. A related motivation for representational theories of consciousness is that an account of intentionality seems more easily given in naturalistic terms, such as in causal theories whereby mental states are understood as representing outer objects in virtue of some reliable causal connection. The idea is that if consciousness can be explained in representational terms and representation can be understood in physical terms, then there is the promise of a naturalistic theory of consciousness. Most generally, however, representationalism can be defined as the view that the phenomenal properties of experience (that is, the "qualia" or "what it is like" of experience) can be explained in terms of the experiences' representational properties. For example, when I look at the blue sky, what it is like for me to have a conscious experience of the sky is simply identical with my experience's representation of the blue sky, and the property of "being blue" is a property of the representational object of experience.

It should be noted that the precise relationship between intentionality and consciousness is itself an ongoing area of research with some arguing that genuine intentionality actually presupposes consciousness in some way (Searle 1992; Siewart 1998; Horgan and Tienson 2002; Pitt 2004). If this is right, then it wouldn't be possible to *reduce* consciousness to intentionality as representationalists desire to do. But representationalists insist instead that intentionality is explanatorily prior to consciousness (Tye 2000; Carruthers 2000; Gennaro 1995; Gennaro 2012, ch. 2). Indeed, representationalists typically argue that consciousness requires intentionality but not vice versa. Few, if any, today hold Descartes' view that mental states are *essentially* conscious and that there are no unconscious mental states.[1]

2 First-Order Representationalism

A first-order representational (FOR) theory of consciousness is one that attempts to explain conscious experience primarily in terms of world-directed (or first-order) intentional states. The two most cited FOR theories are those of Fred Dretske (1995) and Michael Tye (1995, 2000), though there are many others as well. Tye's theory is the focus of this section. Like other FOR theorists, Tye holds that the representational content of my conscious experience is identical with the phenomenal properties of experience. Tye and other representationalists often use the notion of the "transparency of experience" in support for their view (Harman 1990). This is an argument based on the phenomenological first-person observation that when one turns one's attention away from, say, the blue sky and onto one's experience itself, one is still only aware of the blueness of the sky (Moore 1903). The experience *itself* is not blue, but rather one "sees right through" one's experience to its representational properties, and there is nothing else to the experience over and above such properties.

Despite some ambiguity in the notion of transparency (Kind 2003), it is clear that not all mental representations are conscious, and so the key question remains: What distinguishes conscious from unconscious mental states (or representations)? Tye defends what he calls "PANIC theory." The acronym "PANIC" stands for poised, abstract, non-conceptual, intentional content. Tye holds that at least some of the representational content in question is non-conceptual (N), which is to say that the subject can lack the concept for the properties represented by the experience in question, such as an experience of a certain shade of red that one has never seen before. The exact nature, or even existence, of non-conceptual content of experience is itself a highly debated issue in philosophy of mind (Gunther 2003; Gennaro 2012, ch. 6). But conscious states clearly must have "intentional content" (IC) for any representationalist. Tye also asserts that such content is "abstract" (A) and so not necessarily about particular concrete objects. This qualification is needed to handle cases of hallucinations where there are no concrete objects at all. Perhaps most important for mental states to be conscious, however, is that such content must be "poised" (P), which is an importantly functional notion. Tye explains that the

> key idea is that experiences and feelings... stand ready and available to make a direct impact on beliefs and/or desires. For example...feeling hungry...has an immediate cognitive effect, namely, the desire to eat...States with non-conceptual content that are not so poised lack phenomenal character [because]...they arise too early, as it were, in the information processing.
>
> *(Tye 2000: 62)*

One frequent objection to FOR is that it cannot explain all kinds of conscious states. Some conscious states do not seem to be "about" or "directed at" anything, such as pains, itches, anxiety, or after-images, and so they would be non-representational conscious states. If so, then conscious states cannot generally be explained in terms of representational properties (Block 1996). Tye responds that pains and itches do represent, in the sense that they represent parts of the body. After-images and hallucinations either misrepresent (which is still a kind of representation) or the conscious subject still takes them to have representational properties from the first-person point of view. Tye (2000) responds to a whole host of alleged counter-examples to FOR. For example, with regard to conscious emotions, he says that they "are frequently localized in particular parts of the body...For example, if one feels sudden jealousy, one is likely to feel one's stomach sink... [or] one's blood pressure increase" (2000: 51). Tye believes that something similar is true for fear or anger. Moods, however, seem quite different and not localizable in the same way. But, still, if one feels, say, elated, then one's overall conscious experience is changed.[2]

Others use "inverted qualia" arguments against FOR. These are hypothetical cases where behaviorally indistinguishable individuals have inverted color perceptions of objects, such as person A visually experiences a lemon in the same way that person B experiences a ripe tomato, likewise for all yellow and red objects. If it is possible that there are two individuals whose color experiences are inverted with respect to the objects of perception, we would have a case of *different* phenomenal experiences with the *same* represented properties. The strategy is to think of counter-examples where there is a difference between the phenomenal properties in experience and the relevant representational properties in the world. These objections can perhaps be answered by Tye (e.g. in Tye 2000) and others in various ways, but significant debate continues. Moreover, intuitions dramatically differ as do the plausibility and value of these so-called "thought experiments."

A more serious objection to Tye's theory might be that what seems to be doing most of the work on his account is the functional-sounding "poised" notion, and thus he is not really explaining phenomenal consciousness in *entirely* representational terms (Kriegel 2002). It is also unclear how a disposition can confer *actual* consciousness on an otherwise unconscious mental state. Carruthers, for example, asks: "How can the mere fact that an [unconscious state] is now in a position to have an impact upon the...decision-making process [or beliefs and desires] confer on it the subjective properties of feel and 'what-it-is-likeness' distinctive of phenomenal consciousness?" (2000: 170).[3]

3 Higher-Order Representationalism

Recall the key question: What makes a mental state a *conscious* mental state? There is also a tradition that has attempted to understand consciousness in terms of some kind of *higher-order* awareness and this intuition has been revived by a number of contemporary philosophers (Armstrong 1981; Rosenthal 1986, 1997, 2002, 2005; Lycan 1996, 2001; Gennaro 1996, 2004a, 2012). The basic idea is that what makes a mental state M conscious is that it is the object of a higher-order representation (HOR). A HOR is a "meta-psychological" state, that is, a mental state directed at another mental state. So, for example, my desire to write a good chapter becomes conscious when I am (non-inferentially) "aware" of the desire. Intuitively, it seems that conscious states, as opposed to unconscious ones, are mental states that I am "aware of" being in. So conscious mental states arise when two (unconscious) mental states are related in a certain way, namely, that one of them (the HOR) is directed at the other (M).

This intuitively appealing claim is sometimes referred to as the Transitivity Principle (TP):

(TP) A conscious state is a state whose subject is, in some way, aware of being in the state.

Conversely, the idea that I could be having a conscious state while totally *unaware* of being in that state seems very odd (if not an outright contradiction). A mental state of which the subject is completely unaware is clearly an *un*conscious state. For example, I would not be aware of having a subliminal perception and thus it is an unconscious perception. Any theory that attempts to explain consciousness in terms of higher-order states is known as a "higher-order representational theory of consciousness." It is best initially to use the more neutral term "representation," because there are many versions of higher-order theory depending upon how one characterizes the HOR itself.

4 Higher-Order Thought (HOT) Theories

The two main kinds of HOR theory are higher-order thought (HOT) and higher-order perception (HOP). HOT theorists, such as David Rosenthal, think it is better to understand the HOR as a thought containing concepts. HOTs are treated as cognitive states involving some kind of conceptual component. HOP theorists hold that the HOR is a perceptual or experiential state of some kind (Lycan 1996), which does not require the kind of conceptual content invoked by HOT theorists. Although HOT and HOP theorists agree on the need for a HOR theory of consciousness, they do sometimes argue for the superiority of their respective positions (Rosenthal 2004; Lycan 2004). I personally favor a version of the HOT theory of consciousness for the reasons discussed here and elsewhere (Gennaro 1996, 2012). HOT theory is arguably well motivated by the Transitivity Principle and offers a reasoned way to differentiate conscious and unconscious mental states. It may not currently be the best strategy to *directly* reduce consciousness to neurophysiology, but not necessarily because of the usual objections to materialism having to do with the "hard problem" or "explanatory gap" (Gennaro 2012, chs. 2 and 4).

There is something like TP in premise 1 of Lycan's (2001) more general argument for HOR. The entire argument runs as follows:

1 A conscious state is a mental state whose subject is aware of being in it.
2 The "of" in 1 is the "of" of intentionality; what one is aware of is an intentional object of the awareness.
3 Intentionality is representational; a state has a thing as its intentional object only if it represents that thing.

Therefore, 4 Awareness of a mental state is a representation of that state (from 2, 3).
Therefore, 5 A conscious state is a state that is itself represented by another of the subject's mental states (1, 4).

The intuitive appeal of premise 1 leads to the final conclusion – (5) – which is just another way of stating HOR. Another interesting rationale for HOR, and HOT theory in particular, is as follows (based on Rosenthal 2004: 24): A non-HOT theorist might still agree with HOT theory as an account of *introspection* or *reflection*, namely, that it involves a *conscious* thought about a mental state. This seems to be a common sense definition of introspection that includes the notion that introspection involves conceptual activity. It also seems reasonable to hold that when a mental state is unconscious, there is no HOT at all. But then, it stands to reason that there should be something in between those two cases, that is, when one has a first-order (i.e. world-directed) conscious state. So what is in between having no HOT at all and having a conscious HOT? The answer is an unconscious HOT, which is precisely what HOT theory says. In addition, this can neatly explain what happens when there is a shift from a first-order conscious state to an introspective state: an unconscious HOT becomes conscious (more on this below).

HOT theorists also hold that one must become aware of the lower-order (LO) state *noninferentially*. We might suppose, say, that the HOT must be caused noninferentially by the LO state to make it conscious. The point of this condition is mainly to rule out alleged counterexamples to HOT theory, such as cases where I become aware of my unconscious desire to kill my boss because I have consciously inferred it from a session with a psychiatrist, or where my envy becomes conscious after making inferences based on my own behavior. The characteristic *feel* of such a conscious desire or envy may be absent in these cases, but since awareness of them arose via conscious inference, the higher-order (HO) theorist accounts for them by adding this noninferential condition.

A common initial worry to HOR theories is that they are circular and lead to an infinite regress. It might seem that HOT theory results in circularity by defining consciousness in terms of HOTs, that is, we should not explain a concept by using that very same concept. It also might seem that an infinite regress results because a conscious mental state must be accompanied by a HOT, which, in turn, must be accompanied by another HOT *ad infinitum*. However, the standard reply is that when a conscious mental state is a first-order world-directed state, the HOT is not itself conscious; otherwise, circularity and an infinite regress would follow. When the HOT is itself conscious, there is a yet higher-order (or third-order) thought directed at the second-order state. In this case, we have *introspection*, which involves a conscious HOT directed at an inner mental state. When one introspects, one's attention is directed back into one's mind. For example, what makes my desire to write a good chapter a conscious *first-order* desire is that there is an unconscious HOT directed at the desire. In this case, my conscious focus is directed at my computer screen, so I am not consciously aware of having the HOT from the first-person point of view. When I *introspect* that desire, however, I then have a conscious HOT (accompanied by a yet higher, third-order, HOT) directed at the desire itself (see Rosenthal 1986, 1997). Thus, what seems to be an objection is really mainly a request to clarify some further details of the theory (see Figure 8.1).

There are several other objections to HOT theory: First, some argue that various animals (and even infants) are not likely to have the conceptual sophistication required for HOTs, and so that would render animal (and infant) consciousness very unlikely (Dretske 1995; Seager 2004). Are cats and pigs capable of having complex HOTs such as "I am in mental state M"? Although most who bring forth this objection are not higher-order theorists, Peter Carruthers (1989, 2000) is one HOT theorist who actually embraces the conclusion that (most) animals do not have phenomenal consciousness. However, it can be argued that the HOTs need not be as sophisticated as it might initially appear and there is other ample comparative neurophysiological evidence supporting the conclusion that animals have conscious mental states (Gennaro 1993, 1996). Most HOT theorists do not want to accept the absence of animal or infant consciousness as a consequence of holding the theory.

The debate on this issue has continued over the past two decades,[4] but to give one example, Clayton and Dickinson and their colleagues have reported demonstrations of memory for time in scrub jays (Clayton, Bussey, and Dickinson 2003: 37). Scrub jays are food-caching birds, and when they have food they cannot eat, they hide it and recover it later. Because some of the food is preferred but perishable (such as crickets), it must be eaten within a few days, while other food (such as nuts) is less preferred but does not perish as quickly. In cleverly designed experiments using these facts, scrub jays are shown, even days after caching, to know not only *what* kind of food was *where* but also *when* they had cached it (see also Clayton, Emery, and Dickinson 2006). This strongly suggests that the birds have some degree of self-concept (or "I-concept"), which can figure into HOTs. That is, such experimental results seem to show that scrub jays have episodic memory, which involves a sense of self over time. Further, many crows and scrub

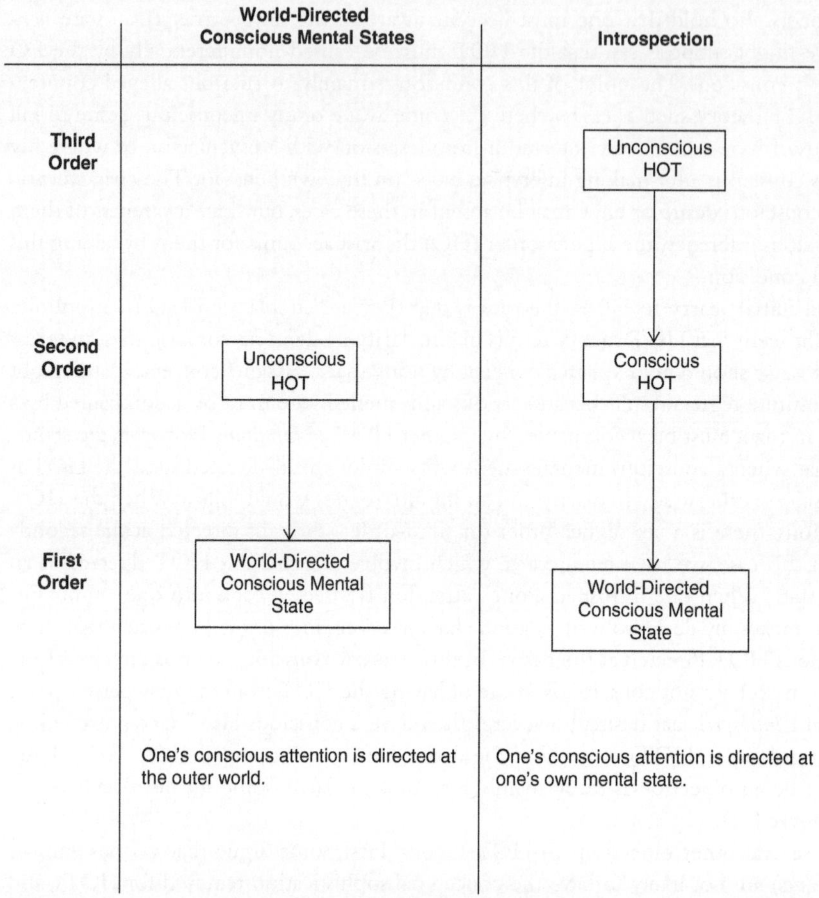

Figure 8.1 The Higher-Order Thought (HOT) Theory of Consciousness

jays return alone to caches they had hidden in the presence of others and recache them in new places (Emery and Clayton 2001). This suggests that they know that *others* know where the food is cached, and thus, to avoid having their food stolen, they recache the food. So it seems that these birds can even also have some concepts of other minds.

A second objection has been called the "problem of the rock" (Stubenberg 1998) and is originally due to Alvin Goldman (1993). When I think about a rock, it is obviously not true that the rock becomes conscious. So why should I suppose that a mental state becomes conscious when I think about it? This objection forces HOT theorists to explain just how adding a HOT changes an unconscious state into a conscious one. There have been, however, a number of responses to this kind of objection (Rosenthal 1997; Van Gulick 2000, 2004; Gennaro 2005, 2012, ch. 4). A common theme is that there is a principled difference in the *objects* of the thoughts in question. For one thing, rocks and similar objects are not mental states in the first place and so HOT theorists are trying to explain how a *mental state* becomes conscious.

Third, the above sometimes leads to an objection akin to Chalmers' (1995) "hard problem." It might be asked just how exactly any HOR theory really explains the subjective or phenomenal aspect of conscious experience. How or why does a mental state come to have a first-person qualitative "what it is like" aspect by virtue of a HOR directed at it? A number of overlapping responses have emerged in recent years. Some argue that this objection misconstrues the more modest purpose of (at least, their) HOT theories. The claim is that HOT theories are theories of consciousness only in the sense that they are attempting to explain what differentiates conscious from unconscious states, that is, in terms of a higher-order awareness of some kind. A full account of "qualitative properties" or "sensory qualities" (which can themselves be unconscious) can be found elsewhere in their work and is independent of their theory of consciousness (Rosenthal 1991; Lycan 1996). Thus, a full explanation of phenomenal consciousness does require more than a HOR theory, but that is no objection to it as such. It may also be that proponents of the hard problem unjustly raise the bar as to what would count as a viable explanation of consciousness, so that any reductionist attempt would inevitably fall short (Carruthers 2000).

My own response to how HOTs explain conscious states has more to do with the rather Kantian idea that the concepts that figure into the HOTs are necessarily presupposed in conscious experience (Gennaro 2012, ch. 4; 2005). The basic idea is that first we receive information via our senses or the "faculty of sensibility." Some of this information will then rise to the level of unconscious mental states but these mental states do not become conscious until the more cognitive "faculty of understanding" operates on them via the application of concepts. We can arguably understand this concept application in terms of HOTs directed at first-order states. Thus, I consciously experience (and recognize) the red barn *as a red barn* partly because I apply the concepts "red" and "barn" (in my HOTs) to my basic perceptual states. If there is a real hard problem, it may have more to do with explaining concept acquisition and application (Gennaro 2012, chs. 6 and 7). It is important to notice, however, that this kind of solution is unlike reductionist accounts in neurophysiological terms and so is immune to Chalmers' main criticism of those theories. For example, there is no problem about how a specific brain activity "produces" conscious experience, nor is there an issue about any allegedly mysterious *a priori* or *a posteriori* connection between brains and consciousness. The issue instead is how HOT theory is realized in our brains.

A fourth and very important objection to HO approaches is the question of how any of these theories can explain cases where the HO state might *misrepresent* the LO mental state (Byrne 1997; Neander 1998; Levine 2001; Block 2011). After all, if we have a representational relation between two states, it seems possible for misrepresentation or malfunction to occur. If it does, then what explanation can be given by the HOT theorist? If my LO state registers a *red* percept and my HO state registers a thought about something *green* due to some odd neural misfiring, then what happens? It seems that problems loom for any answer given by a HOT theorist. For example, if a HOT theorist takes the option that the resulting conscious experience is reddish, then it seems that the HOT plays no role in determining the qualitative character of the experience. On the other hand, if the resulting experience is greenish, then the LO state seems irrelevant. Rosenthal and Weisberg hold that the HO state determines the qualitative properties even when there is no LO state at all, which are called "targetless" or "empty" HOT cases (Rosenthal 2005, 2011; Weisberg 2008, 2011).[5]

My own view is that no conscious experience results in the above cases because it is difficult to see how, even according to HOT theory, a sole unconscious HOT can result in a conscious state (Gennaro 2012, 2013). I think that there must be a conceptual match, complete or partial, between the LO and HO state in order for a conscious state to exist in the first place. Weisberg and Rosenthal argue that what really matters is how things seem to the subject and, if we can explain that, we've explained all that we need to. But somehow the HOT *alone* is now all that matters. Doesn't this defeat the very purpose of HOT theory, which is supposed to explain a conscious mental state in terms of the *relation* between *two states*? Moreover, HOT theory is supposed to be a theory of first-order state consciousness, that is, the *lower-order state* is supposed to be the conscious one. So, I hold that misrepresentations cannot occur between M and HOT *and* still result in a conscious state (Gennaro 2012, 2013).[6]

Let us return briefly to the claim that HOT theory can help to explain how one's conceptual repertoire can transform our phenomenological experience. Concepts, at minimum, involve recognizing and understanding objects and properties. Having a concept C should also give the concept possessor the ability to discriminate instances of C and non-Cs. For example, if I have the concept 'tiger' I should be able to identify tigers and distinguish them from other even fairly similar land animals. Rosenthal invokes the idea that concepts can change one's conscious experience with the help of several nice examples (2005: 187–188). For example, acquiring various concepts from a wine-tasting course will lead to different experiences from those enjoyed before the course. I acquire more fine-grained wine-related concepts, such as "dry" and "heavy," which in turn can figure into my HOTs and thus alter my conscious experiences. As is widely held, I will literally have different qualia due to the change in my conceptual repertoire. As we acquire more concepts, we have more fine-grained experiences and thus we experience more qualitative complexities. A botanist will likely have somewhat different perceptual experiences than I do when we are walking through a forest. Conversely, those with a more limited conceptual repertoire, such as infants and animals, will have a more coarse-grained set of experiences.[7]

5 Dispositional HOT Theory

Carruthers (2000) thinks that it is better to treat HOTs as dispositional states instead of the standard view that the HOTs are actual, though he also understands his "dispositional HOT theory" to be a form of HOP theory (Carruthers 2004). The basic idea is that the consciousness of an experience is due to its *availability* to HOT. So, "conscious experience occurs when perceptual contents are fed into a special short-term buffer memory store, whose function is to make those contents available to cause HOTs about themselves" (Carruthers 2000: 228). Some first-order perceptual contents are available to a higher-order "theory of mind mechanism," which transforms those representational contents into conscious contents. Thus, no actual HOT occurs. Instead, according to Carruthers, some perceptual states acquire a dual intentional content; for example, a conscious experience of yellow not only has a first-order content of "yellow," but also has the higher-order content "seems yellow" or "experience of yellow." Thus, he calls his theory "dual-content theory." Carruthers makes interesting use of so-called "consumer semantics" in order to fill out his theory of phenomenal consciousness. The content of a mental state depends, in part, on the powers of the organisms who "consume" that state, for example, the kinds of inferences the organism can make when it is in that state.

Carruthers' dispositional theory is criticized by those who, among other things, do not see how the mere disposition toward a mental state can render it conscious (Rosenthal 2004; Gennaro 2004b, 2012). Recall that a key motivation for HOT theory is the TP. But the TP clearly lends itself to an *actualist* HOT theory interpretation, namely, that we *are* aware of our conscious states and not aware of our unconscious states. As Rosenthal puts it, "being disposed to have a thought about something doesn't make one conscious of that thing, but only potentially conscious of it" (2004: 28). Thus it is natural to wonder just how dispositional HOT theory *explains* phenomenal consciousness, that is, how a *dispositional* HOT can render a mental state *actually* conscious.

Carruthers is, to be fair, well aware of this objection and attempts to address it in some places (such as Carruthers 2005: 55–60). He again relies on consumer semantics in an attempt to show that changes in consumer systems can transform perceptual contents. But one central problem arguably remains: dual-content theory appears vulnerable to the same objection raised by Carruthers against FOR. On both views, it is difficult to understand how the functional or dispositional aspects of the respective theories can yield actual conscious states (Jehle and Kriegel 2006).

6 Higher-Order Perception (HOP) Theory

David Armstrong (1968, 1981) and William Lycan (1996, 2004) have been the leading HOP theorists in recent years. Unlike HOTs, HOPs are not thoughts and can have at least some non-conceptual content. HOPs are understood as analogous to outer perception. One standard objection to HOP theory, however, is that, unlike outer perception, there is no distinct sense organ or scanning mechanism responsible for HOPs. Similarly, no distinctive sensory quality or phenomenology is involved in having HOPs whereas outer perception always involves some sensory quality. Lycan concedes the disanalogy but argues that it does not outweigh other considerations favoring HOP theory (Lycan 1996: 28–29; 2004: 100). Lycan's reply might be understandable but the objection remains a serious one nonetheless. After all, this represents a major difference between normal outer perception and any alleged inner perception.

Lycan (2004: 101–110) presents several reasons to prefer HOP theory to HOT theory. For example, he urges that consciousness, and especially active introspection, of our mental states is much more like perception than thought because perception allows for a greater degree of voluntary control over what areas of our phenomenal field to make conscious. But one might argue against Lycan's claim that HOP theory is superior to HOT theory by pointing out that there is an important *nonvoluntary* or *passive* aspect to perception not found in thought (Gennaro 2012, ch. 3). The perceptions in HOPs are too passive to account for the dynamic interaction between HORs and first-order states. While it is true that many thoughts do occur nonvoluntarily and somewhat spontaneously, introspective thoughts (i.e. conscious HOTs) can be controlled voluntarily at least as well as conscious HOPs. We often actively search our minds for information, memories, and other mental items. In any case, what ultimately justifies treating HORs as thoughts is the application of concepts to first-order states (Gennaro 1996: 101; 2012, ch. 4).

Lycan has actually recently changed his mind and no longer holds HOP theory, mainly because he thinks that *attention* to first-order states is sufficient for an account of conscious states and there is little reason to view the relevant attentional mechanism as *intentional* or as *representing* first-order states (Sauret and Lycan 2014). Armstrong and Lycan had indeed previously often spoken of HOP "monitors" or "scanners" as a kind of attentional mechanism but now it seems that "...leading contemporary cognitive and neurological theories of attention are unanimous in suggesting that attention is not intentional" (Sauret and Lycan 2014: 365). They cite Prinz (2012), for example, who holds that attention is a psychological process that connects first-order states with working memory. Sauret and Lycan explain that "attention is the mechanism that enables subjects to become aware of their mental states" (2014: 367) and yet this "awareness of" is a non-intentional selection of mental states. Thus, Sauret and Lycan (2014) find that Lycan's (2001) argument, discussed above, goes wrong at premise 2, namely, that the "of" mentioned in premise 1 is perhaps more of an "acquaintance relation," which is non-intentional. Unfortunately, Sauret and Lycan do not present a worked out theory of acquaintance and it is doubtful that the acquaintance strategy is a better alternative (see Gennaro 2015). Such acquaintance relations would presumably be understood as somehow "closer" than the representational relation. But this strategy is at best trading one difficult problem for an even deeper puzzle, namely, just how to understand the allegedly intimate and non-representational "awareness of" relation between HORs and first-order states. It is also more difficult to understand such "acquaintance relations" in the context of a reductionist approach. Indeed, acquaintance is often taken to be unanalyzable and simple, in which case it is difficult to see how it could explain anything, let alone the nature of conscious states.

7 Hybrid and Self-Representational Accounts

A final cluster of representationalist views holds that the HOR in question should be understood as *intrinsic* to an overall complex conscious state. This is in contrast to the standard view that the HOR is *extrinsic* to (that is, entirely distinct from) its target mental state. Rosenthal's view about the extrinsic nature of the HOR has come under attack in recent years and thus various hybrid representational theories can be found in the literature. One motivation for this trend is some dissatisfaction with standard HOR theory's ability to handle some of the objections addressed above. Another reason is renewed interest in a view somewhat closer to the one held by Franz Brentano (1874/1973) and others, normally associated with the phenomenological tradition (Husserl 1913/1931; Sartre 1956; Smith 2004; Textor 2006). To varying degrees, these theories have in common the idea that conscious mental states, in some sense, represent *themselves*. Conscious states still involve having a thought about a mental state but just not a distinct mental state. Thus, when one has a conscious desire for a beer, one is also aware that one is in that very state. The conscious desire both represents the beer and itself. It is this "self-representing" that makes the state conscious and is the distinguishing feature of such states.

These theories are known by various names. For example, my own view is actually that, when one has a first-order conscious state, the (unconscious) HOT is better viewed as intrinsic to the target state, so that we have a complex conscious state with parts (Gennaro 1996, 2006, 2012). I call this the "wide intrinsicality view" (WIV) and argue, for example, that Jean-Paul Sartre's theory of consciousness can also be understood in this way (Gennaro 2002, 2015). On the WIV, first-order conscious states are complex states with a world-directed part and a meta-psychological component. Conscious mental states can be understood as brain states, which are combinations of passively received perceptual input and presupposed higher-order conceptual activity directed at that input. Robert Van Gulick (2004, 2006) has also explored the related alternative that the higher-order state is part of an overall global conscious state. He calls these states "HOGS" (Higher-Order Global States) where a lower-order unconscious state is "recruited" into a larger state, which becomes conscious, partly due to the "implicit self-awareness" that one is in the lower-order state.

This approach is also forcefully advocated by Uriah Kriegel in a series of papers (beginning with Kriegel [2003] and culminating in Kriegel [2009]). He calls it the "self-representational theory of consciousness." To be sure, the notion of a mental state representing itself or a mental state with one part representing another part is in need of further explanation. Nonetheless, there is agreement among these authors that conscious mental states are, in some important sense, reflexive or self-directed.

Kriegel (2006, 2009) interprets TP in terms of a ubiquitous (conscious) "peripheral" self-awareness (or "mine-ness"), which accompanies all of our first-order focal conscious states. Not all conscious "directedness" is attentive and so we should not restrict conscious directedness to that which we are consciously focused on. If this is right, then a first-order conscious state can be both attentively outer-directed and inattentively inner-directed. Still, there are problems with this approach. For example, although it is true that there are degrees of conscious attention, the clearest example of genuine "inattentive" consciousness is outer-directed awareness in one's peripheral visual field. But this obviously does not show that any such inattentional consciousness is *self-directed* when there is outer-directed consciousness, let alone at the very same time. Also, what is the evidence for such self-directed inattentional consciousness? It is presumably based on phenomenological considerations, but, for what it's worth, I have to confess that I do not find such ubiquitous inattentive self-directed "consciousness" in my first-order experience. It does not seem to me that I am consciously aware (in any sense) of my own *experience* when I am, say, consciously attending to a movie or putting together a bookcase. Even some who are otherwise very sympathetic to Kriegel's phenomenological approach find it difficult to believe that "pre-reflective" (inattentional) self-awareness always accompanies conscious states (Siewart 1998; Zahavi 2004; Smith 2004). None of these authors are otherwise sympathetic to HOT theory or reductionist approaches to consciousness.[8]

8 HOT Theory and the Brain

One interesting recent area of emphasis has been on how HOR and self-representationalism might be realized in the brain. After all, most representationalists think that their accounts of the structure of conscious states are realized in the brain (even if it will take some time to identify all the neural structures). To take one question: do conscious mental states require widespread brain activation, or can at least some be fairly localized in narrower areas of the brain? Perhaps most interesting is whether or not the prefrontal cortex (PFC) is required for having conscious states (Gennaro 2012, ch. 9). Kriegel (2007; 2009, ch. 7) and Block (2007) argue that, according to the higher-order and self-representational view, the PFC is required for most conscious states. But even though it is very likely true that the PFC is required for the more sophisticated *introspective* states (or conscious HOTs), this would not be a problem for HOT theory because it doesn't require introspection for first-order conscious states (Gennaro 2012, ch. 9).

Is there evidence of conscious states without PFC activity? Yes. For example, Rafael Malach and colleagues show that when subjects are engaged in a perceptual task, such as absorbed in watching a movie, there is widespread neural activation but little PFC activity (Grill-Spector and Malach 2004; Goldberg, Harel, and Malach 2006). Although some other studies do show PFC activation, this is mainly because subjects are asked to *report* their experiences. Also, basic conscious experience is not decreased entirely even when there is extensive bilateral PFC damage or lobotomies (Pollen 2003). It seems that this is also an advantage for HOT theory with regard to the problem of animal and infant consciousness. If another theory *requires* PFC activity for *all* conscious states and HOT theory does not, then HOT theory is in a better position to account for animal and infant consciousness, since it is doubtful that infants and most animals have the requisite PFC activity.

One might still ask: Why think that unconscious HOTs can occur outside the PFC? If we grant that unconscious HOTs can be regarded as a kind of "pre-reflective" self-consciousness, then we might for example look to Newen and Vogeley (2003) for some answers. They distinguish five levels of self-consciousness from "phenomenal self-acquaintance" and "conceptual self-consciousness" up to "iterative meta-representational self-consciousness." The majority of their discussion is explicitly about the neural correlates of what they call the "first-person perspective." Citing numerous experiments, they point to various "neural signatures" of self-consciousness. The PFC is rarely mentioned and then usually only with regard to more sophisticated forms of self-consciousness. Other brain areas are much more prominently identified, such as the medial and inferior parietal cortices, the temporoparietal cortex, and the anterior cingulate cortex and the posterior cingulate cortex.[9]

9 Brief Summary

The primary focus of the chapter is on representational theories of consciousness, which attempt to reduce consciousness to mental representations rather than directly to neural states. Examples of this popular approach are first-order representationalism (FOR), which attempts to explain conscious experience primarily in terms of world-directed (or first-order) intentional states, and higher-order representationalism (HOR), which holds that what makes a mental state M conscious is that it is the object of some kind of HOR directed at M. Objections to each view were raised and some responses were offered. In addition, some hybrid and self-representational approaches were also critically discussed. The overall question that should be answered by any of these theories is: What makes a mental state a conscious mental state?

Notes

1 Some related literature along these lines has been growing quickly with frequent reference to "phenomenal intentionality" (Kriegel 2013) and "cognitive phenomenology" (Bayne and Montague 2011; Chudnoff 2015). For my own take on this issue, see Gennaro (2012, ch. 2).
2 For a more recent exchange on the representational content of moods, see Kind (2014) and Mandelovici (2014).
3 For other versions of FOR, see Kirk (1994), Byrne (2001), and Droege (2003). See Chalmers (2004) for an excellent discussion of the dizzying array of possible representationalist positions.
4 See Carruthers (2000, 2005, 2008) and Gennaro (2004b, 2009, 2012, chs. 7 and 8).
5 For some other variations on HOT theory, see Rolls (2004), Picciuto (2011), and Coleman (2015).
6 In the end, I argue for the much more nuanced claim that "Whenever a subject S has a HOT directed at experience e, the content c of S's HOT determines the way that S experiences e (provided that there is a *full or partial* conceptual match with the lower-order state, *or* when the HO state contains more specific or fine-grained concepts than the LO state has, *or* when the LO state contains more specific or fine-grained concepts than the HO state has, *or* when the HO concepts can combine to match the LO concept)" (Gennaro 2012: 180). The reasons for these qualifications are discussed at length in Gennaro (2012, ch. 6).
7 In Gennaro (2012, ch. 6), I argue that there is a very close and natural connection between HOT theory and conceptualism. Chuard (2007) defines conceptualism as the claim that "the representational content of a perceptual experience is fully conceptual in the sense that what the experience represents (and how it represents it) is entirely determined by the conceptual capacities the perceiver brings to bear in her experience" (Chuard 2007: 25).
8 For others who hold some form of the self-representationalism, see Williford (2006) and Janzen (2008). Some authors (such as Gennaro [2012]) view their hybrid position to be a modified version of HOT theory and Rosenthal (2004) has called it "intrinsic higher-order theory." I have argued against Kriegel's view at length in Gennaro (2008) and Gennaro (2012, ch. 5).
9 See Kozuch (2014) for a nice discussion of the PFC in relation to higher-order theories.

References

Armstrong, D. (1968) *A Materialist Theory of Mind*, London: Routledge and Kegan Paul.
Armstrong, D. (1981) "What Is Consciousness?" In *The Nature of Mind*, Ithaca, NY: Cornell University Press.
Bayne, T., and Montague, M. (eds.) (2011) *Cognitive Phenomenology*, New York: Oxford University Press.
Block, N. (1996) "Mental Paint and Mental Latex," *Philosophical Issues* 7: 19–49.
Block, N. (2007) "Consciousness, Accessibility, and the Mesh between Psychology and Neuroscience," *Behavioral and Brain Sciences* 30: 481–499.
Block, N. (2011) "The Higher Order Approach to Consciousness Is Defunct," *Analysis* 71: 419–431.
Brentano, F. (1874/1973) *Psychology From an Empirical Standpoint*, New York: Humanities.
Byrne, A. (1997) "Some Like It HOT: Consciousness and Higher-Order Thoughts," *Philosophical Studies* 86: 103–129.
Byrne, A. (2001) "Intentionalism Defended," *Philosophical Review* 110: 199–240.
Carruthers, P. (1989) "Brute Experience," *Journal of Philosophy* 86: 258–269.
Carruthers, P. (2000) *Phenomenal Consciousness*, Cambridge: Cambridge University Press.
Carruthers, P. (2004) "HOP over FOR, HOT Theory," In Gennaro (2004a).

Carruthers, P. (2005) *Consciousness: Essays from a Higher-Order Perspective*, New York: Oxford University Press.

Carruthers, P. (2008) "Meta-Cognition in Animals: A Skeptical Look," *Mind and Language* 23: 58–89.

Chalmers, D. (1995) "Facing Up to the Problem of Consciousness," *Journal of Consciousness Studies* 2: 200–219.

Chalmers, D. (ed.) (2002) *Philosophy of Mind: Classical and Contemporary Readings*, New York: Oxford University Press.

Chuard, P. (2007) "The Riches of Experience," In R. Gennaro (ed.) *The Interplay Between Consciousness and Concepts*, Exeter, UK: Imprint Academic. (This is also a special double issue of the *Journal of Consciousness Studies* 14 (9–10).)

Chudnoff, E. (2015) *Cognitive Phenomenology*, New York: Routledge.

Clayton, N., Bussey, T., and Dickinson, A. (2003) "Can Animals Recall the Past and Plan for the Future?" *Nature Reviews Neuroscience* 4: 685–691.

Clayton, N., Emery, N., and Dickinson, A. (2006) "The Rationality of Animal Memory: Complex Caching Strategies of Western Scrub Jays," In S. Hurley and M. Nudds (eds.) *Rational Animals?* New York: Oxford University Press.

Coleman, S. (2015) "Quotational Higher-Order Thought Theory," *Philosophical Studies* 172: 2705–2733.

Dretske, F. (1995) *Naturalizing the Mind*, Cambridge, MA: MIT Press.

Droege, P. (2003) *Caging the Beast*, Philadelphia and Amsterdam: John Benjamins Publishers.

Emery, N., and Clayton, N. (2001) "Effects of Experience and Social Context on Prospective Caching Strategies in Scrub Jays," *Nature* 414: 443–446.

Gennaro, R. (1993) "Brute Experience and the Higher-Order Thought Theory of Consciousness," *Philosophical Papers* 22: 51–69.

Gennaro, R. (1995) "Does Mentality Entail Consciousness?" *Philosophia* 24: 331–358.

Gennaro, R. (1996) *Consciousness and Self-Consciousness: A Defense of the Higher-Order Thought Theory of Consciousness*, Amsterdam and Philadelphia: John Benjamins.

Gennaro, R. (2002) "Jean-Paul Sartre and the HOT Theory of Consciousness," *Canadian Journal of Philosophy* 32: 293–330.

Gennaro, R. (ed.) (2004a) *Higher-Order Theories of Consciousness: An Anthology*, Amsterdam and Philadelphia: John Benjamins.

Gennaro, R. (2004b) "Higher-Order Thoughts, Animal Consciousness, and Misrepresentation: A Reply to Carruthers and Levine," In Gennaro (2004a).

Gennaro, R. (2005) "The HOT Theory of Consciousness: Between a Rock and a Hard Place?" *Journal of Consciousness Studies* 12 (2): 3–21.

Gennaro, R. (2006) "Between Pure Self-Referentialism and the (Extrinsic) HOT Theory of Consciousness," In U. Kriegel and K. Williford (2006).

Gennaro, R. (2008) "Representationalism, Peripheral Awareness, and the Transparency of Experience," *Philosophical Studies* 139: 39–56.

Gennaro, R. (2009) "Animals, Consciousness, and I-thoughts," In R. Lurz (ed.) *Philosophy of Animal Minds*, New York: Cambridge University Press.

Gennaro, R. (2012) *The Consciousness Paradox: Consciousness, Concepts, and Higher-Order Thoughts*, Cambridge, MA: The MIT Press.

Gennaro, R. (2013) "Defending HOT Theory and the Wide Intrinsicality View: A Reply to Weisberg, Van Gulick, and Seager," *Journal of Consciousness Studies* 20 (11–12): 82–100.

Gennaro, R. (2015) "The 'of' of Intentionality and the 'of' of Acquaintance," In S. Miguens, G. Preyer, and C. Morando (eds.) *Pre-Reflective Consciousness: Sartre and Contemporary Philosophy of Mind*, New York: Routledge Publishers.

Goldberg, I., Harel, M., and Malach, R. (2006) "When the Brain Loses Its Self: Prefrontal Inactivation during Sensorimotor Processing," *Neuron* 50: 329–339.

Goldman, A. (1993) "Consciousness, Folk Psychology and Cognitive Science," *Consciousness and Cognition* 2: 264–282.

Grill-Spector, K. and Malach, R. (2004) "The Human Visual Cortex," *Annual Review of Neuroscience* 7: 649–677.

Gunther, Y. (ed.) (2003) *Essays on Nonconceptual Content*, Cambridge, MA: MIT Press.

Harman, G. (1990) "The Intrinsic Quality of Experience," In J. Tomberlin (ed.) *Philosophical Perspectives*, 4, Atascadero, CA: Ridgeview Publishing.

Horgan, T., and Tienson, J. (2002) "The Intentionality of Phenomenology and the Phenomenology of Intentionality," In Chalmers (2002).

Husserl, E. (1913/1931) *Ideas: General Introduction to Pure Phenomenology (Ideen au einer reinen Phänomenologie und phänomenologischen Philosophie*, Translated by W. Boyce Gibson, New York: MacMillan.

Janzen, G. (2008) *The Reflexive Nature of Consciousness*, Amsterdam and Philadelphia: John Benjamins.

Jehle, D. and Kriegel, U. (2006) "An Argument Against Dispositional HOT Theory," *Philosophical Psychology* 19: 462–476.

Kind, A. (2003) "What's so Transparent about Transparency?" *Philosophical Studies* 115: 225–244.

Kind, A. (2014) "The Case Against Representationalism About Moods," In U. Kriegel (ed.) *Current Controversies in Philosophy of Mind*, New York: Routledge Press.

Kirk, R. (1994) *Raw Feeling*, New York: Oxford University Press.

Kozuch, B. (2014) "Prefrontal Lesion Evidence Against Higher-Order Theories of Consciousness," *Philosophical Studies* 167: 721–746.

Kriegel, U. (2002) "PANIC Theory and the Prospects for a Representational Theory of Phenomenal Consciousness," *Philosophical Psychology* 15: 55–64.

Kriegel, U. (2003) "Consciousness as Intransitive Self-Consciousness: Two Views and an Argument," *Canadian Journal of Philosophy* 33: 103–132.

Kriegel, U. (2006) "The Same Order Monitoring Theory of Consciousness," In U. Kriegel and K. Williford (2006).

Kriegel, U. (2007) "A Cross-Order Integration Hypothesis for the Neural Correlate of Consciousness," *Consciousness and Cognition* 16: 897–912.

Kriegel, U. (2009) *Subjective Consciousness*, New York: Oxford University Press.

Kriegel, U. (2013) *Phenomenal Intentionality*, New York: Oxford University Press.

Levine, J. (2001) *Purple Haze: The Puzzle of Conscious Experience*, Cambridge, MA: MIT Press.

Lycan, W. (1996) *Consciousness and Experience*, Cambridge, MA: MIT Press.

Lycan, W. (2001) "A Simple Argument for a Higher-Order Representation Theory of Consciousness," *Analysis* 61: 3–4.

Lycan, W. (2004) "The superiority of HOP to HOT," In R. Gennaro (ed.) *Higher-Order Theories of Consciousness: An Anthology*, Amsterdam: John Benjamins.

Mandelovici, A. (2014) "Pure Intentionalism about Moods and Emotions," In U. Kriegel (ed.) *Current Controversies in Philosophy of Mind*, New York: Routledge Press.

Moore, G. E. (1903) "The Refutation of Idealism," In G. E. Moore (ed.) *Philosophical Studies*, Totowa, NJ: Littlefield, Adams, and Company.

Neander, K. (1998) "The Division of Phenomenal Labor: A Problem for Representational Theories of Consciousness," *Philosophical Perspectives* 12: 411–434.

Newen, A. and Vogeley, K. (2003) "Self-Representation: Searching for a Neural Signature of Self-Consciousness," *Consciousness and Cognition* 12: 529–543.

Picciuto, V. (2011) "Addressing Higher-Order Misrepresentation with Quotational Thought," *Journal of Consciousness Studies* 18 (3–4): 109–136.

Pitt, D. (2004) "The Phenomenology of Cognition, Or, What Is It Like to Think That P?" *Philosophy and Phenomenological Research* 69:1–36.

Pollen, D. (2003) "Explicit Neural Representations, Recursive Neural Networks and Conscious Visual Perception," *Cerebral Cortex* 13: 807–814.

Prinz, J. (2012) *The Conscious Brain*, New York: Oxford University Press.

Rolls, E. (2004) "A Higher Order Syntactic Thought (HOST) Theory of Consciousness," In Gennaro (2004a).

Rosenthal, D.M. (1986) "Two Concepts of Consciousness," *Philosophical Studies* 49: 329–359.

Rosenthal, D.M. (1991) "The Independence of Consciousness and Sensory Quality," *Philosophical Issues* 1: 15–36.

Rosenthal, D.M. (1997) "A Theory of Consciousness," In N. Block, O. Flanagan, and G. Güzeldere (eds.) *The Nature of Consciousness*, Cambridge, MA: MIT Press.

Rosenthal, D.M. (2002) "Explaining Consciousness," In D. Chalmers (ed.) *Philosophy of Mind: Classical and Contemporary Readings*, New York: Oxford University Press.

Rosenthal, D.M. (2004) "Varieties of Higher-Order Theory," In R. Gennaro (ed.) *Higher-Order Theories of Consciousness: An Anthology*, Amsterdam and Philadelphia: John Benjamins.

Rosenthal, D.M. (2005) *Consciousness and Mind*, New York: Oxford University Press.

Rosenthal, D.M. (2011) "Exaggerated Reports: Reply to Block," *Analysis* 71: 431–437.

Sartre, J. (1956) *Being and Nothingness*, New York: Philosophical Library.
Sauret, W., and Lycan, W. (2014) "Attention and Internal Monitoring: A Farewell to HOP," *Analysis* 74: 363–370.
Seager, W. (2004) "A Cold Look at HOT Theory," In R. Gennaro (ed.) *Higher-Order Theories of Consciousness: An Anthology*, Philadelphia and Amsterdam: John Benjamins.
Searle, J. (1992) *The Rediscovery of the Mind*, Cambridge, MA: MIT Press.
Siewart, C. (1998) *The Significance of Consciousness*, Princeton: Princeton University Press.
Smith, D.W. (2004) *Mind World: Essays in Phenomenology and Ontology*, Cambridge, MA: Cambridge University Press.
Stubenberg, L. (1998) *Consciousness and Qualia*, Philadelphia and Amsterdam: John Benjamins Publishers.
Textor, M. (2006) "Brentano (and Some Neo-Brentanians) on Inner Consciousness," *Dialectica* 60: 411–432.
Tye, M. (1995) *Ten Problems of Consciousness*, Cambridge, MA: MIT Press.
Tye, M. (2000) *Consciousness, Color, and Content*, Cambridge, MA: MIT Press.
Van Gulick, R. (2004) "Higher-Order Global States (HOGS): An Alternative Higher-Order Model of Consciousness," In R. Gennaro (ed.) *Higher-Order Theories of Consciousness: An Anthology*, Amsterdam and Philadelphia: John Benjamins.
Van Gulick, R. (2006) "Mirror Mirror—Is That All?" In U. Kriegel and K. Williford (2006).
Weisberg, J. (2008) "Same Old, Same Old: The Same-Order Representation Theory of Consciousness and the Division of Phenomenal Labor," *Synthese* 160: 161–181.
Weisberg, J. (2011) "Misrepresenting Consciousness," *Philosophical Studies* 154: 409–433.
Williford, K. (2006) "The Self-Representational Structure of Consciousness," In Kriegel and Williford (2006).
Zahavi, D. (2004) "Back to Brentano?" *Journal of Consciousness Studies* 11 (10–11): 66–87.

Related Topics

Materialism
Consciousness in Western Philosophy
Consciousness and Intentionality
Consciousness and Conceptualism
Consciousness and Attention
Animal Consciousness

9
THE GLOBAL WORKSPACE THEORY

Bernard J. Baars and Adam Alonzi

1 Introduction

A global workspace (GW) is a functional hub of signal integration and propagation in a large population of loosely coupled agents, on the model of "crowd" or "swarm" computing, using a shared "blackboard" for posting, voting on, and sharing hypotheses, so that multiple experts can make up for each others' limitations. Crowd computation has become a major technique for web commerce as well as scientific problem-solving.

In the 1970s Allen Newell's Carnegie-Mellon team showed that a GW architecture was able to solve a difficult practical problem, the task of identifying 1,000 normally spoken words in a normally noisy and distorting acoustical environment, including the many challenges of phonemic and syllabic encoding of slow analogue sound resonances interrupted by fast transients, produced by the inertial movements of many different vocal tracts, each with its distinctive acoustical resonance profile beginning with vocal, soft tissue, nasal, and labiodental turbulence, each with overlapping "coarticulation" of phonemic gestures, with its own idiosyncratic speech styles and dialects, all in an acoustical environment with its own mix of sound-absorbing, masking and echoing surfaces. In real speech this difficult signal identification task is also organized in lexical and morphemic units, with real-world referents, with unpredictable and ambiguous grouping, syntactic, semantic, pragmatic, intonational and emotional organization. Newell's HEARSAY system was able to identify more than 90% of the spontaneous words correctly, even without modern formant tracking, a newer and more effective technique.

HEARSAY was one of the first success stories for the new concept of parallel-distributed architectures, now often called "crowdsourcing" or "swarm computing." The most important point here is the surprising effectiveness of expert crowds using GW-mediated signaling, when none of the individual experts could solve the posted problem by themselves.

One of today's leading speech recognition systems, Apple's SIRI, is still making use of web-based crowdsourcing to identify poorly-defined syllables in numerous languages and dialects, spoken by many different voices in acoustically noisy spaces. SIRI also learns to predict the speaker's vocal tract to improve its detection score. It is still imperfect, but it is commercially viable.

Based on Newell's work, Baars (1988) demonstrated the surprisingly close empirical match between the well-known "central limited capacity" components of the brain associated with consciousness and the notion of a global workspace architecture. The resulting GW theory of conscious cognition has been widely adopted and developed, showing some 15,000 citations since Baars (1988).

A new wave of neuroscience evidence shows that the extended cortex – neo, paleo, and thalamus – can support a dynamic, mobile, context-sensitive and adaptive GW function. Many regions of the cortex support conscious experiences, which can be reported with high accuracy, and which generally compete with each other for limited momentary capacity. However, regions like the cerebellum and the dorsal stream of cortex do not enable conscious contents directly.

Modern computation came of age using very clean electrical signals, digitally distinctive and easy to translate into programming code. The first programs used logical, arithmetic and other symbolic operations that came from centuries of mathematics. "Shannon information" was well-defined and relatively easy to implement in practice, and the mathematical Turing Machine supported formal proofs that almost any determinate function could be implemented by TMs.

The challenge for HEARSAY was quite different from the standard problems of classical computation, and much more biological in spirit, because the real sensory world is not digital but analogue, with poorly-defined stimuli, actions and salience boundaries, many of which must be defined "top-down" based on prior knowledge.

The natural world is not engineered to avoid catastrophic events like head injuries and microparasites; modern humans live in highly protected environments, with none of the pitfalls and dangers we encounter when running over an unimproved natural landscape with poor visual conditions. In contrast, ancient cemeteries often show very high rates of broken human bones and other physical damage, often inflicted by other humans.

Modern buildings use parallel and orthogonal ceilings, floors and walls, making perceptual size estimation and action prediction much easier. Their acoustical properties are typically clean and predictable. Conscious distractions are radically reduced. In the last century the spread of sanitary engineering alone has reduced infectious diseases and doubled the human lifespan.

The world in which our ancestors evolved was fundamentally different. Computational architectures built to deal with unpredictable, high risk events are therefore more biologically realistic. Humans may be among the most adaptable species in the animal kingdom, as shown by the fact that Homo Sapiens has colonized an immense diversity of econiches around the globe in the last 30–40,000 years, beginning with a genetically small and homogeneous "founder population" in north-east Africa some 46,000 years ago.

As they spread out of Africa, humans occupied many hostile environments, using a toolkit that included flint cutting tools, hand axes, hunting bows and flint-tipped arrows, projectile weapons, cooperative hunting and fishing, woven and animal skin clothing sown with bone needles, woven reed matting, and effective social organization. Because the descendants of African founder population were able to rapidly colonize the coastal regions of the Old World, including Australia and New Zealand, it is believed that humans understood practical water travel, using reed bundles and rafts, wood, animal bladders, paddled canoes and sailboats that are still in widespread use today.

In a broad sense, all human biocultural adaptation involves cortex, and novel problems require *conscious* cortical regions and networks, like the ones you are using in this moment. The conscious regions of the cortico-thalamic (CT) system give us the gateway for learning and problem-solving. The proposed reason for the efficiency of conscious cortex in the face of novel challenges is its ability to recruit entirely new coalitions of expert sources to "concentrate" on a single unpredictable question.

The mammalian neocortex is roughly 200 million years old. At a basic level, the cortex is a highly flexible swarm computation architecture, although its frontal half also supports executive functions.

The prefrontal cortex (PFC) interacts with the entire sensory and motor brain, with biocultural motivation and emotions, and appetitive drivers ranging from nutrition to reproductive pheromones.

Emotion theorists have pointed out that "emotions" are dramatic fragments that use a kind of narrative case grammar. We don't just feel "anger," but we experience "anger" toward some perceived violator of the perceiver's social boundaries, such as the murder of a socially protected child. To set the balance right again, the emotional actor often engages in compensatory actions, from an act of protection or revenge, to a negotiated compensation for the loss and humiliation. Thus, emotional acts can easily be strung into entire interpersonal narratives of the kind we have in dreams: A norm-violating provocation followed by just retribution is one very common example of a narrative theme, often seen in ancient epics.

Cooperation and planning are important skills largely made possible by the prefrontal cortex. Experiential hippocampal memory (called "episodic") may record every conscious event. Biological examples of swarm computation are extremely common. Eusocial animals (like ants, naked mole rats, and termites) and slime mold colonies (like *P. polycephalum*, which can solve the famously difficult Traveling Salesman Problem using locally emergent parallel-interactive processing) are prime examples (Jones and Adamatsky 2014). Varieties of swarm computation, including mixed cases of swarm and executive computation, are therefore very common. With the emergence of language, humans learned how to implement executive computation, as in playing chess and calculating arithmetic; however, such sequential computation may be rather recent (approximately 100,000 years ago).

2 Consciously Mediated Processing in the Cortex

Functional specialization of cortical regions was controversial until Broca's and Wernicke's language areas were discovered in the 1800s. The cortex does both swarm and sequential symbolic computation. Using high spatiotemporal resolution imaging tools, we can see individual neurons performing tasks, sometimes phase-locked to population oscillations. The primary projection areas of the senses and motor systems are functional hierarchies, which signal bidirectionally, not strictly top-down or bottom-up. Sometimes single functional neuronal members of a hierarchy can be mobilized by conscious neurofeedback.

Learning throughout the brain appears to occur by the Hebbian rule: "neurons that fire together, wire together." Learned inhibition may occur the same way, using inhibitory (GABA-ergic) neurons. New functional groups are therefore constantly being created, decomposed and reorganized. Neurofeedback signaling is a powerful and general method to induce neuronal learning, using conscious feedback stimuli (tones, flashing lights, etc.). However, there is no evidence that unconscious neurofeedback leads to novel learning. This suggests that learning is consciously mediated, as shown in the case of associative conditioning. Baars (1988) describes how the GWT hypothesis can show how conscious (global) neurofeedback can recruit local neuronal groups to acquire control over local target activity.

Neurons and neuronal cell assemblies can be defined as "expert agents" when they respond selectively to input or output. Conscious experiences may therefore reflect a GW function in the brain. The cortex has many anatomical hubs, but conscious percepts are unitary and internally consistent at any given moment. This suggests that a brain-based GW capacity cannot be limited to only one anatomical hub. Rather, a consciousness-supporting GW should be sought in a mobile, dynamic and coherent binding capacity – a functional hub – for neural signaling over multiple networks.

Two research groups have found conscious (rather than unconscious) visual processing high in the visual hierarchy, including the inferotemporal cortex (IT), superior temporal sulcus

(STS), medial temporal lobe (MTL), lateral occipital complex (LOC) and the PFC. In hearing, Heschel's gyrus seems to involve a consciousness-supporting neuronal hub, and in interoceptive feelings, like nausea and "gut emotions," the anterior insula seems to be involved. External touch is probably mediated by area S1 (the somatosensory homunculus), and related sensory body maps, and the corresponding motor areas influence voluntary movement in various subtle ways.

The theater metaphor is ancient and is associated with more than one theory of consciousness. In GWT focal consciousness acts as the bright spot on the stage, which is directed by the spotlight of attention. The bright spot is surrounded by a "fringe" of vaguely conscious events (Mangan 1993). The stage corresponds to "working memory," the immediate memory system in which we talk to ourselves, visualize places and people, and make plans. Information from the bright spot is globally distributed to two classes of complex unconscious processors: those in the shadowy audience, who primarily receive information from the bright spot; and unconscious contextual systems that shape events within the bright spot, who act "behind the scenes." One example of such a context is the unconscious philosophical assumptions with which we tend to approach the topic of consciousness.

Cross-model conscious integration is extremely common, and is presumably mediated by parietal regions, but the prefrontal cortex is also a "hub of many sensory hubs," intimately connected with the others, and it is difficult to rule out a PFC function in any conscious or voluntary experience. Conscious feelings of knowing (FOKs) are vividly illustrated by Wilder Penfield's (1975) long series of open-brain surgeries on epileptic patients, which found that both sides of the prefrontal lobe (medial and lateral) are involved in feelings of effort, such as tip-of-the-tongue. Tip-of-the-tongue experiences, and their accompanying FOKs, can be induced by asking for the technical names of familiar facts. The question "What are two names for flying dinosaurs?" may elicit strong FOK. Subjects who cannot recall those names still choose accurately and quickly between "pterodactyl" and "brontosaurus." Semantic knowledge may be fully primed in tip-of-the-tongue (TOT) states, before the lexical form of the missing words can be recalled. Such FOK commonly occur when we have compelling and accurate expectations and intentions. They are not limited to language.

Our general hypothesis is that the cortical connectome (the enormous mass of myelinated long-distance fibers emerging from pyramidal cells in the neocortex, paleocortex, and thalamus) supports GW functions: That is, the ability to integrate multiple incoming signals into coherent spatiotemporal coalitions, and to "broadcast" the output signals to activate and recruit large functional cell assemblies in pursuit of some high-level goal. Recent cortical network maps using Diffuse Tractography Imaging (DTI) show classical features of large-scale networks, including small-world organization, optimal signaling efficiency, and robust functioning in the face of local damage.

In humans and macaques, the CT complex underlies reportable conscious percepts, concepts, FOKs, visual images and executive functions. While subcortical areas are sometimes claimed to specify conscious contents, the human evidence is slight and disputed. However, basal ganglia can feed back to cortex via a posterior thalamic pathway, and the thalamus is obviously involved in all cortical input-output signaling. In the case of corticofugal signals (e.g. vocalization, voluntary eye movements, corticospinal tracts, corticovagal output), conscious signaling comes from muscular output leading to sensory input, as in the famous example of the articulatory-auditory feedback loop.

Because cortex and thalamus are so densely interleaved as to constitute a single functional system, we will refer here to the CT system as a whole. CT pathways permit constant reentrant signaling, so that multiple spatiotopic maps, internal topographical representations, can sustain or inhibit each other. The CT system resembles an enormous metropolitan street plan, in which one can travel from any street corner to any other. Almost all cortico-cortical and cortico-thalamic links are bidirectional, so that the normal signaling mode in the CT system is not unidirectional, but resonant. This basic fact has many implications.

Global workspace theory follows the historic distinction between the "focus" of experience vs. the largely implicit background of experience. Extensive evidence shows that visual and auditory consciousness flows from the respective sensory surfaces to frontoparietal and particularly prefrontal regions. The CT core is a great mosaic of multi-layered two-dimensional neuronal arrays. Each array of cell bodies and neurites projects to others in topographically systematic ways. Since all CT pathways are bidirectional, signaling is "adaptively resonant" (reentrant). In this complex, layered two-dimensional arrays are systematically mirrored between cortex and thalamus, region by region.

The CT nexus appears to be the most parallel-interactive structure in the brain, allowing for efficient signal routing from any neuronal array to any other. This connectivity is different from other structures that do not directly enable conscious contents, like the cerebellum. The cerebellum is organized in modular clusters that can run independently of each other, in true parallel fashion. But in the CT core any layered array of cortical or thalamic tissue can interact with any other, more like the worldwide web than a server farm.

CT pathways run in all canonical directions and follow small-world organization, so that each array is efficiently linked to many others. The entire system acts as an oscillatory medium, with markedly different global regimes in conscious and unconscious states. Global workspace dynamics interprets the traditional distinction between the "object" and "ground" of experiences as a directional flow between the moment-to-moment focus of conscious experience vs. the implicit background and sequelae of focal contents. The proposed directionality of broadcasting suggests a testable distinction with information integration theory and dynamic core theory.

3 Dynamic GW vis-à-vis Other Theoretical Proposals

We can widely divide current theories into philosophical and empirically based ones. Some of the philosophical theories are now generating testable hypotheses. Empirical theories can be divided into "localist" vs. "local-global" types. There are no exclusively global theories, since no one denies the evidence for local and regional specialization in the brain. Philosophical theories typically aim to account for subjective experiences or "qualia," a notoriously difficult question. Recently some philosophical perspectives, like "higher order theory" (HOT), have also generated testable proposals about the involvement of brain regions like the PFC. However, brain imaging experiments (e.g. Dehaene and Naccache 2001) have long implicated the frontoparietal cortex in subjective experience.

It is not clear at this time whether philosophically based theories generate novel, testable predictions. However, efforts are underway to test HOT theories. In general, claims to explain subjective qualia are still debated. Zeki (2001) makes the localist claim that conscious percepts of red objects involve "micro-conscious" activation of cortical color regions (visual areas V3/V4). However, most empirical theories combine local and global activities, as briefly discussed above. It is still possible that momentary events may be localized for 100 milliseconds or less, and that full conscious contents emerge over some hundreds of milliseconds. The Dynamic GW theory is a specific version of the "dynamic core" hypothesis proposed by Edelman and Tononi (2000) and, in somewhat different forms, by Edelman (1989) and others. Dynamic Global Workspace theory implies a directional signal flow from binding to receiving coalitions. For each conscious event there is a dominant source and a set of receivers, where the propagated signal is interpreted, used to update local processes, and refreshed via reentrant signaling to the source (Edelman 1989). Conscious sensations arise in a different center of binding and propagation than "feelings of knowing" like the TOT experience, as demonstrated by brain imaging studies (Maril et al. 2001). Directional broadcasting of bound conscious contents is one testable distinction from other proposals (Edelman et al. 2011). Supportive evidence has been reported by Doesburg et al. (2009) and others. Other theories, like Tononi's mathematical measure of complexity, phi, seem less directional (Edelman and Tononi 2000). Llinas and Pare (1991) have emphasized the integration of specific and nonspecific thalamocortical signaling, and Freeman et al. (2003) have developed a conception of hemisphere-wide signaling and phase changes. Nevertheless, current local-global theories are strikingly similar. Whether major differences will emerge over time is unclear.

4 Dynamic Global Workspace as a Local-Global Theory

In 1988, GW theory suggested that "global broadcasting" might be one property of conscious events. Other proposed properties were:

1 Informativeness, that is, widespread adaptation to the novelty of the reportable signal, leading to general habituation (information reduction) of the news contained in the global broadcast. The evidence now supports widespread neuronal updating to novel input.
2 Internal consistency of conscious contents, because mutually exclusive global broadcasts tend to degrade each other. This is a well-established feature of conscious contents, first observed in the nineteenth century and replicated many thousands of times. Binocular rivalry is one well-known example.
3 Interaction with an implicit self-system. Baars (1988) proposed that the observing self is coextensive with implicit frames that shape the objects of consciousness. One major kind of access that has been discussed since Immanuel Kant is the access of the "observing self" to the contents of consciousness. Lou et al. (2010) have shown that self-related brain regions like the precuneus and midline structures from the PAG to orbitofrontal cortex may be mobilized by conscious sensory contents. Baars (1988) proposed that self-other access is a specific variety of framing (contextualizing), and that it is a necessary condition for conscious contents.
4 One of the driving questions of GW theory is how the limited capacity of momentary conscious contents can be reconciled with the widespread access enabled by conscious contents. Why is the conscious portion of an otherwise massively parallel-distributed system a limited and serial process? Would our ancestors not have benefited from the ability to competently perform several tasks at once? A stream of consciousness integrates disparate sources of information, but it is limited to a "single internally consistent content at any given moment" (Baars 1988). The Oxford English Dictionary dedicates 75,000 words to the various definitions of "set." However, a native speaker will, while reading or listening, know almost immediately in what sense the word is being used. We can rapidly detect errors in phonology, syntax, semantics, and discrepancies between a speaker's stated and true intentions, but are not necessarily conscious of how this is done The workspace makes sense of novel and ambiguous situations by calling upon unconscious "expert" processors (see Figure 9.1).

Because almost all neural links in the CT system are bidirectional, reentrant signaling from receivers to broadcasting sources may quickly establish task-specific signaling pathways, in the same way that a fire department might locate the source of a community-wide alarm, and, then,

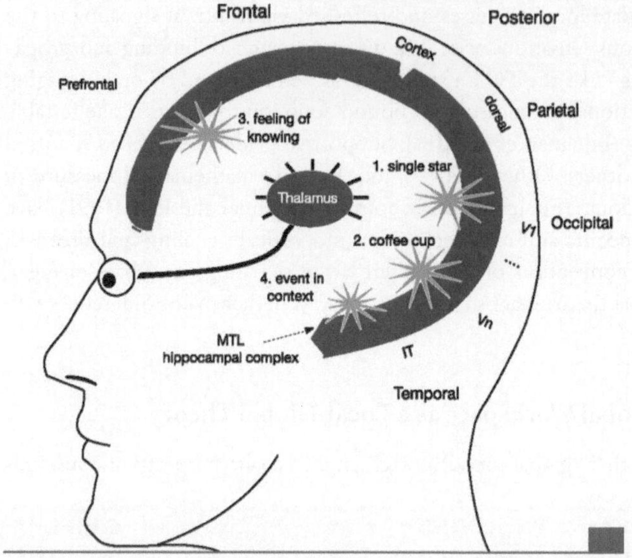

Figure 9.1 Examples of Possible Binding and Broadcasting in the Cortico-Thalamic Core

communicate in a much more task-specific way. Current evidence suggests brief broadcasts, as suggested by the 100 ms conscious integration time of different sensory inputs.

Figure 9.1 shows four examples of possible binding and broadcasting in the CT core (starburst icons). Cortical area V1 and the lateral geniculate nucleus (LGN) – the visual thalamus – can be conceived as two arrays of high-resolution bright and dark pixels, without color. The sight of a single star on a dark night may therefore rely heavily on V1 and its mirror array of neurons in LGN. V1 and LGN interact constantly, with bidirectional signal traffic during waking. The sight of a single star at night reveals some surprising features of conscious vision, including spatial context sensitivity, as in the classical autokinetic effect: single points of light in a dark space begin to wander long subjective distances in the absence of spatial framing cues. The autokinetic effect is not an anomaly, but rather a prototype of decontextualized percepts (Baars 1988). A large literature in perception and language shows scores of similar phenomena, as one can demonstrate by looking at a corner of a rectangular room through a reduction tube that excludes external cues. Any two- or three-way corner in a carpentered space is visually reversible, much like the Necker Cube and the Ames trapezoid. Such local ambiguities exist at every level of language comprehension and production (Baars 1988; Shanahan and Baars 2005).

The dorsal stream of the visual cortex provides egocentric and allocentric "frames" to interpret visual events in nearby space. These parietal frames are not conscious in themselves, but they are required for visual objects to be experienced at all (Goodale and Milner 1992). Injury to the right parietal cortex may cause the left half of visual space to disappear, while contralesional stimulation, like cold water in the left ear, may cause the lost half of the field to reappear. Thus, even a single dot of light in a dark room reveals the contextual properties of conscious perception. Ambiguity and its resolution is a universal need for sensory systems in the natural world, where ambiguity is commonly exacerbated by camouflage, deceptive signaling, distraction, unpredictable movements, ambushes, sudden dangers and opportunities, darkness, fog, light glare, dense obstacles, and constant utilization of cover by predators and prey (Bizley et al. 2012).

Conscious percepts plausibly involve multiple "overlays," like map transparencies, which can be selectively attended. The sight of a coffee cup may involve an object overlaid by color, texture, and reflectance, combining information from LGN, V1, V2, V3/V4, and IT (Crick and Koch 1990). Active cells in those arrays may stream signals across multiple arrays, cooperating, and competing to yield a winner-take-all coalition. Once the winning coalition stabilizes, it may "ignite" a broadcast to other regions.

Conscious vision is strikingly flexible with respect to level of analysis, adapting seamlessly from the sight of a single colored dot to the perception of a dotted (pointillist) painting. An account of conscious vision must therefore explain how a local dot can be perceived in the same visual display as a Georges Seurat painting.

To identify a single star at night, because the highest spatial resolution is attained in the retina, LGN, and V1, the visual cortex must be able to amplify neuronal activity originating in LGN-V1 through attentional modulation. For coffee cups and faces, the relative activity of IT and the fusiform gyrus must be increased. It follows that binding coalitions of visual activity maps can bring out the relative contribution of different feature levels, even for the same physical stimulus (Itti and Koch 2001).

5 Bidirectional Pathways and Adaptive Resonance

Because CT pathways are bidirectional they can support "reentrant signaling" among topographically regular spatial maps. The word "resonance" is often used to describe CT signaling (Wang 2001). It is somewhat more accurate than "oscillation," which applies to true iterative patterns like sine waves. Edelman and coworkers prefer the term "reentry," while others like to use "adaptive resonance." We will use the last term to emphasize its flexible, selective, and adaptive qualities.

Adaptive resonance has many useful properties, as shown in modeling studies like the Darwin autonomous robot series, where it can account for binding among visual feature maps, a basic property of visual perception (Izhikevich and Edelman 2008). Edelman has emphasized that reentry (adaptive resonance) is not feedback, but rather evolves a selectionist trajectory that can search for solutions to biologically plausible problems. Grossberg and others have developed adaptive resonance models for cortical minicolumns and layers.

6 Broadcasting: Any-to-Many Signaling

A few ants can secrete alarm pheromones to alert a whole colony to danger, an example of any-to-many broadcasting among insects. In humans the best-known example is hippocampal-neocortical memory storage of memory traces in the neocortex by way of the hippocampal complex (Nadel et al. 2000; Ryan et al. 2001). Memories of conscious episodes are stored in millions of synaptic alterations in the neocortex (Moscovitch et al. 2005). Computer users are familiar with global memory searches, which are used when specific searches fail. The CT system may enable brain-based global memory searches. "Any-to-many" coding and retrieval can be used to store and access existing information (Nadel et al. 2000; Ryan et al. 2010). It is also useful for mobilizing existing automatisms to deal with novel problems.

Notice that "any-to-many" signaling does not apply to the cerebellum, which lacks parallel-interactive connectivity, or to the basal ganglia, spinal cord, or peripheral ganglia. Crick and Koch have suggested that the claustrum may function as a GW underlying consciousness (Crick and Koch 2005). However, the claustrum, amygdala, and other highly connected anatomical hubs seem to lack the high spatiotopic bandwidth of the major sensory and motor interfaces, as shown by the very high-resolution of minimal conscious stimuli in the major modalities. On the motor side, there is extensive evidence for trainable voluntary control over single motor units and more recently, for voluntary control of single cortical neurons (Cerf et al. 2010). The massive anatomy and physiology of cortex can presumably support this kind of parallel-interactive bandwidth. Whether structures like the claustrum have that kind of bandwidth is doubtful.

We do not know the full set of signaling mechanisms in the brain, and any current model must be considered provisional. Neural computations can be remarkably flexible, and are, to some degree, independent of specific cells and populations. John et al. (2001) has argued that active neuronal populations must have dynamic turnover to perform any single brain function, like active muscle cells. Edelman and Tononi (2000) and others have made the same point with the concept of a dynamic core. GW capacity as defined here is not dependent upon the mere existence of anatomical hubs, which are extremely common. Rather, it depends upon a dynamical capacity, which operates flexibly over the CT anatomy, a "functional hub," so that activated arrays make up coherent "coalitions."

The global neuronal workspace has been used to model a number of experimental phenomena. In a recent model, sensory stimuli mobilize excitatory neurons with long-range cortico-cortical axons, leading to the genesis of a global activity pattern among workspace neurons. This class of models is empirically linked to phenomena like visual backward masking and in attentional blindness (Dehaene and Changeux 2005).

Franklin et al. (2012) have combined several types of computational methods using a quasi-neuronal activation-passing design. High-level conceptual models such as LIDA (Snaider, McCall, and Franklin 2011) can provide insights into the processes implemented by the neural mechanisms underlying consciousness, without necessarily specifying the mechanisms themselves. Although it is difficult to derive experimentally testable predictions from large-scale architectures, this hybrid architecture approach is broadly consistent with the major empirical features discussed in this article. It predicts, for example, that consciousness may play a central role in the classic notion of cognitive working memory, selective attention, learning, and retrieval.

7 Global Chatting, Chanting, and Cheering

Spontaneous conscious mentation occurs throughout the waking state, reflecting repetitive themes described as "current concerns." Conscious mentation is also reported when subjects are awoken from Rapid Eye Movement (REM) dreams and even from slow-wave sleep. The last may reflect waking-like moments during the peaks of the delta wave (Valderrama et al. 2012).

Global brain states can be compared to a football crowd with three states: "chatting," "chanting," and "cheering." Chatting describes the CT activity of waking and REM dreams. It involves point-to-point conversations among spatial arrays in the CT system, which can have very high S/N ratios, though they appear to be random when many of them take place at the same time. Like a football stadium with thousands of coordinated local conversations that are not coordinated globally, the average global activity is a low-level crowd roar, seemingly random, which appears to be fast and low in amplitude.

Nevertheless, as we will see, direct cortical recordings show phase-coupled chatting in the CT core appears to underlie specific cognitive tasks. Thus, chatting activity gives the misleading appearance of randomness *en masse,* but it is in fact highly organized in a task-driven fashion. Because sports arenas show the same properties, the arena metaphor provides us with a useful reminder.

Chanting shows coordinated start-stop crowd activity, about once a second over a prolonged period of time, like the "buzz-pause" rhythm of billions of neurons in the CT core, which results in global delta waves. Chanting sounds like chatting at the peak of the delta wave, followed by simultaneous pausing, which interrupts all conversations at the same time (Massimini et al. 2005).

Finally, a stadium crowd may cheer when a team scores a goal or makes an error. This corresponds to an "event-related" peak of activity. In the brain, the event-related potential (ERP) occurs when a significant or intense stimulus is processed, causing a stereotypical wave pattern to sweep through the brain.

8 Feature and Frame Binding

In GWT frames (previously called contexts) can be thought of as groups of specialists dedicated to processing input in particular ways. As we have seen, there are frames for perception and imagery (where they help shape qualitative experiences), as well as in conceptual thought, goal directed activities and the like (where they serve to access conscious experiences). One of the primary functions of consciousness is to evoke contexts that shape experiences. Some challenges to a dominant frame are more noticeable than others. Consider the following from Eriksen and Mattson (1981):

1 How many animals of each kind did Moses bring on the Ark?
2 In the Biblical story, what was Joshua swallowed by?
3 What is the nationality of Thomas Edison, inventor of the telephone?

While some subjects noticed errors with one or all of these statements, most did not. When asked directly, subjects showed they knew the answers, but it took more severe violations (e.g "How many animals of each kind did Nixon bring on the Ark?") for the majority to see any issues.

Visual features are stimulus properties that we can point to and name, like "red," "bounded," "coffee cup," "shiny," etc. Feature binding is a well-established property of sensory perception. There is much less discussion about what we will call "frame-binding," which is equally necessary, where "frames" are defined as visual arrays that do not give rise to conscious experiences, but which are needed to specify spatial knowledge within which visual objects and events become conscious. Powerful illusions like the Necker Cube, the Ames trapezoidal room, the railroad (Ponzo) illusion are shaped by unconscious Euclidean assumptions about the layout of rooms, boxes, houses, and roads.

The best-known brain examples are the egocentric (coordinate system is centered around the navigator) and allocentric (oriented on something other than the navigator) visuotopic arrays of the parietal cortex. When damaged on the right side, these unconscious visuotopic fields cause the left half of objects and scenes to disappear, a condition called hemi-neglect. Goodale and Milner have shown that even normal visuomotor guidance in near-body space may be unconscious. In vision the dorsal "framing" stream and "feature-based" ventral stream may combine in the medial temporal cortex (MTL) (Shimamura 2010). Baars (1988) reviewed extensive evidence showing that unconscious framing is needed for normal perception, language comprehension and action planning. In sum, normal conscious experiences need both traditional feature binding and frame-binding (Shanahan and Baars 2005).

9 Perceptual Experiences vs. Feelings of Knowing (FOKs)

This Dynamic GW theory figure shows an occipital broadcast (which must mobilize parietal egocentric and allocentric maps as well) evoking spatiotopic activity in the prefrontal cortex, which is known to initiate prefrontal activation across multiple tasks demanding mental effort (Duncan and Owen 2000), and suggests that sensory conscious experiences are bound and broadcast from the classical sensory regions in the posterior cortex, while voluntary effort, reportable intentions, feelings of effort, and the like, have a prefrontal origin, consistent with brain imaging findings.

These findings suggest a hypothesis about sensory consciousness compared to "fringe" FOK, feelings of effort, and reportable voluntary decisions. These reportable but "vague" events have been discussed since William James (1890) who gave them equal importance to perceptual consciousness. Functional magnetic resonance imaging (fMRI) studies show that they predominantly involve prefrontal regions, even across tasks that seem very different.

Because of the small-world connectivity of white matter tracts, different integration and distribution hubs may generate different global wave fronts. The sight of a coffee cup may involve an infero-temporal hub signaling to other regions, while the perception of music may emerge from Heschel's gyrus and related regions. Reportable experiences of cognitive effort might spread outward from a combined dorsolateral prefrontal cortex (dlPFC)/anterior cingulate cortex (ACC) hub.

10 Conscious Events Evoke Widespread Adaptation or Updating

What is the use of binding and broadcasting in the CT system? One function is to update numerous brain systems to keep up with the fleeting present. GW theory suggested that consciousness is required for non-trivial learning (i.e., learning that involves novelty or significance) (Baars 1988). While there are constant efforts to demonstrate robust unconscious learning, after six decades of subliminal vision research there is still little convincing evidence. Subliminal perception may work with known chunks, like facial expressions, but while single-word subliminal priming appears to work, Baars (1988) questioned whether novel two-word primes would work subliminally. The subliminal word pair "big house" might prime the word "tall," while "big baby" might not, because it takes conscious thought to imagine a baby big enough to be called tall. In general, the more novelty is presented, the more conscious exposure is required.

It follows that the Dynamic GW theory should predict widespread adaptive changes after conscious exposure to an event. That is indeed the consensus for hippocampal-neocortical memory coding (Nadel et al. 2012). However, the hippocampal complex is not currently believed to enable conscious experiences. Nevertheless, episodic memory is by definition "memory for conscious events." Conscious events trigger wide adaptation throughout the CT system, and in subcortical regions that are influenced by the CT system. Episodic, semantic, and skill (procedural) processing all follow the same curve of high-metabolic processing to novel, conscious learning followed by a drastic drop in conscious access and metabolic BOLD (blood-oxygen-level dependent) activity after learning.

11 Voluntary Reports of Conscious Events

Conscious contents are routinely assessed by voluntary report, as we know from 200 years of scientific psychophysics. Yet the reason for that fact is far from obvious. Any theory of consciousness must ultimately explain the basic fact that we can voluntarily report an endless range of conscious contents, using an endless range of voluntary actions. Voluntary control is one kind of consciously mediated process. As we learn to ride a bicycle for the first time, each movement seems to come to consciousness. After learning, conscious access drops even as BOLD activity in the CT core declines. We postulate that conscious involvement is necessary for non-trivial acquisition of knowledge and skills, and that the period of conscious access enables permanent memory traces to be established.

While "verbal report" is the traditional phrase, reports do not have to be verbal – any voluntary response will work. Broca's aphasics who cannot speak can point to objects instead. Locked-in (paralyzed) patients, who seem to be comatose, can learn to communicate by voluntary eye movements. Thus "verbal report" should be called "accurate, voluntary report," using any controllable response. Voluntary actions can point to objects and events. A "match to sample" task is commonly used to indicate the similarity of two conscious events, and to specify just noticeable differences. Pointing occurs naturally when mammals orient to a novel or significant stimulus. Children develop pointing abilities using "shared attention" in early childhood.

For simplicity's sake let's assume conscious contents emerge in posterior cortex and voluntary actions emerge in frontal and parietal cortex. We can ask the question in The Dynamic GW theory terms: how is a posterior "binding and broadcasting" event transformed into a frontally controlled action? These facts raise the question of how accurate signal transmission occurs between sensory arrays and frontal executive control. In the case of pointing to a single star on a dark night, the physical minimum of light quanta in the retina can be amplified and transmitted to prefrontal cortex, which can control the movement of a single finger to point to the star.

Even more remarkably, single neurons in the temporal cortex have been shown to be fired at will in surgical patients using intracranial electrodes, providing only that conscious sensory feedback is given during training (Cerf et al. 2010). Thus, the physical minimum to the eye can accurately translate into "any" voluntarily controlled single cell, used as a sensory pointer. Given a million foveal cells for input, and perhaps billions of cortical cells for output, "any-to-any" mapping in the brain can involve remarkably large numbers. With accurate psychophysical performance in both tasks, the signal-to-noise ratio from receptor to effector cell can approach the physical limit. This precision needs explanation in terms of conscious input and voluntary control.

This also suggests an explanation for the standard index of voluntary report. When we report a star on a dark night, posterior broadcasting may lead to frontal binding and ultimately a frontal broadcast. Frontoparietal regents are driven by posterior sensory projections when they become conscious. Because of the striking similarities of spatiotopic coding in frontal and posterior cortices, we can imagine that sensory consciousness can also trigger a new binding, and broadcast an event in the frontal cortex. Voluntary action is therefore an extension of GW dynamics.

Conscious contents enable access to cognitive functions, including sense modalities, working memory, long-term memories, executive decisions and action control. Executive regions of the frontoparietal cortex gain control over distributed unconscious functions. Animals live in a world of unknowns, surrounded by dangers and opportunities that may be fleeting, hidden, camouflaged, surprising, deceptive, and ambiguous. Conscious brains may have evolved to cope with such unknowns (Baars 1988, 2002). Newell and colleagues built the first GW architecture to perform acoustical word recognition, at a time when that task was largely underdetermined (Newell 1990). Their solution was to build a computational architecture, a blackboard model, which would allow many incomplete sources to compete and cooperate to resolve some focal ambiguity. The result was remarkably successful for its time in recognizing nearly 1,000 ordinary words spoken in normal acoustical spaces, complete with hard echoing surfaces, mumbling speakers, and soft, absorbent surfaces, background noises, and the like. Speech recognition is now handled with improved formant tracking, but even today, if semantic unknowns arise in a spoken word stream, a GW architecture may be useful to find the answer. We have no semantic algorithms that interpret word ambiguities across many domains, the way humans routinely do. Baars and Franklin (2003) used GW theory to propose that consciousness enables access between otherwise separate knowledge sources.

GW architectures can also "call" fixed automatisms. For example, in speech recognition word ambiguity may be resolved by a known syntactic rule. A global broadcast of the ambiguous word may recruit routines whose relevance cannot be known ahead of time. We have referred to this as contextualization or frame binding (Baars 1988; Shanahan and Baars 2005). The "frame problem" is a recognized challenge in artificial intelligence and robotics, but it applies equally to living brains. Briefly stated, it is an effort to explain how a "cognitive creature with many beliefs about the world" can regularly update them while remaining "roughly faithful to the world" (Dennett 1978). In GWT this conundrum is solved through the invocation of unconscious context-sensitive and context-shaping processors.

12 Concluding Remarks

The main use of a GW system is to solve problems which any single "expert" knowledge source cannot solve by itself – problems whose solutions are underdetermined. Human beings encounter such problems in any domain that is novel, degraded, or ambiguous. This is obvious for novelty: if we are just learning to ride a bicycle, or to understand a new language, we have inadequate information by definition. Further, if the information we normally use to solve a known problem becomes degraded, determinate solutions again become indeterminate.

What may not be so obvious is that there are problems that are inherently ambiguous, in which all the local pieces of information can be interpreted in more than one way, so that we need to unify different interpretations to arrive at a single, coherent understanding of the information. But there are numerous biological examples of densely vegetated fields and forests that harbor so many hiding places for animals and birds that there is in principle no way to make the visual scene predictable. Many wet jungle regions also have very loud ambient sounds produced by insects, frogs and birds, so that the noise level exceeds the signal emanating from any single individual animal. This situation also applies to the famous human cocktail party effect, where we can understand conversations despite a negative signal-to-noise ratio. Clearly biological sensory systems can thrive in such noisy environments, perhaps using top-down predictions and multimodal signal correlations. Standard sensory studies in humans and animals have generally neglected this ecologically realistic scenario.

Conscious learning is often involved in decomposing such complex signal environments, as in the case of human music conductors, for example, who can rapidly pinpoint wrong notes. In these cases, top-down learning of musical patterns and entire large-ensemble scores is involved, but talented experts spend a lot of conscious time on the learning process, and their spectacular performances do not contradict our observations about the many functions of conscious thought.

References

Baars, B. J. (1988) *A Cognitive Theory of Consciousness*, New York: Cambridge University Press.

Baars, B. J. (2002) "The conscious access hypothesis: origins and recent evidence," *Trends in Cognitive Science* 6: 47–52.

Baars, B. J., and Franklin, S. (2003) "How conscious experience and working memory interact," *Trends in Cognitive Science* 7: 166–172.

Bizley, J. K., Shinn-Cunningham, B. G., and Lee, A. K. (2012) "Nothing is irrelevant in a noisy world: sensory illusions reveal obligatory within and across modality integration," *Journal of Neuroscience* 32: 13402–13410.

Cerf, M., Thiruvengadam, N., Mormann, F., Kraskov, A., Quiroga, R. Q., Koch, C., Fried, I. (2010) "On-line, voluntary control of human temporal lobe neurons," *Nature* 467: 1104–1108.

Crick, F., and Koch, C. (1990) "Towards a neurobiological theory of consciousness," *Seminars in Neuroscience* 2: 263–275.

Crick, F., and Koch, C. (2005) "What is the function of the claustrum?" *Philosophical Transactions of the Royal Society B*. 360: 1271–1279.

Dehaene, S., and Changeux, J. P. (2005) "Ongoing spontaneous activity controls access to consciousness: a neuronal model for inattentional blindness," *PLoS Biology* 3: 0920–0923.

Dehaene, S., and Naccache, L. (2001) "Towards a cognitive permanence of consciousness: basic evidence and a workspace framework," *Cognition*. 79: 1–37.

Dennett, D. (1978) *Brainstorms*, Cambridge, MA: MIT Press.

Doesburg, S. M., Green J. J., McDonald J. J., and Ward L. M. (2009) "Rhythms of consciousness: binocular rivalry reveals large-scale oscillatory network dynamics mediating visual perception," *PLoS ONE* 4: 1–14.

Duncan, J., and Owen, A. M. (2000) "Common regions of the human frontal lobe recruited by diverse cognitive demands," *Trends in Neuroscience* 23: 475–483.

Edelman, G. M. (1989) *The Remembered Present: A Biological Theory of Consciousness*, New York: Basic Books Inc.

Edelman, G. M., Gally, J. A., and Baars, B. J. (2011) "Biology of consciousness," *Frontiers in Psychology* 2: 4.

Edelman, G. M., and Tononi, G. (2000) *A Universe of Consciousness: How Matter Becomes Imagination*, New York: Basic Books Inc.

Erickson, T. D., and Mattson, M. E. (1981) "From words to meaning: A semantic illusion," *Journal of Verbal Learning and Verbal Behavior* 20: 540–551.

Franklin, S., Strain, S., Snaider, J., McCall, R., and Faghihi, U. (2012) "Global workspace theory, its LIDA model and the underlying neuroscience," *Biologically Inspired Cognitive Architectures* 1: 32–43.

Freeman, W. J., Burke, B. C., and Holmes, M. D. (2003) "A periodic phase resetting in scalp EEG of beta-gamma oscillations by state transitions at alpha-theta rates," *Human Brain Mapping* 19: 248–272.

Goodale, M. A., and Milner, A. D. (1992) "Separate visual pathways for perception and action," *Trends in Neuroscience* 15: 20–25.

Itti, L., and Koch, C. (2001) "Computational modelling of visual attention," *Nature Reviews Neuroscience* 2: 194–203.

Izhikevich, E. M., and Edelman, G. M. (2008) "Large-scale model of mammalian thalamocortical systems," *Proceedings of the United States National Academy of Science* 105: 3593–3598.

James, W. (1890) *The Principles of Psychology*, New York: Holt.

John, E. R., Prichep, L. S., Kox, W., Valdés-Sosa, P., Bosch-Bayard, J., Aubert, E., Tom., M. di Michele, F., and Gugino, L.D. (2001) "Invariant reversible QEEG effects of anesthetics," *Conscious Cognition* 10: 165–183.

Jones, J., and Adamatzky, A. (2014) "Computation of the travelling salesman problem by a shrinking blob," *Natural Computing* 13: 1–16.

Llinas, R. R., and Pare, D. (1991) "Of dreaming and wakefulness," *Neuroscience* 44: 521–535.

Lou, H. C., Luber, B., Stanford, A., and Lisanby, S. H. (2010) "Self-specific processing in the default network: a single-pulse TMS study," *Experimental Brain Research* 207: 27–38.

Mangan, B. (1993) "Taking phenomenology seriously: the fringe and its implications for cognitive research," *Consciousness and Cognition* 2: 89–108.

Massimini, M., Ferrarelli, F., Huber, R., Esser, S. K., Singh, H., and Tononi, G. (2005) "Breakdown of cortical effective connectivity during sleep," *Science* 309: 2228–2232.

Moscovitch, M., Rosenbaum, R. S., Gilboa, A., Addis, D. R., Westmacott, R., Grady, C., McAndrews, M.P., Levine, B., Black, S., Winocur, G., and Nadel, L. (2005) "Functional neuroanatomy of remote episodic, semantic and spatial memory: a unified account based on multiple trace theory," *Journal of Anatomy* 207: 35–66.

Nadel, L., Samsonovich, A., Ryan, L., and Moscovitch, M. (2000) "Multiple trace theory of human mem ory: computational, neuroimaging, and neuropsychological results," *Hippocampus* 10: 352–368.

Newell, A. (1990) *Unified Theories of Cognition*, Cambridge, MA: Harvard University Press.

Penfield, W. (1975) *The Mystery of the Mind: A Critical Study of Consciousness and the Human Brain*, Princeton: Princeton University Press.

Ryan, L., Lin, C. Y., Ketcham, K., and Nadel, L. (2010) "The role of medial temporal lobe in retrieving spatial and nonspatial relations from episodic and semantic memory," *Hippocampus* 20: 11–18.

Ryan, L., Nadel, L., Keil, K., Putnam, K., Schnyer, D., Trouard, T., and Moscovitch, M. (2001) Hippocampal complex and retrieval of recent and very remote autobiographical memories: evidence from functional magnetic resonance imaging in neurologically intact people. *Hippocampus* 11: 707–714.

Shanahan, M., and Baars, B. (2005) "Applying global workspace theory to the frame problem," *Cognition* 98: 157–176.

Shimamura, A. P. (2010) "Hierarchical relational binding in the medial temporal lobe: the strong get stronger," *Hippocampus* 20: 1206–1216.

Snaider, J., McCall, R., and Franklin, S. (2011) "The LIDA Framework as a General Tool for AGI," *Fourth Conference on Artificial General Intelligence.* Mountain View, CA: Springer Lecture Notes in Artificial Intelligence.

Valderrama, M., Crépon, B., Botella-Soler, V., Martinerie, J., Hasboun, D., Alvarado-Rojas, C., Baulac, M., Adam, C., Navarro, V., and Le Van Quyen, M. (2012). "Human gamma oscillations during slow wave sleep," *PLoS ONE* 7: 1–14.

Wang, X. J. (2001) "Synaptic reverberation underlying mnemonic persistent activity," *Trends in Neuroscience* 24: 455–463.

Zeki, S. (2001) "Localization and globalization in conscious vision," *Annual Review of Neuroscience* 24: 57–86.

Related Topics

Materialism
Representational Theories of Consciousness
The Multiple Drafts Model
The Information Integration Theory
The Intermediate Level Theory of Consciousness
The Attention Schema Theory of Consciousness
The Neural Correlates of Consciousness
The Biological Evolution of Consciousness

10
INTEGRATED INFORMATION THEORY

Francis Fallon

Integrated Information Theory (IIT) combines Cartesian commitments with claims about engineering that it interprets, in part by citing corroborative neuroscientific evidence, as identifying the nature of consciousness. This borrows from recognizable traditions in the field of consciousness studies, but the structure of the argument is novel.

IIT takes certain features of consciousness to be unavoidably true. Rather than beginning with the neural correlates of consciousness (NCC) and attempting to explain what about these sustains consciousness, IIT begins with its characterization of experience itself, determines the physical properties necessary for realizing these characteristics, and only then puts forward a theoretical explanation of consciousness, as identical to a special case of information instantiated by those physical properties. "The theory provides a principled account of both the quantity and quality of an individual experience… and a calculus to evaluate whether a physical system is conscious" (Tononi and Koch 2015).

1 The Central Claims[1]

IIT takes Descartes very seriously. Descartes located the bedrock of epistemology in the knowledge of our own existence given to us by our thought. "I think, therefore I am" reflects an unavoidable certainty: one cannot deny one's own existence as a thinker (even if one's particular thoughts are in error). For IIT, the relevance of this insight lies in its application to consciousness. Whatever else one might claim about consciousness, one cannot deny its existence.

IIT takes consciousness as primary. What does consciousness refer to here? Before speculating on the origins or the necessary and sufficient conditions for consciousness, IIT gives a characterization of what consciousness means. The theory advances five axioms intended to capture just this. Each axiom articulates a dimension of experience that IIT regards as self-evident. They are as follows:

First, following from the fundamental Cartesian insight, is the axiom of *existence*. Consciousness is real and undeniable; moreover, a subject's consciousness has this reality intrinsically; i.e. it exists from its own perspective.

Second, consciousness has *composition*. In other words, each experience has structure. Color and shape, for example, structure visual experience. Such structure allows for various distinctions.

Third, the axiom of *information*: the way an experience is distinguishes it from other possible experiences. An experience specifies; it is specific to certain things, distinct from others.

Fourth, consciousness has the characteristic of *integration*. The elements of an experience are interdependent. For example, the particular colors and shapes that structure a visual conscious state are experienced together. As we read these words, we experience the font-shape and letter-color inseparably. We do not have isolated experiences of each and then add them together. This integration means that consciousness is irreducible to separate elements. Consciousness is unified.

Fifth, consciousness has the property of *exclusion*. Every experience has borders. Precisely because consciousness specifies certain things, it excludes others. Consciousness also flows at a particular speed.

In isolation, these axioms may seem trivial or overlapping. IIT labels them axioms precisely because it takes them to be obviously true. IIT does not present them in isolation. Rather, they motivate postulates.[2] Each axiom leads to a corresponding postulate identifying a physical property. Any conscious system must possess these properties. The postulates include:

First, the existence of consciousness implies a system of mechanisms with a particular cause-effect power. IIT regards existence as inextricable from causality: for something to exist, it must (be able to) make a difference to other things, and vice versa. (What would it even mean for a thing to exist in the absence of any causal power whatsoever?) Because consciousness exists from its own perspective, the implied system of mechanisms must do more than simply have causal power; it must have *cause-effect power upon itself*.

Second, the compositional nature of consciousness implies that its system's mechanistic elements must have the capacity to *combine*, and that those combinations have cause-effect power.

Third, because consciousness is informative, it must *specify*, i.e. distinguish one experience from others. IIT calls the cause-effect powers of any given mechanism within a system, its cause-effect repertoire. The cause-effect repertoires of all the system's mechanistic elements taken together, it calls its cause-effect structure. This structure, at any given point, is in a particular state. In complex structures, the number of possible states is very high. For a structure to instantiate a particular state is for it to specify that state. The specified state is the particular way that the system is making a difference to itself.

Fourth, consciousness's integration into a unified whole implies that the system must be *irreducible*. In other words, its parts must be interdependent. This in turn implies that every mechanistic element must have the capacity to act as a cause for the rest of the system and to be affected by the rest of the system. If a system can be divided into two parts without affecting its cause-effect structure, it fails to satisfy the requirement of this postulate.

Fifth, the exclusivity of the borders of consciousness implies that the state of a conscious system must be *definite*. In physical terms, the various simultaneous subgroupings of mechanisms in a system have varying cause-effect structures. Of these, only one will have a maximally irreducible cause-effect structure (called the *maximally irreducible conceptual structure*, or MICS). Others will have smaller cause-effect structures, at least when reduced to non-redundant elements. Precisely this – the MICS – is the conscious state.

IIT accepts the Cartesian conviction that consciousness has immediate, self-evident properties, and outlines the implications of these phenomenological axioms for conscious physical systems. This characterization does not exhaustively describe the theoretical ambition of IIT. The ontological postulates concerning physical systems do not merely articulate necessities (or even sufficiencies) for realizing consciousness; the claim is much stronger than this. IIT *identifies* consciousness with a system's having the physical features that the postulates describe. Each conscious state is a MICS, which just is and can only be a system of irreducibly interdependent physical parts whose causal interaction constitutes the integration of information.

An example may help to clarify the nature of IIT's explanation of consciousness. Our experience of a cue ball integrates its white color and spherical shape, such that these elements are inseparably fused. The fusion of these elements constitutes the structure of the experience: the experience is composed of them. The nature of the experience informs (about whiteness and spherical shape) in a way that distinguishes it from other possible experiences (such as of a blue cube of chalk). This is just a description of the phenomenology of a simple experience (perhaps necessarily awkward, because it articulates the self-evident). Our brain generates the experience through neurons physically communicating with one another, in systems linked by cause-effect power. IIT interprets this physical communication as the integration of information, according to the various constraints laid out in the postulates. The neurobiology and phenomenology converge.

Indeed, according to IIT, the physical state of any conscious system *must* converge with phenomenology; otherwise the kind of information generated could not realize the axiomatic properties of consciousness. We can understand this by contrasting two kinds of information. First, Shannon information: When a digital camera takes a picture of a cue ball, the photodiodes operate in causal isolation from one another. This process does generate information; specifically, it generates observer-relative information. That is, the camera generates the information of an image of a cue ball for anyone looking at that photograph. The information that is the image of the cue ball is therefore relative to the observer; such information is called Shannon information. Because the elements of the system are causally isolated, the system does not make a difference to itself. Accordingly, although the camera gives information to an observer, it does not generate that information for itself. By contrast, consider what IIT refers to as *intrinsic* information: unlike the digital camera's photodiodes, the brain's neurons do communicate with one another through physical cause and effect; the brain does not simply generate observer-relative information, it integrates *intrinsic* information. This information from its own perspective *just is* the conscious state of the brain. The physical nature of the digital camera does not conform to IIT's postulates and therefore does not have consciousness; the physical nature of the brain, at least in certain states, does conform to IIT's postulates, and therefore does have consciousness.

To identify consciousness with such physical integration of information constitutes a bold and novel ontological claim. Again, the physical postulates do not describe one way, or even the best way, to realize the phenomenology of consciousness; the phenomenology of consciousness is one and the same as a system having the properties described by the postulates. It is even too weak to say that such systems "give rise to" or "generate" consciousness. Consciousness is fundamental to these systems in the same way as mass or charge is basic to certain particles.

IIT's conception of consciousness as mechanisms systematically integrating information through cause and effect lends itself to quantification. The more complex the MICS, the higher the level of consciousness: the corresponding metric is *phi*. IIT points to certain cases as illustrating this relation, thereby providing corroborative evidence of its central claims. For example, deep sleep states are less experientially rich than waking ones. IIT predicts, therefore, that such sleep states will have lower *phi* values than waking states. For this to be true, analysis of the brain during these contrasting states would have to show a disparity in the systematic complexity of non-redundant mechanisms. In IIT, this disparity of MICS complexity directly implies a disparity in the amount of conscious integrated information (because the MICS is identical to the conscious state). The neuroscientific findings bear out this prediction.

IIT cites similar evidence from the study of patients with brain damage. For example, we already know that among vegetative patients, there are some whose brain scans indicate that they can hear and process language: when researchers prompt such patients to think about playing tennis, e.g., the appropriate areas of the brain become activated. Other vegetative patients do not respond this way. Naturally, this suggests that the former have a richer degree of consciousness than the latter. When analyzed according to IIT's theory, the former have a higher *phi* metric than the latter; once again, IIT has made a prediction that receives empirical confirmation. IIT also claims that findings in the analysis of patients under anaesthesia corroborate its claims.

In all these cases, one of two things happens. First, as consciousness fades, cortical activity may become less global. This reversion to local cortical activity constitutes a loss of integration: the system no longer is communicating across itself in as complex a way as it had. Second, as consciousness fades, cortical activity may remain global, but become stereotypical, consisting in numerous redundant cause-effect mechanisms, such that the informational achievement of the system is reduced: a loss of information. As information either becomes less integrated or becomes reduced, consciousness fades, which IIT takes as empirical support of its theory of consciousness as integrated information.

2 Quantifying Consciousness: Measuring *Phi*

IIT strives, among other things, not just to claim the existence of a scale of complexity of consciousness, but to provide a theoretical approach to the precise quantification of the richness of experience for any conscious system. This requires calculating the maximal amount of integrated information in a system: the system's *phi* value can be expressed numerically (at least in principle). It is important to note that not every system with phi has consciousness. A sub- or super-system of an MICS may have phi, but will not have consciousness. A closer look at the digital photography example affords particularly apt illustrations of some of the basic principles involved in quantifying consciousness.

First, a photodiode exemplifies integrated information in the simplest way possible. A photodiode is a system of two elements, which together render it sensitive to two states only: light and dark. After initial input from the environment, the elements communicate input physically with one another, determining the output. So, the photodiode is a two-element system that integrates information. A photodiode not subsumed in another system of greater phi value is the simplest possible example of consciousness.

This consciousness, of course, is virtually negligible. The photodiode's experience of light and dark is not rich in the way that ours is. The level of information of a state depends upon its specifying that state as distinct from others. The repertoire of the photodiodes allows only for the most limited differentiation ('this' vs. 'that'), whereas the repertoire of a complex system such as the brain allows for an enormous amount of differentiation. Even our most basic experience of darkness distinguishes it not only from light, but from shapes, colors, etc.

Second, as noted in Section 1, a digital camera's photodiodes' causal arrangement neatly exemplifies the distinction between integrated and non-integrated information. Putting to one side that each individual photodiode integrates information (as simply as possible), those photodiodes do not take input or give output *to one another*, so the information does not get integrated across the system. For this reason, the camera's image is informative to us, but not to itself. So, each isolated photodiode has integrated information in the most basic way, and would therefore have the lowest possible positive value of phi. The camera's photodiodes taken as a system do not integrate information and have a phi value of zero.

In order to measure the level of consciousness of a system, IIT must describe the amount of its integrated information. This is done by partitioning the system in various ways.[3] If the digital camera's photodiodes are partitioned (say, by dividing the abstract model of its elements in half) no integrated information is lost, because all the photodiodes are in isolation from each other, and so the division does not break any connections. If no logically possible partition of the system results in a loss of connection, the conclusion is that the system does not make a difference to itself. So, in this case, the system has no phi.

Systems with phi will have connections that will be lost by some partitions and not by others. Some partitions will sever from the system elements that are comparatively low in original degree of connectivity to the system; in other words, elements whose (de)activation has few causal consequences upon the (de)activation of other elements. A system where all or most elements have this property will have low phi.

The lack of strong connectivity may be the result of relative isolation, or locality (an element not linking to many other elements, directly or indirectly) or from stereotypicality (where the element's causal connections overlap in a largely redundant way with the causal connection of other elements). A system whose elements are connected more globally and non-redundantly will have higher phi. These descriptions apply, for example, to the cortical activity of sleep and wake states, respectively (see Section 1 above).

A partition that not only separates all elements that do not make a difference to the rest of the system (for reasons of either isolation or redundancy) from those that do make a difference, but also separates those elements whose lower causal connectivity decreases the overall level of integration of the system from those that do not, will thereby have picked out the MICS, which according to IIT is conscious. The degree of that consciousness, its phi, depends upon its elements' level of causal connectivity. This is determined by how much information integration would be lost by the least costly further partition, or, in other words, how much the cause-effect structure of the system would be reduced by eliminating the least causally effective element within the MICS.

3 What IIT's Central Claims Imply

No controversy attaches to the observation that humans experience varying degrees of consciousness. As noted, consciousness decreases during sleep, for example. IIT implies that brain activity during this time will generate either less information or less integrated information, and interprets experimental results concerning cortical activity as bearing this out. By contrast, the cerebellum, which has many neurons, but neurons that are not complexly interconnected and so do not belong to the MICS, does not generate consciousness.

More controversial is the issue of non-human consciousness. IIT counts among its merits that the principles it uses to characterize human consciousness can apply to non-human cases. On IIT, consciousness happens when a system makes a difference to itself at a physical level: elements causally connected to one another in a re-entrant architecture integrate information, and the subset of these with maximal causal power is conscious. The human brain offers an excellent example of re-entrant architecture integrating information, capable of sustaining highly complex MICSs, but nothing in IIT limits the attribution of consciousness to human brains only.

Mammalian brains share similarities in neural and synaptic structure: the human case is not obviously exceptional. Other, non-mammalian species demonstrate behavior associated in humans with consciousness. These considerations suggest that humans are not the only species capable of consciousness. IIT makes a point of remaining open to the possibility that many other species may possess at least some degree of consciousness. At the same time, further study of non-human neuroanatomy is required to determine whether and how this in facts holds true. As mentioned above, even the human cerebellum does not have the correct architecture to generate consciousness, and it is possible that other species have neural organizations that facilitate complex behavior without generating high phi. The IIT research program offers a way to establish whether these other systems are more like the cerebellum or the cerebral cortex in humans. Of course, consciousness levels will not correspond completely to species alone. Within conscious species, there will be a range of phi levels, and even within a conscious phenotype, consciousness will not remain constant from infancy to death, wakefulness to sleep, and so forth.

IIT claims that its principles are consistent with the existence of cases of dual consciousness within split-brain patients. In such instances, on IIT, two local maxima of integrated information exist separately from one another, generating separate consciousness. IIT does not hold that a system need have only one local maximum, although this may be true of normal brains; in split-brain patients, the re-entrant architecture has been severed so as to create two. IIT also takes its identification of MICSs (through quantification of phi) as a potential tool for assessing other actual or possible cases of multiple consciousness within one brain.

Such claims also allow IIT to rule out instances of aggregate consciousness. The exclusion principle forbids double-counting of consciousness. A system will have various subsystems with phi value, but only the local maxima of phi within the system can be conscious. A normal waking human brain has only one conscious MICS, and even a split-brain patient's conscious systems do not overlap but rather are separate. One's conscious experience is precisely what it is and nothing else. All this implies that, for example, the USA has no superordinate consciousness in addition to the consciousness of its individuals. The local maxima of integrated information reside within the skulls of those individuals; the phi value of the connections among them is much lower.

Although IIT allows for a potentially very wide range of degrees of consciousness and conscious entities, this has its limits. Some versions of panpsychism attribute mental properties to even the most basic elements of the structure of the world, but the simplest conscious entity admitted on IIT to be conscious would have to be a system of at least two elements that have cause-effect power over one another. Otherwise no integrated information exists. Objects such as rocks and grains of sand have no phi (whether in isolation or heaped into an aggregate), and therefore no consciousness.

IIT's criteria for consciousness are consistent with the existence of artificial consciousness. The photodiode, because it integrates information, has a phi value; if not subsumed into a system of higher phi, this will count as local maximum: the simplest possible MICS or conscious system. Many or most instances of phi and consciousness may be the result of evolution in nature, independent of human technology, but this is a contingent fact.

IIT's basic arguments imply, and the IIT literature often explicitly claims, certain important constraints upon artificial conscious systems. Often technological systems involve feed-forward architecture that lowers or possibly eliminates phi, but if the system is physically re-entrant and satisfies the other criteria laid out by IIT, it may be conscious. In fact, according to IIT, we may build artificial systems with a greater degree of consciousness than humans.

At the level of hardware, computation may process information with either feed-forward or re-entrant architecture. In feed-forward systems, information gets processed in only one direction, taking input and giving output. In re-entrant systems, which consist of feedback loops, signals are not confined to movement in one direction only; output may operate as input also.

IIT interprets the integration axiom (the fourth axiom, which says that each experience's phenomenological elements are interdependent) as entailing the fourth postulate, which claims that each mechanism of a conscious system must have the potential to relate causally to the other mechanisms of that system. By definition, in a feed-forward system, mechanisms cannot act as causes upon those parts of the system from which they take input. A purely feed-forward system would have no phi, because although it would process information, it would not integrate that information at the physical level. One implication for artificial consciousness is immediately clear: Feed-forward architectures will not be conscious. Even a feed-forward system that perfectly replicated the behavior of a conscious system would only *simulate* consciousness. Artificial systems will need to have re-entrant structure to generate consciousness.

Furthermore, re-entrant systems may still generate very low levels of phi. Conventional CPUs have transistors that only communicate with several others. By contrast, each neuron of the conscious network of the brain connects with thousands of others, a far more complex re-entrant structure, making a difference to itself at the physical level in such a way as to generate much higher phi value. For this reason, brains are capable of realizing much richer consciousness than conventional computers. The field of artificial consciousness, therefore, would do well to emulate the neural connectivity of the brain.

Still another constraint applies, this one associated with the exclusion (fifth) postulate. A system may have numerous phi-generating subsystems, but according to IIT, only the network of elements with the greatest cause-effect power to integrate information (the maximally irreducible conceptual structure, or MICS) is conscious. Re-entrant systems may have local maxima of phi, and therefore small pockets of consciousness. Those attempting to engineer high degrees of artificial consciousness need to focus their design on creating a large MICS, not simply small, non-overlapping MICSs. If IIT is correct in placing such constraints upon artificial consciousness, deep convolutional networks such as GoogLeNet and advanced projects like Blue Brain may be unable to realize (high levels of) consciousness.

4 Selected Objections

Space prohibits even a cursory description of alternative interpretations of consciousness, as the variety of chapters in this volume alone evidences. Even an exhaustive account of the various objections that have been levelled explicitly at IIT is not possible (nor necessarily desirable) here. What follows will be partial in this sense and in the sense that it reflects the author's opinion of the more serious challenges to IIT.[4]

First, the objection from functionalism: According to functionalism, mental states, including states of consciousness, find explanation by appeal to function. The nature of a certain function may limit the possibilities for its physical instantiation, but the function, and not the material details, is of primary relevance (Dennett 1991, 2005). IIT differs from functionalism on this basic issue: on IIT, the conscious state is identified with the way in which a system embodies the physical features that IIT's postulates describe.

Their opposing views concerning constraints upon artificial consciousness nicely illustrate the contrast between functionalism and IIT. For the functionalist, any system that functions identically to, for example, a conscious human, will by definition have consciousness. Whether the artificial system uses re-entrant or feed-forward architecture is a pragmatic matter. It may turn out that re-entrant circuitry more efficiently realizes the function, but even if the system incorporates feed-forward engineering, so long as the function is achieved, the system is conscious. IIT, on the other hand, expressly claims that a system that performed in a way completely identical to a conscious human, but that employed feed-forward architecture, would only simulate, but not realize consciousness. Put simply, such a system would operate as if it were integrating information, but because its networks would not take output as input, would not actually integrate information at the physical level. The difference would not be visible to an observer, but the artificial system would have no conscious experience.

Those who find functionalism unsatisfactory often take it as an inadequate account of phenomenology: no amount of description of functional dynamics seems to capture, for example, our experience of the whiteness of a cue ball. Indeed, IIT entertains even broader suspicions. Beginning with descriptions of physical systems may never lead to explanations of consciousness. Rather, IIT's approach begins with what it takes to be the fundamental features of consciousness. These self-evident, Cartesian descriptors of phenomenology then lead to postulates concerning their physical realization; only then does IIT connect experience to the physical.

This methodological respect for Cartesian intuitions has a clear appeal, and the IIT literature largely takes this move for granted, rather than offering outright justification for it. In previous work with Edelman, Tononi discusses machine-state functionalism, an early form of functionalism that identified a mental state entirely with its internal, 'machine' state, describable in functional terms (Edelman and Tononi 2000). Noting that Putnam, machine-state functionalism's first advocate, came to abandon the theory (because meanings are not sufficiently fixed by internal states alone), Tononi rejects functionalism generally. More recently, Koch (2012: 92) describes much work in consciousness as "models that describe the mind as a number of functional boxes," where one box is "magically endowed with phenomenal awareness." (Koch confesses to being guilty of this in some of his earlier work.) He then points to IIT as an exception.

Functionalism is not receiving a full or fair hearing in these instances. Machine-state functionalism is a 'straw man': contemporary versions of functionalism do not commit to an entirely internal explanation of meaning, and not all functionalist accounts are subject to the charge of arbitrarily attributing consciousness to one part of a system. The success or failure of functionalism turns on its treatment of the Cartesian intuitions we all have that consciousness is immediate, unitary, and so on. Rather than taking these intuitions as evidence of the unavoidable truth of what IIT describes in its axioms, functionalism offers a subtle alternative. Consciousness indeed seems to us direct and immediate, but functionalists argue that this 'seeming' can be adequately accounted for without positing a substantive phenomenality beyond function. Functionalists claim that the seeming immediacy of consciousness receives sufficient explanation as a set of beliefs (and dispositions to believe) that consciousness is immediate. The challenge lies in giving a functionalist account of such beliefs: no mean feat, but not the deep mystery that non-functionalists construe consciousness as posing. If functionalism is correct in this characterization of consciousness, it undercuts the very premises of IIT.

These considerations relate to the debate concerning access and phenomenal consciousness. Function may be understood in terms of access. If a conscious system has cognitive access to an association or belief, then that association or belief is conscious. In humans, access is often taken to be demonstrated by verbal reporting, although other behaviors may indicate cognitive access. Functionalists hold that cognitive access exhaustively describes consciousness (Cohen and Dennett 2012). Others hold that subjects may be phenomenally conscious of stimuli without cognitively accessing them. IIT may be interpreted as belonging to the latter category.

Interpretation of the relevant empirical studies is a matter of controversy. The phenomenon known as 'change blindness' occurs when a subject fails to notice subtle differences between two pictures, even while reporting thoroughly perceiving each. Dennett's version of functionalism, at least, interprets this as the subject not having cognitive access to the details that have changed, and moreover as not being conscious of them. The subject overestimates the richness of his or her conscious perception. Certain non-functionalists claim that the subject does indeed have the reported rich conscious phenomenology, even though cognitive access to that phenomenal experience is incomplete. Block (2011), for instance, holds this interpretation, claiming that "perceptual consciousness overflows cognitive access." On this account, phenomenal consciousness may occur even in the absence of access consciousness.

IIT's treatment of the role of silent neurons aligns with the non-functionalist interpretation. On IIT, a system's consciousness grows in complexity and richness as the number of elements that could potentially relate causally within the MICS grows. Such elements, *even when inactive*, contribute to the specification of the integrated information, and so help to fix the phenomenal nature of the experience. In biological systems, this means that silent but potentially active neurons matter to consciousness.

Such silent neurons are not accessed by the system. According to IIT, these non-accessed neurons still contribute to consciousness. As in Block's non-functionalism, access is not necessary for consciousness. On IIT, it is crucial that these neurons could potentially be active, so they must be accessible to the system. Block's account is consistent with this in that he claims that the non-accessed phenomenal content need not be inaccessible. Koch, separately from his support of IIT, takes the non-functionalist side of this argument in Koch and Tsuchiya (2007); so do Fahrenfort and Lamme (2012); and for a functionalist response to the latter, see Cohen and Dennett (2011, 2012).

Non-functionalist accounts that argue for phenomenal consciousness without access make sense given a rejection of the functionalist claim that phenomenality may be understood as a set of beliefs and associations, rather than a Cartesian, immediate phenomenology beyond such things. If, on the other hand, access can explain phenomenality, then the appeal to silent neurons as – despite their inactivity – having a causal bearing on consciousness, becomes as unmotivated as it is mysterious.

Another important distinction between functionalism and IIT lies in their contrasting ontologies. Functionalist explanations of consciousness do not augment the naturalistic ontology in the way that IIT does. Any account of consciousness that maintains that phenomenal experience is immediately first-personal stands in tension with naturalistic ontology, which holds that even experience in principle will receive explanation without appeal to anything beyond objective, or third-personal, physical features. As noted (see Section 3), among theories of consciousness, those versions of panpsychism that attribute mental properties to basic structural elements depart perhaps most obviously from the standard scientific position. Because IIT limits its attribution of consciousness to particular physical systems, rather than to, for example, particles, it constitutes a somewhat more conservative position than panpsychism. Nevertheless, IIT's claims amount to a radical reconception of the ontology of the physical world.

IIT's allegiance to a Cartesian interpretation of experience from the outset lends itself to a non-naturalistic interpretation, although not every step in IIT's argumentation implies a break from standard scientific ontology. IIT counts among its innovations the elucidation of integrated information, achieved when a system's parts make a difference intrinsically, to the system itself. This differs from observer-relative, or Shannon, information, but by itself stays within the confines of naturalism: for example, IIT could have argued that integrated information constitutes an efficient functional route to realizing states of awareness.

Instead, IIT makes the much bolder claim that such integrated information (provided it is locally maximal) is *identical* to consciousness. The IIT literature is quite explicit on this point, routinely offering analogies to other fundamental physical properties. Consciousness is fundamental to integrated information, in the same way as it is fundamental to mass that space-time bends around it. The degree and nature of any given phenomenal feeling follow basically from the particular conceptual structure that is the integrated information of the system. Consciousness is not a brute property of physical structure per se, as it is in some versions of panpsychism, but it is inextricable from physical systems with certain properties, just as mass or charge is inextricable from (some) particles. So, IIT is proposing a striking addition to what science admits into its ontology.

The extraordinary nature of the claim does not necessarily undermine it, but it may be cause for reservation. One line of objection to IIT might claim that this augmentation of naturalistic ontology is non-explanatory, or even ad hoc. We might accept that biological conscious systems possess neurology that physically integrates information in a way that converges with phenomenology (as outlined in the relation of the postulates to the axioms), without taking this as sufficient evidence for an identity relation between integrated information and consciousness. In response, IIT advocates might claim that the theory's postulates give better ontological ground than functionalism for picking out systems in the first place.

A second major objection to IIT comes in the form of a reductio ad absurdum argument. The computer scientist Scott Aaronson (2014a) has compelled IIT to admit a counterintuitive implication. Certain systems, which are computationally simple and seem implausible candidates for consciousness, may have values of phi higher even than those of human brains, and would count as conscious on IIT. The IIT response has been to accept the conclusion of the reductio, but to deny the charge of absurdity. Aaronson's basic claim involves applying phi calculation. Advocates of IIT have not questioned Aaronson's mathematics, so the philosophical relevance lies in the aftermath.

IIT refers to richly complex systems such as human brains, or hypothetical artificial systems, in order to illustrate high phi value. Aaronson points out that systems that strike us as much simpler and less interesting will sometimes yield a high phi value. The physical realization of an expander graph (his example) could have a higher phi value than a human brain. A graph has points that connect to one another, making the points vertices and the connections edges. This may be thought of as modelling communication between points. Expander graphs are 'sparse' – having not very many points – but those points are highly connected, and this connectivity means that the points have strong communication with one another. In short, such graphs have the right properties for generating high phi values. Because it is absurd to accept that a physical model of an expander graph could have a higher degree of consciousness than a human being, the theory that leads to this conclusion, IIT, must be false.

Tononi (2014) responds directly to this argument, conceding that Aaronson has drawn out the implications of IIT and phi fairly, even ceding further ground: a two-dimensional grid of logic gates (even simpler than an expander graph) would have a high phi value and would, according to IIT, have a high degree of consciousness. Tononi has already argued that a photodiode has minimal consciousness; to him, accepting where Aaronson's reasoning leads is just another case of the theory producing surprising results. After all, science must be open to theoretical innovation.

Aaronson's rejoinder (2014b) challenges IIT by arguing that it implicitly holds inconsistent views on the role of intuition. In his response to Aaronson's original claims, Tononi disparages intuitions regarding when a system is conscious: Aaronson should not be as confident as he is that expander graphs are not conscious. Indeed, the open-mindedness here suggested seems in line with the proper scientific attitude. Aaronson employs a thought-experiment to draw out what he takes to be the problem. Imagine that a scientist announces that he has discovered a superior definition of temperature and has constructed a new thermometer that reflects this advance. It so happens that the new thermometer reads ice as being warmer than boiling water. According to Aaronson, even if there is merit to the underlying scientific work, it is a mistake for the scientist to use the terms 'temperature' or 'heat' in this way, because it violates what we mean by those terms in the first place: 'heat' means, partly, what ice has less of than boiling water. So, while IIT's phi metric may have some merit, it is not in measuring consciousness degree, because 'consciousness' means, partly, what humans have and expander graphs and logic gates do not have.

One might, in defense of IIT, respond by claiming that the cases are not as similar as they seem, that the definition of heat necessitates that ice has less of it than boiling water and that the definition of consciousness does not compel us to draw conclusions about expander graphs' non-consciousness (strange as that might seem). Aaronson's argument goes further, however, and it is here that the charge of inconsistency comes into play. Tononi's answer to Aaronson's original reductio argument partly relies upon claiming that facts such as that the cerebellum is not conscious are totally well-established and uncontroversial. (IIT predicts this because the wiring of the cerebellum yields a low phi and is not part of the conscious MICS of the brain.) Here, argues Aaronson, Tononi is depending upon intuition, but it is possible that although the cerebellum might not produce our consciousness, it may have one of its own. Aaronson is not arguing for the consciousness of the cerebellum, but rather pointing out an apparent logical contradiction. Tononi rejects Aaronson's claim that expander graphs are not conscious because it relies on intuition, but here Tononi himself is relying upon intuition. Nor can Tononi here appeal to common sense, because IIT's acceptance of expander graphs and logic gates as conscious flies in the face of common sense.

It is possible that IIT might respond to this serious charge by arguing that almost everyone agrees that the brain is conscious, and that IIT has more success than any other theory in accounting for this, while preserving many of our other intuitions (that animals, infants, certain patients with brain damage, and sleeping adults all have dimmer consciousness than adult waking humans, to give several examples). Because this would accept a certain role for intuitions, it would require 'walking back' the gloss on intuition that Tononi has offered in response to Aaronson's reductio. Moreover, Aaronson's arguments show that such a defense of the overall intuitive plausibility of IIT will face difficult challenges.

5 Conclusion

IIT has a good claim to being the most strikingly original theory of consciousness in recent years. Any attempt to gloss it as a variant of Cartesian dualism, materialism, or panpsychism will obfuscate much more than it illuminates. The efforts of its proponents, especially Tononi and Koch (and their respective research centers) continue to secure its place in the contemporary debate. IIT's novelty notwithstanding, attempts to assess it return us to very familiar ground: its very premises take for granted a highly embattled set of Cartesian principles, and its implications – despite its advocates' protests to the contrary – arguably violate both parsimony and intuition. Its fit with certain empirical evidence suggests that the phi measurement may have scientific utility, but it is far from clear that this implies that IIT has succeeded in identifying the nature of consciousness.

Notes

1 Tononi and Koch (2015) outlines the basics; Oizumi, Albantakis, and Tononi (2014) gives a more technical introduction; see also Tononi (2006, 2008).
2 Tononi (2015) adopts the position that the move from the axioms to the postulates is one of inference to the best explanation, or abduction.
3 This is pragmatically impossible for systems with as many components as the human brain, so an ongoing issue within IIT involves refining approximations of these values.
4 It would be remiss to neglect any mention of Searle's (2013a, 2013b) critique of IIT, but as the response from Koch and Tononi (2013) makes very clear, the objection does not succeed.

References

Aaronson, S. (2014a) "Why I Am Not an Integrated Information Theorist (or, the Unconscious Expander)," [Stable web log post]. May 21. Retrieved from *Shtetl-Optimized*, http://scottaaronson.com/blog. Accessed July 27, 2016.

Aaronson, S. (2014b) "Giulio Tononi and Me: A Phi-nal Exchange," [Stable web log post]. May 30, June 2. Retrieved from *Shtetl-Optimized*, http://scottaaronson.com/blog. Accessed July 27, 2016.
Block, N. (2011) "Perceptual Consciousness Overflows Cognitive Access," *Trends in Cognitive Science* 15: 567–575.
Cohen, M., and Dennett, D. (2011) "Consciousness Cannot be Separated from Function," *Trends in Cognitive Science* 15: 358–364.
Cohen, M., and Dennett, D. (2012) "Response to Fahrenfort and Lamme: Defining Reportability, Accessibility and Sufficiency in Conscious Awareness," *Trends in Cognitive Science* 16: 139–140.
Dennett, D. (1991) *Consciousness Explained*, New York: Little, Brown and Co.
Dennett, D. (2005) *Sweet Dreams: Philosophical Obstacles to a Science of Consciousness*, London: A Bradford Book, The MIT Press.
Edelman, G., and Tononi, G. (2000) *A Universe of Consciousness: How Matter Becomes Imagination*, New York: Basic Books.
Fahrenfort, J., and Lamme, V. (2012) "A True Science of Consciousness Explains Phenomenology: Comment on Cohen and Dennett," *Trends in Cognitive Science* 16: 138–139.
Koch, C. (2012) *Consciousness: Confessions of a Romantic Reductionist*, Cambridge, MA: The MIT Press.
Koch, C., and Tsuchiya, N. (2007) "Phenomenology without Conscious Access is a Form of Consciousness without Top-Down Attention," *Behavioral and Brain Sciences* 30: 509–510.
Koch, C., and Tononi, G. (2013) "Can a Photodiode be Conscious?" *New York Review of Books* (3/7/13).
Oizumi, M., Albantakis, L. and Tononi, G. (2014) "From the Phenomenology to the Mechanisms of Consciousness: Integrated Information Theory 3.0," *PLOS Computational Biology* 5: 1–25.
Searle, J. (2013a) "Can Information Theory Explain Consciousness?" *New York Review of Books* (1/10/2013).
Searle, J. (2013b) "Reply to Koch and Tononi," *New York Review of Books* (3/7/13).
Tononi, G. (2008) "Consciousness as Integrated Information: A Provisional Manifesto," *Biology Bulletin* 215: 216–242.
Tononi, G. (2014) "Why Scott Should Stare at a Blank Wall and Reconsider (or, the Conscious Grid)," [Stable web log post]. May 30. Retrieved from *Shtetl-Optimized*, http://scottaaronson.com/blog. Accessed July 27, 2016.
Tononi, G. (2015) "Integrated Information Theory," *Scholarpedia* 10: 4164. http://www.scholarpedia.org/w/index.php?title=Integrated_information_theory&action=cite&rev=147165. Accessed June 27, 2016.
Tononi, G., and Koch, C. (2015) "Consciousness: Here, There and Everywhere?" *Philosophical Transactions of the Royal Society, Philosophical Transactions B* 370, DOI:10.1098/rstb.2014.0167.

Related Topics

Materialism
Dualism
Idealism, Panpsychism, and Emergentism
Biological Naturalism and Biological Realism
Robot Consciousness
Animal Consciousness

11
THE MULTIPLE DRAFTS MODEL

Francis Fallon and Andrew Brook

The phrase "Multiple Drafts Model" (MDM) refers to Daniel Dennett's alternative to a Cartesian model of mind and, in many cases, serves as a synecdoche for Dennett's general theory of consciousness. According to Dennett, clinging to the Cartesian conception of mind involves unwarranted assumptions and precludes a properly naturalistic understanding of mind, including experience. Providing an alternative, on the other hand, opens up the possibility of genuinely explaining consciousness.

Dennett makes a very strong, but also counterintuitive, case, which accounts for its divided reception. Largely thanks to Dennett, Cartesianism, in the sense relevant here, has become a charge most philosophers and scientists would hasten to avoid. This acceptance of Dennett's negative case has not translated into a general acceptance of his positive theory (although the latter has been influential upon many). A clear explanation of the basic principles should show that giving up certain tempting and familiar beliefs is of a piece with demystifying consciousness.

1 The Cartesian Model

Descartes famously – or perhaps infamously – defended substance dualism, the claim that the mental and the physical belong to different realms of existence. (The mental is directly known and not extended in space; the physical is not directly known and has extension.) With some exceptions, more modern theories of mind reject this dualism, claiming that both the mental and the non-mental belong to the same realm of existence (monism). Of these, many propose some version of naturalism. According to naturalism, the mental exists but in principle finds explanation by reference to the natural world.

Throughout his long career, Dennett has advocated explaining mental phenomena as continuous with the workings of the natural, physical world. In other words, he endorses naturalistic principles. Still, even within a broadly naturalistic paradigm, some discussions of consciousness share certain features with Descartes' account. Descartes believed that mechanical events in the brain were unconscious until they passed through the pineal gland, the "turnstile" of consciousness. The idea that consciousness will involve disparate, non-conscious elements all "coming together" in one place (not necessarily the pineal gland) and at one time has appeal. One might hold such a position without committing to any dualism. So, Dennett terms this "Cartesian materialism."

Why should non-conscious mental events have to unite in one time and place in the brain in order to rise to consciousness? A popular metaphor for experience depicts it as a play or movie unfolding in the brain. This implies an internal viewer, who is watching the show. Dennett describes this as the Cartesian Theater, complete with an audience. Such an audience would be a homunculus, i.e., an agent within the experiencing person. This metaphor does not offer genuine explanation. If you experience because you take various visual and auditory percepts into your brain, and they remain unconscious until they unite in an inner theater and are received by a homunculus audience, the question remains: What would allow this homunculus to have one unitary experience of the various elements that have just debuted "on stage"? The only recourse would involve a further regress, where these pass once again into the homunculus's own "brain," where there exists a further theater and audience, ad infinitum.

Now, a dualist might want to insist on a special meeting place for the mental and physical substances,[1] and so would have a motivation for positing a Cartesian Theater, but is there a motivation for the monistic naturalist to posit any such place? Intuitively, it feels like consciousness is unitary. All in one moment we see a cloud pass across the sky, and hear a flock of birds take flight from nearby trees. We focus on one bird for an instant, simultaneously taking in its outline, the backdrop of the sky, and the sound of its cawing. Moreover, it seems that experience proceeds in one single stream – as a storm approaches, one experiences a lightning bolt across the sky, followed by a crash of thunder overwhelming a car alarm, after which comes a cascade of rain soon joined by a gust of wind, and then in the very next moment a combination of some or all of these. At any point, it seems, we are experiencing certain elements at once, and these points together make up our stream of consciousness.

We take this single, unified stream of experiences to be very rich: to include, e.g., detailed vision out to the edges of our visual field, the sound of many individual raindrops pelting the ground, etc. Even if we forget almost immediately where the clouds were, if the lightning bolt began at the western or eastern side of the sky, or if there were more than ten audible raindrops per second, there is a fact of the matter about just what we were experiencing at any given point.

Put briefly, then, intuition motivates even some naturalists to commit to a Cartesian Theater: consciousness seems like a unified stream of experiences proceeding past the "audience" within us. We have seen, though, that this move is non-explanatory, viciously regressive even. So, we have to give up our intuition (or give up on explaining consciousness).

> The brain is headquarters, the place where the ultimate observer is,[2] but there is no reason to believe that the brain itself has any deeper headquarters, any inner sanctum, arrival at which is the necessary or sufficient condition for experience. In short, there is no observer inside the brain.
>
> *(Dennett 1991: 106)*

Given our habitual comfort with Cartesian Theater metaphor, this may appear to remove a useful option, but now we know better, that the promise of explanation via inner movies and audiences is a false one. Instead, this latest step has removed a constraint upon explanation. What does explain consciousness will not have to conform to our (prior) intuitions (although it would help if it could explain their existence).

2 Multiple Drafts

Having let go of the requirement that mental events must pass through a central processing area in order to achieve consciousness implies that consciousness does not have to consist of a seamless stream of unified experiences, even if it seems that way. Immediately, this appears to present a paradox. How could consciousness really be one way and yet seem another way? Isn't consciousness precisely in the seeming? Doesn't the subject have direct access to it, and so infallibility concerning it?

These questions cannot all receive satisfactory answers right away. Dennett knows this, noting in his central expression of MDM that making it a "vivid" and "believable alternative" to the Cartesian Theater "will be the hardest part of the book" (1991: 114). A temporary (and rather unsatisfactory) general answer might note that these questions all reflect deep intuitions, and recall, from the end of Section 1 just above, that explaining intuitions, even while not necessarily granting them authority, should in principle suffice. Only once we pull apart the mechanisms of various "seemings" can we assess their claim upon the reality of our mental lives.

To say that there does not have to be one single stream of consciousness is to say, in other words, that there does not have to be one single, authoritative narrative that makes up consciousness. The brain, in cooperation with senses, registers multiple stimuli, but does not need to re-process those registrations into a final copy for "publication." In Dennett's words:

> Feature detections or discriminations *only have to be made once*. That is, once a particular "observation" of some feature has been made, by a specialized, localized portion of the brain, the information content thus fixed does not have to be sent somewhere else to be rediscriminated by some "master" discriminator. In other words, discrimination does not lead to a re-*presentation* of the already discriminated feature for the benefit of the audience in the Cartesian Theater – for there is no Cartesian Theater.
>
> *(1991: 113)*

This describes a disjointed process, in tension with our belief in a stream of consciousness. Indeed, "this stream of contents is only rather like a narrative because of its multiplicity; at any point in time there are multiple 'drafts' of narrative fragments at various stages of editing in various places in the brain" (Dennett and Akins 2008).

An example will help illustrate how MDM and the Cartesian Model differ in their implications for assessing experience (1991: 137–8; 2008). Intently reading in a study (perhaps having found shelter from the storm), you observe the person sitting across from you look up, and just then you become aware – seemingly for the first time – that the grandfather clock has been chiming. Before the other person looked up at it, this had not come to your attention. You then find yourself able to count, retrospectively, the (three) chimes before you had become aware (at the fourth chime).

What has happened? Were you conscious of the chimes all along, and then became "extra" aware of them? Were you unconsciously registering the chimes, and then called them forth once prompted by an environmental stimulus? Nothing at the level of introspection answers these questions definitively. Mechanisms in the brain will have registered the chimes, possibly in different ways, but why should an examination of these speak with authority to exactly when one became conscious, since introspection will be incapable of confirming one way or the other?

Only on a Cartesian model do these questions require answers, and so only on a Cartesian model does the apparent inability to settle them pose a problem. On MDM, because one single official draft does not proceed through time along a continuous line, there does not have to be a fact of the matter about these issues of timing. Were you to insist that there must be a fact of the matter, this would introduce the strange category of objective facts about your awareness, of which facts you yourself are unaware. Instead – and this is a crucial point – the privileged status of consciousness is conferred *retroactively* upon (even very recent) memories when stimuli prompt us to attend to them.

This, then, is how MDM respects the powerful appearance of a single, "official" draft. Our conviction about the existence of a master narrative does not reflect its existence in the first instance, but is in the event a creation of (sometimes non-veridical) retrospective assembly of various perceptual fragments and associations. What at first may strike us as paradoxical becomes merely (but deeply) counterintuitive: there is no unitary stream of consciousness, but there are drafts whose availability for recall supply the material for the ad hoc manufacture (upon prompting) of linear narratives, and this regular capacity for spinning such yarns makes it *seem* – even in their absence – as though one linear stream of consciousness exists.

Maybe this is too quick: maybe there *was* a fact of the matter of your consciousness. Maybe you were aware of the chimes in real time, but forgot them almost as quickly. Alternatively, the initial chimes were registered unconsciously and then introduced into consciousness later, on a time-delay. Both these interpretations preserve the intuition that one single continuous draft of consciousness exists. The next section addresses this issue.

3 Orwellian and Stalinesque Streams of Consciousness

Dennett speaks to both of these possible interpretations directly. He maintains that while a mental event may bear description as conscious or non-conscious, "it is a confusion… to ask *when it becomes conscious*" (1991: 113). The argument claims not just the non-necessity of there being a fact of the matter concerning such precise timing, but the incoherence of requiring such facts.

It will help to address the terminological distinctions. Imagine the following case of false memory. You remember seeing a woman in a hat at yesterday's party (even though there was no woman in a hat). If you had no initial experience of the woman in the hat, and then after the party you misremembered (or someone surgically implanted a false memory, for that matter), then the chronology is similar to the first interpretation of the chimes case. In both cases, something happens after the fact of our conscious experience to alter our memory of it. Dennett calls such instances Orwellian, because in Orwell's *1984*, the Ministry of Truth rewrites history (1991: 117–18).

The other possibility works pre-emptively. For example, you saw other people in hats at this party, non-consciously perceived a woman without a hat, and quickly afterward, in your single authoritative draft of consciousness, fused hat-wearing with the experience of seeing the woman. Here the chronology is similar to the second interpretation of the chimes, because unconscious registrations are introduced into consciousness with some slight delay. Dennett terms such cases Stalinesque, after the staged trials that took place under Stalin (1991: 119).

Empirical evidence may support one or the other interpretation at a comparatively macro timescale. For instance, if you mentioned the hatlessness of the woman at the party yesterday, but today (mis)remember her as having worn a hat, this would suggest the Orwellian interpretation, where your consciousness was over-written. It seems natural to think that this should apply even at a micro timescale, which of course is what any theory of a single stream of consciousness expects.

Dennett uses a thought experiment to show that at the micro timescale, things change (1991: 117-19). This time, imagine that a woman with long hair runs past you. One second later, the memory of a woman with short hair and glasses contaminates the memory of the long-haired woman, and you believe that you had a visual experience of a long-haired woman with glasses. The Orwellian interpretation suggests itself: you experienced the woman without glasses run past, but then your brain "wrote over" that experience almost immediately. The Stalinesque will work too: "your subterranean earlier memories of that woman with the eyeglasses could just as easily have contaminated your experience on *the upward path*," so that the one authoritative stream of consciousness included only the experience of a woman with glasses running by. No way of determining the truth of the single stream of consciousness makes itself available here. Introspection is blind to the causal mechanisms at work, and unlike in the earlier example, where someone might remind you of having mentioned a hatless woman yesterday (thereby giving the Orwellian interpretation support), there is no further way to settle the matter, "leaving no fact of the matter about whether one is remembering mis-experiences or mis-remembering experiences" (1998: 135). There is nothing unsettling about this on MDM, because unlike Cartesian models it denies the existence of one "official" draft of consciousness.

Empirical experimentation bears out the point. Dennett discusses Kolers' "color phi phenomenon." In this experiment, subjects are shown a red dot (A) at one place on a screen, rapidly followed by a blank screen, and then a green dot (B) on another part of the screen. The experiences involve movement and change of a single spot: "Subjects report seeing the color of the moving spot switch in midtrajectory from red to green" (1991: 120). The Orwellian gloss on the Kolers experiment posits an accurate conscious experience, immediately obliterated and replaced by the midtrajectory shift report: AB, quickly forgotten, replaced with ACDB (where C and D are intermediary imagined spots), which gets reported. The Stalinesque interpretation posits something like a "slack loop of film," allowing for editing and censoring, before consciousness takes place. This has the subject inserting CD preconsciously, so that the whole sequence of color conscious events is ACDB.

> So here's the rub: we have two different models of what happens in the color phi phenomenon.... [B]oth of them are consistent with whatever the subject says or thinks or remembers. Note that the inability to distinguish these two...does not just apply to the outside observers.
>
> *(1991: 122–3)*

Whether cases like this phenomenon have Orwellian or Stalinesque origins would have to have an "answer if Cartesian materialism were true...even if we – and you – could not determine it retrospectively by any test" (1991: 119). On a model of consciousness where there is a strict, non-smeared sequence of events streaming past a conscious homunculus, or entering and exiting a stage in a Cartesian Theater, there would be a fact of the matter about the origins, on any time scale. We may, perhaps through neuroscientific progress, find answers. "But this is just where the reasons run out... [T]here is no behavioural reaction to a content that *couldn't* be a merely unconscious reaction" (124). Focusing on one or another mental event of brain processing as the moment of consciousness "has to be arbitrary," because:

> [T]here are no functional differences that could motivate declaring all prior stages and revisions to be unconscious or preconscious adjustments, and all subsequent emendations to the content (as revealed by recollection) to be post-experiential memory contaminations. The distinction lapses in close quarters.
>
> *(126)*

The problem for the Cartesian model therefore runs deeper than an epistemological shortcoming awaiting empirical resolution: nothing can settle the question of the "true" stream of consciousness, because there isn't one. A distinction in which the truth or falsity of the two sides of the distinction makes no difference is not a basis for an explanation of any kind.[3]

4 "Fame in the Brain" and Probes

In introducing MDM, Dennett describes it as "a first version of the replacement" for the image of mind suggested by the Cartesian model (1991: 111). Since then, he has not abandoned the principles of MDM, but he has augmented it with an alternative metaphor. The original MDM "did not provide... a sufficiently vivid and imagination-friendly antidote to the Cartesian imagery we have all grown up with, so... I have proposed what I consider to be a more useful guiding metaphor: 'fame in the brain' or 'cerebral celebrity'" (2005: 136; see 1998, pp. 137–9, for an early treatment of this metaphor).

The Cartesian model encourages us to think of consciousness as a play (or movie) in the mind, viewed by an audience in a Cartesian Theater within the brain. The tempting notion of a single stream of consciousness fits this well: one single series of conscious states, much like the frames that make up a television show. MDM denies that otherwise unconscious contents travel to a central processing place, where each finds its place in a queue to form the stream of consciousness. Instead, unconscious contents compete with each other for "fame." Not all people can be famous, so the process of becoming famous is competitive. Both fame and consciousness are "not precisely dateable" (Dennett and Akins 2008; for the classic Dennettian analysis of the implications of states of consciousness taking time to come into existence, see Dennett and Kinsbourne 1992). Section 3 above showed why this holds for consciousness, and gaining fame similarly defies exact chronology, even if it can be assessed at a comparatively macro timescale. Moreover, each "is only retrospectively determinable since it is constituted by its sequelae" (Dennett and Akins 2008).

Even if this metaphor does not encourage us to think of consciousness as a medium of representation, like television or theater, might it accidentally rely on a homunculus to decide "fame"? Understanding that the nature of the fame in question commits Dennett to no such fallacy requires returning to a "crucial point" noted in Section 2 above. The privileged status of consciousness is conferred *retroactively* upon (even very recent) memories when stimuli prompt us to attend to them.

Following Dennett, we have been citing instances where attention plays a role in the generation of consciousness. While this indicates an overlap with attentional theories of consciousness, Dennett does not seem to require attention per se. The crucial requirement for conferring consciousness is the involvement of one or more "probes". A probe can be "whatever event in the brain happens to boost some aspect of the current content-fixations into prominence. *In the simplest case*, a probe is a new stimulus that draws attention..." (Dennett and Akins 2008, emphasis added). Because Dennett's examples of probes involve attention, we will continue to feature it centrally.[4] To return to the chimes case, when someone else looked up at the clock, this prompted you to consider the number of chimes – a case of probing mental contents. This drew attention to just-registered sounds. Without this attention, they would not have gained any "fame"; they would have registered as temporary micro-drafts, but without any probing would have remained unnoticed, never rising to prominence. In this context, it makes sense to quote more completely a passage cited in Section 2 above:

> [A]t any point in time there are multiple drafts of narrative fragments at various stages of editing in various places in the brain.... Probing...produces different effects, producing different narratives – and these are narratives: single versions of a portion of 'the stream of consciousness'.
>
> *(Dennett and Akins 2008)*

Because of the probe, these partial drafts become available for further judgments, which may include the retroactive framing of these elements as part of a seamless stream of unitary experiences (which Dennett sometimes calls "retrospective coronation").

Consciousness comes about when mental contents get noticed. Such notice, or fame, depends upon the actualization of available judgments. No re-presentation to an experiencing homunculus enters into the explanation, nor does it incorporate any reliance upon properties qualitatively distinct from discriminative judgments. "Consciousness, like fame, is not an *intrinsic* property, and not even just a *dispositional* property; it is a phenomenon that requires some actualization of the potential" (2005: 141). Only its prominence in cognition – and not a further special quality – makes a mental content conscious. "[T]his is not the prominence, the influence or clout, those contents would have had anyway in the absence of the probe" (Dennett and Akins 2008).

Section 3 explained that requiring an exact moment for consciousness misses an essential truth about experience, that no one definitive chronology of consciousness exists, because it is temporally "smeared" among multiple drafts. The preceding discussion of probes shows that certain (portions of) drafts win competitions for fame, get noticed, and earn judgment as fitting into one single stream. Consider the familiar question of whether you were conscious during your commute home. At first it might seem as though you were not, but upon trying, you find that you recall a number of details. Must you have been conscious of them all along? You certainly registered these in a way that disposed you, upon probing, to recall them. It also stands to reason that more temporally local probes would have resulted in at least as detailed recall. The question is ill-posed. Succinctly put, "A temporally punctate event need not make the transition from unconsciously discriminated to consciously experienced in a temporally punctate moment." In other words,

> We can expect to find, and time the onset of, necessary conditions for fame in the brain... but when sufficient conditions ripen slowly and uncertainly over longer periods of time, identifying these onsets of necessary conditions as the onset of consciousness is at best arbitrary and misleading.
>
> *(Dennett and Akins 2008)*

The dispositions are necessary for entering what one takes to be the stream of consciousness, but are insufficient to count as consciousness without a subsequent boost in content-fixation (as in attention), exemplified by an ability to report these things (veridically or not) to yourself, upon probing, which probing may happen almost in real time, or at quite a delay.

5 The Ontology of Consciousness

Descartes' dualism gives us the most obvious case of claiming different realms of existence for the mental and the physical. As noted, most philosophers and scientists reject dualism in favour of naturalism, but the question of how to explain the mental by reference to nature persists. In particular, the endurance of the consciousness debate stems from its seeming to *be* a different kind of thing from material or arrangements and functions of matter. Even among those who claim common allegiance to naturalism, then, the *ontology* of consciousness remains controversial.

Dennett, sensitive to this, introduces his MDM only after articulating a methodological approach he calls heterophenomenology (1991: 71–78). This approach maintains strict neutrality with respect to the ontological status of experiential (phenomenological) components. Recall the questions posed near the beginning of Section 2 above: How could consciousness really be one way and yet seem another way? Isn't consciousness precisely in the seeming? Doesn't the subject have direct access to it, and so infallibility concerning it? Heterophenomenology begins by making no assumptions about the answers to these questions. It refuses to take for granted that the intuitive responses to these are correct, that intuitions are generally infallible or fallible, or even that these questions are posed unambiguously. The proper methodology is the most cautious: examining the empirical evidence and determining what conclusions it allows.

Returning to the Kolers phenomenon illustrates how one may begin neutrally and proceed to a defense of a particular ontology. As a matter of empirical fact, no changing spots (CD) exist in the middle of the screen. Strictly, then, the subject does not see such a spot, although she may sincerely insist upon having seen such spots. A tension exists, then, between the subject's reports and the empirical evidence. One way to attempt to resolve this, without discounting the subject's authority concerning her experience, maintains that her experience does in fact include referents for the spots CD: phenomenal units, dubbed "qualia." The term "qualia" is, by itself, ontologically neutral. Sometimes it simply serves as a placeholder, covering the various elements in experience, however they might receive characterization or explanation. More typically, however, "qualia" refers to inner, intrinsic, irreducible "bits" of consciousness. This characterization holds important implications: If the components of consciousness are inner, intrinsic, and irreducible, then they are impervious to explanation by reference to an objective, or "third-person," ontology. This rules out any standard scientific explanation of first-person, subjective experience.

Heterophenomenology might admit the logical possibility of such a position, but denies that there is reason to grant it truth. If the proponent of this robust understanding of qualia – Dennett terms such thinkers "qualophiles" – defends her claim on the grounds of its intuitive nature, this simply begs the question concerning the authority of our intuitions. It follows, then, that if an empirical, third-person explanation is available, and moreover can satisfactorily address our intuitions, we should prefer it.

On MDM, the individuated, qualic event "spot changing color in the middle of the screen" is not irreducible. That is – in principle at least – reference to mechanisms can account for the subject's conviction that she saw such a change in spots mid-screen. Mechanisms of perception, association, and memory all work in parallel in the subject's brain. The stimuli include only two spots (A and B), and we cannot assume inner, irreducible CD spots. The experiment itself requires the subject to attend, and therefore serves as a probe. Given these stimuli and the probe, the subject engages in a rapid retroactive synthesis of multiple parallel, non-conscious drafts. This gives rise to a non-veridical, although sincere, judgment that in the middle of the screen a spot changed from red to green (see Dennett [1988] for the classic treatment of the claim that we do not need a notion of ineffable, irreducible qualia; see also [1991: 369–411]).

This respects the subject's conviction about the changing spots. It really seems to her that they existed, in the place and order she reports. That this seeming consists in non-veridical judgment is no denial of that. She has infallibility about how it seems – which is to say that she has authority about what her judgments are – but her judgments themselves are fallible, and in this case are false. At the same time, the MDM explanation has not posited any special objects in its ontology that stand beyond the reach of a standard naturalistic vision. "Conscious experiences *are real events* occurring in the real time and space of the brain, and hence they are clockable and locatable *within the appropriate limits of precision for real phenomena of their type*" (1998: 135, emphases added). The appropriate limits preclude very fine-grained and irreducible, serial qualic events, such as spots C and D: "I *am* denying that there are [qualia]. But… I wholeheartedly agree that there seem to be qualia" (1991: 372).

Dennett routinely describes naturalism about the mental as requiring that each mental phenomenon receive explanation by reference to simpler mechanisms, ultimately bottoming out at the mechanical level of description. Excising irreducible qualia from MDM's ontology is necessary to such an approach. "There is no reality of conscious experience independent of the effects of various vehicles of content on subsequent action (and hence, of course, on memory)" (1991: 132).[5] Those cognitive events that influence action at least have a chance of disposing us to judge them as parts of our stream of consciousness; those that "die on the vine" (i.e. do not influence action) cannot. Probes generate prominence, determining which of the multiple drafts receive retrospective coronation as conscious.

No one particular homunculus decides what content is prominent (just as one person's regard does not confer fame upon another). This role is discharged throughout the brain. Many subpersonal mechanisms underpin the judgments at the personal level that constitute our conviction of having a unified stream of consciousness with particular and seemingly irreducible or contents. Prominent mental content may exert influence upon a variety of actions; among these, the clearest demonstrations come in the form of verbal reports. This is not due to such reports' infallibility (as we have seen, they are fallible in one, direct, sense), nor to verbalization's residence in one privileged conscious arena (MDM has denied anything holding the place of a Cartesian Theater from the very beginning). Rather,

> The personal level of explanation is defined by the limits of our abilities to respond to queries about what we are doing and why... A reported episode or nuance, current or recollected, has left the privacy of the subpersonal brain...
>
> *(Dennett and Akins 2008)*

Just as the life of an organism is explained ultimately by reference to non-living parts, the person is explained by interplay at the subpersonal level; consciousness is explained by the functional roles of non-conscious mental content. Details of how such function itself finds explanation at the mechanical level is a matter of ongoing empirical research.

6 Situating MDM

The introductory passage of this chapter noted that "Multiple Drafts Model" can refer to Dennett's overall theory of consciousness, and what followed linked MDM to the more recent "fame in the brain" metaphor, as well as to the methodological approach of heterophenomenology. Dennett has hewed closely to the core principles of MDM for decades, augmenting it without altering the fundamental arguments, and applying it with varied emphases to suit different contexts. Throughout, his arguments concerning consciousness have enjoyed a high profile: to give an indicative overview of the field, even a brief volume on consciousness would need to include a discussion of MDM. Because MDM challenges familiar assumptions about consciousness, and also because it fits a certain scientific worldview, it has generated an enormous body of literature – hundreds of papers' and several books' worth – both sympathetic and critical. Situating it in the broader discussion in a limited space will have to sacrifice precision for balance of coverage.

By now, it goes without saying that MDM stands at odds with dualism. Perhaps it should go without saying that it stands opposed to eliminative materialism, the position that consciousness strictly merits no ontological status. Dennett has eschewed this association all along, but it is still a matter of some controversy (Fallon, forthcoming). Very recently, for example, Dennett felt the need to offer clarification anew: "Consciousness exists, but just isn't what some folks think it is" (2017: 223).

Those who read Dennett's restrictions on the ontology of consciousness as too reductive accuse him of "explaining consciousness away." Section 5 gave the reasons why denying ontological status to irreducible qualia may not amount to denying that consciousness exists: that things (really) *seem* how – but don't necessarily *exist* in just the way that – they seem. This defense has failed to satisfy many. Among such critics are some of the most influential figures in philosophy of mind, and among their arguments are some of the most famous thought experiments in contemporary philosophy of any kind, themselves very durable and appearing in countless discussions in the literature. (Often the original versions predate MDM.) Uniting all of these is the conviction that Dennett's MDM "leaves something out."

Ned Block has consistently criticized Dennett's theory for being overly cognitive, failing to account for essentially non-cognitive experiences or elements of experience. He maintains a separation between phenomenal consciousness, a domain arguably coextensive with qualia, and access consciousness. Dennett's functional theory has the resources to treat the latter, but the former altogether eludes the explanatory net of MDM. Block (1990) presses his point through the inverted qualia thought experiment, which has several iterations. The basis of each is the intuition that you could see green wherever I see red, and the two of us could function in identical ways. Therefore, function does not exhaust phenomenal experience.

David Chalmers (1996) argues that nothing currently known to science about matter or its arrangement in the brain logically implies experience. We cannot tell why physical systems such as ours could not operate as they do, while remaining "in the dark," i.e. without generating the experience we enjoy. He makes use of a zombie thought experiment: we can imagine a complete physical and functional replica of a human being that has no interior life at all, so current physics and neuroscience cannot account for experience. He entertains the possibility of an augmented, future science that identifies fundamental experiential (or proto-experiential) properties in the physical world.

Another well-known thought experiment that casts doubt on physicalism, and so applies to Dennett, comes from Frank Jackson (1982): Mary is a scientist who has a complete knowledge of the objective facts about color – surface reflectance, visual cortices, conventions of naming, etc. She is confined to a black and white laboratory her entire life until, one day, she is released into the outside world and experiences color for the first time. She has learned something new, which was unavailable to her earlier, despite her expertise about the third-person facts. So, the physical facts do not suffice to explain subjective experience. Similarly, Thomas Nagel (1974) argues that knowing third-person facts about a bat would not suffice for us to understand "what it is like" (subjectively, experientially) to be a bat.

John Searle denies that Dennett has captured the "special causal powers of the brain" that produce consciousness, but is optimistic about future science doing so. His Chinese Room thought experiment (1980), the most written-about thought experiment in the history of philosophy, challenges not just Dennett, but every non-biological form of materialism on a fundamental level, because it concerns the origin of intentionality (or aboutness, upon which accounts of meaning rely). Basically, Searle imagines someone who, like him, knows no Chinese, working in a large room rigged with complex symbolic input-output instructions. When Chinese characters are fed into the room (input), the person uses the instructions (program) to select the appropriate Chinese characters to send out of the room (output). The worker could be an Anglophone monoglot, and the instructions could be all in English. From the outside, though – if everything is set up appropriately – it would seem as though the person inside understood Chinese. Programmatic input-output relations appropriate to the external world therefore do not suffice to ground true meaning. Any mental model that confines itself to describing such functional dynamics leaves something out.

Even this cursory and partial exposition of some of the livelier objections to MDM shows their intuitive appeal. Dennett responds to each of them in numerous contexts (1991, 2005, inter alia).[6] The responses are complex and, again, counterintuitive. One argument thematic among Dennett's responses holds that these thought experiments are merely "intuition pumps," designed to exploit existing intuitions rather than providing good grounds for them. Nevertheless, their intuitive appeal gives the anti-MDM camp a distinct rhetorical edge. The reader should bear in mind that while the anti-MDM arguments typically claim allegiance to naturalism, their references to future science and special, as-yet unknown, causal powers of the brain reveal their uneasy fit with a standard scientific worldview. In this, sometimes less visible, sense, Dennett's MDM has its own intuitive appeal. It lies beyond the scope of this chapter to resolve this clash of intuitions; the objective here has been to clarify Dennett's case, the better for the reader to assess it.

This treatment of objections to MDM should not overshadow its alliances. MDM's rejection of a central Cartesian Theater fits well with Bernard Baars' "global workspace" model (1988). Higher-order theories of consciousness such as David Rosenthal's Higher-Order Thought (HOT) theory (2005 and many earlier works) explain consciousness as arising when mental contents themselves become objects of (ipso facto) higher-order mental states. Unsurprisingly, this thoroughly cognitive model receives a sympathetic hearing in Dennett's work. Along similar lines, and from an evolutionary perspective, Antonio Damasio (1999) explains consciousness as the organism registering, as a stimulus, itself in the act of perceptual change (see Dennett 1993: 920 for more commonalities between Damasio and Dennett). Jesse Prinz (2004) conceives of consciousness similarly to Damasio, and, like Dennett, assigns attention a crucial role. If anything, Dennett is more optimistic about the explanatory reach of the latter two projects than their authors.

More recently, Dennett has enthusiastically endorsed Andy Clark's explanation (2013) of how the brain seems to project phenomenal properties out into the world. The "projection" metaphor glosses a functional process that can be elaborated scientifically. The organism is "designed to deal with a set of [Gibsonian] affordances, the 'things' that matter," and this "*Umwelt* is populated by two R&D processes: evolution by natural selection and individual learning" (Dennett 2017: 165–6). Feedback (probabilistic, Bayesian, feedback) from how top-down guesses do against bottom-up incoming data determines what in the environment becomes salient. A lack of feedback would mark the absence of prediction error (data confirm the top-down guesses) and would work as confirmation (167–9; see also Clark 2013). The affordances we experience result from such processes. One might redescribe this, in the language of MDM, as conceiving of the brain as oriented by evolutionary and developmental pressures to probe its own activity, where unlikelier events win competition for attention from such probes.

The end of Section 5 noted that ascertaining the mechanical realization of the relevant functional processes is a matter of ongoing empirical research. Any model that acknowledges that mental content does not require re-presentation of mental content to an audience in a Cartesian Theater to become conscious – that is, eschews conceiving of consciousness as a single temporal stream of consecutive conscious events, and accounts for the construction of an apparent stream of consciousness retroactively – will hold consistent with MDM's central principles. "A wide variety of quite different specific models of brain activity could qualify as multiple drafts models of consciousness if they honored its key propositions" (Dennett and Akins 2008; see Dennett 2005: 133–42 for a selection of neuroscientific models consistent with MDM). By explaining our intuitions about experience, without granting them ultimate authority, the MDM secures the viability of contemporary scientific research into consciousness.

Notes

1 That a non-extended substance should have the property of being locatable in extended space is, of course, paradoxical. This observation lies at the heart of the general rejection of Cartesian dualism.

2 Even the brain's status as location of the observer is contingent. As Dennett (1981) notes, if one's brain and body were separated, with lines of communication between the two maintained through radio connection, and the brain kept alive in a vat while the body went on a remote mission, one's point of view would be the sensory contacts of the body with its surrounding stimuli, and not the vat.
3 This move has smacked of verificationism to many commentators. Dennett's response has mostly been to accept the charge but deny its force. While we do not have room for a full discussion of this response, it is not clear that there is much wrong with his variety of verificationism (what he once referred to as "urbane verificationism") (Dennett 1991: 461–2; see also Dennett 1993: 921–2, 930n; Dahlbom 1993; and Ross and Brook 2002, Introduction).
4 Of course, if mechanisms other than attention – perhaps less deliberate or guided, less compelled by stimulation, or more sub-personal than attention – can serve as probes, this should be explicated. Depending upon one's sympathies, this point can be regarded as a complaint about Dennett's account or as a research question motivated by it. The same can be said for any lack of specificity concerning the kinds of memory relevant to consciousness.
5 This is a particularly loaded sentence. For an explication of the "vehicles" at issue, see Brook (2000). For a discussion of the ontology of consciousness as including phenomenological *effects*, easily mistaken for inner *causes* of phenomenal experience, see Chapter 14 of Dennett (2017); see also Dennett (2007). Fallon (forthcoming) argues that these claims support a "realist" interpretation of Dennett on consciousness.
6 Dennett's numerous comments (1991) on Fodor's "language of thought" (LOT) account (1975) nicely encapsulate his positive arguments concerning intentionality. See also the exchange between Rey (1994) and Dennett (1994). Dennett (1993: 925–8) gives one succinct response to the "inverted qualia" arguments.

References

Baars, B. (1988) *A Cognitive Theory of Consciousness*, Cambridge: Cambridge University Press.

Block, N. (1990) "Inverted Earth," *Philosophical Perspectives* 4: 53–79.

Brook, A. (2000) "Judgments and Drafts Eight Years Later." In D. Ross, A. Brook, and D. Thompson (eds.) *Dennett's Philosophy: A Comprehensive Assessment*, Cambridge, MA: MIT Press.

Chalmers, D. (1996) *The Conscious Mind: In Search of a Fundamental Theory*, New York: Oxford UP.

Clark, A (2013) "Whatever Next? Predictive Brains, Situated Agents, and the Future of Cognitive Science," *Behavioral and Brain Sciences* 36: 181–204.

Dahlbom, B. (1993) "Editor's Introduction." In B. Dahlbom (ed.) *Dennett and His Critics*, Oxford and Cambridge, MA: Blackwell Publishers.

Damasio, A. (1999) *The Feeling of What Happens: Body and Emotion in the Making of Consciousness*, New York: Harcourt, A Harvest Book.

Dennett, D. (1981) "Where Am I?" In D. Hofstadter and D. Dennett (eds.) *The Mind's I: Fantasies and Reflections on Self and Soul*. New York: Basic Books.

Dennett, D. (1988) "Quining Qualia." In A. Marcel and E. Bisiach (eds.) *Consciousness in Contemporary Science*, Oxford and New York: Oxford University Press.

Dennett, D. (1991) "The Message Is: There Is No Medium," *Philosophy and Phenomenological Research* 53: 919–931.

Dennett, D. (1993) *Consciousness Explained*, Boston: Little, Brown.

Dennett, D. (1994) "Get Real," *Philosophical Topics* 22: 505–568.

Dennett, D. (1998) *Brainchildren*, Cambridge, MA: MIT Press.

Dennett, D. (2005) *Sweet Dreams*, Cambridge, MA: MIT Press.

Dennett, D. (2007) "Heterophenomenology Reconsidered," *Phenomenology and the Cognitive Sciences* 6: 247–270.

Dennett, D. (2017) *From Bacteria to Bach and Back: The Evolution of Minds*, New York: W.W. Norton & Company.

Dennett, D., and Akins, K. (2008) "Multiple Drafts Model," *Scholarpedia* 3: 4321. http://www.scholarpedia.org/article/Multiple_drafts_model. Accessed 17 April 2017.

Dennett, D., and Kinsbourne, M. (1992) "Time and the Observer: The Where and When of Consciousness in the Brain," *Behavioral and Brain Sciences* 15: 183–201.

Fallon, F. (forthcoming) "Dennett on Consciousness: Realism without the Hysterics," *Topoi*.

Fodor, J. (1975) *The Language of Thought*, Scranton, PA: Crowell.
Jackson, F. (1982) "Epiphenomenal Qualia," *Philosophical Quarterly* 32: 127–136.
Nagel, T. (1974) "What Is It Like to Be a Bat?" *The Philosophical Review*, LXXXIII, 435–450.
Prinz, J. (2004) *Gut Reactions: A Perceptual Theory of Emotion*, New York: Oxford University Press.
Rey, G. (1994) "Dennett's Unrealistic Psychology," *Philosophical Topics* 22: 259–289.
Rosenthal, D. (2005) *Consciousness and Mind*, Oxford and New York: Oxford University Press.
Ross, D., and Brook, A. (2002) "Introduction." In Don Ross and Andrew Brook (eds.) *Daniel Dennett*. Cambridge and New York: Cambridge University Press.
Searle, J. (1980) "Minds, Brains, and Programs," *Behavioral and Brain Sciences* 3: 417–458.

Related Topics

Dualism
Materialism
The Global Workspace Theory
Representational Theories of Consciousness
Consciousness and Attention

12
THE INTERMEDIATE LEVEL THEORY OF CONSCIOUSNESS

David Barrett

Jesse Prinz (2011, 2012) advocates what he calls the Attended Intermediate Representation (hereafter 'AIR') theory of consciousness. To understand Prinz's view, it is easiest to begin with the overall method he employs to deliver a theory. He attempts first to answer the question of where in the processing of information in the brain that consciousness arises. The focus on 'intermediate' gives away the answer to this question; he emphasizes the intermediate level of processing as the locus of consciousness. It is representations at this intermediate level that constitute the content of conscious experiences. Once one knows where the conscious states pop up in the processing, one can then employ a further method shared by psychologists and neuroscientists: compare cases where these intermediate level activations occur with and without consciousness, then look for differences elsewhere that could be responsible for the variation in consciousness. Here the 'attention' part of the theory comes to the fore. Prinz contends that a major reason why intermediate level activations can occur without consciousness is that subjects with these activations lack attention. Two subjects can thus process the same information through intermediate areas, but, if one attends to the stimuli responsible for those activations and the other does not, only the former will be conscious of those stimuli. Conjoin the two parts and you have the overall view: conscious states are AIRs.

In this entry, I will review the impressive evidence Prinz presents for his theory, explaining both the arguments he makes for the locus of conscious states in brain processing and also the role of attention in making those states conscious. Most of this evidence concerns what can be called the 'psychological correlates of consciousness,' since the evidence concerns attention, representations, and information processing, which are all psychological notions. Prinz (2013) also offers a theory about the neural implementation of the psychological correlates. Given the speculative nature of this part of Prinz's view, however, I will focus here exclusively on the psychological side of his view. Finally, after laying out Prinz's view and the evidence he cites in favor of it, I will offer a critical voice about his theory.

1 Where Does Consciousness Arise in Neural Processing?

Let us begin with the question of where consciousness arises. What is the 'intermediate' level, what are these 'intermediate representations'? The most intensely studied sense modality is vision, so it is easiest to answer to these questions by looking at what we know about vision.

More specifically, we can investigate *visual object recognition*. Though Marr's (1982) treatment of this process is somewhat out of date, it can still provide clear answers to our questions. The idea is to understand the representations that play a part in his construal of this particular process, and then locate them on a low, intermediate, and high scale. When we locate the representations that are where consciousness arises, we can then move on to the big question of what happens to them to make them conscious.

At the beginning stages of visual object recognition, the primary visual cortex (V1) receives stimulation from the retina and thalamus to form a conglomerate of mental representations of very local features. V1 houses cells that represent edges, lines, and vertices that pop up in very specific areas of the visual field. The composite of these features yields what Prinz helpfully calls something approaching a two-dimensional pixel map. This map only gives information about these very specific properties of very specific locations in visual space. Marr calls this the 'primal sketch'. The representations that make up the primal sketch are *low level* representations. Next comes the processing that yields the '2.5 D sketch,' where figure is separated from ground, surface information is calculated, and a coherent, vantage point-dependent representation is formed of the object(s) found in the visual field. The representations of items here are considered *intermediate level* representations. Finally, the last stage of processing involves a 3D representation of an object, which is formed by a collection of basic shapes like cubes and cylinders. These representations are supposed to represent their objects regardless of perspective. Their main job is for classification and categorization. The 2.5 D sketch is no good for categorization because a shift in perspective yields different representations. There is nothing invariant in the intermediate processing, then, to classify objects as the same through shifts in vantage point. Hence the need for these *high level* representations to complete visual object recognition.

The details of Marr's account have largely stood up to the test of time. Lower level visual areas are perhaps more capable than he realized, there are a wealth of areas that independently process particular features at the 2.5 D level (including color, form, and motion), and the higher-level primitives he hypothesized have not been confirmed. Nevertheless, there is much neurological and psychological evidence to back up his story. Where does consciousness seem to arise in this process, however? Certainly not at the 2-D, pixel map stage. We are conscious of integrated wholes, distinct from their backgrounds. Certainly, also not the perspective-independent, 3-D representational stage. We experience objects from a particular perspective; when they, or we, move, our visual experience of those objects changes. The answer to our question seems to be: consciousness arises at the intermediate level.

Is there any evidence to substantiate this story besides the loose allusion to Marr's fairly successful theory of visual object recognition? Prinz cites three lines of evidence, which are generally used to support all theories about information processing in the brain: cell recordings from neurons in the brains of monkeys, fMRI studies about neural activation in humans, and neuropsychological studies from human patients with brain damage. Beginning with the first line, recordings from cells in the brain areas that correspond to the distinct levels show that only intermediate level neurons fire reliably in response to what monkeys are conscious of. Lower level cells, located in V1, for example, fire in response to two distinct colors that are presented rapidly in succession, despite it being well known that such a presentation of colors is experienced as a *mixture* of the colors. V1 cells also show no activity during color afterimages. Higher-level cells show the same lack of sensitivity to what monkeys are consciously experiencing. Cells in the inferotemporal cortex, those found at the back end of one of the major visual streams in the brain, respond to the same objects regardless of size, orientation, position, and left/right reversal. Evidently changes in these parameters make a huge difference to our conscious visual experiences (and presumably, those of the studied monkeys), yet they make none to the higher-level cells.

Moving to fMRI brain scans in humans, we find much the same pattern of results. During Rapid Eye Movement (REM) sleep, when we do sometimes have visual experiences, activations are found in V3 (an intermediate level visual brain area) but not in V1. Color afterimages associate with activity in the intermediate levels, but not in V1. The interpretation of bistable figures—such a figure can be experienced in two different ways, like how a Necker cube can be experienced as facing up or facing down—from subjects correlated with activity in intermediate levels. And, finally, illusory colors and illusory contours seem to invoke only intermediate areas. This heightened activity in intermediate areas, correlated with our visual experiences, suggests strongly that it is it is the intermediate level of processing where consciousness arises.

Humans with behavioral deficits also provide interesting evidence for Prinz's intermediate representation (IR) hypothesis. Supposing it is the intermediate level of processing that houses conscious representations, we should expect three findings: (1) that damage to early visual areas will largely destroy visual consciousness (since damage here means information does not get to intermediate levels); (2) that damage to intermediate levels means a total loss of visual consciousness (since the areas where the conscious arises will have been destroyed); and (3) that damage to higher-level areas will not destroy consciousness (since the processing for conscious experiences occurs earlier in the sequence). There is much evidence for (1). Many people have had damage to V1, which resulted in blindness. There is still the phenomenon of 'blindsight' (Weiskrantz 1986), where individuals with V1 damage still retain the ability to navigate their environments and retain some small level of visual acuity, but it is well known that this occurs through subcortical projections from the retina to higher levels that bypass the usual route through V1. Since blindsighters have no visual experiences (it is *blind*sight, after all), this does not bother the intermediate level hypothesis. Evidence for (2) exists in abundance, as well. Since the intermediate level in vision is fractionated into different areas that process specific information, there should exist specific kinds of blindnesses to correspond to the different processing areas. One thus finds patients with brain damage who have a form of color blindness known as achromatopsia, those with a motion blindness called akinetopsia (where one experiences the world visually as a series of still frames), and those with something like form blindness called apperceptive agnosia (who, for instance, cannot accurately copy the shapes of pictures they see). The evidence for (3) also exists. What one would predict from Prinz's position is that damage to the higher levels would lead only to inability to recognize and classify objects, not necessarily to experience them from particular perspectives. This is exactly what happens in patients. They are said to suffer from associative agnosia: they can see objects and draw them faithfully, but without any ability to recognize what the object is.

To move beyond vision for a moment, there is ample evidence for the intermediate hypothesis in other sensory modalities. Physiologically the brain areas that support audition and touch are organized hierarchically, just as the areas that support vision. There is a primary auditory cortex and a primary somatosensory cortex, and later processing areas just as with vision ('the belt' for audition, and 'S2'—the touch analogue for V2). More interestingly, one also finds the same sorts of behavioral deficits, as with vision, in these other sense modalities. Damage to the lower levels renders patients deaf or unable to feel. With the primary somatosensory cortex, information about different body parts is processed in different areas, so usually there are deficits for sensation in distinct areas of the body. In the auditory stream, at intermediate levels, one finds deafness just for particular kinds of sounds. As one finds brain damage in these areas, one finds the same apperceptive/associative distinction as with vision. Some auditory deficits leave patients with the inability to recognize sounds, although they are able to match any pairs of sounds as the same. The same kind of deficits are found in touch, as well; patients are able to match two different objects by how they feel, but are totally unable to categorize those objects (as, say, cubes or cylinders). These patients, then, suffer from an associative agnosia: they are able to experience the world consciously, but cannot classify these experiences as belonging to this category or that. This is strong evidence for the existence of an intermediate level of processing in these modalities, and evidence also for consciousness arising at that intermediate level.

2 Attention Is the Mechanism by Which IRs Become Conscious

One main worry for Prinz about the intermediate level hypothesis is that there is ample evidence of activations at the intermediate level and beyond that does not coincide with conscious experience. For example, Pessiglione et al. (2007) show the existence of subliminal motivation; we can be motivated by stimuli of which we are unaware. These motivation-relevant structures are in the forebrain, well after the processing of visual stimuli (at the intermediate level). Berti et al. (1999) describe a subject with brain damage who can make accurate judgments about whether objects he holds in his hands are the same or different, but who, due to the damage, has no tactile experiences in one hand. The ability to make these comparative judgments, as we have seen, is associated with intermediate level structures. Since we have good evidence for intermediate level processing in the absence of consciousness, we have reason to believe that IRs are not *sufficient* for consciousness. This is not necessarily a big surprise. So far, we have only been concerned with figuring out where conscious states can be found in the information processing going on in the brain, not necessarily with trying to theorize about what conditions or features *constitute* consciousness.

To begin this extra theoretical work, there is a simple strategy to follow: compare those cases where IRs are unconscious with those cases where IRs are conscious and see what the difference is. Whatever difference we find will give us interesting clues about the further constitutive question. Not only will we (already) know *where* consciousness arises, we might also find out *how* these IRs become conscious. For example, if we can find cases of people who are blind to stimuli that other people usually see (or deaf to stimuli other people normally hear, etc.), so long as the former process information through the intermediate level the same as the latter, we will be in a perfect position to implement the strategy.

One locus of evidence that fits the strategy comes from *unilateral neglect*. In this condition, patients usually suffer from damage to the right inferior parietal cortex. Given the wiring of the brain, visual information that comes in from the left visual field is processed by the right side of the brain (and vice versa)—even through to the parietal cortex. As a result, these patients have a visual deficit for items in their left visual field. For example, if you present a subject with unilateral neglect a series of lines and ask her to bisect the line exactly in the middle, she will invariably bisect those lines with a mark that is far closer to the right end of the line than the middle. Or, even more strangely, suppose you present such a subject with a series of pictures, where the right side of each picture is the same front of a horse, while the left side of the picture varies—sometimes it is the back of a cow, sometimes the back of a bicycle, and sometimes the (normal) back of a horse. A patient with unilateral neglect will experience them as all the same; the left side of the pictures and their obvious differences are invisible to such patients. Yet, and this part is crucial, there is strong evidence that the invisible stimuli are being processed right on through the intermediate level. Not only do these subjects fail to see what most people typically see, but they appear to process these unseen stimuli in ways that others process stimuli that are seen. To obtain evidence for all the processing, let us stick with the horse example. When asked which of these identical horses seems the most real, patients typically select the correct horse (the one with the horse back end). Clearly, then, something like visual object recognition is taking place. And we already know that process goes straight through intermediate levels to the high level representations.

Hence, we have examples of subjects who process stimuli in totally normal ways, but who are quite blind to those stimuli. They are the perfect implementation of our strategy; we need only consider the differences between them and us to have evidence on which to build a theory of consciousness. What, then, does the right inferior parietal cortex do? It is most associated with the control and allocation of *attention*. Unilateral neglect is understood to be an attentional deficit—as opposed to a strictly visual deficit. We have, then, an extra piece in the puzzle about consciousness: perhaps conscious states are those IRs that are attended to.

Before running with that theory, it would be best to consider extra evidence from people with intact brains. Considered alone, the neurological damage in patients with unilateral neglect is not the most convincing. It could easily be that the damage to the parietal cortex damages attention *and* some other functional capacity (or capacities). This other functional capacity might make a difference to what makes IRs conscious, or it might not. But the unilateral neglect patients will not be able to tell us. So, it is premature to conclude that consciousness is attended IRs purely on the basis of their evidence.

Luckily, there is much known about attention and awareness in people without brain damage. Two conditions are noteworthy: *attentional blink* and *inattentional blindness*. In the first case, when two stimuli are presented in close enough succession (the second stimulus occurring before ~450 milliseconds after the first), the first captures attention while the second stimulus is not noticed. Though Prinz does not mention it, there is evidence that the attentional blink occurs in the auditory and tactile modalities and can even occur cross-modally—that is, a visual target can cause such a 'blink' for an auditory stimulus presented quickly enough after the visual target, or vice versa (see Soto-Faraco et al. 2002). Inattentional blindness occurs when attention is paid to a particular stimulus, rendering other stimuli in (say) the visual field invisible. The classic example is of a person in a gorilla costume walking through a group of people passing around a basketball (see Simons and Chabris 1999). Subjects are tasked with counting the number of passes, attention to which makes some subjects (though not all, unsurprisingly) unaware of the gorilla. Since they are aware of the number of passes, and since the person in the gorilla suit walks straight through the middle of the other people passing the ball, it is clear that the subjects could be conscious of the gorilla. It is as if they have a neglect for anything but the passes of the ball. Since these attention-related phenomena occur in all people, it provides stronger evidence for our working hypothesis that attention is the mechanism that makes our IRs conscious.

It is worth pointing out that these bits of evidence give us suggestive reason to think attention is important for consciousness, but they are somewhat limited. They show that attention is *necessary* for consciousness—in particular, that a lack of attention means a lack of consciousness—but they do not show that attention is sufficient for consciousness. If we can find evidence that attention paid to stimuli renders those stimuli conscious, we would have a strongly supported theory. Prinz is quick to point out that such evidence exists. A phenomenon called 'pop out' is suggestive. When we look at a set of objects that are almost entirely uniform, save for one object that stands out as different, we become quickly conscious of the dissonant object. It is thought, Prinz reports, that attention is grabbed by that object. If so, it is a case of attention paid to an object bringing that object into consciousness. Posner (1980) is also famous for first deploying an experimental method, which is named for him (the so-called 'Posner cuing paradigm'). In these experiments, accuracy in judgments about, and speed of detection for, objects is enhanced by a cue that precedes the target object. In those cases where the cue appears in a location different from the target object (an 'invalid' cue), those same capacities are diminished. Again, it is supposed that the cue acts to grab attention. Once attention is allocated to the area, it is available to process the stimuli that appear in that location. There is also the 'cocktail party effect,' where we are able to pick out our name being said in the din of many conversations, whose contents we would never otherwise be conscious of. It is thought that salient information like this automatically grabs attention. The supposition is that it is another case of attention to a stimulus creating consciousness of that stimulus.

We have, then, the most basic formulation of Prinz's AIR theory of consciousness: consciousness arises when and only when intermediate level representations are modulated by attention. I call it a 'basic' formulation, because it does not go as far as possible in distancing Prinz's view from the views of others. Many hold that attention is important for consciousness, from neuroscience-oriented theorists (Baars 1988; Crick and Koch 1990) to philosophers of the higher-order representation theories of consciousness (Lycan 1996; Rosenthal 2005). To distinguish Prinz's view, it is important to understand what he has in mind by 'attention'. Once we are clear on what it means to modulate IRs by attention, we will have the full theory of consciousness.

What Prinz seizes on is the thought that when attention is paid, information flows through the system in a different way than it otherwise would. Again, Prinz uses the same methodological line as when determining the mechanism that makes IRs conscious: compare cases of stimuli to which attention is paid with stimuli two which attention is not paid and see what the differences are. We already know that processing can occur deep in the brain in the absence of consciousness, and the same is true for unattended stimuli, as well. In cases of binocular rivalry (where each eye is presented with a different stimulus, but we only visually experience the winner of the rivalry between the processing of the stimuli), we know that attention is the main determinant of what we visually experience. Yet the unattended, unconscious stimuli can still cause priming. So, processing for these stimuli proceed to the higher levels. Yet one important difference between the loser and the winner of the rivalry is that only the former is available for executive processes. We can, that is, report about the stimulus, reason about it, remember it for as long as we like, and examine it in detail. The loser can activate semantic networks in the brain (a high-level of processing) but it is unavailable for these kinds of processes. We cannot report about those stimuli or remember them. We have, then, an interesting difference between attended stimuli and unattended stimuli: the former, but not the latter, are available to these executive processes. Of course, there is already a psychological mechanism thought to be responsible for these other processes called *working memory* (see Baddeley 2007). Working memory is where information can be controlled by the subject for various purposes—for memory, for action guidance, for report, etc. To support this connection between attention and working memory, Prinz offers two pieces of evidence. First, there are studies that show that when two shapes are laid on top of each other, and subjects are asked to focus only on one of them, it is only the attended shape that is recalled after a short delay interval. The attention to that shape seems to have made it available to working memory. Second, there is evidence that as working memory is filled (with distractor tasks, for instance), cases of inattentional blindness increase. Apparently as working memory capacity diminishes, so does our attentional capacity (and thus consciousness of stimuli right in the middle of our sensory fields).

The simplest explanation of these results would be to hold that attention and availability to working memory are identical. This is the explanation Prinz favors. What attention *is*, then, is the processing of stimuli that makes the representations of them available to working memory and executive processes. A nice feature of this hypothesis about the nature of attention is that it explains what 'top-down' and 'bottom-up' attention have in common. Top-down attention is where we voluntarily allocate attention on whatever objects or features we are interested in. Bottom-up attention is where objects or their features automatically, non-voluntarily grab our attention. It could have turned out that these were simply two different processes that were mistakenly given the same name (as if there were something in common between them). In the case of top-down attention, where, for instance, we are searching for a particular person in a crowd, what happens is that the location of our target makes that target conscious. We can now report about what we see and begin to take steps towards whatever actions we like. In the case of bottom-up attention, where, for instance, a stimulus pops out of the background, again the stimulus becomes immediately conscious to us in a way that allows executive-style manipulations. In either case, then, it appears what attention is really doing is making our stimuli available to working memory. This unification of what could easily be disparate neural processes is a nice implication of Prinz's view. The simplicity it implies is a further reason to believe that attention just is the process by which information becomes available to working memory.

Putting all of the foregoing together, then, we finally have Prinz's full, unique theory of consciousness:

> *Prinz's AIR Theory of Consciousness:*
> Consciousness arises when and only when intermediate level representations undergo changes that allow them to become available to working memory.[1]

Start with the evidence he provides for the contents of conscious experience, mix that with the evidence Prinz has for attention being the mechanism that makes these contents conscious, then add in his theory about what attention is, and you have the full AIR theory.[2]

3 A Critical Appraisal of Prinz's View

To provide a brief review of the main assertions and to make evaluation more orderly, we can chop up Prinz's overall view into the following succession of claims:

1 Consciousness arises at the intermediate level of processing.
2 Consciousness arises when and only when we attend.
3 Attention is the process by which information becomes available to working memory.
4 Consciousness arises when and only when intermediate level representations undergo changes that allow them to become available to working memory.

From here we can evaluate each claim in turn.

Claim 1

Claim (1) is a view about the contents of conscious experience. For Prinz, what we are conscious of is always construed perceptually. This is a result of his identification of intermediate level, perspective-dependent, and detail-filled representations as the exclusive stuff that populates conscious experience. Thus, there cannot be, for example, any conscious experience that goes along with *cognitive states* (see Pitt 2004; Siegel 2006) or *the self* (see Kriegel 2005). This is why Prinz spends multiple chapters in the middle of his book (2012) arguing against the idea of 'cognitive phenomenology' or 'phenomenology of the self.' Yet the bare existence of these topics, along with their defenders, shows that it is philosophically controversial to locate consciousness exclusively at the intermediate level of processing.

Additionally, Wu (2013) claims that body-centered representations—of the kind it seems reasonable to suppose Prinz would categorize as the perspective-dependent IRs that can become conscious—are localized in the ventral interparietal area. This area is one which Prinz claims to count as part of high-level visual processing. Furthermore, Wu also notes evidence from an apperceptive agnosic, who cannot see shapes and objects, but who can see textures and colors. Her visual experience must be mostly like seeing an animal camouflaged against the background; color and texture is uniform and consciously experienced, but there is no individuation of the boundaries of the animal itself. These object/shape representations are the kinds that Prinz identifies as populating consciousness, yet the evidence shows this woman has damage to the lateral occipital complex, an area Prinz identifies as a part of high-level visual processing.

Prinz can always choose to redraw the lines between low, intermediate, and high-level (visual) processing. Especially given the reliable difference across modalities between associative and apperceptive agnosia, Prinz has some justification for holding out hope for a better remapping. It is out of the scope of the chapter to do justice to the literature on the contents of consciousness, so I leave any controversy surrounding (1) to the side.

Claim 2

Claim (2) is a hotly contested claim in philosophy (Mole 2008; Wu 2014), psychology (Kentridge et al. 2008) and neuroscience (Lamme 2003). For the most part, scrutiny is reserved for the sufficiency claim. That is, many researchers discuss evidence against the idea that all cases of attention imply cases of consciousness; there can be, according to this crowd, attention in the absence of consciousness. Below I will focus only on the sufficiency claim, but it is worth mentioning the necessity claim does come under attack, too. Some think that there can be consciousness in the absence of attention (see Block 2013; van Boxtel et al. 2011). Setting the necessity claim aside, what are the reasons for thinking that attention is not sufficient for consciousness?

The most popular evidence for this claim comes from Kentridge et al. (2008). There they review evidence of a patient with blindsight, who is totally blind to particular locations in his visual fields, but can still make above-chance forced judgments about stimuli that appear in those locations. When a visible cue is presented (to the hemifield which is still conscious), the ability of the patient in question to make discriminations about stimuli in the blind visual field is enhanced. More precisely, when the target stimulus is *validly* cued (that is, when the cue indicates correctly where the stimulus will appear) there is enhanced performance compared to an *invalid* cue. This is an example of the famous Posner experimental paradigm I mentioned above, which is thought to be a measure of attentional capture (by the cue). Hence the evidence for attention (focused on the target in the blind hemifield) without consciousness (of that target).

Prinz's response is that this experiment only captures spatial attention, so that attention is allocated only a particular location in space and not any particular object. The enhancement in performance is explained by a shift in gaze, which results from that capture of spatial attention. Given the shift in gaze, more receptors will be allocated on the target. Whatever remaining processing resources left in the patient's V1 can then act on whatever is found in the gazed-at location. When the cue is misleading with respect to the target, gaze shifts away from where the target is located, where fewer receptors in the periphery will mean less processing of the target. This ultimately means attenuated task performance.

Norman et al. (2013), however, respond to exactly this line of criticism of the evidence from the original blindsight patient. They reprise the Posner technique, this time using invalid cues that appear in different locations with respect to the cue. In either case, the target appears equidistant from the (invalid) cue. This should mean, if Prinz is right, that performance will be the same in either case. The same amount of (diminished) processing power will be available to either kind of invalidly cued target. Yet this is not what Kentridge and his colleagues find. Though the invalidly cued targets are the same distance away from the bad cue, they appear in separate invisible rectangles. One of the targets appears in the *same* invisible rectangle as the invalid cue; the other target appears in a *different* invisible rectangle from the invalid cue. What they find is that performance for the cases where the invalid cue is in the same rectangle is enhanced, compared to cases where the invalid cue is not in the same rectangle as the target. They interpret these results as showing that *object-based* attention is captured in these cases. Attention is captured by the invisible rectangles and this accounts for the facilitating effect when the invalid cue appears in the same rectangle as the cue. Prinz's gaze-shifting explanation cannot account for these results. No matter where the invalidly cued target appears, it is the same distance from the invalid cue; so, the same amount of processing power should be available in both cases. It appears that attention in the absence of consciousness (in this case, of rectangles) is possible.

Claim 3

One would have noticed, I think, the comparative lack of evidence that Prinz presents for his view of what attention is compared to the evidence he provides for (1) and (2). Unsurprisingly, then, claim (3) can come under heavy attack. Since Prinz could, in principle, abandon his view about what attention is and still maintain the details of his theory—whether we would still want to call it a true AIR theory or not—I will be somewhat brief in my remarks here.

To begin, there is a vast literature about attentional effects in V1. These so-called 'early stage' effects seem to show that attention is not what makes information available to working memory, since the information that is being manipulated at these early stages is never anything that enters working memory (it cannot be reported on, stored for later use, etc.). Prinz responds that these modulations might not properly count as *attentional* modulations because they might merely reflect back-propagations from intermediate level processes. It could be, then, that attention does only affect processing at the intermediate level, which then itself alters the early stage processing. Moving on, there is also much evidence for 'late stage' (post-perceptual) attentional modulation. As Awh et al. (2006) describe it, there is even perhaps attentional modulation for items *already encoded in* working memory. Additionally, Awh and colleagues show evidence that spatial attention is the mechanism for maintaining information in visual working memory. Much like covert rehearsal is thought to maintain information in phonological working memory, spatial attention might refresh visual traces in storage.

If any of these hypotheses capture something important about attention, then clearly it is not correct to say that attention is simply the mechanism that makes information available to working memory. In the first case, the information never makes it to working memory; in the other two the information that is modulated is already in working memory. Yet Prinz is free to drop this claim (3) about attention. His main focus is on theorizing about consciousness, not attention. So, what ultimately counts is the evidence in favor of, or against, claim (4). If it turns out that Prinz has no grand, unifying account of attention to offer, it need not show that his theory of consciousness is unacceptable. It is to this theory that we now turn.

Claim 4

I am aware of no evidence against the idea that information that is conscious must also be information that is available to working memory. I will instead focus on the idea that there is information available to working memory that is not conscious. Accordingly, I am attacking the sufficiency of working-memory-available information for consciousness. More specifically, I think there is evidence of information *already encoded* in working memory that is not conscious. If I am right, then it follows that there can be information merely available to working memory that is not conscious. Prinz's theory would, then, be in trouble.

This evidence comes from Hassin et al. (2009) and Soto, Mantyla, and Silvanto (2011). Focusing on the former study, they present subjects with disks that are either 'full' or 'empty.' The task is simple: identify whether the disk that appears on the screen is full or empty. Disks appear in sets of five that can either follow a pattern (like a zig-zag, pattern, for instance), follow a broken pattern, where four disks follow the pattern and the fifth disrupts it, follow a control condition, or finally follow a totally random condition. Subjects are not informed of the patterns. The variable of interest is, of course, the reaction times for the discrimination task. Predictably, the researchers found that reaction times were higher for control and random conditions versus the pattern condition. The interesting result is that reaction times for the broken pattern condition were significantly higher than in the pattern condition. This supports the hypothesis that subjects were extracting information about the pattern and using it to aid in the discrimination task. By multiple post-test measures, subjects were unaware of the presence of the pattern. So, it appears that subjects were unconsciously using an extracted pattern to enhance their performance on the full-or-empty task. Since this is the kind of executive work expected of working memory, this all counts as evidence for encoded items in working memory that are unconscious. Hence it is evidence against Prinz's AIR theory of consciousness.

Prinz's response to this evidence is to implicate another memory store besides working memory. Instead of the information about patterns being stored there, he offers the alternative theory that the information is stored in what is called 'fragile visual short-term memory,' a modality specific, high capacity storage unit. If the information is stored there, and can explain the increased performance in the pattern condition, then there is no evidence from the Hassin study that unconscious information can be encoded in working memory. There is a swift, and to my mind decisive, response to Prinz's alternative theory. One extra experiment run by Hassin and colleagues utilized the same experimental setup, but altered the stimuli. In place of a visual pattern, they substituted an algebraic pattern (e.g., 2, 4, 6, 8; or 1, 3, 5, 7). Again, the same results from they pattern and broken pattern conditions obtained (and so, too, the unawareness of the patterns in post-task examinations). Since the patterns here were algebraic rather than visual, the fragile visual short-term memory store can offer no explanation for the performance. It appears, then, that there can be unconscious information encoded in working memory. Since that information is stored there, it must have been, at some point, available to working memory, too. The result is that information (intermediate level representations, or whatever you like) can be available to working memory without being conscious. This is exactly the wrong result for Prinz's theory of consciousness.

4 Conclusion

What, in the final analysis, should we make of Prinz's AIR theory? I think the empirical details do not bear out his main contentions very well, including crucially the heart of the theory (claim 4). Nevertheless, there is much of value in the evidence he presents for his theory. The methodology he employs comparing cases of consciousness with cases of no consciousness is a very natural idea. Pursuing that method, it might be of value to examine the neural processes of the Hassin subjects and compare them with neural processing involving conscious patterns. Perhaps there are differences that would offer clues about the nature of consciousness, much like hemispatial neglect patients give us reason to implicate attention. It is also very tempting to suppose attention is necessary for consciousness. The evidence Prinz offers for the claim is very strong and, to my mind, gives us good reason to believe attention will play an important role in the ultimately correct theory of consciousness. These are virtues of Prinz's view that anyone hoping to construct that correct theory would do well to consider.

Notes

1 One might have wondered by now why it is *availability* to working memory that matters rather than information being actually encoded there. Here are three quick reasons Prinz offers. First, we can sustain frontal lobe damage (where working memory is housed) without suffering from any deficits in awareness. Second, experience is too complex for working memory's capacities. We can experience, say fifteen items on a computer monitor, but the exact number would never be encoded in working memory. We could not report how many items were on the screen. Third, most famously, our ability to discriminate colors vastly outstrips our ability to recall them. We can separate out millions of different shades of colors, but give us a delay period and we could never recall which exact colors we had experienced. Working memory thus seems to encode higher level representations, not the intermediate level representations we have reason to believe are conscious.

2 Lack of space does not permit full examination of the point, but it is interesting to note that some of the evidence Prinz provides for claim (2)—especially for the necessity of attention for consciousness—can come under fire. Chapter 5 of Wu (2014) presents an interesting discussion of inattentional blindness. He argues that one could explain the findings of the experiments that support inattentional blindness along the lines of inattentional *agnosia*, or inattentional *apraxia*. In either case, one could be fully conscious of the unreported targets. On either interpretation, the evidence for the necessity of attention is undermined. The same explanation could work, I take it, for the attentional blink, undermining that bit of evidence, as well.

References

Awh, E., Vogel, E. and Oh, S. (2006) "Interactions between attention and working memory," *Neuroscience* 139: 201–208.

Baars, B. (1988) *A Cognitive Theory of Consciousness*, Cambridge: Cambridge University Press.

Baddeley, A. (2007) *Working Memory, Thought and Action*, Oxford: Oxford University Press.

Berti, A. Oxbury, S., Oxbury, J., Affanni, P., Umilta, C., and Orlandi, L. (1999) "Somatosensory extinction for meaningful objects in a patient with right hemispheric stroke," *Neuropsychologia* 37: 333–343.

Block, N. (2013) "The grain of vision and the grain of attention," *Thought: A Journal of Philosophy* 1: 170–184.

Crick, F., and Koch, K. (1990) "Towards a neurobiological theory of consciousness," *Seminars in Neuroscience* 2: 263–275.

Hassin, R., Bargh, J., Engell, A., and McCulloch, K. (2009) "Implicit working memory," *Consciousness and Cognition* 18: 665–678.

Kentridge, R., de-Wit, L., and Heywood, C. (2008) "What is attended in spatial attention?" *Journal of Consciousness Studies* 15: 105–111.

Kriegel, U. (2005) "Naturalizing subjective character," *Philosophy and Phenomenological Research* 71: 23–57.

Lamme, V. (2003) "Why visual attention and awareness are different," *Trends in Cognitive Sciences* 7: 12–18.

Lycan, W. (1996) *Consciousness and Experience*, Cambridge: Cambridge University Press.

Marr, D. (1982) *Vision*, San Francisco, CA: Freeman.

Mole, C. (2008) "Attention and consciousness," *Journal of Consciousness Studies* 15: 86–104.

Norman, L., Heywood, C., and Kentridge, R. (2013) "Object-based attention without awareness," *Psychological Science* 24: 836–843.

Pessiglione, M., Schmidt, L., Draganski, B., Kalisch, R., Lau, H., Dolan, R., and Frith, C. (2007) "How the brain translates money into force: a neuroimaging study of subliminal motivation," *Science* 316: 904–906.

Pitt, D. (2004) "The phenomenology of cognition, or, what is it like to think that P?" *Philosophy and Phenomenological Research* 69: 1–36.

Posner, M. (1980) "Orienting of attention," *Quarterly Journal of Experimental Psychology* 32: 3–25.

Prinz, J. (2011) "Is attention necessary and sufficient for consciousness?" in C. Mole, D. Smithies, and W. Wu (eds.) *Attention: Philosophical and Psychological Essays*, Oxford: Oxford University Press.
Prinz, J. (2012) *The Conscious Brain: How Attention Engenders Experience*, New York: Oxford University Press.
Rosenthal, D. (2005) *Consciousness and Mind*, New York: Oxford University Press.
Siegel, S. (2006) "Which properties are represented in perception?" in T. Gendler and J. Hawthorne (eds.) *Perceptual Experiences*, Oxford: Oxford University Press.
Simons, D., and Chabris, C. (1999) "Gorillas in our midst: sustained inattentional blindness for dynamic events," *Perception* 28: 1059–1074.
Soto, D., Mantyla, T., and Silvanto, J. (2011) "Working memory without consciousness," *Current Biology* 21: 912–913.
Soto-Faraco, S., Spence, C., Fairbank, K., Kingstone, A., Hillstrom, A., Shapiro, K. (2002) "A crossmodal attentional blink between vision and touch," *Psychonomic Bulletin and Review* 9: 731–738.
van Boxtel, J., Tsuchiya, N., and Koch, C. (2011) "Consciousness and attention: on sufficiency and necessity," *Frontiers in Psychology* 1–13.
Weiskrantz, L. (1986) *Blindsight: A Case Study and Implications*, Oxford: Oxford University Press.
Wu, W. (2013) "The conscious brain: how attention engenders experience, by Jesse Prinz," *Mind* 122: 1174–1180.
Wu, W. (2014) *Attention*, London: Routledge.

Related Topics

Consciousness and Attention
Representational Theories of Consciousness
The Global Workspace Theory
The Attention Schema Theory of Consciousness
Consciousness and Psychopathology

13
THE ATTENTION SCHEMA THEORY OF CONSCIOUSNESS

Michael S. Graziano

Over the past several years, my colleagues and I outlined a novel approach to understanding the brain basis of consciousness. That approach was eventually called the Attention Schema Theory (AST) (Graziano 2010; Graziano and Kastner 2011; Graziano 2013; Graziano 2014; Kelly at al. 2014; Webb and Graziano 2015; Webb, Kean, and Graziano 2016; Webb et al. 2016). The core concept is extremely simple. The brain not only uses the process of attention to focus its resources onto select signals, but it also constructs a description, or representation, of attention. The brain is a model builder – it builds models of items in the world that are useful to monitor and predict. Attention, being an important aspect of the self, is modeled by an attention schema.

The hypothesized attention schema is similar to the body schema. The brain constructs a rough internal model or simulation of the body, useful for monitoring, predicting, and controlling movement (Graziano and Botvinick 2002; Holmes and Spence 2004; Macaluso and Maravita 2010; Wolpert et al. 1995). Just so, the brain constructs a rough model of the process of attention – what it does, what its most basic properties are, and what its consequences are.

In the theory, the internal model of attention is a high-level, general description of attention. It lacks a description of the physical nuts and bolts that undergird attention, such as synapses, neurons, and competing electrochemical signals. The model incompletely and incorrectly describes the act of attending to X as, instead, an ethereal, subjective awareness of X. Because of the information in that internal model, and because the brain knows only the information available to it, people describe themselves as possessing awareness and have no way of knowing that this description is not literally accurate.

Although AST may seem quite different from other theories of consciousness, it is not necessarily a rival. Instead, I suggest it is compatible with many of the common, existing theories, and can add a crucial piece that fills a logical gap. Most theories of consciousness suffer from what might be called the metaphysical gap. The typical theory offers a physical mechanism, and then makes the assertion, "and then subjective awareness happens." The bridge between a physical mechanism and a metaphysical experience is left unexplained. In contrast, AST has no metaphysical gap, because it contains nothing metaphysical. Instead its explanation arrives at the step, "And then the machine claims that it has subjective awareness; and its internal computations consistently and incorrectly loop to the conclusion that this self-description is literally accurate." Explaining how a machine computes information is a matter of engineering, not a matter of metaphysics. Even if many of the steps have not yet been filled in, none present a fundamental, scientifically unapproachable mystery.

In this chapter, I summarize AST and then discuss some of the ways it might make contact with three specific approaches to consciousness: higher-order thought, social theories of consciousness, and integrated information. This chapter does not review the specific experimental evidence in support of AST, described in other places (Kelly at al. 2014; Webb and Graziano 2015; Webb, Kean, and Graziano 2016; Webb et al. 2016). Instead it summarizes the concepts underlying the theory.

1 Awareness

AST posits a specific kind of relationship between awareness and attention. Explaining the theory can be difficult, however, because those two key terms have an inconvenient diversity of definitions and connotations. The next few sections, therefore, focus on explaining what I mean by "awareness" and "attention."

When people say, "I am aware of X," whatever X may be – a touch on the skin, an emotion, a thought – they typically mean that X is an item within subjective experience, or in mind, at that moment in time. This is the sense in which I use the term in this chapter. To be aware is to have a subjective experience.

The term is also sometimes used in another sense: If someone asks, "Are you aware that paper is made from trees?" you might say, "Of course I am." You are aware in the sense that the information was available in your memory. But by the definition of the word that I use in this chapter, you were not aware of it while it was latent in your memory. You became aware of it – had a subjective experience of thinking it – when you were reminded of the fact, and then you stopped being aware of it again when it slipped back out of your present thought.

A third, less common use of the word, "objective awareness," is found in the scientific literature (Lau 2008). The essential concept is that if the information gets into a person's brain and is processed in a manner that is objectively measurable in the person's behavior, then the person is "objectively aware" of the information. In this sense, one could say, "My microwave is aware that it must stop cooking in thirty seconds." Objective awareness has no connotation of an internal, subjective experience.

In this chapter, when I use the term awareness, I do not mean objective awareness. I also do not mean something that is latent in memory. I am referring to the moment-by-moment, subjective experience. Some scholars refer to this property as "consciousness." Some, in an abundance of zeal, call it "conscious awareness." In this chapter, for simplicity, I will use the term "awareness." One can have awareness of a great range of items, from sensory events to abstract thoughts.

The purpose of AST is to explain how the human brain claims to have so quirky and seemingly magical a property as an awareness of some of its information content. This problem has sometimes been called the "hard problem" of consciousness (Chalmers 1996).

2 Attention

The term "attention" has even more meanings and interpretations than "awareness." Here, I will not be able to give a single definition, but will describe the general class of phenomenon that is relevant to AST.

First, I will clarify what I do *not* mean by attention. A typical colloquial use of the term conflates it with awareness. In that colloquial use, awareness is a graded property – you are more vividly aware of some items than others – and the items of which you are most aware at any moment are the items within your attention. This meaning is close to William James' now famous definition of attention (James 1890): "It is the taking possession of the mind, in clear and vivid form, of one out of what seems several simultaneously possible objects or trains of thought. Focalization, concentration of consciousness are of its essence." In this intuitive approach, attention is part of subjective experience. It is a subset of the conscious mind. If the content of awareness is the food spread at a banquet, attention refers specifically to the food on the plate directly in front of you. However, that is not what I mean by attention.

In this chapter, I use the term "attention" to refer to a mechanistic process in the brain. It can be defined independently of any subjective experience, awareness, or mind. Attention is the process by which some signals in the brain are boosted and therefore processed more deeply, often at the expense of other competing signals that are partially suppressed. Attention is a data-handling process. It can be measured in a great variety of ways, including through faster reaction times and greater accuracy in understanding, remembering, and responding to an attended item.

Many different kinds of attention have been described by psychologists (for review, see Nobre and Kastner 2014). Psychologists have distinguished between overt attention (turning the head and eyes toward a stimulus) and covert attention (focusing one's processing on a stimulus without looking directly at it). Psychologists have also distinguished between bottom-up, stimulus-driven attention (such as to a flashing light) and top-down, internally driven attention (such as looking for a friend in a crowd). Other categorizations include spatial attention (enhancing the sensory signals from a particular location in space) and object attention (enhancing the processing of one object over another, even if the two are superimposed on each other at the same spatial location). One can direct visual attention, auditory attention, tactile attention, and even multisensory attention. It has been pointed out that people can focus attention on specific abstract thoughts, beliefs, memories, or emotions, events that are generated in the brain and that are not directly stimulus-linked (Chun et al. 2011).

One of the most influential perspectives on attention is a neuroscientific account called the biased competition model (Desimone and Duncan 1995; Beck and Kastner 2009). In that account, the relevant signals – whether visual, auditory, or anything else – are in competition with each other. The competition is driven ultimately by synaptic inhibition among interconnected neurons. Because of this inhibition, when many signals are in competition, one will tend to rise in strength and suppress the others. That competition is unstable – shifting from one winner to another, from one moment to the next, depending on a variety of influences that may tip or bias the competition. The biasing influences include bottom-up, stimulus-driven factors (such as the brightness of a stimulus) and top-down, internally generated factors (such as a choice to search a particular location). The biased competition model provides a neuronal mechanism that explains how some signals become enhanced at the expense of others.

Attention is clearly a complex, multifaceted process. It is probably best described as many different processes occurring at many levels in the brain, applied to many information domains. Yet there is a common thread among these many types of attention. Throughout this chapter, when I use the term attention, I am referring to the selective enhancement of some signals in the brain over other competing signals, such that the winning signals are more deeply processed and have a greater impact on action, memory, and cognition.

3 Comparing Awareness to Attention

The relationship between awareness and attention has been discussed many times before (e.g. Koch and Tsuchiya 2007; Lamme 2004). A variety of theories of consciousness emphasize that relationship (e.g. Prinz 2012). In AST, one specific kind of relationship is hypothesized. To better explain that proposed relationship, in this section I list eight similarities and two differences between attention and awareness. The subsequent section will discuss why that list of similarities and differences suggests a specific kind of relationship between attention and awareness.

Similarity 1: Both involve a target. You attend *to* something. You are aware *of* something.

Similarity 2: Both involve a source. Attention is a data-handling operation performed by the processing elements in a brain. Awareness implies an "I," an agent who is aware.

Similarity 3: Both are selective. Only some of the available information is attended at any one time, or enters awareness at any one time.

Similarity 4: Both have an uneven, graded distribution, typically with a single focus. While attending mostly to A, the brain can spare some attention for B, C, and D. One can be most intently aware of A and a little aware of B, C, and D.

Similarity 5: Both imply deep processing. Attention is when an information processor devotes computing resources to a selected signal and thereby arrives at a deeper or more detailed encoding of it. Awareness implies an intelligence seizing on, being occupied by, knowing or experiencing something.

Similarity 6: Both imply an effect on behavior and memory. When the brain attends to something, the enhanced neural signals have a greater impact on behavioral output and memory. When the brain does not attend to something, the neural representation is weak and has relatively little impact on behavior or memory. Likewise, when you are aware of something, by implication you can choose to act on it and are able to remember it. When you are unaware of something, by implication, you probably fail to react to it or remember it.

Similarity 7: Both operate on similar domains of information. Although most studies of attention focus on vision, it is certainly not limited to vision. The same signal enhancement can be applied to signals arising in any of the five senses – to a thought, to an emotion, to a recalled memory, or to a plan to make a movement, for example. Just so, one can be aware of the same range of items. Generally, if you can in principle direct attention to it, then you can in principle be aware of it, and *vice versa*.

Similarity 8: Not only *can* attention and awareness apply to the same item, they almost always do. Here the relationship is complex. It is now well established that attention and awareness can be dissociated (Hsieh et al. 2011; Jiang et al. 2006; Kentridge et al. 2008; Koch and Tsuchiya 2007; Lambert 1988; Lambert et al. 1999; Lamme 2004; McCormick 1997; Norman et al. 2013; Tsushima et al. 2006; Webb, Kean, and Graziano 2016). A great many experiments have shown that people can pay attention to a visual stimulus, in the sense of processing it deeply, and yet at the same time have no subjective experience of the stimulus. They insist they cannot see it. This dissociation shows that attention and awareness are not the same. Awareness is not merely "what it feels like" to pay attention. Arguably, this point could be labeled "Difference 1" rather than "Similarity 8." However, the dissociation between attention and awareness should not be exaggerated. It is surprisingly difficult to separate the two. The dissociation seems to require either cases of brain damage, or visual stimuli that are extremely dim or masked by other stimuli, such that they are near the threshold of detection. Only in degraded conditions is it possible to reliably separate attention from awareness. Under most conditions, awareness and attention share the same target. What you attend to, you are usually aware of. This almost-but-not-quite registration between awareness and attention plays a prominent role in AST.

Awareness and attention are so similar that it is tempting to conclude that they are simply different ways of measuring the same thing, and that the occasional misalignment is caused by measurement noise. However, I find at least two crucial difference that are important in AST.

Difference 1: We know scientifically that attention is a process that includes many specific, physical details. Neurons, synpases, electrochemical signals, ions and ion channels in cell membranes, a dance of inhibitory and excitatory interactions, all participate in the selective enhancement of some signals over others. But awareness is different: we describe it as a thing that has no physical attributes. The awareness stuff itself isn't the neurons, the chemicals, or the signals – although we may think that awareness arises from those physical underpinnings. Awareness itself is not a physical thing. You cannot push on it and measure a reaction force. It is a substanceless, subjective feeling. In this sense, awareness, as most people conceptualize it, is metaphysical. Indeed, the gap between physical mechanism and metaphysical experience is exactly why awareness has been so hard to explain.

Difference 2: Attention is something the brain demonstrably *does* whereas awareness is something the brain *says that it has*. Unless you are a neuroscientist with a specific intellectual knowledge, you are never going to report the state of your actual, mechanistic attention. Nobody ever says, "Hey, you know what just happened? My visual neurons were processing both A and B, and a competition ensued in which lateral inhibition, combined with a biasing boost to stimulus A, caused..." People do not report directly on their mechanistic attention. They report on the state of their awareness. Even when people say, "I'm paying attention to that apple," they are typically using the word "attention" in a colloquial sense, not the mechanistic sense as I defined it above. In the colloquial sense of the word, people typically mean, "My conscious mind is focusing on that apple; it is uppermost in my awareness." Again, they are reporting on the state of their awareness, not on their mechanistic process of attention.

In summary, awareness and attention match point-for-point in many respects. They seem to have similar basic properties and dynamics. They are also tightly coupled in most circumstances, becoming dissociated from each other only at the threshold of sensory performance. But attention is a physically real, objectively measurable event in the brain, complete with mechanistic details, whereas awareness is knowledge that can be reported, and we report it as lacking physical substance or mechanistic details.

This pattern of similarities and differences suggests a possible relationship between attention and awareness: awareness is the brain's incomplete, detail-poor description of its own process of attention. To better grasp what I mean by this distinction between attention (a physically real item) and awareness (a useful if incomplete description of attention), consider the following examples. A gorilla is different from a written report about gorillas. The book may contain a lot of information, but is probably incomplete, perhaps even inaccurate in some details. An apple is different from the image of an apple projected onto your retina. An actual clay pipe is not the same as Magritte's famous oil painting of a pipe that he captioned, "This is not a pipe." The next section describes this hypothesized relationship in greater detail.

4 Analogy to the Body Schema

To better explain the possible relationship between attention and awareness, I will use the analogy of the body and the body schema (Graziano and Botvinick 2002; Holmes and Spence 2004; Macaluso and Maravita 2010; Wolpert et al. 1995). Imagine you close your eyes and tell me about your right arm – not what you know intellectually about arms in general, but what you can tell about your particular arm, at this particular moment, by introspection. What state is it in? How is it positioned? How is it moving? What is its size and shape? What is the structure inside? How many muscles do you have inside your arm and how are they attached to the bones? Can you describe the proteins that are cross-linking at this moment to stiffen the muscles?

You can answer some of those questions, but not all. General information about the shape and configuration of your arm is easy to get at, but you can't report the mechanistic details about your muscles and proteins. You may even report incorrect information about the exact position of your arm. The reason for your partial, approximate description is that you are not reporting on your actual arm. Your cognitive machinery has access to an internal model, a body schema, that provides incomplete, simplified information about the arm. You can report some of the information in that arm schema. Your cognition has access to a repository of information, an arm model, and the arm model is simplified and imperfect.

My point here is to emphasize the specific, quirky relationship between the actual arm and the arm schema. In AST, the relationship between attention and awareness is similar. Attention is an actual physical process in the brain, and awareness is the brain's constantly updated model of attention.

Suppose you tell me that you are aware of item X – let's say an apple placed in front of you. In AST, you make that claim of awareness because you have two closely related internal models. First, you have an internal model of the apple, which allows you to report the properties of the apple. You can tell me that it's round, it's red, it's at a specific location, and so on. But that by itself is not enough for awareness. Second, you have an internal model of attention, which allows you to report that you have a specific kind of mental relationship to the apple. When you describe your awareness of the apple – the mental possession, the focus, the non-physical subjective experience – according to AST, that information comes from your attention schema, a rough, detail-poor description of your process of attention.

5 Why an Attention Schema Might Cause a Brain to Insist That It Has Subjective Awareness – and Insist That It Isn't Just Insisting

Suppose you play me for a fool and tell me that you are literally an iguana. In order to make that claim, you must have access to that information. Something in your brain has constructed the information, "I am an iguana." Yet that information has a larger context. It is linked to a vast net of information to which you have cognitive access. That net of information includes much that you are not verbalizing to me, including the information, "I'm not really an iguana," "I made that up just to mess with him," "I'm a person," and so on. Moreover, that net of information is layered. Some of it is at a cognitive level, consisting of abstract propositions. Some of it is at a linguistic level. Much is at a deeper, sensory or perceptual level. You have a body schema that informs you of your personhood. Your visual system contains sensory information that also confirms your real identity. You have specific memories of your human past.

But, suppose I am cruelly able to manipulate the information in your brain, and I alter that vast set of information to render it consistent with the proposition that you are an iguana. Your body schema is aligned to the proposition. So is the sensory information in your visual system, and the information that makes up your memory and self-knowledge. I remove the specific information that says, "I made that up just to mess with him." I switch the information that says, "I am certain this is not true," to its opposite, "I'm certain it's true." Now how can you know that you are not an iguana? Your brain is captive to the information it contains. Tautologically, it knows only what it knows. You would no longer think of your iguana identity as hypothetical, or as mere information at an intellectual level. You would consider it a ground truth.

Now we can explain the widespread human conviction that we have an inner, subjective experience. In AST, the attention schema is a set of information that describes attention. It does not describe the object you are attending to – that would be a different schema. Instead it describes the act of attention itself. Higher cognition has a partial access to that set of information, and can verbally report some of its contents.

Suppose you are looking at an apple and I ask you, "Tell me about your awareness of the apple – not the properties of the apple, but the properties of the awareness itself. What is this awareness you have?" Your cognitive machinery, gaining access to the attention schema, reports on some of the information within it. You answer, "My mind has taken hold of the apple. That mental possession empowers me to know about the apple, to remember it for later, to act on it."

"Fair enough," I say, "but tell me about the physical properties of this awareness stuff." Now you're stuck. That internal model of attention lacks a description of any of the physical details of neurons, synapses, or competing signals. Your cognition, reporting on the information available to it, says, "The awareness itself has no physically describable attributes. It just is. It's a non-physical essence located inside me. In that sense, it's metaphysical. It's the inner, mental, experiential side of me."

The machine, based on an incomplete model of attention, claims to have a subjective experience.

I could push you further. I could say, "But you're just a machine accessing internal models. Of course, you're going to say all that, because that's the information contained in those internal models." Your cognition, searching the available internal models, finds no information that matches that description. Nothing in your internal models say, "This is all just information in a set of internal models." Instead, you reply, "What internal models? What information? What computation? No, simply, there's a me, there's an apple, and I have a subjective awareness of the apple. It's a ground truth. It simply exists."

This is a brain stuck in a loop, captive to the information available to it.

AST does not explain how the brain generates a subjective inner feeling. It explains how a brain *claims* to have a subjective inner feeling. In this theory, there is no awareness essence that arises from the functioning of neurons. Instead, in AST, the brain contains attention. Attention is a mechanistic, data-handling process. The brain also constructs an incomplete and somewhat inaccurate internal model, or description, of attention. On the basis of that internal model, the brain insists that it has subjective awareness – and insists that it is not just insisting. That general approach, in which awareness does not exist as such, and our claim to have awareness can be cast in terms of mechanistic information processing, is similar to the general approach proposed by Dennett (1991).

In AST, awareness is not merely an intellectual construct. It is an automatic, continuous, fundamental construct about the self, to which cognition and language have partial access.

6 Three Ways in Which the Theory Remains Incomplete

AST is underspecified in at least three major ways, briefly summarized in this section.

First, if the brain contains an attention schema, which of the many kinds of attention does it model? There are many overlapping mechanisms of attention, as noted in an earlier section of this chapter. These mechanisms operate at many levels, from the lowest sensory processing levels to the highest levels of cognition. If the brain has an attention schema, does it model only one type of attention? Many types? Are there many attention schemas, each modeling a different mix of attention mechanisms? In its current provisional form (Graziano 2013; Webb and Graziano 2015), the theory posits that a single attention schema models an amalgam of all levels of attention. In that view, the reality of attention is a complex and layered process, but the attention schema depicts it in a simplified manner as a single amorphous thing – an awareness.

A second way in which AST is not yet fully specified concerns the information content of the attention schema. It is extremely difficult to specify the details of an information set constructed in the brain. In the case of the body schema, for example, after a hundred years of study, researchers have only a vague understanding of the information contained within it (Graziano and Botvinick 2002; Holmes and Spence 2004; Macaluso and Maravita 2010; Wolpert et al. 1995). It contains information about the general shape and structure of the body, as well as information about the dynamics of body movement. In the case of the attention schema, if the brain is to construct an internal model of attention, what information would be useful to include? Perhaps basic information about the properties of attention – it has an object (the target of attention); it is generated by a subject (the agent who is attending); it is selective; it is graded; it implies a deep processing of the attended item; and it has specific, predictable consequences on behavior and memory. Perhaps the attention schema also includes some dynamic information about how attention tends to move from point to point and how it is affected by different circumstances. The fact is, at this point, the theory provides very little indication of the contents of the attention schema. Only future work will be able to fill in those details.

The third way in which AST is underspecified concerns the functions of an attention schema. Why would such a thing evolve? A range of adaptive functions are possible. For example, an attention schema could in principle be used for controlling one's own attention (Webb and Graziano 2015; Webb, Kean, and Graziano 2016). By analogy, the brain constructs the internal model of the arm to help control arm movements (e.g. Haith and Krakauer 2013; Scheidt et al. 2005; Wolpert et al. 1995). It is a basic principle of control engineering (Camacho and Bordons Alba 2004).

A possible additional function of an attention schema is to model the attentional states of other people (Kelly et al. 2014; Pesquita et al. 2016). The more a person attends to X, the more likely that person is to react to X. Modeling attention is therefore a good way to predict behavior. By attributing awareness to yourself and to other people, you are in effect modeling the attentional states of interacting social agents. You gain some ability to predict everyone's behavior including your own. In this way, an attention schema could be fundamental to social cognition.

7 Higher-Order Thought

The higher-order thought theory, elaborated by Rosenthal, is currently one of the most influential theories of consciousness (Lau and Rosenthal 2011; Rosenthal 2005; Gennaro 1996, 2012). I will briefly summarize some of its main points and note its possible connection to AST.

Consider how one becomes aware of a visual stimulus such as an apple. In the higher-order thought theory, the visual system constructs a sensory representation of the apple. Higher-order systems in the brain receive that information and re-represent the apple. That higher-order re-representation contains the extra information that causes us to report not only the presence of the apple, but also a subjective experience.

The higher-order thought theory is a close cousin of AST because of its focus on representation and information. The theory, however, focuses on the representation of the item (such as the apple in the example above) that is within awareness, in contrast to AST which focuses on the representation of the process of attention.

Higher-order thought theory is surprisingly compatible with AST. In the combination theory, the brain constructs a representation of the apple. It also constructs a representation of attention – the attention schema. A higher-order re-representation combines the two. That higher-order representation describes an apple to which one's subjective awareness is attached. Given that higher-order representation, the system can make two claims. First, it can report the properties of the apple. Second, it can report a subjective awareness associated with the apple. By adding an attention schema to the mix, we add the necessary information for the machine to report awareness – otherwise, the machine would have no basis for even knowing what awareness is or concluding that it has any. In this perspective, AST is not a rival to the higher-order thought theory. Instead, the two approaches synergize and gain from each other.

8 Social Attribution of Awareness

Recently, Prinz (2017) outlined a view of consciousness termed import theory. In that perspective, humans first develop the ability to model the mind states of others and then turn that ability inward, attributing similar mind states to themselves. This explanation of conscious mind states, invoking social cognition, has been proposed before many times in different forms, including in the earliest descriptions of AST (Graziano 2013), but Prinz presents the view in a particularly clear and compelling manner.

One of the strengths of the import theory is that it covers a broad range of mind states, all of which compose what most people colloquially think of as consciousness. You can attribute emotions, thoughts, goals, desires, beliefs, and intentions to other people. Just so, you can attribute the same range of mind states to yourself. The theory therefore addresses a rich world of consciousness that is often ignored in discussions of sensory awareness.

However, the theory has the same metaphysical gap as so many other theories contain. It addresses the content of awareness, but it does not address how we get to be aware of it. You may attribute an emotional state to another person, and you may attribute the same emotional state to yourself. But why do you claim to have a subjective *experience* of that emotion? It is not enough for the brain, computer-like, to build the construct, "I am happy." Humans also report a subjective experience of the happiness, just as they report a subjective experience of many other items. Import theory, by itself, does not explain the subjective experience. This point is not meant as a criticism of the theory. It is a valuable theory – but the specific question of awareness may lie outside its domain.

AST may be able to fill that gap. In AST, when we attribute awareness to another person, we are modeling that person's state of attention. When we attribute awareness to ourselves, we are modeling our own state of attention. By adding an attention schema to the system, we add information that allows the brain to know what awareness is in the first place and to claim that it has some, or that someone else has some. Note that, strictly speaking, AST does not explain how people *have* subjective awareness. It explains how people *insist* that they have it and insist that it's real and that they're not just insisting.

I do not mean to take a strong stand here on import theory, for or against. It is possible that people develop the ability to model the mind states of others first and then import that to the self. It is also possible that people develop the capacity of self-modeling first and then export it outward to others. Maybe both are true. Only more data will be able to untangle those possibilities. My point here is that, whichever perspective one prefers, AST makes a useful addition.

A skeptical colleague might wonder, "Why focus on attention, when the brain contains so many different processes? Decisions, emotions, moods, beliefs – all of these are a part of consciousness. Yes, surely the brain constructs a model of attention, but doesn't it also construct models of all its other internal processes?" Indeed, the brain probably does construct models of many internal processes, and all of those models are worthy of scientific study. The reason AST highlights attention is that an attention schema answers one crucial, focused question that was thought to be unanswerable. It explains how people claim to have a subjective experience of anything at all. Because of the narrow specificity of AST, it can be added as a useful component to a great range of other theories.

9 Networked Information

Many theories and speculations about awareness share an emphasis on the widespread networking or linking of information around the brain. Two prominent examples are the Integrated Information Theory (Tononi 2008) and the Global Workspace Theory (Baars 1988; Dehaene 2014).

The essence of the Integrated Information Theory is that if information is integrated to a sufficient extent, which may be mathematically definable, then subjective awareness of that information is present (Tononi 2008). Awareness is what integrated information feels like. The Global Workspace Theory has at least some conceptual similarities (Baars 1988; Dehaene 2014). You become subjectively aware of a visual stimulus, such as an apple, because the representation of the apple in the visual system is globally broadcasted and accessible to many systems around the brain. Again, the widespread sharing of information around the brain results in awareness. Many other researchers have also noted the possible relationship between awareness and the binding, integration, or sharing of information around the brain (e.g. Crick and Koch 1990; Damasio 1990; Engel and Singer 2001; Lamme 2006).

Of all the common theories of consciousness in the cognitive psychology literature, this class of theory most obviously suffers from a metaphysical gap. To explain an awareness of item X, these theories focus on the information about X and how that information is networked or integrated. The awareness is treated as an adjunct, or a symptom, or a product, of the information about X. But once you have information that is integrated, or that is globally broadcasted, or that is linked or bound across different domains, why would it take the next step and enter a state of subjective awareness? Why is it not just a pile of integrated information without the subjective experience? What is the actual awareness stuff and how does it emerge from that state of integration?

Another way to put the question is this: Suppose you have a computing machine that contains information about an apple. Suppose that information is highly-integrated – color, shape, size, texture, smell, taste, identity, all cross-associated and integrated in a massive brain-wide representation. I can understand how a machine like that might be able to report the properties of the apple, but why would I expect the machine to add to its report, "And by the way, I have a subjective, internal *experience* of those apple properties"? What gave the machine the informational basis to report a subjective experience?

The metaphysical gap has stood in the way of these theories that depend on networked information. And yet the conundrum has a simple solution. Add AST to the integrated information account, and you have a working theory of awareness. If part of the information that is integrated globally around the brain consists of information about awareness, about what awareness is, what its properties are, about how you yourself are aware and what in specific you are aware of – if the machine contains an attention schema – then it is equipped to talk about awareness in all its subtle properties and to make the claim that it has those properties. If the machine lacks information about awareness, then logically it cannot claim to have any.

Note that not only is AST a useful addition to the integrated information perspective, but the relationship works both ways. AST depends on integrated information. It does not work as a theory without the widespread networking of information around the brain. In AST, to be aware of an apple, it is not enough to construct an attention schema. The attention schema models the properties of attention itself. The brain must also construct an internal model of the apple and an internal model of the self as a specific agent. All three must be integrated across widely divergent brain areas, building a larger internal model. That overarching, integrated internal model contains the information: there is a you as an agent with a set of specific properties, there is an apple with its own set of specific properties, and at this moment the you-as-agent has a subjective awareness of the apple and its properties. Only with that highly networked information is the brain equipped to claim, "I am aware of the apple." Without the widespread integration of information around the brain, that overarching internal model is impossible, and we would not claim to possess awareness. Thus, even though AST and the integrated information approach rest on fundamentally different philosophical perspectives, they have a peculiarly close, symbiotic relationship.

10 The Allure of Introspection

Before Newton's publication on light (1671), the physical nature of color was not understood. White light was assumed to be pure and colored light to be contaminated. One could say the hard problem of color was this: how does white light become scrubbed clean of contaminants? That hard problem, alas, had no answer because it was based on a physically incoherent model of color and light. The model was not merely a mistaken scientific theory. It was the result of millions of years of evolution working on the primate visual system, shaping an efficient and simplified internal model of reflectance spectrum. Finally, after Newton's insights, it became possible to understand two crucial items. First, white light is actually a mixture of all colors. Second, the model we all automatically construct in our visual systems is simplified and in some respects wrong.

The same issues, I suggest, apply to the study of awareness. Our cognitive machinery gains partial access to deeper internal models, including an attention schema. On the basis of that information, people assert with absolute confidence that they have physically incoherent, magicalist properties. Gradually, as science has made progress over hundreds of years, some of the more obviously irrational assertions have fallen away. Most scientists accept there is no such thing as a ghost. A mysterious energy does not emanate from the eyes to affect other objects and people. Most neuroscientists reject the dualist notion of mind and brain, the notion most famously associated with Descartes (1641), in which the machine of the brain is directed by the metaphysical substance of the mind.

Some of the assertions of magic, however, remain with us in subtle ways. Almost all theories of consciousness rest on a fundamental assumption: we have an inner subjective experience. The experience is not itself a physical substance. It cannot be weighed, poked, or directly measured. You cannot push on it and measure a reaction force. Instead it is a non-physical, side-product – the "what-it-feels-like" when certain processes occur in the brain. The challenge is to explain how the functioning of the brain results in that private feeling.

This perspective has framed the entire field of consciousness studies from the beginning. Yet, I argue it is as futile as the attempt to explain how white light becomes purified of contaminants. It is predicated on false assumptions. As long as we dedicate ourselves to explaining how the brain produces subjective experience, a property we know about only by our cognition accessing our internal models, we will never find the answer. As soon as we step away from the incorrect assumptions, and realize that our evolutionarily built-in models are not literally accurate, we will see that the answer to the question of consciousness is already here.

The heart of AST is that the brain is a machine: it processes information. When we claim to have a subjective experience, and swear on it, and vociferously insist that it isn't just a claim or a conclusion – it's real, dammit – this output occurs because something in the brain computed that set of information. It is a self-description. The self-model is unlikely to be entirely accurate or even physically coherent. As in the case of color, the brain's models tend to be efficient, simplified, useful, and not very accurate on those dimensions where accuracy would serve no clear behavioral advantage. People do not have a magic internal feeling. We have information that causes us to insist that we have the magic. And explaining how a machine computes and handles information is well within the domain of science.

References

Baars, B. J. (1988) *A Cognitive Theory of Consciousness*, New York: Cambridge University Press.

Beck, D. M. and Kastner, S. (2009) "Top-down and bottom-up mechanisms in biasing competition in the human brain," *Vision Research* 49: 1154–1165.

Camacho, E. F. and Bordons Alba, C. (2004) *Model Predictive Control*, New York: Springer.

Chalmers, D. (1996) *The Conscious Mind*, New York: Oxford University Press.

Chun, M. M., Golomb, J. D. and Turk-Browne, N. B. (2011) "A taxonomy of external and internal attention," *Annual Review of Psychology* 62: 73–101.

Crick, F. and Koch, C. (1990) "Toward a neurobiological theory of consciousness," *Seminars in the Neurosciences* 2: 263–275.

Damasio, A. R. (1990) "Synchronous activation in multiple cortical regions: a mechanism for recall," *Seminars in the Neurosciences* 2: 287-296.

Dehaene, S. (2014) *Consciousness and the Brain*, New York: Viking.

Dennett, D. C. (1991) *Consciousness Explained*, Boston: Little, Brown, and Co.

Descartes, R. (1641) "Meditations on first philosophy," in J. Cottingham, R. Stoothoff, and D. Murdock (trans.) *The Philosophical Writings of Rene Descartes*, Cambridge: Cambridge University Press.

Desimone, R. and Duncan, J. (1995) "Neural mechanisms of selective visual attention," *Annual Review of Neuroscience* 18: 193–222.

Engel, A. K. and Singer, W. (2001) "Temporal binding and the neural correlates of sensory awareness," *Trends in Cognitive Sciences* 5: 16–25.

Gennaro, R. (1996) *Consciousness and Self-Consciousness: A Defense of the Higher-Order Thought Theory of Consciousness*, Philadelphia, PA: John Benjamins Publishing.

Gennaro, R. (2012) *The Consciousness Paradox: Consciousness, Concepts, and Higher-Order Thoughts*, Cambridge, MA: The MIT Press.

Graziano, M. S. A. (2010) *God, Soul, Mind, Brain: A Neuroscientist's Reflections on the Spirit World*, Fredonia: Leapfrog Press.

Graziano, M. S. A. (2013) *Consciousness and the Social Brain*, New York: Oxford University Press.

Graziano, M. S. A. (2014) "Speculations on the evolution of awareness," *Journal of Cognitive Neuroscience* 26: 1300–1304.

Graziano, M. S. A. and Botvinick, M. M. (2002) "How the brain represents the body: insights from neurophysiology and psychology," in W. Prinz and B. Hommel (eds.) *Common Mechanisms in Perception and Action: Attention and Performance XIX*, Oxford: Oxford University Press.

Graziano, M. S. A. and Kastner, S. (2011) "Human consciousness and its relationship to social neuroscience: a novel hypothesis," *Cognitive Neuroscience* 2: 98–113.

Haith A. M. and Krakauer, J. W. (2013) "Model-based and model-free mechanisms of human motor learning," in M. Richardson, M. Riley, and K. Shockley (eds.) *Progress in Motor Control: Advances in Experimental Medicine and Biology*, Vol. 782, New York: Springer.

Holmes, N. and Spence, C. (2004) "The body schema and the multisensory representation(s) of personal space," *Cognitive Processing* 5: 94–105.

Hsieh, P., Colas, J. T. and Kanwisher, N. (2011) "Unconscious pop-out: attentional capture by unseen feature singletons only when top-down attention is available," *Psychological Science* 22: 1220–1226.

James, W. (1890) *Principles of Psychology*, New York: Henry Holt and Company

Jiang, Y., Costello, P., Fang, F., Huang, M. and He, S. (2006) "A gender- and sexual orientation-dependent spatial attentional effect of invisible images," *Proceedings of the National Academy of Sciences U. S. A.* 103: 17048–17052.

Kelly, Y. T., Webb, T. W., Meier, J. D., Arcaro, M. J. and Graziano, M. S. A. (2014) "Attributing awareness to oneself and to others," *Proceedings of the National Academy of Sciences U. S. A.* 111: 5012–5017.

Kentridge, R. W., Nijboer, T. C. and Heywood, C. A. (2008) "Attended but unseen: visual attention is not sufficient for visual awareness," *Neuropsychologia* 46: 864–869.

Koch, C. and Tsuchiya, N. (2007) "Attention and consciousness: two distinct brain processes," *Trends in Cognitive Sciences* 11: 16–22.

Lambert, A. J., Beard, C. T. and Thompson, R. J. (1988) "Selective attention, visual laterality, awareness and perceiving the meaning of parafoveally presented words," *Quarterly Journal of Experimental Psychology: Human Experimental Psychology* 40A: 615–652.

Lambert, A., Naikar, N., McLachlan, K. and Aitken, V. (1999) "A new component of visual orienting: Implicit effects of peripheral information and subthreshold cues on covert attention," *Journal of Experimental Psychology: Human Perception and Performance* 25: 321–340.

Lamme, V. A. (2004) "Separate neural definitions of visual consciousness and visual attention; a case for phenomenal awareness," *Neural Networks* 17: 861–872.

Lamme, V. A. (2006) "Towards a true neural stance on consciousness," *Trends in Cognitive Sciences* 10: 494–501.

Lau, H. (2008) "Are we studying consciousness yet?" in L. Weiskrantz and M. Davies (eds.) *Frontiers of Consciousness: Chichele Lectures*, Oxford: Oxford University Press.

Lau, H. and Rosenthal, D. (2011) "Empirical support for higher-order theories of consciousness," *Trends in Cognitive Sciences* 15: 365–373.

Macaluso, E. and Maravita, A. (2010) "The representation of space near the body through touch and vision," *Neuropsychologia* 48: 782–795.

McCormick, P. A. (1997) "Orienting attention without awareness," *Journal of Experimental Psychology: Human Perception and Performance* 23: 168–180.

Newton, I. A. (1671) "Letter of Mr. Isaac Newton, Professor of the Mathematicks in the University of Cambridge; Containing His New Theory about Light and Colors: Sent by the Author to the Publisher from Cambridge, Febr. 6. 1671/72; In Order to be Communicated to the Royal Society," *Philosophical Transactions Royal Society* 6: 3075–3087.

Norbre, K. and Kaster, S. (2014) *The Oxford Handbook of Attention*, New York: Oxford Univeristy Press.

Norman, L. J., Heywood, C. A. and Kentridge, R. W. (2013) "Object-based attention without awareness," *Psychological Science* 24: 836–843.

Prinz, J. J. (2012) *The Conscious Brain*, New York: Oxford University Press.

Prinz, W. (2017) "Modeling self on others: an import theory of subjectivity and selfhood," *Consciousness and Cognition* 49: 347–362.

Pesquita, A., Chapman, C. S. and Enns, J. T. (2016) "Humans are sensitive to attention control when predicting others' actions," *Proceedings of the National Academia of Science U. S. A.* 113: 8669–8674.

Rosenthal, D. (2005) *Consciousness and Mind*, New York: Oxford University Press.

Scheidt, R. A., Conditt, M. A., Secco, E. L. and Mussa-Ivaldi, F. A. (2005) "Interaction of visual and proprioceptive feedback during adaptation of human reaching movements," *Journal of Neurophysiology* 93: 3200–3213.

Tononi, G. (2008) "Consciousness as integrated information: a provisional manifesto," *Biological Bulletin* 215: 216–242.

Tsushima, Y., Sasaki, Y. and Watanabe, T. (2006) "Greater disruption due to failure of inhibitory control on an ambiguous distractor," *Science* 314: 1786–1788.

Webb, T. W., Kean, H. H. and Graziano, M. S. A. (2016) "Effects of awareness on the control of attention," *Journal of Cognitive Neuroscience* 28: 842–851.

Webb, T. W., and Graziano, M. S. A. (2015) "The attention schema theory: a mechanistic account of subjective awareness," *Frontiers in Psychology* 6, article 500, doi:10.3389/fpsyg.2015.00500.

Webb, T. W., Igelström, K., Schurger, A. and Graziano, M. S. A. (2016) "Cortical networks involved in visual awareness independently of visual attention," *Proceedings of the National Academy of Sciences U. S. A.* 113: 13923–13928.

Wolpert, D. M., Ghahramani, Z. and Jordan, M. I. (1995) "An internal model for sensorimotor integration," *Science* 269: 1880–1882.

Related Topics

Consciousness and Attention
The Intermediate Level Theory of Consciousness
Representational Theories of Consciousness
The Global Workspace Theory
The Information Integration Theory
The Neural Correlates of Consciousness

14

BIOLOGICAL NATURALISM AND BIOLOGICAL REALISM

Antti Revonsuo

> Consciousness is a real, natural biological phenomenon, produced by and realized in higher-level neurophysiological processes going on inside the brain.

This thesis is the shared core of two closely related theories of consciousness, *Biological Naturalism* (BN) and *Biological Realism* (BR). Biological Naturalism has been formulated and defended by John Searle ever since the 1990s in numerous writings, especially in his seminal book *The Rediscovery of the Mind* (1992). Biological Realism was put forward by the present author in Revonsuo (2006) *Inner Presence: Consciousness as a Biological Phenomenon.*

Although the two biological approaches to consciousness share a lot of metaphysical ground, there are also significant differences between them. Whereas BN is presented in the context of the philosophy of mind, BR is put forward in the context of the modern empirical science of consciousness as a proposal for a metaphysical basis for that science.

In this chapter I will first summarize the main principles of Biological Naturalism, followed by a summary of Biological Realism. After that, I will analyze some of their similarities and differences. In the final sections, I will contrast the biological approach represented by BN and BR with another currently influential approach: information theories of consciousness, especially the Information Integration Theory.

1 Biological Naturalism

In the study of consciousness, the role of philosophy is, according to Searle, to get us to the point at which we can start to have systematic scientific knowledge about consciousness (Searle 1998). But the history of the philosophical mind-body problem, with its traditional categories and conflicts such as "dualism" vs "materialism," is unhelpful in this endeavour, because it involves a series of philosophical confusions. Consequently, the study of consciousness is an area of science in which scientific progress is blocked by philosophical error, says Searle (1998).

Searle presents BN as an approach to consciousness that, to begin with, invites us to forget about the history of the mind-body problem in philosophy. Instead, we should go back to square one, to build an approach that respects all the facts about consciousness but avoids the traditional philosophical categories. Thereby we will have avoided also the typical philosophical confusions and pitfalls about consciousness that the traditional philosophy of mind suffers from.

BN starts from the facts that we know beyond any reasonable doubts. Searle says that BN is just "scientifically sophisticated common sense" applied to consciousness (Searle 2007). So, what are the facts about consciousness? Searle summarizes BN as a set of four theses (Searle 2004: 113):

1 *Realism:* Consciousness is a real phenomenon in the real world. Consciousness really exists, in its own right, in the physical world. Its existence cannot be denied. Any theory that gets rid of consciousness by eliminating or reducing it to something else, rejects this realist thesis about consciousness and therefore rejects the most undeniable fact we know about consciousness.
2 *Neurophysiological causation and sufficiency:* Consciousness is entirely caused by lower-level neurophysiological processes in the brain. The causally sufficient conditions for any conscious phenomenon are in the brain.
3 *Neurophysiological realization:* Consciousness is a higher-level feature of the brain system. It exists at a level higher than the level of neurons or synapses; individual neurons cannot be conscious.
4 *Mental causation:* Consciousness has causal powers. Our voluntary behaviors are causally driven by our conscious mental states.

In addition to these four fundamental theses, BN needs a definition and description of "consciousness." Searle's definition of consciousness says that *consciousness consists of one's states of awareness or sentience or feeling* (Searle 2007). The definition also points to the conditions under which the phenomenon is to be found in the world: "Conscious states are those states of awareness, sentience, or feeling that begin in the morning when you wake from a dreamless sleep and continue throughout the day until you fall asleep or otherwise become 'unconscious' (Dreams are also a form of consciousness)" (Searle 2007: 326).

The three essential features of consciousness are (1) *qualitative character* or 'what-it-feels-like,' (2) *ontological subjectivity*, and (3) global *unity*. All conscious states are qualitative in the sense that having them *feels like something*. Conscious states cannot exist without their qualitative character. All conscious states are *subjective* in the sense that they exist only when experienced by a human or animal subject, some "I" whose conscious experiences they are. The subjectivity of consciousness is *ontological*, meaning that it is *a special mode of existence* that only conscious phenomena possess. The ontological subjectivity of consciousness prevents the ontological reduction of consciousness to any purely objective phenomena, such as neuronal firings. All momentary conscious phenomena are parts of *a single unified conscious field*. Although our consciousness involves many kinds of qualitative experiences at any given moment – say the text I see on the computer screen, the music I hear in the background, the fresh breeze of cold air I feel coming from the window, and the softness of the carpet under my bare feet – all these qualitatively different contents of consciousness are experienced as happening simultaneously within one unified field by one subject of experience. In a nutshell, for Searle, consciousness is *unified qualitative subjectivity*.

2 The Explanation of Consciousness According to Biological Naturalism

One and the same physical system can have different levels of description that are not competing or distinct; they are simply different levels within a single unified causal system. In this completely non-mysterious way, also the brain has many different levels of description. Higher-level properties of a system can be causally explained by the lower-level or the micro-properties of the same system. Conscious states are thus *causally reducible* to neurobiological processes. Searle says that *"they have no life of their own; causally speaking, they are not something "over and above" neurobiological processes"* (Searle 2004: 113). It is, claims Searle, a fact established by an overwhelming amount of evidence that all of our conscious states are caused by brain processes. Recognizing this fact, and the other theses of BN, amounts to a solution (or a dissolution) of the traditional mind-body problem in philosophy, claims Searle. It is the duty of the biological sciences and the neurosciences to take over and figure out exactly *how* the causal mechanisms between brain processes and consciousness work. The philosophers should just get out of the way.

3 Critical Remarks about Biological Naturalism

It is however not immediately obvious whether the core ideas of BN amount to an internally coherent explanation. How is it possible to simultaneously hold the two claims: "consciousness is causally reducible to the brain" and "consciousness is ontologically subjective; therefore, consciousness is not ontologically reducible to the brain"? Searle explicitly does hold them: "You can do a causal reduction of consciousness to its neuronal substrate, but that reduction does not lead to an ontological reduction because consciousness has a first-person ontology" (Searle 2004: 123).

Searle tries to explain the difference between causal and ontological reduction. By causal reduction Searle means that, causally speaking, consciousness *owes its existence* to the underlying lower-level brain processes that causally bring it about. Although neuroscience doesn't have it yet figured out, there is a full causal, neurophysiological explanation as to where, when, and how conscious states are causally brought about in the brain; so far, we just lack the details of that explanation.

Searle appears to accept the two main components of the supervenience relationship between consciousness and the brain (although he is somewhat reluctant to use the concept of supervenience): there can be no difference in conscious states without a corresponding difference in the underlying brain states (the covariance principle), and the conscious states owe their existence to the underlying brain states (the principle of ontological dependency). He also accepts that BN represents emergentism:

> If we define emergent properties of a system of elements as properties which can be explained by the behavior of the individual elements, but which are not properties of elements construed individually, then it is a trivial consequence of my view that mental properties are emergent properties of neurophysiological systems.
>
> *(Searle 1987: 228)*

What, then, is the "ontological subjectivity" or "first-person ontology" that escapes ontological reduction, and why does it escape it? Searle explains it along the following lines. Objective physical phenomena that can be ontologically reduced to their causal base have two types of properties, surface properties (how they *appear* to human observers) and underlying causal properties (how they *"really" are*, independent of human observation). Causal reduction leads to ontological reduction in the case of third-person objective phenomena (such as the physical explanation of visible light) because we get rid of the surface properties (how visible light looks like to us in our conscious perception), as they are not really properties of the physical phenomenon at all, but of our observations, mere appearances. But in the case of consciousness, *consciousness is identical to its appearance*; its appearance is identical with the subjective ontology of consciousness. The appearance of consciousness is its essential feature; and a thing cannot be deprived of its essential features while still preserving the same ontology. Thus, the appearance of consciousness cannot be carved off, and still preserve some underlying "real" consciousness. Consciousness ceases to be real consciousness if its essential features are carved off. The reality of consciousness is its appearance, and therefore the way we practise ontological reduction in other cases just does not work in the case of consciousness.

At first glance, the contrast between first-person consciousness and its third-person causal basis would thus seem to imply an ontological dualism of some sort. Yet, Searle argues that the ontological irreducibility of consciousness has no deep metaphysical implications. Rather, he claims it is a trivial consequence of how we define "reduction" and of what we find the most interesting features of consciousness. He says that the irreducibility of consciousness does not reveal a deep metaphysical asymmetry as to how conscious experiences relate to their causes.

Searle at one point even goes so far as to admit that (Searle 2004: 120–121) if we wanted, we *could* carve off the surface properties of consciousness and redefine it in terms of its underlying neural causes, thereby conducting an ontological reduction. But the price we would pay for that is we would lose the vocabulary to talk about the surface properties of consciousness, and subsequently, we would lose the purpose of having any concept of consciousness at all. We would still need *some* kind of vocabulary to talk about the surface features of consciousness, because precisely those features of consciousness are the ones that *we care most about,* and which are *of most interest to us,* says Searle (2004).

Searle's attempt to avoid the looming metaphysical asymmetry is not entirely convincing. At this crucial point, Searle is trying to have his cake and eat it too. First, he is trying to *have* his cake: According to his own account, consciousness is real and essentially consists of unified qualitative subjectivity. Its first-person ontology is its essential feature; its very mode of existence. That's why it is ontologically irreducible to any objective physical phenomenon. But then he suddenly also tries to *eat* his cake by denying that the first-person ontology is in any way metaphysically asymmetrical with the third-person ontology of the causal basis of consciousness. There is no ontological breach in the world between brain and consciousness; it is just our definitional practises and the trivial pragmatics of reduction that make it awkward to ontologically reduce consciousness to brain processes, but we *could* do it at least in principle (Searle 2004: 120–121). Going through with the ontological reduction of consciousness just would not serve our *interests* very well, because then we would have difficulties in finding words to describe the features of consciousness that most *interest us*, features that we most *care* about.

At this point, Searle's line of argument loses its credibility. Suddenly, the essential, defining ontological features are treated as accidental features of consciousness that are important only relative to our interests and to our descriptive vocabulary. This move implies that after all, consciousness as unified qualitative subjectivity was not the definitive characterization of a phenomenon ontologically and metaphysically different from third-person physical phenomena. Instead, when we describe conscious states in subjective and qualitative terms (their surface features), we just happen to pick up some accidental features of consciousness that *happen to interest us* and that *we happen to care about.*

By this line of argumentation, Searle seems to paint himself into a corner that he has often warned others about: He warns against confusing the *intrinsic* features of the world with its *observer-relative* features, or descriptions that are merely relative to someone's interests. Now he seems to be guilty of exactly that mistake. Or rather, he first treats unified qualitative subjectivity as intrinsic features of consciousness that define an ontologically and metaphysically distinct phenomenon irreducible to a third-person basis, but then, when ontological subjectivity starts to sound like a metaphysical breach in the brain between consciousness and neurophysiology, Searle treats it (or its surface features, which also happen to be its essential features) as an observer- or interest-relative description whose surface features we just happen to be interested in or to care about, but those features are metaphysically nothing special and imply no ontological asymmetry.

Searle leaves room for this kind of fiddling, because his account is ambiguous about the ontological status of consciousness and about the ontological status of higher levels of neurophysiology. If a higher macro-level has no ontological status different from its micro-level basis, then the higher-level merely constitutes *a level of description* for our practical purposes, *not* an *ontological* level of organization in the world itself. Typically, when Searle talk about levels, he only talks in terms of "levels of description":

> The fact that the brain has different levels of description is no more mysterious than that any other physical system has different levels of description.
>
> *(Searle 2007: 328)*

But to have a level of description in our vocabulary when talking about the brain in neuroscientific terms does *not* entail that there is a corresponding *ontological* level in reality. Perhaps the levels of description just happen to be convenient tools for our scientific practises. At any rate, Searle leaves it open whether the levels he talks about are ontologically real, existing out there in the physical world, independent of our descriptions, or are they only levels of description that serve human purposes, but no such levels "really" exist in the physical reality of the brain.

His characterization of consciousness suffers from a similar ambiguity about ontological status. On the one hand, it seems obvious that he is committed to the view that consciousness is ontologically and metaphysically different from any third-person, objective, physical phenomena. But on the other hand, when he needs to explain *why* the ontological reduction of consciousness is impossible, he ends up denying that consciousness is ontologically or metaphysically in any way special. Should we carry out the ontological reduction of consciousness to its neurophysiological basis, all we will lose is a convenient vocabulary, a level of description. But this betrays Searle's own definitions of the essential ontological features of consciousness. Unified qualitative subjectivity defines the fundamental *ontology* of consciousness, *not just a level of description*.

BN thus fails to offer a coherent account of how the first-person ontology of consciousness is related to the third-person ontology of neurophysiology. Searle suggests that BN solves (or dissolves) the philosophical mind-body problem, but this turns out to be a mere promissory note. Significant philosophical problems remain. Searle however never directly addresses them.

The famous problems known as the Explanatory Gap (Levine 1983) and the Hard Problem (Chalmers 1996) are precisely the types of inescapable philosophical problems that any biological (or emergent physicalist) theory of consciousness must face. They arise *after* we commit ourselves to something like BN, and they arise precisely at the interface between the first-person ontology of consciousness and the third-person ontology of neurophysiology.

It is too early to hand the explanation of consciousness from philosophers to neuroscientists. The interaction between philosophy and neuroscience has in fact been going the other way around recently. Some leading neuroscientists who used to be firmly committed to something like BN in the 1990s, such as Giulio Tononi (then working with Gerald Edelman) and Christof Koch (then working with Francis Crick), have recently turned away from biologically based metaphysics for consciousness (Koch 2012). Because the Explanatory Gap and the Hard Problem remain unsolved and no imaginable solutions are in sight, neuroscientists may not find biological approaches such as BN convincing any longer. The failure to directly address these problems significantly weakens the case for BN.

4 Biological Realism

I have presented Biological Realism in its full formulation in Revonsuo (2006) (and brief summaries in Revonsuo 2010, 2015). The basic thesis of BR is that "Consciousness is a real, natural, biological phenomenon." As it is "real," it exists in physical space and time, and it cannot be eliminated away or reduced to anything else. It is a "natural," not supernatural or metaphysically outlandish phenomenon. It is a biological phenomenon, existing among other biologically based phenomena. It follows that the task of explaining consciousness falls to the biological sciences, especially cognitive neuroscience that has defined itself as "the biology of the mind."

Therefore, if there is going to be a unified science of consciousness, any such research program must be anchored to the biological sciences. The explanatory framework of the biological sciences is the proper framework in which the explanation of consciousness will be sought. By contrast, in the philosophy of mind, the models of explanation are typically drawn from the basic physical sciences. The reductive unity of the physical sciences and the Deductive-Nomological (D-N) model of explanation are typically taken for granted as the paradigms of scientific explanation. The D-N model assumes that all scientific explanations and theories should look like our best theories in physics: mathematically expressed exact laws of nature that accurately predict and explain the behavior of physical entities. Theories describing higher-level macroscopic entities (liquid water, ice) can be logically derived from and reduced back to the fundamental laws in physics (first to theories describing H_2O molecules, then to hydrogen and oxygen atoms, and finally to microphysics and quantum physics).

But explanation in the biological sciences does not typically follow the reductive D-N model, and there are only few exact, mathematically described explanatory laws in biology. Recent work in the philosophy of neuroscience has come up with a different model of explanation for the life sciences: the multilevel mechanistic model, or *multilevel explanation* (see Bechtel and Richardson 1992; Craver 2007, 2016). According to this model, complex physical systems such as biological organisms consist of multiple levels of organization. These levels are real, ontological levels in nature. Different phenomena (such as synapses, neurons, neural networks, the whole central nervous system) reside at different levels of complexity. Higher-level phenomena are constituted by lower-level phenomena but not reducible to them, because the higher-level phenomena have causal powers at their own level of organization that go beyond those of the lower-level phenomena. A synapse or a single neuron does not have the same causal powers as the whole central nervous system does.

All of the above applies to the explanation of complex biological phenomena in general. According to BR, we now simply need to place consciousness into this framework of explanation. In the multilevel framework, consciousness as a biological phenomenon constitutes a higher level of neurophysiological or neuroelectrical organization in the brain. Consciousness can thus be reconceptualized as the *phenomenal level of organization* in the brain. Next, the multilevel explanatory model of consciousness should be constructed.

In general, a multilevel mechanistic model first describes a phenomenon in detail at its own level, and then explains it by specifying its micro-level constitution, its origin (or causal history), and its functional roles (or causal powers) in the world. All this can be achieved by placing the phenomenon into the center of a causal-mechanical network that has several different explanatory dimensions (see Figure 14.1). The full explanation of a phenomenon requires multiple levels of description and three different directions of explanation.

First, the essential features of consciousness itself should be described accurately, at their own level, so that we will have precisely identified what the target phenomenon (the *explanandum*)

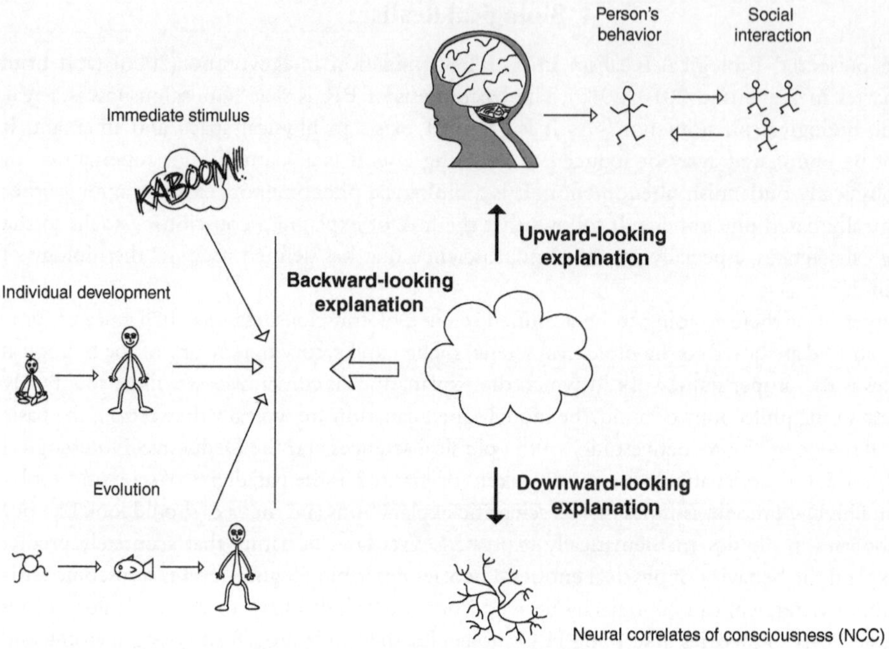

Figure 14.1 The Multilevel Framework

is that we are trying to explain. For example, if we wish to explain perceptual consciousness in wakefulness and dream experiences during sleep, we need to first have accurate descriptions of the phenomenology of both perceptual and dream experiences. How do we visually experience colors, objects, places, and faces in these two states?

Then we can start to fill in the different directions of explanation that "surround" the phenomenon in the causal-mechanical network. The *downward-looking (or constitutive) explanation* describes the lower-level nonconscious neural mechanisms that directly underlie consciousness in the brain. We can also call them *the constitutive mechanisms of consciousness*. As "constitution" implies ontological dependency, this set of neural mechanisms is bound to be a much narrower subset of mechanisms than the mere neural *correlates* of consciousness. For example, the constitutive mechanism of color experiences both during dreaming and wakefulness likely includes the visual cortical area V4. When it is sufficiently activated, we have visual color experiences, whether they are dreams, hallucinations, or visual perceptions. When area V4 is deactivated or destroyed, experiences of color become impossible. Without V4, we can have achromatic visual experiences only.

The *upward-looking explanation* describes the higher-level causal or functional role that consciousness plays in the whole brain, the whole person, and especially in guiding behavior. For example, color vision supports our ability to perceptually detect and discriminate from each other differently colored visual objects such as traffic lights and signs, and quickly adjust our behavior accordingly.

The backward-looking explanation moves backwards in time, tracing the causal chain of events that resulted in or causally modulated consciousness. This explanation can look to the immediate past when explaining how a preceding stimulus resulted in a conscious experience, or to the individual's past, describing how conscious experience emerged and changed during individual development from new-born baby to an adult, or backwards to the evolutionary past, describing how human consciousness emerged during evolution, or how any type of consciousness at all emerged during the evolutionary history of life on the planet.

In the case of color vision, the immediate physical stimulus might be the red traffic light lit up in your visual stimulus field and activating the neural pathways from the eyes to the visual cortex to area V4, until the phenomenal experience of redness emerges into perceptual consciousness. The developmental explanation would explicate how the human visual brain matures early in life to detect and discriminate various colored stimuli, and how this development fails in color-blind individuals. The evolutionary explanation explains the ultimate origins of the ability of humans to see colors: why are we not achromatic creatures? During evolutionary history, there were selective survival advantages in quickly and accurately recognizing colored objects, such as ripe fruits and berries, and poisonous dangerous animals, such as snakes and hornets, among green and brown leaves and grass. This led to highly accurate color vision being selected for in our ancestors and many other primate species before humans.

When we have all these dimensions of explanation covered without gaps, we will have a full mechanistic, multilevel biological explanation (of visual color experiences, for example).

Consciousness is a higher level of biological organization. The problem is that, at the moment, there are too many missing levels in the constitutive explanation; too many gaps in our scientific knowledge, between what we know about the neural levels in the brain and the conscious levels in the mind to yet be able to connect them smoothly within the multilevel model. The Explanatory Gap between consciousness and the brain follows from our ignorance of the intermediate levels, not from any fundamental metaphysical or epistemic inability to explain consciousness.

According to BR, pure phenomenal consciousness is the basic level of consciousness. It is a unified, spatial field or sphere where the qualities of experience come into existence (I thus agree with Searle's unified field theory). Phenomenal consciousness is most likely based on large-scale neuroelectrical activities and recurrent interactions in cortico-cortical and thalamo-cortical networks.

BR (like BN) takes consciousness as a holistic, spatial phenomenon by its fundamental nature. In BR, the spatiality of consciousness is lifted to a special status among the features of consciousness: it is the very feature that crosses from third-person physical ontology to the first-person qualitative ontology, as it has one foot in both realms. The pure spatiality of consciousness, the phenomenal space or field, cannot be experienced as such - it does not *in itself* constitute a content of experience or include a phenomenal character. It is the level of organization that *mediates* between the nonconscious, purely neural levels, and the conscious phenomenal levels in the brain. Thus, it could be called the *sub-phenomenal space*. It is the system at the interface between the phenomenal and nonphenomenal realms that reveals itself only indirectly, in the fact that *all* the phenomenal qualities that we *do* experience *always appear to be spatially organized within a single unified overall context*, the *world-for-me*.

The sub-phenomenal space must be activated for us to be in the conscious state – in the state where all kinds of qualitative experiences are enabled. When it is not activated, as in a coma or in dreamless sleep, we are in an unconscious state, and no experiences are possible. When it is partially damaged, as in unilateral spatial neglect, no experiences are possible in the compromised parts of sub-phenomenal space. Moreover, any direct awareness of a space that is missing from experience is impossible to reach for the neglected patient. The unified field of consciousness always seems like a complete spatial world for the subject, even if it fails to represent some part of the external stimulus space because the corresponding phenomenal space itself is missing. The sub-phenomenal level constitutively supports phenomenal qualities: they can only appear within it. Outside the sub-phenomenal space, qualities of experience do not and cannot exist.

According to BR, consciousness is something very simple. In its barest essence, phenomenal consciousness constitutes an *inner presence* – the simple presence or occurrence of experiential qualities, that is. No separate "self" or "I" or "subject" is required; no representing, no intentionality, no language, no concepts; only the sub-phenomenal space in which self-presenting phenomenal qualities may come into existence and realize inner presence. At higher levels of phenomenal organization, the qualities form complex phenomenal entities or organized bundles of self-presenting properties ("virtual objects") that we typically experience in conscious perception and in vivid dream experiences. Some of the bundles constitute our body-image, others the phenomenal objects we perceive. The entire phenomenal level, when well-organized, constitutes what I call a *world-simulation*: a simulated world, or a virtual reality in the brain.

There is no separate subject or self who "has" or "observes" the experiences, or who inhabits the virtual world. What we normally call the "self" is the body-image in the center of the simulated world, and what we call the "subject" is simply the overall system of self-presenting qualities that forms the phenomenal level in our brain. Thus, any particular experience is "had" by the "subject" simply because "having" reduces to "being a part of" the phenomenal level. Your momentary total experience simply consists of all the qualities that are simultanously present within the sphere of phenomenality. It is your subjective world, the *world-for-you*. You are both a part of the world (you as the "self" embedded within a body-image and visual perspective), and the whole world (you as the subject whose experiences constitute all the present contents of the sphere).

The phenomenal level and the "subject" thus refer to the same entity: they both are simply the sum of spatiotemporally connected phenomenality in the brain; the totality of self-presenting qualitative patterns that are spatially connected and temporally simultaneous in the brain. Therefore, the concept of a subject, as something separate from the phenomenal experiences themselves, is superfluous. In addition to the interconnected self-presenting qualities, no notion of a subject is necessary. The notion of a "self," by contrast, applies to most experiences, but it is also possible to have selfless and bodiless experiences where even the perspectivalness and the egocentricity of experience disappears. When this happens, the experience is fundamentally one; an experience of ego-dissolution, oneness and unity, or of being one with the world; the separation between a self and a world is gone. It could be called, not a being-*in*-the-world, but rather, a *being-the-world* experience. Mystical experiences and altered states of consciousness are sometimes associated with this sort of experiential unity.

In BR, the problem of explaining the emergence of consciousness and closing the Explanatory Gap boils down to the problem of understanding the constitutive relationships between the lower nonconscious or sub-phenomenal levels and the phenomenal level. Will an unbridgeable Explanatory Gap between them remain?

Dainton (2004) agrees that the idea of a sub-phenomenal, physical space that is the constitutive level for consciousness might build a bridge across the Explanatory Gap:

> For Revonsuo... our experiences inherit their spatial characteristics... from a physical field of a kind which is not... phenomenal in nature. This at least narrows the Explanatory Gap, and does so while minimizing the risk of panpsychism. It may well be that our brains generate coherent spatially extended fields. If these fields are... imbued with localized patterns of phenomenal properties by neural activity, we have a direct link between phenomenal and physical space. In fact, we have an identity: phenomenal space *is* physical space, albeit field-filled physical space. Of course, there is still a good deal to be explained: how, exactly, does a physical field come to carry or be imbued with phenomenal properties as a consequence of neural activity?

> Even so, progress has been made, the gap between the phenomenal and the physical is less wide than it was.
>
> *(Dainton 2004: 19)*

So, the first step to solve the Explanatory Gap is taken by marrying phenomenal space with physical space. The second step is to explain what the modulations of this field are and how they constitute the qualitative contents of phenomenal consciousness. A serious challenge for the explanation of consciousness crystallizes in the notion of "presence" and "self-presentation":

> Although biological self-presentation appears for us to be a magical feat, perhaps it is no more magical than biological self-replication...there may be particular biological mechanisms that render a biological process *present-for-itself*...
>
> ...the problem of understanding phenomenal consciousness seems to boil down to the problem of mechanistically modeling "self-presentation." Do we have any idea why some levels of biological organization may "feel" or "sense" their own existence whereas others have no means for sensing any existence at all? ... Most physical phenomena exist but in the dark, hidden even from themselves. They are not present for themselves and nothing is present for them. Somehow, for a physical or biological system to sense its own existence... It must make an appearance to itself, in order to create its own, self-contained, inner presence: the world-for-me. The "self-sensing" capability might be the result of the system being connected to itself in a particular way at the lower levels of organization, which would support a special type of global unity at the higher level. Every part of the system should become present to every other part simultaneously, to create their spatial co-presence in the same phenomenal world... This kind of neural architecture might be found in the thalamo-cortical loops... The integrated sphere of neuroelectrical flow may thus become present-for-itself, a world-for-itself.
>
> *(Revonsuo 2006: 360–361)*

5 Biological Realism at Work in the Study of Consciousness

Biological Realism guides the research in several ways. One of the implications of BR deals with the objective measurement of consciousness and the problem of accessing other conscious minds. According to BR, the current methods of cognitive neuroscience are not sufficient to measure consciousness, because they do not deliver data from the higher levels of organization in the brain where phenomenal consciousness is realized. Thus, we cannot "see" the phenomenal level in the brain via any brain scanning methods that are currently available. However, there is no reason why the objective measurement of consciousness should be impossible in principle. What we need is first, research methods that retrieve signals directly from the phenomenal level and its constituents in the brain. Secondly, we also need more sophisticated technology to understand and model the data. As phenomenal qualities only have an existence inside the phenomenal level, the brain imaging data that captures conscious experiences should be presented to observers within their own phenomenal level, by making the observers' phenomenal level *simulate* the state and contents of the phenomenal level of the observed subject. Observation of phenomenal consciousness, the patterns of qualities in someone's phenomenal level in one conscious brain, thus is simply *a shared simulation* that will run in the observers' phenomenal level and recreates similar patterns of qualities in each observer's phenomenal level. Two separate worlds-for-me become one shared world-for-us by the observer's consciousness closely mirroring the subject's consciousness. This renders the "public observation" of anyone's consciousness feasible, as anybody can log into the simulation and personally witness or live through the what-it-is-likeness of the observed person's phenomenal level. I call this the Dream-Catcher method of future consciousness science (Revonsuo 2006).

The world-simulation view of consciousness has consequences for empirical research as well. It has led to a new definition of dreaming as a world-simulation in the brain during sleep. The world-simulation concept of dreaming, in turn, has led to new ideas about the function of dreaming, called simulation theories. The first simulation theory is known as the Threat Simulation Theory (Revonsuo 2000), and it argues that dreaming (especially bad dreams and nightmares) is an ancient evolutionary program in the brain for the repeated, automatically programmed simulation of dangerous situations in a safe place, in order to rehearse important survival skills by facing dangers believed to be real while the dream experience takes place. The theory is testable and there is already considerable empirical evidence supporting it (Valli and Revonsuo 2009). Other simulation theories include the Protoconsciousness Theory (Hobson 2009) and the Social Simulation Theory (Revonsuo, Tuominen and Valli 2016a, 2016b).

Another empirical topic that BR has some relevance for is the conceptual and empirical distinction between *phenomenal* and *access* consciousness. According to the BR conception of consciousness as a simple inner presence of qualities, phenomenal consciousness is independent of access. This conceptual distinction has led to an empirical line of research where we have presented evidence for the empirical separability of the electrophysiological correlates of phenomenal visual consciousness from those of access consciousness (see e.g. Koivisto and Revonsuo 2010; Railo, Koivisto, and Revonsuo 2011).

6 Biological Realism in Relation to Informational Theories of Consciousness

Recently, a major shift away from biological theories to informational theories of consciousness has taken place. Influential philosophers (such as David Chalmers) and leading neuroscientists (such as Giulio Tononi and Christof Koch) metaphysically anchor consciousness to *information* rather than to biology. Interestingly, this shift seems to be motivated by the perceived inability of the biological approach to rise to the challenges of the Hard Problem and the Explanatory Gap. Koch (2012) explicitly confesses that he has switched from materialism to informational theories because he cannot see how phenomenal consciousness could emerge from neural processes.

The Integrated Information Theory (IIT) (Tononi 2008) is currently the most influential informational theory of consciousness. The core thesis of IIT says that consciousness is integrated information. Integrated information is a property defined by the internal causal interconnectedness of a system. The amount of integrated information possessed by any physical system can be quantified as its phi-value (Tononi and Koch 2015). A phi greater than zero means that the system is conscious, and the degree or quantity of its consciousness is expressed as its phi-value. The conscious human brain has perhaps the highest phi-value of any physical system, but even simple physical and nonbiological systems have phi-values above zero.

It is important to be clear that the biological and the informational research program have very different views on the fundamental nature of consciousness. Even though it is difficult to give a general definition of "information," one thing about the ontology of information of any type is clear: information consists of *abstract* patterns, *realized in* or *carried by* physical systems.

Information is not a concrete physical entity like a DNA molecule, a neuron, or an action potential is. Information has *no physical or biological essence*. Information is a second-order property. Its identity is defined at the abstract level; therefore, it is not ontologically type-identical with any particular physical properties.

Consequently, information theories of consciousness, such as IIT, easily lead to panpsychism (because almost anything can "carry" information) or at the very least, to ultra-liberal multiple realizability of consciousness (because almost any causally connected system can realize an amount of integrated information that is above zero). *Panpsychism* is the idea that each and every fundamental physical entity is coupled with some kind of elementary mental properties: everything physical has at least a degree of consciousness. *Multiple realizability* is the idea that the mind or consciousness has no physical essence, but can be carried by and realized in radically different types of physical systems (as long as they realize the appropriate abstract patterns of e.g. computation or information processing). Thus, computers and robots may have consciousness as long as their processing units carry the appropriate patterns of information for conscious processing.

Information itself is non-material and abstract, but it is easy to confuse its abstractness with the concrete physical complexity of the physical vehicles that "carry" this information. Information is easily masquerading as a higher-level physical property, but ontologically it is no such thing. Information is a second-order, formal property, not a higher-level concrete physical property. There is no such higher level of physical organization in the brain, where "information" emerges out of non-informational physical phenomena, and where this information then forms the constitutive basis of yet higher physical levels. Because of its abstractness, information can exist at (or be carried by) any arbitrary physical level. If consciousness consists of information, there is nothing particularly biological about the fundamental nature of consciousness.

Informational theories – like their close relatives, functionalist and computationalist theories – posit an abstract metaphysical domain as being the fundamental ontological nature of consciousness. Information, causal roles, computations, algorithms: their essence resides in the world of abstract forms.

But any theory that identifies consciousness with an abstract metaphysical domain "realized" by concrete physical entities pays a high price. Second-order properties inherit all their causal powers from their first-order physical realizers; they have no causal powers of their own. Abstract entities constituted by second-order properties like "information" or "computation" as such *have no causal powers of their own in the physical world*; they have no effects on anything; rather it is the concrete material, physical entities or processes, "realizing" the abstract patterns, that are causally efficacious (see e.g. Jaegwon Kim's well-known arguments on this in Kim 1998).

Thus, informational theories have two unacceptable consequences: First, they typically assign consciousness to all sorts of extremely simple or otherwise unlikely physical systems (such as photodiodes, bacteria, iPhones, etc.). The empirical evidence and testability for such claims is nil, and the intuitive plausibility even less. Secondly, they rob consciousness of any causal powers in the physical realm. If consciousness consists of information or anything else in the abstract metaphysical domain, it is doomed to be epiphenomenal.

By contrast, the biological approaches BN and BR assign consciousness to the concrete metaphysical domain of higher-level physical and biological phenomena. Such phenomena have a concrete spatiotemporal emergent structure and possess concrete causal powers of their own. The biological approach rejects panpsychism, but allows multiple realizability within narrow limits. The brains of different animal species can support consciousness although the lower-level neurophysiological basis may be slightly different.

7 Conclusions

BN and BR argue that consciousness is a higher-level of physical organization in the physical world, a concrete emergent biological phenomenon that supervenes on lower-level neural activities in the brain but cannot be reduced to them. Consciousness forms its own level of phenomenal organization in the brain, a level of spatially unified qualitative subjectivity. This level is constituted by concrete physical (perhaps complex neuroelectrical) phenomena, located in and unfolding across the physical space and time inside the brain. Consciousness is not built out of abstract entities or second-order properties such as computations, algorithms, or information. For unified qualitative subjectivity to emerge, the highly specific biological conditions inside the brain are required. Consciousness is not likely to be found in physical systems completely unlike the brain. As a higher-level physical phenomenon, consciousness possesses causal powers of its own, manifested in consciously guided behaviors.

Compared to the large extent of their shared ground, the differences between BN and BR are only minor. One difference however is that BN appears to not acknowledge the challenges of the Explanatory Gap and the Hard Problem. BR by contrast takes them to be serious but not insurmountable anomalies for the cognitive neuroscience of consciousness (Revonsuo 2015).

Other philosophers who have recently defended a position close to BR include O'Brien and Opie (2015). They argue that conscious experiences are emergent physical structures just like molecules and cells, and thereby the biological approach avoids two mortal pitfalls that have long plagued the philosophy of mind. Firstly, they avoid the reductionistic pitfall that leads to *microphysicalism* (the belief that only the bottom level of elementary physics really exists and everything, including consciousness, reduces to that level). Secondly, they avoid the functionalist, information-theoretic, computationalist pitfall that identifies consciousness with second-order abstract entities that have no causal powers of their own in the physical world (epiphenomenalism) and the implausible idea that consciousness can exist or be realized in nearly all physical systems (panpsychism).

Biological Naturalism and Biological Realism place consciousness where it belongs: As a real higher-level physical phenomenon in the brain, with special features and causal powers of its own, just like any other higher level biological phenomena. The biological approach avoids falling into the traps of epiphenomenalism and panpsychism, but must face the Explanatory Gap and the Hard Problem. BN and BR remain optimistic that understanding consciousness as a biological phenomenon will in the future close the gap between subjective consciousness and objective brain activity.

References

Bechtel, W. and Richardson, R.C. (1993) *Discovering Complexity: Decomposition and Localization as Strategies in Scientific Research*, Princeton, NJ: Princeton University Press.

Chalmers, D.J. (1996) *The Conscious Mind*, Oxford: Oxford University Press.

Craver, C. (2007) *Explaining the Brain*, New York: Oxford.

Craver, C. (2016) "Levels," in T. Metzinger and J. Windt (eds.) *Open MIND: Philosophy and the Mind Sciences in the 21st Century*, Vol. 1; 305–330, Cambridge, MA: MIT Press.

Dainton, B. (2004) "Unity in the void: reply to Revonsuo," *Psyche* 10 (1): 1–26.

Hobson, A.J. (2009) "REM sleep and dreaming: towards a theory of protoconsciousness," *Nature Reviews Neuroscience* 10: 803–813.

Kim, J. (1998) *Mind in a Physical World*, Cambridge, MA: MIT Press.

Koch, C. (2012) *Consciousness: Confessions of a Romantic Reductionist*, Cambridge, MA: MIT Press.

Koivisto, M. and Revonsuo, A. (2010) "Event-related brain potential correlates of visual awareness," *Neuroscience and Biobehavioral Reviews* 34: 922–934.

Levine, J. (1983) "Materialism and qualia: the explanatory gap," *Pacific Philosophical Quarterly* 64: 354–361.

O'Brien, J.P. and Opie, G.J. (2015) "The structure of phenomenal consciousness," in S.M. Miller (ed.) *The Constitution of Phenomenal Consciousness*, Amsterdam: John Benjamins.

Railo, H., Koivisto, M. and Revonsuo, A. (2011) "Tracking the processes behind conscious perception: a review of event-related potential correlates of visual consciousness," *Consciousness and Cognition* 20: 972–983.

Revonsuo, A. (2000) "The reinterpretation of dreams: an evolutionary hypothesis of the function of dreaming," *Behavioral and Brain Sciences* 23: 877–901.
Revonsuo, A. (2006) *Inner Presence: Consciousness as a Biological Phenomenon*, Cambridge, MA: MIT Press.
Revonsuo, A. (2010) *Consciousness: The Science of Subjectivity*, Hove and London: New York: Routledge, Psychology Press.
Revonsuo, A. (2015) "Hard to see the problem?," *Journal of Consciousness Studies* 22 (3-4): 52–67.
Revonsuo, A., Tuominen, J. and Valli, K. (2016a) "The avatars in the machine: dreaming as a simulation of social reality," in T. Metzinger and J. Windt (eds.) *Open MIND: Philosophy and the Mind Sciences in the 21st Century*, Vol. 2: 1295–1322, Cambridge, MA: MIT Press.
Revonsuo, A., Tuominen, J. and Valli, K. (2016b) "The simulation theories of dreaming: how to make theoretical progress in dream science," in T. Metzinger and J. Windt (eds.) *Open MIND: Philosophy and the Mind Sciences in the 21st Century*, Vol. 2: 1341–1348, Cambridge, MA: MIT Press.
Searle, J.R. (1987) "Minds and brains without programs," in C. Blakemore and S. Greenfield (eds.) *Mindwaves*, Oxford: Blackwell.
Searle, J.R. (1992) *The Rediscovery of the Mind*, Cambridge, MA: MIT Press.
Searle, J.R. (1998) "How to study consciousness scientifically," in S. Hameroff, A. Kaszniak, and A. Scott (eds.) *Toward a Science of Consciousness II: The Second Tucson Discussions and Debates*, Cambridge, MA: MIT Press.
Searle, J.R. (2004) *Mind: A Brief Introduction*, New York: Oxford.
Searle, J.R. (2007) "Biological naturalism," in M. Velmans and S. Schneider (eds.) *The Blackwell Companion to Consciousness*, Oxford: Blackwell.
Tononi, G. and Koch, C. (2015) "Consciousness: here, there and everywhere?" *Phil. Trans. R. Soc. B* 370: 20140167.
Valli, K. and Revonsuo, A. (2009) "The threat simulation theory in the light of recent empirical evidence: a review," *The American Journal of Psychology* 122: 17–38.

Related Topics

Materialism
Dualism
Idealism, Panpsychism, and Emergence
Information Integration Theory
Consciousness and Dreams
The Unity of Consciousness
The Biological Evolution of Consciousness

15
SENSORIMOTOR AND ENACTIVE APPROACHES TO CONSCIOUSNESS

Erik Myin and Victor Loughlin

1 Introduction

What is it like to have a sensation of red, or to consciously see a blue car parked in the street? On established philosophical understandings of the relation between the mental and the physical, these questions concern how it is possible for brain states or inner representations to give rise to phenomenal feel.

According to the sensorimotor approach to perceptual experience, the pressing philosophical questions about phenomenal feel are answerable only if it is recognized first that such experience essentially is "something we do, not something that happens in [us]" (O'Regan and Noë 2001b: 80). That is, if it is understood that having perceptual experience is fundamentally a matter of engaging with our environments in certain ways. Forgetting that perceptual awareness is something we do and instead aiming for an understanding of perceptual experience in terms of inner neural or representational events only invites, insist sensorimotor theorists, further unsolvable problems about how these events give rise to consciousness.

This chapter will be devoted to unpacking the sensorimotor thesis that experience is something we do, and explicating how it helps to deal with the philosophical problem of consciousness. The key to understanding the sensorimotor position, so we propose, is to recognize it as a form of identity theory. Like the early mind/brain identity theorists, the sensorimotor approach holds that the solution to the philosophical problem of phenomenal experience lies in realizing that phenomenal experience is identical with something which, while at first sight might seem different, turns out not to be different after all. Like the classical identity theorists, sensorimotor theorists reject the claim that identities can and need to be further explained once identification is made. Sensorimotor theorists consequently oppose the idea that there is a genuine scientific issue with the identity relation between experience and what perceivers do. However, unlike other identity positions, the identification proposed by the sensorimotor approach is wide. That is, conscious experience is identified, not with internal or neural processes, but instead with bodily (including neural) processes in spatially and temporally extended interactions with environments.

However, if experience is identified with doing, there is a further issue about what are the conditions needed for the appropriate doings to be possible. In Mind-Life Continuity (MLC) Enactivism (Thompson 2007), it has been argued that consciousness can occur only when and where the organization of life is present. After briefly discussing the relation of the sensorimotor approach with MLC Enactivism, we will also compare the sensorimotor approach with Radical Enactivism, according to which basic perception is contentless and argue that considerations of coherence should push the sensorimotor approach to endorse Radical Enactivism. Adding to replies to other standard criticisms of the sensorimotor approach, which we give earlier in the chapter, we will end by showing how our construal of the sensorimotor approach to consciousness can be used to reject the criticism often made against sensorimotor theory, namely that by invoking the environment in its account of consciousness, the sensorimotor theory deepens, rather than overcomes, the philosophical problem of consciousness.

2 Sensorimotor Sensation and Perception

The sensorimotor approach to perceptual experience is built on the idea that "experience is something we do, not something that happens in [us]." But what exactly does this mean?

Consider the having of a sensation. Having a visual sensation of red, so the approach holds, is a matter of perceptually engaging with the environment. But such an engagement only constitutes the perceiver's visual experience if the perceiver is sensitive, adapted or attuned to particular sensorimotor contingencies. Sensorimotor contingencies are lawful patterns in the way stimulation changes, including the lawful ways in which stimulation for a perceiver changes as a function of the perceiver's bodily movement. In the case of light and vision, for example, sensorimotor contingencies concern the ways in which light interacts with objects, with other light, and with perceivers. The sensorimotor contingencies typical for red thus include the lawful ways in which light of a particular constitution gets reflected by particular surfaces, how the reflection changes when the constitution of the light changes, how the reflection differs along different angles of perception and how the reflected light differentially affects receptors on a perceiver's retina.

This reveals that the sensorimotor approach construes the having of a red sensation in terms of bodily interaction with certain surfaces (or, occasionally, lights) in ways that are adapted to or attuned to the relevant sensorimotor contingencies. A perceiver's being attuned to such contingencies shows, for example, in the fact that she still has the same experience when only the illumination but not the surface changes—the phenomenon known as color constancy. Another example of attunement is when the same color is experienced when the perceiver moves and the surface comes to stimulate a different part of the retina. The paradigm case of experiencing red, so the sensorimotor contingency approach holds, is thus one in which the agent perceptually engages with an object in its environment in ways that are appropriately sensitive to the sensorimotor contingencies typical for red objects.

The sensorimotor approach accounts for the quality of sensory modalities as a whole in the same way in which it accounts for the quality of particular sensations. That is, what gives visual experience the quality of seeing, as different from hearing, is that seeing is a specific way of interacting with the environment, subject to its own particular sensorimotor contingencies. Closing your eyes will interrupt your vision but not your hearing, for example. Standing on your head will invert your visual experience but not your auditory experience.

The sensorimotor approach consequently offers a recipe by which to analyze any qualitative aspect of perceptual experience. Such sensorimotor analysis proceeds by characterizing the specific kind of interaction that the experience is to be identified with. Consider the perceptual experience of objects. Typically, when one perceives an object, one has only a partial view of it. Nevertheless, one's experience relates to the complete object, not to only the fragment that is currently in view. The sensorimotor approach explains that one relates to the whole object through one's sensitivity to the changes in stimulation that would happen if one were to move with respect to the object. So, for example, one will not be surprised by how one's visual experience changes when one moves around the object. Also, if one would grasp the object, one's hand aperture would be appropriate to the orientation of the unseen parts of the object. As with the case of sensation, conscious perceptual experience is understood in terms of what perceivers do and can do if and when they engage with their environments, in a way that is adapted to the relevant sensorimotor contingencies.

Crucially, sensorimotor theorists claim that understanding sensation and perception as doings holds decisive advantages over explaining sensation and perception in terms of internal neural or representational events. Sensorimotor theorists acknowledge that neural processes are involved when, say, a sensation of red is felt. Still, they insist that the conscious quality of having the sensation cannot be adequately understood in terms of such processes. The same position has also been adopted with respect to inner representational events. Sensorimotor theorists reject the idea that the phenomenology of being perceptually related to unfaced parts of an object can be explained in terms of the activation of internal mental representations that 'stand for' these parts. O'Regan and Noë (2001a: 939–940) illustrate their stance by commenting on an extensive list of contemporary proposals for the mechanisms alleged to explain the generation of consciousness. These include, in their formulation, "a 'commentary' system situated somewhere in the fronto-limbic complex (taken to include the prefrontal cortex, insula and claustrum; cf. Weiskrantz 1997: 226)"; "coherent oscillations in the 40–70 Hz range, which would serve to bind together the percepts pertaining to a particular conscious moment" (Crick and Koch 1990); "a quantum process in neurons' microtubules" (Hameroff 1994); and "reentrant signaling between cortical maps" (Edelman 1989). O'Regan and Noë claim all these examples raise the following issue:

> A problem with proposals of this kind is that they do little to elucidate the mystery of visual consciousness (as pointed out by, for example, Chalmers 1996). For even if one particular mechanism — for example, coherent oscillations in a particular brain area — were proven to correlate perfectly with behavioral measures of consciousness, the problem of consciousness would simply be pushed back into a deeper hiding place: the question would now become, why and how should coherent oscillations ever generate consciousness? After all, coherent oscillations are observed in many other branches of science, where they do not generate consciousness. And even if consciousness is assumed to arise from some new, previously unknown mechanism, such as quantum-gravity processes in tubules, the puzzle still remains as to what exactly it is about tubules that allows them to generate consciousness, when other physical mechanisms do not.
>
> *(O'Regan and Noë 2001a: 939–940)*

This passage shows that O'Regan and Noë object to a number of proposals to understand consciousness in terms of specific inner (neural) processes. It also offers their grounds for such rejection, namely that all such proposals invite the further question as to why the particular inner process proposed gives rise to, or generates, consciousness (O'Regan 2011: 97 raises the same point). However, if this criticism of internalist approaches to consciousness is correct, then one may wonder why the sensorimotor approach is not itself susceptible to a similar critique. For why is the idea that the qualitative aspects of sensation and perception should be understood as doings not itself vulnerable to the worry that there is a gap between, on the one side, consciousness, and on the other side, doings? If there is such a gap, then one can ask: why should engaging with the environment perceptually give rise to consciousness at all? Also, one can ask: why should *this* particular doing or action generate *this* particular sensation and/or perception?

In what follows, we will show how the sensorimotor approach provides the means to tackle this criticism and to deal with these questions.

3 Sensorimotor Identity

Getting a grip on how the sensorimotor approach to experience allows us to answer questions about an alleged gap between experience and doing requires that we first clarify what exactly is the proposed account of experience, and its relation to doing. However, this is complicated by the fact that the canonical writings in which the sensorimotor approach has been expressed leave room for more than one interpretation (O'Regan and Noë 2001a, b). This has not gone unnoticed by commentators such as Gennaro (2017), who wonders:

> What exactly is the view? Sometimes it is unclear. On the one hand, it often sounds like a stronger identity or constitutive claim is being made about the relationship between sensorimotor skills and consciousness. "Perceptual experience... *is* an activity of exploring the environment drawing on knowledge of sensorimotor dependencies and thought" (Noë 2004: 228) and "perceptual experience *just is* a mode of skilful exploration of the world" (Noë 2004: 194). Again: "Visual experience is simply not generated [in the brain] *at all*. Experience is not the end product of some kind of neural processing" (O'Regan 2011: 65). On the other hand, there are many examples of a much more modest causal or dependency claim. "I have been arguing that, *for at least some experiences*, the physical substrate [vehicle] of the experience *may* cross boundaries, implicating neural, bodily, and environmental features" (Noë 2004: 221) and "experiencing a raw feel *involves* engaging with the real world" (O'Regan 2011: 112).
>
> *(Gennaro 2017: 85–86)*

We propose to resolve this possible lack of clarity by taking the sensorimotor proposal that experience is doing to be an *identity claim,* similar in some respects but dissimilar in others, to the identity claims made by the classical mind/brain identity theorists Ullin Place and Jack Smart (Place 1956; Smart 1959). This reading, so we will attempt to show, allows for the most viable form of sensorimotor theory, and is consistent with the bulk of the canonical sensorimotor writings.

According to this reading, sensorimotor theorists, like the classical mind/brain identity theorists, propose that the solution to the mind/body problem lies in *identifying* what might seem like otherwise different relata. And as was the case for classical identity theorists, this enables sensorimotor theorists to declare that there are no further issues concerning the relation between the mind and the body, since identities do not stand in need of further explanation. However, in contrast to classical mind/brain identity theorists, sensorimotor theorists propose that sensations and perceptions should be identified with, not brain processes, but instead wide, environment-involving activities (see also Hutto and Myin 2013, ch. 8; Myin 2016).

Let us first quickly address the idea that is shared by the mind/brain identity theorists and the sensorimotor approach, namely that identities don't stand in need of explanation. According to this idea, asking for an explanation of why E happens when C happens only makes sense if E and C are not identical. In such a case, one may wonder how the occurrence of C makes possible the happening of E. For example, C might be a mechanism that produces E. By contrast, if C and E are identical, then the question as to why E occurs when C occurs becomes the question as to why E occurs when E occurs, or why C occurs when C occurs. In other words, once C and E are understood to be identical, then it no longer makes any sense to wonder why E occurs when C occurs.

Of course, even in the case of an identity, one might think the question about how C gives rise to E does make sense if one is not aware of the identity between C and E. One could be puzzled, for example, about why Clark Kent's footprints where at the same spot as Superman's footprints. One might then conjecture that perhaps the two cooperated and then ponder about their motives for doing so. But even in this case, the only genuine questions involving distinct relata and how they relate concern conceptions of Superman and Clark Kent, not Clark Kent and Superman themselves. After all, Clark Kent and Superman remain one and the same person even if you are entirely ignorant of this fact and mistakenly take your thoughts about Superman to be about someone different from your thoughts about Clark Kent.

Summing up: the central claim of the sensorimotor approach, we propose, is that perceptual experience is identical to a bodily activity, within which sensitivity, adaptivity, or attunement to sensorimotor contingencies is displayed. Our interpretation is in line with the beginning of O'Regan and Noë's landmark paper:

> We propose that seeing is a way of acting. It is a particular way of exploring the environment. (…). The experience of seeing occurs when the organism masters what we call the governing laws of sensorimotor contingency.
>
> *(O'Regan and Noë 2001a: 939)*

We have emphasized that identifying experience with activity leads to the conclusion that further questions as to how experience and activity are related no longer make sense. This is congruent with the way in which O'Regan and Noë compare their account with developments in physics. They write:

> In understanding the epistemological role of the present theory, an analogy can be made with the situation facing nineteenth-century physicists, who were trying to invent mechanisms by which gravitational or electrical forces could act instantaneously at a distance. To solve this problem, Faraday developed the idea of a field of force, which was, according to Einstein, the single most important advance in physics since Newton (cf. Balibar 1992). But, in fact, the idea of a field of force is not a theory at all, it is just a new way of defining what is meant by force. It is a way of abandoning the problem being posed, rather than solving it. Einstein's abandoning the ether hypothesis is another example of how advances can be made by simply reformulating the questions one allows oneself to pose.
>
> *(O'Regan and Noë 2001a: 949)*

However, as noted earlier, while appealing to the same abstract logic of identity, the identity claim at the center of the sensorimotor approach fundamentally differs from the sort of identity claim made by classical identity theorists. For the sensorimotor approach identifies perceptual experience, not with neural processes, but rather with bodily activity. Indeed, with respect to the identification of the mental and the neural, sensorimotor theorists take a diametrically opposed position to classical identity theory.

Classical identity theorists have claimed that the identification of the mental with the neural settles issues of how the physical generates the experiential (see in particular Smart 1959). Contrarily, sensorimotor theorists posit that such classical identification does exactly the opposite: it invites unsolvable generation issues. According to sensorimotor theorists, the reason such unsolvable problems arise is that the identification proposed by the classical theorists is wrong headed: experience is identified, not with what it is identical with, i.e. embodied activity, but instead with what is only a necessary condition for it, i.e. neural processes. According to this sensorimotor critique, one can't identify a property displayed by a system with the activity of part of that system, even when the systemic property always involves that part's contribution. If one does make this mistaken identification, then unsolvable problems arise, that is, problems that are logically or conceptually flawed.

Invoking an analogy, O'Regan and Noë argue that one should not single out the beating heart as the "biological correlate" for life. Such a move would, they state, invite the problem of how the beating heart, all by itself, can *generate* life. Yet even though the beating heart is quite clearly necessary for its owner's life:

> (n)either the beating heart, nor any other physiological system or process, or even the sum of all them, generate life. Being "alive" is just what we say of systems that are made up in this way and that can do these sorts of things.
>
> *(O'Regan and Noë 2001a: 1018)*

The same goes for the thinking about perceptual experience in terms of neural correlates. While neural events form a necessary condition for such experience, they nonetheless provide the wrong "targets" for making identifications, since identifying neural events with experience will always invite the further question: how can such neural events, all by themselves, generate experience? By contrast, identifying experience with doings prevents any such generation question from arising. The reason is that this identification is in fact the right one to make: doings are precisely the "sorts of things," to paraphrase from the quote above, that perceivings are. Engaging in particular doings simply *is* what it is to have perceptual experience.

That an identification of experience with doings makes more sense than an identification with internal happenings can be further argued for by comparing and contrasting answers to the following question: what distinguishes perceptual experience from other kinds of experiences, such as imagination, or thought?

Consider perceiving first. When you see an object, your movements will bring into view different parts of the object. Closing your eyes will interrupt your seeing of it. Moreover, when something suddenly changes, as when the color on part of the object would suddenly change to a very different color, this would draw your attention to the spot on the object where the change has happened. Now consider imagination or thought. When you visualize an object in imagination, or simply think about that object, neither your bodily movements nor changes that can happen to the object impact in the same way. For example, the real-life counterpart to your imagined object might be annihilated and yet you can still continue to imagine that object or think about that object.

This contrast between perceiving and imagining or thinking reveals that perceptual experience has a profile that can be characterized in terms of "bodiliness" and "grabbiness" (see O'Regan and Noë 2001b; O'Regan, Myin and Noë 2005a,b). The ways you move your body will affect what you perceive (bodiliness) and changes in the object you perceive will grab your attention, such that you will perceive those changes (grabbiness). Contrarily, visually imagining or simply thinking about that same object has no such bodiliness or grabbiness. According to the sensorimotor approach, it is this difference in sensorimotor profile that ensures that perceptual experience has the specific quality of being perceptual, or has "perceptual feel."

Note that bodiliness and grabbiness are characteristics of the interaction between perceivers and their environments, that is, characteristics of doings. Bodiliness and grabbiness concern how the perceiver's activity affects the environment (and thereby further activity) and how the environment affects the perceiver's activity. By tracing the conscious quality of "being perceptual" back to bodiliness and grabbiness, sensorimotor theorists analyze the quality in terms of specific ways of doings. As before, sensorimotor theorists argue that such an analysis holds advantages over one which invokes internal (neural or representational) factors.

For example, suppose that there is some neural correlate typical of interactions, which involve bodiliness and grabbiness. Call this neural correlate N. Further suppose that N never occurs in cases of imagining or thinking. The sensorimotor theorist will argue that N nonetheless offers an inferior explanation for the feel of perceiving (versus the feel of imagining or thinking) than an explanation in terms of bodiliness and grabbiness. For N raises the sort of generation question posed before: why does N give rise to the quality of "being perceptual"? A satisfactory answer to that question, insists the sensorimotor theorist, must require invoking N's role in interactions that have bodiliness and grabbiness. But then one is back to the sensorimotor position.

The same holds true for explanations in terms of internal representations. It might be proposed that some experiences are "perceptual" because they carry a label, mark or code. Some brain events, in other words, get tagged as being perceptual and the tag should be understood as the representation "this is perceptual". However, without a detailed and convincing story about how such labels actually work, little ground is gained by invoking such a representational story. For while a possible explanans is pointed at, this explanans is tailor made to have exactly those properties that provide the explanation. Yet the fact that the explanans has these properties is all we know about it: what has been invoked, that is, is an unexplained explainer.

Moreover, even if a story about how the label in question represents could be told, such that our hitherto unexplained explainer would then be explained, such a story would have to mention bodiliness and grabbiness, or other interactive factors, which are characteristic of being perceptual. For what makes perception perceptual is what the label represents. In other words, even under these conditions, we end up very close to the sensorimotor theory, and need to invoke its explanantia or something very much like them.

4 Objections and Replies

Despite these arguments in favor of the sensorimotor approach, many opponents have rejected the sensorimotor approach without giving it due theoretical consideration. For such opponents, the imagined or actual existence of vivid perceptual-like experience in imaginations, dreams, hallucinations, or through direct stimulation of the brain provides simple but conclusive empirical proof that the sensorimotor identification of experience with doings is mistaken (see Gennaro 2017: 86, for a formulation of this worry; Block 2005; Prinz 2006). Such phenomena are taken to run counter to the identification of experiences with doings because they are possible without any movement at all. Indeed, a completely paralyzed person could have them. Apart from not demanding movement, such experiences are not voluntary: they happen to us, whether we want it or not. This provides an additional reason, claim some, for concluding that these experiences can't be doings.

The sensorimotor theorist will deny however that phenomena like dreams run counter to the identification of experience and doings, and instead argue that the sensorimotor approach contains resources to explain the particular characteristics of dreams—characteristics that are in fact left unexplained by rival approaches. The key to getting a sensorimotor grip on phenomena like dreams is to point out that they are, like all other experiences, embodied and embedded. Dreams are embodied in the sense that they are dreamt by persons, who are bodies, and it is the same body that perceives and acts during the day that then dreams during the night. Evan Thompson in a recent book which treats of enactivism and, inter alia, dreaming, invokes an ancient image from the Indian *Upanishads*: "like a great fish swimming back and forth between the banks of a wide river, we journey between waking and dreaming. The image hints of deeper currents beneath the surface while allowing for intermediate areas and eddies where waking and dreaming flow into each other" (Thompson 2015: 110).

Moreover, what people dream is only partly contingent. People dream of their mothers, brothers, and friends, in ways determined by their unique personal relations to them, and in situation types the dreamers desire or fear. Dreams are moulded by personal feelings, anxieties and preferences. They take place, and can only be understood when they are considered against the background of a person's "active life" (Noë 2004: 231).

If dreams thus are embedded in a personal situation, then they are tied to the specifics of the immediately occurring surrounding circumstances as well. Dreamers hear the fire wagon speeding by when their alarm sets off, and, if Nietzsche is to be believed, "the man who ties two straps around his feet, for example, may dream that two snakes are winding about his feet" (Nietzsche 1878/1986, section one, aphorism 13). Of course, interaction with the environment in dreams is severely restricted: after all, we have closed our eyes and don't see our surroundings. Sensorimotor theorists have argued however that this restriction of perceptual interaction in dreaming holds the key to understanding the particular characteristics and dynamics of dreams.

For example, the lack of perceptual interaction between the dreamer and its environment might explain why whole series of bizarre changes can be experienced when dreaming. The perceptual experience of a horse, unless one is at the movies or so, won't turn into the perceptual experience of a cat because the flow of stimulation from a horse remains the flow of stimulation from a horse, even if it (the horse), or the perceiver, moves. But when experience is only minimally conditioned by such flow, nothing then stands in the way of such a transformation (Noë 2004: 213–214; O'Regan, Myin and Noë 2005b: 62–64; O'Regan 2011: 66; Dennett 1991, ch. 1).

The claim that perception-like experience is possible without movement, be it in dreams, hallucinations, paralysis, or just when standing still, only runs against the sensorimotor idea that experience is a doing if we further assume that all doings involve movement. But this assumption is false. For it is a mistake to confuse doings with moving or making movements. In fact, one can do very specific things by arresting any movement. Think about obeying a police officer's order to stand still, or what a statue artist does to make money. Interestingly, it seems people only don't act out their dreams because they are physiologically prevented from doing so—their muscles being temporarily and selectively paralyzed during Rapid Eye Movement (REM) sleep, by known neurophysiological mechanisms (Brooks and Peever 2012). In that sense, dreams can be considered doings whose movements are prevented from occurring—and in rare occasions, when the physiological mechanisms sub-serving the prevention fail to function, people do, with much danger to themselves and their surroundings, in fact act out their dreams (Howell and Schenck 2015).

Another reason why dreams, or hallucinations—a fortiori when these are imagined to be induced by directly stimulating the brain (as in the classic studies by Walter Penfield 1975; for comparable experiments using transcranial magnetic stimulation, see Hallet 2000)—can be considered to run against the sensorimotor view of experiences as doings is that they occur involuntarily. They seemingly "happen to us," rather than "something we do." As such, they are on a par with a much wider class of perceptual, or perception-related experiences, including bodily feelings such as pains, twinges or itches, or sensations of the sensory modalities like vision, hearing or smell. We have to do nothing, apart from keeping our eyes open, to receive perceptual impressions from the world. And pain strikes us, often very much against what we want. This leads to the question as to how this apparently passive nature of such experiences can be reconciled with the sensorimotor idea of perception as a doing.

It is important to realize, however, that many of our doings are provoked, rather than intended with conscious premeditation. Still they are things we do. Consider for example a person that swears when he accidentally hits his thumb while hammering a nail in a beam, or a person that shouts, "take care," or "watch out!" when a teammate comes running into him when playing sports. This swearing and shouting are things the person does, despite not being planned, or wanted. What makes them a person's doings are rather that they are learned reactions, arising from, and grounded in this person's history. Moreover, they take place as part of the person's interactions with his environment. They are reactions to a specific situation. Yet, as the examples of shouting, "take care," or "watch out!" clearly show, they can be forward-looking and anticipatory.

A promising lead to follow for sensorimotor theorists, so we think, would be to view sensory experiences in analogy to such acts. Feeling pain, feeling the tactile sensation of being stroked by a feather or seeing red could then be seen as adaptive anticipatory bodily reactions of an organism to specific kinds of environmental offerings. These anticipatory reactions are grounded in evolutionary history, but they also form part of a person's or organism's contextualized engagement with their current situation, in a way that is sensitive to "cognitive, emotional and evaluative contributions" (Ben-Zeev 1984). Such analysis can be fruitfully applied to pain. Aaron Ben-Zeev, for example, cites Melzack, in order to underscore the personal nature of pain:

> The psychological evidence strongly supports the view of pain as a perceptual experience whose quality and intensity are influenced by the unique past history of the individual, by the meaning he gives to the pain-producing situation and by his 'state of mind' at the moment… In this way pain becomes a function of the whole individual, including his present thoughts and fears as well as his hopes for the future.
>
> *(Melzack 1973: 48)*

The claim that pain is situationally and personally sensitive is further indicated by the finding that as many as 37% of the patients arriving at an emergency clinic reported a period, normally of about an hour but in some cases lasting up to nine hours, of absence of the experience of pain after the injury—a finding lending support to the fact that athletes and soldiers sometimes succumb to serious injury, but they report being unaware of the pain until the end of the competition or battle (Beecher 1956). The picture of pain as purely passive, that is, as an impersonal event an organism simply "undergoes" as a result of inflicted damage consequently appears fundamentally flawed. Moreover, pain is anticipatory: it sometimes already happens before damage occurs. That is, rather than being invariably a reaction to actual tissue damage, pain also occurs whenever there is the threat of tissue damage (Melzack 1996; Moseley 2007; Wall 1999). In those cases, it seems pain's evolutionary rationale is to steer the organism away from activity that will inflict damage. The anticipatory character of pain is also discernible at the neural level. It has been shown that nociceptive neurons in area 7b of the monkey brain respond with increasing strength to temperatures between 47 and 51°C, which is just below the level at which tissue damage occurs (Dong et al. 1994).

Though further work is of course needed, conceiving of sensory experience—both sensation and perception—along these lines seems both promising and congruent with the existing sensorimotor literature (see O'Regan and Noë 2001a,b; some of the points made in Myin and Zahidi, in press). Sensory awareness of red becomes an anticipatory embodied interaction pattern provoked by and specific to environmental conditions or sensorimotor contingencies, which prepares and disposes the perceiver to interact in ways appropriate to how the conditions or sensorimotor contingencies have varied in the past, for example, as a function of movement. Similarly, perceiving a particular object is an embodied anticipatory interaction, forged and attuned by situations in which the same sensorimotor contingencies have occurred again and again—seeing the frontside of a cube is being ready to deal with its hidden sides, for example.

In fact, it is now possible to say how a sensorimotor identity theory answers the question posed in the opening phrase of this chapter, namely what it is like to see red. It is to be identical to a creature that shows a phylo and ontogenetically acquired interaction pattern adapted to the circumstances that forged such reaction. Moreover, it is equally possible to see how the sensorimotor approach accounts for perception-like awareness in circumstances in which the environmental part of the normal environment is missing, as in dreams. In such cases, parts of the interactive pattern occur and make it seem to the subject that perception occurs. But such perception-like experiences are different; that is, they lack the solidity of genuine perceptual experiences, because they are not directly regimented by environmental regularities.

5 Other Enactive Approaches

"Enactivism" is a term that encompasses a wide variety of approaches to mind and experience. These various approaches all share the view that action and interaction are at the basis of all (human and animal) mentality. The enactive nature of experience is, for example, central to the particular brand of enactivism proposed and defended by, among others, Francesco Varela and Evan Thompson. According to Mind-Life Continuity (MLC) Enactivism, as we'll call it, living beings have unique organizational properties, "and the organizational properties distinctive of mind are an enriched version of those fundamental to life. Mind is life-like, and life is mind-like" (Thompson 2007: 128). For MLC Enactivism, this principle is true of all living beings, from language-using creatures such as ourselves, right down to single cellular organisms, such as bacteria. There is thus a deep-seated continuity between mind and life. To be alive is to have a mind, albeit in the case of single cellular organisms a very primitive one.

This raises the question as to what extent MLC Enactivism is compatible with the sensorimotor approach, or if in fact it runs entirely counter to it. Answering this question depends upon those conditions needed for the doings to occur, doings that, according to the sensorimotor approach, experiences are identical to. If only living beings can engage in activity that deserves to be called a "doing" (in the sense which the sensorimotor approach uses this term), then the sensorimotor approach is a de facto brand of MLC Enactivism. Alternatively, if nonliving systems, for example artificial agents that do not share the organizational properties typical of life, are capable of such doings, then the sensorimotor approach is not compatible with MLC Enactivism. In any case, MLC enactivists have drawn attention to the fact that the sensorimotor contingencies that shape consciousness do not occur as free-floating patterns, but are rather regularities in the embodied interactions of living organisms with their environments. Moreover, they have argued that in order to provide a more complete treatment of consciousness, the sensorimotor approach "needs to be underwritten by an enactive account of selfhood or agency in terms of autonomous systems" (Thompson 2005: 417; Di Paolo, Buhrmann and Barandarian 2017 for providing an account of agency).

Hutto and Myin (2013, 2017) have also defended a view on enactivism, which they term Radical Enactivism. They have argued that many forms of cognition exist which do not involve content, where content is defined in terms of the having of truth or accuracy conditions. Hutto and Myin's proposal runs counter to ideas about perception and cognition that have become standard in philosophy and cognitive science, such as that perception always involves representing the world to a subject in a way in which the world is or could be. Hutto and Myin object that we currently don't have reasons to endorse the idea that cognition and perception always involve contentful representation. Moreover, they hold that we don't need to appeal to such contentful representations anyway, since perception, and the way in which perception interacts with other forms of cognition, can be explained without invoking content (see Hutto and Myin 2017, ch. 7).

To assess whether sensorimotor enactivism should (as argued in Hutto 2005 and Hutto and Myin 2013, ch. 2; see also Loughlin 2014) embrace Radical Enactivism, it is helpful to consider why Radical Enactivism rejects content in the case of basic perception. Radical Enactivism opposes invoking content in such cases because it rejects unexplained explainers. If one wants to invoke content in characterizing perception, and in explaining the role perception plays in further cognitive activities, then one should have a story about how content comes about and how it has effects qua content. Crucially, such a story must be about content, and not about something else—for example correlation, or isomorphism—that is merely stipulated to be content.

Recall the above account of the sensorimotor stance on internal representations as a means to explain perceptual phenomenology. Representations were rejected because they either contained an unexplained explainer or could be assimilated to the sensorimotor approach. As such, Radical Enactivism and the sensorimotor approach both oppose the invoking of representations for the same reason, namely because they are assigned the role of explainers yet they themselves are not explained.

However, while the sensorimotor approach and Radical Enactivism both reject internal representations, they do so in different contexts. Sensorimotor theorists reject representations when they are proposed to explain consciousness. Radical enactivists reject representations when they are proposed to explain cognition. Theoretically, it might be possible for a sensorimotor theorist to reject representations for explaining consciousness, while still holding on to representations for the sake of explaining cognition. Yet though such a position is theoretically possible, it is only plausible if there are good reasons to hang on to representations for cognition. Prima facie, it might seem that explaining an organism's sensitivity to sensorimotor contingencies might provide such a reason. But an organism's sensitivity to sensorimotor contingencies means nothing more than that its engagements with the environment are adapted to the fact that certain sensorimotor regularities occur. Explaining such adaptation simply requires appealing to the regularities themselves, an organism's adaptation to them, and how they are underwritten by bodily and neural changes. In her recent book on vision, Nico Orlandi states this point very clearly:

> The embedded view understands internal biases of the visual system as neurophysiological responses to environmental pressure that perform a certain function, not as representations. (…) It favors explanations that make essential reference to the environmental conditions under which vision occurs, and under which it evolved. We see edges when exposed to discontinuities in light intensity because edges are the typical environmental causes of such discontinuities—and because they are advantageous for us to see. We don't know anything, either implicitly or explicitly, about these environmental contingencies, prior to studying vision.
>
> *(Orlandi 2014: 102–103)*

The fact that representations are not required to explain sensitivity to sensorimotor contingencies further supports our proposal that the most coherent and theoretically elegant option for the sensorimotor theorists is to endorse Radical Enactivism (a conclusion reached on different grounds in Silverman, forthcoming; see also Di Paolo, Buhrmann and Barandiaran, ch. 2).

6 Conclusion

We have construed the sensorimotor approach to perceptual consciousness as proposing that episodes of perceptual awareness are identical to the engagement of organisms with their environments in ways that are sensitive, adaptive or attuned to sensorimotor contingencies. However, the fact that such doings are wide, or environment-involving, has been used to promote a prominent criticism of the sensorimotor approach to consciousness, namely that such an approach is deeply flawed. For example, Prinz (2006) has claimed that if is it indeed hard to understand how neural correlates can generate phenomenal experience, then not only is it no advance to appeal to bodily interaction with an environment, it actually compounds our original problem. For whereas previously we needed to explain how neural correlates generate experience, now we need to explain how neural correlates *plus* bodily interaction with an environment generates phenomenal experience. Going outside the head to explain experience is deemed by Prinz as nothing more than a "fool's errand."

Yet our construal of the sensorimotor approach demonstrates why this criticism is misplaced. For the sensorimotor approach, in our view, is precisely not the claim that brain plus body and environment generates experience. It is instead the claim that perceptual experience is something people (and animals) do. This is to identify experience with doing and so prevent possible generation questions from arising in the first place. We can thus clear away the sorts of unsolvable issues that have previously dogged investigations into consciousness (notably the Hard Problem of Consciousness), and thereby target those genuine empirical and theoretical issues that consciousness can raise. That they can lead us away from principally intractable problems and so clear the road for real progress reveals, so we propose, the true merit of sensorimotor and enactive approaches to consciousness.

Acknowledgments

We are grateful to Rocco Gennaro, Kevin O'Regan and Farid Zahnoun for very helpful comments on a previous version of this chapter. The research of the authors is supported by the Research Foundation Flanders (FWO), projects G048714N 'Offline Cognition' and Victor Loughlin's postdoctoral fellowship 12O9616N, 'Removing the Mind from the Head: A Wittgensteinian Perspective,' as well as the DOCPRO3 project 'Perceiving affordances in natural, social and moral environments' of the BOF Research Fund of the University of Antwerp.

References

Balibar, F. (1992) *Einstein 1905. De L'éther aux quanta*, Paris: Presses Universitaires de France.

Beecher, H.K. (1956) "Relationship of significance of wound to pain experienced," *Journal of the American Medical Association* 161: 1609–1613.

Ben-Zeev, A. (1984) "The passivity assumption of the sensation-perception distinction," *The British Journal for the Philosophy of Science* 35: 327–343.

Block, N (2005) "Review of Alva Noë, *Action in Perception*," *Journal of Philosophy* 102: 259–272.

Brooks, P. J. and Peever, J. H. (2012) "Identification of the transmitter and receptor mechanisms responsible for REM sleep paralysis," *Journal of Neuroscience* 32: 9785–9795.

Chalmers, D. J. (1996) *The Conscious Mind: In Search of a Fundamental Theory*, New York: Oxford University Press.

Crick, F. and Koch, C. (1990) "Toward a neurobiological theory of consciousness," *Seminars in the Neurosciences* 2: 263–275.

Dennett, D. C. (1991) *Consciousness Explained*, Boston, MA: Little, Brown and Co.

Di Paolo, E., Buhrmann, T. and Barandiaran, X. (2017) *Sensorimotor Life: An Enactive Proposal*, New York: Oxford University Press.

Dong, W.K., Chudler, E.H., Sugiyama, K., Roberts, V.J. and Hayashi, T. (1994) "Somatosensory, multisensory and task-related neurons in cortical area 7b (PF) of unanesthetized monkeys," *Journal of Neurophysiology* 72: 542–564.

Edelman, G. M. (1989) *The Remembered Present*, New York: Basic Books.

Gennaro, R (2017) *Consciousnes*, London: Routledge.
Hallet, M. (2000) "Transcranial magnetic stimulation and the human brain," *Nature* 406: 147–150.
Hameroff, S. R. (1994) "Quantum coherence in microtubules: A neural basis for emergent consciousness?" *Journal of Consciousness Studies* 1 (1): 91–118.
Howell, M.J. and Schenck, C.H. (2015) "Rapid eye movement sleep behavior disorder and neurodegenerative disease," *JAMA Neurology* 72: 707–712.
Hutto, D. (2005) "Knowing what? Radical versus conservative enactivism," *Phenomenology and the Cognitive Sciences* 4: 389–405.
Hutto, D. and Myin, E. (2013) *Radicalizing Enactivism: Basic Minds without Content*, Cambridge, MA: The MIT Press.
Hutto, D. and Myin, E. (2017) *Evolving Enactivism: Basic Minds Meet Content*, Cambridge, MA: The MIT Press.
Loughlin, V. (2014) "Sensorimotor knowledge and the radical alternative," in J. Bishop and A. Martin (eds.) *Contemporary Sensorimotor Theory, Studies in Applied Philosophy, Epistemology and Rational Ethics*, New York: Springer.
Melzack, R. (1973) *The Puzzle of Pain*, New York: Basic Books.
Melzack, R. (1996) "Gate control theory: On the evolution of pain concepts," *Pain Forum* 5: 128–138.
Moseley, G.L. (2007) "Reconceptualising pain according to its underlying biology," *Physical Therapy Reviews* 12: 169–178.
Myin, E. (2016) "Perception as something we do," *Journal of Consciousness Studies*, 23 (5-6): 80–104.
Myin, E. and Zahidi, Z. (in press) "Sensations," *Routledge Encyclopedia of Philosophy* Online.
Nietzsche, F. (1878/1986) *Human, All Too Human: A Book for Free Spirits*. Trans. R. J. Hollingdale, Cambridge University Press, 1996.
Noë, A. (2004) *Action in Perception*, Cambridge, MA: The MIT Press.
O'Regan, J. K. (2011) *Why Red Doesn't Sound Like a Bell: Understanding the Feel of Consciousness*, New York: Oxford University Press.
O'Regan, J.K., Myin, E., and Noë, A. (2005a) "Sensory consciousness explained (better) in terms of bodiliness and grabbiness," *Phenomenology and the Cognitive Sciences* 4: 369–387.
O'Regan, J.K., Myin, E., and Noë, A. (2005b) "Skill, corporality and alerting capacity in an account of sensory consciousness," *Progress in Brain Research* 150: 55–68.
O'Regan, J.K. and Noë, A. (2001a) "A sensorimotor account of vision and visual consciousness," *Behavioral and Brain Sciences* 24: 939–1031.
O'Regan, J.K. and Noë, A. (2001b) "What it is like to see: A sensorimotor theory of perceptual experience," *Synthese* 129: 79–83.
Orlandi, N. (2014) *The Innocent Eye: Why Vision is Not a Cognitive Process*, New York: Oxford University Press.
Place, U.T. (1956) "Is consciousness a brain process?," *British Journal of Psychology*, 47: 44–50.
Penfield, W. (1975) *The Mystery of the Mind: A Critical Study of Consciousness and the Human Brain*, Princeton, NJ: Princeton University Press.
Prinz, J. (2006) "Putting the brakes on enactive perception," *Psyche* 12 (1): 1–19.
Silverman, D. (in press) "Bodily skill and internal representation," *Phenomenology and the Cognitive Sciences*.
Smart, J.J.C. (1959) "Sensations and brain processes," *Philosophical Review* 68: 141–156.
Thompson, E. (2007) *Mind in Life: Biology, Phenomenology, and the Sciences of the Mind*, Cambridge, MA: Harvard University Press.
Thompson, E. (2015) *Waking, Dreaming, Being: Self and Consciousness in Neuroscience, Meditation, and Philosophy*, New York: Columbia University Press.
Wall, P.D. (1999) *Pain: The Science of Suffering*, London: Weidenfeld & Nicolson.
Weiskrantz, L. (1997) *Consciousness Lost and Found: A Neuropsychological Exploration*, Oxford University Press.

Related Topics

Materialism
Biological Naturalism and Biological Realism
The Neural Correlates of Consciousness
Consciousness and Action
The Biological Evolution of Consciousness
Consciousness and Dreams

Further Reading

Di Paolo, E, Buhrmann, T and Barandiaran, X. (2017) *Sensorimotor Life: An Enactive Proposal*, New York: Oxford University Press. (Combines Sensorimotor and Mind/Life Continuity Enactivism, in a manner congenial to Radical Enactivism too.)

Hutto, D. and Myin, E. (2013) *Radicalizing Enactivism: Basic Minds without Content,* Cambridge, MA: The MIT Press. (The Radical Enactivist manifesto.)

Noë, A. (2004) *Action in Perception,* Cambridge MA: The MIT Press. (Alva Noë's further elaboration of the sensorimotor contingency approach.)

O'Regan, J. K. (2011) *Why Red Doesn't Sound Like a Bell: Understanding the Feel of Consciousness*, New York: Oxford University Press. (Kevin O'Regan's further elaboration of the sensorimotor contingency approach.)

O'Regan, J.K. and Noë, A. (2001a) "A sensorimotor account of vision and visual consciousness," *Behavioral and Brain Sciences*, 24: 939–1031. (The classic original statement of the sensorimotor contingency approach.)

Thompson, E. (2007) *Mind in Life: Biology, Phenomenology, and the Sciences of the Mind*, Harvard University Press. (An extensive presentation and defence of Mind/Life Continuity Enactivism.)

16
QUANTUM THEORIES OF CONSCIOUSNESS

Paavo Pylkkänen

> …quantum consciousness theory offers not just a solution to the mind-body problem, or additionally, to the nature of life and of time… And it does not just solve the Agent-Structure and Explanation-Understanding problems, or explain quantum decision theory's success in predicting otherwise anomalous behavior. What the theory offers is all of these things and more, and with them a unification of physical and social ontology that gives the human experience a home in the universe. With its elegance… comes not just extraordinary explanatory power, but extraordinary meaning, which at least this situated observer finds utterly lacking in the classical worldview. … I hope I have given you reason to suspend your belief that we really are just classical machines, and thus to suspend your disbelief in quantum consciousness long enough to try assuming it in your work. If you do, perhaps you will find your own home in the universe too.
>
> *(Alexander Wendt,* Quantum Mind and Social Science*, 2015: 293)*

1 Introduction

Much of contemporary philosophy of mind and cognitive neuroscience presupposes that the physical framework to use when explaining mind and consciousness is the framework of classical physics (and neurophysiological and/or computational processes embedded in this framework); it is typically assumed that no ideas from quantum theory, or relativity theory, are needed. Of course, not all theories of consciousness are trying to reduce conscious experience to mechanistic physical interactions at the neural level, but this tacit commitment to the classical physics of Newton and Maxwell introduces a strong mechanistic element into contemporary theorizing about consciousness, at least whenever the theories make a reference to physical processes.

One could argue that much of mainstream consciousness studies is an attempt to "domesticate" the radically non-mechanistic and experiental features of conscious experience by trying to force them to fit into the mechanistic framework (cf. Ladyman and Ross 2007: 1–2). Some researchers are happy to assume that people are just very complicated machines, or even (philosophical) zombies – machines who think they are conscious, while in fact they are just walking computers, with no such exotic features as qualia, subjectivity, experiencing and the like. Others feel that consciousness remains unexplained rather than explained by these mechanistic explanatory attempts, but even they cannot come up with a view that coherently unites conscious experience and physical processes. Thus some kind of uneasy dualism of the mental and the physical (whether acknowledged or not) often looms in those theories of consciousness that take conscious experience seriously.

However, it has been known since the early 20th century that classical physics provides a very limited, albeit useful, description of the physical world. Classical physics fails completely in certain important domains; at high speeds and with large masses we need special and general theories of relativity (respectively), and at the atomic level, we need quantum theory. Because of experimentally detected features, such as the indivisibility of the quantum of action, wave-particle duality and non-locality (to be briefly explained below), it can be argued that quantum theory requires a holistic rather than a mechanistic picture of reality. The mechanistic world of relatively independent objects that we find in everyday experience is then a special, limiting case that arises from a more fundamental dynamical ground, in which wholeness prevails (Bohm 1980, ch. 7; Pylkkänen 2007).

It is widely agreed that conscious experience has dynamical and holistic features. Could it be that these features are in some way a reflection of the dynamic and holistic quantum physical processes associated with the brain that could underlie (and make possible) the more mechanistic neurophysiological processes that contemporary cognitive neuroscience is measuring? If so, these macroscopic processes would be a kind of shadow, or amplification of the results of quantum processes at a deeper, possibly pre-spatial level where our minds and conscious experience essentially live and unfold (cf. Penrose 1994). The macroscopic, mechanistic level is of course necessary for communication, cognition and life as we know it, including science; but perhaps the experiencing (consciousness) of that world and the initiation of our actions takes place at a more subtle, non-mechanical level of the physical world, which quantum theory has begun to discover (Bohm and Hiley 1993: 176–180).

In this chapter, after a brief historical introduction to quantum theory, we will see that the theory opens up some radically new ways of thinking about the place of mind and consciousness in nature. This need not (at least not always) deny what the other theories of consciousness are saying, but also complement them. At the very least, a quantum perspective will help a "classical" consciousness theorist to become better aware of some of the hidden assumptions in his or her approach. Given that consciousness is widely thought to be a "hard" problem (Chalmers 1996), its solution may well require us to question and revise some of our assumptions that now seem to us completely obvious. This is what quantum theory is all about – learning, on the basis of scientific experiments, to question the "obvious" truths about the nature of the physical world and to come up with more coherent alternatives.

2 Quantum Theory: A Brief Introduction

Quantum theory originated at the turn of the 19th and 20th centuries when Planck and Einstein were studying certain experiments in which matter exchanged energy with the electromagnetic field (this section relies mostly on Bohm 1984: 70–84 and Ney 2013). Classical physics assumed that matter is composed of bodies that move continuously (determined by Newton's laws), while light consists of waves in the electromagnetic field (determined by Maxwell's equations). This implies that matter and light should exchange energy in a continuous fashion. However, to explain the photoelectric effect (in which a beam of light ejects electrons from within a metal), Einstein postulated in 1905 that light transmits energy to matter in the form of small indivisible particles or "quanta". Planck had a few years earlier postulated the existence of such quanta when explaining black-body radiation; thus, the theory was to be called "quantum theory."

The above did not, however, mean that the wave nature of light that had been experimentally detected already in 1801 in Young's two-slit interference experiment was given up. On the contrary, the energy of a "particle" of light was given by the famous Planck-Einstein equation $E = hf$, where h is Planck's constant and f is the frequency of the light. Thus, the energy of a *particle* of light depends on the frequency of the *wave aspect* of the same light. Light thus has *both* wave *and* particle properties, and this somewhat paradoxical feature is called wave-particle duality.

Quantization of energy was also postulated in Bohr's 1913 model of the atom, to explain the discrete spectra emitted by a gas of, say, hydrogen. In this model a hydrogen atom consists of a proton in the nucleus, and an electron orbiting it. Bohr postulated that only certain energy levels are allowed for the electron, and when the electron jumps from a higher to a lower level, it emits a quantum of light with $E = hf$. Conversely, in order to jump from a lower to a higher level it needs to absorb a quantum of a suitable energy. A limited number of allowed energy levels implies a limited number of possible jumps, which in turn gives rise to the discrete spectral lines that had been observed.

It became possible to explain the discrete (quantized) energies of atomic orbits when de Broglie postulated in 1923 that atomic particles have a wave associated with them (Wheaton 2009). This implies that wave-particle duality applies to all manifestations of matter and energy, not just to light. In an enclosure, such as when confined within an atom, such a wave associated with an electron would vibrate in discrete frequencies (a bit like a guitar string), and if we assume that the Planck-Einstein relation $E = hf$ holds for de Broglie's waves, then discrete frequencies imply discrete energy levels, as in Bohr's model (Bohm 1984: 76).

Finally, Schrödinger discovered in 1926 an equation that determines the future motion of de Broglie's waves (which are mathematically described by a complex wave function ψ), much in the same way as in classical physics Maxwell's equations determine the future motions of electromagnetic waves. One puzzle was how the wave function ought to be interpreted. Schrödinger was hoping to give it a physical interpretation, but did not manage to do this at the time. Max Born suggested in 1926 that the wave function describes a *probability density* for finding the electron at a certain region. More precisely the probability density ρ at a given region is given by the square of the absolute value of the wave function, or the probability amplitude $| \psi |^2$ in that region, which is known as the Born rule $\rho = | \psi |^2$.

Another important development was Heisenberg's uncertainty principle. If, in a given moment, we want to measure both the position (x) and the momentum (p) of a particle, the uncertainty principle gives (roughly) the maximal possible accuracy $\Delta p \Delta x \geq h$ (Δp is uncertainty about momentum, Δx is uncertainty about position, h is Planck's constant, also known as the quantum of action, where action $h = Et$). This limits what we can know about a particle. But how should we interpret the uncertainty principle? Does the electron always have a well-defined position and momentum, but it is for some reason difficult for us to get knowledge about them at the same time (the epistemic interpretation)? Or does the electron not even have simultaneously a well-defined position and momentum (the ontological interpretation)? (von Wright 1989).

To observe an electron with light, we need at least one light quantum, with the energy $E = hf$. Bohr assumed that such a quantum (or more precisely the quantum of action $h = Et$) is indivisible, and its consequences in each measurement are unpredictable and uncontrollable. Because of such nature of the quantum link in each measurement, Bohr said that the form of the experimental conditions and the meaning of the experimental results are a whole that is not further analyzable. This whole constitutes what Bohr called the "quantum phenomenon." Such wholeness means that the results of experiment cannot be ascribed to the properties of a particle that is assumed to exist independently of the rest of the quantum phenomenon. So Bohr interpreted the uncertainty principle in an ontological sense. We cannot define the state of being of the observed system because this state is inherently ambiguous. Depending on the experimental set-up, we can apply either the concept of position or momentum. But these concepts are *complementary*: incompatible yet both necessary for a full description of the possible quantum phenomena. The situation is very different from that in classical physics (Bohm and Hiley 1993: 13–17; Faye 2014; Plotnitsky 2010; Pylkkänen 2015).

In 1935 Schrödinger drew attention to a curious holistic feature of quantum mechanics, which he called *Verschränkung*, later translated as "entanglement". This played a key role in the 1935 thought experiment by Einstein, Podolsky and Rosen (EPR). Bohr had said that because of the uncertainty principle it is meaningless to talk about an electron as if this had simultaneously a well-defined momentum and position. However, quantum mechanics implies that there are quite generally situations where two systems that interact with each other can become entangled. EPR pointed out that if two such entangled systems are separated from each other, their properties remain correlated in such a way that by measuring the position of a particle A one can obtain information about the position of particle B, and the same for momentum – and according to them this happens, "without in any way directly influencing B." But surely, argued EPR, the particle B must have both a well-defined position and a well-defined momentum already prior to measurement, if an experimenter can choose which one of these she wants to measure (i.e., an experimenter can choose to measure either the position or the momentum of particle A, and in this way [without disturbing B] get information about either the position or the momentum of particle B; so surely particle B must have these properties well-defined, waiting to be revealed?). EPR concluded that quantum theory is incomplete, as it cannot account for the simultaneous existence of the position and momentum of particle B, i.e. properties which they thought that obviously exist.

Bohr's reply to EPR emphasized that we should not, like EPR did, attribute properties to particle B, conceived in isolation from a particular quantum phenomenon involving a particular experimental set-up (see Fine 2016).

But for those physicists who think that quantum theory describes a world that exists independently of the observer, entanglement implies that experimental interventions at subsystem A influence subsystem B *instantaneously*, without any mediating local contact between them. Because relativity requires that signals cannot be transmitted faster than the speed of light, Einstein considered such non-locality "spooky," but experiments seem to imply a non-locality in nature (see Aspect et al. 1982; Bricmont 2016, ch. 4). We will return to the issue of non-locality below in connection with the Bohm interpretation of quantum theory.

A better understanding of some of the above ideas can be obtained by considering the famous two-slit experiment. When classical particles (e.g. bullets) pass through a wall with one or two slits in it, they build up either one or two piles on the detecting screen, depending on whether one or two slits are open. With waves the situation is different. If the size of the slit is roughly the same as that of the wavelength, the wave will bend or *diffract* after it passes through the slit. With two slits open, the diffracted waves from the two slits will meet and *interfere* with each other, giving rise to an interference pattern where areas where the waves add to produce a wave of large amplitude alternate with areas where the waves cancel each other out.

What happens with electrons with two slits open? The electron has typical particle properties such as mass and charge, so physicists expected that it should behave like a little bullet. However, the electrons collectively build up an interference pattern (Tonomura et al. 1989). They appear at the plate one by one at localized points, which suggests that they are particles. But it seems that each individual electron *also* has wave-like properties – for how else could the individual systems "co-operate" to build up an interference pattern? Note that we get an interference pattern even if we send just one electron at a time, so the pattern is not produced by the electrons interacting with each other. (For an entertaining video demonstration of the two-slit experiment, see e.g. Dr. Quantum's lecture on YouTube, "Dr Quantum – Double Slit Experiment". The lecture is an excerpt from the film *What The Bleep: Down The Rabbit Hole*. There is some simplification and interpretation in the demo, but it gives a nice visual illustration of the experiment.)

Let us now see what the different interpretations of quantum theory say about situations like the two-slit experiment, and also consider what kind of theories of mind and consciousness some interpretations have inspired.

3 The Bohr Interpretation

We have already discussed Bohr's views, so I will describe them only briefly here. Bohr said in a minimalist way that we should think of the wave function merely as a mathematical tool, as a part of an algorithm we use to calculate probabilities for the observed results of experiments. So, in the two-slit experiment we can use the Born rule to obtain probabilities for where the spots will appear in the photographic plate. As we have seen, Bohr's interpretation is very subtle and emphasizes the unanalyzable wholeness of a quantum phenomenon. Bohr did suggest that quantum theory could be relevant to understanding biological systems and even the mind (see e.g. Bohr 1934: 99), and his writings inspired others to start thinking about such issues; but as Bohr did not advance a detailed quantum theory of mind or consciousness we will not consider his view here further.

4 Von Neumann's Interpretation: Consciousness Collapses the Wave Function

Other physicists, such as Dirac and von Neumann, assumed that the quantum theory describes quantum reality, saying that the wave function provides the most complete possible description of the so-called "quantum state" of the electron. Bohm and Hiley (1993: 20) provide a succinct description of von Neumann's (1955) view of the quantum state and its relation to the large-scale level where we observe the results of measurement:

> This state could only be manifested in phenomena at a large-scale (classical) level. Thus he was led to make a distinction between the quantum and classical levels. Between them, he said there was a "cut." This is, of course, purely abstract because von Neumann admitted, along with physicists in general, that the quantum and classical levels had to exist in what was basically one world. However, for the sake of analysis one could talk about these two different levels and treat them as being in interaction. The effect of this interaction was to produce at the classical level a certain observable experimental result. … But reciprocally, this interaction produced an effect on the quantum level; that is the wave function changed from its original form ψ to ψ_n, where n is the actual result of the measurement obtained at the classical level. This change has been described as a "collapse" of the wave function. Such a collapse would violate Schrödinger's equation, which must hold for any quantum system. However, this does not seem to have disturbed von Neumann unduly, probably because one could think that in its interaction with the classical level such a system need not satisfy the laws that apply when it is isolated.

So note that two changes take place as a result of the interaction between the quantum level and the classical level. On the one hand there will be an observable effect (e.g. a macroscopic pointer pointing to a given value) at the classical level. On the other hand, it is assumed that at the quantum level the wave function will collapse from what typically is a superposition of many possible states to a single state (a so-called "eigenstate"). Note also that the terms "quantum state" and "wave function" are used interchangeably in the above quote, which is common in the discussion about the quantum theory. In this way of talking, the term "wave function" is taken to refer to the physical quantum field that exists objectively in some sense, and not merely to a piece of mathematics.

However, there is a problem in von Neumann's approach. It is not clear what causes the collapse, because von Neumann thought that the location of the cut between the quantum level and the classical level was arbitrary. He thought that we can in principle include the observed quantum object and the measuring apparatus as part of a single combined system, which has to be treated quantum mechanically (Bohm and Hiley 1993: 20). To bring about the collapse of the wave function of this combined system, we then need to bring in a second measuring apparatus at the classical level to interact with the combined quantum system. But because the place of the cut is arbitrary, even this second apparatus can be included in the combined system, which requires that we introduce yet another classical apparatus, if we want to bring about a collapse, and so on. If we keep going we realize that even the brain of the observer could in principle be included in the combined quantum system. However, at the end of the experiment we experience a definite outcome rather than a complex superposition of possible states, so it seems obvious that a collapse has taken place somehow. But how could the collapse possibly happen anywhere in the physical domain, given that the cut between the quantum and classical levels is arbitrary and can be moved indefinitely? This, essentially, is the (in)famous measurement problem of quantum theory.

Given this problem, von Neumann and Wigner (1961) were led to speculate that it is only when we bring in something non-physical, namely the consciousness of the observer, that we need not apply a non-collapsed wave function ψ and we get the definite outcome (e.g. a spot at a definite location *n*) we observe and can then describe the quantum system with the collapsed wave function ψ_n. This idea that it is only consciousness that can cause the collapse of the wave function and thus account for the well-defined physical reality we find in every-day experience is a historically important suggestion about the role of consciousness in quantum theory (for a critical discussion of von Neumann's and Wigner's ideas, see e.g. Bohm and Hiley 1993: 19–24; see also Stapp 1993).

In recent years, the von Neumann-Wigner approach has been advocated and modified, especially by Henry Stapp. Alexander Wendt (2015) provides a succinct summary of Stapp's (2001) approach:

> Whereas Wigner argued that consciousness causes collapse, Stapp sees the role of the mind here as more passive, as coming to know the answer nature returns to a question. Importantly, the two roles of the mind both involve the brain/mind complex. In contrast to Cartesian dualism, therefore, Stapp's ontology is more like a psycho-physical duality or parallelism, in which every quantum event is actually a pair: a physical event in an entangled brain-world quantum system that reduces the wave function to an outcome compatible with an associated (not causal) psychical event in the mind.
>
> *(Wendt 2015: 84)*

The above implies that the collapse takes place without consciousness playing a causal role. It is not possible here to enter into a detailed analysis of Stapp's view, but Wendt's summary indicates that he has developed the approach in subtle ways (see also Atmanspacher 2015).

5 Penrose and Hameroff: Quantum Collapse Constitutes Consciousness

Later on physicists such as Ghirardi, Rimini and Weber (1986), as well as Diósi (1989) and Penrose (1996) have developed concrete physical models about how the collapse of the quantum state happens objectively, without the consciousness of the observer having to play any role. Typically this type of theory involves introducing a mathematically described mechanism which accounts for the collapse in situations where we expect there to be just one outcome (rather than a number of possibilities typically implied by the description in terms of an uncollapsed wave function that obeys the Schrödinger equation). Thus, in the two-slit experiment we may say – in a somewhat simplified way – that the electron is a wave (described by the wave function) when it moves, but when it interacts with matter in the photographic plate, the wave collapses into a small region with a probability that obeys the Born rule and we observe a definite outcome. While this type of theory aims to show that there is no need for consciousness for there to be definite outcomes, for Penrose and Hameroff a certain kind of quantum collapse *constitutes* moments of conscious experience, and thus plays a key role in their quantum theory of consciousness. Let us now briefly examine this theory.

In his book *The Emperor's New Mind* Penrose was concerned with the physical underpinnings of human mathematical insight or understanding (Penrose 1989). Reflecting upon Gödel's theorem, he was led to propose that human conscious understanding is non-computable. As he wanted to avoid the dualism of mind and matter, the question then became what sort non-computable physical process could underlie mathematical insight. After considering some possibilities, he suggested that the most likely candidate would be a certain kind of collapse or reduction of the quantum state. However, this would not be the usual random collapse of the quantum state (which obeys the Born rule), but rather a more subtle kind of collapse induced by gravity in some circumstances, or what Penrose later called an orchestrated objective reduction – "Orch-Or" for short.[1]

The question then arose concerning where in the brain such a collapse could possibly be taking place. The kind of large-scale coherent quantum states that Penrose needed in his model are fragile, and would, it seemed, be easily destroyed by the so-called environmental decoherence taking place in the warm, wet and noisy environment of the human brain. There should thus be some way in which the coherent quantum states could be protected from decoherence, so that they would survive long enough and then collapse in a suitable way, to properly underlie conscious understanding in the way Penrose's model had proposed. Penrose was aware that Fröhlich (1968) had suggested that there should be vibrational effects within active cells, as a result of a biological quantum coherence phenomenon. These effects were supposed to arise from the existence of a large energy of metabolic energy and should not need a low temperature (Penrose 1994: 352).

Penrose then discovered that the anesthesiologist Stuart Hameroff had suggested that a computation-like action takes place within the microtubules in the cytoskeleton of neurons (Hameroff and Watt 1982; Hameroff 1987). Could such microtubules be a sufficiently protected site in the brain where the kind of large-scale quantum-coherent behavior and collapse, proposed by Penrose to underlie conscious understanding, might happen? Penrose and Hameroff teamed up and proposed in the mid 1990's the Orch-Or theory of consciousness, which today is the best-known quantum theory of consciousness. In a 2014 review article Hameroff and Penrose summarize their proposal:

> consciousness depends on biologically "orchestrated" coherent quantum processes in collections of microtubules within brain neurons, ... these quantum processes correlate with, and regulate, neuronal synaptic and membrane activity, and ... the continuous Schrödinger evolution of each such process terminates in accordance with the specific Diósi-Penrose scheme of "objective reduction" of the quantum state. This orchestrated

> OR activity ("Orch-Or") is taken to result in moments of conscious awareness and/or choice.
>
> *(Hameroff and Penrose 2014: 39)*

Note that this provides a concrete suggestion for a mechanism for how the "quantum mind" could influence (and be influenced by) the large-scale, classical neural processes that mainstream cognitive neuroscience is focusing upon.

There have been many criticisms of the Penrose-Hameroff proposal, often in prestigious scientific journals, for example by Grush and Churchland (1995), Tegmark (2000a and 2000b), Litt et al. (2006), Koch and Hepp (2006), Reimers et al. (2009) and McKemmish et al. (2009). However, Hameroff and Penrose have provided detailed responses to the criticisms, and the theory still remains a live option, albeit an exotic one (for a summary of and references to their replies see their 2014: 66–68; for discussion see Wendt 2015: 102–108).

6 Everett's Many Worlds Interpretation

Yet other physicists have tried to account for the experimental quantum phenomena without postulating a collapse. One radical possibility is to follow Everett (1954) and assume that in each situation where the wave function implies a number of possible outcomes, but we perceive only one outcome (e.g. an electron at point *n*), there is no collapse of the quantum state, but instead the world at a macroscopic level branches into copies so that there is a branch corresponding to each possible outcome. So with two possible outcomes ($x = 1$ or $x = 2$) the world branches into two copies that differ in that in one of them the macroscopic pointer indicates, say, that the electron is at point $x = 1$ (which the observer in that branch sees) and in the other one it is at point $x = 2$ (which the observer in that branch sees), and so on. In the two-slit experiment there are a large number of possible places where the electron can be detected, and correspondingly the world branches into a large number of copies each time an electron is detected (Lewis 2016: 6). While this "many worlds" interpretation may sound very implausible, some physicists find it attractive because they think it best reflects the experimentally verified Schrödinger equation and has also other virtues (Saunders et al. 2010; Wallace 2012). Some researchers have even proposed in the context of the Everett theory that each conscious brain is associated with many minds, where some of the minds follow each branch! (Albert and Loewer 1988; Lockwood 1989, 1996; for discussion see Lewis 2016: 132–133).

7 The Bohm Interpretation: The Wave Function Describes Active Information

Yet another interpretation which avoids the need to postulate a collapse is due to de Broglie (1927) and Bohm (1952 a, b). This assumes that the electron is a particle always accompanied by a new type of quantum field, described by the wave function. We will focus on Bohm and Hiley's (1987, 1993) version of the de Broglie-Bohm interpretation and will call it hereafter "the Bohm theory" (for de Broglie's views, see Bacciagaluppi and Valentini 2009).

In the Bohm theory the field gives rise to a quantum potential, which influences the movement of the particle, besides classical potentials, and in this way gives rise to quantum effects. Let us see how the theory deals with the two-slit experiment. In Figure 16.1 the particles are coming towards us from the two slits. When a particle passes a slit it will encounter the quantum potential which arises from the quantum field that has passed both slits and interfered with itself.

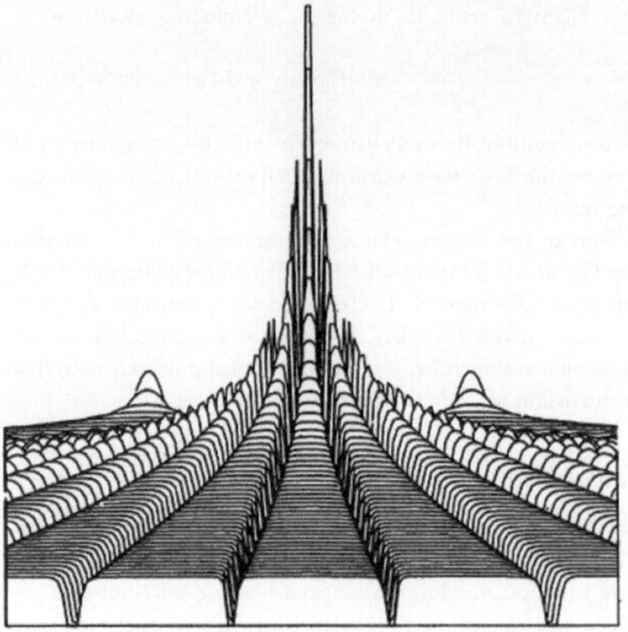

Figure 16.1 Quantum Potential for Two Gaussian Slits (from Philippidis, Dewdney and Hiley 1979). Reprinted with kind permission of Società Italiana di Fisica, copyright (1979) by the Italian Physical Society (https://link.springer.com/article/10.1007%2FBF02743566)

One can think of a potential as a bit analogous to a mountain, so that the quantum potential will, for example, keep the electrons away from areas where it has a high value.

The particles (electrons) have their source in a hot filament, which means that there is a random statistical variation in their initial positions. This means that each particle typically enters the slit system in a different place. The Bohm theory assumes that this variation in the initial positions is typically consistent with the Born rule, so that the theory gives the same statistical predictions as the usual quantum theory. Figure 16.2 shows some possible trajectories that an electron can take after it goes through one of the slits. Which trajectory it takes depends, of course, on which place it happens to enter the slit system. The theory provides an explanation of the two-slit experiment without postulating a collapse of the wave function.

Note that the trajectories in the Bohm theory should be seen as a *hypothesis* about what may be going on in, say, the two-slit experiment. Because of the uncertainty principle we are not able to observe the movement of individual quantum particles. However, there is currently an attempt to experimentally determine the *average trajectories* of atoms by making use of the measurements of so-called weak values (Flack and Hiley 2014). Over the years there have been many criticisms of the de Broglie-Bohm interpretation, but its proponents have been able to provide answers (see Goldstein 2013; Bricmont 2016).

When Bohm re-examined his 1952 theory with Basil Hiley in the early 1980s, he considered the mathematical form of the quantum potential. With classical waves the effect of the wave upon a particle is proportional to the amplitude or size of the wave. However, in Bohm's theory the effect of the quantum wave depends only upon the *form* of the quantum wave, not on its amplitude (mathematically, the quantum potential depends upon the second spatial derivative of the amplitude). Bohm realized that this feature might be revealing something important about

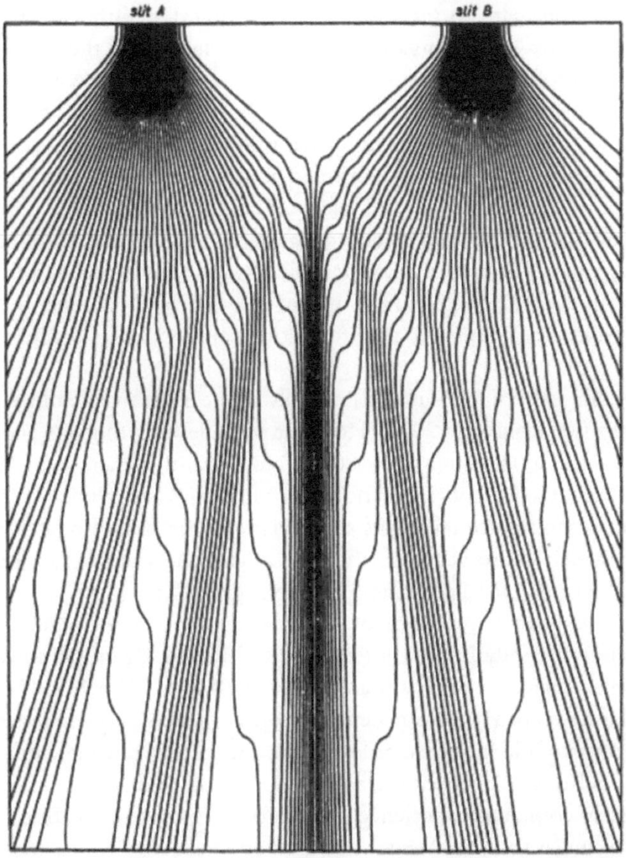

Figure 16.2 Trajectories for Two Gaussian Slits (from Philippidis, Dewdney and Hiley 1979) with kind permission of Società Italiana di Fisica, copyright (1979) by the Italian Physical Society (https://link.springer.com/article/10.1007%2FBF02743566)

the nature of quantum reality. For instead of saying that the quantum wave pushes and pulls the particle mechanically, the mathematics suggests that the form of the quantum field is literally informing the energy of the particle. This is somewhat analogous to the way a radar wave guides a ship on autopilot. The radar wave is not pushing and pulling the ship, but rather the form of the radar wave (which reflects the form of the environment) informs the greater energy of the ship. Analogously, Bohm thought that the quantum field carries information about the form of the environment (e.g. the presence of slits) and this information directs the particle to move in a particular way.

Another puzzling feature in quantum mechanics (and also in Bohm's theory) is that the wave function for a many-body system necessarily lives in a 3N-dimensional configuration space (where N is the number of particles in a system). So for a two-particle entangled system the wave lives in a six-dimensional space, and so on. But how could one possibly give a physical interpretation to such a multidimensional field? This was not a problem for Niels Bohr, because he thought we should not give an ontological interpretation to the wave function in the first place. But approaches that assume that the wave function describes reality have to deal with this issue of multidimensionality (for a discussion, see Ney and Albert 2013).

The idea of active information also helps to make sense of this multidimensionality, for it is common to think that information can be organized multidimensionally. If the essential nature of the quantum field is information, then it is perhaps not such a mystery that it is organized in a multidimensional way. This does not mean that Bohm's suggestion is not exotic – for one thing the Bohmian multidimensional information mediates non-local correlations through the quantum potential. But as was mentioned above, experiments indicate that there exists some kind of quantum non-locality in nature. This seems to create a tension with relativity, according to which it is not possible to signals faster than the speed of light. However, Bohm and Hiley point out that it is not possible to send signals non-locally by modulating the wave function (1993: 282–284). Also, recent research by Walleczek and Grössing (2016) shows how a certain kind of non-local information transfer can be compatible with the theory of relativity.

Bohm and Hiley's proposal about active information has not always been received enthusiastically in the physics community (see e.g. Riggs 2008). However, some leading thinkers take it seriously (e.g. Holland 1993; Smith 2003; Khrennikov 2004). Note also that there exists a more minimalist version of the Bohm theory known as "Bohmian mechanics" which does not give the quantum potential a great significance (and thus usually ignores the notion of active information). (For this approach which has some support among philosophers of physics, see Goldstein 2013; Bricmont 2016; Bell 1987; for a discussion, see Holland 2011.)

Bohm had been interested in the possible relevance of quantum theory to understanding the nature of mind and consciousness already in his 1951 textbook *Quantum Theory*, pointing to some striking analogies between quantum processes and thought (Bohm 1951: 168–172; Pylkkänen 2014). In the 1960s he developed a more general framework for physics, which he called the implicate order. The notion of the implicate order tries to capture the flowing, undivided wholeness of quantum and relativistic phenomena, and Bohm also applied it to describe the holistic and dynamic features of conscious experience, such as time consciousness (Bohm 1980, 1987; Pylkkänen 2007). In a similar vein, he thought that the notion of active information is relevant to understanding the relationship between mind and matter. He proposed that the active information carried by the quantum field could be seen as a primitive mind-like quality of, say, an electron. This sounds like a panpsychist move, but Bohm thought it was obvious that an electron does not have consciousness, and was thus not embracing panpsychism in the traditional sense, which attributes experience to the ultimate constituents of the world (Bohm 1989, 1990; Pylkkänen, forthcoming; cf. Strawson 2006a, b).

How might the above be relevant to the mind-matter problem? Bohm and Hiley suggested that it is natural to extend the quantum ontology (1993: 380). So, just as there is a quantum field that informs the motion of the particle, there could be a super-quantum field that informs the movement of the first-order quantum field, and so on. Bohm speculated that the information in our mental states could be a part of the information contained in this hierarchy of fields of quantum information. This way, the information in our mental states could influence neural processes by reaching the quantum particles and/or fields in a suitable part of the brain (e.g. in synapses or microtubules or other suitable sites, to be revealed by future quantum brain theory). In effect, Bohm was proposing a solution to the problem of mental causation.[2]

8 Explaining Qualia in a Quantum Framework

We have above given a brief introduction to some aspects of quantum theory, as well as to some quantum theories of mind and consciousness. However, the above only gives a small glimpse of the great variety and diversity of such theories. In this section we will approach the question differently, by taking up an essential feature of consciousness, namely *qualia*, and considering how some of the quantum approaches might help to explain them.

Presumably, the most discussed and debated feature of conscious experience is its qualitative character – the blueness of the sky, the taste of chocolate, and similar sensory qualia. Do quantum theories of consciousness have anything to say about qualia? In further developments of their theory, Hameroff and Penrose have introduced an explicitly panpsychist element to it. For they (2014: 49) note that the Diósi-Penrose proposal suggests that "each OR [objective reduction] event, which is a purely physical process, is itself a primitive kind of 'observation,' a moment of 'proto-conscious experience'." They (2014: 72) further elaborate this idea: "…in the Orch-OR scheme, these [non-orchestrated OR] events are taken to have a rudimentary subjective experience, which is undifferentiated and lacking in cognition, perhaps providing the constitutive ingredients of what philosophers call *qualia*." The idea is that the unorchestrated and ubiquitous objective reductions involve proto-qualia, but when such reductions are orchestrated (e.g. in the human brain), then qualia in a full sense emerge. Of course, this idea may sound very speculative and even *ad hoc*; but given that very little can be said about the origin of qualia in the mechanistic classical physical framework of mainstream neuroscience, perhaps one should keep an open mind here.

Also, we saw above how Bohm and Hiley proposed that the wave function describes a field of active information, which can be seen as a primitive mind-like quality of the particle. The idea of quantum theoretical active information is perhaps most naturally seen as proposing that electrons have "proto-cognition" (because of the information aspect) and "proto-will" (because the information is fundamentally active) (cf. Wendt 2015: 139). But in search of a panpsychist solution to the hard problem of consciousness one could also, somewhat similar to Chalmers's (1996) double-aspect theory of information, postulate that Bohmian quantum theoretical active information has proto-phenomenal and proto-qualitative aspects. Such proto-qualia could be the *content* of such active information, a kind of "proto-meaning" that active information has for the electron (cf. Pylkkänen 2007: 244–246). Again, this is very speculative, but the basic idea is that the quantum ontology with its subtle, non-classical properties provides the ground from which qualia in a full sense might emerge, in a suitably organized biological or artificial system.

9 Quantum Biology, Quantum Cognition and Quantum Computation

The attempt to explain mind and consciousness in terms of the quantum theory involves heavy speculation – can we really cross the explanatory gap with a quantum leap? While we may not be able to answer that question in the near future, it is worth noting that in recent years we have seen significant advances in other areas where the ideas and formalisms of quantum theory have been applied to new domains. In biology, it has been shown how quantum effects (e.g. quantum-coherent energy transfer and entanglement) are likely to play a role photosynthesis and avian magnetoreception (Ball 2011; Lambert et al. 2013). Lambert et al. (2013: 16) conclude their review article of quantum biology in *Nature Physics* as follows:

> The fact that there is even the possibility of a functional role for quantum mechanics in all of these systems suggests that the field of quantum biology is entering a new stage. There may be many more examples of functional quantum behavior waiting to be discovered.

These advances in quantum biology, while not giving direct support to quantum brain theory, perhaps make a biologically grounded quantum theory of consciousness seem less inconceivable.

Another area where there has been interesting cutting-edge research is quantum cognition (sometimes also called "quantum interaction"). In recent years a number of researchers have proposed that certain principles and mathematical tools of quantum theory (such as quantum probability, entanglement, non-commutativity, non-Boolean logic and complementarity) provide a good way of modeling many significant cognitive phenomena (such as decision processes, ambiguous perception, meaning in natural languages, probability judgments, order effects and memory; for an introduction, see Wang *et al.* 2013; Pothos and Busemeyer 2013; Busemeyer and Bruza 2012). While quantum cognition researchers are typically agnostic regarding whether there are any significant quantum effects in the neural processes underlying cognition, it can be argued that the success of quantum cognition also provides support for the stronger quantum mind and consciousness programs (Wendt 2015: 154–155).

Finally, there has been significant research in areas such as quantum information, computation and cryptography, providing yet another example where it has been valuable to apply quantum theory to new domains (Bouwmeester et al. 2000).

There are a number of important quantum approaches to mind and consciousness that we have not covered in this short review: There is the quantum field theoretical program that involves a quantum view of memory, going back to Umezawa and Ricciardi (Ricciardi and Umezawa 1967; Jibu and Yasue 1995; Vitiello 2001; Globus 2003; for a succinct account see Atmanspacher 2015). There is also Beck and Eccles's (1992) proposal that synaptic exocytosis can be controlled by a quantum mechanism (see Atmanspacher 2015; Hiley and Pylkkänen 2005). Eccles saw this proposal as opening up a way for the (non-physical) self to control its brain, without violating the energy conservation laws. In a recent development, the physicist Matthew Fisher has given support to a strong version of quantum cognition by proposing that quantum processing with nuclear spin might be operative in the brain (Fischer 2015). There are also interesting approaches that see quantum theory as grounding a double-aspect view of mind and matter and which have been inspired by the ideas of Jung and Pauli (Atmanspacher 2014, 2015).

Many tend to dismiss quantum theories of consciousness as too speculative and implausible. Others, however, hold that it is only through such radical thinking, guided by our best scientific theories, that we will ever make progress with the harder problems of mind and consciousness.

Notes

1 See his 1994 book *Shadows of the Mind* for a detailed exposition of these ideas; for criticisms by a number of commentators as well as Penrose's reply, see the internet journal *Psyche* at http://journalpsyche.org/archive/volume-2-1995-1996/; see also Pylkkö (1998, ch. 4).

2 Bohm (1990); Pylkkänen (1995, 2007, 2017); Hiley and Pylkkänen (2005); for criticisms see Kieseppä (1997a, b) and Chrisley (1997); for a reply see Hiley and Pylkkänen (1997).

References

Aspect, A., Grangier, P. and Roger, G. (1982) "Experimental test of Bell's inequalities using time-varying analyzers," *Physical Review Letters* 49: 1804–1807.

Atmanspacher, H. (2014) "20th century variants of dual-aspect thinking (with commentaries and replies)," *Mind and Matter* 12: 245–288.

Atmanspacher, H. (2015) "Quantum approaches to consciousness," *The Stanford Encyclopedia of Philosophy* (Summer 2015 Edition), E. N. Zalta (ed.), URL = http://plato.stanford.edu/archives/sum2015/entries/qt-consciousness/.

Bacciagaluppi, G. and Valentini, A. (2009) *Quantum Theory at a Crossroads: Reconsidering the 1927 Solvay Conference*, Cambridge: Cambridge University Press.

Ball, P. (2011) "The dawn of quantum biology," *Nature* 474: 272–274, URL= http://www.nature.com/news/2011/110615/full/474272a.html.

Beck, F. and Eccles, J. (1992) "Quantum aspects of brain activity and the role of consciousness," *Proceedings of the National Academy of Sciences* 89 (23): 11357–11361.

Bell, J. (1987) *Speakable and Unspeakable in Quantum Mechanics*, Cambridge: Cambridge University Press.

Bohm, D. (1951) *Quantum Theory*, Englewood Cliffs, NJ: Prentice Hall. Dover edition 1989.

Bohm, D. (1952 a and b) "A suggested interpretation of the quantum theory in terms of 'hidden variables' I and II," *Physical Review* 85 (2): 166–179 and 180–193.

Bohm, D. (1980) *Wholeness and the Implicate Order*, London: Routledge.

Bohm, D. (1984) *Causality and Chance in Modern Physics*, London: Routledge. New edition with new preface. First edition published in 1957.

Bohm, D. (1987) "Hidden variables and the implicate order," in B.J. Hiley and F. D. Peat (eds.) *Quantum Implications: Essays in Honour of David Bohm*, London: Routledge.

Bohm, D. (1989) "Meaning and information," in P. Pylkkänen (ed.) *The Search for Meaning*, Wellingborough: Crucible.

Bohm, D. (1990) "A new theory of the relationship of mind and matter," *Philosophical Psychology* 3: 271–286.

Bohm, D. and Hiley, B. J. (1987) "An ontological basis for quantum theory: I. Non-relativistic particle systems," *Physics Reports* 144: 323–348.

Bohm, D. and Hiley, B. J. (1993) *The Undivided Universe: An Ontological Interpretation of Quantum Theory*, London: Routledge.

Bohr, N. (1934) *Atomic Theory and the Description of Nature*, Cambridge: Cambridge University Press (new edition 2011).

Bouwmeester, D., Ekert, A., and Zeilinger, A. K. (eds.) (2000) *The Physics of Quantum Information: Quantum Cryptography, Quantum Teleportation, Quantum Computation*, Heidelberg and Berlin: Springer.

Bricmont, J. (2016) *Making Sense of Quantum Mechanics*, Heidelberg: Springer.

Busemeyer, J. and Bruza, P. (2012) *Quantum Models of Cognition and Decision*, Cambridge: Cambridge University Press.

Chalmers, D. (1996) *The Conscious Mind: In Search of a Fundamental Theory*, Oxford: Oxford University Press.

Chrisley, R. C. (1997): "Learning in non-superpositional quantum neurocomputers," in Pylkkänen et al. 1997.

Diósi, L. (1989) "Models for universal reduction of macroscopic quantum fluctuations," *Physical Review A* 40: 1165–1174.

Everett, H. III (1957) "'Relative state' formulation of quantum mechanics," reprinted in J. Wheeler and W. Zurek (eds.) (1983) *Quantum Theory and Measurement*, Princeton, NJ: Princeton University Press.

Faye, J. (2014) "Copenhagen interpretation of quantum mechanics," *The Stanford Encyclopedia of Philosophy* (Fall 2014 Edition), E. N. Zalta (ed.), URL = https://plato.stanford.edu/archives/fall2014/entries/qm-copenhagen/.

Fine, A. (2016) "The Einstein-Podolsky-Rosen argument in quantum theory," *The Stanford Encyclopedia of Philosophy* (Fall 2016 Edition), E. N. Zalta (ed.), URL = https://plato.stanford.edu/archives/fall2016/entries/qt-epr/.

Fischer, M.P.A. (2015) "Quantum cognition: The possibility of processing with nuclear spins in the brain," *Annals of Physics* 362: 593–602.

Flack, R. and Hiley, B. J. (2014) "Weak measurement and its experimental realisation," *Journal of Physics: Conference Series* 50. arXiv:1408.5685

Fröhlich, H. (1968) "Long range coherence and energy storage in biological systems," *International Journal of Quantum Chemistry* II: 641–649.

Ghirardi, G. C., Rimini, A. and Weber, T. (1986) "Unified dynamics for microscopic and macroscopic systems," *Physical Review D* 34: 470.

Globus, G. (2003) *Quantum Closures and Disclosures*, Amsterdam: John Benjamins.

Goldstein, S. (2013) "Bohmian mechanics," *The Stanford Encyclopedia of Philosophy*, E. N. Zalta (ed.), URL = http://plato.stanford.edu/archives/spr2013/entries/qm-bohm/

Grush, R. and Churchland, P.S. (1995) "Gaps in Penrose's toilings,'" *Journal of Consciousness Studies*, 2 (1): 10–29.

Hameroff, S. (1987) *Ultimate Computing: Biomolecular Consciousness and Nano-Technology*, Amsterdam: North-Holland.

Hameroff, S., and Penrose, R. (2014) "Consciousness in the universe: A review of the 'Orch OR theory'," *Physics of Life Reviews* 11: 39–78.

Hameroff, S. and Watt, R. (1982) "Information processing in microtubules," *Journal of Theoretical Biology* 98: 549–61.

Hiley, B.J. and Pylkkänen, P. (1997) "Active information and cognitive science: A reply to Kieseppä," in Pylkkänen et al. 1997.
Hiley, B.J. and Pylkkänen, P. (2005) "Can mind affect matter via active information?," *Mind and Matter* 3 (2): 7–26, URL = http://www.mindmatter.de/resources/pdf/hileywww.pdf
Holland, P. (1993) *The Quantum Theory of Motion: An Account of the de Broglie-Bohm Causal Interpretation of Quantum Mechanics*, Cambridge: Cambridge University Press.
Holland, P. (2011) "A quantum of history," *Contemporary Physics* 52: 355.
Jibu, M. and Yasue, K. (1995) *Quantum Brain Dynamics and Consciousness*, Amsterdam: John Benjamins.
Kieseppä, I. A. (1997a) "Is David Bohm's notion of active information useful in cognitive science?" in Pylkkänen et al. 1997.
Kieseppä, I. A. (1997b) "On the difference between quantum and classical potentials – A reply to Hiley and Pylkkänen," in Pylkkänen et al. 1997.
Khrennikov, A. (2004) *Information Dynamics in Cognitive, Psychological and Anomalous Phenomena*, Series Fundamental Theories of Physics 138, Dordrecht: Kluwer.
Koch, C. and Hepp, K. (2006) "Quantum mechanics in the brain," *Nature* 440: 661.
Ladyman, J. and Ross, D. (with Spurrett, D. and Collier, J.) (2007) *Every Thing Must Go: Metaphysics Naturalized*, Oxford: Oxford University Press.
Lambert, N. Chen Y.-N., Cheng, Y.-C., Li, C.-M, Chen, G.-Y. and Nori, F. (2013) "Quantum biology," *Nature Physics* 9: 10–18.
Lewis, P. (2016) *Quantum Ontology: A Guide to the Metaphysics of Quantum Mechanics*, Oxford: Oxford University Press.
Litt, A., Eliasmith, D., Kroon, F., Weinstein, S., and Thagard, P. (2006) "Is the brain a quantum computer?," *Cognitive Science* 30: 593–603.
Lockwood, M. (1989) *Mind, Brain and the Quantum*, Oxford: Blackwell.
Lockwood, M. (1996) "Many minds interpretations of quantum mechanics," *British Journal for the Philosophy of Science* 47: 159–188.
McKemmish, L., Reimers, J., McKenzie, R., Mark, A., and Hush, N. (2009) "Penrose-Hameroff Orchestrated objective-reduction proposal for human consciousness is not biologically feasible," *Physical Review E* 80 (2): 021912-1 to 021912-6.
Ney, A. (2013) "Introduction," in A. Ney and D. Albert (eds.) (2013).
Ney, A. and Albert, D. (eds.) (2013) *The Wave Function: Essays on the Metaphysics of Quantum Mechanics*, Oxford: Oxford University Press.
Penrose, R. (1989) *The Emperor's New Mind*, Oxford: Oxford University Press.
Penrose, R. (1994) *The Shadows of the Mind*, Oxford: Oxford University Press.
Penrose, R. (1996) "Wavefunction collapse as a real gravitational effect," *General Relativity and Gravitation* 28: 581–600.
Philippidis, C., Dewdney, C. and Hiley, B. J. (1979) "Quantum interference and the quantum potential," *Il Nuovo Cimento* 52: 15–28.
Plotnitsky, A. (2010) *Epistemology and Probability. Bohr, Heisenberg, Schrödinger and the Nature of Quantum-Theoretical Thinking*, Heidelberg and New York: Springer.
Pothos, E. M. and Busemeyer, J. R. (2013) "Can quantum probability provide a new direction for cognitive modeling?" *Behavioral and Brain Sciences* 36: 255–327.
Pylkkänen, P. (1995) "Mental causation and quantum ontology," *Acta Philosophica Fennica* 58: 335–348.
Pylkkänen, P. (2007) *Mind, Matter and the Implicate Order*, Berlin and New York: Springer Frontiers Collection.
Pylkkänen, P. (2014) "Can quantum analogies help us to understand the process of thought?" *Mind and Matter* 12: 61–91, URL= http://www.mindmatter.de/resources/pdf/pylkkaenen_www.pdf.
Pylkkänen, P. (2015) "The quantum epoché," *Progress in Biophysics and Molecular Biology* 119: 332–340.
Pylkkänen, P. (2017) "Is there room in quantum ontology for a genuine causal role of consciousness?," in A. Khrennikov and E. Haven (eds.) *The Palgrave Handbook of Quantum Models in Social Science*, London: Palgrave Macmillan UK.
Pylkkänen, P. (forthcoming) "A quantum cure for panphobia," to appear in W. Seager (ed.) *Routledge Handbook of Panpsychism*, London: Routledge.
Pylkkänen, P., Pylkkö, P. and Hautamäki, A. (eds.) (1997) *Brain, Mind and Physics*, Amsterdam: IOS Press.
Pylkkö, P. (1998) *The Aconceptual Mind: Heideggerian Themes in Holistic Naturalism*, Amsterdam: John Benjamins.
Reimers, J., McKemmish, L., McKenzie, R., Mark, A., and Hush, N. (2009) "Weak, strong, and coherent regimes of Frohlich condensation and their applications to terahertz medicine and quantum consciousness," *Proceedings of the National Academy of Sciences* 106: 4219–4224.

Ricciardi, L. and Umezawa. H. (1967) "Brain and physics of many-body problems," *Kybernetik* 4 (2): 44–48.
Riggs, P. (2008) "Reflections on the de Broglie – Bohm quantum potential," *Erkenntnis* 68: 21–39.
Saunders, S., Barrett, J., Kent, A. and Wallace, D. (eds.) (2010) *Many Worlds? Everett, Quantum Theory, & Reality*, Oxford: Oxford University Press.
Smith, Q. (2003) "Why cognitive scientists cannot ignore quantum mechanics?" in Q. Smith and A. Jokic (eds.) *Consciousness: New Philosophical Perspectives*, Oxford: Oxford University Press.
Stapp, H. (1993) *Mind, Matter and Quantum Mechanics*, Berlin: Springer Verlag.
Stapp, H. (2001) "Quantum theory and the role of mind in nature," *Foundations of Physics* 31: 1465–1499.
Strawson, Galen (2006a) "Realistic monism – why physicalism entails panpsychism," *Journal of Consciousness Studies* 13 (10–11): 3–31.
Strawson, Galen (2006b) "Panpsychism? Reply to commentators with a celebration of Descartes," *Journal of Consciousness Studies* 13 (10–11): 184–280.
Tegmark, M. (2000a) "Importance of quantum decoherence in brain processes," *Physical Review E* 61 (4): 4194–4206.
Tegmark, M. (2000b) "Why the brain is probably not a quantum computer," *Information Sciences* 128: 155–179.
Tonomura, A., Endo, J, Matsuda, T., Kawasaki, T. and Ezawa, H. (1989) "Demonstration of single-electron buildup of an interference pattern," *American Journal of Physics* 57: 117–120.
Vitiello, G. (2001) *My Double Unveiled: The Dissipative Quantum Model of the Brain*, Amsterdam: John Benjamins.
von Neumann, J. (1955) *Mathematical Foundations of Quantum Mechanics*, Princeton: Princeton University Press. (First edition in German, *Mathematische Grundlagen der Quantenmechanik*, 1932.)
von Wright, G.H. (1989) "Images of science and forms of rationality," in S. J. Doorman (ed.) *Images of Science: Scientific Practise and the Public*, Aldershot: Gower.
Wallace, D. (2012) *The Emergent Multiverse: Quantum Theory according to the Everett Interpretation*, Oxford: Oxford University Press.
Walleczek, J. and Grössing, G. (2016) "Nonlocal quantum information transfer without superluminal signalling and communication," *Foundations of Physics* 46: 1208–1228.
Wang, Z., Busemeyer, J. R., Atmanspacher, H., and Pothos, E. M. (2013) "The potential of using quantum theory to build models of cognition," *Topics in Cognitive Science* 5: 672–688.
Wheaton, B.R. (2009) "Matter waves," in D. Greenberger, K. Hentschel and F. Weinert (eds.) *Compendium of Quantum Physics: Concepts, Experiments, History and Philosophy*, Berlin: Springer.
Wigner, E. (1961) "Remarks on the mind-body problem," in I.J. Good (ed.) *The Scientist Speculates*, London: Heinemann.

Related Topics

Materialism
Dualism
Idealism, Panpsychism, and Emergentism

Further Reading

Atmanspacher, H. (2015) "Quantum approaches to consciousness," *The Stanford Encyclopedia of Philosophy* (Summer 2015 Edition), E. N. Zalta (ed.), URL – http://plato.stanford.edu/archives/sum2015/entries/qt-consciousness/. (Essential reading for anyone interested in quantum theories of consciousness; see also the many other articles on quantum theory in *The Stanford Encyclopedia*.)
Bohm, D. and Hiley, B.J. (1993) *The Undivided Universe: An Ontological Interpretation of Quantum Theory*, London: Routledge. (An attempt to make quantum theory intelligible which includes accurate descriptions and critical reflections of the views of Bohr, von Neumann, Everett, Ghirardi, Rimini & Weber, Stapp and Gell-Mann & Hartle.)
Polkinghorne, J. (2002) *Quantum Theory: A Very Short Introduction*, Oxford: Oxford University Press. (A remarkably lucid introduction to quantum theory for the uninitiated.)
Wendt, A. (2015) *Quantum Mind and Social Science: Unifying Physical and Social Ontology*, Cambridge: Cambridge University Press. (An insightful and comprehensive review of the revolutionary quantum mind proposals by a leading social scientist.)

PART III

Major Topics in Consciousness Research

17

THE NEURAL CORRELATES OF CONSCIOUSNESS

Valerie Gray Hardcastle and Vicente Raja

1 What Is the Neural Correlate of Consciousness?

At first blush, it seems that explaining what the "Neural Correlate of Consciousness" (NCC) is should be straightforward: it is whatever that happens in our brains when we have a conscious experience that is lacking when we are not having conscious experiences. But this simple answer is misleading. It turns out that there might not be an NCC – even if we adopt a purely materialistic and reductionistic framework for explaining consciousness.

To explain more definitely what NCC references, we must first say a bit about what "consciousness" refers to. Although dualists and materialists disagree with each other on just about everything, they do agree, in general, about the phenomena they are trying to explain. As John Searle explains, consciousness refers to "those states of sentience and awareness that typically begin when we awake from a dreamless sleep and continue until we go to sleep again, or fall into coma or die or otherwise become unconscious" (1997: 5).

Being sentient or aware seems to be a pretty straightforward account of consciousness, but a closer look reveals that it does not fully account for the complexity of conscious phenomena. Intuitively, we would want to say that both of the following are instances of being conscious: being aware of the brown table in front of me as being brown and in front of me, on the one hand, and being alert and ready to interact with the environment, on the other. These two cases are different enough from one another that it might make more sense to understand consciousness as a set of phenomena and not as a unitary thing. At least until we know more about what consciousness is exactly, we should divide conscious phenomena into at least two categories: being aware of a perception and being in a state such that we can have such a perception in the first place.

Following David Chalmers (1998, 2000), we can talk about content states of consciousness and background states of consciousness. Content states of consciousness align more closely with what most philosophers mean when they talk about consciousness. These states are "the fine-grained states of subjective experience that one is in at any given time" (Chalmers 2000: 19). They encompass, for instance, the experience of the sound pattern of a song one is listening to, the experience of the softness of a surface one is touching, or the experience of one's own train of thoughts. These conscious states are specific events in our day-to-day experience and Chalmers calls them "content states," because they are usually differentiated by their content (i.e., the specific sound pattern, the specific haptic feeling, the specific thought, and so on). The content states of consciousness are also sometimes called phenomenal states (Block 1995, 2004) or subjective experiences (Dennett 1993) or qualia (Crane 2000; Jackson 1982). These names all refer to "the way things seem to us" (Dennett 1993: 382). While we may not know what consciousness is exactly, we do know a lot about where and how perception is processed in our brains. Some of these experiences are conscious, so perhaps this is the way in to identifying what it is that happens in our brain when we are conscious.

The background states of consciousness are "overall state[s] of consciousness such as being awake, being asleep, dreaming, being under hypnosis, and so on" (Chalmers 2000: 18). This is what most doctors are referring to when they say that their patient is conscious. These states are the common framework for other more specific conscious states (the content states of consciousness) and can influence the latter. For instance, different background states of consciousness, such as being alert as a healthy individual or being alert but with schizophrenia may affect the way one perceives objects and events in the world. We would expect – indeed, we know – that the brains of patients with schizophrenia are structurally different from the brains of normal controls, and that the two different types of brains can react differently when given the same stimuli. We know some things that are going on in the brain that keeps us alert and oriented – though not as much as we do about perception – and perhaps this too is a way in to identifying what it is about the brain that connects alertness with being aware.

It is clear that the two main categories of conscious states are related to each other in important ways, but that they are also connected to different brain structures. The part of our brain that perceives things in our environment is different from the part of our brain that keeps us oriented to the world. And yet, something must distinguish those neurons or neural firing patterns or brain structures that are aligned with consciousness from those that are not. If we could isolate what that something is, then perhaps we will have found the NCC.

Francis Crick and Christof Koch (1990) were the first ones to discuss the idea of NCCs in the scientific press in a serious fashion (see also Crick 1994; Crick and Koch 1995; Koch 2004; Rees et al. 2002). They started by assuming an unabashedly materialistic position and argued that eventually consciousness has to be explained by something at the neural level. They then assumed that "all different aspects of consciousness ... employ a basic common mechanism or perhaps a few such mechanisms" (1990: 264). That is, they just assumed that whatever consciousness is, it can be explained by a single thing in the brain – the thing that distinguishes conscious brain activity from unconscious brain activity. This is an important and significant presumption, for they were claiming more than consciousness being just correlated with some sort of brain activity (despite the name "neural correlates of consciousness"). They are claiming that understanding the difference between conscious and unconscious phenomena will depend fundamentally on understanding something about the brain.

In their original paper, they argued that consciousness is realized by a group of cortical neurons all firing together in unison at some particular frequency (see also Hardcastle 1995; Singer 1999). This is one possibility for what the mechanism of consciousness might be, but many other proposals have been floated over the years. For example, at about the same time as Crick and Koch were postulating neural oscillations as the correlate for consciousness, Gerald Edelman (1989) hypothesized that consciousness was localized to the thalamocortical and limbic systems; ten years later, in a similar vein, Antonio Damasio (1999) claimed consciousness was to be found in the frontal-limbic nexus.

Edelman distinguishes between two types of immediate consciousness awareness: primary and higher-order. "Primary" consciousness refers to awareness of objects and their properties present in the world around us. "Higher-order" consciousness refers to being aware that we are aware. It is the mental model we have of ourselves as a thinking, experiencing creature. He notes that the interactions of our thalamus and cortex allow us to perceive things in the world around us. When our thalamocortical system is connected to our limbic system, we are then able to assign valances and values to things in the world. Edelman claims that a special reentrant signaling process evolved in our cortical systems that permit us to connect our memories of valued things with incoming perceptions in real time and in parallel across multiple sensory systems. With the advent of this special process, primary consciousness appeared.

Damasio also postulates two types of consciousness: core consciousness and extended consciousness. These map directly onto Edelman's primary and higher-order consciousness. Like Edelman – and unlike Crick and Koch – he does not see consciousness as a single unified phenomenon. Also much like Edelman, Damasio believes that the interactions among the limbic system, the thalamic region, and cortical areas are the correlates for his "core" consciousness. Though he emphasizes different aspects of the structures (less importance is attached to reentrant signaling; more is given to the contributions of the posteromedial cortex), in both views, the thalamocortical system is key to understanding the neural correlate of primary or core consciousness.

Both Edelman and Damasio conclude that multiple brain regions underlie higher-order or extended consciousness, for our self-models need access to many different memory systems, among other things. Later research suggests that perhaps the "default network" recently uncovered in imaging studies might be a central actor in the neural correlates of our conscious sense of self (Addis et al. 2004; Vogeley and Fink 2003). (The default network [or default mode network, as it is sometimes called] refers to the multiple interconnected regions of the brain that remain active when a person is not thinking about or noticing anything in particular. It is what is active by default, as it were.)

Other proposals for the NCC have included left hemisphere based interpretative processes (Gazzaniga 1988), global integrated fields (Kinsbourne 1988), the extended reticular-thalamic activation system (Newman and Baars 1993), intralaminar nuclei in thalamus (Bogen 1995), neural assemblies bound by NMDA (Flohr 1995), action-prediction-assessment loops between frontal and midbrain areas (Gray 1995), hemostatic processes in the periacqueductal gray region (Panksepp 1998), and thalamically modulated patterns of cortical activation (Llinás 2001). Suffice it to say, there really is no agreement among scientists or philosophers regarding what the NCC might be.

More importantly, as these suggestions have piled up over time, we are beginning to realize that perhaps Crick and Koch were wrong in their initial assumption that there is a single brain mechanism that would account for all of consciousness. For each suggestion really only attempts to explain how the hypothesized brain structure or activity gives rise to some aspect or other of consciousness. For example, global fields or transient synchronous firing assemblies of neurons might indeed underlie individual subjective experiences, but thalamic projections into the cortex could help knit the diverse individual experiences into a single integrated conscious perceptual experience of the world. Left hemisphere interpretative processes could explain our sense of conscious self-awareness over time, and the reticular-activating system could help to account for our background sense of alertness. It is possible that each of these distinct neural theories is true, with each contributing some partial explanation of the full complexity of consciousness.

Despite Crick and Koch's initial assertion, there is no reason to think that consciousness cannot be realized in various locations or by utilizing a number of different mechanisms. Perhaps there is no single NCC, but we should be looking for several different neural mechanisms to account for the full range of conscious phenomena. Instead of the "neural correlate of consciousness," we should seek instead the "neural correlates of consciousness" (the NCCs).

The complexity of what the brain might be doing that differentiates being conscious from not being conscious suddenly becomes enormous. A better research agenda might be to investigate different aspects of conscious awareness, looking for different NCCs that underlie each one.

In hindsight, simply assuming an NCC appears to be theoretical overreach. And yet, even if we adopt the more sophisticated approach for identifying underlying brain activities or processes associated with aspects of consciousness, there are still several troubling problems with this putative research agenda.

2 Some Problems with NCCs

Consciousness is often described as being completely puzzling – McGinn (1991) refers to it as a "residual mystery." Such a description is completely understandable because of consciousness's ill-defined boundaries, its fully subjective character, the vividness of our phenomenological experiences, and, perhaps most importantly, the complete opacity of its purpose. What does being conscious do for us? Or what does it allow us to do? For any particular human activity, it seems that we (or perhaps a computer) could do the same task without being conscious. And indeed, for many human abilities, we do have machines that can mimic them successfully. This hazy status of consciousness in relation to human thought and behavior raises questions regarding its relation to the science of psychology or neurobiology. Here, we look at two of these problems: the *explanatory gap* and the *hard problem of consciousness*.

The "explanatory gap" (Levine 1983; Horgan 1999) is the name given to the inability of physical theories to fully account for the phenomenology of consciousness. Any scientific account of consciousness – any brain theory about the NCC – faces the problem of explaining how the rich Technicolor of conscious phenomenology just is, or is the product of, some non-experiential physical interaction. Saying "the experience of the color red just is coherent oscillations of such-and-such neurons in area V4" says nothing about how these oscillations give rise to the feeling of seeing the color red. While the experience of the color red might be *correlated with* neural oscillations, that is not the same thing as them being reducible to or identical with these oscillations. Any putative reduction of consciousness to some physical interaction seems to leave out the very thing it wants to explain: the conscious experience itself. There is necessarily a gap in any putative scientific account of consciousness and its target of explanation.

And this of course creates problems for people like Crick and Koch, who believe that identifying the NCC will provide a theoretical explanation for what consciousness is. If there is an explanatory gap, it will not really help us to know that coincident oscillations in the brain co-vary with consciousness, for example, because brain oscillations are just too different a thing from conscious mental states. We have no intellectual or theoretical bridge from one to the other. Without such a bridge, knowing about these sorts of co-variations is not going to be explanatorily helpful, and if it is not explanatorily helpful, then it will not provide a good foundation for a scientific theory of consciousness.

Discussions of the explanatory gap come in different degrees of severity. Some think the explanatory gap is a kind of practical inability of our current scientific theories (Dennett 1991; Nagel 1974). In principle, it might be possible to bridge the gap, but our current scientific theories are not ready to do it yet. Others agree that it is in principle possible to bridge the explanatory gap; however, it is impossible for creatures like us to do it (McGinn 1991, 1995; Papineau 1995, 2002; Van Gulick 1985, 2003). Human beings are just lacking the cognitive capacity to close the explanatory gap. Finally, there are those who claim that the explanatory gap is impossible to resolve in principle, that the gap is conceptually unbridgeable (Chalmers 1996, 2005; Jackson 1993). Dualists primarily believe this strongest version.

We can see the hard problem of consciousness (Chalmers 1996, 2010) as the other side of the explanatory gap coin. The explanatory gap refers to the putative impossibility of accounting for subjective experience by physical theories, and the hard problem of consciousness asserts that physical theories cannot account for why subjective experience exists at all. Proponents of the hard problem argue that science cannot account for why some systems are conscious, but others are not. The hard problem of consciousness assumes the veracity of the explanatory gap, in other words.

David Chalmers articulates the issue in the following way:

> What makes the hard problem hard and almost unique is that it goes beyond problems about the performance of functions. To see this, note that even when we have explained the performance of all the cognitive and behavioral functions in the vicinity of experience – perceptual discrimination, categorization, internal access, verbal report – there may still remain a further unanswered question: *Why is the performance of these functions accompanied by experience?*
>
> *(Chalmers 1995: 202, emphasis in original)*

Even if all of science is completed, there will always remain the further question of why some of the systems science has explained are conscious, in addition to whatever else it is they do. Chalmers believes that such question cannot be addressed in physical terms; otherwise, science would have addressed it. Therefore, consciousness has to be something non-physical. Others believe that the hard problem is just that – a hard problem – and it does not make any further claims about the metaphysics of consciousness.

These two problems affect consciousness studies in general. However, the problems are especially relevant for neurobiological approaches to consciousness, and hypotheses about the NCC in particular. The challenge for NCCs raised by these two problems can be understood as operating at two levels. On the one hand, it is possible that no NCC could ever account for the whole phenomena of consciousness. That is to say, after finding the neural correlate for some specific subjective experience, we still will not be able to know why it is that particular experience that feels that particular way. Even if we have a well-constructed and refined theory of the neural correlates of consciousness, it may never completely explain what we really care about with respect to phenomenological experience. On the other hand, NCCs may simply be the wrong approach to explaining consciousness. If the subjective experience of the color red, or any other qualia, is non-physical, then seeking their neural correlates seems to be a task without a purpose.

Most advocates for trying to identify the NCCs have little patience with these two alleged problems (cf., Churchland 1986; Crick and Koch 1990; Hardcastle 1995). They see the arguments as some sort of exaggerated reasoning from oddness that many dualists have, a kind of intellectual hysteria, as it were. In particular, they believe that proponents of the explanatory gap and the hard problem do not understand how science proceeds. A lot of science tackles strange things and many of its explanations are counter-intuitive and, frankly, intellectually unsatisfying. Quantum mechanics can be like this. That some folk now cannot see how a biological theory of consciousness could account for the raw feelings of phenomenology is not a strike against biology or a victory for consciousness mysterians; rather, it says something about those folk. Perhaps their inability to see how 40-Hz oscillations just are a conscious visual experience points to a failure of imagination on their part, and not to a failure of science.

But more importantly, science is in the business of seeking correlations. Smoking is correlated with lung cancer. Physiologists can dig into the chemistry of cigarette smoke and the biological composition of lung tissue to help flesh out this correlation. We learn that benzo(a) pyrene, a chemical found in cigarette smoke, is correlated with damage to DNA, and chromium, another chemical in cigarette smoke, is correlated with benzo(a)pyrene sticking more actively to DNA, and arsenic, yet another chemical in cigarette smoke, is correlated with slower DNA repair processes. But, it is just correlations all the way down. Some of the correlations we call "causes" to emphasize their importance in our ultimate story, but all scientific investigations can ever give us are a series of correlations.

We can turn the series of correlations into an explanatory narrative: the chemicals in cigarette smoke cause cancer because benzo(a)pyrene damages our DNA, while chromium helps benzo(a)pyrene stick to our DNA, which increases the amount of damage done, and arsenic prevents the DNA damaged by benzo(a)pyrene from repairing itself. And it is this narrative of a series of correlations that gives us the satisfying sense of a good explanation. But it is very difficult, in a pre-narrative state, to argue that any possible story about how something comes about is just not going to work for us. We have to start with the correlations and then build from there.

Any science of consciousness follows exactly this pattern. Perhaps we find some correlations in the brain between neural activity and consciousness. Neurophysiologists can then dig into these correlations to learn more precisely what is correlated with what, which we can then turn into an explanatory narrative. There is nothing different about consciousness that makes its science suspect or more difficult. The only thing that is different is that people cannot see how the narrative might go before we have the correlations. But we daresay that that has been true for a lot of things we would like to have had explained: like the plague before we knew about bacteria, or fire before we knew about oxidation. Who could have antecedently imagined germs or invisible chemical reactions before we uncovered correlations that indicated that these things existed? At best, right now, we can say that science should do its work and then later we can see where we stand with respect to moving from NCCs to a story about where consciousness comes from and what it is. It could be that we never will be able to develop an intellectually satisfying neurobiological explanation of consciousness. But even if we cannot, it is unclear, at this stage in the game, whether this would be because consciousness falls outside the realm of what science can explain, or because we failed in our scientific endeavors, or because we do not like what our science is telling us.

3 Embodied Approaches to NCC

A different sort of problem for uncovering NCCs comes from positions that are purely materialistic, but not brain-centered. We are speaking of "embodied cognition" approaches to understanding the mind (Calvo and Gomila 2008; Shapiro 2014). Embodied cognition refers to a wide range of theories of cognition that assume that any explanation of a cognitive process will also have to reference both the body and the environment. As such, embodied cognition approaches challenge brain-exclusive explanations of cognitive phenomena.

Andy Clark and David Chalmers (1998) highlight the relevance of body and environment for cognition using the following demonstration. First, they challenge students to do a difficult multiplication problem (each multiplicand is ten digits long, say) in their heads. This is hard, if not impossible, for most math-literate students to complete successfully. Then, they pose the same challenge again, but this time the students get to use pencil and paper. The task suddenly becomes much easier. What differentiates the two tasks? The math problem is exactly the same. What changes are the resources available to accomplish the assignment. If a pencil, a piece of paper, and a hand to write with are all crucial for accomplishing this cognitive task, then it would seem that any explanation of mathematical ability requires more than simply a description of brain activity. Proponents of embodied cognition believe that these sorts of examples demonstrate that body and environment have to be included in our explanations of the mind, for our minds can only function in our particular bodies interacting with our particular environment. And we need to understand our bodies and our environment in order to understand our minds.

Consciousness seems to be a cognitive phenomenon and, as such, should therefore perhaps also be considered embodied. This means that NCCs alone will not suffice in explaining subjective experience. For defenders of embodied cognition, as much as we have to appeal to a pencil, paper, and a hand in offering a complete explanation of the cognitive process behind multiplication, we would also need to appeal to the relevant aspects of body and environment in order to offer a complete explanation of consciousness. Hence, NCCs might form some part of a scientific theory of consciousness – they could potentially help us understand connections between bodily action and conscious experience – but, in and of themselves, they would never be a complete explanation of the phenomena (see also Hutto and Myin 2013).

For example, we have data concerning the neural correlates of consciousness in action. Increased activity in the posterior parietal cortex, which is tied to our intending to move and our picking out which action we want to select (Tosoni et al. 2008), is correlated with our experiences of motor execution in the environment (Desmurget and Sirigu 2009, 2012). When surgeons gently stimulate Brodman's areas 39 and 40 (which form part of the parietal cortex) in awake and alert patients undergoing brain surgery, the patients report experiencing intentions to move, as well as illusory movements themselves. When the electrical stimulation increases, patients believe they actually have moved, even though there is no neuromuscular activity. In contrast, when surgeons stimulate the premotor region, the area of the brain associated with actual bodily movements, the patients have no conscious awareness of any action, even though they really do move (Desmurget et al. 2009).

Interestingly, our conscious experiences of action seem to be independent of the physical movements themselves. Instead, we become aware of intentions to move just slightly in advance of the movements themselves (cf., Haggard 2005), and these experiences double as our experience of the movement itself (cf., Desmurget and Sirigu 2009, 2012). Conscious awareness then seems to co-occur with intending to move purposefully in the world instead of actually moving. That is, consciousness (or, more accurately, some aspect of consciousness) might be correlated with our planning how to move our bodies in our current environments.

If embodied cognition is the right approach for explaining cognition, then we are left with two questions for explaining consciousness. First, what are the relevant features of body and environment that would be included in a complete theory of consciousness? And second, what kind of descriptions and methodologies should be used to integrate NCCs, the body, and the environment into the one explanation? We explore two different approaches to answering these questions using embodied perspectives: *neurophenomenology* and *extended conscious mind*. The first supports the project of identifying NCCs; the second does not.

Francisco Varela (1996) first proposed the idea of "neurophenomenology," which combines first-person reports of conscious experiences with the neurophysiological approach typically used in NCC research. Neurophenomenology integrates phenomenological and neurophysiological investigations of conscious experience with each other, while at the same time, trying to make explicit the relationship these two methodologies have to each other.

Phenomenology has a long and multifarious history in philosophy, starting with Edmund Husserl (1900, 1913, 1928) in the early 20th century. But the basic idea behind phenomenology is a rigorous examination of the structure of conscious experience, as experienced from a first-person point of view. The expectation is that all first-person experiences have invariant features, things that are common to all conscious experiences. Identifying those stable features is the ultimate goal of a phenomenological investigation.

The "neuro" of neurophenomenology refers to a physiological account of consciousness, the same as the NCC project. However, proponents of neurophenomenology put a twist on the basic NCC assumptions. The classic Crick and Koch approach to the NCCs is unidirectional, moving from neural activity to conscious experience. That is, something about some particular set of neurons (or their interactions) causes consciousness to occur. But neurophenomenologists believe that neural events, which are embedded in bodies and their environments, and conscious states exist in a bidirectional or "reciprocal" relationship (Thompson and Varela 2001: 418). Conscious states emerge from brain-body-world interactions and then they in turn constrain what the brain-and-body can do in its environment.

Neurophenomenology describes NCCs as part of a larger and more complex system that accounts for consciousness as a whole. Using embodied approaches to consciousness, in general, and neurophenomenology, in particular, means that consciousness must be understood in terms of brain-body-environment interactive systems, with each component constraining and being constrained by the others. Any complete explanation of a conscious experience will have to integrate elements across all these different elements, showing how they are all connected to each other. Classic conceptions of NCCs as the things in the brain that are the only things that are correlated with conscious experiences are wrong-headed. NCCs are likely to be included in any explanation of consciousness, but a complete one will have to take in account other aspects of both the body and the environment.

In addition, neurophenomenology requires that the aspects of the body and the environment that play a role in conscious experiences are constitutive parts of the experiences themselves. The core idea is that, unlike more reductive accounts of consciousness, both body and environment are causally relevant for consciousness; both body and environment are part of the mechanisms of consciousness, just as the NCCs are. There is a multi-directional causal relation between NCCs, the body, and the environment, with each affecting the others as the others are also affecting each. A consequence of this view is that how to articulate what the NCCs are is likely to be extended in space and in time, as the usual ways that the body and the environment affect neural firings in the brain is through very particular neurosensory and neuromuscular junctions.

In general, according to proponents of neurophenomenology, the way to understand the underlying physiology of consciousness goes as follows: a localized neural state causally affects large-scale neural dynamics, which then causes the body to move, which impacts the environment, which causes changes in sensory inputs, which affects the large-scale neural dynamics, which also change local neural states. A full explanation of consciousness should account for all these different types of interactions and the multi-directional relations among them.

Perhaps most importantly, though: "phenomenologically precise first-person data produced by employing first-person methods provide strong constraints on the analysis and interpretation of the physiological processes relevant to consciousness" (Lutz and Thompson 2003: 33). That is, we need our subjective descriptions of conscious experiences to help us interpret what is happening inside the brain. Perhaps, too, as we use our first-person descriptions to inform our neuroscience, then the "third-person data produced in this manner might eventually constrain first-person data, so that the relationship between the two would become one of dynamic 'mutual' or 'reciprocal constraints'" (Lutz and Thompson 2003: 33).

The methodology of neurophenomenology is bidirectional as well. On the one hand, using a phenomenological analysis, subjects will provide refined and precise reports of conscious experiences to researchers, which could provide important details that otherwise might be glossed over. Such a practice might describe distinctions between two similar conscious events that would have remained unnoticed without this type of analysis, for example. This then might improve the analysis of the physiological data, as the target of investigation would be clearer. Small differences in EEG results, for example, could gain a new meaning, if small differences in conscious experiences are articulated. On the other hand, a more detailed physiological analysis of the NCCs might constrain additional phenomenological analysis of conscious experiences. For example, by understanding the changes at the neurophysiological scale, subjects might be able to understand their own conscious states better or to see new distinctions among them. So, not only could the phenomenological analysis of conscious events lead to a better understanding of the physiological results, but improved physiological analysis leads to a better phenomenological interpretation of consciousness. This bi-directionality of theory and methodology are the cornerstones of neurophenomenological approaches to understanding conscious experiences.

To take a concrete example: focal epileptic seizures start in specific parts of cortex and then remain confined to that area or spread to other parts of the brain. Where the seizure originates and how it spreads determine the symptoms of the seizure. Often these symptoms include changes in the conscious experiences of the patients. Patients with epilepsy often experience "auras," or sensory hallucinations (usually visual or auditory, though sometimes olfactory or gustatory), at the onset of a seizure. Temporal lobe seizures can also result in the experience of familiarity, or *déjà-vu*. Walter Penfield (1938) discovered that stimulating small areas in the temporal lobe also causes this experience of familiarity. It follows that activity in the relevant area in the temporal lobe is relevant to the experience of familiarity. The local neural event helps account for a global sensation.

The converse appears to be the case as well. We know that both bodily states (like stress or lack of sleep) and the surrounding environment (flashing lights) can trigger seizures in those with epilepsy (Engel 1989). About half of patients with epilepsy experience warning signs (headaches, nausea, irritability) that a seizure in immanent. Scientists are now able to align these symptoms with changes in the global dynamics of brain activity as well as with changes in bodily states and the environment (Le Van Quyen et al. 2001; Le Van Quyen and Petitmengin 2002). Importantly, it appears that some patients can use these experiential and environmental cues to decrease the probability that they will have a seizure by using biofeedback and classical conditioning techniques (Fenwick 1981; Schmid-Schönbein 1998). The patients are using global parameters, including their own insights regarding their conscious experiences, to affect local neural events.

Neurophenomenology seems to articulate the way the science of consciousness actually proceeds. As the examples recounted above show, scientists who study aspects of awareness spend their time measuring behavior and environmental events, as well as brain changes, and trying to account for how they all impact personal experiences. Philosophical analyses of neurophenomenology help clarify the importance of accurate and full first-person descriptions of experiences and, we hope, we will be able to see its influence in seeking NCCs going forward.

The other dominant embodied approach to understanding consciousness is the Embodied Conscious Mind (ECM). Alva Noë is one of its main proponents. He argues that, "for at least some experiences, the physical substrate of the experience may cross boundaries, implicating neural, bodily and environmental features" (2004: 221). The main thesis of ECM is that consciousness itself is not exclusively located at neural level, but it crosses boundaries at the brain-body-environment system.

Notice that this is a more radical position than neurophenomenology, which remains agnostic regarding where consciousness resides exactly. Neurophenomenologists hold that brains, bodies, and environment are all necessary for understanding what consciousness is and how it functions, and that there is a causal bi-directionality between experience and neural states, but most are perfectly comfortable with the idea that particular brain states or activities correlate with particular conscious experiences. It is just that the brain states or activities come about via an interaction with other experiences, bodies, and the environment. ECM, in contrast, holds that the correlates of consciousness itself run outside of the brain. Neurophenomenologists believe some version of NCCs exists, but ECM-ers explicitly deny that there is such a thing as NCCs.

We can understand ECM as a particular instance of the extended mind thesis in general (Clark and Chalmers 1998). Just as pencils, paper, and hands comprise part of the cognitive process of multiplication, so too do bodies and the environment comprise part of conscious experience. From the extended mind perspective, these external (or non-neural) objects constitute, at least in part, the very physical substratum of mental states.

One way to appreciate the ECM perspective is to consider how we project ourselves through objects in the world. A blind man using a cane, for example, does not experience the cane while he taps his way down a sidewalk; rather, he experiences the world at the end of the cane. Similarly, when we write with a pencil (perhaps when we are doing multiplication) we feel the end of the pencil writing on the paper, even though of course our nerve endings do not extend to the end of the writing implement. We are projecting our bodily awareness through the end of the pencil. Or when we walk in shoes, we can feel the pavement below us; we do not feel the inside of the shoes. We project ourselves through our shoes. Proponents of ECM say that the cane, the pencil, and the shoes all become part of our conscious system.

We can measure the edges of conscious projections experimentally. For example, Tony Chemero and his colleagues devised an experiment that forces change in our extended conscious experience (Dotov et al. 2010, 2017). Undergraduates engaged in a simple video game, using a computer monitor and a mouse. At irregular intervals during each trial, the connection between the mouse and the monitor was disrupted. When students were engaged in the video game, they were not aware of the hand-mouse interface per se, but once the connection between mouse and monitor was altered, then the mouse grabbed their attention and they become aware of it. Once the disruption was over and the connection returned to normal, then the awareness of the hand-mouse connection disappeared. Chemero and his colleagues argue that during the normal phase of the task, the mouse was part of the conscious system. During disruption, it was not. Though the details would take us too far afield, they are able to measure changes in underlying behavioral dynamics that reflect the changes in conscious experience. Their point is that we project ourselves into our environment, and in so doing, we consciously experience the edges of our extended cognitive system.

The remaining question for ECM is how this might relate to putative NCCs. Most proponents of ECM (Hurley 1998; Kiverstein and Farina 2012; Loughlin 2012; Manzotti 2011; Ward 2012) accept that the role neural states play in conscious experience is fundamental. (However, some of the radical embodied approaches to consciousness deliberately and explicitly avoid the appeal to brain states as an explanatory tool [e.g., Silberstein and Chemero 2012, 2015].) They take ECM as a theory that may account for some or most conscious experiences although they agree that some other experiences might be purely internal or brain-dependent (like headaches, for example). However, as Pepper (2013) points out, the way in which ECM relates to NCCs is different from the way the extended mind thesis relates to brain states in general. According to the main proponents of the extended mind thesis, external objects constitute mental events when they are used in the same way that we might otherwise just use our brains to achieve the same end. Using a pencil and paper might help us do a multiplication problem, but the pencil and paper are substituting for what we could do with just our brains if we had to. The case of consciousness, however, seems to be slightly different. It is difficult to imagine an extension of a conscious event in functional terms. But, when we project ourselves into the environment, we are not substituting something external for some inner process of consciousness. Conscious experience is not extended into the environment by causally replicating what we might do internally, but the experience itself is constituted by whatever it is we are projecting ourselves through. In this sense, consciousness "extends beyond the brain by its very nature" (Pepper 2013: 100).

If this is the case, then there is no NCC. Conscious experience is not constrained by the sorts of bodies we have and the types of environments we are interacting with. Rather, consciousness is located out in the world just as much as it is located inside the head.

Of course, there are several who flatly disagree with this perspective (e.g., Gennaro 2017; Metzinger 2000; Revonsuo 2000) and argue that proponents of ECM confuse constituency with causal relevancy. That is, if we lost significant portions of our brain, we could thereby lose consciousness. However, if we lost significant portions of our environment, our ability to perceive our environment, our ability to interact with the world, even if we lost significant portions of our body, we could still be fully and richly conscious. We are reminded of the anti-war classic *Johnny Got His Gun* (Trumbo 1939/1994). The point is, some things in the brain-body-environment complex are more relevant for consciousness than others. The search for the NCC is a search for those things most relevant.

Crick and Koch (1990) articulated a very simple vision for how to investigate and understand consciousness: isolate the thing inside the brain that is correlated with experience and you will have identified what consciousness is. Unfortunately, it turns out that whatever story ends up being told about consciousness is going to be much more complicated. Already we can see that there likely is not a single thing that accounts for the wide variety of conscious experiences we have. Probably, we will find many different neural correlates for many different aspects of consciousness. In addition, it seems naïve to believe that we can understand the brain or our minds in isolation from the bodies they are housed in and the environments in which we live. Hence, understanding consciousness is going to at least require matching changes in brain activity with changes in its surroundings and vice versa. At the end of the day, it remains to be seen whether seeking the neural correlates for consciousness is a productive approach for understanding our phenomenal experiences.

References

Addis, D.R., McIntosh, A.R., Moscovitch, M., Crawley, A.P., and McAndrews, M.P. (2004) "Characterizing spatial and temporal features of autobiographical memory retrieval networks: A partial least squares approach," *Neuroimage* 23: 1460–1471.

Block, N. (1995) "On a confusion about a function of consciousness," *The Behavioral and Brain Sciences* 18: 227–287.

Block, N (2004) "Consciousness," In R. Gregory (ed.) *The Oxford Companion to the Mind*, Oxford, UK: Oxford University Press.

Bogen, J. E. (1995) "On the neurophysiology of consciousness: I. An overview," *Consciousness and Cognition* 4: 52–62.

Calvo, P., and Gomila, T. (2008) *Handbook of Cognitive Science: An Embodied Approach*. San Diego, CA: Elsevier.

Chalmers, D. (1996) "Moving forward on the problem of consciousness," *Journal of Consciousness Studies* 4: 3–46.

Chalmers, D. (1998) "On the search for the neural correlate of consciousness," In S. Hameroff, A. Kaszniak, and A. Scott (eds.), *Towards a Science of Consciousness II*, Cambridge, MA: The MIT Press.

Chalmers, D. (2000) "What is a neural correlate of consciousness?" In T. Metzinger (ed.) *Neural Correlates of Consciousness: Empirical and Conceptual Issues*, Cambridge, MA: The MIT Press.

Chalmers, D. (2005) "Phenomenal concepts and the explanatory gap," In T. Alter and S. Walter (eds.) *Phenomenal Concepts and Phenomenal Knowledge: New Essays on Consciousness and Physicalism*, Oxford, UK: Oxford University Press.

Chalmers, D. (2010) *The Character of Consciousness*, Oxford, UK: Oxford University Press.

Churchland, P. (1986) "Some reductive strategies in cognitive neurobiology," *Mind* 95: 279–309.

Clark, A., and Chalmers, D. (1998) "The extended mind," *Analysis* 58: 10–23.

Crane, T. (2000) "The origins of qualia," In T. Crane and S. Patterson (eds.) *History of the Mind Body Problem*, London: Routledge.

Crick, F. (1994) *The Astonishing Hypothesis: The Scientific Search for the Soul*, New York: Charles Scribner's Sons.

Crick, F., and Koch, C. (1990) "Towards a neurobiological theory of consciousness," *Seminars in Neuroscience* 2: 263–275.

Crick, F., and Koch, C. (1995) "Are we aware of neural activity in primary visual cortex?" *Nature* 375: 121–123.

Damasio, A. (1999) *The Feeling of What Happens: Body and Emotion in the Making of Consciousness*, New York: Harcourt Brace.

Dennett, D. (1991) *Consciousness Explained*, Cambridge, MA: The MIT Press.

Dennett, D. (1993) "Quining qualia," In A. Goldman (ed.) *Readings in Philosophy and Cognitive Science*, Cambridge, MA: The MIT Press.

Desmurget, M., and Sirigu, A. (2009) "A parietal-premotor network for movement intention and motor awareness," *Trends in Cognitive Science* 13: 411–419.

Desmurget, M., and Sirigu, A. (2012) "Conscious motor intention emerges in the inferior parietal lobule," *Current Opinion in Neurobiology* 22: 1004–1011.

Desmurget, M., Reilly, K. T., Richard, N., and Szathmari, A. (2009) "Movement intention after parietal cortex stimulation in humans," *Science* 324: 811–813.

Dotov, D., Nie, L., and Chemero, A. (2010) "A demonstration of the transition from readiness-to-hand to unreadiness-to-hand," *PLoS ONE*, 5: e9433.

Dotov, D., Nie, L., Wojcik, K., Jinks, A., Yu, X., and Chemero, A. (2017) "Cognitive and movement measures reflect the transition to presence-at-hand," *New Ideas in Psychology* 45: 1–10.

Edelman, G. M. (1989) *The Remembered Present: A Biological Theory of Consciousness*, New York: Basic Books.

Engel, J. (1989) *Seizure and Epilepsy*, Contemporary Neurology Series. Philadelphia: F.A. Davis Company.

Fenwick, P. (1981) "Precipitation and inhibition of seizures," In: E. Reynolds and M. Trimble (eds.), *Epilepsy and Psychiatry*, London: Churchill Livingstone.

Flohr, H. (1995) "Sensations and brain processes," *Behavioral Brain Research* 71: 157–161.

Gazzaniga, M. S. (1988) *Mind Matters*, Boston, MA: Houghton Mifflin.

Gennaro, R. (2017) *Consciousness*, New York: Routledge.

Gray, J. (1995) "The contents of consciousness: A neuropsychological conjecture," *Behavior and Brain Sciences* 18: 659–722.

Haggard, P. (2005) "Conscious intention and motor cognition," *Trends in Cognitive Sciences* 9: 290–295.

Hardcastle, V. G. (1995) *Locating Consciousness*, Amsterdam, Netherlands: John Benjamins Press.

Horgan, J. (1999) "The undiscovered mind: How the human brain defies replication, medication, and explanation," *Psychological Science* 10: 470–474.

Hurley, S. L. (1998) *Consciousness in Action*, Cambridge, MA: Harvard University Press.

Husserl, E. (1900/1970) *Logical Investigations, Volumes One and Two*. J.N. Findlay (Trans.) London: Routledge and Kegan Paul.

Husserl, E. (1913/1963) *Ideas: A General Introduction to Pure Phenomenology*, W.R. Boyce Gibson (Trans.), New York: Collier Books.

Husserl, E. (1928/1989) *Ideas Pertaining to a Pure Phenomenology and to a Phenomenological Philosophy, Second Book*, R. Rojcewicz and A. Schuwer (Trans.), Boston: Kluwer Academic Publishers.

Hutto, D., and Myin, E. (2013) *Radicalized Enactivism: Basic Minds without Content*, Cambridge, MA: The MIT Press.

Jackson, F. (1982) "Epiphenomenal qualia," *Philosophical Quarterly* 32: 127–136.

Jackson, F (1993) "Armchair metaphysics," In J. O'Leary-Hawthorne and M. Michael (eds.) *Philosophy of Mind*. Dordrecht, Netherlands: Kluwer Books.

Kinsbourne, M. (1988) "An integrated field theory of consciousness," In A. Marcel and E. Bisiach (eds.) *Consciousness in Contemporary Science*, Oxford, UK: Oxford University Press.

Kiverstein, J., and Farina M. (2012) "Do sensory substitution devices extend the conscious mind?" In F. Paglieri (ed.) *Consciousness in Interaction*, Amsterdam, Netherlands: John Benjamins.

Koch, C. (2004) *The Quest for Consciousness: A Neurobiological Approach*, Englewood, CO: Roberts & Company Publishers.

Le Van Quyen, M., and Petitmengin, C. (2002) "Neuronal dynamics and conscious experience: An example of reciprocal causation before epileptic seizures," *Phenomenology and the Cognitive Sciences* 1: 169–180.

Le Van Quyen, M., Martinerie, J., Navarro, V., Baulac, M., and Varela, F. J. (2001) "Characterizing the neurodynamical changes prior to seizures," *Journal of Clinical Neurophysiology* 18: 191–208.

Levine, J. (1983) "Materialism and qualia: The explanatory gap," *Pacific Philosophical Quarterly* 64: 354–361.

Llinás, R. (2001) "Consciousness and the brain: The thalamocortical dialogue in health and disease," *Annals of the New York Academy of Sciences* 929: 166–175.

Loughlin, V. (2012) "Sketch this: Extended mind and consciousness extension," *Phenomenology and the Cognitive Sciences* 12: 41–50.
Lutz, A., and Thompson, E. (2003) "Neurophenomenology: Integrating subjective experience and brain dynamics in the neuroscience of consciousness," *Journal of Consciousness Studies* 10: 31–52.
Manzotti, R. (2011) "The spread mind: Is consciousness situated?" *Teorema* 30: 55–78.
McGinn, C. (1991) *The Problem of Consciousness*, Oxford, UK: Blackwell.
McGinn, C. (1995) "Consciousness and space," *Journal of Consciousness Studies* 2: 220–230.
Metzinger, T. (2000) "Introduction: Consciousness research at the end of twentieth century." In T. Metzinger (ed.) *Neural Correlates of Consciousness: Empirical and Conceptual Issues*, Cambridge, MA: The MIT Press.
Nagel, T. (1974) "What is it like to be a bat?" *Philosophical Review* 83: 435–456.
Newman, J. B., and Baars, B. J. (1993) "A neural attentional model for access to consciousness: A global workspace perspective," *Concepts in Neuroscience* 4: 255–290.
Noë, A. (2004) *Action in Perception*, Cambridge, MA: The MIT Press.
Panksepp, J. (1998) *Affective Neuroscience: The Foundations of Human and Animal Emotions*, Oxford, UK: Oxford University Press.
Papineau, D. (1995) "The antipathetic fallacy and the boundaries of consciousness," In T. Metzinger (ed.) *Conscious Experience*, Thoverton, UK: Imprint Academic.
Papineau, D. (2002) *Thinking about Consciousness*, Oxford, UK: Oxford University Press.
Penfield, W. (1938) "The cerebral cortex in man. I. The cerebral cortex and consciousness," *Archives of Neurology and Psychiatry* 40: 417–442.
Pepper, K. (2013) "Do sensorimotor dynamics extend the conscious mind?" *Adaptive Behavior* 22: 99–108.
Rees, G., Kreiman, G, and Koch, C. (2002) "The neural correlates of consciousness in humans," *Nature Reviews Neuroscience* 3: 261–270.
Revonsuo, A. (2000) "Prospects for a scientific research program on consciousness," In T. Metzinger (ed.) *Neural Correlates of Consciousness: Empirical and Conceptual Issues*, Cambridge, MA: The MIT Press.
Schmid-Schönbein, C. (1998) "Improvement of seizure control by psychological methods in patients with intractable epilepsies," *Seizure* 7: 261–270.
Searle, J. (1997) *The Mystery of Consciousness*, New York: The New York Review of Books.
Shapiro, L. (2014) *The Routledge Handbook of Embodied Cognition*, New York: Routledge.
Silberstein, M., and Chemero, A. (2012) "Complexity and extended phenomenological-cognitive systems," *Topics in Cognitive Science* 4: 35–50.
Silberstein, M., and Chemero, A. (2015) "Extending neutral monism to the hard problem," *Journal of Consciousness Studies* 22: 181–194.
Singer, W. (1999) "Neuronal synchrony: A versatile code for the definition of relations?" *Neuron* 24: 49–65.
Thompson, E., and Varela, F. (2001) "Radical embodiment: Neural dynamics and consciousness," *Trends in Cognitive Sciences* 5: 418–425.
Tosoni, A., Galati, G., Romani, G. L., and Corbetta, M. (2008) "Sensory-motor mechanisms in human parietal cortex underlie arbitrary visual decisions," *Nature Neuroscience* 11: 1446–1453.
Trumbo, D. (1939/1994) *Johnny Got His Gun*, New York: Carol Publishing Group.
Van Gulick, R. (1985) "Physicalism and the subjectivity of the mental," *Philosophical Topics* 12: 51–70.
Van Gulick, R (2003) "Maps, gaps, and traps," In A. Jokic and Q. Smith (eds.) *Consciousness: New Philosophical Perspectives*, Oxford, UK: Oxford University Press.
Varela, F. J. (1996) "Neurophenomenology: A methodological remedy to the hard problem," *Journal of Consciousness Studies* 3: 330–350.
Vogeley, K., and Fink, G.R. (2003) "Neural correlates of the first-person perspective," *Trends in Cognitive Science* 7: 38–42.
Ward, D. (2012) "Enjoying the spread: Conscious externalism reconsidered," *Mind* 121: 731–751.

Related Topics

Sensorimotor and Enactive Approached to Consciousness
Materialism
Representational Theories of Consciousness

18
CONSCIOUSNESS AND ATTENTION

Wayne Wu

1 Introduction

This review will summarize work relevant to four questions

1 Is attention necessary for consciousness?
2 Is attention sufficient for consciousness?
3 Does attention alter the character of consciousness?
4 How does attention give us access to consciousness?

Remember that when we say that attention is necessary for consciousness, we mean that if a subject S is not attending to X, then S is not conscious of X, or equivalently, S's being conscious of X implies (requires) that S is attending to X. When we say that attention is sufficient for consciousness, we mean that if the subject attends to X, this implies that the subject is conscious of X. Attention is enough for consciousness. The relevant senses of "attention" and "consciousness" will now be specified.

2 What Is Attention? What Is Consciousness?

A challenge to assessing our questions is to fix what attention and consciousness are. After all, it is difficult to talk clearly about how the two are related if the relata are unclear. Let us begin with attention, which has been actively studied in cognitive science but has only recently become a topic of philosophical research (on philosophical theories, see Mole 2013; Wu 2014). One thing is apparent in looking at the empirical literature on attention: there seems to be a lack of consensus on what it is. Thus, psychologists bemoan the absence of a uniform account of attention. Here is a representative quote:

> In general, despite the ingenuity and subtlety of much of the experimental literature that has been devoted to these two enduring controversies [early versus late selection and automaticity and control in processing], the key concepts (selection, automaticity, attention, capacity, etc.) have remained hopelessly ill-defined and/or subject to divergent interpretations. Little wonder that these controversies have remained unresolved.
> *(Allport 1993: 118)*

For current purposes, it will be enough to provide a sufficient condition for attention that is widely accepted in the empirical literature: if a subject S perceptually selects X to perform a task T, then the subject is perceptually attending to X. The rationale for this proposal is that it is assumed in designing experiments on attention. When one wishes to study attention, say visual attention to a moving object, one needs to ensure that during the study, subjects are attending to the targeted object. To ensure this, experimenters design a task where it is a necessary condition on performing the task correctly that the subject perceptually selects a target, or information from it, to guide task performance. If the task is designed correctly, then proper task performance entails appropriate perceptual selection and thus, appropriate perceptual attention. For current purposes, we can understand this sufficient condition as identifying the forms of attention of primary interest in cognitive science (this is not a surprise given that it is assumed in experimental design). A broader characterization of attention expands from focusing on common experimental tasks to actions. If we expand the sufficient condition to encompass all action and endorse the necessary condition, we have the following definition of attention: attention to X just is selection of X for action. Nevertheless, for current discussion, we need only the sufficient condition.

What of consciousness? Ned Block (Block 1995) distinguished between *access* and *phenomenal* consciousness. Access consciousness, at root, concerns the use of information by the subject. Intuitively, to be access conscious of X is to be able to use X in some way. Indeed, attention, as given in our sufficient condition, embodies access for action. Block himself spoke of access for the sake of *rational* control of action, thus limiting the type of informational access that qualified as conscious in the relevant sense.

Our focus, however, will be on phenomenal consciousness, but as scientists have noted, this notion is not well-defined. A salient attempt at a definition was given by Thomas Nagel, when he suggested that a state is (phenomenally) conscious if and only if there is something it is like for the subject to be in that state. The problem is that the definition is no more illuminating than the elusive notion of "what it is like" for the subject. As a scientist might complain, how does one "operationalize" that definition to allow it to guide empirical study of consciousness?

Empirical work, however, can proceed so long as one can track the phenomenon in question. At this point, philosophers and sympathetic scientists will rely on introspection: one can track consciousness because one can access what it is like for one in experience, and this access is just *introspection*. So, we can assess claims about the relation between consciousness and attention by drawing on introspection to track consciousness and the empirical sufficient condition to fix when attention is present and, with some additional assumptions, when attention is absent.

One important distinction that we will largely ignore concerns the different targets of attention, as in the visual case when we speak about attention to objects, spatial locations or features. This introduces important distinctions that any complete analysis of the relation between attention and consciousness must confront, but we shall focus on their interrelations at a more abstract level of analysis, namely in terms of selecting targets for tasks and whether such selection is necessary and/or sufficient for consciousness.

3 Is Attention Sufficient for Consciousness?

One central issue in the empirical literature concerns whether attention and consciousness are the same process (Koch and Tsuchiya 2007). The identity is false if one can occur without the other, so we can investigate whether there can be selection for task without phenomenal consciousness and vice versa. The empirical sufficient condition allows us to draw on experimental paradigms to *test* whether attention and consciousness are tightly correlated. For example, if we can demonstrate that subjects are attending using a concrete attentional paradigm where selection for task is operative, and yet show that subjects are not conscious, then we will have shown a case where attention to X is *not* sufficient for consciousness of X. It then follows that attention and consciousness are not the same.

Are there counterexamples to the sufficiency of attention for consciousness? Blindsight patients provide a possible instance. These subjects have damage to primary visual cortex leading to hemianopia in the contralateral visual field, namely a blind field. They report not being able to see stimuli in that portion of the visual field, but strikingly, when forced to guess about stimulus properties in that blindfield, their perceptual reports show above chance accuracy (Weiskrantz 1986). Hence, they exhibit visually guided response in an area of purported blindness, hence blindsight. This ability is likely mediated by subcortical visual pathways that reengage cortical vision in a way that supports the observed perceptual discrimination behavior (Schmid and Maier 2015). While questions have been raised as to whether cases like blindsight present *phenomenal* blindness (Phillips 2016), let us assume with most theorists that blindsighters are phenomenally blind in the relevant part of the visual field. Can we then show that they can attend to the objects within the blind field?

Given the sufficient condition, we need to locate a task where appropriate task performance requires selection of, and hence attention to, X. One standard paradigm is spatial cueing (Posner 1980). In a standard version, a subject is asked to detect visual targets that are flashed on the screen peripheral to the point of fixation, the point on which subjects must keep their eyes fixed. During the task, the subject maintains fixation while attempting to detect targets that appear in the periphery. During the interstimulus interval period before the flashing of the target, a cue will appear, either a central (symbolic) cue at the point of fixation such as an arrow pointing to a peripheral location or a peripheral cue that occurs at the possible target location. Cues can be valid or invalid, that is, they can appear where the target subsequently appears (valid) or does not appear (invalid). During an experiment, the ratio of valid/invalid cues is often in the range of 80/20, so cues carry information about the location of the target (for a discussion of other psychological paradigms, see Wu (2014, ch. 1)).

Where attention is engaged, a standard observation is that relative to a neutral condition, valid cues lead to faster response times and/or greater accuracy, while invalid cues lead to slower response times and/or greater inaccuracy. If visual attention were a spotlight, the idea would be that valid cues draw the spotlight to the location of a future target facilitating target detection while invalid cues draw the spotlight away, leading to decrements in performance, say slower reaction times, due to having to reset the spotlight (such metaphors like the spotlight should be taken with many grains of salt). Thus, increases in reaction time and/or accuracy *during performance of target detection in this paradigm* are a signature of visual spatial attention. This provides a case of selection for task that we can use to fix the deployment of attention.

We can now combine spatial cueing with blindsight: Do blindsighters show spatial cueing effects of the sort associated with spatial attention? Bob Kentridge and colleagues demonstrated this with the blindsight patient GY who showed spatial cueing effects to targets in his blindfield (Kentridge et al. 1999). Later work reproduced similar results with normal subjects by inducing blindsight-like responses using techniques such as visual masking which makes targets "invisible" (Kentridge 2011). The results seemingly demonstrate cases where attention and consciousness come apart, namely where attention to a location is not sufficient to induce consciousness.

Earlier, I noted that we should keep track of the "kind" of attention at issue, and in the visual domain, whether attention is directed to locations, features or objects. Those distinctions are relevant since the previous paradigm is typically understood as a test of *spatial* attention, yet blindsight in the first case is the claim that subjects do not consciously perceive the stimuli whose features they can reliably report when forced to guess. That is, blindsight concerns object or feature perception. Thus, one can argue that the case of spatial cueing in blindsight does not provide a counterexample to the claim that if one is attending to an object, one is conscious of that object (Prinz 2011). After all, we have *spatial* attention and failure of *object* consciousness. However, recall that the task is a target detection task that is facilitated by a cue, so attention to objects is plausibly present. How else could the subjects make the appropriate report? So, blindsight does provide a case of object attention (detection of targets) along with blindness to those objects (for a demonstration of an object attention effect in blindsight patients, see Norman et al. 2013).

Does this mean that attention never gives rise to consciousness? That is a trickier claim to assess. We will consider two cases. The first is whether attention can alter consciousness, say when one shifts spatial attention thereby causing changes in conscious experience. We will consider that possibility in Section 5. The other case is the claim that attention makes consciousness possible. This idea can be unpacked in light of the claim that attention is necessary for consciousness, such that when one is not attending to a stimulus, one is thereby not conscious of it. If attention is like a gate, then perhaps when one then shifts attention to the stimulus one is conscious of it. If the latter claim is true, then in that context, attention can be sufficient for consciousness by making it come on the scene.

4 Is Attention Necessary for Consciousness?

Call the claim that attention is necessary the *Gatekeeping Thesis*:

> ***Gatekeeping:*** one is perceptually conscious of X only if one perceptually attends to X.
> *(where perception is in the same modality)*

Why think that this thesis is true? It might seem that consciousness and attention are tightly yoked because to report on (introspect) consciousness, we need to attend.

Is there evidence for Gatekeeping? It is widely thought that a wealth of empirical evidence supports it. Given that Gatekeeping expresses a necessary condition, there is a clear prediction: if we can find a context where attention to X can be or is disrupted, then consciousness of X can be or is disrupted. For example, if one can manipulate attention by pulling it away from X, one will thereby eliminate consciousness of X if attention gates consciousness. This would lead to *inattentional blindness.* Let us consider two putative sources of empirical evidence.

The first case involves paradigms where subjects are asked to do an attentionally demanding task that is directed at Y in the presence of X where Y≠X. The idea is that given the widespread view that attention is capacity limited (you can't attend to *everything*), an appropriately demanding task directed at Y will remove the possibility of attending to X. In effect, task demands distract the subject away from X. A famous example is an experiment conducted by Daniel Simons and co-workers, where they presented subjects with a video of two groups of players, one group dressed in white shirts, the other in black shirts, each group passing a basketball amongst themselves. Subjects were tasked with counting the number of passes by the white shirted players (notice that this invokes the empirical sufficient condition to direct attention to the ball by making it task relevant). At a certain point, a person dressed in a gorilla suit walks through the scene, turns and pounds its chest, and walks off. About 50% of subjects fail to notice the gorilla, i.e. do not report the gorilla's presence (Simons and Chabris 1999). Here, it seems that without attention to the gorilla, subjects are not conscious of the gorilla.

A second case involves neuropsychological patients. Subjects who suffer strokes, often in parietal cortex, can acquire hemispatial neglect. There are many ways of testing for neglect, but the basic symptom is that subjects seem to be unaware of the side of space contralateral to the brain lesion (typically right-side lesions lead to the neglect of the left side of space). Strikingly, patients with left hemispatial neglect fail to eat food on the left side of their dinner plate or fail to detect objects on the left side of a sheet of paper. It then seems that subjects are strikingly unconscious of items to their left. Neglect is thought to be due to an inability to attend to the relevant side of visual space (Corbetta and Shulman 2011), so again, neglect suggests failures of consciousness linked to the absence of attention.

Theorists conclude that both cases exemplify inattentional blindness, but let us spell out the reasoning. Recall that we need to experimentally secure the absence of attention. In the gorilla experiment, this is achieved by manipulating attention to distract the subject away from the gorilla. So, inattention is achieved methodologically through task demands. In the case of neglect, inattention is a result of brain damage. Let us grant that attention to the relevant objects are missing in these two conditions. We must now establish that consciousness is absent. How? Here, we rely on introspective reports, or indeed their absence, as a sign of the conscious state of the individual. In the case of the gorilla, the relevant report that is absent is in fact a perceptual report: subjects fail to report the gorilla. Let us treat that failure as a surrogate of a plausible additional failure to introspect and detect a visual experience as being of a gorilla. Similar points arise for neglect patients who fail to report stimuli present in the neglected part of their conscious field. We could further probe subjects as to whether they are aware of anything beyond the items they report (as was done in Mack and Rock 1998), and perhaps subjects will explicitly deny being aware of anything out of the ordinary. Given a failure to generate reports of experiencing relevant stimuli or an explicit denial that anything odd is seen, we infer that subjects are not visually aware of the relevant targets and hence are blind to them. So, we have inattention and blindness, and it might then seem plausible that inattention explains the blindness, namely that it is because we remove attention that blindness results. Attention then would be necessary for visual consciousness.

In the visual case, Gatekeeping can be understood as holding that (spatial) attention defines the extent of the conscious visual field, so that objects that are not in the area targeted by spatial attention are effectively outside the visual field. In that sense, they could just as well be located behind the head even though they are right before the eyes. In the gorilla experiment, while the subject is doing the task and is not attending to the gorilla, the subject is blind to the gorilla. This blindness is temporary in that when one directs the subject's attention to the gorilla, the subject immediately recognizes it. In effect, such shifts of attention to the gorilla will bring the gorilla within the conscious visual field, thus making the gorilla an object of visual awareness. So, in this context, attending to the gorilla is sufficient for consciousness of the gorilla.

Does the evidence noted earlier support the Gatekeeping Thesis? The standard inference from data provided by inattentional blindness experiments and by spatial neglect do not support the Gatekeeping Thesis despite widespread assumptions that they do. Theorists have failed to notice this because they have failed to be clear about what attention is. Recall that we take attention to be selection for task, so in the case of the gorilla, selection for task is directed towards the basketball. The basketball, as the task relevant object, is the object of attention. To test Gatekeeping, we must insure the absence of attention to the gorilla, so if that condition is satisfied, the subject is not attending to the gorilla. The question then emerges: *Why should the subject report the gorilla if the subject isn't attending to it?* To report on an X, one needs to attend to it, to select it to guide report capacities. If I ask you to name the objects in a picture, you will scan each one, and when your eyes lock on, you are then in a position to report the object. Without that perceptual selection, there is no reason for an object to prompt a response. Thus, the very methodology used to demonstrate inattentional blindness undercuts the proposed result, for to test Gatekeeping, the subject must not attend to the gorilla, but that condition *guarantees* that the subject will not report the gorilla since the necessary selective capacity for report is *distracted*. So, the experimental design ensures the failure to report or the design fails because the gorilla does capture attention. In fact, those are the observed results, and they are *consistent* with the subject being consciously aware of the gorilla. In other words, the experimentally imposed distraction is sufficient to explain failure to report in subjects whose attention is not captured by the gorilla.

The same point holds for those individuals suffering from hemispatial neglect: a neurological basis for failure of attention also insures that one cannot deploy the needed capacities for reporting objects. Indeed, if one observes the pattern of a neglect patient's eye movements across a picture where some item X is located in the neglected side of space (e.g. the left side of the picture), one will notice that the eye effectively never crosses the midline of the picture as defined by the body midline (Karnath 2015). Indeed, if one observes the posture of neglect patients, their head is always oriented away from the neglected side of space. So, in a clear sense, the neglect patient never looks over to the side of space where X is and a fortiori, never looks at X (fixes eyes on it). So, overt attention is never directed at X and if overt attention follows covert attention, then the subject never attends to the neglected side of space. Is it any wonder that one will not report X? The failure to look at and attend to X is sufficient to explain the failure to report even if the subject is conscious of X.

There is then a general problem for assessing Gatekeeping, since the relevant experiment apparently cannot be done. A crucial component in the experimental strategy is eliminating a form of attention to assess effects on consciousness. The problem is that in lieu of an adequate definition of consciousness, we empirically track consciousness by attention in introspection, so the experiment undercuts the possibility of tracking consciousness or its absence. This does not show that Gatekeeping is false, but it does undercut a wealth of empirical evidence that is thought to support the position.

Let us then consider the alternative to Gatekeeping, namely *Overflow:*

> ***Overflow:*** a subject can be conscious of X without attending to X.
>
> *(Block 2007)*

Can we empirically demonstrate Overflow? Again, we confront limits set by attention: we must determine that the subject is conscious of X despite having attention directed away from X. Yet as before, we track consciousness by introspective attention. This means that to test Overflow, we must induce conditions where attention is not directed to X thereby undercutting the very access we need to track consciousness. It seems that given the central role attention plays in introspection, we are not in a position to empirically assess either Gatekeeping or Overflow. Some think that Overflow is thus untestable (Cohen and Dennett 2011), but as we have seen, the same problem accrues to Gatekeeping.

Clarity on these issues requires clarity on the concept of attention. Let me note a recent study that is claimed to demonstrate that consciousness overflows attention. Christof Koch and co-workers have done experiments that they argue demonstrates consciousness in the "near absence" of attention (Li et al. 2002). Such a thesis would not, of course, demonstrate the falsity of Gatekeeping though it raises a host of important issues. Can there be different amounts of attention? If a Gatekeeping Thesis is reformulated to consider different amounts of attention, does that mean there will be different amounts of consciousness? What would talk of different amounts of consciousness mean? Clearly, some further conceptual work is needed to clarify these issues.

It might seem obvious that there can be more or less attention, but what precisely does that mean? It would be good in this domain to not rely on intuitions but draw on analyses that are as precise as possible, and in the case of attention, we draw on the empirical sufficient condition. In that case, there is one clear notion of amount of attention that we can formulate at the outset, namely the amount of selection for action with respect to the targets of attention within a specific task context. For example, if one is dealing with many objects as opposed to a few, then the amount of attention can be fixed by the number of objects selected, and here there are clear limits to the number of objects subjects can attend to. With such specifications in place, we can then deal with claims about consciousness in the "near absence" of attention. On quantity of attention, it is up to researchers to specify what the relevant measurement is. That said, the critical question addressed by our formulation of Gatekeeping is whether the loss of attention results in the loss of consciousness, so on that point, near absence of attention is not sufficient to address the issue with which we began. Yet in every case, we face the original problem: the assessment of consciousness requires attention, and to the extent that subjects report that they are aware of objects "outside of attention" through their behavior, that behavior itself implicates attention and undercuts the core claim.

These are troubling results in that we seem to be unable to empirically support Overflow or Gatekeeping. Still, there might be reasons to query the severity of Gatekeeping, since it implies blindness without attention in the visual case. Blindness must be the absence of visual consciousness, but this seems both odd and severe. If a gorilla is standing behind you, then you are in a clear sense blind to it in that you have no visual experience of it. Now, as the gorilla walks around to come before your eyes, imagine that your attention is fully locked onto some other object so that no attention is directed at the gorilla. Is it plausible that the gorilla is phenomenally absent as when it was standing behind you? Let us imagine that you attend to the gorilla momentarily but ignore it (you know it is your friend dressed up as a gorilla and expect him to be moving about). Your shifting attention to it brings it into consciousness, but now you go back to attending to other matters. Does the gorilla somehow literally disappear before your very eyes, a *phenomenal hole* in the fabric of the visual field?

The idea that attention leads to blindness seems severe given that there is an alternative that seems plausible. When attention is removed from the gorilla, the gorilla does not disappear but becomes less in focus. A similar effect is achieved when you foveate the gorilla and then saccade to another object, putting the gorilla in peripheral vision, where it appears like an indistinct black blob. At that point, the gorilla remains in consciousness but no longer appears as a gorilla but rather as a black shape. The idea then is that attention puts things, metaphorically, in focus. Again, we are not in a position to establish what we might call *inattentional agnosia* or perhaps *inattentional blurriness* since that would require attention. Yet, this picture has what seems like an advantage, that the issue is not the absence of consciousness in the absence of attention but the absence of typical clarity that attention brings. Put another way, a middle ground position is to acknowledge that attention changes the character of consciousness without gating it.

5 Does Attention Affect Consciousness?

Well, certainly. The idea bruited in the last section is that attention "puts things in focus." We can put this slightly more precisely by saying that attention sharpens representations, something that we will unpack in a moment. Let us first consider a case where shifts of attention do seem to change consciousness. Figure 18.1 is an illusion discovered by Peter Tse (redrawn based on Tse 2005):

Maintain fixation on any of the dots but shift attention between disks. Notice anything different about how the disks appear to you? To many people, the attended disk looks darker than the unattended disks.

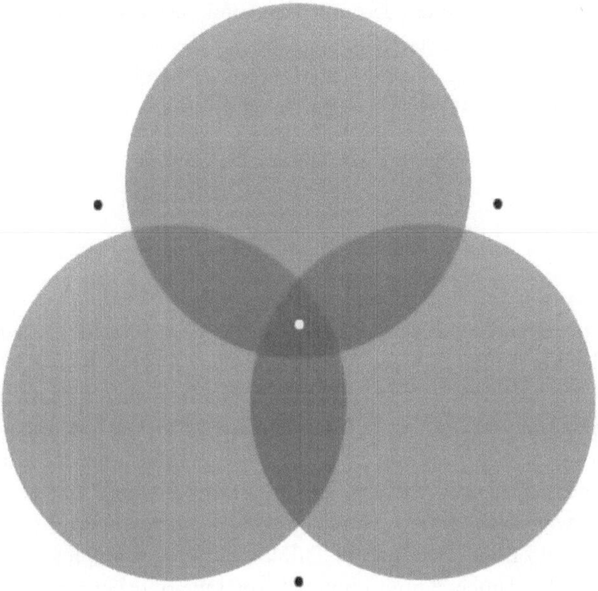

Figure 18.1 Illusion by Peter Tse (adapted from Tse 2005)

The common idea of an attentional spotlight as a characterization of attention suggests that one effect of attending to X is altering the representations of X. For example, it might intuitively seem that when we attend to X, say in vision, we have a clearer view of it. Attention changes the quality of perception. We must be careful with introspection, however, for in the visual case, clarity of vision depends on fixating the item of interest so that it stimulates the fovea, the area of the retina that provides for the highest spatial acuity. While cognitive scientists take moving the eye to foveate objects to count as *overt* attention, it is not clear that foveation should be equated with attention since one can pay attention "out of the corner of one's eye" while maintaining focus on an object at the center of one's visual field.

At the neural level, attending to objects is associated with a variety of neural responses that seem to suggest changes in representation. For example, visual attention can increase the strength of neural signaling (gain modulation), sharpen selection as when neural spatial fields contract around targeted objects, or sharpen contrast representations (contrast gain). Do these neural effects have upshots in visual consciousness? Work by Marisa Carrasco has probed this possibility (Carrasco et al. 2004). Carrasco and co-workers asked subjects to detect the orientation of Gabor patches, i.e. luminance contrast gradients. In one experiment, subjects maintained fixation on a central cross while they reported on the orientation of a targeted Gabor which could be tilted either to the left or the right. Two Gabors were presented at the periphery, one to the left of fixation, the other to the right. The target was defined as "the Gabor that appeared of the highest contrast." In this way, the subjects had to perform two tasks, discriminating which of two possible targets had the higher contrast appearance relative to the other, and then, reporting the orientation of that target. In effect, the first task probes how the Gabor's appeared to the subject. The additional factor in the experiment was to use spatial cueing to direct attention to one of the two Gabor patches. Carrasco provided evidence that when attention was deployed to a Gabor, the contrast appeared to increase. This suggests that attention can alter conscious appearances perhaps by altering underlying neural properties (for counterarguments, see Schneider and

Komlos (2011)). Carrasco's group has demonstrated similar effects for size and color (Fuller and Carrasco 2006; Gobell and Carrasco 2005).

There are limits to attention's effects, as can be seen in the phenomenon of visual crowding (Whitney and Levi 2011). Visual crowding can be demonstrated in the following display:

+	**X**
+	**AXA**

Fix your eyes on the "+" and try to attend the "X" in the periphery. In the first line, you can still make out the "X." In the second line, you cannot since the "A"s that flank the "X" crowd it. The current views about crowding conceive of the flankers as disrupting feature integration and it is plausible that when the visual system fails to integrate features, it fails to construct a coherent representation of objects (Whitney and Levi 2011). One might then think that the necessary neural object representations will not form and thus, that we should not be able to see objects in conditions of crowding. Indeed, in many natural scenes, crowding in the periphery occurs given the natural clutter of our environment. Think of walking through a park or reading a text. Crowding identifies a fundamental limit on visual representation, but it is also resistant to attention (Intriligator and Cavanagh 2001). It is not clear that attention can even dissect the crowded letter but even spatial attention to the area of crowding cannot lead to an escape from it.

6 Attention and Introspection?

We began our discussion by noting that we do not need to define consciousness to study it. We just need a way to track it. This tracking capacity is provided by introspection, which deploys a type of attention or focus on the properties of consciousness. Yet how does attention work in introspection?

One possibility raised by the last section is that in attending to consciousness, we might thereby change its character. That is "observation" of conscious states changes the very state observed (again note the Carrasco results discussed earlier; this possibility was noted early on by Hill (1991)). One question then would be whether introspective attention could give us an undistorted view of consciousness. But set aside that concern and focus on a pressing question: what exactly is introspection?

A common idea is that of inner focus: when we introspect our conscious experiences, it is as if we turn our attention inwards to an internal feature of our minds. For example, Brie Gertler writes:

> By focusing your attention on the phenomenal quality of [a sensation], you can come to know something about your current experience. Philosophers generally agree on this much.
>
> *(Gertler 2012)*

Putting a different spin on the idea, William Lycan writes:

> When we attend to our own mental states, it feels like that is just what we are doing: focusing our internal attention on something that is there for us to discern.
>
> *(Lycan 2003)*

The problem is that philosophers do not typically say more in terms of the psychological details of what introspection is as a psychological capacity. What would it be to have this capacity?

What is the form of attention referred to here? The challenge is to say something about this capacity that helps us understand consciousness itself.

In recent years, some philosophers have pressed the question concerning introspective reliability (Schwitzgebel 2011) leading to a skepticism about introspection of consciousness. Others have suggested that introspection does not provide a fruitful method in the empirical investigation of experience (Irvine 2012). In response, philosophers have attempted to calibrate attention (Bayne and Spener 2010; Spener 2015). What remains missing is a psychologically realistic account of what introspective attention involves. When such an account is provided, we can then put ourselves in a better position to understand introspective attention and hopefully, thereby understand when introspection is reliable and when it is not.

Lycan, as we saw, clearly thinks that when we introspect on how introspection of consciousness works, it appears that introspection of consciousness involves a form of internal attention. We can literally focus on our internal states. Still, not everyone finds this when they introspect. Thus, Harman writes:

> When Eloise sees a tree before her, the colors she experiences are all experienced as features of the tree and its surroundings. None of them are experienced as intrinsic features of her experience. Nor does she experience any features of anything as intrinsic features of her experiences. And that is true of you too. There is nothing special about Eloise's visual experience. When you see a tree, you do not experience any features as intrinsic features of your experience. Look at a tree and try to turn your attention to intrinsic features of your visual experience. I predict you will find that the only features there to turn your attention to will be features of the presented tree.
>
> *(Harman 1990: 667)*

Harman's point is that when we attend in introspecting, our attention does not seem to be internally directed but rather points outward to the world. Such *transparency* accounts of introspection have been developed where there is no internally directed introspective capacity (Dretske 1995). The distinction of how attention is deployed in introspection divides between two conceptions of consciousness. On one, the phenomenal is in a sense external, so that in focusing on the qualitative aspects of conscious experience, our attention is directed outwards. On the other, the phenomenal is in a sense internal, so that in focusing on consciousness, attention is directed inwards.

The point is that our conception of how we access consciousness is not independent of our conception of what consciousness is or consists of. We might have hoped for a more neutral yet substantive characterization of introspection, beyond the common invocation of attention to consciousness. Yet, the conception of attention as deployed in introspection is divided by a border that also divides metaphysical views about consciousness. In that sense, introspection is no less controversial than consciousness. This opens up the possibility that investigation into the nature of introspection might have a role to play in helping us assess theories of consciousness.

7 Conclusion?

There is no doubt that attention has an intimate relation to consciousness. Attention provides for our distinctive access to consciousness, and when it is disrupted, so is our ability to introspect what consciousness is like. At the same time, attention guides our actions, which are often influenced and controlled by what we perceive, and in that link, it can exert its influence, perhaps bringing items to awareness, changing how we experience them, all within the limits and parameters that are fixed by our brains.

We have learned much about attention in recent years, and deploying and modulating attention has played a central role in casting light on consciousness. Nevertheless, there remain a variety of questions of which we shall emphasize three:

1 Can we find an experimental way to assess the debate between Overflow and Gatekeeping, namely whether attention is necessary for some aspect of consciousness?
2 How precisely does attention affect the character of consciousness?
3 How does attention control our access to consciousness in introspection?

These questions have both an empirical and philosophical character, and the issue of the relation between attention and consciousness offers an opportunity for genuine interdisciplinary work involving cognitive science and philosophy.

References

Allport, A. (1993) "Attention and control: have we been asking the wrong questions? A critical review of twenty-five years," in D. E. Myer and S. Kornblum (eds.) *Attention and Performance XIV: Synergies in Experimental Psychology, Artificial Intelligence, and Cognitive Neuroscience* pp. 183–218. Cambridge, MA: MIT Press.

Bayne, T., and Spener, M. (2010) "Introspective humility," *Philosophical Issues*, 20: 1–22.

Block, N. (1995) "On a confusion about the function of consciousness," *Behavioral and Brain Sciences*, 18: 227–247.

Block, N. (2007) "Consciousness, accessibility, and the mesh between psychology and neuroscience," *The Behavioral and Brain Sciences*, 30: 481–499.

Carrasco, M., Ling, S., and Read, S. (2004) "Attention alters appearance," *Nature Neuroscience*, 7: 308–313.

Cohen, M. A., and Dennett, D. C. (2011) "Consciousness cannot be separated from function," *Trends in Cognitive Sciences*, 15: 358–364.

Corbetta, M., and Shulman, G. L. (2011) "Spatial neglect and attention networks," *Annual Review of Neuroscience*, 34: 569–599.

Dretske, F. (1995) *Naturalizing the Mind*, Cambridge, MA: MIT Press.

Fuller, S., and Carrasco, M. (2006) "Exogenous attention and color perception: performance and appearance of saturation and hue," *Vision Research*, 46: 4032–4047.

Gertler, B. (2012) "Renewed acquaintance," in D. Smithies and D. Stoljar (eds.) *Introspection and Consciousness*, New York: Oxford University Press.

Gobell, J., and Carrasco, M. (2005) "Attention alters the appearance of spatial frequency and gap size," *Psychological Science*, 16/8: 644–651.

Harman, G. (1990) "The intrinsic quality of experience," *Philosophical Perspectives, 4 Action Theory and Philosophy of Mind*, 31–52. Atascero.

Hill, C. S. (1991) *Sensations: A Defense of Type Materialism*, Cambridge: Cambridge University Press.

Intriligator, J., and Cavanagh, P. (2001) "The spatial resolution of visual attention," *Cognitive Psychology*, 43: 171–216.

Irvine, E. (2012) "Old problems with new measures in the science of consciousness," *British Journal for the Philosophy of Science*, 63: 627–648.

Karnath, H.-O. (2015) "Spatial attention systems in spatial neglect," *Neuropsychologia*, 75: 61–73.

Kentridge, R. W. (2011) "Attention without awareness: a brief review," in C. Mole, D. Smithies, and W. Wu (eds.) *Attention: Philosophical and Psychological Essays*, Oxford University Press: New York.

Kentridge, R. W., Heywood, C. A., and Weiskrantz, L. (1999) "Attention without awareness in blindsight," *Proceedings of the Royal Society London B*, 266: 1805–1811.

Koch, C., and Tsuchiya, N. (2007) "Attention and consciousness: two distinct brain processes," *Trends in Cognitive Sciences*, 11: 16–22.

Li, F. F., VanRullen, R., Koch, C., and Perona, P. (2002) "Rapid natural scene categorization in the near absence of attention," *Proceedings of the National Academy of Sciences of the United States of America*, 99: 9596–9601.

Lycan, W. G. (2003) "Perspectival representation and the knowledge argument," in Q. Smith and A. Jokic (eds.) *Consciousness: New Philosophical Perspectives*, Oxford: Oxford University Press.

Mack, A., and Rock, I. (1998) *Inattentional Blindness*, Cambridge, MA: MIT Press.
Mole, C. (2013) *Attention Is Cognitive Unison: An Essay in Philosophical Psychology*, Oxford: Oxford University Press.
Norman, L. J., Heywood, C. A., and Kentridge, R. W. (2013) "Object-based attention without awareness," *Psychological Science*, 24: 836–843.
Phillips, I. (2016) "Consciousness and criterion: On block's case for unconscious seeing," *Philosophy and Phenomenological Research*, 93: 419–451.
Posner, M. I. (1980) "Orienting of attention," *The Quarterly Journal of Experimental Psychology*, 32/1: 3–25.
Prinz, J. (2011) "Is attention necessary and sufficient for consciousness," in C. Mole, D. Smithies, and W. Wu (eds.) *Attention: Philosophical and Psychological Essays*, New York: Oxford University Press.
Schmid, M. C., and Maier, A. (2015) "To see or not to see—thalamo-cortical networks during blindsight and perceptual suppression," *Progress in Neurobiology*, 126: 36–48. DOI:10.1016/j.pneurobio.2015.01.001
Schneider, K. A., and Komlos, M. (2011) "Attention alters decision criteria but not appearance: a reanalysis of Anton-Erxleben, Abrams, and Carrasco (2010)," *Journal of Vision*, 11/13: 1–10.
Schwitzgebel, E. (2011) *Perplexities of Consciousness*, Cambridge, MA: MIT Press.
Simons, D. J., and Chabris, C. F. (1999) "Gorillas in our midst: sustained inattentional blindness for dynamic events," *Perception*, 28/9: 1059–1074.
Spener, M. (2015) "Calibrating introspection," *Philosophical Issues*, 25/1: 300–321.
Tse, P. U. (2005) "Voluntary attention modulates the brightness of overlapping transparent surfaces," *Vision Research*, 45: 1095–1098.
Weiskrantz, L. (1986) *Blindsight: A Case Study and Implications*, Oxford: Clarendon Press.
Whitney, D., and Levi, D. M. (2011) "Visual crowding: a fundamental limit on conscious perception and object recognition," *Trends in Cognitive Sciences*, 15: 160–168.
Wu, W. (2014) *Attention*, London: Routledge.

Related Topics

Materialism
The Intermediate Level Theory of Consciousness
The Attention Schema Theory of Consciousness
Consciousness and Psychopathology

19
CONSCIOUSNESS AND INTENTIONALITY

David Pitt

A mental state is conscious just in case there is something it is like to be in it. The properties in virtue of which there is something it is like to be in a mental state are *phenomenal properties*, or *qualia*. A mental state is intentional just in case it is *about* something, and thereby has truth or veridicality conditions. The feature of an intentional state in virtue of which it has these properties is called its *intentional content*. In analytic philosophy of mind there was for many years a consensus that consciousness and intentionality are properties of metaphysically exclusive kinds. Conscious qualitative states, such as visual, auditory and olfactory experiences, do not, *per se*, have intentional content; and intentional states, such as thoughts, beliefs, desires and intentions do not, *qua intentional*, have phenomenal properties. To be sure, perceptual states such as seeing a dog or hearing it bark are perceptions *of* dogs and barks, and thereby have intentional content. But their intentionality was typically taken to be determined by causal relations between perceiver and perceived, and not by any intrinsic qualitative features they might have. And though thoughts, beliefs and desire may be conscious, whatever qualitative features might be associated with thinking, believing and desiring were taken to be irrelevant to their intentional content. In general, the phenomenal character of conscious states was seen as having no essential connection to their intentional contents.

Consciousness is extremely difficult (some think impossible) to explain within the naturalist framework that has prevailed in analytic philosophy of mind for most of the twentieth century, and into the twenty-first. Intentionality, on the other hand, insofar as it is a phenomenon that is not essentially tied to consciousness, was seen to be more tractable, and various theories grounding it in or reducing it to natural relations between the brain and the world it represents were proposed and developed. Philosophers working on intentionality, both perceptual and cognitive, felt they could safely ignore the vexing problem of the naturalization of consciousness.

More recently, however, this consensus has begun to weaken, as naturalistic theories of intentionality have faced problems that a growing number of philosophers believe are due to their failure to take conscious qualitative experience into account. These philosophers have argued that intentionality is essentially an experiential phenomenon, and, as such, cannot be reductively explained unless consciousness can – however problematic this may be for the naturalistic program in philosophy of mind. They have taken a stance reminiscent of classical phenomenology, which "brackets" the relation of experience to the world in order to study it on its own terms. These *analytic* phenomenologists tend to bracket the relation between experience and the brain, pursuing a phenomenal theory of intentionality free from, as Charles Siewert (2011: 242) so memorably put it, "the tyrannizing anxieties and ambitions of mind-body metaphysics." While not ignoring the metaphysical problem of consciousness, these analytic phenomenologists insist that reductive explanation is not the only project one might profitably pursue in the study of consciousness.

1 Causal-Informational Psychosemantics

Fred Dretske was set to be the Darwin of intentionality. His insight that causal relations, insofar as they carry *information* about the occurrence of the events they relate, establish a kind of proto-intentionality, is profound. It is the kind of idea – intuitive, simple and powerful – we all wish we had thought of (and wonder why we didn't).[1] Though not yet what *we* have, this proto-intentionality is sufficiently *like* it to get us a conceptual foot in the seemingly unopenable door between this aspect of mind and our physical constitution. Dretske's idea promised to show how it is *possible* that a highly sophisticated and puzzling aspect of our mental nature could arise from simple beginnings, by entirely natural processes.

In the 1980s and '90s there was, understandably, a great deal of excitement among analytic philosophers of mind over this idea. Jerry Fodor went as far as to suggest that (modulo a syntactic solution to Frege's Puzzle) "Turing and Dretske have between them solved the mind/body problem" (Fodor 1994: 56). Turing showed how a physical thing could *reason*, and Dretske showed how a physical thing could *represent*. The philosophical project of naturalizing the mind, of bringing it within the scope of the kind of empirical methodology that led to such spectacular successes in our understanding of the world, seemed to be, if not complete, at least halfway there.

The view has the added benefit of building a connection between thought and its objects into the very nature of representational content. Concepts are individuated by the object(s) or property instantiation(s) whose presence is lawfully causally correlated with their occurrence, and thus acquire their contents and their extensions simultaneously.

There was (as Dretske and Fodor were always well aware) still the problem of consciousness to be addressed. Causal relations *per se* do not seem to be sufficient to bring about conscious experience, or even some kind of proto-conscious experience. Qualia freaks would have to await their own Darwin. But the other half of the mind-body problem was, from a philosophical point of view, in its essential outlines, thought to have been solved.

Philosophy being philosophy, of course there were dissenters all along. In particular, there have been those, such as John Searle and Galen Strawson, who have long insisted that *genuine* intentionality (what *we* have) is essentially a conscious, experiential phenomenon. Searle has argued for what he calls the "connection principle" (Searle 1992), according to which a mental state cannot have fine-grained intentional content (what he calls "aspectual shape") unless it is either conscious or potentially conscious,[2] and Strawson (1994) has argued for the essential experientiality of mentality in general, and of conceptual intentionality in particular. According to these theorists, resources sufficient for constructing an information processor are not sufficient for constructing a mind, since information *per se* is not conscious, and consciousness is required for genuine intentionality. Another important defender of this idea is Charles Siewert (1998).

Causal-informational theorists have, unsurprisingly, resisted this claim. If true, it would short-circuit their naturalistic explanation of intentionality, since at present there is no adequate naturalistic account of conscious experience (and some would argue that there can never be one). Fodor even pronounced commitment to an essential link between intentionality and conscious experience "intellectual suicide."[3] But, as we will see, it is a position that has recently been gaining adherents in analytic philosophy of mind, who so far appear to have remained intellectually above ground.

In spite of their promise, causal-informational theories face internal difficulties – the most persistent of which have been problems of indeterminacy. There is Quine's Problem, which arises out of what may be called *causal superimposition*; the Disjunction Problem, which arises out of what may be called *causal spread*; and the Stopping Problem, which arises out of what may be called *causal depth*. In all of these cases, there are multiple candidates for content determiner/extension, and no obvious way to choose among them derivable from the basic machinery of the theory.

Quinean examples of indeterminacy of radical translation (Quine 1960) can be taken to show that for any property that is a candidate for determining the content of a concept (the meaning of a term), there are indefinitely many other simultaneously instantiated (superimposed) properties that cannot be teased apart causally. Any instantiation of *rabbithood*, for example, is also, necessarily, an instantiation of *undetached-rabbit-parts-hood*, *rabbit-stage-hood*, and indefinitely many other properties. Assuming that these properties are distinct, they are candidates for distinct contents for the meaning of 'rabbit' (and the concept [mental representation] RABBIT). (Names for concepts are here written in SMALL CAPS, and names of properties in *italics*.) Given that these properties are (at least physically) necessarily instantiated by the same things, there can be no lawful relations between mental states and *one* of them that are not also lawful relations between mental states and *all* of them. Hence, a causal-informational theory cannot, at least *prima facie*, assign one of them as the content of RABBIT. There is by the theory's lights no fact of the matter about which of these properties is content-determinative of the concept RABBIT (the term 'rabbit').

Though Quinean examples can be taken as entailing *indeterminacy* of content, they can also be viewed as entailing massive *disjunctiveness* of content. On this construal, the content of RABBIT would be *rabbithood or undetached-rabbit-parts-hood or rabbit-stage-hood or* .... In this case there would be a fact of the matter about what the content of a given concept is, but it would be, counterintuitively, open-endedly disjunctive. This is problematic because, as Fodor has often pointed out (e.g., Fodor 1987), there ought to be psychological generalizations that apply to mental states in virtue of their content. However, in keeping with the naturalistic project, such laws would be *causal* (or otherwise nomological). But natural laws typically are not formulated in terms of disjunctive properties, which do not in general constitute *natural kinds*.

Dretske (1981) himself recognized this problem (named the "Disjunction Problem" in Fodor 1984), which arises from the fact that there are causal correlations between the occurrence of mental representations and the presence of a wide range of things (property instantiations) that are, intuitively, not in the extension of those representations. Thus, though there may be a law-like regularity between horses (instantiations of *horsehood*) and occurrences of the concept HORSE, such relations *also* hold between HORSE occurrences and indefinitely many other things: *donkeys on dark nights, zebras in the mist, merest ripples in horse-infested waters,*[4] ... – anything that might cause one to think, correctly or incorrectly, (e.g.) LO, A HORSE! Thus, for HORSE (or any empirical concept), there is a *spread* of different property instantiations (by distinct objects) sufficient for its tokening, and, hence, by the theory's lights, sufficient for determining its content. But HORSE cannot mean *all* of these indefinitely many things. And the reasons for resisting a *disjunctive* content are the same here as they were in the causal superimposition cases.

Indeed, though this is not always remarked upon, one could just as well construe this as a problem of indeterminacy: there is, consistent with the resources of the theory, no fact of the matter about which one of the indefinitely many causally correlated property instantiations determine a concept's content.

Another problem (named the "Stopping Problem" in Strawson 2008) first arises when the causal relations that are supposed to establish content hold between mental states and *distal* objects.[5] Thus, causal relations to cows – instantiations of *cowhood* – are supposed to constitute a mental representation the concept COW. But there are *also* causal relations between occurrences of COW and any link in the causal chain between cows and COWS. These include links *within* the perceptual system, such as *bovine retinal images, bovine olfactory bulb stimulations, bovine visual or olfactory cortex activation patterns*, etc.,[6] as well as links between retinal images (or other sensory-organ representations) and cows – such as *cow reflections, cow shadows, cow breezes*, ... There are also less obvious candidates, like *photons reflected from a cow, the cow's parents, distant ancestor bovine species, ..., the Big Bang*. All of these can lay equal claim to inclusion in the causal chain leading to tokenings of COW, although, obviously, the vast majority of them are not plausible candidates for being (or determining) the content or extension of the concept COW.

The causal chains connecting concept tokenings to their content-conferring property instantiations are *deep*, involving a densely packed series of property instantiations (events) as links. And while we may find it impossible to take seriously candidates such as objects or events in the distant past, or property instantiations undetectable by us, if *all* we have at our disposal is causal relations, it is not obvious what principled reasons there could be for excluding any of them. And if there is no way to prune away the unwanted causes, then we are faced, as with the other problematic cases, with the invidious choice between indeterminacy and massive disjunction.[7]

And there are other apparent problems, as well: How are causal theories to explain the contents of mathematical, logical and other concepts, whose referents are abstract, causally-inert objects? Or the contents of concepts of non-existent objects?

Causal-informational theorists have expended considerable effort and ingenuity in the search for a solution to these problems (see e.g. Dretske 1988, 1995; Fodor 1987, 1990; Millikan 1984, 1989; Neander 1995; Papineau 1998; Prinz 2002; Rupert 1999, to cite just a few examples from a very large literature). Some see a solution in teleology – the evolved function of representation-producing mechanisms; though there are residual indeterminacy problems for such views (see Fodor 1990). Others appeal to causal-inferential relations among mental representations (see Block 1986; Field 1977; Harman 1973, 1987; Loar 1981; and McGinn 1982 for foundational statements of the view). These "conceptual-," "functional-," or "inferential-role" theories are typically integrated with Dretske-style accounts in constructing "two-factor" (internal and external, "narrow" and "wide") theories of content. These theories have their own technical difficulties, arising from their prima facie commitment to meaning holism (see e.g. Fodor and Lepore (1992). (An intuitive objection to such views is that inferential relations among concepts are determined by their contents, not vice versa.) But it would not be accurate to say that naturalistic approaches of these kinds are defunct.[8]

2 Phenomenal Intentionality

Other philosophers have proposed that in order to solve these problems – or, even better, to avoid them entirely – causal relations should be replaced with (or at the very least supplemented by) qualitative features of experience as determiners of content. Searle and Strawson have already been mentioned as early analytic proponents of an experience-based approach to intentionality.[9] Searle (1987) responds to Quinean indeterminacy; and Strawson addresses the Stopping Problem in his 2008. It has also been argued that phenomenology can solve the Disjunction Problem (Pitt 2009; Horgan and Graham 2012).

The shared idea is that what our concepts are concepts *of* is what we *take them to be of*, where *taking* is a manner of *experiencing*. What HORSE means is what *we* mean by it; and what we mean is experiential, and introspectively available to us. We *know*, from a first-person perspective, that the extension of HORSE is horses, and not *horse-part-fusions* or *zebras in the mist* or *equine retinal arrays*, ...[10] And we know this in this way because conceptual contents (and thought contents) are *experiential* in nature.

Searle calls the experiential content of a concept its "aspectual shape." Strawson (1994) calls it "understanding experience." Siewert (2011) speaks of "phenomenal thought."[11] It has lately come to be known as "cognitive phenomenology" (Pitt 2004; Bayne and Montague 2011a; Chudnoff 2015a; see Strawson 1986 for an early use of this term). Without claiming that everyone who subscribes to this view agrees about the nature of conceptual experience and its relation to intentional mental content (some theorists claim that it does not determine content at all [Siewert 1998], some say it constitutes only an internally determined component of content [Horgan and Kriegel 2008; Strawson 2008], while others reject the idea that content should be factored into internally and externally determined components [Pitt 2013]), we can say that there is a shared commitment to the thesis that genuine conceptual intentionality of the kind we have is essentially an experiential phenomenon. Without experience (which for most philosophers means without *consciousness*) there can be no mental representation with the fineness of grain or selectivity that our thoughts and concepts display.

Apart from its value as a prophylactic (or cure) for Indeterministic Disjunctivitis, conceptual phenomenology has been recommended on independent grounds.

One common form of argument is from *phenomenal contrast*. In one kind of case, we are invited to compare the experience of hearing discourse in a language that is understood to the experience of discourse in a language that is not understood (Strawson 1994: 5–6). In another, we are invited to consider *changes* in our own conscious occurrent thought (Siewert 1998: 275–278). In yet another, we are to imagine an individual who lacks all sensory, emotional, algedonic, etc., experience, yet who can still *think*, and consider what it is like for this individual to reason mathematically (Kriegel 2015: 56–62). In all cases, it is argued that there is a phenomenal difference, a difference in what it's like for the thinker, and, further, that this is not a difference in familiar kinds of phenomenology, such as that of verbal or auditory imagery, emotional tone, etc. It is then concluded that there is an irreducible, distinctively *cognitive* kind of experience that accompanies (or constitutes) thinking, differences in which account for the experiential contrasts.[12]

Phenomenal contrast arguments are vulnerable to competing claims about what the contrast between experiences with and without understanding actually consists in. What proponents attribute to a difference in cognitive phenomenology, critics maintain is a difference in auditory, visual, emotional, or some other more familiar kind of phenomenology. Such positions are bolstered by claims of a lack of introspective evidence in the objector's own experience for the existence of such *sui generis* cognitive phenomenology.[13] Disputes over what is phenomenally manifest in introspection are notoriously difficult (though not impossible) to adjudicate. This has led some to doubt whether the phenomenal contrast strategy is the best way to try to establish the existence of cognitive phenomenology. (Sacchi and Voltolini [2016] offer a version of the contrast argument that, they claim, does not rely on introspection.)

A different sort of approach, due to Strawson, focuses on the *significance* or *value* of conscious experience in general, and of conscious thought in particular. Strawson (2011) argues that our conscious experience would be significantly *less interesting* if it did not include an experience of thinking. If thoughts were just unconscious subpersonal computational states, our conscious mental lives would be drastically impoverished. We would have no experience of *grasping* truths, of *wondering* why something is the case, of *realizing* and *solving* problems, etc.

Another type of argument for cognitive phenomenology appeals to a particular kind of self-knowledge we are capable of. Pitt (2004) argues that it is possible to know, consciously, introspectively and non-inferentially, what one is consciously occurrently thinking, and that this would not be possible if thought (and conceptual) contents were not immediately present in consciousness. Just as one can know in this way that one is hungry, hearing a trumpet or tasting ashes, because there is something it is like to be in these states, one can know *that* one is thinking, and *what* one is thinking, because there is something it is like to think, and what it is like to think thoughts with different contents is phenomenally different. Conscious occurrent thoughts could not be introspectively distinguished from other kinds of conscious states, and from each other, in this way if they were not phenomenally individuated. Moreover, since it is possible to have auditory or visual experience of linguistic expressions without thinking what they mean, or thinking anything at all, this individuative phenomenology cannot be the phenomenology of inner speech or visualization.

Pitt (2009) argues, further, that this cognitive kind of phenomenology *is* cognitive intentional content. To consciously think that three is a prime number is to consciously token a maximally determinate cognitive-phenomenal type which *is* the proposition that three is a prime number. (Just as to be in a maximally determinate pain state is to token a maximally determinate pain type.)

Pitt (2011) offers another argument for cognitive phenomenology, based upon the claim that conscious states, as such, are individuated phenomenologically. That is, what distinguishes conscious states of different kinds is their different kinds of phenomenal character. Conscious sensory states, such as visual, auditory and olfactory experiences, are distinguished by, respectively, visual, auditory and olfactory phenomenology, each a *sui generis* kind of experiential quality. And conscious sensory states within a given modality are, as such, individuated by different determinate phenomenologies within their respective determinable phenomenal kinds. Pitt argues that conscious thought, qua conscious, is individuated in the same way as other kinds of conscious experience, as are distinct thoughts within the cognitive experiential modality. Hence, there must be a proprietary, distinctive and individuative phenomenology of occurrent conscious thought.

Perceptual states are also intentional. In their various modalities, they represent to us the world around us, providing information about the existence and states of the things with which we interact. And they can be more or less accurate, veridical or not. What is the role of consciousness in the intentionality of perception? Obviously, conscious perceptual experiences must be conscious. But what role do the phenomenal properties apparent in conscious experience play in determining the intentional content of a perceptual state – what it is a perception *of*? On what can be called the Pure Causal View, they play no role whatever. A perceptual state is a representation of an object or property (instantiation) if and only if it is caused by that object or property. Whatever qualitative properties may be consciously apparent determine, at best, only how accurately or inaccurately a perceptual state represents, not whether or not it represents. Toward the other end is what Montague (2016) calls the Matching View, according to which there is a (probably vague) limit to how badly a perceptual state can misrepresent its cause before it ceases to be a perception of it.

Most (if not all) philosophers would agree that a causal relation between token perceptual states and specific objects or properties is *necessary* for genuine perception. No state not caused by an elephant is a perception *of* an elephant. The role of causation with respect to perceptual states is thus different from its role with respect to cognitive (conceptual) states. In the latter case, we want to allow that token concepts can be *of* things that are *not* their token causes. A token concept ELEPHANT should be a concept *of* elephants (have elephants in its extension), no matter what causes it, and whether or not it was caused by any external thing or property. But a token perceptual state cannot be a perception of an elephant unless it is caused by an elephant. Because of this difference, the Disjunction Problem does not arise for perceptual states. Perceptions of elephants cannot be caused by hippos-in-the-mist or large grey rocks, or by nothing at all.

Quine's problem also does not arise for perceptual states, since, for example, a perceptual state caused by an elephant is also caused by an elephant-stage and a sum of undetached elephant parts, etc. The *conceptual* distinctions do not seem to be relevant to what is being perceived in the way they are relevant to what is being thought about.

But the Stopping Problem does arise. Any state caused by an *F* is also caused by other links in the causal chain leading to the occurrence of the state. A visual perception of an elephant is caused by the elephant; but it is also caused by whatever caused the elephant, the photons reflected from the elephant, the firing of cells in the retina, the lateral geniculate nuclei and the primary visual cortex, etc. – none of which we would want to say the experience is *of*.

The Matching View has a straightforward solution to this problem: the visual experience one has upon looking at an elephant is not an experience of any of these other causes because it does not *resemble* any of them. This is analogous to the cognitive-phenomenal solution to the Quine and Disjunction Problems for conceptual representations – the concepts RABBIT and RABBIT-STAGE, HORSE and COW-IN-THE-MIST, are introspectively distinguishable cognitive experiences. What it is like to think that something is a rabbit is different from what it is like to think that it is a rabbit-stage.

Some philosophers (e.g. Evans 1982 and Dretske 2006) have argued that in order for a state to be a perception of an *F*, it must not just be caused by an *F*, but also enable the perceiver to *locate* or *track* the *F*. And this might seem to be enough to solve the perceptual Stopping Problem, since the state of perceiving an elephant does not provide the perceiver information about the location of the elephant's ancestors, the photons bouncing off it, the perceiver's retina, or parts of the perceiver's brain. Moreover, since on this account the state itself need not (at least for Dretske) be phenomenally conscious, it need not *resemble* its cause in any robust sense. And even if it is acknowledged that perceptual states are typically conscious, and that conscious states have (or present) qualitative properties, one may allow that these properties establish whether or not the state resembles its cause, but still deny that resemblance is necessary for genuine perception.

Montague insists, however, that there are limits to how badly a conscious perceptual state can misrepresent its cause before it is disqualified as a perception of it. On her Matching View, a perceptual state "must represent a sufficient number of [an] object's properties correctly in order for it to be true that one [perceives] it" (Montague 2016: 156). On this view, an experience that in no way resembles an elephant cannot be a perception of the elephant that caused it.

The intuitions on both sides are respectable. On the one hand, it seems reasonable to say that an experience caused by an *F* is a perception of that *F* no matter how unlike its cause it is – just as it seems reasonable to say that a photograph is of an *F* if it was photons bouncing off the *F* that were responsible for its production (cf. Evans 1982, 78), no matter how little it resembles its cause; or a painting is a painting of an *F* if the artist intended it to be a painting of an *F*, no matter how little it might resemble the *F* (cf. modern art).

On the other hand, if we consider things from the perspective of the representation itself, it seems reasonable to say that resemblance *is* required. No one shown a picture of an elephant would take it to be a picture of a pink Cadillac, or vice versa. And no one would take a completely blank image to be a photograph of either an elephant or a pink Cadillac. Moreover, it seems entirely natural to say that an image with the appropriate properties is an image of an elephant, whether or not it resulted from causal interaction with one, and somewhat perverse to say that such an image is not an image of an elephant, because it was not caused by one.

These intuitions are not inconsistent. There is a perfectly good sense of 'a perception *of* an *F*' on which it means *a perception caused by an F*, and an equally good sense on which it means *a perception resembling an F*. The latter sense is commonly marked out with the phrase 'perception *as of* an *F*' (or of an *F as* an *F*). A perception *of* an *F* (like a photograph or picture of an *F*) may or may not also be a perception *as of* an *F*. Being caused by an *F* does not entail, and is not entailed by, resembling an *F*. A state caused by an elephant could resemble virtually anything, or nothing at all; and a state resembling an elephant could be caused by virtually anything, or nothing at all. (Additionally, the former sense may be used in reference to a perception [or photograph or painting] of a *particular F*, the latter in reference to a perception [or photograph or painting] of a *typical F*, though none in particular.)

However, if the issue is the intentionality of perceptual experience itself, then it is arguable that the latter sense of 'perception of' is more appropriate. For the content of perceptual experience *as one has it* is constituted by its phenomenal character. Perceivers do not have direct access to external causal relations between objects and their perceptions of them. And if the *role* of perception is to inform perceivers of the existence and states of external objects, then complete misrepresentation of its external cause should disqualify an experience as a genuine perception, since such an experience would be (more or less) *useless* to the perceiver.

Dretske and others (e.g. Dretske 1995; Harman 1990; Lycan 1996; Tye 2000) have proposed extensions of the casual-informational theory to give a naturalistic account of the qualitative properties apparent to us in perceptual experience. Such "reductive representationalist" views (see Chalmers 2004 for terminological clarification) attempt to explain the phenomenology of perception in terms of causal-informational representation of objectively instantiated phenomenal properties. The yellowness one might mention in describing what it is like to see a ripe banana, for instance, is a property of the banana, not one's experience of it. And it is easy to see how this account could be used to solve the Stopping Problem for perception: a perceptual state represents the thing whose phenomenal properties are apparent to the perceiver. However, this "qualia externalism" (see Byrne and Tye 2006) faces serious problems in accounting for dreams, illusions and hallucinations (Thompson 2008; Pitt 2017). (Moreover, it is far from obvious how externalist theories of this kind could solve the indeterminacy problems for cognitive states. See Byrne 2011, 2008 and Pitt 2011.)

3 Conclusion

There is a common point to be made about the role of phenomenology in determining conceptual and perceptual intentionality (content). A theory that takes causal-informational relations between representation and represented to be sufficient to determine the content of the representation (what the representation is about) will encounter indeterminacy/disjunction problems that cannot be solved in purely causal-informational terms. The diagnosis offered by advocates of phenomenal intentionality is that such difficulties are bound to arise whenever the intrinsic properties of representations are ignored. Such properties have an essential role in both determining representational contents and making them available to the thinker or perceiver. If thought and perception are to establish useful and accurate representational connections between conscious thinker-perceivers and their worlds, it must be apparent to them *what* is being represented and *how* it is being represented, and how a thing is represented must sufficiently resemble (accurately characterize) what it represents. In consciousness, appearance is, necessarily, phenomenal. Nothing can appear to a thinker-perceiver without appearing in some *way* or other, and the ways of appearing are constituted by phenomenal properties. And nothing can be accurately and (hence) usefully conceived or perceived unless the way it appears to the thinker-perceiver is the way it *is*. In spite of the fact that consciousness and phenomenality stubbornly resist naturalistic explanation, no theory of intentionality can afford to ignore them.

Notes

1 See Dretske (1981, 1988, 1995). C.B. Martin had a different, but also inspired, idea when he noticed that the relation between dispositions and their manifestations can also be seen as a kind of proto-intentionality. Dispositional states are directed at, indicate, or point to, their manifestations. See Martin (2008).
2 See Fodor and Lepore (1994) and Gennaro 2012 (sec. 2.3.1) for critical discussion of Searle's connection principle. Searle (1984) also objected to the idea that Turing solved the naturalization problem for reasoning, arguing that rule-governed symbol-manipulation without understanding (for Searle, a form of experience) is not thinking.
3 I am not aware of this remark appearing in print. I have heard Fodor say it, and Strawson reports it in 2008.
4 Cf. Fodor (1990).
5 As Strawson notes, this is a common problem for causal theories generally (e.g., the causal theory of perception, to be discussed below).
6 As pointed out also in Sterelny (1990), Antony and Levine (1991), Adams and Aizawa (1997) and Fodor (1990).
7 The Stopping Problem has also been called the "horizontal disjunction problem." The three problems discussed here are really versions of a general problem that we might call the "Problem of Causal Proliferation."
8 Philosophical views are rarely, if ever, definitively defunct. What usually happens is that people get bored with their problems and move on to something else. Often enough, old views get resurrected once new ones become stale.
9 In the Phenomenological tradition, the experiential nature of intentionality is taken to be self-evident.
10 Indeed, as has often been pointed out, if we could make no such distinctions as those between the contents *rabbit* and *rabbit-stage*, indeterminacy and disjunction would not appear to us to be *problems* at all. Of course, Quine famously denied that, after all, there is a difference between *rabbit* and *rabbit-stage* for the radical translator. But, as Searle (1987) argued, this strains credibility (to say the least). It seems more plausible to see Quinean indeterminacy as a *reductio* of empiricist semantics.
11 See also the discussion of the experience of thinking in Siewert (1998, ch. 8).
12 Other proponents of this kind of argument include Horgan and Graham (2012), Horgan and Tienson (2002), Moore (1962), Peacocke (1998) and Siewert (1998, 2011).
13 See, e.g., Carruthers and Veillet (2011), Chudnoff (2015b), Koksvik (2015), Levine (2011), Pautz (2013), Prinz (2011) and Tye and Wright (2011).

References

Adams, F. R. and Aizawa, K. (1997) "Fodor's Asymmetric Causal Dependency Theory and Proximal Projections," *Southern Journal of Philosophy* 35: 433–437.

Antony, L. and Levine, J. (1991) "The Nomic and the Robust," in B. M. Loewer and G. Rey (eds.), *Meaning in Mind: Fodor and His Critics*, Oxford: Blackwell.

Bayne, T. and Montague, M. (eds.) (2011) *Cognitive Phenomenology*, Oxford: Oxford University Press.

Bayne, T. and Montague, M. (2011a) "Cognitive Phenomenology: An Introduction," in Bayne and Montague (2011).

Block, N. (1986) "Advertisement for a Semantics for Psychology," *Midwest Studies in Philosophy* 10, Minneapolis, MN: University of Minnesota Press.

Byrne, A. (2008) "Knowing That I Am Thinking," in A. E. Hatzimoysis (ed.), *Self-Knowledge*, Oxford: Oxford University Press.

Byrne, A. (2011) "Transparency, Belief, Intention," *Aristotelian Society Supplementary Volume* 85: 201–221.

Byrne, A. and Tye, M. (2006) "Qualia Ain't in the Head," *Noûs* 40: 241–255.

Carruthers, P. and Veillet, B. (2011) "The Case Against Cognitive Phenomenology," in Bayne and Montague (2011).

Chalmers, D. (2004) "The Representational Character of Experience," in B. Leiter (ed.), *The Future for Philosophy*, Oxford: Oxford University Press.

Chudnoff, E. (2015a) "Phenomenal Contrast Arguments for Cognitive Phenomenology," *Philosophy and Phenomenological Research* 90: 82–104.

Chudnoff, E. (2015b) *Cognitive Phenomenology*, London: Routledge.

Dretske, F. (1981) *Knowledge and the Flow of Information*, Cambridge, MA: MIT Press.
Dretske, F. (1988) *Explaining Behavior*, Cambridge, MA: MIT Press.
Dretske, F. (1995) *Naturalizing the Mind*, Cambridge, MA: MIT Press.
Dretske, F. (2006) "Perception Without Awareness," in T. S. Gendler and J. Hawthorne (eds.), *Perceptual Experience*, Oxford: Oxford University Press.
Evans, G. (1982) *The Varieties of Reference*, Oxford: Oxford University Press.
Field, H. (1977) "Logic, Meaning and Conceptual Role," *Journal of Philosophy* 69: 379–409.
Fodor, J. A. (1984) "Semantics, Wisconsin Style," *Synthese* 59: 231–250.
Fodor, J. A. (1987) *Psychosemantics*, Cambridge, MA: MIT Press.
Fodor, J. A. (1990) "A Theory of Content," in *A Theory of Content and Other Essays*, Cambridge, MA: MIT Press.
Fodor, J. A. (1994) *The Elm and the Expert*, Cambridge, MA: MIT Press.
Fodor, J. A. and Lepore, E. (1992) *Holism: A Shopper's Guide*, Oxford: Blackwell
Fodor, J. A. and Lepore, E. (1994) "What *Is* the Connection Principle?" *Philosophy and Phenomenological Research* 54: 837–845.
Gennaro, R. (2012) *The Consciousness Paradox*, Cambridge, MA: MIT Press.
Harman, G. (1973) *Thought*, Princeton, NJ: Princeton University Press.
Harman, G. (1987) "(Nonsolipsistic) Conceptual Role Semantics," in E. Lepore (ed.), *New Directions in Semantics*, London: Academic Press.
Harman, G. (1990) "The Intrinsic Quality of Experience," *Philosophical Perspectives* 4: 31–52.
Horgan, T. and Graham, G. (2012) "Phenomenal Intentionality and Content Determinacy," in R. Schantz (ed.), *Prospects for Meaning*, Berlin: de Gruyter.
Horgan, T. and Kriegel, U. (2008) "Phenomenal Intentionality Meets the Extended Mind," *The Monist* 91: 347–373.
Horgan, T. and Tienson, J. (2002) "The Intentionality of Phenomenology and the Phenomenology of Intentionality," in D. Chalmers (ed.), *Philosophy of Mind: Classical and Contemporary Readings*, Oxford: Oxford University Press.
Kriegel, U. (ed.) (2013) *Phenomenal Intentionality: New Essays*, Oxford: Oxford University Press.
Kriegel, U. (2015) *The Varieties of Consciousness*, Oxford: Oxford University Press.
Koksvik, O. (2015) "Phenomenal Contrast: A Critique," *American Philosophical Quarterly* 52: 321–334.
Levine, J. (2011) "On the Phenomenology of Thought," in Bayne and Montague (2011).
Loar, B. (1981) *Mind and Meaning*, Cambridge, MA: Cambridge University Press.
Lycan, W. G. (1996) *Consciousness and Experience*, Cambridge, MA: MIT Press.
McGinn, C. (1982) "The Structure of Content," in A. Woodfield (ed.), *Thought and Object*, Oxford: Oxford University Press.
Martin, C. B. (2008) *The Mind in Nature*, Oxford: Clarendon Press.
Millikan, R. (1984) *Language, Thought and Other Biological Categories*, Cambridge, MA: MIT Press.
Millikan, R. (1989) "Biosemantics," *Journal of Philosophy* 86: 281–297.
Montague, M. (2016) *The Given*, Oxford: Oxford University Press.
Moore, G. E. (1962) "Propositions," in *Some Main Problems of Philosophy*, New York: Collier Books: 66–87.
Neander, K. (1995) "Misrepresenting and Malfunctioning," *Philosophical Studies* 79: 109–114.
Papineau, D. (1998) "Teleosemantics and Indeterminacy," *Australasian Journal of Philosophy* 76: 1–14.
Pautz, A. (2013) "Does Phenomenology Ground Mental Content?" in Kriegel (2013).
Peacocke, C. (1998) "Conscious Attitudes, Attention, and Self-Knowledge," in C. Wright, B. C. Smith, and C. Macdonald (eds.), *Knowing Our Own Minds*, Oxford: Oxford University Press.
Pitt, D. (2004) "The Phenomenology of Cognition, or, *What Is It Like to Think That P?*", *Philosophy and Phenomenological Research* 69: 1–36.
Pitt, D. (2009) "Intentional Psychologism," *Philosophical Studies* 146: 117–138.
Pitt, D. (2011) "Introspection, Phenomenality and the Availability of Intentional Content," in Bayne and Montague (2011).
Pitt, D. (2013) "Indexical Thought," in U. Kriegel (2013).
Pitt, D. (2017) "The Paraphenomenal Hypothesis," *Analysis* 77: 735–741.
Prinz, J. (2002) *Furnishing the Mind: Concepts and Their Perceptual Basis*, Cambridge, MA: MIT Press.
Prinz, J. (2011) "The Sensory Basis of Cognitive Phenomenology," in Bayne and Montague (2011).
Quine, W.V. O. (1960) *Word and Object*, Cambridge, MA: MIT Press.
Rupert, R. (1999) "The Best Test Theory of Extension: First Principles," *Mind and Language* 14: 321–355.
Sacchi, E. and Voltolini, A. (2016) "Another Argument for Cognitive Phenomenology," *Rivista Internazionale di Filosofia e Psicologia* 7: 256–263.

Searle, J. (1984) *Minds, Brains and Science*, Cambridge, MA: Harvard University Press.
Searle, J. (1987) "Indeterminacy, Empiricism, and the First Person," *Journal of Philosophy* 84: 123–146.
Searle, J. (1992) *The Rediscovery of the Mind*, Cambridge, MA: MIT Press.
Siewert, C. (1998) *The Significance of Consciousness*, Princeton, NJ: Princeton University Press.
Siewert, C. (2011) "Phenomenal Thought," in Bayne and Montague (2011).
Sterelny, K. (1990) *The Representational Theory of Mind*, Oxford: Blackwell.
Strawson, G. (1986) *Freedom and Belief*, Oxford: Oxford University Press.
Strawson, G. (1994) *Mental Reality*, Cambridge, MA: MIT Press.
Strawson, G. (2008) "Real Intentionality 3: Why Intentionality Entails Consciousness," in *Real Materialism and Other Essays*, Oxford: Oxford University Press.
Strawson, G. (2011) "Cognitive Phenomenology: Real Life," in Bayne and Montague (2011).
Thompson, B. (2008) "Representationalism and the Argument from Hallucination," *Pacific Philosophical Quarterly* 89: 384–412.
Tye, M. (2000) *Consciousness, Color and Content*, Cambridge: MIT Press.
Tye, M. and Wright, B. (2011) "Is There a Phenomenology of Thought?" in Bayne and Montague (2011).

Related Topics

Representational Theories of Consciousness
Consciousness and Conceptualism
Dualism
Materialism

20

CONSCIOUSNESS AND CONCEPTUALISM

Philippe Chuard

Consciousness takes many forms. There's the visual awareness of a yellow school bus, the auditory consciousness of its engine's roar, the olfactive experience of its exhaust fumes, etc. There's also the cognitive awareness of objects, facts, and states-of-affairs when thinking of Budapest or of the French elections, when endorsing the classical properties of logical consequence. Not to mention other kinds, including emotions, bodily sensations, etc. These forms of consciousness differ in a variety of respects: whether they are tied to a specific sense organ (if at all), their distinctive phenomenology (what "it is like" to be in such-and-such conscious psychological state), the sorts of things they make us aware of (colors, sounds, smells, cities, social phenomena, or logical properties, etc.), and the different functions they occupy in our psychological lives (their contribution to rational thought-processes, to the acquisition of evidence and knowledge, how they can lead to action and behavior, etc.).

One significant contrast in this regard concerns the broad divide between *sensory* forms of conscious awareness and purely *cognitive* ones—between thinking, believing, judging, supposing, on the one hand, and seeing, hearing, touching, smelling, etc., on the other. Such contrast involves differences along several dimensions too (in phenomenology, functional role, etc.). But one crucial dimension is the manner in which sensory perception and cognition make us consciously aware of objects, features, situations, facts, etc. Whereas thoughts and beliefs seem to be essentially tied with concepts, sensory consciousness may not be, one suggestion goes.[1]

According to this "nonconceptualist" approach, sensory perception is separate from—and need not involve—conceptualization: the deployment of concepts marks a later stage in the cognitive process, one that causally depends on sensory consciousness rather than "permeates" or "suffuses" it entirely.[2] By contrast, *conceptualism* maintains that, just as thoughts and beliefs essentially depend upon the use and possession of concepts to represent what they do, perceptual experiences cannot make us aware of objects and features in our environment without conceptual identification and categorization.[3] Conceptualists needn't deny there's a divide between perception and cognition: such a divide, they say, has little to do with the presence or absence of conceptualization—in fact, it is essential to the interaction between sensory consciousness and cognition that such divide be bridged by conceptualization on both sides, conceptualists argue, in what appears to be one main motivation for the view.[4]

At the heart of this disagreement lies a series of related questions about the nature of sensory perception and the distinctive kind of consciousness it constitutes. For instance:

i Should all forms of consciousness (i.e., all types of conscious psychological states) be modelled on the specific kind of cognitive awareness underpinning conscious thoughts and beliefs?

ii Is there a purely sensory level of consciousness and, if so, how does it differ from, and interact with, cognitive states like judgments and beliefs that typically accompany conscious perception?

iii Does the phenomenology of conscious sensory states essentially differ with respect to how things appear to us in perception from how we think about and recognize perceived items in our environment?

Answers to such questions largely divide conceptualists and non-conceptualists, with wide-ranging implications for a host of related issues in the philosophy of mind and epistemology, many of which I'll have to ignore, unfortunately: e.g., how animals and young infants perceive and know their environments,[5] what contribution conscious perception makes to the acquisition of concepts,[6] and how it serves as a source of evidence and knowledge.[7]

I begin with some of the background assumptions informing the dispute (Section 1), and then review some of the considerations advanced against conceptualism (Sections 2–3). I'll limit myself to arguments that explore features specific to the distinctive kind of consciousness characteristic of sensory experiences.[8]

1 Conceptual Content?

What are concepts? And what roles do they play in our psychology? The starting point of the dispute between conceptualists and non-conceptualists resides in a number of related platitudes—largely shared on both sides[9]—about how concepts are connected with thoughts and beliefs.

First, concepts can be *possessed* (or not) by psychological subjects.[10] And *which* concepts subjects do possess sets constraints on *which* beliefs and thoughts they are capable of thinking. Without the concept of an *ARMADILLO*,[11] Fred can't think about armadillos as such. Second, concepts seem essentially *representational*, serving to classify different *kinds* or *categories* of things. The concept *KANGAROO* allows one to think about kangaroos, just as *RED* serves to think about red things. This representational function is *compositional*: single concepts can combine into more complex strings of concepts—e.g., the concept of a *RED KANGAROO*, or the complex representation that *KANGAROOS AREN'T ARMADILLOS*—some of which are assessable for truth or falsity. And the ability to combine and recombine different concepts appears central to the ability to entertain new thoughts. Finally, concepts have an *inferential* function: at least for some inferences, how good the inference is might depend in part on some of the logical or semantic relations between the concepts used, as when one infers that *SKIPPY IS A MARSUPIAL* from the belief that *SKIPPY IS A KANGAROO*.[12]

It's quite natural, then, to view concepts and their possession as inevitably linked with distinctive *psychological and cognitive capacities* underpinning thoughts and beliefs. Concepts are crucial to the ability of *understanding* thoughts, including new thoughts a subject hasn't entertained before, as well as the ability to draw certain inferences. By representing *kinds* of things, concepts scaffold the ability to *identify* certain individuals as belonging to some category or other and to *discriminate* individuals from distinct kinds. Such capacities likely serve to determine what it takes to possess a given concept—that's not to say, note, that the lists of conceptual features and capacities just mentioned are exhaustive.

For conceptualists, the functions concepts play in our psychology aren't limited to thoughts and beliefs: the realm of the conceptual extends to perception and sensory awareness too. Conceptualism thus involves a commitment to:

c_1: subject *S* is sensorily conscious of object/event/fact/property *x* *only if* *S* possesses some concept c for *x*.

Sensory awareness requires concept-possession, yet c_1 isn't enough: after all, *S* might possess more than one concept applicable to the kangaroo in front of them (e.g., the concepts *KANGAROO, SKIPPY'S MATE, SMELLY PEST, MACROPUS RUFUS*, etc.). That *S* merely possesses *some* concept leaves unspecified *how* it is exactly that *S* currently conceptualizes the kangaroo in front of them. What matters, then, is *which* concept *S* *deploys* in a given perceptual experience:

c_2: *S* is sensorily conscious of *x* *only if* *S* applies some specific concept *c* to *x* in *S*'s conscious experience of *x*.

Where applying a concept may be tantamount, in this context, to *identifying* (correctly or not) *x* as falling under *c*. But even that may not suffice. Suppose *S* looks at a farmyard filled with cows, ducks, kangaroos, cats, and armadillos, and conceptually identifies each animal merely *as an ANIMAL*, nothing more, using the very same concept for all the different animals *S* sees. There's a sense in which the concepts deployed by *S* fail to capture everything *S* is visually aware of, including the diverse animals and their visible differences. If conceptualism aims to account for how we are sensorily aware of things and features in our environment, in the sense that conscious awareness depends for a crucial part on conceptualization, a more demanding constraint is needed:

c_3: *S* is sensorily conscious of *x* and *y* and of some difference between *x* and *y* *only if S identifies x as* c and *y as* c^*, where c and c^* are different concepts (concepts of different objects/properties, etc.).

Otherwise, concepts deployed in experience might fail to match what one is in fact sensorily aware of. Conceptualism, so construed, amounts to the suggestion that *what* we are consciously aware of in sensory perception is a function of *which concepts* are deployed in perception. Perhaps, other conditions need to be added to this minimal set of conceptualist requirements.

Against this, the weakest form of nonconceptualism denies that conditions c_1 to c_3 apply to *everything* we perceptually experience: if, typically, we conceptually identify in experience the objects, events, and features we are sensorily aware of, it's possible to fail to do so. A more radical version rejects, not just that conceptual identification be necessary for sensory awareness, but that there's any conceptualization in sensory consciousness at all: we might of course conceptually identify most of what we perceive in beliefs and thoughts *based on* our conscious experiences, although such conceptualization occurs in a later stage, causally downstream of sensory consciousness *per se*.

In sketching the nature of the disagreement between conceptualism and nonconceptualism, note, I haven't even yet mentioned the notion of "conceptual content," let alone that of a "proposition".[13] In part, this is because the platitudes about concepts and their connections to various capacities used in thoughts and beliefs, and whether such connections extend to sensory consciousness, is really what the dispute is mostly about, it appears. The notion of "content" is often associated with theories of propositional contents, which treat the latter as some sort of abstract objects, and disagree about the metaphysical nature of what composes these abstract contents—be it concepts considered as abstract objects themselves (Fregean contents), or physical objects and properties arranged in sets (Russellian contents), or possible worlds in sets thereof (e.g., Stalnaker 1998).[14] Unless such theories can shed genuine light on the platitudes we started with about concepts and their connections with thoughts and beliefs in our psychological lives, however, it's not all that clear why metaphysical concerns about the nature of certain abstract objects have much to contribute to issues about the place and function of sensory consciousness in our cognitive architecture.

Still, how to spell out the terms of the dispute has become a somewhat contentious issue, of late. It's now common to draw a distinction between two ways of understanding the dispute: as (i) having to do with a certain kind of content, *conceptual content*, which is composed of concepts, and whether such content can be found both in thoughts and beliefs as well as perceptual experiences—or whether the content of the latter is of a different, non-conceptual, kind—or, rather, as (ii) being about what it takes to be in certain *types* of psychological *states, conceptual states*, and whether concept-possession is as necessary for perception as it is for thoughts and beliefs.[15] One concern is that many arguments take as their explicit target the view that perceptual experiences have a conceptual content (content conceptualism), but end up, if successful at all, discarding at most the view that such experiences are conceptual states (state conceptualism): there is, that is, a recurrent but illegitimate shift between content conceptualism and state conceptualism, the worry goes. Relatedly, it may seem unfortunate that, being so often cashed out in terms of content conceptualism, the disagreement remains closed to theorists who deny some of the starting assumptions, in particular, the idea that mental contents are composed of concepts at all (Stalnaker 1998).

Phrasing the dispute so as to allow diverse theorists to partake should be desirable, undoubtedly. In this light, it might seem as though the platitudes about concepts and their connections to thoughts and beliefs, which we started with, align nicely with the construal behind state conceptualism—especially the idea that one's concepts and conceptual abilities constrain which thoughts and beliefs one can entertain. However, some of these platitudes treat concepts as representations, we saw, which can combine into more complex conceptual representations: this seems to constitute a *conceptual content* of a sort which many could accept—especially if treated as a concrete psychological kind for the purpose of psychological explanation, rather than a constituent of abstract propositions.[16] Why demand more, exactly?

In addition, the equivocation from content conceptualism to state conceptualism seems avoidable: if *only* conceptual psychological states *require* possession of the relevant concepts to understand, entertain, think, differentiate, and believe the contents associated with such states, it looks as though, based on the platitudes unearthed earlier, conceptual states involve, by virtue of their content, specific combinations of concepts. If a subject lacks some concept necessary for understanding the content of the psychological state they are in, it then seems legitimate to infer that such a psychological state isn't entirely conceptual. This means it doesn't come equipped with the fully conceptual representations underpinning conceptual states—a claim about the type of semantic features (at least some of them) associated with the state in question.[17]

So, what is it about sensory awareness that is supposed to be significantly different from the sort of conceptual awareness essentially at play in thoughts and beliefs, to the effect that the former needn't depend on the possession and deployment of concepts?

2 Fineness of Grain

Examples like the gray patches in Figure 20.1 suggest that sensory awareness can be quite fine-grained, in the sense that we can be sensorily aware of highly, if perhaps not perfectly, *determinate* properties (colors in this case) and some of the specific differences between them.[18] If conceptualism is true and conditions c_1 to c_3 hold, subjects who visually discriminate all the shades in Figure 20.1 should be deploying *different* color concepts to identify each shade.[19]

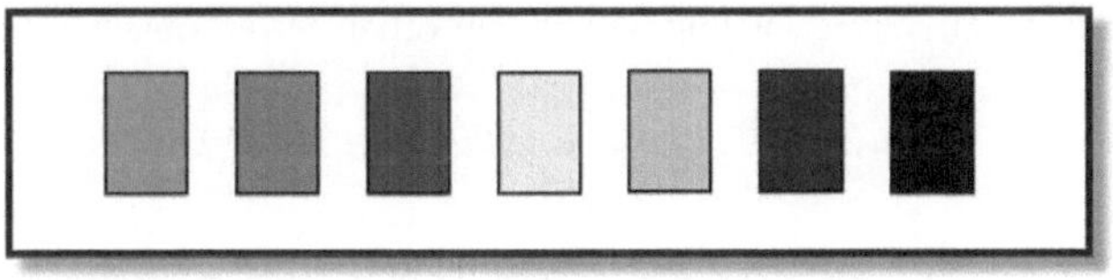

Figure 20.1 Fineness of Grain Example

Yet, the argument goes, it seems many subjects who can visually discriminate these shades might have difficulties conceptually identifying the colors in question (Evans 1982: 229). Not just that they lack words to express such concepts, but it seems as though few subjects possess for each of the specific shades in Figure 20.1 the sort of conceptual representations typically used to think about red and yellow (say). But, if it's possible for subjects to lack the corresponding concepts even when visually discriminating the shades in question, it should follow that some of the conditions conceptualism imposes upon sensory awareness—including c_3—don't in fact apply.

In response, conceptualists typically reject the assumption that subjects lack the relevant concepts:

> There is an unacceptable assumption behind this line of argument, that concepts necessarily correspond with entirely context-independent classification of things, … This restriction unacceptably rules out any appeal to context-dependent demonstrative concepts, though—concepts associated with expression like 'that shade of red', or 'just that large in volume,' …
>
> *(Brewer 1999: 171)*[20]

That is, in line with c_3, subjects could deploy different *demonstrative concepts* to discriminate conceptually the chromatic differences they are sensorily aware of—where concepts are *demonstrative* at least in that the perceptual context helps fix what the concept picks out, aided in part by the subject's ability to narrow down what is picked out in the context.

This appeal to demonstrative concepts has invited various objections. One question is how demonstrative concepts pick out the specific features they do. If a subject deploys distinct demonstrative concepts—*THIS* and *THAT*—for the two shades on the left of Figure 20.1, say, how is it that *THIS* picks out the shade on the outmost left rather than the one on its right—the "differentiation problem" (see Raffman 1995)? Relatedly, what else is needed to guarantee that *THIS* picks out a specific shade of gray rather than the shape of the patch, its location, size, the color of the background, etc.—the "supplementation problem"? Perhaps, *THIS* can be combined with some non-demonstrative concept like *SHADE* to pick out the color rather than some other property of the object. But as Peacocke (2001: 245–250) points out, this suggestion doesn't help answer the differentiation problem: why *THIS SHADE* picks out the color of the patch on the extreme left, rather than that of an adjacent patch.

Brewer proposes that perceptual attention is key: "concepts figuring in experiential contents do not simply pop up from nowhere" but "are provided directly by [the subject's] attentional relations with the particular things around him" (1999: 185). Accordingly,

> … determinacy of reference is secured by the supplementation of the bare demonstrative "that", by the subject's actual attention to the *color* of the object in question, as opposed to its shape or movement, say, where this is a neurophysiologically enabled

> relation between the subject and *that property*, as opposed to any other, of the object which he is perceiving.
>
> *(Brewer 2005: 224; also 1999: 172–3, 187ff., 226)*

In light of this, Adina Roskies (2010) has asked what it takes to form *new* demonstrative concepts. If demonstrative concepts "are to be understood as a mental analogue of these more familiar linguistic demonstratives" like "this" and "that" (Roskies 2010: 120) so that attention can fill the role for demonstrative concepts that demonstrative gestures play when communicating with demonstrative expressions (2010: 121), "the act of focusing attention must be intentional," Roskies (2010: 122) argues, since demonstrations are (2010: 120). But then, any intentional shift in attention must exploit the content of the relevant conscious perceptual experience, Roskies continues: to selectively focus on some element of one's visual field, the element in question must already be consciously available in perception. However, if such perceptual contents serve in intentionally directing attention, and the latter determines how demonstrative concepts pick out their referents, the perceptual contents in question cannot be demonstrative or conceptual, on pain of circularity, she concludes (2010: 123). That's why conceptualist appeals to demonstrative concepts must ultimately draw upon the non-conceptual content of experience, the suggestion goes.[21]

It looks as if there's room for conceptualists to resist Roskies' argument. Note that, in the passage cited above, Brewer seems to hint that, on his proposal, attention operates upon subpersonal representations—I assume this is what he means by "a neurophysiologically enabled relation" (Brewer 2005: 224)—rather than conscious (hence personal) perceptual contents, in determining the semantic features of demonstrative concepts, *once attention has shifted*.[22] Roskies' argument, however, concerns the causal process leading to attentional shifts (Roskies 2010: 128). Conceptualists could retort that, before it is attended, a shade of gray might be experienced in a less fine-grained, and less determinate, manner: meaning it could be conceptualized *via* coarser-grained concepts.[23] Once such a shade is selectively attended, however, its perceptual representation is more fine-grained as a result, and so is the demonstrative concept grounded in such attentional focus. Hence, even if attentional shifts causally depend upon earlier perceptual representations of the target, the latter can still be conceptual, provided they involve different coarser-grained concepts. No circularity here, and none either in how the more fine-grained demonstrative concept is semantically determined by subpersonal processes underlying selective attention, provided the contents of earlier experiences before the shift aren't responsible for this part of the explanation.[24]

A different worry relates to the *possession-conditions* for demonstrative concepts. For Sean Kelly (2001b), if demonstrative concepts really *are* concepts of *kinds* like colors, they must satisfy the *re-identification constraint*: if a subject *S* possesses a concept *c*, *S* must be able to *re-identify* distinct instances of *c as being the same* (e.g., the same color), consistently and reliably (2001b: 406). Yet, it's not uncommon for subjects to fail to recognize the fine-grained chromatic shades they previously discriminated (Raffman 1995; Dokic and Pacherie 2001; Kelly 2001b). Imagine Figure 20.1 is a color chart found in a hardware store: you might have chosen one of the shades as the ideal color to repaint your wheelbarrow with. If you then accidentally drop the chart, pick it up again along with the other charts that fell, you may not quite remember which color you had just decided upon (based on the color alone, rather than its location on a chart or other). If, before dropping the chart, you were able to think demonstratively of the color you chose, you are now unable to re-identify it as such only a few seconds later. Hence, the argument goes, the demonstrative concept used earlier wasn't really a concept for a specific kind of color, if re-identification is necessary for possessing demonstrative concepts (Kelly 2001b: 411; also Jacob and Jeannerod 2003: 25; Smith 2002: 111; and Tye 2005).

One complication owes to the different re-identification constraints available (Chuard 2006a). It's one thing to identify distinct objects—e.g., x at t and y at t+n—as falling under the same concept: here, re-identification is nothing but repeated identification of instances of a concept. It's another to identify y as falling under concept c while remembering x in such a way as to identify both x and y as being relevantly the same in that both fall under c. The latter kind of re-identification is more demanding: it involves explicit memory of past encounters with other instances, together with comparative judgments based on such memories. It's not clear we always meet such a requirement, even when it comes to non-demonstrative concepts: if my memory is rather poor in recalling past encounters with a specific type (e.g., hexagon), this needn't undermine my ability to identify instances of that type. The less demanding notion of "re-identification" (as mere repeated identification) is unproblematic in this respect. Yet Kelly's argument presupposes that the subject doesn't just identify a determinate shade, but that they also recognize the color thus identified as the same as one previously identified: it appears to rest on the more demanding constraint.

Nor is it even clear, in fact, whether the less demanding re-identification constraint applies to *demonstrative concepts* (see Chuard 2006a; Coliva 2003). As Brewer suggests in the passage cited earlier, demonstrative concepts are meant to be *context-dependent* tools for categorizing objects and their properties. And part of the context involves how one perceptually attends to the relevant samples. In this sense, possession of such concepts may be quite fickle: demonstrative concepts can be thought of as *disposable classificatory devices*, to be used in a given context, but not beyond. Accordingly, if there's any change in the perceptual scene, or if the perceiver shifts the focus of their attention, etc., such contextual changes may suffice to ground *distinct* demonstrative concepts (for the same sample, even). Hence, a subject may not be able to identify a color as falling under the same demonstrative concept twice, simply because the first demonstrative concept deployed isn't available the second time around. This needn't imply that demonstrative concepts aren't concepts, or that they aren't concepts of kinds. Just like other concepts, they can serve in conceptual discriminations or inferences (within a context), etc. But there's no obvious reason to assume that re-identification ought to have a special status when it comes to the possession-conditions of different types of concepts.[25]

Finally, a worry about the *extension* of demonstrative concepts (Dokic and Pacherie 2001; Kelly 2001a). Samples of highly similar but *distinct* colors can be arranged so as to be perceptually indiscriminable (despite their fine-grained differences): patch *a*'s color may be visually indiscriminable from patch *b*'s, which is indiscriminable from *c*'s, even though *a* and *c* are visually *discriminable*, owing to their greater chromatic difference—perceptual indiscriminability is non-transitive, that is.

Under the conceptualists' proposal, *b*'s color falls under the demonstrative concept $\textit{THIS}_b$ formed when attending to *b*'s specific shade. And since *b*'s color is perceptually indiscriminable from *a*'s, it seems *a* should also fall under the demonstrative concept $\textit{THIS}_b$—being indiscriminable, they might be conceptually identified in the same way, one might think. But if *a* falls under $\textit{THIS}_b$, *c* should too, since *c*'s color is also indiscriminable from *b*'s. Hence, *a*, *b*, and *c*, fall in the extension of $\textit{THIS}_b$. However, the color of *a* and *c* are discriminable, which means that $\textit{THIS}_b$ applies, not just to *distinct* colors, but to visually *discriminable* ones, even though $\textit{THIS}_b$ was supposed to be a fine-grained concept of a determinate shade. Demonstrative concepts aren't fine-grained enough, after all (Dokic and Pacherie 2001: 195; see also Peacocke 1992: 83; Martin 1992: 757).

Conversely, since *b* is chromatically indiscriminable from both *a* and *c*, it seems it should also fall under the respective demonstrative concepts formed when attending to *a*, $\textit{THIS}_a$, and when attending to *c*, $\textit{THIS}_c$. Which means that *b*'s determinate color falls under at least three distinct demonstrative color concepts: distinct, since $\textit{THIS}_a$ and $\textit{THIS}_c$ pick out different, and discriminable, colors. Demonstrative concepts are too fine-grained then (Dokic and Pacherie 2001: 195). Consequently, if more than one such demonstrative concept applies to *b*'s color, it follows that, were a perceiver to look at *b* and deploy all the relevant concepts (i.e., THIS_a, THIS_b, and THIS_c), a "uniformly colored" object like *b* might "present[…] more than one shade at a given time to a given observer" (Dokic and Pacherie 2001: 195), in a manner which does "not correspond to any phenomenological differences" (Dokic and Pacherie 2001: 197).

These concerns, however, rest largely on additional assumptions conceptualists needn't grant (Chuard 2007a; Pelling 2007). To begin with, there's the assumption that chromatic indiscriminability should suffice for two colored samples to fall under the same demonstrative color concept. True, demonstrative concepts are supposed to be concepts of highly determinate shades, so that two samples had better be perceptually indiscriminable to fall in the extension of the same concept. But a necessary condition isn't a sufficient one, especially if demonstrative concepts are indeed contextually tied to the particular samples attended when deploying such concepts.[26]

As for the consequence that deploying different demonstrative concepts for the same shade might lead to differences in experiences that are in fact not there, it seems to involve a simple confusion regarding conceptualism itself. If C_3 requires that differences in sensory awareness must depend upon matching conceptual differences, the converse requirement—that different concepts suffice to ground differences in sensory awareness—isn't, strictly speaking, implied by or required for conceptualism (Chuard 2007a).

In sum, demonstrative concepts have proven a useful tool for conceptualists: one which helps escape a host of related objections about fine-grained experiences, and sheds some light on what concepts might be deployed in experience and how.

3 Informational Richness

Look at a crowded city street and you become sensorily aware of a great many things: different objects interspersed through your field of vision (people, cars, street signs, buildings and trees, etc.), many of their visible properties and relative arrangement. Sensory consciousness can be a source of *rich* information about your environment.

This isn't to say all perceptual experiences, let alone all visual experiences, are like that: fixate on a blank sheet of paper close-up and, though you might be aware of quite a few things about the paper and its parts, the information you thereby access pales in comparison to that available in other situations. Nor is it to say that *every* object or perceivable feature in a perceptual scene is in fact available in your experience of that scene. Finally, the rich information conveyed in some experience may or may not be fine-grained: imagine seeing a crowded street through a very thick fog, when the shape and colors of the passers-by are all blurred in the fog (Chuard 2007b).

That sensory experiences can be informationally rich isn't really at issue between conceptualists and non-conceptualists (see Brewer 1999: 240–241; McDowell 1994: 49). What is, is whether such informational richness raises a difficulty for conceptualism about the deployment of concepts in perception: whether perceivers can simultaneously deploy sufficiently many concepts to match the amount of information conveyed through sensory awareness to ensure that such rich information is entirely conceptualized—not whether subjects *possess* the relevant concepts. The question is whether conditions C_1 to C_3 hold for *every* perceived object or property in a given experience, or whether, owing to capacity limitations, some object or feature amidst such rich information might be left unconceptualized.

Earlier attempts to exploit the informational richness of sensory consciousness against conceptualism (Dretske 1993; Martin 1992) relied largely on the idea that conceptualization is constrained by what perceivers *notice*. Since it seems possible that a subject fails to notice someone's thin moustache or a pair of cufflinks in a drawer, yet it also seems possible for subjects in those situations to later remember experiencing these unnoticed elements, this suggests that, if they can be recalled in memory, what is unnoticed in a scene must in fact have been perceived. Whether these considerations really threaten conceptualism, however, hangs mostly on whether there really is no conceptualization outside of what perceivers notice, and it's not altogether clear how this question could be settled—conceptualists might well grant that not everything that is conceptualized in experience ends up being thought about, let alone believed.

Since at least Dretske (1981), much attention in this context has been devoted to Sperling's (1960) work on iconic memory.[27] Drawing on earlier results that, when presented with arrays of twelve letters (three rows of four letters each) for relatively short intervals (less than 15ms), subjects can correctly report the identity and location of a little more than four letters in the display. Sperling proceeded to show that performance improved when subjects were prompted to provide partial reports after being cued to attend one of the three rows of letters by a tone: importantly, the cue occurred after the stimulus offset (i.e., once the array of letters had been masked). In these partial reports, subjects were able on average to correctly report the identity and location of a little more than three letters out of four in the cued row.

The standard interpretation of Sperling's results assumes that, if subjects can access the identity and location of three of the letters in a given row when it is cued, this holds not just for the cued row but for the other two as well, suggesting that sensory consciousness is informationally rich, at least to the extent that the identity and location of about nine of the twelve letters in the display must be consciously perceived. Yet, on average, subjects can only report not more than four or five letters, because, the standard explanation goes, not more than four or five letters can be stored in short-term memory. Accordingly, sensory consciousness is informationally richer than what subjects can conceptualize and store in memory, and then verbally report. So much the worse for conceptualism.

One difficulty with this interpretation concerns the assumption that Sperling's subjects were sensorily aware of the identity and location of about nine letters out of twelve in the display. This assumption apparently relies on extrapolating how much is consciously perceived of the two uncued rows by projecting from the subjects' partial reports of the cued row, and then adding up the estimated results from such extrapolation (roughly, three letters for each of the three rows). The thought might be that, since the cue occurs *after* the display has disappeared from view, it couldn't retrospectively affect how each row is perceived (Fodor 2007; Tye 2005).[28]

Even so, the cue serves to draw attention to the letters in the cued row, and this may affect how much information about the letters' identity and location in that row can be retrieved and become consciously available. There's no guarantee, in other words, that just as much information from the other two uncued rows is in fact made available in conscious experience. A visual experience may well carry information about the overall display, *albeit determinably*, without explicit representation of the identity or specific shape of each letter at each location (Phillips 2011: 402).[29]

Moreover, it's unclear to what extent evidence about how many letters subjects *can report* provides evidence about how many letters are *conceptually identified* in sensory consciousness. For one thing, Sperling's results might reveal limitations on the processing mechanisms needed for *reportability*, rather than on conceptual identification (for discussion, see Pashler 1998)—even if what is reported has been conceptualized, the converse may not hold.

4 What Next?

This brief discussion concentrated on just two considerations advanced against conceptualism, and both fail: once the relevant details about concepts, their possession and deployment, are worked out, we can see that conceptualism has the resources to deal with those features which, at first sight, appeared to separate sensory awareness from conscious thoughts and beliefs. At best, such considerations might have helped develop conceptualism, by sorting out which commitments conceptualists really need.

But note how these objections were assessed in relative isolation from one another. What would happen if we considered them jointly—after all, it seems many experiences can be both fine-grained *and* rich in information? The question is whether the commitments conceptualists take on board to address one objection cohere with those needed to address another.

The fine-grained differences between the seven shades in Figure 20.1 seem immediately available as soon as you set your eyes on the whole figure. Presumably, you must deploy seven distinct demonstrative concepts, one for each shade, *if* conceptualism is true. And for these demonstrative concepts to pick out the specific shade they do, your attention must have selected one shade when forming the concept in question. At this point, various empirical questions arise about how subjects selectively attend to each shade in such a case. Presumably, there are some constraints on selective attention—including processing and temporal constraints on the mechanisms at play, especially if the process of briefly scanning the seven shades is serial or not. But if subjects can't attend separately to each of the seven shades more or less simultaneously, they may not be able to form fine-grained demonstrative concepts for each; at least not straight away. Hence, there might well be fine-grained differences between some such shades, which subjects are sensorily aware of at a time without being able to conceptualize them at that time, even demonstratively—because their attention is currently allocated to some of the other shades in Figure 20.1.

At this point, conceptualists might well resort to the idea that attention functions like a fluid, quantities of which can be distributed across several items in a scene. That is, if attention can be *divided*, a perceiver might focus her divided attention *simultaneously* on all the gray samples in Figure 20.1: rather than process each shade serially, attentional mechanisms may be deployed in parallel for each of the seven shades. However, divided attention—and the parallel processing that seems to underlie it—is usually observed in cases where stimulus attributes involve highly salient *coarse-grained* differences (see Pashler 1998: 218 for a review). When the differences are fine-grained, there is evidence that attentional mechanisms slow down and operate separately on each stimulus, as it were.

Perhaps, conceptualists will balk at the suggestion that a conscious experience of Figure 20.1 can be so fine-grained that almost all chromatic differences between the seven shades are immediately available. Perhaps, they might borrow the suggestion (Mandik 2012; Phillips 2014) that some of the specific shades, and some of their chromatic differences, are consciously available only in a *determinable*, and coarse-grained, manner. This would mean that the impression that chromatic differences between the seven shades are immediately available when glancing at them must be illusory: even if all shades appear at the center of one's visual field, and despite the fact that subjects typically don't seem remotely aware of any shift—from determinable to determinate—in how colors appear in experience, let alone any changes in conceptualization of these shades, from coarse-grained to fine-grained demonstrative concepts. All this to suggest, in any case, that conceptualists aren't entirely out of the woods when it comes to the fineness of grain and informational richness of experience.[30]

There's another, more general, matter: namely, how conceptualists propose to account for the distinctive phenomenology of sensory consciousness. Suppose, as intentionalists have it, that what it's like to be in a specific sensory experience, phenomenologically speaking, has a lot to do with how things appear in such an experience—as Brewer (1999: 156) seemed to allow.[31] Presumably, the distinctive phenomenology of a visual experience of a yellow school bus differs noticeably from that of an auditory experience of its engine, and both contrast phenomenologically from beliefs one might have about the bus in question.

But if the phenomenology of sensory experiences is essentially determined by their contents, it seems it isn't quite possible to strictly believe the exact same content one experiences, contrary to conceptualists' insistence (Brewer 1999; Speaks 2005): at least not literally, as a claim about the very contents of experiences and beliefs (rather than, say, the objects and features so represented). Given intentionalism, if their contents really are the same, why isn't their phenomenology the same too?[32]

Unless, of course, the conceptual contents in question somehow lose their essential connection with some distinctive sensory phenomenology in the transition from sensory consciousness to belief—does this mean there are two kinds of conceptual contents, after all, one perceptual, the other doxastic? Either way, conceptualists here face an additional explanatory burden, in accounting for how this is preferable to the non-conceptualists' proposal that such differences in phenomenology map onto different kinds of contents.

5 Concluding Remarks

Whether sensory consciousness crucially depends or not on what subjects conceptually identify in experience remains quite open, at present. Some progress on psychological theories of concepts and how conceptual capacities are used in thought and perception could help a great deal at this juncture. On the other hand, unless there are compelling reasons for conceptualism, the development of a detailed non-conceptualist view might hold the promise of providing a more integrated explanation of some of the aspects of sensory consciousness surveyed here: even if conceptualists can address all such objections, it remains to be seen how stable the resulting conceptualist view will turn out.[33]

Notes

1 For alternate ways the perception-cognition divide can be drawn, see, e.g., Beck (forthcoming), Crane (1992), Orlandi (2014), Smith (2002), Vision (1997).

2 For instance, Bermúdez (1998, 2009), Crane (1992, 2001), Dretske (1981, 1993), Evans (1982), Fodor (2007, 2008), Martin (1992), Peacocke (1986, 1989, 1992, 2001), Pylyshyn (2003), Raftopoulos (2009), Schmidt (2011), and Tye (1995, 2000, 2005).

3 Conceptualists include Brewer (1999, 2005), McDowell (1994, 1998), Sedivy (1996), Strawson (1992), and Gennaro (2012).

4 For the merits of that argument, see references in note 7. Conceptualism also aligns with higher-order thought theories of consciousness: Gennaro (2012).

5 Beck (2012), Bermúdez (2007b), and Peacocke (2001).

6 Gennaro (2012: chs. 6–7), Peacocke (2001), Roskies (2008, 2010).

7 See Brewer (1999, 2005), Heck (2000), McDowell (1994), Millar (1991), Peacocke (1998, 2001, 2006), Pryor (2005), Roessler (2011), Sosa (1997, 2003), Wright (1998, 2002).

8 I also ignore interesting issues about cognitive penetration in perception (Macpherson 2015), and whether there are relevant structural differences between perception and conceptual thought (Crane 1992; Fodor 2007, 2008; Heck 2007; Matthen 2005a; Toribio 2011). For reviews of different aspects of the dispute, see also Bermúdez and Cahen (2015), Crane (2001), Chuard (2009), Laurence and Margolis (2012), Schmidt (2011), Siegel (2016), Toribio (2007), Wright (2015).

9 See Laurence and Margolis (2012: 293).

10 Even if most concepts can be expressed linguistically, caution is due to separate concepts from their linguistic expressions.

11 I follow the standard convention of capitalizing names of concepts, as in my concept of *CONCEPT*.

12 For theories of concepts, see, e.g., Fodor (1998), Laurence and Margolis (1999, 2007), Prinz (2002).

13 Nor have I explicitly relied on the assumption that perceptual experiences have representational contents. In principle at least, it seems possible for those who profess to reject such an assumption (i.e., naïve realists for whom perception is only a direct relation to perceived objects) to agree that sensory awareness requires concept-possession in the way characterized by c_1–c_3. For useful discussion of whether experiences have representational contents, see Siegel (2012, 2016).

14 For how such views apply to perceptual content: Siegel (2016).

15 Byrne (2005), Crane (1992, 2001), Heck (2000), Speaks (2005), Stalnaker (1998).

16 In a different vein, Bermúdez (2007a, 2009; Bermúdez and Cahen 2015) emphasizes the difficulties in making sense of conceptual states independently of their contents being conceptual.

17 See also Toribio (2006) for more discussion, as well as Crowther (2006), Duhau (2011), Hanna and Chadha (2011), Laurier (2004), and Schmidt (2011).

18 If in doubt, imagine the patches instantiate very different shades of green—unfortunately not available here.

19 Compare Peacocke (1986, 1989) on fine-grained spatial perception.

20 Also, McDowell (1994: 56ff). Whether combinations of coarser-grained non-demonstrative concepts—involving comparative concepts such as *BRIGHTER THAN*—suffice as a response, is developed in Mandik (2012)—compare Noë (2004: ch. 6) for a different suggestion. It's not entirely obvious, however, how combinations of non-demonstrative and comparative concepts might accurately capture *all* relevant differences between an experience of Figure 20.1 and another (simultaneous or not) of the same colored patches but spatially arranged differently, for instance.

21 Compare Peacocke (1998, 2001) and Heck (2000) for similar arguments, where the latter is particularly concerned with non-veridical experiences. In response, Bengson, Grube, and Korman (2011) have suggested there is a relation of direct and non-contentful sensory awareness which could fix the referents of demonstrative concepts. For conceptualists to resort to such a relation, however, seems tantamount to acknowledging there are *two kinds* of sensory awareness, one constrained by concepts, and one which isn't. See also Brewer's (2005: 222–223) response and, for more on some of these questions, Dickie (2016).

22 Roskies (2010: 129) rejects this appeal to subpersonal mechanisms: this is where the *intentionality* of attentional shifts becomes relevant, requiring resources at the personal level, she insists.

23 See Mandik (2012).

24 Compare Speaks (2005: 386) on Heck's (2000: 492) version of the argument. See Gennaro (2012: chs. 6 and 7) for another response.

25 For alternative approaches, see Hanna and Chadha (2011); Matthen (2005b). Veillet (2014) raises the further worry that Kelly's objection, if it worked, would undermine certain assumptions about the conceptual content of *demonstrative beliefs*, which non-conceptualists typically rely upon.

26 Being so highly context-dependent, demonstrative concepts might be so fine-grained that differences in lighting conditions, background, etc., which affect the appearance of a chromatic shade (Kelly 2001a), suffice to give way to new demonstrative concepts. Compare Peacocke (1998, 2001).

27 See Coltheart (1980), Pashler (1998), and Phillips (2011) for useful surveys.

28 For skepticism about this point, and the suggestion that the mechanism behind Sperling's results is postdictive, see Phillips (2011: 393–394); compare Pashler (1998).

29 For discussions of recent variants of Sperling's experiment, see Block (2007, 2011) and Phillips (2011, 2016).

30 See also Chuard (2006b) and Pelling (2008).

31 For instance, Byrne (2001), Chalmers (2010), Crane (2001), Dretske (1995), Harman (1990), and Tye (1995, 2000).

32 See Smith (2002: ch. 3), Tye (1995, 2000).

33 All my gratitude to Rocco Gennaro and Jennifer Matey for helpful suggestions.

References

Beck, J. (2012) "The Generality Constraint and the Structure of Thought," *Mind* 121: 563–600.

Beck, J. (forthcoming) "Marking the Perception-Cognition Dependence: The Criterion of Stimulus-Dependence," *Australasian Journal of Philosophy*.

Bengson, J., Grube, E., and Korman, D. (2011) "A New Framework for Conceptualism," *Noûs* 45: 167–189.

Bermúdez, J. L. (1998) *The Paradox of Self-Consciousness*, Cambridge, MA: MIT Press.

Bermúdez, J. L. (2007a) "What is at Stake in the Debate on Nonconceptual Content?" *Philosophical Perspectives* 21: 55–72.

Bermúdez, J. L. (2007b) *Thinking without Words*, Oxford: Oxford University Press.

Bermúdez, J. L. (2009) "The Distinction between Conceptual and Nonconceptual Content," in A. Beckermann and B. McLaughlin (eds.) *The Oxford Handbook of Philosophy of Mind*, Oxford: Oxford University Press.

Bermúdez, J. L. and Cahen, A. (2015) "Nonconceptual Mental Content," *The Stanford Encyclopedia of Philosophy*. [https://plato.stanford.edu/entries/content-nonconceptual/]

Block, N. (2007) "Consciousness, Accessibility and the Mesh between Psychology and Neuroscience," *Behavioral and Brain Sciences* 30: 481–548.

Block, N. (2011) "Perceptual Consciousness Overflows Cognitive Access," *Trends in Cognitive Science* 15: 567–575.

Brewer, B. (1999) *Perception and Reason*, Oxford: Oxford University Press.

Brewer, B. (2005) "Do Sense Experiential States Have Conceptual Content?" in E. Sosa and M. Steup (eds.) *Contemporary Debates in Epistemology*, Oxford: Blackwell.

Byrne, A. (2001) "Intentionalism Defended," *The Philosophical Review* 110: 199–240.

Byrne, A. (2005) "Perception and Conceptual Content," in E. Sosa and M. Steup (eds.) *Contemporary Debates in Epistemology*, Oxford: Blackwell.

Chalmers, D. (2010) "The Representational Character of Experience," in D. Chalmers, *The Character of Consciousness*, Oxford: Oxford University Press.

Chuard, P. (2006a) "Demonstrative Concepts without Re-identification," *Philosophical Studies* 130: 153–201.

Chuard, P. (2006b) *From Concepts to Appearances: A Critical Evaluation of Conceptualism*, Ph.D. Dissertation, Australian National University.

Chuard, P. (2007a) "Indiscriminable Shades and Demonstrative Concepts," *The Australasian Journal of Philosophy* 85: 277–306.

Chuard, P. (2007b) "The Riches of Experience," in R. Gennaro (ed.) *The Interplay between Consciousness and Concepts*, Special Issue of *The Journal of Consciousness Studies* 14 (9–10): 20–42.

Chuard, P. (2009) "Non-conceptual Content," in T. Bayne, A. Cleeremans, and P. Wilken (eds.) *The Oxford Companion to Consciousness*, Oxford: Oxford University Press.

Coliva, A. (2003) "The Argument from the Finer-Grained Content of Color Experiences," *Dialectica* 57: 57–70.

Coltheart, M. (1980) "Iconic Memory and Visible Persistence," *Perception and Psychophysics* 27: 183–228.

Crane, T. (1992) "The Non-Conceptual Content of Experience," in T. Crane (ed.) *The Contents of Experience*, Cambridge: Cambridge University Press.

Crane, T. (2001) *The Elements of Mind*, Oxford: Oxford University Press.

Crowther, T. (2006) "Two Conceptions of Conceptualism and Nonconceptualism," *Erkenntnis* 65: 245–276.

Dickie, I. (2016) *Fixing Reference*, Oxford: Oxford University Press.

Dokic, J., and Pacherie, É. (2001) "Shades and Concepts," *Analysis* 61: 193–202.

Dretske, F. (1981) *Knowledge and the Flow of Information*, Cambridge, MA: MIT Press.

Dretske, F. (1993) "Conscious Experience," *Mind* 102: 263–283.

Dretske, F. (1995) *Naturalizing the Mind*, Cambridge, MA: MIT Press.

Duhau, L. (2011) "Perceptual Nonconceptualism: Disentangling the Debate Between Content and State Nonconceptualism," *European Journal of Philosophy* 22: 358–370.

Evans, G. (1982) *The Varieties of Reference*, Oxford: Oxford University Press.

Fodor, J. (1998) *Concepts: Where Cognitive Science Went Wrong*, Oxford: Oxford University Press.

Fodor, J. (2007) "The Revenge of the Given," in B. McLaughlin and J. Cohen (eds.) *Contemporary Debates in Philosophy of Mind*, Oxford: Blackwell.

Fodor, J. (2008) *LOT 2: The Language of Thought Revisited*, Oxford: Oxford University Press.

Gennaro, R. (2012) *The Consciousness Paradox*, Cambridge, MA: MIT Press.

Hanna, R. and Chadha, M. (2011) "Non-conceptualism and the Problem of Perceptual Self-Knowledge," *European Journal of Philosophy* 19: 184–223.

Harman, G. (1990) "The Intrinsic Quality of Experience," *Philosophical Perspectives* 4: 31–52.

Heck, R. (2000) "Non-conceptual Content and the 'Space of Reasons'," *The Philosophical Review* 109: 483–523.

Heck, R. (2007) "Are There Different Kinds of Content?" in B. McLaughlin and J. Cohen (eds.) *Contemporary Debates in Philosophy of Mind*, Oxford: Blackwell.

Jacob, P., and Jeannerod, M. (2003) *Ways of Seeing*, Oxford: Oxford University Press.
Kelly, S. D. (2001a) "The Non-Conceptual Content of Perceptual Experience: Situation Dependence and Fineness of Grain," *Philosophy and Phenomenological Research* 62: 601–608.
Kelly, S. D. (2001b) "Demonstrative Concepts and Experience," *The Philosophical Review* 110: 397–420.
Laurence, S., and Margolis, E. (1999) "Concepts and Cognitive Science," in E. Margolis and S. Laurence (eds.) *Concepts: Core Readings*, Cambridge: MIT Press.
Laurence, S., and Margolis, E. (2007) "The Ontology of Concepts: Abstract Objects of Mental Representations?" *Noûs* 41: 561–593.
Laurence, S., and Margolis, E. (2012) "The Scope of the Conceptual," in E. Margolis, R. Samuels, and S. Stich (eds.) *The Oxford Handbook of Philosophy of Cognitive Science*, Oxford: Oxford University Press.
Laurier, D. (2004) "Nonconceptual Contents vs Nonconceptual States," *Grazier Philosophische Studien* 68: 23–43.
McDowell, J. (1994) *Mind and World*, Cambridge: Harvard University Press.
McDowell, J. (1998) "Having the World in View: Sellars, Kant, and Intentionality (The Woolbridge Lectures 1997)," *Journal of Philosophy* 95: 431–491.
Macpherson, F. (2015) "Cognitive Penetration and Nonconceptual Content," in J. Zeimbekis and A. Raftopoulos (eds.) *The Cognitive Penetrability of Perception*, Oxford: Oxford University Press.
Mandik, P. (2012) "Color-Consciousness Conceptualism," *Consciousness and Cognition* 21: 617–631.
Martin, M. G. F. (1992) "Perception, Concepts and Memory," *Philosophical Review* 101: 745–763.
Matthen, M. (2005a) *Seeing, Doing, and Knowing*, Oxford: Oxford University Press.
Matthen, M. (2005b) "Visual Concepts," *Philosophical Topics* 33: 207–233.
Millar, A. (1991) *Reasons and Experiences*, Oxford: Oxford University Press.
Noë, A. (2004) *Action in Perception*, Oxford: Oxford University Press.
Orlandi, N. (2014) *The Innocent Eye*, Oxford: Oxford University Press.
Pashler, H. (1998) *The Psychology of Attention*, Cambridge, MA: MIT Press.
Peacocke, C. (1986) "Analogue Content," *Proceedings of the Aristotelian Society: Supplementary Volume* 60: 1–17.
Peacocke, C. (1989) "Perceptual Content," in J. Almog, J. Perry, and H. Wettstein (eds.) *Themes from Kaplan*, Oxford: Oxford University Press.
Peacocke, C. (1992) *A Study of Concepts*, Cambridge, MA: MIT Press.
Peacocke, C. (1998) "Non-Conceptual Content Defended," *Philosophy and Phenomenological Research* 57: 381–388.
Peacocke, C. (2001) "Does Perception Have a Nonconceptual Content?" *The Journal of Philosophy* 98: 239–264.
Peacocke, C. (2006) *The Realm of Reason*, Oxford: Oxford University Press.
Pelling, C. (2007) "Conceptualism and the (Supposed) Non-Transitivity of Colour Indiscriminability," *Philosophical Studies* 134: 211–234.
Pelling, C. (2008) "Concepts, Attention, and Perception," *Philosophical Papers* 37: 213–242.
Phillips, I. (2011) "Perception and Iconic Memory," *Mind and Language* 26: 381–411.
Phillips, I. (2016) "No Watershed for Overflow: Recent Work on the Richness of Consciousness," *Philosophical Psychology* 29: 236–249.
Prinz, J. (2002) *Furnishing the Mind: Concepts and Their Perceptual Basis*, Cambridge, MA: MIT Press.
Pryor, J. (2005) "There Is Immediate Justification," in E. Sosa and M. Steup (eds.) *Contemporary Debates in Epistemology*, Oxford: Blackwell.
Pylyshyn, Z. (2003) *Seeing and Visualizing: It's Not What You Think*, Cambridge, MA: MIT Press.
Raffman, D. (1995) "On the Persistence of Phenomenology," in T. Metzinger (ed.) *Conscious Experience*, Munich: Imprint Academic Verlag.
Raftopoulos, A. (2009) *Cognition and Perception*, Cambridge, MA: MIT Press.
Roessler, J. (2011) "Perceptual Attention and the Space of Reasons," in C. Mole, D. Smithies, and W. Wu (eds.) *Attention: Philosophical and Psychological Essays*, Oxford: Oxford University Press.
Roskies, A. (2008) "A New Argument for Nonconceptual Content," *Philosophy and Phenomenological Research* 76: 633–659.
Roskies, A. (2010) "'That' Response Doesn't Work: Against a Demonstrative Defense of Conceptualism," *Noûs* 44: 112–134.
Schmidt, E. (2011) *Modest Nonconceptualism: Epistemology, Phenomenology, and Content*, Springer.
Sedivy, S. (1996) "Must Conceptually Informed Perceptual Experience Involve Non-Conceptual Content?" *Canadian Journal of Philosophy* 26: 413–431.

Siegel, S. (2012) *The Contents of Visual Experience*, Oxford: Oxford University Press.
Siegel, S. (2016) "The Contents of Perception," *Stanford Encyclopedia of Philosophy* [https://plato.stanford.edu/entries/perception-contents/].
Smith, A. D. (2002) *The Problem of Perception*, Cambridge: Harvard University Press.
Sosa, E. (1997) "Mythology of the Given," *History of Philosophy Quarterly* 14: 275–286.
Sosa, E. (2003) "Beyond Internal Foundations to External Virtues," in L. Bonjour and E. Sosa (eds.) *Epistemic Justification*, Oxford: Blackwell.
Speaks, J. (2005) "Is There a Problem about Nonconceptual Content?" *The Philosophical Review* 114: 359–398.
Sperling, G. (1960) "The Information Available in Brief Visual Presentations," *Psychological Monographs* 74: 1–29.
Stalnaker, R. (1998) "What Might Non-conceptual Content Be?" in E. Villanueva (ed.) *Philosophical Issues: Concepts* 9: 339–352.
Strawson, P. F. (1992) *Analysis and Metaphysics*, Oxford: Oxford University Press.
Toribio, J. (2006) "State versus Content: The Unfair Trial of Perceptual Nonconceptualism," *Erkenntnis* 69: 351–361.
Toribio, J. (2007) "Nonconceptual Content," *Philosophy Compass* 2/3: 445–460.
Toribio, J. (2011) "Compositionality, Iconicity, and Perceptual Nonconceptualism," *Philosophical Psychology* 24: 177–193.
Tye, M. (1995) *Ten Problems of Consciousness*, Cambridge: MIT Press.
Tye, M. (2000) *Consciousness, Color, and Content*, Cambridge: MIT Press.
Tye, M. (2005) "Nonconceptual Content, Richness, and Fineness of Grain," in T. Gendler and J. Hawthorne (eds.) *Perceptual Experience*, Oxford: Oxford University Press.
Veillet, B. (2014) "Belief, Re-identification, and Fineness of Grain," *The European Journal of Philosophy* 22: 229–248.
Vision, G. (1997) *Problems of Vision*, Oxford: Oxford University Press.
Wright, C. (1998) "McDowell's Oscillation," *Philosophy and Phenomenological Research* 57: 395–402.
Wright, C. (2002) "Human Nature," in N. Smith (ed.) *Reading McDowell: On Mind and World*, London: Routledge.
Wright, W. (2015) "Nonconceptual Content," in M. Matthen (ed.) *The Oxford Handbook of Philosophy of Perception*, Oxford: Oxford University Press.

Related Topics

Consciousness and Intentionality
Consciousness and Attention
Representational Theories of Consciousness
Multisensory Consciousness and Synesthesia

21
CONSCIOUSNESS, TIME, AND MEMORY

Ian Phillips

To do full justice to our conscious perceptual experience, mention must be made of our awareness of succession, duration and change. Even the banal experience as I look out of my office window includes leaves and branches nodding in the wind, cars moving down the road, their lights blinking successively on-and-off as they turn, and now a bus pausing for a moment at its stop before moving on.[1] Many theorists have suggested that our capacity to see such happenings, and more generally to perceive temporal aspects of reality, depends essentially on memory. Here I explore (and ultimately defend) this putative connection.

I begin by motivating the idea that memory must be involved in our temporal consciousness via the notorious slogan that a succession of experiences is not, in and of itself, an experience of succession (Section 1). This leads to the introduction of a traditional memory theory and, by way of objections, its replacement by a more refined version (Section 2). This refined theory distinguishes between ordinary recollective memory, and a form of memory often called "retention," which is held to be distinctively implicated in temporal experience. In Section 3, I discuss how these theories relate to Dainton's influential cinematic/retentional/extensional trichotomy of models of temporal consciousness. Here, I suggest that, *contra* certain contemporary theorists, there are grounds for thinking that some form of memory is involved in all variants of the class of models which Dainton calls retentional—a claim I later extend also to all extensionalist models. In Section 4, I introduce a further issue, namely whether retentions can occur in the absence of prior experience of the retained contents. Many contemporary retentionalists insist they can. However, it is striking to note that Husserl (historically, the most influential retentionalist) denies that possibility. In Section 5, I suggest that appreciating Husserl's version of retentionalism threatens to subvert Dainton's distinction between retentional and extensional models. More importantly, it helps pinpoints what is really at issue between theorists of temporal experience, namely whether the temporal structure of experience itself is implicated in explaining our consciousness of time.

1 Motivating Memory

One source of the idea that temporal experience essentially depends on memory begins with the Kantian principle that a mere succession of experiences is insufficient for an experience of succession. Or as James puts it in his celebrated discussion of time consciousness: "*A succession of feelings, in and of itself, is not a feeling of succession*" (1980: 628–629; likewise, Husserl 1991: 12–13). If this principle is read simply as saying that *some* successions of experiences do not compose experiences of succession, it is beyond reproach. Successions of experiences enjoyed by *different subjects* do not compose experiences of succession. Even within a subject, not all successive experiences compose experiences of succession. If, early this morning, I hear a robin's tuneful warble, and, late this evening, a nightingale's strident *jug jug jug*, I will not thereby enjoy an experience of these sounds as successive.

In this light, we might ask: when do successive experiences compose experiences of succession?[2] A natural suggestion is that successive experiences compose experiences of succession when a single subject enjoys them close enough together in time. This suggestion is firmly and widely dismissed in the literature. Temporal proximity is widely held to be obviously insufficient for experiences to compose an experience of succession.[3] Instead, it is commonly insisted that experiences of succession must somehow involve a *unified* apprehension of the successive elements. On this point, James quotes Volkmann who, he suggests, "has expressed the matter admirably":

> successive ideas are not yet the idea of succession ... If idea A follows idea B, consciousness simply exchanges one for another ... if A and B are to be represented *as occurring in succession* they must be *simultaneously represented.*
>
> *(1875: §87, cited in James 1890: 629)*

Notice how Volkmann is here, in effect, denying that successive experiences *ever* compose an experience of succession. Instead, the requirement that an experience of succession demands the *unification* of the successive elements is taken to require that the elements be presented *at one and the same moment.* This widespread commitment has subsequently been labelled the *Principle of Simultaneous Awareness*, or PSA (Miller 1984: 109). Another important early proponent is Lotze, who writes: "In order for this comparison in which *b* is known as later to occur, it is surely again necessary that the two representations *a* and *b* be the absolutely simultaneous objects of a knowing that puts them in relation and that embraces them quite indivisibly in a single indivisible act" (1879: 294, cited by Husserl 1991: 21).[4] Again, we find here a denial that successive experience ever composes experience of succession.

The PSA helps us understand why memory might be thought an essential requirement for temporal experience. Consider an experience of two sounds. For these two sounds to be experienced as successive, they must—according to the PSA—be experienced together and so simultaneously. But, one might think, they cannot both be *heard* simultaneously, for then we would hear the two notes as "a chord of simultaneous tones, or rather a disharmonious tangle of sound" (Husserl 1991: 11), or as Brough nicely puts it (in his introduction to Husserl 1991: xxxv) as an "instantaneous tonal porridge." It must instead be that, when we hear the later sound, the earlier sound is simultaneously presented *in memory*.

A view of this kind—call it the traditional memory theory—can be found in Reid who holds that "the motion of a body, which is a successive change of place, could not be observed by the senses alone without the aid of memory" (1785: 326). What happens in Reid's view is this: "We see the present place of the body; we remember the successive advance it made to that place: The first can then only give us a conception of motion, when joined to the last" (Reid 1785: 327). Husserl finds a related view in Brentano's early work on temporal experience. Interestingly, Husserl notes: "As a consequence of his theory, Brentano comes to deny the perception of succession and change" (1991: 14). Arguably the same is true for Reid who observes "that if we speak strictly and philosophically, no kind of succession can be an object either of the senses, or of consciousness" (1785: 325–326). Reid however does seem to think that we are *aware* of succession, it is just that this awareness is not strictly speaking *perceptual* (see Falkenstein 2017: 48–49).

2 Problems for the Traditional Memory Theory

The traditional memory theory faces serious difficulties. One difficulty is local to theories which think of memory as distinguished from perception only in causal origin, as apparently Brentano once did (1874/1973: 316; quoted in Miller 1984: 105). On this view, it is hard to see how a satisfactory solution has been offered to the tonal porridge objection above, since on such a view there will be no *intrinsic, phenomenological* difference between a case of simultaneously hearing two sounds, and a case of hearing one whilst remembering the other. However, we need not endorse this way of thinking of the relation between perception and memory. For example, we might follow Martin in holding that episodic memory is "the representational recall of... an experiential encounter" (2001: 270) with a particular event or object, whereas perception involves the genuine presentation of such particulars to the mind. In this way, we can insist on a phenomenological difference between simultaneously hearing A and B, and simultaneously remembering A whilst hearing B.[5]

A more recent and general criticism of the traditional memory theory is offered by Tye (see also Lockwood 1989).

> Consider … hearing the sequence of musical notes, do, re, mi, in rapid succession. [According to the memory theory] … first, one experiences do; then one experiences re in conjunction with a short-term phenomenal memory of having just heard do; then finally one has an experience of mi, along with a short-term phenomenal memory of having just heard re.
>
> Patently, however, this won't do. One has an experience of do *followed* by re *followed* by mi; and this experienced temporal sequence has not been explained. It does not help to add that when one experiences mi, one has a short-term phenomenal memory of having just heard do followed by re. For one can only remember having just heard do followed by re, if one has experienced do followed by re; and it is precisely this experience of succession, of do's being followed by re, that the appeal to memory is supposed to explain. Moreover, no account at all has been offered of the experience of re followed by mi.
>
> *(Tye 2003: 87–88)*

Though superficially convincing, on reflection it is unclear how forceful Tye's argument really is. Let us begin with the simple case of hearing two notes: *do* followed by *re*. The memory theorist's account of this experience is, as Tye says, the following: one first experiences *do*, then one experiences *re* in conjunction with a short-term phenomenal memory of having just heard *do*. In the case where one hears three notes, one's experience unfolds further: one next experiences *mi* in conjunction with a short-term phenomenal memory of having just heard *re*, and further in conjunction (we might add) with a short-term phenomenal memory of having just had a short-term phenomenal memory of having just heard *do*. Tye objects: "one can only remember having just heard do followed by re, if one has experienced do followed by re; and it is precisely this experience of succession, of do's being followed by re, that the appeal to memory is supposed to explain." But one *has* experienced *do* followed by *re*, and this *was* explained—by appeal to our having an experience of *do* followed by an experience of *re* in conjunction with a short-term phenomenal memory of having just heard *do*. Tye further objects that "no account at all has been offered of the experience of re followed by mi." But, again, an account *has* been given in terms of hearing *re*, and then hearing *mi* in conjunction with a short-term phenomenal memory of having just heard *re*.

A simpler objection ultimately undoes the traditional memory theory. This objection is that the theory cannot distinguish between perceiving succession and merely perceiving *that* succession has occurred. As Broad famously notes, "to see a second-hand moving is quite a different thing from 'seeing' that an hour-hand has moved" (1923: 351; also Locke 1690/1975: II.xiv.11; Russell 1927: 281; Dainton 2008b: 619–621; Hoerl 2017: 174). Likewise, to hear a succession of sounds as such is quite a different thing from hearing *that* a succession of sounds has occurred. Yet it is obscure what resources the traditional memory theory has to mark the distinction. For, plainly, I can see the present position of the hour-hand whilst simultaneously recalling its earlier position, without yet enjoying an experience of those positions as successive. Likewise, I can hear a sound whilst recalling some earlier sound, without yet being aware of those sounds as successive. When one sees the present position of an hour-hand and recalls its different earlier position, one is thereby in a position to know change has occurred. In such a case, we talk of seeing *that* change has occurred, where this means knowing or being in a position to know, on a perceptual basis, a certain fact about change. The difficulty for the memory theory is that none of this suffices for *seeing change* (i.e. the event or process of change itself).

The standard response to this concern is to distinguish two forms of memory, one variously called *primary, elementary*, or *fresh memory*, or *retention*; the other *secondary memory* or *recollection*. Perceiving change is then said to require the involvement of primary memory, whereas secondary memory at most affords knowledge that change has occurred. On this primary form of memory James comments:

> what elementary memory makes us aware of is the *just* past. The objects we feel in this directly intuited past differ from properly recollected objects. An object which is recollected … is one which has been absent from consciousness altogether, and now revives anew. It is brought back, recalled, fished up, so to speak, from a reservoir in which … it lay buried and lost from view. But an object of primary memory is not thus brought back; it never was lost; its date was never cut off in consciousness from that of the immediately present moment. In fact, it comes to us as belonging to the rearward portion of the present space of time, and not to the genuine past.
>
> *(1890: 646–647)*[6]

Before exploring this alleged form of memory further, it is worth pausing to consider the path we have taken and how our two memory theories relate to the now standard way of carving up the contemporary landscape of positions due to Dainton.

3 Dainton's Trichotomy of Models: Cinematic, Retentional and Extensional

Dainton influentially carves up the landscape of positions regarding temporal awareness in terms of three distinct models (see esp. Dainton 2000, 2017b). First there are "cinematic models," according to which change experience is analysable into a sequence of instantaneous or near-instantaneous sensory atoms, each individually bereft of dynamic content.[7] Second, there are "retentional models." On such models whilst experiences of change can be analysed into a sequence of instantaneous or near-instantaneous sensory atoms, these atoms *do* possess temporally-extended contents (i.e. they individually present goings on over a period of time, as such). Finally, there are "extensional models" according to which "our episodes of experiencing are themselves temporally-extended, and are thus able to incorporate change and persistence in a quite straightforward way" (Dainton 2017b).

How do the traditional and refined memory theories introduced above relate to this three-part framework? It is natural to think of the traditional memory theory as a form of cinematic theory. Notice that the cinematic theorist, as Dainton is thinking of her, grants that we are aware of change. (She is not in his terminology an "anti-realist.") To accommodate this, whilst cleaving to her claim that change experience can be analysed into a series of atoms each individually lacking temporal content, the cinematic theorist has unsurprisingly looked to memory (see the discussion of Reid above, and, for a contemporary defence, Chuard [2011, 2017]).

It is equally natural to think of the refined memory theory as a form of retentionalism. Primary memory, after all, is not conceived of as a separate act but as contributing to the content of a complex perceptual episode. It should be recognized, however, that many contemporary retentionalists make no mention of memory in their accounts. Indeed, some explicitly deny it any role. Instead, they simply attribute representational contents, which concern extended periods of time to experiences. And further, in deliberate contrast to extensionalism, deny that the intrinsic temporal features of experience have any direct explanatory connection to their conscious character. Content does all the work. In this light, should we consider the primary memory theory as simply one form of retentionalism, or is memory in fact implicit in all such accounts?

Consider Lee (2014), who defends a view he calls "atomism." Lee's view is arguably a form of retentionalism, but Lee resists that label because he denies that memory or retention plays any part in his view, which simply appeals to temporally-extended contents to make sense of temporal experience. Furthermore, Lee (2014: 6) gives four reasons for thinking that we should eschew talk of retention or memory. First, he suggests that the contents of temporal experience need not be tensed at all (i.e. represent events as past, present and future as opposed to simply standing in B-theoretic relations of earlier or later-than; see Hoerl 2009). Second, he thinks temporal experience "might involve just one kind of conscious perceptual experience, not differentiated between 'retention' and 'perception'." Third, he thinks that temporal experience need not retain "contents from immediately past experiences." That is, "a temporally-extended content could include—perhaps exclusively—information about events that were not presented in any previous experiences." Finally, he notes the plausible involvement of prediction, and so presumably of forward-looking contents in temporal experience.

It is unclear how serious the second and last of these concerns are. The primary memory theorist conceives of retentional awareness as an aspect of a single kind of perceptual state (it is for Husserl, for example, a "dependent moment" of a perceptual act[8]). Thus, they need not disagree that temporal experience involves "just one kind" of perceptual experience, albeit one with multiple aspects. The retentional theorist may equally include forward-looking aspects as amongst these different aspects. Indeed, Husserl's account of temporal consciousness involves a *three-fold* intentionality, comprising retention, *now*-awareness and (forward-looking) protention. Temporal experience may then count all-at-once as a form of memory, and of perception, and of anticipation.

What about Lee's objections that the contents of experience might be tenseless, and that temporal contents might include aspects that have not featured in any earlier experience? Do these tell against the involvement of memory? In making that claim, Lee implicitly invokes two constraints on what it is for a state to count as a state of memory. A past-awareness constraint, viz. that memory states must present their content as past; and a previous awareness constraint, viz.

that (perceptual) memory states must have contents, which have previously figured in perceptual awareness.

In earlier work I have suggested, following Martin (2001), that the fundamental unifying feature common to all forms of memory is that they are all ways of preserving past psychological success. Secondary memory or recollection is plausibly thought of as the preservation of past apprehension or acquaintance (or more precisely, the preservation of an associated ability). Primary memory, however, is, in James' words "not thus brought back; it never was lost; its date was never cut off in consciousness from that of the immediately present moment. … it comes to us as belonging to the rearward portion of the present space of time, and not to the genuine past" (1890: 646–647). Consider awareness of two notes *do* and *re*. Suppose one hears *re* in a different manner depending on whether one hears it as part of a succession or not. We might then suppose that hearing *re* involves primary memory insofar as it involves hearing *re* in a particular way, namely as succeeding on from *do*. This modification of one's manner of awareness plausibly counts as a form in which a psychological success (namely awareness of *do*) can be preserved. Moreover, in itself, it does not commit us to the idea that *do* is presented *as past* as opposed simply to *re* being heard as succeeding on from *do*.

Reconstructing his argument, it may nonetheless seem that Lee is right to find an inconsistency between the following three claims:

i Temporal experience essentially involves memory.
ii Memory essentially involves the preservation of past psychological success.
iii Temporal experience can occur independently of the preservation of past psychological success.

In the above example, for example, it would surely be problematic to claim that an awareness of *re* as succeeding on from *do* counted as a form of memory if *do* had never been heard. However, the tension might seem to be straight-forwardly resolved by weakening claim (i) to read: temporal experience essentially involves memory *or apparent memory*. This weakened claim is arguably sufficient to constitute a genuine memory theory, and might seem capable of accommodating the kind of case which Lee has in mind where one has an experience with temporal extended contents despite no such contents appearing in any earlier experience. For all Lee says, then, there are grounds for thinking that some form of memory is involved in all variants of the class of models that Dainton calls retentional.

Lee's discussion raises an interesting question, however, namely whether we should in fact admit the possibility (as Lee does) of courses of experience where one set of extended contents bears no relation to previous experiential contents. As I now discuss, this possibility is precisely rejected by Husserl in his own discussion of primary memory. Its exploration serves to raise a doubt about the distinctness of retentional and extensional accounts. It also reveals what is, I suggest, fundamentally at issue between different theorists of temporal awareness.

4 Retention and Prior Awareness

In discussing primary memory, a dominant concern of Husserl's is to distinguish primary memory from any form of weak or faded perception.

> The reverberation of a violin tone is precisely a feeble present violin tone and is absolutely different from the retention of the loud tone that has just passed. The echoing itself and after-images of any sort left behind by the stronger data of sensation, far from

> having to be ascribed necessarily to the essence of retention, have nothing at all to do with it.
>
> *(1991: 33)*

Husserl is also explicit that retention is not to be thought of in terms of representation or phantasy (i.e. imagination).[9] Husserl further embraces the Jamesian idea that retention does not involve a new act of consciousness. Rather, "primary memory … extends the *now*-consciousness" (47). Indeed, he offers a Jamesian metaphor to illustrate, characterizing "primary memory or retention as a comet's tail that attaches itself to the perception of the moment" (37).

However, focussing on such negative points, some critics have complained: "Husserl tells us what retention is not, and what it does, but provides no explanation as to how it accomplishes this" (Dainton 2000: 156). Can Husserl answer this objection? We began with a question about when a succession of experiences composes an experience of succession. It might be thought that we have lost track of this thought. Indeed, as we saw, theorists from Lotze and Volkmann through to Lee (and likewise other retentionalists such as Tye [2003] and Grush [2005, 2007]) embrace the idea that we could have an experience of succession without a succession of experiences at all. All we need is a single episode with suitable contents representing goings on over a stretch of time as such, and quite irrespective of its own temporal structure (be it momentary or otherwise). On such a conception there is no obvious reason to deny that such experiences can occur entirely in independence, or indeed in isolation, from one another. Indeed, various theorists take this as a positive virtue. For it provides the freedom for the past directed contents of later experiences to revise how things were originally presented in the light of new information, a thought made particular use of by Grush and Tye in discussing postdictive phenomena.[10]

The same might appear true for Husserl. That is, it might at first seem that all Husserl requires for an experience of succession is a single episode, which has both *now*-awareness and retentional awareness as aspects. Yet this is not Husserl's view. Husserl holds that there is an "epistemic" distinction between primary and secondary memory (1991: §22). In particular, he holds that retentional consciousness is "absolutely certain," writing: "If I am originally conscious of a temporal succession, there is no doubt that a temporal succession has taken place and is taking place" (51). Husserl is clear here that he does not mean that there can be no illusions or hallucinations in respect of temporal perception. He acknowledges the possibility that "no [objective] reality corresponds" to the appearances in question (51–52). What he means is that awareness of a temporal succession *guarantees* that a succession of appearances (i.e. experiences) has occurred, be these veridical or otherwise (ibid., see also 35). On this view experience of succession *does* require successive experience, for there is a constitutive connection between one's current experience (here in particular its retentional component) and one's past experience. *One could not be experiencing the way one presently is, were one not to have experienced a certain way in the past.*[11]

On Husserl's view then it is not after all possible simply to have isolated acts of temporal awareness. Nor is it possible to have the kind of revisions which Lee and Grush propose. What appear to be imposed are certain coherence constraints on the way that experience can unfold over time.[12] In this way, Husserl's conception of retention is not well-captured simply in terms of the contemporary thought that contents do all the work. Instead, for him, explaining temporal experience requires appeal to the idea of a sequence of experiences unfolding over time and standing in complex relations to one another. This is further brought out by the fact that, for Husserl, momentary phases of awareness are considered abstractions from an on-going flow of experience, and not independent, self-contained episodes. As Husserl puts it:

> This continuity [of constantly changing modes of temporal orientation] forms an inseparable unity, inseparable into extended sections that could exist by themselves and inseparable into phases that could exist by themselves, into points of the continuity. The parts that we single out by abstraction can exist only in the whole running-off; and this is equally true of the phases, the points that belong to the running-off continuity.
>
> *(1991: 29)*

These elements in Husserl's account, which following Hoerl (2013a) we might call "externalist," are also arguably found in O'Shaughnessy's richly suggestive discussion, in which he argues that in temporal experience "present experience must both unite with and depend upon past experience." He continues:

> This means that the past must in some sense be co-present with the present, and such a co-presence is a mode of remembering. Doubtless it is a developmentally early form of memory, to be supplemented later by additional less primitive ways of relating to one's past, notably cognitive modes. What in effect we are concerned with here is the tendency on the part of experience and its given objects to unite across time to form determinate wholes.
>
> *(2000: 56)*

Here, O'Shaughnessy suggests that our awareness of change (in his view, essential to all conscious experience) involves a constitutive dependence of present experience on recently past experience. Furthermore, this—we are told—suffices for such present experience to count as a primitive form of memory (see Phillips 2010: 193–194). Of course, this returns us to the idea of primary memory as distinct from recollection. It also provokes a question as to how retentionalism so conceived really differs from extensionalism. This is the topic of the next and final section.

5 Extensionalism

Extensionalism was introduced above as the view that "our episodes of experiencing are themselves temporally-extended, and are thus able to incorporate change and persistence in a quite straightforward way" (Dainton 2017b). But how is extensionalism so conceived supposed to contrast with cinematic and retentional models? Lee is not alone in complaining here that Dainton's definition is in fact "a claim that … all parties to the debate … can and should accept" (2014: 3). For Lee this is because he thinks it "very plausible" both that "*all* experiences are realized by extended physical processes" (5) and further that "experiences have the *same* timing as their realizers" (3). Consequently, it is equally true that on his exclusively content-focused, atomist view, "our episodes of experiencing are themselves temporally-extended."

What Lee misses here is the word "thus" in Dainton's definition (see Hoerl 2013a: 397). For on Lee's atomism there is no direct explanatory connection between experience's temporal extension and its content. In contrast, Dainton precisely holds that there is such a connection. Recall our starting point: the idea that a succession of experiences is not (at least in itself) an experience of succession. Dainton's extensionalist agrees that any model of temporal experience which works only with momentary apprehensions is unsustainable, no matter how closely one packs the experiences. And he agrees for the familiar sounding reason that "the required synthesis or combination is entirely lacking" (Dainton 2008b: 623). However, Dainton does not unpack this unity requirement in terms of the PSA (the requirement, recall, that for us to enjoy an experience of succession, the successive elements must be presented *at one and the same moment*). Instead, he invokes a "phenomenal binding principle," the principle that awareness of change requires "each brief phase of a stream of consciousness [to be] *phenomenally bound* to the adjacent (co-streamal) phases" (2000: 129). This binding requires adjacent co-streamal phases to be co-conscious. Co-consciousness, for Dainton, is a "primitive experiential relationship" (131), which holds between our experiences both at times and across time. Dainton's extensionalism thus not only involves the denial that the unity required for experience of succession should be conceived of in terms of simultaneity.[13] It also appeals essentially to relations holding between phases of experience occurring at different times. As a result, Dainton keeps hold of the claim that experiences of succession require successions of experiences, ones properly co-conscious with one another.[14] Here is a point of real disagreement with Lee's atomist.

At this juncture, I suggest we find the most fundamental divide between theorists of temporal consciousness. This divide turns on whether a theorist sees the unfolding of experience itself as having explanatory bearing on the possibility of temporal experience. On the one side of this divide are those for whom experiences of succession do not involve successive experiences at all. Traditional such views hold that temporal experiences are instantaneous events which nonetheless present us with temporally-extended goings on. Contemporary such views, like Lee's, hold that temporal experiences are brief-lived events whose intrinsic temporal structure is irrelevant to their phenomenal character, which is determined solely by their temporally-extended contents. On the other side of the divide are those who insist that it is only *because* our experience is a process, which unfolds in time, that it can acquaint us with the temporal structure of reality as it does.

If we divide the landscape in this way, however, theorists who we might initially conceive of as rivals, namely extensionalists such as Dainton, and retentionalists such as Husserl and O'Shaughnessy, do not obviously disagree on substance. All agree that experience of succession requires successive experience and so insist on an explanatory connection between the unfolding temporal structure of experience and its contents. They thereby depart from theorists such as Tye, Grush and Lee who reject this connection. Furthermore, whilst Dainton does not conceive of extensionalism in terms of memory, it is arguable that extensionalism does in fact implicate memory in temporal experience. This is because one can reasonably consider the relation of co-consciousness, which Dainton invokes as unifying earlier and later phases of experience, as constitutive of a form of memory.[15] More generally, on extensionalist views, the nature of one's current experience is not independent of past psychological successes (i.e. previous phases of experience). As we saw above, this arguably suffices for memory to be in play.

6 Conclusions and Further Issues

Discussion so far has revealed that, though traditional memory theories are untenable, the idea that memory is involved in all temporal experience can in fact be sustained across the accounts of temporal experience which we have considered in detail. This includes not only Husserl's retentionalism, but contemporary views that deny any role for memory such as Lee's atomism, and also extensionalist views. We have also seen that Dainton's partition of the landscape of positions on temporal experience into three camps masks a deeper and rather different dividing line between theorists. This more fundamental divide concerns whether or not an explanatory connection obtains between the unfolding of experience itself and its capacity to present us with change and succession. Or put another way: whether experience of succession requires successive experience.

Recognition of this more fundamental divide prompts various critical issues for future investigation. But, above all, we need to ask what motivates the thought that there is an explanatory connection between the unfolding of experience itself and its capacity to present us with change and succession. Insofar as there is no such connection between the spatiality of experience (if that notion is even coherent) and its capacity to present us with spatial features, what makes time special (if it is)? Some theorists have proposed that a connection between the temporal structure of experience and the temporal features it presents to us best articulates how experience seems to us on pre-theoretic reflection, and so can rightly be considered the proper starting point for theorizing about experience. See here, in particular, Phillips (2014a, b) on what he calls the naïve view of temporal experience.[16] Others have argued for a deep connection between views of temporal experience and views in the metaphysics of perception more generally. In particular, Hoerl (2013b, 2017) and Soteriou (2010, 2013) suggest that the idea of an explanatory connection between the unfolding of experience itself and its capacity to present us with change and succession goes hand-in-glove with relational or naïve realist views of perception. Conversely, they suggest that atomist views such as Lee's, Tye's and Grush's are the product of a more general representationalism about perception. These are important ideas, and merit further serious scrutiny.

Notes

1 Even awareness of an entirely unchanging scene arguably involves awareness of the continual unfolding of experience itself. As O'Shaughnessy writes, "Even when experience does not change in type or content it still changes in another respect: it is constantly renewed, a new sector of itself is then and there taking place. This is because experiences are events or processes, and each momentary new element of any given experience is a further happening or occurrence" (2000: 42). For development and discussion see Soteriou (2013: ch. 6). Cf. Husserl: "Even the perception of an unchanging object possesses in itself the character of change" (1991: 239).

2 Compare here the so-called Special and General Composition Questions explored in van Inwagen (1990). These focus on the conditions under which objects compose something. The Special Composition Question asks what relations must hold amongst some objects for them to compose *something or other*. The General Composition Question asks rather what relations must hold between a whole and some objects when those objects compose that whole. In the present context we might ask (in line with the General Composition Question): what relations must hold between an experience of succession and some experiences when those experiences compose that experience of succession? Note that if one thinks of the fundamental units of experience as extended stretches as opposed to moments (see note 14), one will abjure the corresponding "Special" Question given its presupposition that one can say when composition occurs independently of facts about the nature of what is composed.

3 It is sometimes questioned whether proximity is even necessary. Tye (2003: 106, following Dainton 2000: 131), for instance, imagines he is experiencing a scale *do-re-mi* but "just as I finish hearing re, God *instantaneously* freezes all my internal physical states as well as all physical processes in my surrounding environment for … five years … and then unfreezes them *instantaneously*." In this case, Tye suggests that I do (presumably, *ceteris paribus*) experience the succession. See below note 12.

4 Husserl himself traces the principle back to Herbart. For other citations and critical discussion thereof see Phillips (2010), Hoerl (2013b) and Rashbrook-Cooper (2013).

5 See Phillips (2010: 201, fn. 21). See also note 9.

6 The "space of time" here is James' "specious present." This is one of a number of occasions on which James construes the specious present in terms of memory.

7 See Chuard (2011, 2017) for recent defence of such a model. For critical discussion see my critique of the "zoëtrope conception" (after James 1890: 200) in Phillips (2011a).

8 See here Brough's introduction to Husserl (1991: §B, esp. xxxix).

9 Husserl is here supposing that perception on the one hand, and memory and imagination on the other, involve distinct kinds of conscious acts, the former being a case of presentation, the latter cases of representation. As already mentioned, a contemporary picture which develops this kind of distinction can be found in Martin (2001).

10 See Grush (2005, 2007) and Tye (2003) for this suggestion; see also Dennett and Kinsbourne (1992). For critical discussion see Dainton (2008a), Hoerl (2013b), and Phillips (2011b, 2014a).
11 See Hoerl (2013a) for an extended examination of this aspect of Husserl's view.
12 See Phillips (2010). It is a nice question whether putting the thought in terms of coherence constraints allows for the kind of possibility envisaged by Dainton and Tye of a freeze or gap in experience (see above note 3) consistent with the kind of necessity which Husserl has in mind.
13 Dainton inherits this view from Foster (1979, 1982). But, as he discusses in detail in later work (Dainton 2017a), the view probably is first articulated by Stern (1897/2005).
14 An important issue which arises here is how we should conceive of the relations between brief phases of the stream of consciousness. For example, do the phases connected by such relations have independent existence, or are they better thought of as dependent for their existence and nature on the extended stretch of experience of which they are parts? Put another way, what are the fundamental units of experience: moments or extended stretches?
15 Might Dainton be open to this suggestion? Consider this passage in a discussion of Bergson who he suggests holds a form of extensionalism. "There is one consideration which could be taken to point in precisely the opposite direction. When attempting to characterize *durée* Bergson often suggests that *memory* is involved. In … *Duration and Simultaneity* … he tells us that even in the briefest of physical events there will be "a memory that connects" their earlier and later phases. … A case can be made, however, for holding that in these contexts Bergson's "memory" is simply the unifying relation, which connects the earlier and later phases of a single episode of *durée*" (2017c: 104, fn. 10).
16 Phillips embraces a more precise claim about the relation between the temporal structure of experience itself and the temporal goings on it presents to us, which he calls naïve inheritance. This is the claim that that for any temporal property apparently presented in experience, our experience itself possesses that temporal property. For critical discussion see Watzl (2013) and Frischhut (2014). For a reply to Watzl, see Phillips (2014c).

References

Brentano, F. (1874/1973) *Psychology from an Empirical Standpoint*, A. C. Rancurello, D. B. Terrell, and L. L. McAllister (trans.), London: Routledge.

Chuard, P. (2011) "Temporal Experiences and Their Parts," *Philosophers' Imprint* 11: 1–28.

Chuard, P. (2017) "The Snapshot Conception of Temporal Experiences," In I. Phillips (ed.) *The Routledge Handbook of Philosophy of Temporal Experience*, London: Routledge.

Dainton, B. (2000, 2nd edition 2006) *Stream of Consciousness: Unity and Continuity in Conscious Experience*, London: Routledge.

Dainton, B. (2008a) "Sensing Change," *Philosophical Issues* 18: 362–384.

Dainton, B. (2008b) "The Experience of Time and Change," *Philosophy Compass* 3/4: 619–638.

Dainton, B. (2017a) "William Stern's *Psychische Präsenzzeit*," In I. Phillips (ed.) *The Routledge Handbook of Philosophy of Temporal Experience*, London: Routledge.

Dainton, B. (2017b) Temporal Consciousness. *The Stanford Encyclopedia of Philosophy* (Fall 2017 Edition), E. N. Zalta (ed.), https://plato.stanford.edu/archives/fall2017/entries/consciousness-temporal/

Dainton, B. (2017c) "Bergson on Temporal Experience and *Durée Réelle*," In I. Phillips (ed.) *The Routledge Handbook of Philosophy of Temporal Experience*, London: Routledge.

Dennett, D. C. and Kinsbourne, K. (1992) "Time and the Observer: The Where and When of Consciousness in the Brain," *Behavioral and Brain Sciences* 15: 183–200.

Falkenstein, L. (2017) "Hume on Temporal Experience," In I. Phillips (ed.) *The Routledge Handbook of Philosophy of Temporal Experience*, London: Routledge.

Foster, J. (1979) "In *Self*-Defence," In G. F. Macdonald (ed.) *Perception and Identity*, London: Macmillan.

Foster, J. (1982) *The Case for Idealism*, London: Routledge and Kegan Paul.

Frischhut, A. M. (2014) "Diachronic Unity and Temporal Transparency," *Journal of Consciousness Studies* 21(7-8): 34–55.

Grush, R. (2005) "Internal Models and the Construction of Time: Generalizing from *State* Estimation to *Trajectory* Estimation to Address Temporal Features of Perception, including Temporal Illusions," *Journal of Neural Engineering* 2: 209–218.

Grush, R. (2007) "Time and Experience," In T. Müller (ed.) *Philosophie der Zeit*, Frankfurt: Klosterman.

Hoerl, C. (2009) "Time and Tense in Perceptual Experience," *Philosophers' Imprint* 9: 1–18.

Hoerl, C. (2013a) "Husserl, The Absolute Flow, and Temporal Experience," *Philosophy and Phenomenological Research* 86: 376–411.
Hoerl, C. (2013b) "A Succession of Feelings, in and of Itself, Is Not a Feeling of Succession," *Mind* 122: 373–417.
Hoerl, C. (2017) "Temporal Experience and the Philosophy of Perception," In I. Phillips (ed.) *The Routledge Handbook of Philosophy of Temporal Experience*, London: Routledge.
Husserl, E. (1991) *On the Phenomenology of the Consciousness of Internal Time* (1893–1917), J. B. Brough (ed. and trans.), Dordrecht: Kluwer.
James, W. (1890) *The Principles of Psychology*, New York: Dover.
Lee, G. (2014) "Temporal Experience and the Temporal Structure of Experience," *Philosophers' Imprint* 14: 1–21.
Locke, J. (1690/1975) *An Essay Concerning Human Understanding*, P. H. Nidditch (ed.), Oxford: Oxford University Press.
Lockwood, M. (1989) *Mind, Brain and the Quantum*, Oxford: Blackwell.
Lotze, R. H. (1879) *Metaphysik*, Leipzig: Hirzel.
Martin, M. G. F. (2001) "Out of the Past: Episodic Recall as Retained Acquaintance," In C. Hoerl and T. McCormack (eds.) *Time and Memory*, Oxford: Oxford University Press.
Miller, I. (1984) *Husserl, Perception and Temporal Awareness*, Cambridge, MA: MIT Press.
O'Shaughnessy, B. (2000) *Consciousness and the World*, Oxford: Clarendon Press.
Phillips, I. (2010) "Perceiving Temporal Properties," *European Journal of Philosophy* 18: 176–202.
Phillips, I. (2011a) "Indiscriminability and the Experience of Change," *The Philosophical Quarterly* 61: 808–827.
Phillips, I. (2011b) "Perception and Iconic Memory," *Mind and Language* 26: 381–411.
Phillips, I. (2014a) "The Temporal Structure of Experience," In V. Arstila and D. Lloyd (eds.) *Subjective Time: The Philosophy, Psychology, and Neuroscience of Temporality*, Cambridge, MA: MIT Press.
Phillips, I. (2014b) "Experience Of and In Time," *Philosophy Compass* 9: 131–144.
Phillips, I. (2014c) "Breaking the Silence: Motion Silencing and Experience of Change," *Philosophical Studies* 168: 693–707.
Rashbrook-Cooper, O. (2013) "An Appearance of Succession Requires a Succession of Appearances," *Philosophy and Phenomenological Research* 87: 584–610. (Originally published under the name "Rashbrook.")
Reid, T. (1785) *Essays on the Intellectual Powers of Man*, Edinburgh: J. Bell and G. G. J. and J. Robinson.
Russell, B. (1927) *The Analysis of Matter*, London: Routledge.
Soteriou, M. (2010) "Perceiving Events," *Philosophical Explorations* 13: 223–241.
Soteriou, M. (2013) *The Mind's Construction: The Ontology of Mind and Mental Action*, Oxford: Oxford University Press.
Stern, L. (1897/2005) "Mental Presence-Time," N. De Warren (trans.). In C. Wolfe (ed.) *The New Yearbook for Phenomenology and Phenomenological Research*, London: College Publications.
Tye, M. (2003) *Consciousness and Persons*, Cambridge, MA: MIT Press.
Van Inwagen, P. (1990) *Material Beings*, Ithaca, NY and London: Cornell University Press.
Volkmann, W. F. (1875) *Lehrbuch der Psychologie* (2nd ed.), Cöthen: Otto Schulze.
Watzl, S. (2013) "Silencing the Experience of Change," *Philosophical Studies* 165: 1009–1032.

Related Topics

The Unity of Consciousness
Consciousness, Personal Identity, and Immortality
Consciousness and Attention

22

CONSCIOUSNESS AND ACTION

Shaun Gallagher

Consciousness happens before, during and after action. I'll argue that for a complete and correct understanding of action, one needs to know precisely what role consciousness plays across these different temporal parameters. This is especially true in regard to the question of whether consciousness is epiphenomenal – something that accompanies action but plays no causal role in action. I'll argue that insofar as consciousness is characterized by intentionality, it should not be regarded as epiphenomenal.

1 Consciousness before Action

We can start by thinking of one delineated action, although in many cases it is difficult to individuate actions; actions are often embedded in larger projects; and any one action may be composed of several more basic actions. There are debates, for example, about the nature of "basic" actions, i.e., actions that are not composed of (or caused by) other actions (Danto 1963; Goldman 1970; Hornsby 2013). A reaching or a grasping might be considered a basic action. When we add these actions together, we may get a more complex action of, e.g., picking up a cup; and that may be part of a more complex action that we would define as clearing up the dining table. Let's take a medium sized complex action as our example – something more than a basic action (e.g., reaching for something), but something less than a large complicated process (e.g., getting married). We can settle for a rather boring non-adventurous example – preparing a cup of tea.

It may be that every morning after I wake up, one action among various others that I perform is making myself a cup of tea. In my case, this is not something automatic, even in the general sense that it is a habitual event. I confess that I usually find myself checking email before I get to the event in question. In such cases, what often happens is that I am sitting at my computer immersed in the action of checking email. I am not unconscious at this point. I may in fact be conscious of a particular email content that I need to respond to; and I may be carefully deciding how to construct my response. As I engage in this process, I may also start to realize how much time this is taking, and then become aware of the fact that I have not yet had my tea. This may motivate a decision either to keep at it, or to get up and prepare some tea.

Two points about this seem important. First, before my action of preparing a cup of tea, I am engaged in a complex conscious life. If I seemingly wake up to the fact that I have not yet made my tea, this is not literally a waking up from a non-conscious state. I've already done that earlier in the morning, and I've been conscious of something or other throughout the various actions with which I have been engaged since. Second, my awareness of something is a motivating factor in moving me into the action of preparing my tea. It's not simply that I am aware of the fact that I have not yet made my tea. It's rather a more positive feeling that I should have some tea. I notice that I am thirsty or in need of some sustenance. This is a bodily feeling; quite literally, a gut feeling rather than a consciousness of an objective fact. The phenomenology, prior to my action of preparing tea, is what motivates me to get up to start that process.

All of this may seem obvious and rather tedious to even state. I do so only because it is often the case that in scientific experiments addressing the question of consciousness one gets the impression that one begins with a non-conscious 0-point and then asks what brings on consciousness. For example, in the famous Libet experiments, the question is: When do I become conscious of the decision or the urge to flick my finger or wrist? The results of Libet's experiment suggest that we become conscious of the decision or urge only after some 500–850 milliseconds of brain activity (the "readiness potential") that seems to correlate with preparation for that specific action, and this suggests that consciousness doesn't play a role in causing the action, at least until approximately 150 milliseconds before motor activation (Libet 1985, 1992). Libet and everyone else know, however, that the subject in these experiments is conscious well before the action and throughout the experiment. The instructions given to the subject assume he is conscious, and in fact they direct the subject to consciously think about the kind of abstract and artificial action (the basic action of flicking a finger or wrist) that he is called on to perform. It's not at all clear that this prior consciousness, which itself involves not only some patterns of neural activation, but also intentional aspects (in the sense of intentionality as consciousness *of* the instructions), or what Roepstorff and Frith (2004) call 'top-top-down' socially contextualized factors, does not activate further processes in the brain that ultimately lead to the readiness potential, which then seemingly accompanies motor preparation for the specific action in question.[1] It's clear that there is consciousness relevant to the action long (relatively speaking) before this specific action occurs.

Moreover, in the Libet experiment, consciousness of the instructions clearly motivates the subject to perform the specific action he is instructed to perform. The subject would not sit there and flick his wrist whenever he wanted, without those instructions. So even if he is not conscious of his precise decision to flick his wrist prior to the specific brain activation that corresponds to the specific action of flicking his wrist, it doesn't seem right to claim that the subject's previous consciousness plays no role in the action (see Gallagher 2017a).

We come back to the example of my growing gustatory awareness that I have not yet made my tea. Can we not say that this awareness does play some role in my now getting up in order to prepare my tea? Clearly it plays a motivating role. The notion of motivation is not equivalent to a strong, billiard-ball conception of causality. But it seems clear that if I am asked why I started to make this move at this point, I would say something like, "Because I just became aware that I did not yet have my tea." Or, "Because I felt the urge." What did I become aware of? Not the physiological processes that define thirst, but perhaps some interoceptive sensations of thirst; or maybe just the thought that it was getting late and I had to move on to other things, like getting my tea. There is not only consciousness before action, but in some cases that consciousness motivates action.

I started with the example of preparing tea, rather than with the example of preparing a tea party. As we all know from our experience of tea parties, preparing a tea party involves some planning, and typically such planning is conscious and deliberate. In this regard, we often find ourselves consciously deliberating and forming prior intentions to act in some specific way. This clearly involves a consciousness before action that seems to have some effect on action, otherwise we would never say that people sometimes act in a way inconsistent with their intentions, nor would we ask them their reasons for acting as they did. I take this idea that we sometimes consciously form prior or distal intentions that motivate later action to be uncontroversial.

2 Consciousness during Action

Clearly, I do not fall into an unconscious state when I start to prepare my tea. This is not the same as saying that I am fully conscious of everything I'm doing. I may in fact be operating from habit, and many of my movements may be non-conscious. I reach for the kettle, clearly without a focused attention on the movements of my arm, or my grasp, and perhaps not even with a focused attention on the kettle. I may still be thinking about that last email, and that may be taking all of my attention. The fact that I am not conscious of my movements (reaching and grasping) is the circumstance of ordinary motor control features of action. In the Libet experiments, the subject may be conscious specifically of the flicking of his hand – but that's because the subject is in an abstract and artificial situation, where he is asked to focus on precisely these things. Even in this case, as in every case, however, the subject is not aware of the neural components of the motor control process. In almost all cases of intentional action, the agent is not aware of the precise specifics of bodily movement – joint angles, muscle tension, etc. As many experiments have shown, kinaesthesis (movement sense) and proprioception (postural sense) are attenuated, i.e., pushed into the outer margins of consciousness to the point where an agent can say that she is aware that she is moving, but not precisely how she is moving. Moving, in this case, is like posture. Unless we are practicing a kind of mindfulness or Alexander Technique (where one's focus is on one's situated posture), we are not usually aware of our posture or of how we are moving. Nonetheless, as ecological psychology shows, I do have a minimal awareness of, for example, the fact that I am sitting, or standing, or walking, or running (e.g., Gibson 2014).

Accordingly, I may not be aware of the details of my movement as I am engaged in preparing my tea – much of the process, such as the shaping of my fingers as I grasp the kettle, may be non-conscious (Jeannerod 1997), and many other aspects of the process may be shunted off into the margins of my consciousness, such as transformations in my posture, my walking around grabbing the tea and the mug and so forth. The latter are things that I can become aware of, e.g., if something suddenly goes wrong and I spill some water, or burn myself. But even without being aware of such things, in some sense I am aware that I am preparing my tea. There is a certain level of action that not only describes best what I am doing, but to which I am consciously attuned.

What's the nature of this conscious attunement? Again, the idea is not that at every moment I am fully attending to what I am doing. Indeed, I may be doing a number of different things across the same span of time – making tea, thinking about my email, responding to my wife's (previous or ongoing) instructions about not making a mess. This may be distributed across bouts of interim consciousness. Right now, I'm attending to the amount of water to put into the kettle and asking my wife if she wants a cup. Setting the kettle on the stovetop, I start thinking about that email as I get my cup from the cupboard. What action am I engaged in? If you ask me, I'll respond, "I'm making some tea," rather than:

1 I'm exercising my neurons.
2 I'm moving my body thus and so.
3 I'm reaching and grasping.
4 I'm being careful not to make a mess.
5 I'm reconsidering the email I just sent.

It's unlikely that I would answer 1, 2, or 3 – since I'm not conscious or fully conscious of such things. Depending on my understanding of your question, and perhaps prior bits of conversation, however, I might answer 4 or 5. If my wife asks me what I'm doing I might indeed say, "I'm being careful not to make a mess." But, generally speaking, if out of the blue you asked what I was doing, I would say, "I'm making some tea." This suggests that at some level, when I am acting, and barring any other unusual happenings, I am generally aware of my actions as I am engaged in them. I may not be aware of a lot of details – that I am tilting the kettle thus and so, for example – but I am at least minimally aware that I am in the process of making tea, even if I am thinking hard about the email, or talking to my wife at the same time. I'm not only conscious during my action, I am, at some level, conscious of my action.

Again, two points are important. First, consciousness during action may or may not involve awareness of specific details as I engage in action. Another appeal to the experimental literature may make this clear. In experiments that are specifically about consciousness one might be led to think that the subject is simply unconscious as they are engaged in a particular task. For example, in experiments on blindsight, subjects are conscious as they perform the various tasks asked of them. They are not conscious of a specific thing – an object, or location, or orientation, etc. – in the visual modality; but they are conscious of what they are thinking; and they are conscious in other sensory modalities. They hear the experimenter's instructions, for example. And they are aware of what they are doing in terms of the general task. They are asked to indicate the location of some object that they cannot see, for example. They are conscious that they are being forced to guess at the location. So, they are aware of specific things concerning the action when they are engaged in the action, and they are aware of what they are doing, even if they are not aware of how they are doing it, i.e., how they are getting the right answers. If we take the performance of the blindsight subject in such experiments to be a case of non-conscious visual perception, there is, nonetheless, a good deal of consciousness involved in their action.

Second, what I am conscious of, or to what degree I am conscious of it, whether it is my specific action or something else, may have some impact upon my action. Motor control processes that I am not conscious of, such as processes described in terms of forward control models that operate at the subpersonal neuronal level, do have impact on my action performance (Wolpert and Flanagan 2001; Wolpert, Doya and Kawato 2003). But intentional actions, and even high-level expert performances, are not reducible to just such non-conscious processes.

Dreyfus (2005), however, has famously argued that expert performance, when the expert is in-the-flow of performance, is mindless. Even if there are some aspects of action of which we are not aware – again, for example, the details of how we do an action, as well as subpersonal motor control aspects – he would go further and suggest that anything like reflective consciousness would necessarily interfere with expert performance. We may need to attend to our movements as we learn new kinds of action, such as a new dance; and there may also be cases in which we attend to our bodily movements in unusual circumstances (e.g., when we are on a ledge and being careful about how we move). But when we are engaged in expert performance, Dreyfus would even rule out thinking or reflecting about different aspects of the environment. The downhill skier, for example, would not be thinking even about the snow or about upcoming conditions on the hill when she is engaged in expert skiing (Gallagher 2016).

A phenomenological study by Høffding (2015), however, suggests that Dreyfus is wrong. Høffding studied the expert musical performance of members of the Danish String Quartet. In terms of being conscious of certain aspects of their playing, or of other things unrelated to their playing, as they played in the quartet, Høffding showed variation across different circumstances. In some cases, the musicians paid no explicit attention to what the others were doing. They may be attending to the music itself, letting their movements and playing of instruments be entrained by the music. In that regard, they are not aware of their movements, but they are definitely aware of the music. In other cases, however, they may be much more explicitly conscious of the other players or of their own performance.

When observing chamber ensemble musicians playing, one very often sees intentional communication in the form of body language or facial expressions – cues in the form of winks, blinks and laughs. Sometimes, there might be an element disturbing the unfolding of the performance and this may call for compensatory strategies (Salice, Høffding, and Gallagher 2017).

In the latter case, one musician may focus his attention on one of the other musicians in order to match the other's movements; this involves a conscious "trying to be together" or trying to achieve explicit coordination. A third type of situation is where there is an affective consciousness that involves a "feeling of intimate trust in the situation." Høffding describes this as a complex state that involves auditory perception, affectivity, interoception and intersubjective proprioception. All of these factors may remain in marginal consciousness; the players may be minimally aware of these different aspects of their experience. At the same time, this adds up to a conscious sense that they are playing well together, and in some cases, they are surprised by where the process goes. One of the musicians, Fredrik, describes it this way:

> there were a couple of times where I was surprised by where we were going.... Suddenly we find ourselves in a tempo we hadn't planned for at all, but we couldn't have done otherwise, because the preceding notes leading into it, they had laid the ground for it. And then you cannot get out of it.
>
> *(cited in Salice, Høffding and Gallagher 2017)*

This is not an unconscious surprise; it's an aesthetic feeling that emerges in the action process, and in contrast to ongoing conscious expectations of where the performance would go.

Such aesthetic consciousness is a pleasurable experience that may emerge in other types of performance. In dance, for example, there are numerous detailed aspects of movement that are not attended to and that do not come to consciousness. But there is often an aesthetic consciousness – a pleasure that is in the movement, occurring during movement, associated with affective touch and affective proprioception. Cole and Montero (2007) have explained the neurophysiological underpinnings of such experiences that may include the experience of effortlessness.

> One significant aspect of this experience is the feeling of effortlessness, of the body moving almost on its own without the need of conscious direction. When absorbed in movement there may even be what might be described as a loss of self, a feeling that, at least as a locus of thought, one hardly exists at all. And of course the best performances are those where one is not thinking about the steps at all but is rather fully immersed in the experience of moving itself.
>
> *(Cole and Montero 2007: 304)*

This description captures two aspects of this experience at once: (1) that there are some things of which one is not conscious (the details of moving, which do not need conscious direction), and (2) that there is a feeling of effortlessness that comes along with this kind of movement. Even deep immersion in performance is not entirely non-conscious.

It is also the case that consciousness during action may actually derive from subpersonal processes that occur prior to action. As Anthony Marcel puts it, "awareness of voluntary action appears to derive from a stage later than intention [formation] but earlier than movement itself"

(2003: 48). We may, for example, be aware of the accuracy or inaccuracy of our movement just as we are making it, and without the power to change it.

> If this is so, awareness of the accuracy of movement may precede feedback of that movement's accuracy and, then some of the harmonious feelings associated with accurate movement may derive from its accurate elaboration and selection within the brain as well as, or even rather than, from peripheral sensory feedback resulting from the movement.
>
> *(Cole and Montero 2007: 306)*

3 The Sense of Agency

Consciousness during action may be more complex than already indicated if it also involves a sense of agency. This has become a controversial issue, however. Do we have, in addition to a sense of what we are doing, and a variable consciousness of our surroundings, and possibly an aesthetic feeling of pleasure that derives from kinaesthetic and control factors, also a sense of agency that is separable from these other aspects?

Elsewhere, I distinguished between the sense of ownership (SO) for movement or action, and the sense of agency (SA), and argued that both of these experiences, even if difficult to distinguish in everyday intentional action, are dissociable in the case of involuntary movement (Gallagher 2000; 2012). In that case, if you push me from behind, I can easily say that *I* am the one moving, or *my* body is moving (SO), but that I was not the agent of that movement (no SA). Phenomenologists thus argue that in voluntary action the agent has both SO and SA for the action (Gallagher and Zahavi 2012), and that these are non- or pre-reflective, implicit aspects of our consciousness of action. It's important to distinguish between pre-reflective SO and SA, and retrospective reflective judgments about ownership and agency (Vosgerau and Newen 2007). The phenomenological claim is that SO and SA are non-reflective, occurring at a first-order level of consciousness.

Mike Martin (1995) first used the term 'sense of ownership' in connection with an experience of one's spatial boundaries. According to Martin, "when one feels a sensation, one thereby feels as if something is occurring within one's body" (1995: 267). This is not a matter of explicit judgment, as if I were experiencing a free-floating sensation, concerning which I needed to judge its spatial location as falling within my body boundaries. Rather, as Martin argues, the experience of location is an intrinsic feature of the sensation itself. This experience just is the SO for one's body as a whole, so that I have SO for particular body parts only as being parts of that whole body (Martin 1995: 277–278). SO is not a quality in addition to other qualities of experience, but "already inherent within them" (1995: 278). I take Martin's concept of SO to be close to, if not identical to, the phenomenological concept to the extent that SO for one's action is an intrinsic aspect of proprioceptive and kinaesthetic experiences of bodily movement (Gallagher 2017b).

José Bermúdez (2011; in press), however, in a critical discussion of SO, in contrast to Martin, rejects the idea that SO is a "special phenomenological relation" (Martin 1995: 267), although he accepts the importance of "boundedness." He denies that there is a positive first-order (non-observational) phenomenology of ownership or feeling of 'mineness.' In contrast to this "inflationary" conception, which he attributes to phenomenologists like Merleau-Ponty (he also cites Gallagher 2005; DeVignemont 2007, 2013), he offers a deflationary account. "On a deflationary conception of ownership the SO consists, first, in certain facts about the phenomenology of bodily sensations and, second, in certain fairly obvious judgments about the body (which we can term judgments of ownership)" (2011: 162). His deflationary view is that an explicit experience of ownership only comes up when we turn our reflective attention to our bodily experience and attribute that experience to ourselves. But that involves adding something that's not there to begin with. "There are facts about the phenomenology of bodily awareness (about position sense, movement sense, and interoception) and there are judgments of ownership, but there is no additional feeling of ownership" (2011: 166).

For the phenomenologists, however, to say that SO is an intrinsic aspect of proprioception and kinaesthesia is to agree that it is not an additional or independent feeling, but rather, a sense "already inherent within" the phenomenology of bodily sensations. On the phenomenological view, and in contrast to Bermúdez, this intrinsic aspect is pre-reflective in the sense that one has this intrinsic experience of ownership without having to make a reflective judgment about ownership. This can be read in the deflationary way, so that the phenomenologists can agree that there is no additional feeling of ownership, or "perfectly determinate 'quale' associated with the feeling of myness" (Bermúdez 2011, 165), independent of the proprioceptive and kinaesthetic sensations. In fact, this implicit self-experience is precisely what makes first-person bodily (proprioceptive, kinaesthetic) awareness itself (i.e., prior to any judgment) a form of self-consciousness. It's what puts the 'proprio' in proprioception (Gallagher and Trigg 2016).

Parallel to Bermúdez's argument about SO, Thor Grünbaum (2015) has offered a detailed critique of the notion of SA, drawing a similar conclusion, namely that there is no separate and distinct pre-reflective SA that acts as the basis for a judgment about agency. Grünbaum's analysis is focused on accounts of SA that make it the experiential product of comparator mechanisms involved in motor control. Although he doesn't deny the possibility that a comparator mechanism may be involved in motor control, he challenges the idea that such subpersonal mechanisms can generate a distinct experience of agency. Specifically, he interprets the claim that SA is generated in this way to mean that SA is intention-free. That is, on such accounts, SA is generated even if the agent has not formulated a prior, personal-level intention to act in a certain way. Reaching for my tea as I answer my email does not require that I consciously deliberate and form a plan to do so. Even if it does not involve the formation of a prior intention, however, it nonetheless counts as an intentional action and certainly involves a motor intention and a present intention (or intention-in-action) (see Pacherie 2006, 2007). In this respect, *pace* Grünbaum, it's not clear that SA can be characterized as intention-free.[2]

A more important challenge to the comparator model, however, involves the characterization of SA found in a number of experiments. Grünbaum, for example, indicates a problem like this in Daprati et al. (1997). They ask subjects to perform a hand movement and to monitor it on a computer screen, which shows either their own hand movement, or a hand movement made by someone else. Subjects are then asked whether what they saw was their action or not, i.e., whether they had a sense of self-agency for the movement.[3] The results show that subjects mistook the actions of the other's hand as their own in about 30% of the cases; schizophrenic subjects with a history of delusions of control and/or hallucinations misjudged 50–77% of the cases. In this experiment, the subject is in fact moving his hand in each task. Grünbaum thinks there is a confusion here about SA. To get clear on what the problem with this experiment is, I suggest it is helpful to look at a similar experiment by Farrer and Frith (2002), who, in contrast to Daprati et al., employ the distinction between SO and SA. In their similar experimental design, the subject is asked to control a cursor on a computer screen using a joystick, and then to judge whether the event on the screen was caused by their action or not. Again, in each case the subject moved the joystick. Farrer and Frith claim this allows for a controlled dissociation between SO and SA. They argue that SA varied depending on whether subjects felt they were in control of what was happening on the computer screen, while SO remained constant since "subjects were requested to execute an action during all the different experimental conditions" (2002: 597).

Grünbaum claims that in such experiments, rather than reporting SA for their movement (given that the subjects moved their hand in each case, SA would seemingly remain constant across all trials), subjects were simply reporting differences in what they were monitoring on the screen. An alternative interpretation, however, is that the pre-reflective SA is more complex than an experience of movement generated in motor control processes. Rather, the experiments make sense if one allows that SA involves at least two aspects – one having to do with the control of bodily movement in action, and one having to do with the intentional aspect of the action, i.e., what the action is about, or what it accomplishes in the world. Grünbaum is correct that these experiments did not eliminate a confounding between SO and the first motoric aspect of SA, even if they did eliminate it for the second intentional aspect of SA. To get clarity on these issues, however, the distinction between the two aspects of SA should not be implicitly assumed (as in Farrer and Frith 2002), but explicitly stated (as in Gallagher 2005, 2012; Haggard 2005; Kalckert and Ehrsson 2012).

As I've argued elsewhere (Gallagher and Trigg 2016), even if Grünbaum were right about comparator mechanisms not generating SA, SA may still be generated in our perceptual monitoring of what our actions are accomplishing in the world. I note that Langland-Hassan (2008) raised worries similar to Grünbaum's about the positive phenomenology of SA, but concluded that the phenomenology of agency is "one that is embedded in all first-order sensory and proprioceptive phenomenology as diachronic, action-sensitive patterns of information; it does not stand apart from them as an inscrutable emotion" (392). This is exactly right, and consistent with the phenomenological and the deflationary views that regard both SA and SO as intrinsic features of the bodily experience of action. SA may be no less intrinsic even if we include the intentional aspect of action. In this regard, Grünbaum may be right about the problems with the comparator model of SA (for a more embodied and situated conception of SA, see Buhrmann and Di Paolo 2017; Gallagher 2012). Indeed, there are good reasons to question whether comparator models of motor control offer the best explanation of SA (see, e.g., Langland-Hassan's [2008] explanation in terms of filtering models; also see Friston 2011; Synofzik, Vosgerau and Newen 2008).

4 Consciousness after Action

After these considerations about consciousness before and during action, let me introduce what I take to be a helpful threefold distinction concerning time scales. These are distinctions that derive from studies by Pöppel (1988, 1994) and Varela (1999), which distinguish between

1 The *elementary* time scale measured in milliseconds.
2 The *integrative* time scale measured in the seconds that comprise the specious present (James 1890).
3 The *narrative* time scale measured across any period that extends beyond the specious present.

The elementary time scale of 10–100 milliseconds corresponds to non-conscious subpersonal processes that underpin actions and motor control processes such as forward models, and what Pacherie (2006) calls motor intentions. The integrative time scale refers to a duration in which these subpersonal processes are integrated and result in an identifiable basic action or the conscious awareness of meaningful movement. The specious present concept found in James tends to be measured in a .5- to 5-second duration; in the phenomenological tradition it is explicated in terms of the intrinsic temporal (retentional-protentional) structure of consciousness (Gallagher 1998; Gallagher and Zahavi 2012). Roughly, anything above that time scale is regarded as involving reflective consciousness, recollection and the possibility of narrative description.

Clearly, the time scales related to consciousness are the integrative and narrative ones. Consciousness prior to action falls on either the narrative time scale in the form of deliberation and distal intention formation (see, e.g., Velleman 2005) or the integrative time scale in the form of the activity that may be involved in our first example of preparing tea. I don't have to formulate a large narrative about making tea; I don't usually engage in the formation of prior intentions about making myself a cup of tea. Planning a tea party would involve such narrative-level processes, however. Consciousness during action falls primarily on the integrative time scale, since my awareness of what I am doing falls within the immediacy of the ongoing action. Here, however, we see some issues pertaining to the individuation of action. Cleaning up after the tea party might be regarded as one action composed of a set of simpler actions, and any of them might extend beyond the upper limits of a specious present. Some actions, accordingly, extend into the narrative time scale. Still, as I am conscious of such an action during the action, my immediate first-order (e.g., perceptual) consciousness extends only across the seconds of the specious present (involving working memory but not recollection) before it requires the enaction of a new and different kind of consciousness, i.e. episodic memory, to hold onto previous parts or elements of the action. More generally, second-order, reflective consciousness involves the narrative time scale. In this respect, consciousness after action tends to be framed on this time scale.

On the narrative time scale, I can retrospectively reflect, attribute, and report on my actions. On this time scale, I can evaluate and give reasons for my actions. None of this changes the physical nature of my actions, although such procedures can change the interpretation or the meaning of my actions. This is an important consideration, however, since what my actions mean, to me or to others, is in some sense part of what my actions are.

In a relatively simple action, such as preparing tea, there may not be much to interpret or evaluate. Yet, if I consistently prepare my tea in such a way that I make a mess that others have to live with; or if I consistently invite my wife to join me for a cup of tea; or if I consistently fail to invite my wife to join me for a cup of tea, etc., becoming conscious of my past actions, perhaps in conversation with my wife, may make my actions something different than what I or others considered them to be. Physically, one might say, they are what they are. But actions are certainly more than what they physically are. What a particular action is, depends on what that action means. We've known this at least since the time Socrates explained his disappointment with Anaxagoras in Plato's *Phaedo*. Sequencing through a series of physical movements does not make an action what it is. Rather, what an action means in a particular pragmatic or social context determines what it is, and this is what a narrative consciousness captures in any giving of account. A purely physical, or even neurophysiological, description tells us nothing about *what* the action is, even if it explains how it might be possible. What an action is depends in a real sense on context (or intentionality), the agent's intention, and how the agent and others interpret the action.

One finds clear examples of such complications in legal contexts. A physical description of the action by itself does not determine whether the action is one of homicide, manslaughter, or accident. Arguably, what an action *is* is determined by the agent's conscious intention prior to and during the action, and sometimes by retrospective judgments made after the action. All of these factors have to do with consciousness – consciousness before, during, and after action. Take away consciousness at any of these points and what we consider to be an action may no longer be an action, or may be a different sort of action than an action that involves consciousness.

5 Conclusion

Epiphenomenalism is the idea, explicated by Shadworth Holloway Hodgson in 1870, that in regard to action, the presence of consciousness doesn't matter, since it plays no causal role. One can think of this in terms that place neural events in charge of action: neural events form an autonomous causal chain that is independent of any accompanying conscious mental states. The epiphenomenalist holds that neural events cause bodily movement and consciousness, but consciousness cannot cause neural events or bodily movement. Consciousness is thus incapable of having any effect on the nervous system. We could, for example, build a non-conscious robot that would prepare my tea, and it is also possible that aspects of my initial motivation, e.g., my thirst, are themselves reducible to non-conscious processes that launch the larger action process. Consciousness, on this view, may keep us informed about what is going on in broad and general terms, and it can act as a "dormant monitor" (Jeannerod 2003: 162), allowing us to know what we are doing, but it plays no role in the causal processes that move or motivate us. We, as conscious animals, are seemingly just along for the ride.

In contrast to this epiphenomenal view, I've suggested that in a variety of contexts, consciousness prior to action is motivating for action; conscious attunement to events or to other people in the environment during action may have an effect on action; and even consciousness after action may redefine what that action is. More generally, the extent to which consciousness before, during, and after action involves intentionality – that is, to the extent that consciousness is directed to the world, or to the action itself – whether in terms of intentional planning, or in terms of a marginal perceptual monitoring of what one is doing, or in terms of retrospective evaluation, consciousness may motivate, or guide, or interpret action in a way that makes the action what it is. In these regards, it is difficult to regard consciousness as epiphenomenal.

Acknowledgements

Research on this chapter was supported by the author's Anneliese Maier Research Award from the Humboldt Foundation, and by an Australian Research Council (ARC) grant, Minds in Skilled Performance (DP170102987).

Notes

1 But see Schurger, Sitt and Dehaene (2012) for a dissenting discussion on the nature of the readiness potential.
2 In addition, some comparator models include the idea that there is a functional element in the system that counts as an intention; this subpersonal intention is compared to efference copy or sensory input from the movement to facilitate motor control (e.g., Frith 1992; Wolpert and Flanagan 2001).
3 Here I ignore the fact that in asking the subjects to report on their experience Daprati et al. are introducing a judgment about agency over and above SA. The same applies to the experiment by Farrer and Frith (2002), discussed below. In both cases, the assumption is simply that the judgment of agency is a veridical report on the sense of agency. This is a complication Grünbaum does not discuss.

References

Bermúdez, J. L. (2011) "Bodily awareness and self-consciousness," in S. Gallagher (ed.) *Oxford Handbook of the Self*, Oxford: Oxford University Press.

Bermúdez, J. L. (2017) "Ownership and the space of the body," in F. de Vignemont and A. Alsmith (eds.) *The Subject's Matter*, Cambridge, MA: MIT Press.

Buhrmann, T., and Di Paolo, E. (2017) "The sense of agency – a phenomenological consequence of enacting sensorimotor schemes," *Phenomenology and the Cognitive Sciences* 16: 207–236.

Cole, J., and Montero, B. (2007) "Affective proprioception," *Janus Head* 9: 299–317.
Danto, A. C. (1963) "What we can do," *Journal of Philosophy* 60: 435–445.
Daprati, E., Franck, N., Georgieff, N., Proust, J., Pacherie, E., Dalery, J., and Jeannerod, M. (1997) "Looking for the agent: An investigation into consciousness of action and self-consciousness in schizophrenic patients," *Cognition* 65: 71–86.
Dreyfus, H. L. (2005) "Overcoming the myth of the mental: How philosophers can profit from the phenomenology of everyday expertise," *Proceedings and Addresses of the American Philosophical Association* 79: 47–65.
Farrer, C., and Frith, C. D. (2002) "Experiencing oneself vs. another person as being the cause of an action: The neural correlates of the experience of agency," *Neuroimage* 15: 596–603.
Friston, K. (2011) "Embodied inference: Or 'I think therefore I am, if I am what I think'," in W. Tschacher and C. Bergomi (eds.) *The Implications of Embodiment: Cognition and Communication*, Exeter: Imprint Academic.
Frith, C. D. (1992) *The Cognitive Neuropsychology of Schizophrenia*, Hillsdale, NJ: Lawrence Erlbaum Associates.
Gallagher, S. (1998) *The Inordinance of Time*, Evanston, IL: Northwestern University Press.
Gallagher, S. (2000) "Philosophical conceptions of the self: Implications for cognitive science," *Trends in Cognitive Sciences* 4: 14–21.
Gallagher, S. (2005) *How the Body Shapes the Mind*, Oxford: Oxford University Press.
Gallagher, S. (2012) "Multiple aspects in the sense of agency," *New Ideas in Psychology* 30: 15–31.
Gallagher, S. (2016) "The practice of thinking: Between Dreyfus and McDowell," in T. Breyer (ed.) *The Phenomenology of Thinking*, London: Routledge.
Gallagher, S. (2017a) *Enactivist Interventions: Rethinking the Mind*, Oxford: Oxford University Press.
Gallagher, S. (2017b) "Self-defense: Deflecting the deflationary and eliminativist critiques of the sense of ownership," *Frontiers in Human Neuroscience*. 8: 1612. https://doi.org/10.3389/fpsyg.2017.01612
Gallagher, S., and Zahavi, D. (2012) *The Phenomenological Mind*, 2nd ed. London: Routledge.
Gallagher, S., and Trigg, D. (2016) "Agency and anxiety: Disorders of the minimal self," *Frontiers in Neuroscience* 10: 1–12.
Gibson, J. J. (2014) *The Ecological Approach to Visual Perception* (Classic edition), Psychology Press.
Goldman, A. I. (1970) *A Theory of Human Action*, Princeton, NJ: Princeton University Press.
Grünbaum, T. (2015) "The feeling of agency hypothesis: A critique," *Synthese* 192: 3313–3337.
Haggard, P. (2005) "Conscious intention and motor cognition," *Trends in Cognitive Sciences* 9: 290–295.
Høffding, S. (2015) "A Phenomenology of Expert Musicianship," PhD Thesis. Department of Philosophy. University of Copenhagen.
Hodgson, S. (1870) *The Theory of Practice*, London: Longmans, Green, Reader, and Dyer.
Hornsby, J. (2013) "Basic activity," *Aristotelian Society Supplementary* 87: 1–18.
James, W. (1890) *Principles of Psychology*, 2 vols. New York: Dover, 1950.
Jeannerod, M. (1997) *The Cognitive Neuroscience of Action*, Oxford: Blackwell Publishers.
Kalckert, A., and Ehrsson, H. H. (2012) "Moving a rubber hand that feels like your own: A dissociation of ownership and agency," *Frontiers in Human Neuroscience* 6: 40.
Langland-Hassan, P. (2008) "Fractured phenomenologies: Thought insertion, inner speech, and the puzzle of extraneity," *Mind and Language* 23: 369–401.
Libet, B. (1985) "Unconscious cerebral initiative and the role of conscious will in voluntary action," *Behavioral and Brain Sciences* 8: 529–566.
Libet, B. (1992) "The neural time-factor in perception, volition, and free will," *Revue de Métaphysique et de Morale* 2: 255–72.
Marcel, A. (2003) "The sense of agency: Awareness and ownership of action," in J. Roessler and N. Eilan (eds.) *Agency and Self-Awareness*, Oxford: Oxford University Press.
Martin, M.G.F. (1995) "Bodily awareness: A sense of ownership," in J. L. Bermúdez, T. Marcel, and N. Eilan (eds.) *The Body and the Self*, Cambridge, MA: MIT Press.
Pacherie, E. (2006) "Towards a dynamic theory of intentions," in S. Pockett, W. P. Banks, and S. Gallagher (eds.) *Does Consciousness Cause Behavior? An Investigation of the Nature of Volition*, Cambridge, MA: MIT Press.
Pacherie, E. (2007) "The sense of control and the sense of agency," *Psyche* 13: 1–30.
Pöppel, E. (1988) *Mindworks: Time and Conscious Experience*, Boston: Harcourt Brace Jovanovich.
Pöppel, E. (1994) "Temporal mechanisms in perception," *International Review of Neurobiology* 37: 185–202.
Roepstorff, A., and Frith, C. (2004) "What's at the top in the top-down control of action? Script-sharing and "top-top" control of action in cognitive experiments," *Psychological Research* 68: 189–198.

Salice, A., Høffding, S., and Gallagher, S. (2017) "Putting plural self-awareness into practice: The phenomenology of expert musicianship," *Topoi* OnLine First: 1–13.
Schurger, A., Sitt, J. D., and Dehaene, S. (2012) "An accumulator model for spontaneous neural activity prior to self-initiated movement," *Proceedings of the National Academy of Sciences* 109 (42) E2904–E2913.
Synofzik, M., Vosgerau, G., and Newen, A. (2008) "Beyond the comparator model: A multifactorial two-step account of agency," *Consciousness and Cognition* 17: 219–239.
Varela, F. J. (1999) "The specious present: A neurophenomenology of time consciousness," in J. Petitot, F. J. Varela, B. Pachoud, and J.-M. Roy (eds.) *Naturalizing Phenomenology: Issues in Contemporary Phenomenology and Cognitive Science*, Stanford: Stanford University Press.
Velleman, J. D. (2005) "Self as narrator," in J. Christman and J. Anderson (eds.) *Autonomy and the Challenges to Liberalism: New Essays*, New York: Cambridge University Press.
Vosgerau, G., and A. Newen. (2007) "Thoughts, motor actions, and the self," *Mind and Language* 22: 22–43.
Wolpert, D. M., and Flanagan, J. R. (2001) "Motor prediction," *Current Biology* 11: R729–R732.
Wolpert, D. M., Doya, K., and Kawato, M. (2003) "A unifying computational framework for motor control and social interaction," *Philos Trans R Soc Lond B Biol Sci* 358: 593–602.

Related Topics

Materialism
Dualism
Consciousness, Free Will, and Moral Responsibility
Sensorimotor and Enactive Approaches to Consciousness
Consciousness, Time, and Memory
Consciousness and Psychopathology

23
CONSCIOUSNESS AND EMOTION

Demian Whiting

> If we fancy some strong emotion, and then try to abstract from our consciousness of it all the feelings of its characteristic bodily symptoms, we find that we have nothing left behind, no 'mind-stuff,' out of which the emotion can be constituted…
>
> *(William James 1884: 193)*

> It is surely of the essence of an emotion that we should feel it, i.e. that it should enter consciousness.
>
> *(Sigmund Freud 1950: 109–110)*

What is the relationship between emotion and consciousness? Is consciousness needed for emotion? Might it even be the case that emotion is required for consciousness? This latter idea is an interesting one. So, we might wonder what it would be like to be a creature that has no emotion at all, no pleasure occasioned by things when they go right, no sorrow when things go wrong, no fear occasioned by threats to oneself or others. Certainly, it would be nothing like what it is like to be a normal human being. Perhaps there are people who experience very little emotion, but, if so, such cases are extremely rare. Even psychopaths have some emotion despite lacking the deeper emotional responses that characterize a fully functioning and flourishing human life. Nevertheless, it is not obvious that a creature that lacked emotion could have no conscious experience. Surely, there are other forms of consciousness that such a creature could undergo if not emotional consciousness. What about perceptual consciousness, or the consciousness that comes with thinking a thought or working out a puzzle? Possibly, what it is like to think a thought or to visually perceive something would be very different without accompanying emotional consciousness. So, perhaps, such mental activity would lack a certain vibrancy or color, so to speak, if there were no accompanying emotion. But, it is unclear what the justification would be for saying that there cannot be consciousness without emotion.

Much more plausible, we might think, is the idea that consciousness is needed for emotion. Indeed, it might seem obvious that emotions are conscious and necessarily or constitutively so. After all, we speak about emotions as feelings and feelings are felt, are they not? This claim that emotions are conscious is the idea to be explored in this chapter. More specifically, I address three sorts of questions: First, what is meant by saying that emotions are conscious? Second, why does it matter whether emotions are conscious? For instance, does saying that emotions are conscious have metaphysical or epistemological implications? And third, is it the case that emotions are always conscious? What about longstanding or persistent emotions, such as an enduring fear of heights or love for another person? Also, can there be episodic or occurrent emotional states that are not conscious?

1 What Is Meant by Saying That Emotions Are Conscious?

In this chapter, 'consciousness' will be understood as referring to phenomenal consciousness or, in other words, the what-it-is-likeness of being in or occupying a certain mental state (Nagel 1974). It follows that to say that emotions are conscious is to say at a minimum that there is something that it is like for us to undergo emotion, or that emotions have a characteristic feel or phenomenology. For instance, it might be held that fear has an edgy feel, and anger has an irritable or hot-headed feel. That being said, note that the claim that emotion feels a certain way doesn't on its own tell us anything about the character of emotional phenomenology. To be sure, that claim is consistent with holding that emotional phenomenology takes the form of bodily sensation or feeling. For instance, about fear and anger, William James writes:

> What kind of an emotion of fear would be left, if the feelings neither of quickened heartbeats nor of shallow breathing, neither of trembling lips nor of weakened limbs, neither of goose-flesh nor of visceral stirrings, were present, it is quite impossible to think. Can one fancy the state of rage and picture no ebullition of it in the chest, no flushing of the face, no dilatation of the nostrils, no clenching of the teeth, no impulse to vigorous action, but in their stead limp muscles, calm breathing, and a placid face? The present writer, for one, certainly cannot.
>
> *(James 1884: 194)*

According to James, then, what it is like to undergo an emotion is what it is like to undergo or experience certain bodily changes. James may or may not be right, but the claim that emotions feel a certain way need not rule out other views, however. So, perhaps, emotional phenomenology has a cognitive or perceptual-like character. For instance, perhaps what it is like to feel afraid is what it is like to judge or perceive something as fearsome or dangerous (e.g. Solomon 1993; Döring 2007; Poellner 2016). Or, perhaps, emotional phenomenology has a desire-like character. For instance, perhaps what it is like to feel afraid is what it is like to want to run away or to avoid the object feared (e.g. Maiese 2011). Or, perhaps, emotional phenomenology is composite in nature, comprising cognitive, sensory, and conative elements (e.g. Kriegel 2012). Or then again, perhaps emotional phenomenology is not described properly by any of the views just outlined; for perhaps emotional phenomenology is altogether distinctive and sui generis, unlike any other kind of consciousness.

The claim that emotions feel a certain way is consistent also with two further ideas. To begin with, it is consistent with the view that the what-it-is-likeness of emotion is a non-constitutive or extrinsic property of emotion. On this view, although emotions feel a certain way, emotions are not constituted by how they feel, in the same way that cuts and bruises, for instance, are generally taken not to be constituted by the way they feel to us. So, perhaps emotions are states of bodily arousal that feel a certain way when undergone, but – as with such things as cuts and bruises – are to be characterized entirely independent of how they feel. Call this the 'non-constitution view.' But, the claim that emotions feel a certain way is consistent as well with a stronger and potentially more interesting view, namely the idea that emotions are constituted by their characteristic phenomenology or way of feeling. Call this the 'constitution view'.

According to this stronger view, there is no distinction to be had between the emotion and the way the emotion feels to us.

Is there a way of deciding between the constitution and non-constitution view? Considerations relevant to answering that question might be conceptual and/or empirical. Conceptual considerations draw on what our idea of emotion might tell us about emotion. That we talk about emotions as feelings might provide some support for the constitution view, since arguably it is part of our concept of feeling that feelings are felt and necessarily so. Another conceptual consideration in favor of the constitution view might be that although we say emotions are conscious, we don't similarly say that cuts and bruises are conscious, even though cuts and bruises feel a certain way when we experience them. For instance, cuts and bruises can be painful, and they also look a certain way when viewed in normal lighting conditions. Why don't we say that cuts and bruises are conscious? One reason might be that saying something is conscious is saying more than merely that thing feels a certain way, namely it is saying that thing comprises the way it feels, or has consciousness built into it, so to speak.

Empirical considerations, on the other hand, resort to observational data, our introspective observations of the emotions that we undergo, for instance. Here is how one simple argument for the constitution view grounded in the experience might go, then. Next time you undergo an emotion – say, an episode of fear or anxiety – introspectively attend to the emotion and hold what you are attending to in mind. Then introspectively attend to how the emotion feels and hold what you are attending to in mind. Finally, having done both of these things, let it be asked: when moving from the emotion to how the emotion felt did you experience a shift in attention? Did you find yourself attending to one thing when attending to the emotion and another thing when attending to how the emotion felt? Plausibly, it might be held that you experienced no change in attention at all, that what you found yourself attending to when attending to the emotion was exactly what you found yourself attending to when attending to the characteristic feel of the emotion. And if that is the case, then it seems to constitute quite a strong argument – again, one rooted in first-person observation or experience of emotion, note – for the view that emotions are indistinguishable from their characteristic ways of feeling.

2 Why It Matters Whether Emotions Are Conscious

If the what-it-is-likeness of emotion were a non-constitutive property of emotion, then the fact of emotion having a characteristic feel would tell us little about emotion and the kinds of things that emotions might and might not be. Of course, if emotions are constituted by something other than how they feel, then the fact of emotion feeling a certain way might still hold interest for other reasons. For instance, the existence of the phenomenology would provide us with evidence for the existence of the emotion and the emotion type to which it belongs. This might prove useful when trying to navigate the world, for example. For instance, knowing something about the emotions we are undergoing – that we are anxious or happy or angry, say – might tell us something about how we are faring, and how our behaviors might need to be modified for us to get on better. Also, if we know on the basis of their characteristic feel that we are undergoing certain emotions, then we will be in a position to share that knowledge with other people, and this might serve various personal and social goals.

However, if emotions are constituted by their characteristic feel or phenomenology, then such a fact about emotion will tell us important truths about emotion and the sorts of things that emotions might and might not be. For instance, such a fact about emotion would mean that emotions are conscious through and through, and that for an emotion to be is for an emotion to be felt or experienced by us.

Moreover, if emotions are how they feel, then that will create serious difficulties for the view that emotions are neural or bodily states, which do not seem to have consciousness built into them. Sometimes the point is made that it is conceivable that bodily or neural activity take place, but without consciousness being present, as in a zombie world, for instance (see Chalmers 1996). But whatever we think of 'conceivability arguments' (and it is sometimes held that what is conceivable is not always a good test of what is actual or possible), a simple but effective argument can be given for the view that neural or bodily states do not consist of how they might feel to us when we experience them.

In a nutshell, the argument is that if neural or bodily states comprise how they feel, then their natures would be evident to us. This is because the way something feels – for instance, the edgy feel of fear – is evident to us. Indeed, there is no part of an edgy feel (say) that is not apparent to us; fear's edginess is evident to us in its entirety, in other words. But, as with other physical phenomena, neural or bodily states have natures – namely, atomic natures – that are not evident to us, even in part. Although we can theorize about the atomic makeup of a given neural or bodily state, we cannot know the atomic constitution in question by directly observing it. Therefore, bodily or neural states are by their nature distinct from the way they feel, and it follows that emotions cannot be bodily or neural states if emotions are not distinct from the way they feel. And, note, that it would be no good to reply that bodily or neural states might be the way they feel because the way they feel might have a nature that is not evident to us. If something is a property that it has and that is apparent to us – for instance, if a bodily or neural state is the way it feels to us – then the nature of the property with which the thing is identical will be evident to us as well. This is because the thing and its property are one and the same, and, therefore, their natures, the properties composing them, will be the same also.

To say that emotions are not bodily states isn't to rule out the idea that emotions require a body, however. Indeed, there is ample empirical evidence to show that bodily activity or change underlies and shapes the emotions that we feel (Damasio 1999; Prinz 2004; Maiese 2011). For instance, it has been shown that fear, anger, joy, sadness, surprise, and disgust (the so-called six basic emotions) can be distinguished by patterns of autonomic activity (Levenson et al. 1992; Levenson 2003). And, again, other studies have identified areas of the brain that are implicated in emotion. For instance, the amygdala has been linked with fear (LeDoux 2000), and amygdala hyperactivation has been found to be associated with anxiety disorders (Garfinkel and Liberzon 2009). Moreover, it has seemed to many emotion theorists, from William James onwards, that emotions just are feelings or sensations of bodily movement or activity – that fear, for instance, is the feeling or sensation of quickened heartbeats, weakening limbs, and trembling lips – or, at least, that emotions are experienced by us as occupying parts of the body (e.g. James 1884; Prinz 2004; Whiting 2011; Maiese 2011). Such considerations lend support to the idea that emotions require a body. If emotions are constituted by their characteristic phenomenology, then there may not be a relationship of identity between emotion and bodily activity, but there is reason to think there is a very close relationship all the same.

If emotions are constituted by the way they feel, then this would have important implications for the epistemology as well. In particular, that fact about emotion would entail that attention to the phenomenology of emotion will be inescapable if we wish to know something about emotion's intrinsic properties or character. In that case, attending to other observable data, such as brain scans and observable bodily changes, would be of limited value insofar as getting to know the nature of emotion is concerned – though, again, such forms of empirical inquiry might tell us about the physical structures that underlie emotion, and out of which emotion may emerge somehow.

Also of limited value would be the language that we employ about emotion. Indeed, if the way to get to know about emotion is by attending to what it is like to undergo emotion, then we will have reason to accept everyday talk about emotion only if such talk is vindicated by the experience of emotion. As a way of illustration, notice that we commonly talk about emotions being *about* or *of* things. For instance, I might say that I am frightened of a dog or frustrated about the weather. Everyday talk suggests that emotions are intentional or object-directed mental states, in the same way that thoughts, desires and perceptual states are commonly taken to be intentional mental states. But is such talk actually borne-out by the phenomenology of emotion? Although some emotion theorists answer in the affirmative (e.g. Tye 2008; Montague 2009; Maiese 2011; Kriegel 2012; Lutz 2015), there is reason to be sceptical. This is because it might be argued that what we find when we attend to the phenomenology of emotion are 'raw' feels with no representational or intentional character of their own (see Whiting 2011). For instance, consider what it is like to undergo a fear sensation in the stomach region. To be sure, when we undergo fear thoughts about the object exciting the fear are never far away and those thoughts have an object-directed character or feel. But the fear itself? Plausibly, it might be held that attention to the phenomenology shows us that such a state is 'blind,' thus promising to vindicate Hume's claim that emotions are 'original existences' with no representational quality of their own (Hume 1969). In this respect, it might be thought that emotions more closely resemble pain sensations, for instance, which also seem to fail to phenomenally manifest an object-directed character.

If emotions comprise their phenomenology, then the only reply available to someone who thinks that emotions are intentional, but accepts that emotions don't manifest an object-directed character, would be to claim that the intentional properties of emotion are extrinsic or non-constitutive properties of emotion. This might be the view of someone who thinks that intentional relations are types of causal relations, for instance. On such a view, a state of fear (say) might be nothing other than a bodily sensation (say), but have intentional content by virtue of standing in some causal relation with what it represents or is about. On such a view, the representational properties of an emotion are not of the nature of emotion, the emotion itself being nothing more than a bodily feeling or sensation.

Now, of course, even supposing that the intentional properties of a mental state are extrinsic in this sense, this would provide us with reason only to question whether emotion lacks such properties. It would not be reason to reject the idea that attending to the phenomenology is crucial for getting to know the nature of emotion or emotions of different types, which pertains to those properties that are intrinsic to emotion. But, moreover, there is good reason to think that intentional properties are felt properties in fact, and, therefore, would be evident to us in the experience of emotion if emotions are truly intentional. Indeed, introspection makes compelling the idea that the intentional properties of a bona-fide representational mental state are phenomenally manifest in the mental state. For example, the representing of Paris as being the capital of France is a property that is manifest in the thought that Paris is the capital of France, and the presentation of a red object in front of me is very much part of the phenomenology of a visual experience of a red object in front of me (see Chalmers 2004; Horgan and Tienson 2002; Kriegel 2012; Whiting 2012; Whiting 2016). If it is the case that intentional properties are phenomenally manifest properties, then emotions will lack intentional properties if emotions do not phenomenally manifest such properties.

But, is it really the case that emotions fail to phenomenally manifest intentional properties? Here, various reasons might be given for answering in the negative. To begin with, one might appeal to a form of 'strong representationalism,' according to which the phenomenal character of a mental state is identical to the representational content of the mental state (e.g. Tye 2000, 2008).

The worry with this way of arguing, however, is that it risks begging the question. If emotions do not manifest intentional properties, then surely the right thing to say is that emotion's phenomenal character *cannot* be identical to emotion's representational content, as emotion would then seem to lack any such content.

A more promising view, perhaps – and one that might be congenial to some strong representationalists as well – would be to hold that even if emotions are not experienced as being directed at things outside the body, nevertheless emotions are experienced as being about the body or its parts. That might be the view of someone who thinks that emotions are sensations or perceptions of bodily change or activity (see, for instance, James 1888; Prinz 2004; Tye 2000, 2008). And arguably, this is the most promising line for someone to adopt, given the close relationship that clearly exists between emotion and the body. That being said, however, an opposing view would be to maintain that although we do indeed experience emotions as taking place or being instantiated in the body, that isn't the same as saying that emotions are experiences of the body.

A final reason to doubt that emotions fail to manifest an intentional character is more concessionary, insofar as it aims to show only that some emotions manifest such a character. For what about so-called 'higher-cognitive' emotions such as pride and guilt, which might obviously seem to involve intentional properties? Thus, isn't pride bound up with ideas of the self and guilt bound up with ideas of personal wrongdoing? Arguably, pride and guilt are bound up with such ideas, but what are we to infer from this? To begin with, we might accept that pride and guilt manifest an intentional character, but disagree that this shows that emotions have such a character. For, perhaps, it shows only that pride and guilt are compound states, comprising non-intentional emotions and mental representations. Alternatively, we could accept that states such as pride and guilt are emotions and nothing more, but deny that attention to the phenomenology shows they have the intentional properties they might be claimed to have. For, perhaps, guilt manifests as a distinctive anxiety-like sensation (say), but qualifies as 'guilt' by virtue of being triggered by thoughts of personal wrongdoing, and, perhaps, pride manifests as a pleasurable feeling (say) but qualifies as 'pride' in virtue of being triggered by thoughts of the self. This would be to allow that pride and guilt are bound up with certain thoughts (for without the triggering thoughts the feelings would not qualify as pride and guilt), while denying that pride and guilt comprise those thoughts, the emotions themselves being objectless feeling states.

3 Unconscious Emotions

Barring some notable exceptions (e.g. Damasio 1999; Prinz 2005a; Winkielman et al. 2005), the idea that emotions are always conscious or felt is a view that has been held by a great many psychologists and philosophers of emotion (e.g. James 1884; Clore 1994; Panksepp 2005; Hatzimoysis 2007; Maiese 2011; Whiting 2011; Deonna and Teroni 2012). Even Freud – who is often credited with showing us the unconscious – can be attributed the view that emotions are felt necessarily and that there are no such things as unconscious emotions (e.g. Freud 1950). But is the view that emotions are always conscious a correct one? There are two sorts of cases that might be offered as counterexamples to that view. First, there are those cases involving emotions sometimes considered to have dispositional natures, such as a fear of heights. Second, there are those cases involving emotions that have occurrent, or episodic natures, but which are held by some emotion theorists to be unconscious. Let us consider each in turn.

Suppose Sebastian has a fear of heights, Joan has loved her partner for many years, and Robert has been angry all day for not getting an expected pay rise. Now, it seems that we can assign these emotions to Sebastian, Joan, and Robert; even when they are not undergoing occurrent fear, or anger, or amorous affect. Thus, we can attribute a fear of heights to Sebastian and love of one's partner to Joan, even when Sebastian and Joan are dreamlessly sleeping, for instance. And, throughout the day we can attribute to Robert an anger for not getting an expected pay rise, including at times in the day when Robert is not *feeling* angry (for instance, when he is working or otherwise preoccupied and not thinking about the pay rise that never materialized).

For this reason, some philosophers have taken these emotional states to have dispositional or non-episodic natures (see, for instance, Lyons 1985; Prinz 2003; Prinz 2004; Goldie 2010). For instance, Sebastian's fear of heights might be considered to be a disposition to undergo occurrent states of fear in the perceived presence of heights; Joan's love for her partner might be taken to be a disposition to undergo occurrent positive or amorous emotions when perceiving or thinking about her partner; and Robert's anger for not getting an expected pay rise might be taken to comprise a tendency throughout the day to get worked up when thinking about the pay rise that failed to materialize (see e.g. Prinz 2003; Naar 2013). Unlike occurrent mental states, which tend to be short-lived and possess a characteristic phenomenology, emotional states such as a fear of heights, love for one's partner, and anger for not getting a pay rise, might seem to be longstanding or persistent mental states that need have no phenomenology of their own. There is nothing that it is like to be afraid of heights or to love one's partner when dreamlessly sleeping, for example. But that might look to be tantamount to saying that these dispositional or longstanding states are emotions that are not conscious.

However, the idea that emotions can be both episodic mental states and the dispositions to undergo such episodic mental states commits us to an unhappy metaphysics. Indeed, it would be to hold that a given emotion type – fear or love or anger, for instance – can have two natures or be more than one thing, metaphysically-speaking, and that looks to be the sort of view we should try to avoid adopting if possible. Therefore, unless we wish to reject the idea that there are episodic or occurrent emotions, we need a way of treating longstanding emotions that acknowledges there are episodic emotions only. And here there are two strategies available to us.

The first is to insist that mental states such as a fear of heights are *emotional dispositions*, not dispositional emotions. Thus, although to fear heights is to be disposed to undergo episodic emotions in the perceived presence of heights, that amounts to saying that a fear of heights is the disposition to undergo emotion, and not an emotion itself (see e.g. Deonna and Teroni 2012). This strategy accepts that mental states, such as a fear of heights or love of one's partner, have dispositional natures, but denies this entails some emotions are dispositional, since the mental states in question are not emotions.

An alternative strategy, however, is to say that such mental states are emotions because it turns out they are episodic emotions and not the dispositions to undergo those emotions as originally supposed (see Whiting, unpublished). This strategy requires us to distinguish between the *emotion* and the *having of the emotion*, and holds that it is the having of the emotion that has a dispositional nature, not the emotion itself. And here like examples are not difficult to find. For instance, many birds have alarm calls that they emit in situations of danger. Moreover, it is true to say of birds that they have alarm calls when they are sleeping and not emitting those calls. Are we to conclude that the alarm calls of birds are the dispositions to emit certain calls or sounds? The answer is negative. Although to have an alarm call might be a disposition to emit certain calls when encountering danger, the calls themselves have episodic, not dispositional, natures.

And likewise, the argument goes, we can agree that what it is to *be* in love with one's partner, or to *be* afraid of heights, or to *be* angry all day for not getting a pay rise, is to be such so as to undergo certain episodic emotions when particular circumstances obtain. So, for instance, if throughout the day Robert doesn't get worked up when reflecting on his failure to get an expected pay rise, then it would seem wrong to describe Robert as being angry that day for failing to get an expected pay rise. However, that is not to say that the emotions themselves have dispositional natures. Indeed, according to the present way of treating these emotional states, the emotions themselves are to be viewed as discrete mental episodes, albeit discrete mental episodes that are undergone when certain circumstances obtain. For instance, Robert's anger for not getting an expected pay rise is to be identified with the angry feelings that Robert undergoes throughout the day when reflecting on his failure to get a pay rise.

Now, regardless of whether we opt for the first strategy or the second strategy just outlined, the important thing to note is that both strategies would entail that mental states, such as a fear of heights or love of one's partner, fail to be counterexamples to the idea that emotions are always conscious. If we adopt the first strategy, then these mental states turn out to be other than emotions. On the other hand, if we adopt the second strategy, then although these mental states turn out to be emotions, they are emotions only because they are episodic, not dispositional, mental states, and we will still lack positive reason for thinking the emotions in question can be anything other than conscious.

The second group of cases that might be offered as counterexamples to the idea that emotions are always conscious might seem to pose a greater threat to that idea, because they pertain to emotional states that almost everyone will from the beginning agree have episodic, not dispositional, natures. Thus, a number of writers have drawn on findings in empirical psychology and elsewhere in support of the view there are unconscious emotions of an episodic kind (see e.g. Kihlstrom et al. 2000; Winkileman et al. 2005; Prinz 2005a; Prinz 2005b; Lane 2007; Smith and Lane 2016). However, as we shall see, it is far from clear the evidence in question does show this.

To begin with, then, some of the findings support only the view that the causes of emotion are not always consciously felt. Consider a study by Robert Zajonc, which found that repeated exposure to subliminal stimuli influenced people's preferences (Zajonc 1980; see also Öhman and Soares 1994). Now, although the findings support the idea that the elicitors of emotion can lie outside conscious awareness, they don't show that emotion itself can be unconscious.

And in other cases, the evidence shows only that people can mislabel or misidentify emotion (see also Deonna and Teroni 2012: 16–17). For instance, Smith and Lane describe the case of a bereaved husband who originally thinks he is angry with what he perceives to be the unfairness of life, but comes to realize he is angry with his spouse for dying (Smith and Lane 2016). But, even supposing that this is a case of misidentified emotion, and not merely a failure to recognize the real cause of an emotion, such a case as this also fails to show that emotions can be unconscious. This is because an emotion that is mislabelled or misidentified is not the same as an emotion that fails to have a characteristic phenomenology.

More puzzling, perhaps, are those cases where there is reason to think emotions have been undergone but which involve people who fail to consciously report on their emotions (Winkielman et al. 2005; Winkielman et al. 2007). Such cases concern the emotions themselves, and not the causes of the emotion, and involve not misidentification, but wholesale unawareness of emotion. Consider a much-discussed experiment, in which participants were subliminally shown happy, angry or neutral faces depending on the group to which they had been assigned (Winkielman et al. 2005). Participants were then asked to rate the pleasantness of a fruit beverage. It was found that participants who had been shown happy faces rated the pleasantness of the beverage higher than those shown angry or neutral faces. It was also found that participants shown happy faces consumed larger amounts of the beverage than those shown angry or neutral faces. Strikingly, however, participants reported no differences in how they felt. For this reason, the study has been taken to show that emotions can be unconscious. The behaviors of participants evidence there being emotions explaining differences in behaviors, but these emotions were not detected by participants (Winkielman et al. 2005; Prinz 2005b; Smith and Lane 2016).

Now, it can be questioned whether the Winkielman et al. study succeeds in showing that participants failed to report accurately on their emotions. As just explained, the justification for thinking that participants underwent changes in emotion is that this is the most plausible explanation for differences in behaviors observed. Winkielman et al. consider one alternative explanation, namely that participants might have cognitively appraised their situations differently when presented with subliminal stimuli, and these cognitive appraisals explained differences in behavior (Winkielman et al. 2005: 133). However, it is unclear whether this is the only possible alternative explanation available. For instance, perhaps the subliminal stimuli were themselves directly responsible for differences in behavior, as might be the case if certain subliminal stimuli have the power to influence or modulate how people are disposed to behave. This would make the receiving of subliminal stimuli, and not an emotion or cognitive appraisal, responsible for the differences in behaviors, and would be consistent with supposing that participants reported on their emotions accurately.

But suppose the explanation that Winkielman et al. give for study participants' behaviors is correct. Does the fact that participants failed to be able to report on their emotions establish the truth of the emotion without consciousness view? It is difficult to see how, since there is nothing to the claim that emotions have a characteristic feel that rules out the idea that people can sometimes fail to register or reflect on their emotions (cp. Lambie and Marcel 2002; Maiese 2011; Deonna and Teroni 2012). After all, few would deny there is something that it is like for young children to undergo emotion, but it would surely be stretching things to say that young children are able to reflect on and form beliefs about their emotions.

Another way to make the point is by distinguishing between two ways in which talk of conscious awareness of emotion might be taken. On the one hand, when we say someone is consciously aware of an emotion we might mean the person is experiencing an emotion. Here conscious awareness is the sort of awareness in virtue of which there is something that it is like to undergo an emotion, or in virtue of which an emotion has a phenomenology. After all, an emotion can have a characteristic phenomenology only insofar as the emotion is experienced or felt by us. But this sense of 'conscious awareness' is to be distinguished from another use of that term, where conscious awareness is a matter of consciously reflecting on, or forming certain beliefs about, an emotion, for instance the belief that one occupies the emotion in question. Here conscious awareness is knowledge of undergoing a certain emotion, a state of fear, for instance.

And therein lies the problem. *Pace* Winkielman et al. why not think that the study results show only that there can be changes in emotion that people can fail to consciously reflect on or form certain beliefs about? Indeed, there are a number of reasons why study participants might have been unable to reflect on changes in felt emotion. For instance, the changes might have been too subtle and/or short-lived to be registered or reflected on (which is consistent with supposing that such changes might nevertheless have had significant effects on behavior, note). And although the authors of the study acknowledge this kind of response to their argument (Winkielman et al. 2005: 132), they say little to remove or mitigate the worry.

There is reason to think the evidence just discussed does not establish the existence of unconscious emotions. Although the evidence might succeed in showing that people can fail to consciously register or reflect on their emotions, there is reason to doubt that the evidence shows there can be emotions that are unfelt or which fail to have a characteristic feel. Still, does the possibility remain that considerations may yet be presented demonstrating that emotions can be unconscious? It is unclear how we can rule out that possibility altogether, I suppose. Nevertheless, one reason for thinking that emotions are always conscious has been alluded to already in this chapter. For suppose we are able to see from our own case that an emotion of a certain type is constituted by its characteristic feel. For instance, suppose we can see by means of introspection that an episode of fear is indistinguishable from its unpleasant edgy feel. But, if that is the case, and an emotion of a certain type is indistinguishable from its characteristic feel, then how can an emotion of that type fail to be conscious, or fail to feel a certain way? To think the emotion in question can be unconscious would be to think that an emotion of that type does and doesn't comprise its phenomenology, which isn't a very cogent position to hold. In other words, then, if emotions are constituted by their characteristic feel, then emotions cannot be unconscious, and the task of seeking to show otherwise will indeed be forever a hopeless one.

Should we conclude that emotions are always conscious, then? Again, only if emotions are constituted by their characteristic feel. Earlier in the chapter, conceptual and empirical considerations were given in support of what I called the 'constitution view.' I will leave it to the reader to judge whether they succeed or not. Certainly, they are likely to need developing. But needless to say, if those considerations fail, then this will be either because the constitution view is not supported by our concept of emotion, or because the constitution view is not supported by our observations of emotion. Thus, if it turns out that our concept of emotion does not support the idea that emotions are constituted by how they feel – either because our concept of emotion supports an alternative view, or because conceptual considerations are unable to establish deep metaphysical truths (thus, perhaps such considerations only ever tell us about our *idea* of emotion) – then the constitution view cannot be endorsed on conceptual grounds. Similarly, if it turns out that introspection fails to support the idea that emotions are constituted by how they feel – either because it supports an alternative view, or because introspection is unable to establish deep metaphysical truths (see, for instance, Schwitzgebel 2008; Schwitzgebel 2011) – then we will lack reason rooted in first-person experience for accepting the constitution view.

For my part, I have defended elsewhere the claim that introspection can deliver us truths about the nature of conscious mental states, including our emotions (Whiting 2016). And as regards the question of whether introspection might support instead the view that emotions don't comprise their phenomenology, again my thinking is that although it is clear that when we observe cuts and bruises (say), we see there to be a gap between the things observed and how they feel to us, we find no such gap in the case of emotions and how they feel to us. And it is for this reason that my money is on there being no such things as unconscious emotions.[1]

Note

1 I am very grateful to Paul Gilbert and Rocco Gennaro for helpful comments on an earlier draft.

References

Chalmers, D. (1996) *The Conscious Mind: In Search of a Fundamental Theory*, New York and Oxford: Oxford University Press.

Chalmers, D. (2004) "The Representational Character of Experience," in B. Leiter (ed.) *The Future for Philosophy*, Oxford: Oxford University Press.

Clore, G. (1994) "Why Emotions Are Never Unconscious," in P. Ekman and R. Davidson (eds.) *The Nature of Emotion: Fundamental Questions*, New York: Oxford University Press.

Damasio, A. (1999) *The Feeling of What Happens: Body and Emotion in the Making of Consciousness*, New York: Harcourt Brace.

Deonna, J., and Teroni, F. (2012) *The Emotions. A Philosophical Introduction*, Oxon: Routledge.

Döring, S. (2007) "Seeing What to Do: Affective Perception and Rational Motivation," *Dialectica* 61: 363–394.

Freud, S. (1950) *Collected Papers* (J. Riviere, Trans. Vol. 4), London: Hogarth Press and The Institute of Psychoanalysis.

Garfinkel, S., and Liberzon, I. (2009) "Neurobiology of PTSD: A Review of Neuroimaging Finding," *Psychiatric Annals* 39: 370–381.

Goldie, P. (2010) "Love for a Reason," *Emotion Review* 2: 61–67.
Hatzimoysis, A. (2007) "The Case Against Unconscious Emotions," *Analysis* 67: 292–299.
Horgan, T., and Tienson, J. (2002) "The Intentionality of Phenomenology and the Phenomenology of Intentionality," in D. Chalmers (ed.) *Philosophy of Mind: Classical and Contemporary Readings*, Oxford: Oxford University Press.
Hume, D. (1969) *A Treatise of Human Nature*, London: Penguin.
James, W. (1884) "What Is an Emotion?" *Mind* 9: 188–205.
Kihlstrom, J., Mulvaney, B., and Tobias, T. (2000) "The Emotional Unconscious," in E. Eich, J. Kihlstrom, G. Bower, J. Forgas, and P. Niedenthal (eds.) *Cognition and Emotion*, New York: Oxford University Press.
Kriegel, U. (2012) "Towards a New Feeling Theory of Emotion," *European Journal of Philosophy* 3: 420–442.
Lambie, J., and Marcel, A. (2002) "Consciousness and the Varieties of Emotion Experience: A Theoretical Framework," *Psychological Review* 109: 219–259.
Lane, R. (2007) "Neural Substrates of Implicit and Explicit Emotional Processes: A Unifying Framework for Psychosomatic Medicine," *Psychosomatic Medicine* 70: 214–231.
LeDoux, J. (2000) "Emotion Circuits in the Brain," *Annual Review of Neuroscience* 23: 155–158.
Levenson, R., Ekman, P., Heider, K., and Friesen, W. (1992) "Emotion and Autonomic Nervous System Activity in the Minangkabau of West Sumatra," *Journal of Personality and Social Psychology* 62: 972–988.
Levenson, R. (2003) "Blood, Sweat, and Fears: The Autonomic Architecture of Emotion," *Annals of the New York Academy of Sciences* 1000: 348–366.
Lutz, A. (2015) "The Phenomenal Character of Emotional Experience: A Look at Perception Theory," *Dialectica* 69: 313–334.
Lyons, W. (1985) *Emotion*, Cambridge: Cambridge University Press.
Maiese, M. (2011) *Embodiment, Emotion, and Cognition*, London: Palgrave MacMillan.
Montague, M. (2009) "The Logic, Intentionality, and Phenomenology of Emotion," *Philosophical Studies* 145: 171–192.
Naar, H. (2013) "A Dispositional Theory of Love," *Pacific Philosophical Quarterly* 94: 342–357.
Nagel, T. (1974) "What Is It Like to Be a Bat?" *Philosophical Review* 83: 435–450.
Öhman, A., and Soares, J. (1994) "'Unconscious Anxiety': Phobic Responses to Masked Stimuli," *Journal of Abnormal Psychology* 103: 231–240.
Panksepp, J. (2005) "Affective Consciousness: Core Emotional Feelings in Animals and Humans," *Consciousness and Cognition* 14: 30–80.
Poellner, P. (2016) "Phenomenology and the Perceptual Model of Emotion," *Proceedings of the Aristotelian Society* 116: 261–288.
Prinz, J. (2003) "Emotions, Psychosemantics, and Embodied Appraisals," in A. Hatzimoysis (ed.) *Philosophy and the Emotions*, Cambridge: Cambridge University Press.
Prinz, J. (2004) *Gut Reactions: A Perceptual Theory of Emotion*, Oxford: Oxford University Press.
Prinz, J. (2005a) "Are Emotions Feelings?" *Journal of Consciousness Studies* 12: 9–25.
Prinz, J. (2005b) "Emotions, Embodiment, and Awareness," in L. Barrett, P. Niedenthal, and P. Winkielman (eds.) *Emotion and Consciousness*, London: Guildford Press.
Solomon, R. (1993) *The Passions: Emotions and the Meaning of Life*, Indianapolis, IN: Hackett Publishing Company.
Schwitzgebel, E. (2008) "The Unreliability of Naïve Introspection," *Philosophical Review* 117: 255–273.
Schwitzgebel, E. (2011) *Perplexities of Consciousness*, Cambridge, MA: MIT Press.
Smith, S., and Lane, R. (2016) "Unconscious Emotion: A Cognitive Neuroscientific Perspective," *Neuroscience and Biobehavioral Reviews* 69: 216–38.
Tye, M. (2000) *Consciousness, Color, and Content*, Cambridge, MA: MIT Press.
Tye, M. (2008) "The Experience of Emotion: An Intentionalist Theory," *Revue Internationale de Philosophie* 243: 25–50.
Whiting, D. (2011) "The Feeling Theory of Emotion and the Object-Directed Emotions," *European Journal of Philosophy* 19: 281–301.
Whiting, D. (2012) "Are Emotions Perceptual Experiences of Value?" *Ratio* 25 93–107.
Whiting, D. (2016) "On the Appearance and Reality of Mind," *Journal of Mind and Behavior* 37: 47–70.
Whiting, D. (unpublished) "The Myth of Dispositional Emotions," Copy available on request.
Winkielman, P., Berridge, K., and Wilbarger, J. (2005) "Emotion, Behavior, and Conscious Experience: Once More without Feeling," in L. Barrett, P. Niedenthal, and P. Winkielman (eds.) *Emotion and Consciousness*, London: Guildford Press.

Winkielman, P., Knutson, B., Paulus, M., and Trujillo, J. (2007) "Affective Influence on Decisions: Moving Towards the Core Mechanisms," *Review of General Psychology* 11: 179–192.
Zajonc, R. (1980) "Feeling and Thinking: Preferences Need No Inferences," *American Psychologist* 35: 151–175.

Related Topics

Dualism
Representational Theories of Consciousness
Consciousness and Intentionality

24
MULTISENSORY CONSCIOUSNESS AND SYNESTHESIA

Berit Brogaard and Elijah Chudnoff

1 Introduction

Suppose you hear your colleague Magdalena speak with someone in the hallway while you are reading a paper on your computer in your office. In the envisaged scenario, you have an auditory experience of the sounds coming from Magdalena's mouth and a visual experience of the graphemes on your computer screen. These two experiences are constituent parts of the total sensory experience you currently have. They are not integrated in any substantial sense. They merely co-exist as constituents of your total experience. That is, the two states are not integrated in a way more substantial than the way any two co-conscious states are integrated into a total experience at a time.

Suppose instead that you are having a conversation with Magdalena. In this case you see her lips move and you hear the sounds that come from them. We can literally say that you *see Magdalena talk*. The integration of your two experiences in this second scenario is different from the mere co-presence of your two experiences in the first. In the first scenario your experiences co-exist as part of your total experience. In the second scenario, your experiences are bound together. How do we account for the difference between the two cases?[1]

Casey O'Callaghan (2008, 2012, 2014, 2015; see also Dainton 2000; Nudds 2001; Bayne 2014; Deroy 2014; de Vignemont 2014a; Briscoe 2016 in press; Bourget 2017) argues that the difference between the first and the second scenario is that in the first scenario the phenomenology of the result of binding your separate experiences can be fully accounted for by appeal to the phenomenology of the individual sensory modalities but that this is not so in the second case. In the second case, he argues, the overall phenomenology reflects that the two experiences are bound together amodally in perceptual faculties that are neither auditory nor visual in nature—for instance, in higher non-sensory regions of the brain, such as the parietal cortex.[2]

In this chapter, we provide an argument for thinking that we can account for the difference in phenomenology between the two cases by appeal to the phenomenology of the individual sensory modalities. We argue that the phenomenology of one type of multisensory experience that goes beyond mere co-consciousness derives exclusively from the individual sensory modalities (for some empirical considerations in favor of a third type of multisensory experience, largely presented by speech perception, see Tuomainen et al. 2005). Call this type of experience "modal multisensory experience."[3] We then argue that another kind of normal multisensory experience that goes beyond mere co-consciousness requires a different treatment. The second type of experience is one where the phenomenology is distinctively multisensory and perceptual, yet amodally integrated. Call this type of experience "amodal sensory experience." This appears to be the kind of multisensory experience O'Callaghan (2012) has in mind. When you perceptually attribute both of the features, *having a coffee look* and *having a coffee smell,* to the dark liquid in your mug, the phenomenology of your experience reflects this type of integration. In the final section of the chapter, we look at the case of synesthesia—a type of atypical integration in which different sensory streams are bound together in unusual ways, for instance, sounds may be bound together with color. We argue that some forms of synesthesia may be helpful in investigating the neural mechanism underlying amodal multisensory binding.

The plan is as follows: In Section 2, we provide an account that draws a distinction among mere co-consciousness and modal and amodal multisensory experience. On this account, modal multisensory experience has a phenomenology that derives from the individual senses and hence lacks the amodal component of multisensory experiences such as that of seeing and holding a tomato. In Section 3, we give reasons for thinking that some cases of integration should be conceived of as instances of modal rather than amodal integration. In Section 4, we compare certain types of synesthesia to amodal multisensory perception and argue that these types of synesthesia may shed light on amodal integration.

2 Modal versus Amodal Integration

The case in which you see Magdalena speak clearly differs from a case in which you have a unified experience visually representing graphemes on your computer screen and auditorily representing Magdalena's voice in the hallway.[4] In the first case you seem to see the event that produces the sound, viz. Magdalena's moving lips. Experiences of this kind are quite different from experiences that attribute features perceived in different sensory modalities to an object, as in the case of visually attributing *being coffee* to the dark liquid in the mug and olfactorily attributing coffee smell to that same liquid; or perceiving the flavor of the Oxtail flatbread by gustatorily, olfactorily, somato-sensorily, thermally and nociceptually attributing features to the flatbread. Although you attribute the smell of coffee to the coffee in front of you, it's not as if you see the coffee smell in any way analogous to the way you hear someone speak.

The case of seeing someone speak should thus be set apart from a case of multisensory experience that is merely about or directed at the same perceptible object or feature, such as a case in which you both hold and see a firm ripe tomato, or see and smell coffee.[5] When you hold and see a tomato, a shape, viz. the common sensible *roundness*, is both seen and felt. When you see and smell coffee, the attributes *being coffee* and *having coffee smell* are both perceptually attributed to the coffee, one by the visual modality and the other by the olfactory modality. In both cases, you are perceptually attributing features to one and the same object but the integration of these attributions into a complete experience cannot be accounted for by appeal to individual sensory modalities. Rather: the integration appears to be amodal: it occurs independently of the mechanisms of the individual sensory modalities.[6] This type of multisensory experience thus seems to have the characteristic that O'Callaghan (2012, 2014, 2015) thinks multisensory experience has. He calls this type of binding "amodal integration." Integration is also closely related to what Tim Bayne and David Chalmers (2003) call "objectual unity."

We will look at the details of the arguments for thinking that the two types come apart below (e.g., seeing someone speak versus seeing and smelling coffee). Suffice it to say at this point that *if* the cases come apart in that the phenomenology of the former (e.g., seeing someone speak) derives fully from the individual senses whereas the phenomenology of the latter (e.g., holding a tomato you also see) does not, then O'Callaghan's amodal view cannot be construed as a general view of multisensory perception. As noted in the previous section, although O'Callaghan does not argue that amodal unification is the only sort of integration that goes beyond mere co-consciousness, he does not distinguish between modal and amodal unification.

The conditional claim made in the previous paragraph raises an interesting question. *If* the phenomenology in the first type of case (e.g., seeing someone speak) derives from the individual senses (viz., from vision and audition), how do we distinguish this type of case from the second type (e.g., seeing and feeling the roundness of the tomato)?

The solution to this problem, we will now argue, is to reconceive of what is actually perceived by the individual senses in the first type of case when, say, we hear a source produce a sound. Our suggestion is that when we perceive, say, sound being produced by a source, the auditory experience attributes audible qualities to an object picked out by a perceptual demonstrative whose reference is anchored to an object, by virtue of that object being visible. For example, the auditory experience attributes sounding like such and such to a lip-moving event picked out by a perceptual demonstrative that refers to the event, by virtue of its presence in vision.

A visual demonstrative is the perceptual equivalent of demonstrative terms that occur in ordinary language, such as "this" and "that." Demonstratives are referential terms that have a referent only when accompanied by a demonstration that successfully picks out an entity or a previously mentioned referent. A demonstration is, for example, a gesture, a glance or a nod in a particular direction or a speaker intention comprehensible by the hearer in the conversational context. When a demonstrative refers back to a previously mentioned referent, as in "John continually scratched his skull. This annoyed Anna," this is also known as "anaphora." In the example we just provided, the anaphoric pronoun "this" refers back to the event *John's scratching of his skull*. In anaphora, the referents of anaphoric pronouns (the anaphor) depend on the referents of the bit of language they are anaphoric on, i.e. the antecedent (or the postcedent in the case of anaphora, such as "It was her own fault that Jamie didn't get to go to the prom."). As we will see, some perceptual demonstratives function in a way analogous to anaphoric pronouns. Perceptual references to objects in different sensory modalities can thus be interdependent in the way that certain linguistic references to objects in different parts of speech are interdependent.

In the case of seeing someone speak, the visual experience provides a visual demonstrative that picks out a speaking or lip-moving event, and the auditory experience attributes audible qualities to it by using the visual demonstrative. By using a visual demonstrative an auditory experience can become dependent on and not just co-conscious with the visual experience.[7]

It may be thought that seeing sound-events is the only example of multisensory experience in which perceptual unification takes place as a result of demonstrative reference being made by one sense and anchored by another. This, however, does not seem to be the case. Suppose you are lifting weights, holding one weight in your right hand. As you bend your arm, the tactile feel of the weight, together with the feeling of how heavy the weight is, attributes qualities to a demonstrative provided by the visual experience of the lifting event. That is, the feeling of exercising effort in lifting a weight consists in tactually and proprioceptively attributing qualities to a seen event, namely the lifting.

Tactile experience itself may very well be multisensory in this sense (Brogaard 2012; de Vignemont and Massin 2015; Briscoe 2016; see Fulkerson 2014 for challenges to the mainstream view that haptic touch is multisensory). Tactile experiences can reasonably be thought to involve not just representations of properties of objects but also properties of the body (Brogaard 2012; Briscoe 2016).[8] Plausibly, you cannot have a tactile experience as of an object being hard without experiencing pressure to the part of your body that does the haptic touching. If you feel a rock press against the palm of your hand, we can take you to have an experience of the palm responding to the hardness of the rock, or alternatively we can take you to have an experience of the hardness of the rock producing a particular sensation in your hand. One aspect of touch thus anchors a tactile demonstrative reference to the rock. Another aspect of touch attributes causing certain bodily sensations in me. And that is what constitutes felt pressure. So, if as some research literature on touch suggests (for discussion see Loomis and Lederman 1986; Jones and Lederman 2006; Fulkerson 2011; Gallace and Spence 2014; Linden 2015), the two aspects of touch involve two different sensory modalities, then this is a case of modal multisensory integration.

Further: on the assumption that emotions are multisensory experiences, it can be argued that they are also integrated by means of perceptual reference. Suppose you fear a particular tiger that bares her sharp teeth at you. Your bodily sensations (e.g., sensations of a quickened heartbeat, sweaty palms and shaky legs) are a response to the tiger's fearfulness (Brogaard 2012; Brogaard and Chudnoff 2016). Your being afraid of a seen tiger consists of attributing the property of causing bodily sensations indicating a threat to your well-being to the seen tiger. Vision allows you to refer to the tiger, and the sensations allow you to attribute properties such as causing sensations indicating threat to your well-being. The overall fearful response just is the act of attributing the properties introduced by the bodily sensation to the object introduced by vision. This type of integration can be cashed out as follows: your visual experience identifies a visual event, viz. the tiger baring her teeth, and the bodily sensation attributes certain qualities, such as causing various events felt in your body, to the visually identified event.

For the case of visual-auditory binding, we can capture the distinctions among co-consciousness, modal integration, and amodal integration as follows:

> *Co-Consciousness*
> Your overall experience has the content: that_v is F and that_h is G [where that_v is a visual demonstrative, F is a visible quality, that_h is an auditory demonstrative reference, and G is an audible quality].
>
> *Modal Integration*
> Your overall experience has the content: that_v is F and that_v is G [where that_v is a visual demonstrative, F is a visible quality, G is an audible quality].
>
> *Amodal Integration*
> Your overall experience has the content: that_v is F and that_h is G and $\text{that}_v = \text{that}_h$ [where that_v is a visual demonstrative, F is a visible quality, that_h is an auditory demonstrative, G is an audible quality, and $\text{that}_v = \text{that}_h$ is an amodally represented identification].

One might wonder how an experience can attribute an audible quality to the referent of a visual demonstrative, thinking, perhaps, that an experience can only attribute audible qualities to referents that are picked out in an auditory manner, say, by an auditory demonstrative or a description built out of audible qualities. However, as noted above in the discussion of anaphora, here we are simply extending a familiar form of representational dependence to representations in different sensory modalities. The familiar form of representational dependence in which one act of reference depends on another act of reference. As noted above, this is a common phenomenon in linguistic representation, for instance, in anaphora and in communication across people. For example, you refer to something because your friend referred to it in her speech.

The phenomenon is also common in mental representation. Say you see a chess piece in a certain position on a chessboard. Then you close your eyes and think about or imagine moving it to another position on the board. Your cognitive or imaginative reference to that particular piece depends on your visual reference to it. It is because you saw *that* piece that your thought or imaginings are about it and not something else.

Before proceeding to our argument for the distinction between modal and amodal multisensory experience, let us consider some potential challenges to this account.

One might argue that vision and audition have different manners of representation (Chalmers 2004): vision represents visually, whereas audition represents auditorily. But multisensory experience does not represent visually or auditorily. It represents amodally (Bourget 2017). So, the phenomenology of modal multisensory experience is not wholly derived from the phenomenology of the individual sensory modalities. Or so the argument goes.

This argument can be resisted, however. Rather than saying that manners of representation change from visual to amodal when the sound is added, it is perfectly plausible to take manners of representation to be additive. When you hear someone speak, your experience represents in a visuo-auditory manner.

A further challenge to the proposed account is that of explaining where in the brain binding takes place if indeed its phenomenology is fully derived from the phenomenology associated with the individual sensory modalities. This challenge can be met. We know from the McGurk effect that seeing lip movements can influence and alter what we hear. The McGurk effect arises when auditory speech cues are presented in synchrony with incongruent visual speech cues (McGurk and MacDonald 1976). For example, when the auditory syllable "ba" is presented in synchrony with a speaker mouthing "ga," subjects typically report hearing "da." We also know from the double-flash illusion that auditory input sometimes influences what we see. The double-flash illusion occurs when the presentation of two brief auditory beeps makes a single flash look like two flashes (Shams et al. 2000). So, just considering visuo-auditory cases for now, the answer to the question of where in the brain this type of binding takes place is likely that it sometimes occurs in the auditory cortex and sometimes in the visual cortex. Whether the integration occurs in visual or auditory areas is likely to depend on what is taken to produce what. When seen lip movements are taken to produce sound in the McGurk illusion, it is likely that the binding takes place in the auditory cortex. When the beeps are taken to produce the flashes in the double-flash illusion, the binding likely takes place in the visual cortex.

A third worry one might have about our proposed account is that it implies that multisensory integration is perceptual. But, it may be argued that multisensory integration is associative or inferential rather than perceptual. This has indeed been the traditional view of multisensory perception (see e.g. Bloom and Lazeron 1988). However, there are numerous empirical considerations in favor of the view that multisensory experience typically is genuinely perceptual and not e.g. associative (for an overview of empirical considerations, see e.g. Giard and Péronnet 1999; Molholm et al. 2002; Klemen and Chambers 2011; Talsma 2015).

Here are two philosophical considerations in favor of the view that multisensory experience is perceptual rather than, say, associative. Ordinary visual experiences that result from stimulation of the individual senses, such as your visual experience of the line drawing of a rectangle in Figure 24.1, possess two interesting characteristics.

Figure 24.1 Line Drawing of a Rectangle. Every part of your experience of the line drawing has presentational phenomenology

One is that your experience does not just represent something as being the case, but is also felt as putting you in touch with its subject matter (see Chudnoff 2014, 2016 for a discussion of this characteristic). One way to understand the idea of subject matter is in terms of truthmakers, where the truthmaker of an experience can be understood as the external mind-independent object in virtue of whose existence or non-existence the content of the experience is true or false (cf. Armstrong 1989: 88). So, here the truthmaker of your experience is the drawing of the rectangle. It is as if your experience makes you directly aware of the drawing of the rectangle. We'll call this characteristic "presentational phenomenology."

In order for an experience to have presentational phenomenology, it is not necessary that we appear to see all aspects of what is presented to us. Consider the following case. You walk down the hallway and see a dog, partially occluded from your field of vision (Figure 24.2).

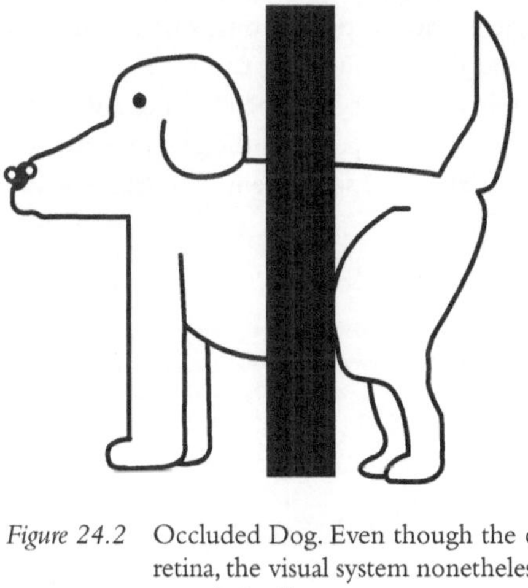

Figure 24.2 Occluded Dog. Even though the occluded parts of the dog do not make an imprint on the retina, the visual system nonetheless generates a complete dog. This is also known as "amodal completion"

In spite of the fact that only the non-occluded parts of the dog reflect light that reaches your retina, it appears to you as if there is a whole dog, not merely a part of a dog.[9] So, your experience of the dog has presentational phenomenology.

Another characteristic of ordinary visual experience is that it is evidence insensitive (understood as a feature of the phenomenology; see Brogaard 2016, in press a, for a discussion of this characteristic). Consider the Müller-Lyer illusion in Figure 24.3 (the figure on the left).

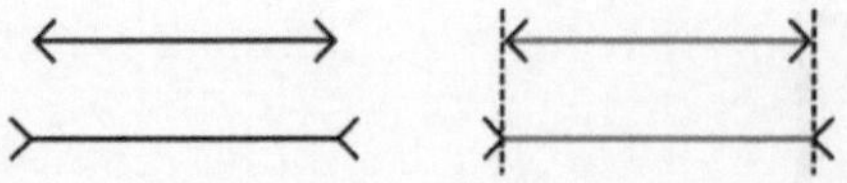

Figure 24.3 The Müller-Lyer Illusion. Even when you learn that the line segments on the left have the same length, they continue to appear as if they have different lengths

The two line segments on the left strongly appear to have different lengths. However, as the marking on the right illustrates, they have exactly the same length. Our knowledge of this fact (our possession of evidence), however, does not change the visual appearance of the line segments on the left. They continue to look as if they have different lengths. This evidence insensitivity is typical of the archetypes of visual experience.

Just like experiences that result from amodal completion can have presentational phenomenology, they can also be evidence insensitive. Consider again the image of the occluded dog in Figure 24.2. Although an occluder obscures your line of sight, you naturally see this as a complete dog. Now, let's remove the occluder (Figure 24.4). The experience produced by the process of amodal completion in Figure 24.2 turned out to be illusory. The dog is lacking its middle part. However, even after it's revealed that there isn't a complete dog behind the occluder, what is presented in Figure 24.2 still appears equally complete. So, the amodally completed experience persists (i.e., the dog looks complete) even when we know that the world is not as it appears to be.

Now, let's consider whether modal and amodal multisensory experience possess the two characteristics: presentational phenomenology and evidence insensitivity. We shall here focus on the modal integration cases, but nothing in what follows hinges on this. Consider once again a case of seeing a source produce a sound. Suppose we see a busboy lose his grip on a stack of plates he is carrying. They hit the tile floor in the restaurant right in front of our table. This results in the loud sound of plates breaking against the tile floor. We can literally hear the plates break. In the envisaged scenario, it would appear that we are in direct conscious touch with the event producing the sound. The multisensory experience of hearing the plates break has an integrated presentational phenomenology that consists partly in the phenomenology of the visual experience that produces the analog of a demonstrative and partly in the phenomenology of the auditory experience that attributes audible qualities to the seen event. The fact that the

Figure 24.4 Incomplete Drawing of a Dog. Even after seeing that there is nothing behind the occluder in this figure, the visual nonetheless still generates a visual experience of a dog when viewing the occluded figure in Figure 24.2

multisensory experience has an integrated presentational phenomenology gives us some reason to think that the phenomenology is perceptual.

Now consider the phenomenon of ventriloquism. We know that the ventriloquist produces the voice of the puppet in his hand. Even so, the voice perceptually appears to come from the puppet's mouth. The appearance that the puppet is speaking is so strong that it persists in spite of our knowledge that this is not so—which is to say, ventriloquism is evidence insensitive. In fact, most of us are taking advantage of the evidence-sensitivity of ventriloquism on a daily basis, when we watch television.

Multisensory experience thus can have a presentational phenomenology and may be evidence insensitive. This indicates that the integration process is perceptual as opposed to inferential or loosely associative.

3 Modal versus Amodal Binding: An Argument

Above, we distinguished a notion of modal integration and made a prima facie case for describing some cases of mutisensory integration in terms of it. The aim of this section is to argue that there are cases of integration that cannot be accounted for in terms of amodal integration but can be explained only given the notion of modal integration.

Modal integration requires that qualities represented because they are perceived in one modality are attributed to an object or event represented because it is perceived in another modality. Amodal integration does not require this sort of dependence; modality 1 attributes qualities to an object because it is perceived in modality 1, and modality 2 attributes qualities to an object because it is perceived in modality 2; the integration, that is the identification of the object presented in modality 1 with the object presented in modality 2, is performed amodally.

To see that the two notions of integration come apart, let us consider some phenomena of referential dependence that amodal integration by itself is unable to account for.[10] Imagine being at a cocktail party whose acoustics disrupt one's ability to hear sounds as coming from specific directions and where everyone has the same voice and everyone is speaking the same words. Despite the unusual conditions you might have a sensory experience as of some specific person speaking. In order for this to be the case, you will need to have some sensory manner of picking out the person, let's stipulate a visual manner. The experience of seeing someone speak cannot be the result of visually referring to a person, aurally referring to a person, and amodally identifying the referents. Since you are in an environment where you cannot pick out individuals by their sounds alone there is no aural reference to any particular person. You can, however, pick out people by their looks, positions, and motions. So, you are able to identify who is saying what by visually referring to a person and aurally exploiting that reference in order to attribute the quality of saying something to him or her. This results in what we call a modally integrated experience as of some specific person speaking.

It should be emphasized that the dependence relation in modal multisensory experience can go in both directions. Suppose you are out jogging one particularly foggy morning. You see a person wave to you from the other side of the street. As it turns out, it is your colleague Magdalena. But the visibility is not good enough for you to identify the speaking event as an event in which your colleague is speaking on the basis of the person's look, posture or gait. You can, however, identify the event as a speaking event by your colleague by the sound of her voice as she shouts "Hey! See you later at work!" In this case the low visibility prevents you from identifying the speaking event as being the event of your colleague speaking. You can, however, identify this event on the basis of your auditory perception of the sound event. Here visual qualities are attributed to a sound event identified by audition.

To summarize: In our first case vision enables you to pick out a person and their speaking motions. But your overall experience involves attributing audible qualities to that person and their speaking motions. Audition alone, however, is not sufficient for this attribution. Audition is dependent on vision in that it attributes the quality of making certain sounds to the seen person and their speaking motions. It does this by making use of a reference to that person and their speaking motions which is supplied by vision. In the third case the dependence goes in the other direction since vision depends on audition for its possession of the further content that the person speaking is one we recognize (e.g., Magdalena).

Amodal integration differs from modal integration in that there is no referential dependence. If you are seeing and holding a tomato, the object you are seeing and holding can be picked out in virtue of how it appears within each sensory modality. You see the tomato as shiny, and your touch identifies the tomato as firm. You do not need vision to confirm that the tomato you see is firm, and you do not need your sense of touch to confirm that the tomato is shiny. What integration accomplishes in this case is the attribution of the two qualities *shiny* and *firm* to one and the same object.

Of course, multisensory perception also attributes common sensibles to objects, for instance, roundness to the tomato. But you can confirm that the tomato is round by sight or touch alone. You don't need both sensory modalities to perceptually establish this. If you are unable to perceive the roundness in one sensory modality, this simply means that roundness is not a common sensible for you. Integration is needed for you to come to have a mutlisensory experience of tomato as round. This suggests that the unitary experiences in the case of amodal multisensory integration are prior temporally to the multisensory experience itself, which is consistent with the integration taking place in higher brain regions, such as the parietal cortex. So, amodal multisensory experience does not require that qualities perceived in one sensory modality are attributed to an event perceived in another sensory modality in order for the integration to occur. Hence, modal and amodal multisensory experience are distinct.

4 Synesthesia

In the previous sections, we have been concerned primarily with ordinary multisensory experience. We should, however, briefly compare ordinary multisensory experience to one of the most common forms of atypical multisensory experience, viz. synesthesia (occurring in 4–7 percent of the population). Synesthesia is a peculiar way of experiencing the world in which internal or external input gives rise to atypical sensations or thoughts (Baron-Cohen et al. 1987; Cytowic 1989; Grossenbacher and Lovelace 2001; Ramachandran and Hubbard 2001b; Rich and Mattingley 2002; Ward 2013). For example, seeing the number 3 printed in black ink may lead to a sensation of copper green, hearing the word "abyss" may flood the mouth with the flavor of minestrone soup and hearing the key of C# minor may elicit a slowly contracting turquoise spiral.

In grapheme-color synesthesia, one of the most common forms of synesthesia, perceiving or thinking about an achromatic grapheme (also known as the "inducer") triggers the sensation or thought (also known as the "concurrent") that the grapheme has a specific color with a highly specific hue, brightness and saturation (Simner et al. 2006). The concurrent images are either projected onto the external world (projector synesthesia) or perceived in the mind's eye (associator synesthesia) (Dixon et al. 2004). In projector synesthesia, the projected concurrent may be seen as instantiated like non-synesthetic colors, as floating above its inducer or as an "afterimage" that floats close to the subject's eyes. In associator synesthesia, the concurrent image is seen internally, much like a visual image retrieved from memory or produced by imagination.

Two key characteristics of synesthesia are (i) automaticity and (ii) stability and consistency over time. Automaticity refers to the observation that synesthetes cannot suppress the

RED RED

Figure 24.5 The Stroop Effect. The word "red" is displayed in the color black (left) and the color green (right—here displayed in gray). It takes longer for subjects to name the ink color of the word "red" when it is printed in green than when it is printed in black or red

association between an inducer and its concurrent. Stability and consistency over time refer to the observation that inducer-concurrent associations are highly stable and consistent in more than 80 percent of cases (Mattingley et al. 2001). Automaticity is supported by research showing that synesthetes are susceptible to Stroop effects (Stroop 1935). The most common Stroop task demonstrates that it takes significantly longer for neurotypical individuals to name the color in which a color word is printed if the color referred to by the word is incongruent with the printed color (see Figure 24.5). Likewise, it takes significantly longer for synesthetes to name the printed color of a grapheme if the synesthetic color induced by the grapheme is incongruent with the printed color (Mattingley et al. 2001).

Consistency and stability over time in grapheme-color and sound-color synesthesia is commonly tested using the synesthesia battery on separate occasions (Eagleman et al. 2007). In the test of grapheme-color synesthesia, a subject is presented with a randomly chosen grapheme, for which she must choose a specific hue, brightness and saturation from a color palette representing over 17.6 million distinct choices. After the subject has repeated the task three times for each grapheme (108 trials; graphemes A–Z and 0–9), the geometric distance among the subject's answers in red, green and blue color space is calculated. Synesthesia requires that the geometric distance falls below a normalized threshold.

Projector synesthesia (where the concurrent is projected out onto the external visual scene) is not always a genuine form of multisensory or multisensory-stream experience. Evidence indicates that some grapheme-color synesthetes have an unusual structural connection between the color area in the brain and the neighboring form area (Ramachandran and Hubbard 2001a; Rouw and Scholte 2007; Jancke et al. 2009; Hanggi et al. 2011). Likewise, there is some evidence to suggest that some sound-color synesthetes have an unusual structural relation between auditory areas and the form area (Zamm et al. 2013). Because a structural relation directly combines two areas of the brain that are not normally directly combined, the synesthetic experiences in structurally induced synesthesia are not best characterized as multimodal but are better characterized as an augmented form of ordinary unimodal perception. The individual sensory pathways are simply mistakenly, or atypically, blended into a single pathway, thus forming an augmented sensory pathway that yields illusory or hallucinatory experiences (e.g., the experience of the musical note D as purple or the experience of a black letter as red).

The more interesting cases of synesthesia for our purposes are cases of associator synesthesia (and perhaps functional cases of projector synesthesia) that are a result of unusual binding in higher areas of the brain, most likely the parietal cortex. On the so-called disinhibited integration model, synesthesia occurs owing to disinhibition of an area in the parietal cortex that is thought to bind information from different senses, causing information from one sensory modality to trigger the projection of information from another modality (Grossenbacher 1997; Armel and Ramachandran 1999; Grossenbacher and Lovelace 2001; Myles et al. 2003). Information from the two sensory sources is then integrated. For example, information about the identity of a grapheme may combine with abnormal color information, giving rise to an experience of an abnormally colored grapheme.

One important piece of evidence cited in favor of this hypothesis comes from a case study in which a patient PH reported seeing visual movement in response to tactile stimuli following

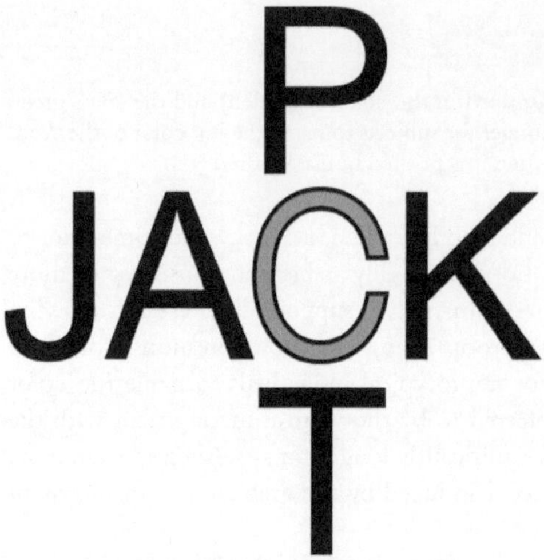

Figure 24.6 Jackpot Figure. Synesthetes interpret the middle letter as a C when it occurs in "Jack" and as an O when it occurs in "pot." The color of their synesthetic experience will depend on which word the grapheme is considered a part of

acquired blindness (Armel and Ramachandran 1999). As PH was blind, he could not have received the information via standard visual pathways. It is plausible that the misperception was a result of disinhibited integration of tactile information and information from the visual motion areas.

Another piece of evidence cited in favor of the disinhibited integration model is the observation that visual context and meaning can influence the phenomenal character of synesthetic experience (Myles et al. 2003; Dixon and Smilek 2005). To illustrate, consider the two words in Figure 24.6. Some grapheme-color synesthetes assign different colors to the shared letter depending on whether they interpret the string of letters as spelling the word "POT" or the word "JACK." For example, a grapheme-color synesthete might have an experience of the shared letter as yellow (**O**) when she reads the word "pot" but have an experience of the letter as pink (**C**) when she reads the word "Jack." One way to explain this phenomenon is that amodal completion of the shared grapheme takes place when the synesthete is reading the word "POT" but not when she is reading the word "JACK." This explanation is consistent with there being a direct structural connection between the form and the color area in the brain. However, the more widely accepted explanation is that it's not the shape actually presented in experience that triggers the color experience, but rather the higher-level property of being a particular grapheme (e.g., *being the grapheme O* or *being the grapheme C*) (Cytowic and Eagleman 2009: 75).

Synesthesia of this second type and ordinary multisensory experience that results from amodal integration both appear to involve higher-level perceptual brain regions (like the parietal cortex) in the integration process, and both types of integration involve attributing features to one and the same object (e.g., *being the musical note D* and *being purple*). Using the example of seeing and holding a firm tomato and hearing the musical note D as purple, we can illustrate the commonalities between the two phenomena as follows:

Normal Amodal Integration

Your overall experience has the content: that_v is a tomato and that_t is firm and $\text{that}_v = \text{that}_t$ [where that_v is a visual demonstrative, and that_t is a tactile demonstrative, and $\text{that}_v = \text{that}_t$ is an amodally represented identification].

Amodal Integration in Sound-Color Synesthesia
Your overall experience has the content: $that_h$ is the musical note D and $that_v$ is purple and $that_h = that_v$ [where $that_h$ is an auditory demonstrative, and $that_v$ is a visual demonstrative, and $that_h = that_v$ is an amodally represented identification].

Now, as noted above, synesthesia also occurs within a single sensory modality combining different sensory streams. In grapheme-color synesthesia, for example, a shape property (e.g., *having the shape of the grapheme 3*) and a color (e.g., *being green*) are visually attributed to a grapheme printed in black. Although this phenomenon is not genuinely *multi*sensory, it nonetheless fits the model. Two visual properties that normally are not integrated are amodally attributed to one and the same object, after being computed separately in separate sensory streams. Using the example of seeing the grapheme 3 as green, we can illustrate this as follows:

Amodal Integration in Grapheme-Color Synesthesia
Your overall experience has the content: $that_{v1}$ has the shape of the grapheme 3 and $that_{v2}$ is purple and $that_{v1} = that_{v2}$ [where $that_{v1}$ is a form area demonstrative, and $that_{v2}$ is a color area demonstrative, and $that_{v1} = that_{v2}$ is an amodally represented identification].

Because of the similarities between synesthesia of the kind under consideration and amodal multisensory experience, research into synesthesia of this type will likely be able to shed light on the process underlying integration in ordinary amodal multisensory experience.

5 Conclusion

We can divide multisensory experiences that go beyond mere co-consciousness (co-consciousness as in the experience of tasting the wine and hearing the siren from the street) into two broad categories. Multisensory experiences—such as feeling the roundness of a tomato through touch and seeing the roundness of the tomato, smelling the Indian curry and seeing it boil or perceiving the flavor of the Oxtail flatbread by gustatorily, olfactorily, somato-sensorily, thermaly and nociceptually attributing features to the flatbread—attribute one or more features to a single object. Experiences of this kind arguably have a phenomenology that reflect that they are integrated amodally. As we have seen, however, not all forms of multisensory perception are amodal in this sense. Some forms are distinctly perceptual and have a phenomenology that derives from the phenomenology of the individual sensory modalities. Seeing someone speak and feeling the rock press against the palm are experiences of this latter kind. Synesthesia is a form of atypical multisensory experience that in some instances involves integration of the first type. Research into this type of synesthesia might thus help shed light on the mechanism underlying amodal integration.[11]

Notes

1 It is widely agreed that there are temporal and spatial congruity constraints on multisensory integration (see e.g. O'Callaghan 2014). If, for example, the visual and audible properties are temporally incongruous you will fail to see a seen event as the one producing the sound. If, for instance, you see a drummer but then hear the drumming sounds only ten seconds later, you will fail to attribute the sound to the drumming. Likewise, if the visual and audible properties are blatantly spatially incongruous, you will fail to see a seen event as the one producing the sound. Suppose, for instance, that you see a person to the left of you move her lips and you also hear corresponding sounds in the distance—far too removed

from the person to be attributable to her. In that case, you will not perceive the person as producing the sounds. We are going to take that for granted in what follows.

2 The idea of the phenomenology deriving exclusively from the phenomenology of the individual sensory modalities is formulated as follows by O'Callaghan (2015: 555): "The phenomenal character of each perceptual episode is exhausted by that which is associated with each individual modality, along with whatever accrues thanks to mere co-consciousness."

3 While amodal experience may seem to be a kind of perception that is cognitively penetrated, most cognitive effects on the integration turn out primarily to be related to attention. Multisensory integration is thus largely accounted for by attentional mechanisms (see Talsma 2015).

4 For simplicity's sake, we shall here assume a representational account of experience according to which the phenomenology of experience (at least typically) reflects a representational content. This is also an assumption made by e.g. O'Callaghan (2012). See also Bourget (2017). Here we are not taking a stance on the question of whether strong representationalism about multisensory experience is feasible (for discussion see e.g. O'Dea 2006; Tye 2007; and Bourget 2017).

5 Bourget (2017) also distinguishes between these two types of multisensory experience (that go beyond mere co-consciousness). However, he argues for a view where the two have different generalized contents. Seeing something produce a sound has a content of the form $\exists x,y(F(x) \wedge G(y) \wedge R(x,y))$, where x and y range over related entities to which different features are attributed. Seeing and feeling a tomato, by contrast, has a content of the form: $\exists x(F(x) \wedge G(x))$. Here different features are attributed to one and the same object.

6 The intermodal interaction can be direct or facilitated by cortico-thalamo-cortical pathways (see Talsma 2015).

7 We can still allow for the possibility that lip reading can produce an experience of meanings (cf. Brogaard 2016). In this case, however, the experience of meanings is not auditory but visual, much like the case of ordinary reading.

8 Bodily sensations (or bodily feelings—a sub-set of the set of interceptive experiences) have not traditionally been construed as sensory experiences. However, one might argue that the modality that produces bodily feelings just is a sensory modality closely related to proprioception, our sense of balance (the vestibular system) and nociception (pain and spice perception), which arguably *are* sensory modalities, unlike intuition and introspection (Macpherson 2011; Schwenkler 2013; Briscoe 2016). Not much hinges on how we settle this issue.

9 We shall set aside the issue of whether we can perceive high-level properties like that of being a dog. Let it be granted for argument's sake that we can perceive such properties. Nothing in what follows hinges on this assumption.

10 For other illustrative examples of cases where the information in one sensory modality cannot be decoded without the assistance of a second sensory modality, see e.g. Talsma (2015). One illuminating example is that of the Swedish chef in *The Muppet Show*. Upon your first encounter with the character, his speech sounds entirely garbled. After multiple other cues (primarily visual) have been presented to you, you realize that the character actually utters English sentences but with an extremely anomalous accent (analogous to sine-wave speech).

11 For comments on this chapter we are grateful to Anna Drozdzowicz, Rocco J. Gennaro, Anders Nes, Sebastian Watzl, the participants in a multisensory perception seminar in Oslo and an audience at a cognitive penetration workshop in Bergen.

References

Armel, K.C., and Ramachandran, V.S. (1999) "Acquired Synesthesia in Retinitis Pigmentosa," *Neurocase* 5: 293–296.

Armstrong, D.M. (1989) *Universals: An Opinionated Introduction*, Boulder: Westview Press.

Baron-Cohen, S., Wyke, M., and Binnie, C. (1987) "Hearing Words and Seeing Colors: An Experimental Investigation of Synesthesia," *Perception* 16: 761–767.

Bayne, T. (2014) "The Multisensory Nature of Perceptual Consciousness," in D. Bennett and C. Hill (eds.) *Sensory Integration and the Unity of Consciousness*, Cambridge, MA: MIT Press.

Bayne, T., and Chalmers, D. J., (2003) "What Is the Unity of Consciousness?" in A. Cleeremans (ed.) *The Unity of Consciousness: Binding, Integration and Dissociation*, Oxford: Oxford University Press.

Bloom, F.E., and Lazeron, A. (1988) *Brain, Mind, and Behavior*, New York: W. H. Freeman and Company.

Bourget, D. (2017) "Representationalism and Sensory Modalities: An Argument for Intermodal Representationalism," *American Philosophical Quarterly* 54: 251–267.

Briscoe, R. E. (2016) "Multisensory Processing and Perceptual Consciousness: Part I," *Philosophy Compass* 11 (2): 121–133.

Briscoe, R. E. (In Press) "Multisensory Processing and Perceptual Consciousness: Part II," *Philosophy Compass*.

Brogaard, B. (2012) "What Do We Say When We Say How or What We Feel?" *Philosophers Imprint* 12 (11), June 2012.

Brogaard, B. (2016) "In Defense of Hearing Meanings," *Synthese* (2016). doi:10.1007/s11229-016-1178-x.

Brogaard, B. (In Press a) *Seeing and Saying*, New York: Oxford University Press.

Brogaard, B. (In Press b) "Knowledge-How and Perceptual Learning," in S. Heatherington and M. Valaris (eds.) *Knowledge in Contemporary Philosophy*, London: Bloomsbury.

Brogaard, B., and Chudnoff, E. (2016) "Against Emotional Dogmatism," *Philosophical Issues*, a supplement to *Nous* 26 1: 59–77

Chalmers, D. J. (2004) "The Representational Character of Experience," in Brian Leiter (ed.) *The Future for Philosophy*, Oxford: Oxford University Press.

Chudnoff, E. (2014) "Review of Tucker (eds.) *Seemings and Justification*," *Notre Dame Philosophical Reviews*.

Chudnoff, E. (2016) "Moral Perception: High Level Perception or Low Level Intuition?" In T. Breyer and C. Gutland (eds.) *Phenomenology of Thinking: Philosophical Investigations into the Character of Cognitive Experiences*, New York: Routledge.

Chudnoff, E. (In Press) "The Epistemic Significance of Perceptual Learning," *Inquiry*.

Cytowic, R.E. (1989) *Synesthesia: A Union of the Senses*, New York: Springer Verlag.

Cytowic, R.E., and Eagleman, D.M. (2009) *Wednesday Is Indigo Blue*, Cambridge, MA: MIT Press.

Dainton, B. (2000) *Stream of Consciousness: Unity and Continuity in Conscious Experience*, New York: Routledge.

Degenaar, M. and Lokhorst, G.J., "Molyneux's Problem," *The Stanford Encyclopedia of Philosophy* (Spring 2014 Edition), E.N. Zalta (ed.), URL = https://plato.stanford.edu/archives/spr2014/entries/molyneux-problem/.

Deroy, O. (2014) "The Unity Assumption and the Many Unities of Consciousness," in D. Bennett and C. Hill (eds.) *Sensory Integration and the Unity of Consciousness*, Cambridge, MA: MIT Press.

de Vignemont, F. (2014) "A Multimodal Conception of Bodily Awareness," *Mind* 123: 989–1020.

de Vignemont, F., and Massin, O. (2015) "Touch." In M. Matthen (ed.) *The Oxford Handbook of the Philosophy of Perception*, Oxford: Oxford University Press.

Dixon, M.J., Smilek, D., and Merikle, P.M. (2004) "Not All Synaesthetes are Created Equal: Projector versus Associator Synaesthetes," *Cognitive, Affective, and Behavioral Neuroscience* 4: 335–343.

Dixon, M.J., and Smilek, D. (2005) "The Importance of Individual Differences in Grapheme-Color Synesthesia," *Neuron* 45: 821–823.

Eagleman, D.M., Kagan, A.D., Nelson, S.S., Sagaram, D., and Sarma, A.K. (2007) "A Standardized Test Battery for the Study of Synesthesia," *Journal of Neuroscience Methods* 159: 139–145.

Fulkerson, M. (2011) "The Unity of Haptic Touch," *Philosophical Psychology* 24: 493–516.

Fulkerson, M. (2014) *The First Sense: A Philosophical Study of Human Touch*, Cambridge, MA: MIT Press.

Gallace, A., and Spence, C. (2014) *In Touch with the Future: The Sense of Touch from Cognitive Neuroscience to Virtual Reality*, Oxford: Oxford University Press.

Giard, M. H., and Péronnet, F. (1999) "Auditory-Visual Integration during Multimodal Object Recognition in Humans: A Behavioral and Electrophysiological Study," *Journal of Cognitive Neuroscience* 11: 473–490.

Grossenbacher, P.G. (1997) "Perception and Sensory Information in Synaesthetic Experience," in S. Baron-Cohen and J.E. Harrison (eds.) *Synaesthesia: Classic and Contemporary Readings*, Malden, MA: Blackwell Publishers.

Grossenbacher, P. G., and Lovelace, C. T. (2001) "Mechanisms of Synesthesia: Cognitive and Physiological Constraints," *Trends in Cognitive Science* 5: 36–41.

Hanggi, J., Wotruba, D., and Jäncke, L. (2011) "Globally Altered Structural Brain Network Topology in Grapheme-Color Synesthesia," *Journal of Neuroscience* 31: 5816–5828.

Jancke, L., Beeli, G., Eulig, C., and Hanggi, J. (2009) "The Neuroanatomy of Grapheme-Color Synesthesia," *European Journal of Neuroscience* 29: 1287–1293.

Jones, L.A., and Lederman, S.J. (2006) *Human Hand Function*, New York: Oxford University Press.

Klemen, J., and Chambers, C. D. (2011) "Current Perspectives and Methods in Studying Neural Mechanisms of Multisensory Interactions," *Neuroscience and Biobehavioral Reviews* 36: 111–133.

Linden, D. J. (2015) *Touch: The Science of Hand, Heart, and Mind*, New York: Penguin Publishing Group.

Loomis, J., and Lederman, S. (1986) "Tactual Perception," in K.R. Boff, L. Kaufman, and J.P. Thomas (eds.) *Handbook of Perception and Human Performance*, New York: Wiley and Sons.
Macpherson, F. (ed.) (2011) *The Senses: Classic and Contemporary Philosophical Perspectives*, Oxford: Oxford University Press.
McGurk, H., and MacDonald, J. (1976) "Hearing Lips and Seeing Voices," *Nature* 264: 746–748.
Mattingley, J.B., Rich, A.N., Yelland, G., and Bradshaw, J.L. (2001) "Unconscious Priming Eliminates Automatic Binding of Colour and Alphanumeric Form in Synaesthesia," *Nature* 410: 580–582.
Molholm, S., Ritter, W., Murray, M. M., Javitt, D. C., Schroeder, C. E., and Foxe, J. J. (2002) "Multisensory Auditory-Visual Interactions during Early Sensory Processing in Humans: A High-Density Electrical Mapping Study," *Cognitive Brain Research* 14: 115–128.
Myles, K.M., Dixonm M.J., Smilek, D., and Merikle, P.M. (2003) "Seeing Double: The Role of Meaning in Alphanumeric-Colour Synaesthesia," *Brain Cognition* 53: 342–345.
O'Callaghan, C. (2008) "Seeing What You Hear: Cross-Modal Illusions and Perception," *Philosophical Issues* 18: 316–338.
O'Callaghan, C. (2012) "Perception and Multimodality," in E. Margolis, R. Samuels, and S. Stich (eds.) *The Oxford Handbook of Philosophy of Cognitive Science*, Oxford: Oxford University Press.
O'Callaghan, C. (2014) "Not All Perceptual Experience Is Modality Specific," in D. Stokes, M. Matthen, and S. Biggs (eds.) *Perception and Its Modalities*, Oxford: Oxford University Press.
O'Callaghan, C. (2015) "The Multisensory Character of Perception," *Journal of Philosophy* 112: 551–569.
O'Dea, J. (2006) "Representationalism, Supervenience, and the Cross-Modal Problem," *Philosophical Studies* 130: 285–295.
Ramachandran, V.S., and Hubbard, E.M. (2001a) "Psychophysical Investigations into the Neural Basis of Synaesthesia," *Proceedings of the Royal Society B: Biological Sciences* 268: 979–983.
Ramachandran, V.S., (2001b) "Synaesthesia: A Window into Perception, Thought and Language," *Journal of Consciousness Studies* 8: 3–34.
Rouw, R., and Scholte, H.S. (2007) "Increased Structural Connectivity in Grapheme-Color Synesthesia," *Nature Neuroscience* 10: 792–797.
Rich, A.N., and Mattingly, J.B. (2002) "Anomalous Perception in Synaesthesia: A Cognitive Neuroscience Perspective," *Nature Reviews Neuroscience* 3: 43–52.
Schwenkler, J. (2013) "The Objects of Bodily Awareness," *Philosophical Studies* 162: 465–472.
Shams, L., Kamitani, Y., and Shimojo, S. (2000) "Illusions: What You See Is What You Hear," *Nature* 408 (6814): 788.
Simner, J., Mulvenna, C., Sagiv, N., Tsakanikos, E., Witherby, S.A., Fraser, C., Scott, K., and Ward, J. (2006) "Synaesthesia: The Prevalence of Atypical Cross-modal Experiences," *Perception* 35: 1024–1033.
Stroop, J.R. (1935) "Studies of Interference in Serial Verbal Reactions," *Journal of Experimental Psychology* 18: 643–662.
Talsma, D. (2015) "Predictive Coding and Multisensory Integration: An Attentional Account of the Multisensory Mind," *Frontiers in Integrative Neuroscience* 9: 19. doi:10.3389/fnint.2015.00019.
Tuomainen, J., Andersen, T. S., Tiippana, K., and Sams, M. (2005) "Audio-Visual Speech Perception Is Special," *Cognition* 96: B13–B22.
Tye, M. (2007) "The Problem of Common Sensibles," *Erkenntnis* 66: 287–303.
Ward, J. (2013) "Synesthesia," *Annual Reviews of Psychology* 64: 49–75.
Zamm, A., Schlaug, G., Eagleman, D. M., and Loui, P. (2013) "Pathways to Seeing Music: Enhanced Structural Connectivity in Colored-Music Synesthesia," *Neuroimage* 74: 359–366.

Related Topics

The Unity of Consciousness
Consciousness and Conceptualism
Consciousness and Attention

25
CONSCIOUSNESS AND PSYCHOPATHOLOGY

Rocco J. Gennaro

This chapter reviews the interdisciplinary field sometimes called "philosophical psychopathology" (Graham and Stephens 1994), which is also related to "philosophy of psychiatry" (Fulford, Thornton, and Graham 2006). I'll focus first on various psychopathologies with special attention to how they negatively impact and distort conscious experience, such as amnesia, somatoparaphrenia, schizophrenia, visual agnosia, autism, and Dissociative Identity Disorder (DID). Many of them are disorders of "self" or "self-awareness" which force us to consider related philosophical problems, such as the problem of personal identity. There are of course many other abnormal conditions not discussed in this chapter. I'll then discuss "philosophy of psychiatry," covering the overlapping topics of psychopathy, mental illness, and moral responsibility. In addition to the work of philosophers, recent interest is also due to the accessible writings of neurologists, most notably Oliver Sacks (starting with his 1987 book), Todd Feinberg (2001), and V.S. Ramachandran (2004). One of the results is the important interdisciplinary interest that has been generated among philosophers, psychologists, and scientists.[1]

Let us begin with a number of disorders that challenge our notion of personal identity and self.

1 Disorders of the Self and Self-Awareness

Philosophers have always been intrigued by disorders of consciousness. Part of the reason is that if we can understand how consciousness goes wrong, then we can better theorize about the normal functioning mind. Locke (1689/1975) discussed the philosophical implications of something like multiple personality disorder (MPD) which is now called "Dissociative Identity Disorder" (DID). He recognized that difficult questions arise: could there be two centers of consciousness in one body? What makes a person the same person over time? These questions are closely linked to the traditional philosophical problem of personal identity. Related questions also arise for various memory disorders, e.g. does consciousness or personal identity require some kind of autobiographical memory or psychological continuity?

DID describes a condition in which a person displays multiple distinct identities (or "alters"), each interacting with the environment in its own way. Each alter has its own pattern of behaving, perceiving, thinking, and pattern of speech. The alters seem to emerge involuntarily and not all alters are even always known to the patient. Thus, there is significant amnesia on the patient's part for periods of time. This would seem to imply that there are really two (or even more) persons and one body, at least at different times, especially if Locke is correct. His account of personal identity through time famously appealed to consciousness and memory. On his view, a later person (P2) is identical to an earlier person (P1), just in case P2's consciousness "can be extended backwards" to P1. This is taken to mean that P2 consciously remembers P1's thoughts and experiences, which is often called the "psychological continuity" account of personal identity. My personhood goes with my consciousness and memory, not necessarily with my body. So, a case of DID would seem to be a case where a single body houses two distinct personalities, often with profoundly different character traits and behavior patterns.

Further, Locke recognized that if there are really two different persons, then it is difficult to make sense of holding one *morally responsible* for the other's actions. Indeed, there is an often-cited link between personal identity and moral responsibility (Kennett and Matthews 2002). One might think of the well-known *Dr. Jekyll and Mr. Hyde* tale of two persons, one good and one evil, inhabiting the same body (from the Robert Louis Stevenson 1886 classic novel). It would seem to be wrong to punish or blame one person for the actions of another person. We are justified in holding a person responsible for some past action only if the person is identical with the person who performed that action. Locke argues that one is justifiably held accountable only for those actions performed by a person to whom one's present consciousness extends. This would apply not only to a person whose alter had committed crimes but perhaps also to, say, an elderly inmate who has lost virtually all memory of committing a crime due to long-term aging.

It should be noted that DID has been, at times, a very controversial diagnosis. When it was called multiple personality disorder (MPD), there was a problem of over-diagnosis especially in the 1980s, perhaps stemming from the publication of the book *Sybil* which, along with the subsequent film, had a major impact on the popular culture of the time. Even in the psychiatric community, there has been a great deal of disagreement about DID. Some argue that DID (and MPD previously) does not really exist at all and point to cases of irresponsible therapists who encouraged their patients to believe (falsely) that they had been abused as children or who were accused of implanting such memories in patients via hypnosis. Significant controversy still surrounds the diagnosis of DID but it remains as a category in the *Diagnostic and Statistical Manual of Mental Disorders*, the DSM-5 (American Psychiatric Association, 2013). Most today hold that DID results from repeated childhood abuse where dissociating is a way to cope with traumatic experiences.

As we saw above, the link between personal identity and memory is a close one, at least according to one prominent theory. So, what about cases where we clearly have a single person but her consciousness is negatively affected by severe memory loss (amnesia)? *Anterograde* amnesia is the loss of short-term memory or impairment of the ability to form new memories through memorization. *Retrograde* amnesia is the loss of pre-existing memories to conscious recollection, well beyond a normal degree of forgetfulness. The person may be able to memorize new things that occur after the onset of amnesia (unlike in anterograde amnesia), but be unable to recall some or all of one's life prior to the onset.

To take one example, Sacks' (1987) patient Jimmie G. didn't even recognize himself in the mirror because he thought he was 20 years younger, having no "episodic memory" of those intervening years. He didn't recognize the doctor each time he came in to see him. Jimmie G. had Korsakoff's syndrome due to long-term heavy drinking. Nonetheless, he retained much of his "procedural memory," that is, memory for how to do certain things or display various skills. This is an example of a "dissociation" between two cognitive abilities, that is, a case where one cognitive function, A, is preserved but another, B, is damaged. In Jimmie G. and other similar cases, procedural memory can not only remain undamaged (such as typing or riding a bicycle), but the patient can also learn new skills (such as following a moving target with a pointer) without having any conscious memory of the previous learning episodes. A "double dissociation" is found when A and B can each function independently of the other.

It is worth mentioning that neither the dualist nor the materialist can easily explain what happens in DID. For example, if we take DID to show that there can be more than one conscious mind associated with a body, then, for example, a substance dualist would be even more hard pressed to explain how more than one nonphysical mind comes to be causally connected to a single body. On the other hand, materialists have the difficulty of explaining how two people, often exhibiting very different personalities and behaviors, can co-exist within a single brain.[2]

Another intriguing abnormality of self and consciousness comes from patients who have undergone a commissurotomy or a "brain bisection" operation (Nagel 1971; Sperry 1984). They were performed decades ago as a last resort to relieve the seizure symptoms of severe epilepsy. During the procedure, the nerve fibers connecting the two brain hemispheres (the corpus callosum) are cut, resulting in so-called "split-brain" patients. So "split-brain cases" are those patients where severing the corpus callosum blocks the inter-hemispheric transfer of, for example, perceptual and motor information. The human retina functions in such a way that the left half of each retina is primarily connected to the left hemisphere of the brain and the right half of each retina is primarily hooked up to the right hemisphere of the brain. The visual system takes information from the left visual field of both eyes to the right hemisphere and information from the right visual field of both eyes to the left hemisphere. Given the crossing over of fibers in the optic chiasm, the effect is that the two sides of the brain reflect opposite sides of the outer world.

In some laboratory conditions, these patients seem to behave as though two "centers of consciousness" are present, each associated with one of the two cerebral hemispheres. Some puzzling results were found in controlled laboratory presentations where patients were forced to look straight ahead without being able to move their heads as usual. Different stimuli were shown to patients' left and right visual fields, revealing striking dissociations of consciousness and behavior. Their left hands literally do not know what their right hands are doing. Shown one stimulus on the left and a different one on the right, each hand will respond appropriately to its respective stimulus but not to the other. Moreover, patients typically showed verbal knowledge only of the stimuli shown to their right field and not to those in their left field, even though their left hand is able to respond appropriately to the left field stimuli. Given the usual location of primary language function to the left hemisphere, the right hemisphere, which receives visual input from the left visual field, remains unable to verbally describe what it sees. During an object recognition task, a subject might report seeing a bottle, due to the left hemisphere, while his left hand (controlled by his right hemisphere) was searching to find a hammer from a group of objects. Patients would report seeing one thing but some of their bodily behavior would indicate that they saw something else.

Split-brain cases often have more to do with "synchronic identity," that is, how many selves are present *at a given time*, as opposed to DID where "diachronic identity" is the main issue, that is, how many selves are present *over time*. Nonetheless, even for split-brain cases, Bayne (2010) for example argues for a "rapid-switching model," in which selves alternate in existence. So only one self exists at a given time although which one exists may switch rapidly and repeatedly over short intervals of time. The rapid-switching hypothesis seems to have the advantage of a kind of "unity thesis," such that all the experiences had by a given human at a given time would remain phenomenally unified.[3]

Somatoparaphrenia is a very strange "depersonalization disorder" and body delusion where one denies ownership of a limb or an entire side of one's body. Anosognosia is a related condition, in which a person who suffers from a disability seems unaware of the existence of the disability. A person whose limbs are paralyzed will insist that his limbs are moving and will become furious when family and caregivers say that they are not. Somatoparaphrenia is usually caused by extensive right hemisphere lesions. Lesions in the temporoparietal junction are common but deep cortical regions (for example, the posterior insula) and subcortical regions (for example, the basal ganglia) are also sometimes implicated (Valler and Ronchi 2009). Anton's syndrome is a form of anosognosia in which a person with partial or total blindness denies being visually impaired, despite medical evidence to the contrary. The patient confabulates, that is, makes up excuses for the inability to see, rationalizing what would seem to be delusional behavior. Thus, the blind person will insist that she can see and stumble around a room bumping into things. Patients with somatoparaphrenia utter some rather stunning statements, such as "parts of my body feel as if they didn't belong to me" (Sierra and Berrios 2000: 160), and "when a part of my body hurts, I feel so detached from the pain that it feels as if it were somebody else's pain" (Sierra and Berrios 2000: 163). It is difficult to grasp what having these conscious thoughts and experiences is like.

Interestingly, the higher-order thought (HOT) theory has been critically examined in light of some psychopathologies because, according to HOT theory, what makes a mental state conscious is a HOT of the form that "I am in mental state M" (Rosenthal 2005; Gennaro 2012). The requirement of an I-reference leads some to think that HOT theory cannot explain or account for some of these depersonalization pathologies. There would seem to be cases where I can have a conscious state and not attribute it to myself but rather to someone else. Liang and Lane (2009) initially argued that somatoparaphrenia threatens HOT theory because it contradicts the notion of the accompanying HOT that "I am in mental state M." The "I" is not only importantly *self-referential*, but essential in tying the conscious state to *oneself* and, thus, to one's *ownership* of M.

Rosenthal (2010) responds that one can be aware of bodily sensations in two ways that, normally at least, go together: (1) aware of a bodily sensation *as one's own*, and (2) aware of a bodily sensation *as having some bodily location*, like a hand or foot. Patients with somatoparaphrenia still experience the sensation as their own but also as having a mistaken bodily location (perhaps somewhat analogous to phantom limb pain where patients experience pain in missing limbs). Such patients still do have the awareness in (1), which is the main issue at hand, but they have the strange awareness in sense (2). So, somatoparaphrenia leads some people to misidentify the bodily location of a sensation as someone else's, but the awareness of the sensation itself remains one's own. Lane and Liang (2010) are not satisfied and counter that Rosenthal's analogy to phantom limbs is faulty and that he has still not explained why the identification of the *bearer* of the pain cannot also be mistaken.

Nonetheless, among other things, we must first remember that many of these patients often deny feeling *anything* in the limb in question (Bottini et al. 2002; Gennaro 2015b: 57–58). As Liang and Lane themselves do point out, patient FB (Bottini et al. 2002), while blindfolded, feels "no tactile sensation" when the examiner would in fact touch the dorsal surface of FB's hand (Liang and Lane 2009: 664). In these cases, it is therefore difficult to see what the problem is for HOT theory at all. But when there really is a bodily sensation of some kind, a HOT theorist might also argue that there are really *two* conscious states that seem to be at odds (Gennaro 2015b). There is a conscious feeling in a limb but also the (conscious) attribution of the limb to someone else. It is also crucial to emphasize that somatoparaphrenia is often characterized as a *delusion* of *belief* under the broader category of anosognosia (de Vignemont 2010; Feinberg 2011). A delusion is often defined as a false belief that is held based on an incorrect (and probably unconscious) *inference* about external reality or oneself that is firmly sustained despite what almost everyone else believes and despite what constitutes incontrovertible evidence to the contrary (Radden 2010). In some cases, delusions seriously inhibit normal day-to-day functioning.

This "doxastic" conception of delusion is common among psychologists and psychiatrists (Bayne and Pacherie 2005; Bortolotti 2009). Beliefs, generally speaking, are themselves often taken to be intentional states integrated with other beliefs. They are typically understood as caused by perceptions that then lead to action or behavior. Thus, somatoparaphrenia is, in some ways, closer to self-deception and involves frequent confabulation.

A related critique of self-representationalism (Kriegel 2009) might also be posed to what Billon and Kriegel (2015) call "subjectivity theories," which says that "there is something it is like for a subject to have mental state M only if M is characterized by a certain mine-ness or for-me-ness. Such theories appear to face certain psychopathological counterexamples: patients appear to report conscious experiences that lack this subjective element" (Billon and Kriegel 2015: 29). Patients with somatoparaphrenia seem to be cases where one has a conscious state without the "for-me-ness" aspect and thus not experienced as one's own. However, Billon and Kriegel counter that "none of the patients that we know of claim feeling *sensations that are not theirs*. Rather, they say that they feel touch in someone else's limb. This does not yet imply that they feel sensations that are not their own—unless it is analytic that one cannot feel one's sensations but in one's own body, which we have phenomenological and empirical reasons to deny" (Billon and Kriegel 2015: 37).[4]

It is worth emphasizing again that many disorders, including somatoparaphrenia, involve delusion or self-deception. A delusion is different than a belief that is based on incorrect information, poor memory, illusion, or other effects of perception. Self-deception is a process of denying or rationalizing away the relevance, significance, or importance of opposing evidence and logical argument. Self-deception involves convincing oneself of a truth (or lack of truth) so that one does not reveal any self-knowledge of the deception. Delusions have received extensive treatment from philosophers in recent years, sometimes in connection with self-deception.[5]

Schizophrenia is a mental disorder characterized by disintegration of thought processes and of emotional responsiveness. It most commonly manifests itself as auditory hallucinations, paranoid or bizarre delusions, or disorganized speech and thinking, and it is accompanied by significant social or occupational dysfunction. Thought insertion, a common symptom of schizophrenia, is the delusion that some thoughts are not one's own and are somehow being inserted into one's mind. In some particularly severe forms of schizophrenia, the victim seems to lose the ability to have an integrated or "unified" experience of her world and self. The person often speaks in an incoherent fashion, doesn't even complete sentences, and is unable to act on simple plans of action. Once again, it is difficult to understand what it is like to consciously experience the world in this way.

Stephens and Graham (2000) suggest that thought insertion should be understood as alienated self-consciousness or meta-representation.[6] They think that schizophrenics make introspective inferential mistakes about the source of inserted thoughts based on delusional background beliefs. Some bodily movements can of course be movements of *my* limbs without counting as actions *of mine* or as caused by me. Perhaps someone else is controlling my movements or they are entirely involuntary such as the physical (motor) tics and vocalizations in Tourette's syndrome. But in these cases, the bodily movements are still self-attributed to the person with the disorder, so something else must be going on to explain attributions to others in thought insertion. If a song spontaneously runs through my mind, I still think of it as an episode in my mind. But it does not count as *my* mental activity in the same way as when I am thinking through a math problem or trying to plan a trip. The latter, but not the former, involves intentional thought that expresses my agency. There seems to be something special going on when I consciously engage in some activity which involves mental effort and voluntariness. Stephens and Graham (2000: 152) call the feeling of *having* a mental state the "sense of subjectivity," and the feeling of *causing* my mental state the "sense of agency." They urge that these two can come apart in unusual cases, so that thought insertion involves the sense of subjectivity without the sense of agency, which also accounts for the curious "passivity experience" of schizophrenics. So attributing thoughts to someone else makes sense because it still must be caused by something or someone.

Gallagher (2000) makes the similar distinction between a "sense of ownership" and a "sense of agency," but, in contrast to Stephens and Graham's (2000) "top-down" approach, argues instead that the primary deficit regarding thought insertion is more of a "bottom-up" problem with the first-person experience itself rather than a self-monitoring abnormality. What happens at the introspective level is not erroneous but rather a correct report of what the schizophrenic actually experiences, that is, thoughts that feel different and externally caused. Gallagher also points to some preliminary neurological evidence which indicates abnormalities in the right inferior parietal cortex for delusions of control.[7]

2 Disorders of Outer Perception

Of course, many psychopathologies of consciousness involve abnormally experiencing the outer world via outer perception. Agnosia, for example, is the loss of ability to recognize objects, persons, sounds, shapes, or smells while the specific sense itself is not necessarily defective and there is no significant memory loss. Focusing on *visual* agnosia, it is known to come in two types: *apperceptive* visual agnosia, cases where "recognition of an object fails because of an impairment in visual perception," and *associative* visual agnosia, cases "in which perception seems adequate to allow recognition, and yet recognition cannot take place" (Farah 2004: 4). The latter is far more interesting to philosophers since it appears to be an instance of having a "normal percept stripped of its meaning" (Teuber 1968). So, for example, a patient will be unable to name or recognize a bicycle. Associative agnosics are not blind and do not have damage to the relevant areas of the visual cortex. In addition, associative agnosics tend to have difficulty in naming tasks and with grouping objects together. Unlike in apperceptive agnosia, there seems to be intact basic visual perception; for example, patients can copy objects or drawings that they cannot recognize, albeit very slowly. But the deficit in associative agnosics is more cognitive than in patients with apperceptive agnosia. Patients will also often see the details of an object but not the whole of the object at a glance. The main point is that the very phenomenal experience of associative agnosics has changed in a way that corresponds to a lack of conceptual deployment. It is, however, important to recognize that associative agnosics still do possess the relevant correct concept because they can apply it via other modalities. For example, a patient might easily identify a whistle by sound instead of sight. Like other disorders, the lack of unity in consciousness is striking for those who suffer from agnosia.

There are other similar psychopathologies resulting in devastating effects on consciousness. Prosopagnosia (also known as "faceblindness" and "facial agnosia") occurs when patients cannot consciously recognize very familiar faces, sometimes even including their own. They do not have a sense of familiarity when looking at another's face but can sometimes make inferences via auditory or other visual cues (e.g. clothes, hair) to compensate. However, skin conductance responses show that there is some kind of emotional arousal when in the presence of a known person. Akinetopsia is the loss of motion perception or visual animation, sometimes called "motion blindness." The visual world seems to come to a standstill or appears more like a sequence of frozen snapshots, such that objects don't really move but appear to "jump" from one place to another. One's visual consciousness is distorted both with respect to temporal sequence and the unity of consciousness. Simultanagnosia occurs when patients can recognize objects in their visual field but only one at a time. They cannot make out the full scene they belong to or make out a whole image out of the details. Outside of a narrow area in the visual field, these patients say they see nothing but an undifferentiated mess. Hemispatial neglect (also called hemiagnosia and hemi-neglect) is a neuropsychological condition in which there is a deficit in attention to and awareness of one side of space. There is typically damage to the right posterior parietal lobe. It is defined by the inability for a person to process and perceive stimuli on one side of the body or environment that is not due to a lack of sensation. In hemi-neglect, one loses all sense of one side of one's body or sometimes half (divided vertically) of everything spatial in one's experience. When asked to draw a clock, patients typically try to jam all of the numbers into the right side of the image.

It is worth mentioning one important launching point for some related work in abnormal psychology and psychopathology, namely, the discovery of "blindsight" (Weiskrantz 1986), which is very frequently discussed in the philosophical literature regarding its implications for consciousness and unconscious visual processing. Blindsight patients are blind in a well-defined part of the visual field (due to specific cortical damage), but yet, when forced, can guess the location or orientation of an object in the blind field with a much higher than expected degree of accuracy (even up to 90% of the time). For example, patients can correctly guess the orientation of lines in circles filled with vertical or horizontal shapes in their blind field. The same goes for some shapes, such as an 'X' or 'O.' The patient is surprised by her success since she tends to think that she is purely guessing without gaining any conscious visual information at all.

3 Autism

There has been increased interest in autism among philosophers and psychologists in recent decades. Autism is a disorder characterized by impaired social interaction and communication, as well as by restricted and repetitive behavior. It is a developmental disorder that affects a child's ability to acquire social skills and engage in social activities. Autism is sometimes thought of as a more serious version of a spectrum of cases called Asperger's syndrome. Researchers also tend to agree that autistic humans have impaired empathizing skills and deception detection. Autistics also exhibit a lack of imagination and an inability to pretend. There is typically a lack of normal eye contact and gaze monitoring, along with a lack of normal social awareness and responsiveness, such as would normally occur when one is embarrassed or sympathetic to another's embarrassment (Hillier and Allinson 2002). So, for example, Baron-Cohen (1995) argues that various mechanisms are deficient in the mind of autistic humans, such as major impairment of what he calls the "Shared Attention Mechanism." Thus, it seems clear that autistic humans have particular difficulty with mind-reading (Baron-Cohen 1995; Frith and Hill 2003; Nichols and Stich 2003). "Mind-reading" is a technical term used in the literature to refer to a set of abilities to discern other people's mental states from their behaviors and from contextual factors. It does not refer to supernatural abilities or telepathy.

Some, however, have claimed that individuals with autism are "mind-blind" in more significant ways and are virtually incapable of mind-reading and perhaps even meta-cognition entirely (Carruthers 1996; Frith and Happé 1999). Given his parallel arguments regarding animals and infants, Carruthers seems committed to the rather startling view that autistic children lack conscious states. He reasons that if autistic subjects lack self-awareness, then autistic individuals should be "as blind to their own mental states as they are to the mental states of others" (Carruthers 1996: 262), and "they lack phenomenally conscious mental states" (Carruthers 2000: 202).

On the other hand, there seem to be numerous cases where autistic subjects engage in deep meditation and prolonged focusing of attention on inner-feelings or images (Gennaro 2012:

257–262). Thus, we have examples where introspective ability is sometimes even *greater* than normal, not to mention the unusual case of Temple Grandin who holds a Ph.D. and is a professor of animal science. So, we might argue that instead of a lack of mind-reading skills negatively impacting one's metacognitive ability, such an intense self-awareness might cause subjects to lack the typical awareness of others. That is, the self-preoccupation of some autistic individuals might even explain their lack of mind-reading skills. Many of the main deficits in question, such as impaired empathizing skills, lack of imagination, and difficulties with joint attention, might actually result from a heightened sense of introspection.[8]

4 Psychopathy

Psychopathy is a mental disorder characterized by a lack of empathy and remorse, shallow emotions, and egocentricity. These abnormalities certainly seem to include deficits of consciousness, such as the inability to show empathy to others or to experience deep emotional connection to others. Psychopathy is sometimes accompanied by "narcissistic personality disorder," which results in a pattern of grandiosity, need for admiration, along with lack of empathy. Psychopaths are unable to feel distress by the perception of others in pain. Although the degree to which someone has the capacity for empathic distress can obviously vary, psychopaths are very different from other people.

Psychopaths have difficulty distinguishing between violating *moral* norms or rules and violating *conventional* norms (Dolan and Fullam 2010). Non-psychopaths understand the difference, say, between beating someone (a moral norm violation) and not responding to a formal invitation (a conventional norm violation). Normal people tend to characterize moral norms as very serious, whereas conventional norms are thought of as dependent on context and authority. Even children begin to grasp the distinction between moral and conventional norms at a very early age. Psychopaths, on the other hand, tend to treat all norms as norms of convention. It is not even clear that psychopaths fully grasp moral concepts.

Psychopaths are often diagnosed using the "Psychopathy Checklist," which includes such personality traits as lack of remorse or guilt, shallow affect, grandiose sense of self-worth, and socially deviant lifestyle (Hare 2003; Malatesti and McMillan 2010). Still, the very category of psychopathy is somewhat controversial within psychiatry. The DSM treats the diagnosis as "antisocial personality disorder," which includes such symptoms as destructive and criminal behavior. For this reason, there is the worry that the diagnosis will be used to excuse such behavior. The philosophical literature on the moral responsibility of psychopaths is enormous (beginning with Murphy 1972), but it is worth noting that some psychopathic personality traits can actually be positive in certain contexts (e.g. high self-confidence and toleration of unfamiliarity or danger). These traits can be found in high achievers in corporate and other respected institutional settings, such as the academic, legal, and medical professions (Babiak 2010).

Despite a lack of empathy and the inability to identify with other's sensory experiences, psychopaths are clearly excellent at times in understanding other minds, especially in more cognitive aspects. After all, psychopaths are often very good at deceiving and manipulating others, which most certainly requires some mind-reading skills. Some serial killers and child-molesters can be cunning and patient in order to gain a victim's trust. Of course, having abnormal mind-reading skills does not automatically lead to psychopathy. As we saw in the previous section, autistic people also have mind-reading difficulties, but they do not share most of the other characteristics of psychopaths.

Of course, whenever there is a mass murder or we become aware of a psychopathic serial killer, we also often wonder about a suspect's mental health and whether it should excuse them from responsibility. One reason for discussions like these is probably because mental illness might have an effect on moral responsibility attributions. Another reason is obviously to help us to prevent similar crimes by looking for "warning signs" in others. But this line of thought also leads some to worry that "[t]o diagnose someone as mentally ill is to declare that the person is entitled to adopt the sick role and that we should respond as though the person is a passive victim of the condition" (Edwards 2009: 80).

Much of this discussion about psychopathy occurs against the backdrop of the perennial problem of free will and determinism for which there is an enormous literature (see e.g. Kane 2011; McKenna and Pereboom 2016). For example, so-called "libertarian" free will says that the core ideas of "could have done otherwise" and "control" over actions are essential for free will and for holding someone morally responsible for an action. If one really couldn't do otherwise, as determinists believe, then how could we blame, punish, or otherwise hold that person morally responsible? The idea, for example, is that if a man were to rob an elderly lady, then he was compelled to do so given his state of mind at that time. But if he really *couldn't* have done otherwise, how can we really hold him morally responsible for the action? So-called "compatibilists," on the other hand, believe that one can still be morally responsible for an action that one cannot avoid at that time (Frankfurt 1969, 1971). They think that the "principle of alternative possibilities" (PAP) is false:

> (PAP) A person is morally responsible for what she does do only if she can do otherwise.

Nonetheless, even a compatibilist believes that there are some situations in which one is *not* morally responsible for an action, such as when one is externally coerced or when one desires to behave in ways that run counter to one's "true self" or true motives. At least some cases of mental illness may fall into this group, such as in obsessive compulsive disorder (OCD) and various addictions. For example, I may strongly desire to take various drugs or drink alcohol but yet wish that I did not have that desire.[9]

Should we then conclude that, say, a psychopathic serial killer shouldn't be punished? If determinism is true, then it may be that we should no longer think of punishment as some kind of retribution based on libertarian free will. Perhaps we should simply focus on deterring others (and the criminals themselves) from committing future crimes. Most people in a society will wish to avoid incarceration and will thus behave accordingly. But for those who still do harm others, incarceration is at least a way to keep them from harming others in the general population. Maybe serial killers and pedophiles really can't help what they do and really aren't morally responsible, but that doesn't mean that we should allow them out on the streets. It is also important to note that to say that a psychopathic murderer is determined does not necessarily mean that he is "legally insane," which is a narrower and technical legal notion. In the United States at least, to be legally insane has more to do with "not understanding the difference between right and wrong," or "not understanding the consequences of one's actions," which is a very high hurdle for the defense to clear.

With regard to consciousness, it is often with respect to emotions where there is significant focus on moral responsibility (Fischer and Ravizza 1998; Brink and Nelkin 2013). So-called "reactive attitude theories" give moral emotions a key role in both attribution and accountability (or responsibility). The term "reactive attitude" was originally coined by Peter Strawson as a way to refer to the emotional responses that arise when we respond to what people do (Strawson 1962). Reactive attitudes are often intense conscious emotional states such as resentment, indignation, disgust, guilt, hatred, love, and shame. So, for example, we get angry at and disgusted by the pedophile murderer and our reaction is even stronger when the victim is someone we know or a family member.

Still, some think that a responsible agent must have conscious access to *moral reasons* along with the ability to understand how such reasons fit together with one's behavior (Fischer and Ravizza 1998). Psychopaths are puzzling for many reasons, for example, they seem to be rational in one sense but also mentally ill at the same time. Reactive attitude theorists have thus argued that psychopaths should be excused from moral responsibility given their difficulty in distinguishing between moral and conventional norms and since they are not properly sensitive to moral reasons (Fischer and Ravizza 1998; Russell 2004). It would therefore be inappropriate to express reactive attitudes toward psychopaths, perhaps analogous to getting angry at a lion for killing someone after escaping from the zoo. However, others do think that psychopaths can and should be held accountable for their actions. Shoemaker (2011), for example, has argued that: "[a]s long as [the psychopath] has sufficient cognitive development to come to an abstract understanding of what the laws are and what the penalties are for violating them, it seems clear that he could arrive at the conclusion that [criminal] actions are not worth pursuing for purely prudential reasons, say. And with this capacity in place, he is eligible for criminal responsibility" (Shoemaker 2011: 119). Shoemaker may be correct with respect to *legal* responsibility, but the main problem for philosophers is whether or not psychopaths are *morally* responsible for their actions.[10]

Some of the related interdisciplinary work in this area is termed "philosophy of psychiatry" and centers around the very nature of mental illness. There are some who argue that our current diagnostic categories, as found in the *Diagnostic and Statistical Manual of Mental Disorders*, the DSM-5 (American Psychiatric Association, 2013), are faulty because, among other things, they are derived only from symptoms rather than underlying physical pathologies (Poland 2014). Genuine mental illnesses are not just symptoms but destructive pathological processes which occur in biological systems. So, some have doubted the very existence of mental illnesses as they are often understood (Szasz 1974). Still, it is generally accepted that mental illnesses are real and involve serious disturbances of consciousness which cause significant impairment in people, sometimes even leading to self-destructive behavior and suicide. The most serious mental illnesses, such as schizophrenia, bipolar disorder, depression, and schizoaffective disorder, are often chronic and can cause serious disability.[11]

5 Brief Summary

This chapter explored the growing and cutting edge interdisciplinary field called "philosophical psychopathology," along with the related "philosophy of psychiatry," which covers the overlapping topics of mental illness, psychopathy, and moral responsibility. Numerous abnormal phenomena were explained with the focus on how they negatively impact consciousness, such as somatoparaphrenia, visual agnosia, and schizophrenia. For example, a number of psychopathologies are commonly viewed as pathologies of self- or body-awareness in some way. Many of these disorders forced us to discuss the importantly related philosophical problems of personal identity, the unity of consciousness, and free will and moral responsibility.[12]

Notes

1 Such as Stephens and Graham (2000), Farah (2004), Feinberg and Keenan (2005), Graham (2013), and Gennaro (2015a).

2 For further discussion on personal identity, see Kind "Consciousness, Personal Identity, and Immortality," this volume, and Kind (2015).

3 See also Schechter (2012) and Brook (2015) for critical discussion of Bayne's view. For much more on split-brain cases and the unity of consciousness, see Schechter (2018) and her chapter in this volume.

For other recent philosophical discussion on the unity of consciousness, see Tye (2003, chapter 5), Cleeremans (2003), Dainton (2008), and Bayne (2008, 2010, chapter 9).

4 Lane (2015) replies not only to Gennaro and Billon and Kriegel, but also clarifies and further develops some of his influential previous work in this area.

5 Radden (2010) and Pliushch and Metzinger (2015). There are also numerous other very strange delusions discussed in the literature, such as *Cotard syndrome*, which is a rare neuropsychiatric disorder in which people hold a delusional belief that they are dead (either figuratively or literally), do not exist, are putrefying, or have lost their blood or internal organs; *Capgras syndrome*, which is a disorder in which a person holds a delusion that a friend, spouse, parent, or other close family member has been replaced by an identical-looking impostor; and *Fregoli Delusion*, the belief that various people met by the deluded subject are actually the same person in disguise.

6 See also Frith (1992), Sabanz and Prinz (2006), Bortolotti and Broome (2009), Parnas and Sass (2011), and Graham (2013: 254–261).

7 There is also a disorder called "body integrity identity disorder" (BIID) whereby an individual has the desire to amputate one or more healthy limbs (First and Fisher 2012). It is not clear what causes BIID, but it may be due to an abnormality of not including the limb in the brain's "body map," located in the right parietal lobe. Ethical issues also arise about whether or not the patient should be able to have a limb amputated (Ryan 2009; Muller 2009).

8 For more on various psychopathologies, see Radden (2004), Hirstein (2005), Bortolotti (2009), Bayne and Fernandez (2009), Bayne (2010), Feinberg (2011), Graham (2013), and Gennaro (2015a). See also McGeer (2004) on autism and self-awareness.

9 For more on consciousness and free will, see Baumeister, Mele, and Vohs (2010) and Caruso (2012).

10 Some philosophers believe that moral responsibility (and free will for that matter) more generally requires consciousness of at least some kind. Given the close connection between acting freely and making conscious decisions, it is not surprising that one might, in turn, hold that consciousness is necessary for moral responsibility as well. Levy (2014), for example, defends what he calls the consciousness thesis, that is, "consciousness of some of the facts that give our actions their moral significance is a necessary condition for moral responsibility" (Levy 2014: 1). For much more on this theme, see Caruso "Consciousness, Free Will, and Moral Responsibility," this volume. For an example of the opposing view, see Sher (2009).

11 For more on mental illness and philosophy of psychiatry, see Fulford, Thornton, and Graham (2006), Graham (2013), Zachar (2014), and Kincaid and Sullivan (2014).

12 Journals such as *Philosophy, Psychiatry, and Psychology, Cognitive Neuropsychiatry*, *Psychopathology*, and *NeuroEthics* have also helped to foster interdisciplinary work on psychopathologies and mental illness. In addition to MIT Press's *Philosophical Psychopathology* book series, Oxford University Press's *International Perspectives in Philosophy and Psychiatry* series and the *Oxford Series in Neuroscience, Law, and Philosophy* are invaluable.

References

American Psychiatric Association. (2013) *Diagnostic and Statistical Manual of Mental Disorders*, 5th edition, Washington, DC: American Psychiatric Association.

Babiak, P., Neumann, C., and Hare R. (2010) "Corporate Psychopathy: Talking the Walk," *Behavioral Sciences and the Law* 28: 174–193.

Baron-Cohen, S. (1995) *Mindblindness*, Cambridge, MA. MIT Press.

Baumeister, R., Mele, A., and Vohs, K. (eds.) (2010) *Free Will and Consciousness: How Might They Work?*, New York: Oxford University Press.

Bayne, T. (2008) "The Unity of Consciousness and the Split-Brain Syndrome," *The Journal of Philosophy* 105: 277–300.

Bayne, T. (2010) *The Unity of Consciousness*, New York: Oxford University Press.

Bayne, T. and Fernandez, J. (eds.) (2009) *Delusion and Self-Deception*, East Sussex, UK: Psychology Press.

Bayne, T. and Pacherie, E. (2005) "In Defence of the Doxastic Conception of Delusion," *Mind and Language* 20: 163–188.

Billon, A. and Kriegel, U. (2015) "Jaspers' Dilemma: The Psychopathological Challenge to Subjectivity Theories of Consciousness," In Gennaro 2015a.

Bottini, G., Bisiach, E., Sterzi, R., and Vallar, G. (2002) "Feeling Touches in Someone Else's Hand," *NeuroReport* 13: 249–252.

Bortolotti, L. (2009) *Delusions and Other Irrational Beliefs*, New York: Oxford University Press.
Bortolotti, L. and Broome, M. (2009) "A Role for Ownership and Authorship in the Analysis of Thought Insertion," *Phenomenology and the Cognitive Sciences* 8: 205–224.
Brink, D. and Nelkin, D. (2013) "Fairness and the Architecture of Responsibility," In D. Shoemaker (ed.) *Oxford Studies in Agency and Responsibility*, volume 1, New York: Oxford University Press.
Brook, A. (2015) "Disorders of Unified Consciousness: Brain Bisection and Dissociative Identity Disorder," In Gennaro 2015a.
Carruthers, P. (1996) "Autism as Mindblindness: An Elaboration and Partial Defence," In P. Carruthers and P. Smith (eds.) *Theories of Theories of Mind*, New York: Cambridge University Press.
Carruthers and P. Smith (2000) *Phenomenal Consciousness*, Cambridge: Cambridge University Press.
Caruso, G. (2012) *Free Will and Consciousness: A Determinist Account of the Illusion of Free Will*, Lanham, MD: Lexington Books.
Cleeremans, A. (ed.) (2003) *The Unity of Consciousness: Binding, Integration and Dissociation*, Oxford: Oxford University Press.
Dainton, B. (2008) *The Phenomenal Self*, New York: Oxford University Press.
de Vignemont, F. (2010) "Body Schema and Body Image – Pros and Cons," *Neuropsychologia* 48: 669–680.
Dolan, M. and Fullam, R. (2010) "Moral/Conventional Transgression Distinction and Psychopathy in Conduct Disordered Adolescent Offenders," *Personality and Individual Differences* 49: 995–1000.
Edwards, C. (2009) "Ethical Decisions in the Classification of Mental Conditions as Mental Illness," *Philosophy, Psychiatry, and Psychology* 16: 73–90.
Farah, M. (2004) *Visual Agnosia*, 2nd ed., Cambridge, MA: MIT Press.
Feinberg, T. (2001) *Altered Egos: How the Brain Creates the Self*, New York: Oxford University Press.
Feinberg, T. (2011) "Neuropathologies of the Self: Clinical and Anatomical Features," *Consciousness and Cognition* 20: 75–81.
Feinberg, T. and Keenan, J. (eds.) (2005) *The Lost Self: Pathologies of the Brain and Identity*, New York: Oxford University Press.
First, M. and Fischer, C. (2012) "Body Integrity Identity Disorder: The Persistent Desire to Acquire a Physical Disability," *Psychopathology* 45: 3–14.
Fischer, J. and Ravizza, M. (1998) *Responsibility and Control: A Theory of Moral Responsibility*, New York: Cambridge University Press.
Frankfurt, H. (1969) "Alternate Possibilities and Moral Responsibility," *Journal of Philosophy* 66: 829–39.
Frankfurt, H. (1971) "Freedom of the Will and the Concept of a Person," *Journal of Philosophy* 68: 5–20.
Frith, C. (1992) *The Cognitive Neuropsychology of Schizophrenia*, East Sussex, UK: Psychology Press.
Frith, C. and Happé, F. (1999) "Theory of Mind and Self-Consciousness: What Is It Like to Be Autistic?" *Mind and Language* 14: 1–22.
Frith, C. and Hill, E. (eds.) (2003) *Autism: Mind and Brain*, New York: Oxford University Press.
Fulford, K., Thornton, T., and Graham, G. (eds.) (2006) *Oxford Textbook of Philosophy and Psychiatry*, Oxford: Oxford University Press.
Gallagher, S. (2000) "Philosophical Conceptions of the Self: Implications for Cognitive Science," *Trends in Cognitive Sciences* 4: 14–21.
Gennaro, R. (2012) *The Consciousness Paradox: Consciousness, Concepts, and Higher-Order Thoughts*, Cambridge, MA: MIT Press.
Gennaro, R. (ed.) (2015a) *Disturbed Consciousness: New Essays on Psychopathology and Theories of Consciousness*, Cambridge, MA: MIT Press.
Gennaro, R. (2015b) "Somatoparaphrenia, Anosognosia, and Higher-Order Thoughts," In Gennaro 2015a.
Graham, G. (2013) *The Disordered Mind: An Introduction into Philosophy of Mind and Mental Illness*, 2nd edition, London: Routledge.
Graham, G. and Stephens, G.L. (eds.) (1994) *Philosophical Psychopathology*, Cambridge, MA: MIT Press.
Hare, R.D. (2003) *The Hare Psychopathy Checklist-Revised*, 2nd edition, Toronto: Multi-Health Systems.
Hillier, A. and Allinson, L. (2002) "Understanding Embarrassment Among Those with Autism: Breaking Down the Complex Emotion of Embarrassment Among Those with Autism," *Journal of Autism and Developmental Disorders* 32: 583–592.
Hirstein, W. (2005) *Brain Fiction: Self-Deception and the Riddle of Confabulation*, Cambridge, MA: MIT Press.
Kane, R. (ed.) (2011) *Oxford Handbook of Free Will*, 2nd edition, New York: Oxford University Press.
Kennett, J. and Matthews, J. (2002) "Identity, Control and Responsibility: The Case of Dissociative Identity Disorder," *Philosophical Psychology* 15: 509–526.
Kind, A. (2015) *Persons and Personal Identity*, Cambridge: Polity Press.

Kincaid, H. and Sullivan, J. (eds.) (2014) *Classifying Psychopathology: Mental Kinds and Natural Kinds*, Cambridge, MA: MIT Press.

Kriegel, U. (2009) *Subjective Consciousness*, New York: Oxford University Press.

Lane, T. (2015) "Self, Belonging, and Conscious Experience: A Critique of Subjectivity Theories of Consciousness," In Gennaro 2015a.

Lane, T. and Liang, C. (2010) "Mental Ownership and Higher-Order Thought," *Analysis* 70: 496–501.

Levy, N. (2014) *Consciousness and Moral Responsibility*, New York: Oxford University Press.

Liang, L. and Lane, T. (2009) "Higher-Order Thought and Pathological Self: The Case of Somatoparaphrenia," *Analysis* 69: 661–668.

Locke, J. (1689/1975) *An Essay Concerning Human Understanding*, P. Nidditch (ed.) Oxford: Clarendon.

McGeer, K. (2004) "Autistic Self-Awareness," *Philosophy, Psychiatry, and Psychology* 11: 235–251.

McKenna, M. and Pereboom, D. (2016) *Free Will: A Contemporary Introduction*, New York: Routledge Press.

Malatesti, L. and McMillan, J. (eds.) (2010) *Responsibility and Psychopathy: Interfacing Law, Psychiatry, and Philosophy*, Oxford: Oxford University Press.

Muller, S. (2009) "Body Integrity Identity Disorder (BIID) – Is the Amputation of Healthy Limbs Ethically Justified?" *American Journal of Bioethics* 9: 36–43.

Murphy, J. (1972) "Moral Death: A Kantian Essay on Psychopathy," *Ethics* 82: 284–298.

Nagel, T. (1971) "Brain Bisection and the Unity of Consciousness," *Synthese* 22: 396–413.

Nichols, S. and Stich, S. (2003) *Mindreading*, New York: Oxford University Press.

Parnas, J. and Sass, L. (2011) "Bodily Awareness and Self-Consciousness," In S. Gallagher (ed.) *The Oxford Handbook of the Self*, New York: Oxford University Press.

Pliushch, I. and Metzinger, T. (2015) "Self-Deception and the Dolphin Model of Cognition," In Gennaro 2015a.

Poland, J. (2014) "Deeply Rooted Sources of Error and Bias in Psychiatric Classification," In H. Kincaid and J. Sullivan (eds.) *Classifying Psychopathology: Mental Kinds and Natural Kinds*, Cambridge, MA: MIT Press.

Radden, J. ed. (2004) *The Philosophy of Psychiatry: A Companion*, New York: Oxford University Press.

Radden, J. (2010) *On Delusion*, Abingdon and New York: Routledge.

Ramachandran, V.S. (2004) *A Brief Tour of Human Consciousness*, London: Pearson Education.

Rosenthal, D. (2005) *Consciousness and Mind*, New York: Oxford University Press.

Rosenthal, D. (2010) "Consciousness, the Self and Bodily Location," *Analysis* 70: 270–276.

Russell, P. (2004) "Responsibility and the Condition of Moral Sense," *Philosophical Topics* 32: 287–305.

Ryan, C. (2009) "Out on a Limb: The Ethical Management of Body Integrity Identity Disorder," *Neuroethics* 2: 21–33.

Sabanz, N. and Prinz, W. (eds.) (2006) *Disorders of Volition*, Cambridge, MA: MIT Press.

Sacks, O. (1987) *The Man Who Mistook His Wife for a Hat and Other Clinical Tales*, New York: Harper and Row.

Schechter, E. (2012) "The Switch Model of Split-Brain Consciousness," *Philosophical Psychology* 25: 203–226.

Schechter, E. (2018) *Self-Consciousness and "Split" Brains: The Minds' I*, New York: Oxford University Press.

Sher, G. (2009) *Who Knew? Responsibility Without Awareness*, New York: Oxford University Press.

Shoemaker, D. (2011) "Psychopathy, Responsibility, and the Moral/Conventional Distinction," *Southern Journal of Philosophy* 49: 99–124.

Sierra, M. and Berrios, G. (2000) "The Cambridge Depersonalisation Scale: A New Instrument for the Measurement of Depersonalisation," *Psychiatry Research* 93: 153–164.

Sperry, R. (1984) "Consciousness, Personal Identity and the Divided Brain," *Neuropsychologia* 22: 611–73.

Stephens, G.L. and Graham, G. (2000) *When Self-Consciousness Breaks*, Cambridge, MA: MIT Press.

Strawson, P. (1962) "Freedom and Resentment," *Proceedings of the British Academy* 48: 1–25.

Szasz, T. (1974) *The Myth of Mental Illness*, New York: Harper and Row.

Teuber, H. (1968) "Alteration of Perception and Memory in Man," In L. Weiskrantz (ed.) *Analysis of Behavioral Change*, New York: Harper & Row.

Tye, M. (2003) *Consciousness and Persons*, Cambridge, MA: MIT Press.

Vallar, G. and Ronchi, R. (2009) "Somatoparaphrenia: A Body Delusion: A Review of the Neuropsychological Literature," *Experimental Brain Research* 192: 533–551.

Weiskrantz, L. (1986) *Blindsight*, Oxford: Clarendon.

Zachar, P. (2014) *A Metaphysics of Psychopathology*, Cambridge, MA: MIT Press.

Related Topics

The Unity of Consciousness
Global Disorders of Consciousness
Consciousness, Free Will, and Moral Responsibility
Consciousness, Personal Identity, and Immortality
Representational Theories of Consciousness
Consciousness and Attention

26
POST-COMATOSE DISORDERS OF CONSCIOUSNESS

Andrew Peterson and Tim Bayne

1 Introduction

Disorders of consciousness can be grouped into two classes: local disorders and global disorders. Local disorders of consciousness involve circumscribed deficits to a particular domain of conscious processing. Blindsight is perhaps the most well-known of all local disorders of consciousness. Patients with blindsight have lesions in the visual cortex that result in a loss of conscious experience for stimuli presented to certain regions of their visual field. Global disorders of consciousness do not involve focal deficits of this kind, but instead involve impairments to the subject's overall state of consciousness—they are domain-general rather than domain-specific.

Some of the most intensely studied global disorders of consciousness involve patients who have undergone a severe brain injury, typically caused by trauma or the loss of oxygen to the brain. Those who sustain the most severe brain injuries will spend some time in a state of unarousable unawareness, also known as the comatose state (Young 2009). After a period of days or weeks, some comatose patients may enter one of two post-comatose states: the Vegetative State (VS) or the Minimally Conscious State (MCS). This chapter is concerned with these post-comatose disorders of consciousness, the use of neuroimaging and electroencephalography to detect preserved consciousness in post-comatose patients, and the ethical challenges that are raised by the detection of consciousness in these patients.

2 The Vegetative and Minimally Conscious States

Clinically, the VS, which is sometimes referred to as "unresponsive wakefulness syndrome" (Laureys et al. 2010), is characterized by wakefulness without awareness. VS patients exhibit a sleep/wake cycle and spontaneously open their eyes, but they do not manifest any behavioral signs of awareness. In the words of the Royal College of Physicians (2003: §2.2), in the VS there should be "no evidence of awareness of self or the environment, no responses to external stimuli of a kind that would suggest volition or purpose (as opposed to reflexes), and no evidence of language expression or comprehension."

Although some patients remain in the VS indefinitely or die, others transition into the MCS (Beaumont and Kenealy 2005). Patients with non-traumatic etiologies (e.g., cardiac arrest) are unlikely to transition to the MCS if they remain in a VS for longer than three months, while patients with traumatic etiologies are unlikely to transition to the MCS if they remain in a VS longer than 12 months (Multi-Society Task Force 1994). In current clinical practice, the transition from the VS to the MCS is marked by the repeated and task-appropriate production of at least one of the following behaviors: sustained visual pursuit, localization of noxious stimuli, and command-following (Giacino et al. 2002). Patients who communicate (e.g., respond correctly to a series of questions) or use objects appropriately (e.g., demonstrating how a coffee mug is used) are taken to have left the MCS and to have entered a state of "full consciousness," albeit a state of consciousness that will typically be characterized by confusion and disorientation (Giacino et al. 2014).

The VS and the MCS can be contrasted with the Locked-In-Syndrome (LIS), a disorder that is caused by focal lesions to the brainstem resulting in widespread paralysis (Bauer et al. 1979). LIS patients resemble post-comatose patients in possessing the capacity for only a very restricted range of behaviors. However, unlike VS or MCS patients, LIS patients have not suffered damage to the cortical structures responsible for awareness. Their capacity for motor responses is severely limited, but they possess a normal suite of rational capacities and are undoubtedly conscious.

Discussion of post-comatose disorders of consciousness has centered on two sets of questions. The first set of questions concerns the ascription of consciousness to these patients. The border between unconsciousness and consciousness in post-comatose disorders has traditionally been taken to coincide with the border between the VS and the MCS; indeed, the very labels for these diagnoses presuppose that only MCS patients have a standing capacity for consciousness. However, this assumption has been undermined over the last decade by neuroimaging and electroencephalographic (EEG) data, which suggest that significant numbers of patients who appear to be in the VS at the bedside—and indeed *are* VS according to current medical guidelines—do retain some form of consciousness. Such patients can be described as "covertly conscious."

The second set of questions concerns the implications of this research for our treatment of post-comatose patients. Should neuroimaging and EEG measures be incorporated into the routine clinical assessment of such patients? If so, how should the information that is gleaned from such tests impact on the medical care that is provided to them? And in what way should the ascription of consciousness to these patients change our view of their moral status?

The remainder of this chapter examines the debates surrounding these two sets of questions: Sections 3 through 6 examine the question of how covert consciousness might be investigated. Section 7 examines the implications of covert consciousness for diagnostic taxonomy, while Section 8 considers some of the ethical issues raised by the discovery of covert consciousness in post-comatose patients.

3 Bedside Behavioral Examination

The question of whether post-comatose patients are conscious has traditionally been addressed by bedside examination. The most widely used examination is the JFK-Coma Recovery Scale-Revised (CRS-R), which tests the capacity of patients to execute a range of behaviors, such as raising an arm in response to command or fixating on a visual stimulus. Individuals who perform well on the CRS-R are regarded as being either minimally conscious or—if they show evidence of the capacity to communicate and use objects in appropriate ways—as having entered a state of full consciousness.

One's attitude toward the CRS-R (and related behavioral examination schedules) will depend on one's view of the relationship between consciousness and behavior. Such schedules carry the implicit assumption that the capacity to produce behaviors of the relevant type is strongly correlated with the presence of consciousness, whereas the incapacity to produce such behaviors is strongly correlated with the absence of consciousness. Call this the *behavioral assumption*.

There are two kinds of problems with the behavioral assumption. The first set of problems is practical: it is often very difficult for clinicians to determine whether a patient possesses the capacity to execute the relevant behaviors. Consider a patient who fails to respond to a request to blink. Although it is possible that the patient fails to blink because she doesn't possess the capacity for consciousness, it is also possible that she fails to blink because she is asleep at the time of examination. The behavioral repertoire of patients depends on their level of arousal, and arousal in post-comatose patients can fluctuate widely throughout a 24-hour period (Candelieri et al. 2011). Alternatively, the patient might have failed to blink because, although conscious, she: was deaf and couldn't hear the instruction; was aphasic and couldn't understand the instruction; or suffered from poor muscle control and couldn't comply with the command. Assuming that the patient is in a VS provides one explanation for why she failed to blink on command, yet this will rarely be the *only* explanation for her failure to blink, and it may not be easy to determine which of the various explanations on offer is the most plausible. Even if the patient does blink, it may be unclear whether the blink was a response to a command or whether it was a purely reflex response that was unrelated to the instruction (Majerus et al. 2005). A general challenge here is that the VS is diagnosed on the basis of an *absence* of evidence of consciousness, thus making it highly susceptible to false negatives.

Although careful and repeated clinical assessment over days or weeks can address some of these problems, it cannot address a second challenge to the behavioral assumption, a challenge that derives from the subjective nature of consciousness. A conscious organism is an organism that has a subjective perspective on the world (Nagel 1974). The clinician, of course, has access only to what a patient can—or, as the case may be, cannot—do. Bedside examinations attempt to bridge the gap between the objective and the subjective by making assumptions about the relationship between consciousness and behavioral capacities, but these assumptions can be challenged.

Most discussion of this issue has focused on the worry that the assumptions that are embedded in the CRS-R are too conservative, and that many of the patients that the CRS-R regards as being in the VS (and thus unconscious) might actually be conscious. (As we shall see, there are good reasons to think that this worry is well-founded.) It is also possible that some of the assumptions that are embedded in the CRS-R are too liberal, and that certain behaviors that are treated by the CRS-R as markers of consciousness might be produced unconsciously. Consider the relationship between consciousness and the control of eye movements. Although the CRS-R treats visual fixation and pursuit as diagnostic of the MCS, it is unclear whether these responses are robust indicators of visual awareness (Spering and Carrasco 2015; Vanhaudenhuyse et al. 2008). In this regard, it is worth noting that the clinical guidelines for the assessment of patients themselves betray a certain amount of uncertainty about the relationship between consciousness and the control of eye movements. Although the CRS-R takes visual fixation and pursuit to be sufficient for a diagnosis of the MCS, the Royal College of Physicians (2003), to whom Britain and much of Europe defers for diagnosis, regards these behaviors as consistent with, albeit atypical manifestations of, the VS.

4 From the Beside to the Scanner: The Command-Following Paradigm

In response to these challenges, researchers have developed methods that aim to detect consciousness on the basis of brain activity. The hope is that such methods could complement behavioral examination, particularly when a patient's clinical presentation is ambiguous. (Indeed, some theorists have even suggested that brain-based methods might eventually displace behaviorally-based methods for assaying consciousness.) Although the discussion of these methods has focused on the question of whether they can be used to support the ascription of consciousness to patients who appear from the bedside to be VS, it is worth noting that they might also be used to show that some MCS patients enjoy a much wider range of conscious states than might be apparent on the basis of their overt behavior.

The first neuroimaging data to provide robust evidence of consciousness in a VS patient were obtained in a functional magnetic resonance imaging (fMRI) study of a 23-year-old woman who had been in a VS for five months (Owen et al. 2006). While lying in the scanner, the patient was given the auditory command to imagine one of two activities—playing tennis or visiting the rooms of her home—for discrete and repeated 30-second intervals. The brain activity preferentially involved in the two tasks (compared with rest) was indistinguishable from that seen in 34 healthy volunteers: the instruction to play tennis was associated with increased activation in the supplementary motor area (SMA), while the spatial navigation instruction was associated with increased activation in the parahippocampal gyrus, posterior parietal lobe, and the lateral premotor cortex (Boly et al. 2007; Owen et al. 2006).

In the decade since the publication of this ground-breaking paper, a number of other command-following studies have been conducted. Some of these studies have followed Owen et al. in using fMRI to assess command-following (Stender et al. 2014; Bardin et al. 2011; Monti et al. 2010), whereas others have used EEG (Goldfine et al. 2011; Cruse et al. 2011). Some of these studies have followed Owen et al. in assessing the capacity of patients to engage in imagery involving spatial navigation or simple motor responses, whereas others have instructed patients to perform tasks that involve attending to certain aspects of their environment. For example, some studies have examined the capacity of VS and MCS patients to attend to instances of particular words (such as their own name) that are presented in an auditory stream. Neural evidence of the capacity for selective attention has been found in some VS (Naci and Owen 2013) and MCS patients (Schnakers et al. 2008; Monti et al. 2015). Interestingly, the capacity of the MCS patients to follow commands in the aforementioned studies was not manifest in their overt behavior, and they had been classified as MCS only because of the low-level behaviors (such as visual fixation and pursuit) that they had shown.

The most striking command-following studies have used neuroimaging as a channel of communication. The first of these studies was conducted by Monti et al. (2010), who asked a VS patient six yes/no autobiographical questions (such as, "Is your father's name Alexander?"). The patient was instructed to engage in either motor imagery or spatial imagery (depending on the trial) to answer "yes" or "no." The patient produced activation indicative of a correct answer in response to five of the six questions. The sixth question elicited no significant activation in the regions of interest. These findings have been replicated in other VS patients (Naci and Owen 2013; Bardin et al. 2011). In the most extensive communication study to date, a patient diagnosed as being in a VS for 12 years answered ten yes/no questions, including "Are you in pain?" (Fernández-Espejo and Owen 2013). Some authors have argued that, given certain technological advances, these methods might allow some post-comatose patients to participate in some medical decisions (Cairncross et al. 2016; Glannon 2016; Peterson et al. 2013a; 2013b; Bendsten 2013).

A recent meta-analysis of command-following studies suggests that approximately 15% of patients who would be regarded as VS on the basis of current clinical assessment schedules appear to be capable of covert command-following—that is, they produce sustained, region-appropriate neural activity in response to commands (Kondziella et al. 2016). The philosophically interesting questions are whether these neural responses ground an ascription of consciousness to these patients and, if so, how that ascription is to be justified.

There are two ways in which one might use a patient's performance in a command-following paradigm to make a case for consciousness. Most fundamentally, one might argue that command-following is itself good evidence of consciousness. Alternatively, one might argue that whether or not command-following is good evidence of consciousness, certain command-following paradigms require the patient to perform cognitive tasks, the execution of which provides good evidence of consciousness. Consider the Monti et al. (2010) study, in which a patient correctly answered questions by engaging in either motor imagery or spatial imagery. There is little doubt that this patient was conscious, but one could plausibly argue that the ascription of consciousness is justified not because the patient was following commands, but because he was answering questions correctly, or because his answers indicated that he had access to autobiographical knowledge.

The interesting question here is whether covert command-following is *itself* robust evidence of consciousness. This is an important question, for only a small fraction of behaviorally non-responsive patients who are able to engage in covert command-following are also able to use this capacity to communicate. Thus, we need to know whether mere command-following provides us with good reason to regard a patient as conscious.

The authors of the original command-following study left little room for doubt regarding their answer to this question:

> [the patient's] decision to cooperate with the authors by imagining particular tasks when asked to do so represents a clear act of intention, which confirmed beyond any doubt that she was consciously aware of herself and her surroundings.
>
> *(Owen et al. 2006: 1402)*

This position has not gone unchallenged. Some commentators have taken issue with the assumption that intentional agency is a marker of consciousness, claiming that "consciousness is univocally probed in humans through the subject's report of his or her own mental states" (Naccache 2006). We will set this challenge to one side, for it sets too high a bar on the ascription of consciousness given that we routinely ascribe consciousness to non-verbal beings, such as pre-linguistic infants and non-human animals. A more serious challenge is that even if intentional agency provides robust evidence of consciousness, command-following studies may not establish that post-comatose patients are capable of intentional agency (Drayson 2014; Davies and Levy 2016; Klein 2017). Different authors develop this objection in slightly different ways, but at the heart of their concerns is the claim that the command-following studies show only that patients are able to act in response to direct perceptual stimulation, whereas genuine intentional agency requires endogenously-generated (that is, stimulus-independent) intentions. Call this the *challenge from intentional agency*.

In responding to this challenge, we might begin by noting that it is not entirely clear that covert command-following *is* purely stimulus-driven. Although the patient's mental imagery is triggered by a command (presumably the patient wouldn't have imagined herself playing tennis unless she had been instructed to do so), the imagery is sustained for 30 seconds, and terminates only when the command, "relax," is given. It seems unlikely that the patient's mental imagery would exhibit this temporal profile if it weren't being appropriately guided by an intention, albeit one that the patient has presumably formed as a result of the instructions.

A second and more fundamental response to the challenge from intentional agency is that the very conception of intentional action that lies behind the challenge is open to dispute. Proponents of the challenge deny that stimulus-driven actions qualify as genuinely intentional. We regard this as an overly restrictive conception of intentional action, and would allow actions that are guided by external stimuli to qualify as genuinely intentional. Consider the kind of unreflective behavior that characterizes much of our daily life, such as picking up a slice of pizza or braking in response to a traffic light turning red. We view such unreflective actions as intentional, despite the fact that they are not guided by endogenous intentions.

Thirdly, skepticism about the degree to which *covert* command-following is a marker of consciousness threatens to spawn a broader skepticism about the degree to which *overt* command-following is a marker of consciousness. If one is unwilling to ascribe consciousness on the basis of covert evidence of command-following, why should one be willing to ascribe it on the basis of overt evidence of command-following?

Fourthly, it is possible to defend the use of command-following studies to ascribe consciousness to patients without appealing to the connection between intentional agency and consciousness. Instead, one might appeal to the content of the command in question. Highly automated responses to familiar stimuli (e.g., braking in response to a red traffic light) may not require consciousness, but appropriate responses to *unfamiliar* and *complex* stimuli probably do. In our view, the instruction to imagine oneself playing tennis or navigating around one's home is both unfamiliar and complex. To comply with these commands one must *understand* them, and it is doubtful that comprehension of the relevant sentences could occur unconsciously.

5 From the Bedside to the Scanner: Passive Paradigms

Methods that require patients to covertly follow commands are cognitively demanding, and it is therefore conceivable that certain patients could be conscious but be unable to pass command-following tests. To identify such patients, methods have been developed that assess consciousness without requiring the patient to perform a task. These methods are called "passive" paradigms.

One such paradigm probes the capacity of patients to respond to stimuli with narrative structure, such as films. In the first demonstration of this method, Naci et al. (2014) presented an engaging, eight-minute Alfred Hitchcock film entitled "*Bang! You're Dead*" to a group of healthy participants and post-comatose patients while recording their brain activity with fMRI. The film's plot involves a young boy who mistakes his uncle's loaded revolver for a toy. In several scenes of the film, the boy spins a single bullet in the revolver (as in a game of Russian roulette) and, taking aim at other characters, pulls the trigger.

Naci et al. (2014) found that neural activity associated with executive processing (such as in the fronto-parietal network) was synchronized across healthy participants while they viewed the film, particularly during suspenseful scenes. This activity was then compared against the brain activity of two VS patients as they viewed the film. One of these patients produced brain activity that was highly synchronized with that of the healthy participants, suggesting that the film had elicited experiences in him akin to those enjoyed by the healthy participants. Astonishingly, this patient had been behaviorally non-responsive for 16 years. In a subsequent study, Naci et al. (2017) found evidence of neural synchronization in similar patients using an engaging sound clip from the film *Taken*.

Another passive paradigm that has been developed for detecting covert consciousness involves the use of EEG to identify neural responses to two kinds of auditory irregularities: local irregularities and global irregularities. Local irregularities occur when a deviant tone follows a repeating standard tone (e.g., four low tones followed by a high tone), whereas global irregularities occur when a deviant sequence of tones follows a repeating standard sequence (e.g., a sequence of five low tones following three iterations of a sequence in which four low tones are followed by a high tone). Evidence from studies involving priming and anesthesia suggests that although local irregularities can be detected unconsciously, consciousness is required for the detection of global irregularities. Crucially, global irregularities are associated with a distinctive brain-response—known as the P300b response—that can be detected using EEG. In light of these findings, a number of research teams have looked for evidence of the P300b response in post-comatose patients. King et al. (2013) found a global effect in 14% of 70 VS patients and 31% of 65 MCS patients, and similar findings have been obtained in a number of other studies (Bekinschtein et al. 2009; Faugeras et al. 2011; Chennu et al. 2013).

The fact that these methods do not require command-following has important implications. Passive paradigms allow for the evaluation of consciousness in patients who may be conscious but are unable to follow commands due to the high cognitive load imposed by command-following tasks. Passive paradigms might also allow for the assessment of particular cognitive capacities that are not recruited by command-following paradigms. The capacity to follow the narrative structure of a film requires a number of cognitive capacities, and information about which patients retain these capacities may be very useful for attempts to improve quality-of-life in patients (Naci et al. 2016).

6 The Puzzle of Behavioral Inactivity

Although debate continues about the evidential force of the data surveyed in the previous two sections, it is increasingly difficult to deny that significant numbers of patients who qualify as VS according to current clinical guidelines are covertly conscious. This finding raises the *puzzle of (behavioral) inactivity*: given that such patients are conscious, why do they not produce purposeful motor behavior (Shea and Bayne 2010)? Locked-in syndrome patients raise no such puzzle, for we know that they are paralyzed. By contrast, VS patients are not paralyzed. Those who ascribe consciousness to putatively VS patients need to explain why they do not produce overt behavior in the way that MCS patients do.

Klein (2017) has suggested that this puzzle might be solved by appealing to the idea that such patients have *motivational* deficits: although they *can* generate motor responses, they are simply not motivated to do so. In defense of this idea, he notes that there is a close anatomical relationship between the VS and akinetic mutism, a condition in which patients do not act unless prompted to do so by an external stimulus. Klein's proposal is helpful in that it might explain why some patients fail to act. However, it cannot explain why those patients who comply with commands to generate mental imagery do not also comply with commands to execute motor responses.

A recent study by Fernández-Espejo et al. (2015)[1] may provide evidence against Klein's proposal. The study examined the connections between the thalamus and primary motor cortex in two post-comatose patients, one of whom was VS according to behavioral criteria but showed evidence against command-following, and one of whom was MCS. The white matter tracts projecting from the thalamus to the motor cortex were damaged in the former patient but not in the latter. It is plausible that these pathways are necessary for producing motor responses but are not required for mental imagery. Damage to these pathways would explain why a patient who was able to sustain covert command-following was overtly unresponsive. Such a patient might have intentions to produce overt behavior, but would be unable to implement those intentions.

7 Validation and Taxonomy

The research surveyed in this chapter promises to reshape how we think about the recovery of consciousness following severe brain injury, for clinicians now have compelling reason to supplement the behavioral assessments of patients with EEG and neuroimaging (Coleman et al. 2009).

However, the translation of these methods to the clinical setting raises a number of challenging epistemic questions. For example, decisions will need to be made about how much evidential weight should be given to covert command-following, the capacity to follow narrative structure, or to detect violations of global regularity when it comes to the ascription of consciousness.

These questions are challenging, but they are not different in kind from those that we face in determining how much weight to place on *behavioral* measures of consciousness. (Consider, for example, the debate regarding visual fixation and pursuit as markers of consciousness, which we noted in Section 3.) Indeed, in certain cases—for example, with respect to covert command-following and overt command-following—these questions are essentially variants of a single question about clinical validation: how do we know that any given assay of consciousness is actually measuring consciousness?

In previous work, we have articulated what we have variously called the "consilience" (Peterson 2016) or "natural kind" approach (Shea and Bayne 2010) to questions of validation. In essence, this approach holds that novel measures of consciousness should be validated on the basis of their fit with other putative measures of consciousness. Greater evidential weight should be given to novel measures that produce evidence that is conciliate with other putative measures, while less evidential weight should be accorded to novel measures that are less strongly associated with other putative measures of consciousness. A certain kind of circularity is inherent in this approach, but this is arguably a form of circularity that is inherent in any attempt to validate novel measurement techniques in domains in which we lack "gold standard" tests. Recently, several research teams have adopted this approach in their own neuroimaging and EEG studies (Sergent et al. 2017; Chennu et al. 2017; Demertzi et al. 2017).

Another way in which neuroimaging and EEG research promises to reshape how we think about the recovery of consciousness following brain injury concerns the very diagnostic categories that are used in neurology (Bayne et al. 2017; Peterson and Bayne 2017). As we have seen, neurologists recognize only two categories in this area: the VS and the MCS. It is now clear that this taxonomic system is overly simple, and that the field would benefit from the development of a more nuanced taxonomy that reflects the spectrum of conscious states and capacities that can be found in post-comatose patients. Some patients exhibit low-level perceptually guided behaviors, such as visual pursuit and fixation. Some patients exhibit command-following behaviors. Some patients appear to be able to follow the plot of a film. Indeed, even further distinctions need to be made within these categories, for some patients have the capacity to engage in overt command-following but lack the capacity to engage in covert command-following, whereas other patients manifest precisely the opposite suite of capacities (Gibson et al. 2014; Monti et al. 2010). Further, the fact that a patient is sensitive to the narrative structure of a film does not imply that she can communicate through neuroimaging (Naci et al. 2014, 2016). Although some of the capacities that are probed by these paradigms are hierarchically related, others are not.

An adequate taxonomy for post-comatose disorders of consciousness should capture these variations in cognitive capacities. The recent introduction by Bruno et al. (2011b) between two types of MCS patients, MCS+ patients and MCS− patients, is a first step in this direction. Bruno et al. (2011b) suggest that the label "MCS+" should be reserved for patients who can follow commands, whereas the label "MCS−" should be reserved for patients who lack this capacity but show other behavioral signs of consciousness, such as visual fixation and pursuit or the ability to localize noxious stimuli. Although these categories have not yet been incorporated into neurological taxonomy, in our view there is every reason for this step to be taken.

An important backdrop to the taxonomic developments in this field concerns a current debate about how best to understand "global" states of consciousness. Many authors embrace a level-based conception of global states, according to which alterations in an individual's global state of consciousness are to be understood in terms of changes in their level of consciousness (e.g. Laureys 2005). Those who endorse this conception typically assume that VS patients have a lower level of consciousness than MCS patients, and that MCS patients in turn have a lower level of consciousness than patients who have returned to "full consciousness." However, a number of authors have argued that this conception of global states of consciousness is problematic (Bayne et al. 2016; Bayne and Hohwy 2016; Klein and Hohwy 2015). For one thing, genuine VS patients (that is, patients who have sleep/wake cycles but no standing capacity for consciousness) are completely unconscious, and thus it is misleading to describe them as possessing any "level" of consciousness. Even when we restrict our attention to post-comatose patients who are conscious, it is far from clear that talk of "levels of consciousness" is helpful. Although many of the cognitive capacities that are associated with consciousness can be graded (for example, one patient might be more responsive to perceptual stimulation than another), it is unclear whether consciousness itself can be graded. To be conscious is to possess a subjective perspective, and that is not a property that seems to admit of degrees.

Instead of assuming that all (conscious) post-comatose patients can be assigned to a single "level of consciousness," we think that a multidimensional approach to consciousness is more likely to do justice to the variation exhibited by post-comatose patients (Bayne et al. 2016; Sergent et al. 2017; Peterson and Bayne 2017). This approach promises to provide a more fine-grained framework that captures the variation in cognitive and behavioral capacities exhibited by post-comatose patients, without suggesting that this variation must correspond to variation in degrees of consciousness. A key question, of course, concerns the nature of the dimensions employed by any such taxonomy. We expect that this issue will be on the agenda of theorists in this field for some time.

8 Ethical Dimensions of Post-Comatose Disorders of Consciousness

We turn now to some of the many ethical issues raised by the findings that have been surveyed in this chapter (Weijer et al. 2014; Fins et al. 2008). What implications might covert consciousness have for the moral and legal protections that we should afford to post-comatose patients? In what ways might it change the attitudes that medical professionals and family members ought to take toward patients? Would the discovery of covert conscious give us a reason to *provide* a patient with life-sustaining treatment, or might it instead give us a reason to *withdraw* life-sustaining treatment?

At the heart of these questions lie issues of moral status. A being has moral status if we owe it moral consideration for its own sake. Violation of a being's moral status results in harm to it. A corpse has no moral status, for although there are moral (and indeed legal) constraints on what can be done to a corpse, we do not owe a corpse moral consideration for its own sake—a corpse cannot be harmed. You, by contrast, can be harmed, and are thus owed moral consideration for your own sake.

What difference—if any—might the ascription of consciousness make to the moral status of post-comatose patients? There is little doubt that the presence of consciousness should make *a* difference to moral status, but there is debate about precisely what kind of difference it makes and why it makes the difference it does.

Let us begin with what we regard as common ground in the literature: things matter to conscious patients in a way in which they don't matter to unconscious patients. A conscious patient is likely to have the capacity to experience pleasurable bodily sensations and perceptual states; they are likely to also have the capacity to experience unpleasant bodily sensations and perceptual states. We should therefore treat patients who are likely to be conscious with care and consideration, attempting to both minimize their chances of having aversive experiences and maximize their chances of having meaningful and positive experiences (Graham et al. 2015). If we remain uncertain about the presence of consciousness in certain patients, it is ethically prudent to treat them as if they were conscious (Peterson et al. 2015). Doing so avoids the various ethical hazards that might arise if we are mistaken about a patient's purported unconsciousness.

The more controversial question is whether the (mere) presence of consciousness implies that patients have what many authors refer to as "full moral status." A creature has full moral status if and only if it has a right to life. If post-comatose patients have full moral status, then we have an obligation not merely to prevent them from suffering, but also to ensure that they are not subject to premature death.

A number of prominent bioethicists have argued that even covertly conscious post-comatose patients lack full moral status. Levy and Savulescu (2009) write:

> A being acquires a full moral status, including the right to life, if its life matters to it; that is, if it is not only momentary experiences that matter—as for the being capable only of phenomenal consciousness—but also an ongoing series of experiences. A full right to life requires that it is not only experiences that matter to one, but also how one's life actually goes; that is, that satisfaction of one's interests matter to one, and this requires very sophisticated cognitive abilities, such as an ability to conceive of oneself as a being persisting through time, to recall one's past, to plan, and to have preferences for how one's life goes.
>
> *(Levy and Savulescu 2009: 367; see also Davies and Levy 2016; Kahane and Savulescu 2009)*

On this view, post-comatose patients have the kind of moral status that is often attributed to many non-human animals: although we ought to take their welfare into account in our decision-making, they lack a right to life, and we need little justification for (painlessly) ending their life. What should we make of this position?

We agree with Levy and Savulescu that the fMRI and EEG data surveyed in this chapter provide little reason to think that post-comatose patients possess the cognitive capacities that they take to be required for full moral status. For example, the fact that a patient might be able to follow commands does not give us reason to think that he or she conceives of him- or herself as a being that persists through time. The ability to correctly answer autobiographical questions arguably provides the strongest evidence for the kind of cognitive capacities that Levy and Savulescu require for full moral status, but even here it is unclear whether this ability involves episodic memory (and thus speaks to the question of moral status) or whether it involves only semantic memory (and thus does not speak to this issue). Indeed, even patients who have emerged from the MCS and are able to communicate may be unable to "recall their past, plan, or have preferences for how their life goes."

However, these considerations show that post-comatose patients lack full moral status only if Levy and Savulescu's account of full moral status is defensible. A full-scale examination of that issue goes well beyond the scope of this chapter, but we do want to draw attention to some of the central challenges facing their account. Most fundamentally, it is at odds with current conceptions of the scope of full moral status, for we routinely afford full moral status to individuals—such as neonates, amnesiacs, and those suffering from advanced dementia—who lack the cognitive capacities that Levy and Savulescu take to be required for full moral status. In other words, Levy and Savulescu's account of full moral status has radically counter-intuitive consequences.

Levy and Savulescu might respond to this challenge in one of two ways. On the one hand, they could bite the bullet, and hold that our current practice of affording full moral status to neonates, amnesiacs, and those suffering from advanced dementia is mistaken. Although this response will have its advocates, we see little prospect of it finding broad favor within the healthcare and bioethics communities, for we do treat these patient groups as though they have full moral status, regardless of the deficiencies they might have in particular cognitive capacities.

An alternative—and to our mind more plausible—response would be to modify their account of full moral status so as to ensure that neonates, amnesiacs, and those suffering from advanced dementia possess full moral status despite their defects. Here one might argue that an account of a creature's moral status must be informed not only by information about its cognitive capacities, but also by information about the social relationships in which it is embedded. If an account of full moral status is modified in this way, then post-comatose patients might indeed qualify as possessing full moral status. After all, brain-damaged patients once enjoyed complex forms of self-consciousness, and they are the natural objects of moral care and concern from families and community-members in the way that neonates, amnesiacs, and those suffering from severe dementia are.

In addition to wrestling with questions of moral status, the bioethics literature has also wrestled with questions relating to quality-of-life in post-comatose patients. What kind of subjective life do such patients have, and is it a life that is in any sense worth living?

Kahane and Savulescu (2009) argue that there is good reason to think that such patients are unlikely to have a life worth living, and thus that it might be in the interests of such patients for their life to be terminated. In support of this position, they note that a number of studies have indicated that the majority of people would prefer not to be given life-sustaining treatment if they were in a non-reversible VS (Frankl et al. 1989; Emmanuel et al. 1991). They go on to suggest that this preference might reflect "recognition of an objective interest in not continuing to exist in a state that has no personal meaning and that could be seen as degrading to one's dignity as a rational being" (Kahane and Savulescu 2009: 15). We are not persuaded that the results of these studies have much relevance for questions about the moral significance of covert consciousness in post-comatose patients, for the individuals who participated in these studies presumably understood by "the vegetative state" a state of complete unconsciousness.

Of more direct relevance to the current discussion are studies of quality-of-life in locked-in syndrome patients, which suggest that such patients enjoy surprisingly high levels of subjective well-being (Bruno et al. 2011a; Lulé et al. 2009; Nizzi et al. 2012). On the basis of these studies, one might be tempted to draw conclusions about subjective well-being in covertly conscious VS patients. However, Kahane and Savulescu point out that there are important differences between the locked-in syndrome patients that are surveyed in these studies and covertly conscious VS patients. The former are capable of some form of communication and agency, whereas the latter (typically) have no capacity for communication or external agency. Kahane and Savulescu conclude that it is far from obvious that the lives of covertly conscious VS patients are worth living, and suggest that their condition is "far worse than that of someone in the worst form of solitary confinement." Terminating the life of such individuals, they suggest, might not be merely permissible but morally required.

Kahane and Savulescu are right to point out the potential problems in generalizing from claims about the subjective well-being in (partially) locked-in patients to that of covertly conscious VS patients. However, we are not convinced that the quality-of-life that they imagine covertly conscious VS patients might have would be as bleak as they suggest. Although such patients will not be able to communicate with those around them (at least not without the aid of neuroimaging), they are likely to be aware of the presence of friends and family members, and they may understand at least some of what is said to them (Graham 2017). Such a life affords some possibilities for meaningful social interaction, albeit in forms that are highly restricted in scope. It is far from obvious to us that such a life is not a life worth living, although we would be the first to admit that this question deserves further attention.

9 Conclusion

This chapter has provided an overview of the state-of-the-art neuroimaging and EEG methods used to detect covert consciousness in post-comatose patients. We have outlined several philosophical controversies surrounding this research, and have sketched preliminary responses to a number of these issues. Although post-comatose disorders of consciousness concern only a relatively narrow subset of brain-injured patients, the scientific study of these patients and related philosophical reflection on the implications of such research promises to benefit not only brain-injured patients but will also contribute to our understanding of the nature of consciousness itself.[2]

Notes

1 Note that although the print version of Klein's paper appeared only in 2017, it was published online in 2015, and prior to the publication of Fernández-Espejo et al. (2015).

2 We are grateful to Neil Levy for comments on an earlier draft of this chapter. We also gratefully acknowledge the support of an Australian Research Council Future Fellowship to Bayne (FT150100266).

References

Bardin, J.C., Fins, J.J., Katz, D.I., Hersh, J., Heier, L.A., Tabelow, K., Dyke, J.P., Ballon, D.J., Schiff, N.D., and Voss, H.U. (2011) "Dissociations between Behavioral and Functional Magnetic Resonance Imaging-Based Evaluation of Cognitive Function after Brain Injury," *Brain* 134: 769–782.

Bauer, G., Gerstenbrand, F., and Rumpl, E. (1979) "Varieties of the Locked-In Syndrome," *Journal of Neurology*, 221: 77–91.

Bayne, T. and Hohwy, J. (2016) "Modes of Consciousness," in W. Sinnott-Armstrong (ed.) *Consciousness after Severe Brain Damage: Medical, Legal, Ethical, and Philosophical Perspectives*. New York: Oxford University Press, pp. 57–80.

Bayne, T., Hohwy, J., and Owen, A. (2016) "Are There Levels of Consciousness?" *Trends in Cognitive Sciences* 20: 405–413.

Bayne, T., Hohwy, J., and Owen, A. (2017) "Reforming the Taxonomy of Disorders of Consciousness," *Annals of Neurology* 82: 866–872.

Beaumont, J. G. and Kenealy, P. M. (2005) "Incidence and Prevalence of the Vegetative and Minimally Conscious States," *Neuropsychological Rehabilitation* 15: 184–189.

Bekinschtein, T.A., Dehaene, S., Rohaut, B., Tadel, F., Cohen, L., and Naccache, L. (2009) "Neural Signature of the Conscious Processing of Auditory Regularities," *Proceedings of the National Academy of Sciences* 106: 1672–1677.

Bendtsen, K. (2013) "Communicating with the Minimally Conscious: Ethical Implications in End-of-Life Care," *AJOB Neuroscience* 4: 46–51.

Boly, M., Coleman, M.R., Davis, M.H., Hampshire, A., Bor, D., Moonen, G., Maquet, P.A., Pickard, J.D., Laureys, S., and Owen, A.M. (2007) "When Thoughts Become Action: An fMRI Paradigm to Study Volitional Brain Activity in Noncommunicative Brain Injured Patients," *Neuroimage* 36: 979–992.

Bruno, M.-A., Bernheim, J.L., Ledoux, D., Pellas, F., Demertzi, A., and Laureys, S. (2011a) "A Survey on Self-Assessed Well-Being in a Cohort of Chronic Locked-In Syndrome Patients: Happy Majority, Miserable Minority," *British Medical Journal Open* 1: e000039.

Bruno, M-A., Vanhaudenhuyse, A., Thibaut, A., Moonen, G., and Laureys, S. (2011b) "From Unresponsive Wakefulness to Minimally Conscious PLUS and Functional Locked-In Syndromes: Recent Advances in our Understanding of Disorders of Consciousness," *Journal of Neurology* 258: 1373–1384.

Cairncross, M., Peterson, A., Lazosky, A., Gofton, T., and Weijer, C. (2016) "Assessing Decision Making Capacity in Patients with Communication Impairments," *Cambridge Quarterly of Healthcare Ethics* 25: 691–699.

Candelieri, A., Cortese, M.D., Dolce, G., Riganello, F., and Sannita, W.G. (2011) "Visual Pursuit: Within-Day Variability in the Severe Disorder of Consciousness," *Journal of Neurotrauma* 28: 2013–2017.

Chennu, S., Norieka, V., Gueorguiev, D., Blenkmann, A., Kochen, S., Ibáñez, A., Owen, A.M., and Bekinschtein, T.A. (2013) "Expectation and Attention in Hierarchical Auditory Prediction," *Journal of Neuroscience* 33: 11194–11205.

Chennu, S., Annen, J., Wannez, S., Thibaut, A., Chatelle, C., Cassol, H., Martens, G., Schnakers, C., Gosseries, O., Menon, D., and Laureys, S. (2017) "Brain Networks Predict Metabolism, Diagnosis and Prognosis at the Bedside in Disorders of Consciousness," *Brain* 140(8): 2120–2132.

Coleman, M.R., Davis, M.H., Rodd, J.M., Robson, T., Ali, A., Owen, A.M., and Pickard, J.D. (2009) "Towards the Routine Use of Brain Imaging to Aid the Clinical Diagnosis of Disorders of Consciousness," *Brain* 132: 2541–2552.

Cruse, D., Chennu, S., Chatelle, C., Bekinschtein, T.A., Fernández-Espejo, D., Pickard, J.D., Laureys, S., and Owen, A.M. (2011) "Bedside Detection of Awareness in the Vegetative State: A Cohort Study," *Lancet* 378: 2088–2094.

Davies, W. and Levy, N. (2016) "Persistent Vegetative State, Akinetic Mutism, and Consciousness," in W. Sinnott-Armstrong (ed.) *Finding Consciousness*. New York: Oxford University Press, pp. 122–136.

Demertzi, A., Sitt, J.D., Sarasso, S., and Pinxten, W. (2017) "Measuring States of Pathological (Un)consciousness: Research Dimensions, Clinical Applications, and Ethics," *Neuroscience of Consciousness* 3(1): nix010.

Drayson, Z. (2014) "Intentional Action and the Post-Coma Patient," *Topoi* 33: 23–31.

Emanuel, L.L., Barry, M.J., Stockle, J.D., Ettelson, L.M., and Emanuel, E.J. (1991) "Advance Directives for Medical Care—A Case for Greater Use," *New England Journal of Medicine* 324: 889–895.

Faugeras, F., Rohaut, B., Weiss, N., Bekinschtein, T.A., Galanaud, D., Puybasset, L., Bolgert, F., Sergent, C., Cohen, L., Dehaene, S., and Naccache, L. (2011) "Probing Consciousness with Event-Related Potentials in the Vegetative State," *Neurology* 77: 264–268.

Fernández-Espejo, D. and Owen, A.M. (2013) "Detecting Awareness after Severe Brain Injury," *Nature Reviews Neuroscience* 14: 801–809.

Fernández-Espejo, D., Rossit, S., and Owen, A.M. (2015) "A Thalamocortical Mechanism for the Absence of Overt Motor Behavior in Covertly Aware Patients," *Journal of the American Medical Association: Neurology* 72: 1442–1450.

Fins, J.J., Illes, J., Bernat, J.L., Hirsch, J., Laureys, S., and Murphy, E. (2008) "Neuroimaging and Disorders of Consciousness: Envisioning an Ethical Research Agenda," *The American Journal of Bioethics* 8: 3–12.

Frankl, D., Oye, R.K., and Bellamy, P.E. (1989) "Attitudes of Hospitalized Patients Toward Life Support: A Survey of 200 Medical Inpatients," *American Journal of Medicine* 86: 645–648.

Giacino, J.T., Ashwal, S., Childs, N., Cranford, R., Jennett, B., Katz, D.I., Kelly, J.P., Rosenberg, J.H., Whyte, J, Zafonte, R.D., and Zasler, N.D. (2002) "The Minimally Conscious State: Definition and Diagnostic Criteria," *Neurology* 58: 349–353.

Gibson, R., Fernández-Espejo, D., Gonzalez-Lara, Kwan, B., Lee, D.H., Owen, A.M., and Cruse, D. (2014) "Multiple Tasks and Neuroimaging Modalities Increase the Likelihood of Detecting Covert Awareness in Patients with Disorders of Consciousness," *Frontiers in Human Neuroscience* 8: 1–9.

Glannon, W. (2016) "Brain-Computer Interfaces in End-of-Life Decision-Making," *Brain-Computer Interfaces* 3: 133–139.

Goldfine, A.M., Victor, J.D., Conte, M.M., Bardin, J.C., and Schiff, N.D. (2011) "Determination of Awareness in Patients with Severe Brain Injury Using EEG Power Spectral Analysis," *Clinical Neurophysiology* 122: 2157–2168.

Graham, M. (2017) "A Fate Worse Than Death? The Well-Being of Patients Diagnosed as Vegetative With Covert Awareness," *Ethical Theory and Moral Practice* 20: 1005–1020.

Graham, M., Weijer, C., Cruse, D., Fernández-Espejo, D., Gofton, T., Gonzalez-Lara, L.E., Lazosky, A., Naci, L., Norton, L., Peterson, A., and Speechley, K.N. (2015) "An Ethics of Welfare for Patients Diagnosed as Vegetative with Covert Awareness," *AJOB Neuroscience* 6: 31–41.

Kahane, G. and J. Savulescu (2009) "Brain-Damaged Patients and the Moral Significance of Consciousness," *The Journal of Medicine and Philosophy* 33: 1–21.

King, J.R., Faugeras, F., and Gramfort, A. (2013) "Single-Trial Decoding of Auditory Novelty Responses Facilitates the Detection of Residual Consciousness," *Neuroimage* 83: 726–738.

Klein, C. (2017) "Consciousness, Intention, and Command Following in the Vegetative State," *The British Journal for the Philosophy of Science* 68: 27–54.

Klein, C. and Hohwy, J. (2015) "Variability, Convergence, and Dimensions of Consciousness," in M. Overgaard (ed.) *Behavioural Methods in Consciousness Research*. Oxford: Oxford University Press.

Kondziella, D., Friberg, C.K., Frokjaer, V.G., Fabricius, M., and Møller, K. (2016) "Preserved Consciousness in Vegetative and Minimal Conscious States: Systematic Review and Meta-Analysis," *Journal of Neurology, Neurosurgery and Psychiatry*, 87: 485–492.

Laureys, S. (2005) "The Neural Correlates of (Un)awareness: Lessons from the Vegetative State," *Trends in Cognitive Sciences* 9: 556–559.

Laureys, S., Celesia, G., Cohadon, F., Lavrijsen, J., Leon-Carrion, J., Sannita, W.G., Sazbon, L., Schmutzhard, E., von Wild, K.R., Zeman, A., and Dolce, G. (2010) "Unresponsive Wakefulness Syndrome: A New Name for the Vegetative State or Apallic Syndrome," *BMC Medicine* 8: 68.

Levy, N. and Savulescu, J. (2009) "Moral Significance of Phenomenal Consciousness," in S. Laureys et al. (eds.) *Progress in Brain Research* 177: 361–370.

Lulé, D., Zickler, C., Häcker, S., Bruno, M.A., Demertzi, A., Pellas, F., Laureys, S., and Kübler, A. (2009) "Life Can be Worth Living in Locked-In Syndrome," in S. Laureys et al. (eds.) *Progress in Brain Research* 177: 339–351.

Majerus, S., Gill-Thwaites, H., Andrews, K., and Laureys, S. (2005) "Behavioral Evaluation of Consciousness in Severe Brain Damage," *Progress in Brain Research* 150: 397–413.

Monti, M.M., Rosenberg, M., Finoia, P., Kamau, E., Pickard, J.D., and Owen, A.M. (2015) "Thalamo-Frontal Connectivity Mediates Top-Down Cognitive Functions in Disorders of Consciousness," *Neurology* 84: 167–173.

Monti, M.M., Vanhaudenhuyse, A., Coleman, M., Boly, M., Pickard, J., Tshibanda, L., Owen, A., and Laureys, S. (2010) "Willful Modulation of Brain Activity in Disorders of Consciousness," *The New England Journal of Medicine* 362: 579–589.

Multi-Society Task Force on Persistent Vegetative State (1994) "Medical Aspects of the Persistent Vegetative State," *New England Journal of Medicine* 330: 1499–1508, 1572–1579.

Naccache, L. (2006) "Is She Conscious?" *Science* 313: 1395–1396.

Naci, L., Cusack, R., Anello, M., and Owen, A. (2014) "A Common Neural Code for Similar Conscious Experiences in Different Individuals," *Proceedings of the National Academy of Sciences* 111: 14277–14282.

Naci, L. and Owen, A.M. (2013) "Making Every Word Count for Nonresponsive Patients," *Neurology* 70: 1235–1241.

Naci, L., Graham, M., Owen, A.M., and Weijer, C. (2016) "Covert Narrative Capacity: A Cross-Section of the Preserved Mental Life in Patients Thought to Lack Consciousness," *Annals of Clinical and Translational Neurology* 1: 61–70.

Naci, L., Sinai, L., and Owen, A.M. (2017) "Detecting and Interpreting Conscious Experiences in Behaviorally Non-Responsive Patients," *NeuroImage* 145(Pt B): 304–313.

Nagel, T. (1974) "What Is It Like to Be a Bat?" *The Philosophical Review* 83: 435–450.

Nizzi, M.-C., Demertzi, A., Gosseries, O., Bruno, M.A., Jouen, F., and Laureys, S. (2012) "From Armchair to Wheelchair: How Patients with a Locked-In Syndrome Integrate Bodily Changes in Experienced Identity," *Consciousness and Cognition* 21: 431–437.

Owen, A.M., Coleman, M.R., Boly, M., Davis, M.H., Laureys, S., and Pickard, J.D. (2006) "Detecting Awareness in the Vegetative State," *Science* 313: 1402.

Peterson, A. (2016) "Clinical Validation and the Science of Consciousness," *Neuroscience of Consciousness* 1–9, doi:10.1093/nc/niw011.

Peterson, A., and Bayne, T. (2017) "A Taxonomy for Disorders of Consciousness That Takes Consciousness Seriously," *AJOB Neuroscience* 8: 153–155.

Peterson, A., Cruse, D., Naci, L., Weijer, C., and Owen, A.M. (2015) "Risk, Diagnostic Error, and the Clinical Science of Consciousness," *Neuroimage: Clinical* 7: 588–597.

Peterson A., Naci, L., Weijer, C., Cruse, D., Fernández-Espejo, D., Graham, M., and Owen, A.M. (2013a) "Assessing Decision Making Capacity in the Behaviorally Non-Responsive Patient with Residual Covert Awareness," *AJOB Neuroscience* 4: 3–14.

Peterson A., Naci, L., Weijer, C., Owen, A.M. (2013b) "A Principled Argument, but Not a Practical One," *AJOB Neuroscience* 4: 52–53.

Royal College of Physicians (2003) "The Vegetative State: Guidance on Diagnosis and Management" [Report of a working party], *Clinical Medicine* 3: 249–254.
Schnakers, C., Perrin, F., Schabus, M., Majerus, S., Ledoux, D., Damas, P., Boly, M., Vanhaudenhuyse, A., Bruno, M.-A., Moonen, G., and Laureys, S. (2008) "Voluntary Brain Processing in Disorders of Consciousness," *Neurology* 71: 1614–1620.
Sergent, C., Faugeras, F., Rohaut, B., Perrin, F., Valente, M., Tallon-Baudry, C., Cohen, L., and Naccache, L. (2017) "Multidimensional Cognitive Evaluation of Patients with Disorders of Consciousness using EEG: A Proof of Concept Study," *NeuroImage: Clinical* 13: 455–469.
Shea, N. and Bayne, T. (2010) "The Vegetative State and the Science of Consciousness," *British Journal for the Philosophy of Science* 61: 459–484.
Spering, M. and Carrasco, M. (2015) "Acting Without Seeing: Eye Movements Reveal Visual Processing Without Awareness," *Trends in Neurosciences* 38: 247–258.
Stender, J., Gosseries, O., Bruno, M.-A., Charland-Verville, V., Vanhaudenhuyse, A., Demertzi, A., Chatelle, C., Thonnard, M., Thibaut, A., Heine, L., Soddu, A., Boly, M., Gjedde, A., and Laureys, S. (2014) "Diagnostic Precision of PET Imaging and Functional MRI in Disorders of Consciousness: A Clinical Validation Study," *Lancet* 384: 514–522.
Vanhaudenhuyse, A., Schnakers, C., Bredart, S., and Laureys, S. (2008) "Assessment of Visual Pursuit in Post-Comatose States: Use a Mirror," *Journal of Neurology, Neurosurgery and Psychiatry* 79: 223.
Weijer C., Peterson, A., Webster, F., Graham, M., Cruse, D., Fernández-Espejo, D., Gofton, T., Gonzalez-Lara, L.E., Lazosky, A., Naci, L., and Norton, L. (2014) "Ethics of Neuroimaging after Serious Brain Injury," *BMC Medical Ethics* 15: 41.
Young, G.B. (2009) "Coma," *Annals of the New York Academy of Sciences* 1157: 32–47.

Related Topics

Consciousness and End of Life Ethical Issues
The Neural Correlates of Consciousness
Consciousness and Psychopathology
The Unity of Consciousness

27

THE UNITY OF CONSCIOUSNESS

Elizabeth Schechter

1 Introduction

I turn my head to gaze out the screen door to my left. It's a pretty scene—grass, rocks, flowers, and woods beyond—but I'm not inspecting it closely. Out of the corner of my eye I see someone crossing the kitchen, and hear someone else walking on the floor above me. Water is running somewhere. I've paused to consider whether I should go to the supermarket before or after lunch. Knowing B___, she will be happy either way. Outside, my dog runs by in a happy white flash, and I realize I've been sitting with one leg bent under me. It's starting to tingle. Shifting position, I decide to get the grocery shopping done with.

Consider this *5-second episode* a window into my conscious experience, encompassing a number of elements. Some are perceptual: I hear someone walking overhead; I become proprioceptively aware of my posture. Some are cognitive: I weigh considerations, make a prediction about someone else's preferences. Some are metacognitive: I recognize that I'm looking outside only absentmindedly. Some elements are agential: I make a decision. Some seem motoric: I'm conscious of repositioning myself in my chair.

We can consider these and other elements of my experience individually, or we can consider the whole of my experience over the course of or at any moment within the episode. *Experiential wholes* are indeed not homogeneous or simple, but incorporate a multitude of *experiential elements.*

Questions about the *unity of consciousness* concern relationships between the individual elements comprised by an experiential whole, between those elements and that whole, and between these experiential phenomena and their experiencing subjects.

These questions can be sorted into six rough categories. First is the *taxonomy question:* if we say that I enjoyed a unified consciousness throughout the 5-second window, what feature or quality of my experience do we mean to indicate? The *analysis question* concerns one particular kind of conscious unity—phenomenal unity—and asks which non-phenomenal relation it is equivalent to. There are *metaphysical questions*, focusing especially on the relata of conscious unity relations. *Mechanism questions* concern the functional and neural bases of conscious unity. Whether and how consciousness can fail to be unified is the *disunity question. Identity questions* concern the relationship between conscious unity and subjects of conscious experience.

2 Taxonomy Question: What We Talk about When We Talk about Unity

Conscious experience has a number of unified and unifying features (Bennett and Hill 2014; Brook and Raymont 2014). Start with the fact that we seem to perceive things as standing in ordered *spatial* relations to each other and to ourselves. Or consider that for any two concurrent elements of one's experiential whole, it seems that one can demonstratively co-refer to them ("I am experiencing *those*"). For that matter, we don't experience the world as a mere jumble of features and occurrences, but as containing coherent objects participating in coherent events. Something seems to promote *mutual consistency* among the various elements of a single experiential whole, in such a way that the experience affords a coherent perspective on the world. Consider *binocular rivalry*. If two images, each of a different object, are presented simultaneously to a subject, one to each eye, the subject is not thereby induced to undergo a single fantastical visual experience of two material objects occupying the very same region of space-time. Typically, instead, subjects spontaneously alternate between seeing first one object in the region, and then the other (see Logothetis 1998 for review). Finally one might argue that the unity of consciousness is essentially tied up with agential and rational unity more broadly (Shoemaker 1996).

On the other hand, it seems to many philosophers that even if your consciousness failed to be unified in any or all of the foregoing senses, it might remain unified in just this one respect: that there would still be *something that it was like* (Nagel 1974)—some *one* thing—to *be* you experiencing whatever you're experiencing.

Conscious unity in this last sense is called *phenomenal unity.* Just as philosophical work on consciousness most often concerns phenomenal consciousness, philosophical work on the unity of consciousness most often concerns phenomenal unity. This is no coincidence. Phenomenality is puzzling, and philosophers love puzzles.

Phenomenal properties are famously difficult to define (see Block 1978: 281; cf. Dennett 1988), other than by ostension, that is, by pointing to examples. Consider this contrastive pair of cases (modified from Fodor 1981). Suppose you open your eyes in a hotel room one morning and in the first moment of consciousness, before you have any thoughts at all, you visually experience the sparsely furnished room, whose walls are all painted *red*. When you return to your hotel that evening you're informed that there was a problem with your room and that your things have all been moved to a new room with the same layout and furnishings and dimensions. The next morning you open your eyes and in the first moment of consciousness, before you have any thoughts at all, you visually experience the sparsely furnished room, whose walls are all painted *green*. Suppose that your old room and your new room are otherwise visually indistinguishable, and consider only your first moment of consciousness, before you think, "What is this unfamiliar *red* room?" on the first morning and "What is this unfamiliar *green* room?" on the second morning. The subjective character of your experience on the first morning would presumably be different from the subjective character of your experience on the second morning. This difference in subjective character is, ostensibly, a *phenomenal* difference, a difference between the phenomenal properties of your experience on those two days.

I will take phenomenal unity to be a *phenomenal* relation between phenomenal properties, in the sense that the phenomenal unity of two elements of experience makes a phenomenal difference to their subject. What makes it difficult to characterize or define phenomenal unity is that the phenomenal difference it makes is so abstract: presumably any two phenomenal properties can have the further property of being phenomenally unified, despite incredible diversity among phenomenal properties and thus among the phenomenal characters of their unified pairs.

This makes it especially tempting to try to articulate what phenomenal unity is by pointing to a pair of contrastive examples, as I did for phenomenal properties above. The most helpful pair of contrastive examples would be a case in which two elements of experience were phenomenally unified and a case in which they were not. The problem is that for various reasons it is not clear that the elements of a subject's experience at a time could ever fail to be phenomenally unified, nor that we can even imagine a case in which they so failed (see Sections 6 and 7).

To characterize phenomenal unity, then, reference is instead often made to there being something that it's like for the subject of a unified consciousness to experience *together* everything they experience, e.g.: "Experiences, when they occur simultaneously, do not occur as phenomenal atoms but have a conjoint phenomenology—there is something it is like to have them *together*" (Bayne 2010: 280; emphasis added). During the five-second episode, say, there wasn't just something it was like for me to see someone walk across the kitchen and something it was like for me to hear someone walking overhead. There was something it was like for me to see someone walk across the kitchen *while* hearing someone walking overhead.

Unfortunately it is very difficult to provide an elaborated description of this experienced "togetherness." On pain of redundancy, it can't just mean that the elements are experienced simultaneously or by the same subject. We might say that if you have a unified consciousness, then everything you experience is experienced as part of *a unified whole*. But a whole *what?* And again—unified *how?*

3 The Analysis Question: What Does Phenomenal Unity Reduce to or Consist In?

The ineffability of phenomenal unity makes it desirable that it should consist in some non-phenomenal relation. The most basic proposal is that phenomenal unity can be understood in terms of the conjunction of the contents of experience, a proposal explored by, among others, Hurley: "if and only if a conscious state with content *p* and a conscious state with content *q* are co-conscious [i.e. unified], then there is a conscious state with content *p and q*" (Hurley 1998: 117; citing Chisolm 1981; see also Tye 2003). This analysis of phenomenal unity is presumably entailed by representationalist accounts of consciousness, whose proponents believe that phenomenal properties reduce to causal-intentional properties.

Another proposal is that phenomenal unity might reduce to or consist in something like *feature binding* (or similar—see Revonsuo 1999), the process whereby multiple features of objects come to be accurately associated with each other in perception. Suppose I am watching my friendly white dog trot toward my wary gray rabbit. To experience this, I must perceive their colors, shapes, and movements—and as it happens, color, shape, and motion are all represented in different regions of the brain and in partial independence of each other; one can lose the ability to perceive one of these features without losing the ability to perceive the others, for instance (e.g. Zihl, Cramon, and Mai 1983). The so-called *binding problem* concerns how the features we experience come to be correctly bound, or associated, in perception—so that I perceive my *rabbit* and not my dog as *gray*, my *dog* and not my rabbit as *trotting*—indeed, so that I perceive colored moving objects at all, as opposed to a mere jumble of color, shape, motion. So, again, one proposal is that phenomenal unity consists in this kind of representational integration.

The philosophical literature contains several other potential analyses of the phenomenal unity relation. Nagel referred to our assumption that "for elements of experience…occurring simultaneously or in close temporal proximity, the mind which is their subject can also experience the simpler relations between them if it attends to the matter" (Nagel 1971: 407; emphasis omitted; and see again Tye 2003). We could pull from this the proposal that the phenomenal unity of elements of experience consists in their *access unity*, which is an extension of the concept of *access consciousness* (Block 1995). Very roughly, a mental state is said to be access conscious when its subject can introspect or report it and use it in reasoning and the rational guidance of action. Access conscious states can be contrasted with the multitude of mental states we know about only from experimental or theoretical work demonstrating that they play roles in the non-rational guidance of our behavior and automatic responses, or in determining what we consciously experience. So, two experiential elements are *access unified* if they (or their contents) can *jointly* figure in reasoning. Perhaps this is what phenomenal unity amounts to.

Marks meanwhile suggested that two experiences "belong to the same unified consciousness only if they are [or could be] known, by introspection, to be simultaneous" (Marks 1981: 13). We might pull from this an analysis of phenomenal unity into *awareness unity*, an extension of the concept of *higher-order awareness* (Rosenthal 1986). According to higher-order theories of consciousness, a mental state is conscious in virtue of its (occurrent or dispositional) relation to a second mental state which is about it, or which takes it as its content. So, two elements of experience are *awareness unified* if they are or are disposed to be made the common object of a single higher-order mental state.

Although each of these candidate analyses of phenomenal unity picks out an interesting respect in which human consciousness is unified, none is widely agreed to offer an analysis of phenomenal unity. The trouble is that some philosophers believe that phenomenal unity could conceivably exist in the absence of any other kind of conscious unity: access unity, awareness unity, feature-binding, and perhaps even the conjunction of contents (see discussion in Bayne 2010). Indeed, philosophers have sometimes concluded that the phenomenal unity of experiential elements admits of no analysis—that it is a basic relation (Dainton 2000). Although most philosophers seem reluctant to accept this, the literature on the analysis question consists largely of arguments rejecting various candidates: not spatial or representational unity, not object or introspective unity, not access or awareness unity. (For taxonomies and critical discussions, see Bayne and Chalmers 2003; Tye 2003.)

The central difficulty concerns the notion of the phenomenal generally. It may not be essential to the concept of phenomenal unity that it should have any particular causal profile. Of course, this might prompt suspicion about the coherence of the concept of phenomenal unity (just as there has been about the concept of the phenomenal, generally; e.g. Dennett 1988, Church 1995).

Debates about phenomenal unity meanwhile inherit all of the controversies about phenomenal consciousness generally. For instance, suppose that there is no such thing as *cognitive phenomenology*: that coming to believe something, or judging or doubting something, or making a decision, have no intrinsic phenomenal characters—unlike, say, conscious seeing or conscious hunger. If this is right, then an account of phenomenal unity can simply ignore the *propositional attitudes* (beliefs, judgments, doubts, decisions—mental states that can be rational or irrational). This in turn makes it more plausible that phenomenal unity should reduce to a kind of *spatial unity* of (the contents of) experience. This analysis is less plausible where propositional attitudes are concerned, however. Thus, uncertainty about the contents of phenomenal consciousness affects the plausibility of competing accounts of phenomenal unity.

4 Metaphysical Questions: Conscious Unity Relations at a Time and over Time

Every experience soon fades from consciousness. Most are never recalled. This makes it unlikely that all of one's experiences stand in any relation to each other than mere succession. On the other hand, it seems possible that all of one's experiences at a given moment are substantively unified in some respect. It is thus standard for philosophers to distinguish between conscious unity *at a given time*, or synchronically, and conscious unity *over* time, or diachronically. Many metaphysical questions about conscious unity concern these temporal concepts.

Conscious experience has both objective and subjective temporal features. A given experience occurs at a particular time and has a particular duration, and these are objective temporal properties. Some experiences meanwhile seem to their subject to be momentary or to drag on, to come before or after others, and these are subjective temporal properties. Naturally objective and subjective temporal features are connected but not identical. For instance, consciousness takes time, so an event that I experience as happening now, in the *subjective present*, may in fact have happened in the very recent *objective past*.

Discussions of conscious unity relations between experiences *at a moment in time* often don't specify whether the moment referred to is one of objective or subjective time. That is, is synchronic conscious unity a relation between all the objectively simultaneous experiences of a subject? Or between all the happenings that the subject's experience presents to him as occurring simultaneously?

Many philosophers believe that we don't experience the present moment as strictly instantaneous but rather as a very brief interval; the term "specious present" is used to refer to the subjective present when it is conceptualized in this way. (The term is associated with William James 1890, though contemporary usages don't line up neatly with his.) It is tempting to conceptualize the subjective present as an interval in order to account for the difference between *successive experiences* and *experiences of succession*. Imagine watching someone dance under a strobe light: you have first a visual experience of the dancer in one position, then a visual experience of darkness, then a visual experience of the dancer in a new position. Your extended experience, of repeatedly seeing the dancer *having moved*, will be different from an experience of seeing a *dancer moving*, in ordinary light. Or imagine hearing a piano key hit repeatedly, versus hearing a key pressed and held. It is tempting to cast these differences in temporal terms. In the first kind of case, you perceive change (or stasis) across intervals of experiencing, but not within any one interval. In the second kind of case, you perceive change (or stasis) within an interval.

A philosopher employing the concept of the specious present for this purpose may view himself as offering an account of very short-term diachronic unity (Dainton 2000). Indeed, within the momentary interval of the specious present, phenomenal unity might even be *transitive*, such that two elements of experience that are both unified with a third must also be unified with each other. But if phenomenal unity were transitive throughout any arbitrary interval of time, then what it's like for, say, me to blow out candles on my 40th birthday cake, necessarily depends upon whether or not I experienced blowing out candles on my 4th birthday cake. This seems implausible. It may then be that for a subject to have a phenomenally unified consciousness over any extended period of time is simply for him continuously to have a consciousness that is phenomenally unified within the specious present (Dainton 2000; Dainton and Bayne 2005).

Metaphysical questions about conscious unity often concern the relata of conscious unity relations. The literature on conscious unity often takes it to be a relation between *experiences*, but this is not without controversy, since there is no universal agreement about the identities of experiences. According to the "many-in-one" position, my experiential whole is an experience, and so are the numerous elements it contains (Bayne 2010; Lockwood 1989). According to the "only-one" position, only the whole of my experience is itself an experience; its elements are not (Tye 2003). Finally, there is a "many-only" position, according to which I have a multitude of experiences at every moment that are unified without being somehow incorporated into a single overarching experience.

This disagreement concerns the individuation of experiences. Philosophical works on conscious unity mainly individuate experiences using a "tripartite" conception (Bayne 2010) that identifies them as particular phenomenal characters of particular subjects at particular times (Dainton 2000). That is, given a subject and a time, the tripartite conception individuates experiences on the basis of their phenomenal characters. This rules out certain possibilities, like the possibility that a single subject might have at a given moment in time multiple experiences with the same character (cf. Schechter 2013). The tripartite conception does not yield a single way of identifying experiences, however.

One might suppose that the tripartite conception rules out the only-one view: how could my experience of my white dog be the very same experience as my experience of my gray rabbit, given the phenomenal difference between experienced whiteness and experienced grayness? But it begs the question against the only-one view to read "experienced whiteness" as "experienced whiteness *alone*, in the absence of any experienced grayness." According to the only-one view, an experience of my dog can equally be an experience of my rabbit, in the same way that my photograph of my dog can equally be a photograph of my rabbit—by being a photograph of them both! In itself, the tripartite conception of experiences is consistent with this position.

It is also consistent with the many-only view. It is true that my experiential whole, at any moment, has a phenomenal character, *W*, that differs from the phenomenal character of any one of my experiences, and also from the mere (disunified) sum of phenomenal characters associated with each of those experiences. The many-only view accepts this, however: it simply denies that *W* is the character of an experience. According to the many-only view, *W* is a *unity* of the phenomenal characters of my multitude of experiences, without itself being an experience.

Naturally, the tripartite account is also consistent with the many-in-one view. In fact, one intuitive metaphysics of the phenomenal unity relation casts it as a mereological relation between experiences. Bayne has offered this kind of account: two experiences are unified in virtue of being *subsumed* by another experience, with an experiential whole subsuming all (other) experiences of its subject (Bayne 2010).

The differences between these accounts must be subtle. Even the only-one view allows multiplicity somewhere, if only in the multiplicity of things that I experience in the world. Even the many-only view allows that a subject's consciousness is unitary in the sense that she has, say, only a single perspective or stream of consciousness. It is therefore unclear what hinges upon whether we identify as *experiences* experiential wholes, experiential elements, or both. Perhaps some objective basis for individuating experiences could be drawn from empirical generalizations about, say, working memory capacity. This kind of possibility has not been much explored by philosophers of conscious unity, however, who may believe that since experience is essentially subjective, a subject should be able to tell precisely on the basis of her own experience just how many experiences she has (but see Hurley 1998; cf. Bayne 2010).

5 The Mechanism Question: How to Explain Conscious Unity

Further questions about conscious unity concern its broadly causal explanation, including its neural basis. Answers depend upon the kind of conscious unity at issue. Suppose that we understand conscious unity to be the product of feature binding. Then an answer to the *mechanism question* consists of an articulation of the mechanism of feature binding: say, two elements are unified when spatial attention selects both of them as concerning goings-on at a single location in space, thus "tagging" their contents, for further processing, as being features of a common object (Treisman 1998; Treisman and Gelade 1980). If one instead takes conscious unity to be a kind of access unity, then two elements may be unified when their contents are jointly broadcast by attention into the "global workspace" (Baars 1988), or when they are simultaneously encoded in working memory. If conscious unity is taken to be a kind of unified awareness of experiences, then two experiential elements may be unified when their contents become the object of a single higher-order state.

On the other hand, if conscious unity is taken to be phenomenal unity, and if phenomenal unity is not analyzable into any non-phenomenal relation, then it's not immediately clear how to explain it. Here again there exists a parallel between phenomenal unity and what's often said to be true of phenomenality generally: that any causal accounting of it may leave an *explanatory gap* (Levine 1983).

One preliminary mechanism question about phenomenal unity is whether it even requires explanation in addition to whatever causal accounting is offered for phenomenal consciousness generally, or whether, instead, the mechanism of phenomenal unity is just the mechanism of phenomenal consciousness. Bayne (2010) uses the terms "atomism" and "holism" to refer to the former and to the latter possibility, respectively.

Holism is a more sensible proposal with respect to synchronic conscious unity. Any kind of extended diachronic conscious unity will presumably require mnemonic integration not necessary for experience itself. On the other hand, it seems possible that the mechanism of consciousness is also the source of its unity at a given moment.

Neither holism nor atomism need be true of *all* kinds of conscious unification. Consider *multisensory integration*—a kind of perceptual binding—as it occurs when, say, one experiences an audiovisual recording designed to induce the so-called *McGurk effect* (McGurk and MacDonald 1976). In one such video, a sound reel of an actor speaking the syllables *ba-ba* plays over a visual recording of him mouthing the syllables *ga-ga*. To the perceiver, however, the man seems to be saying *da-da*. The perceiver does not hear one thing, see another, and then struggle to work out what the speaker is actually saying. Rather, what the speaker is saying is worked out by the perceiver's perceptual systems, prior to conscious experience.

Multisensory integration and perceptual binding more generally have sometimes been proposed to integrate what would otherwise be disunified conscious contents (Bartels and Zeki 1998), and they do clearly contribute to the *coherence unity* of consciousness, creating for the experiencing subject a coherent world of objects and events. Thus atomism could be true for coherence unity. On the other hand these mechanisms of coherence unity are not well suited to explain phenomenal unity. Multisensory integration means that in the perceiver's actual experience, there is not a *ga-ga* thing that it is like to *see* the speaker speak unified with a *ba-ba* thing that it is like to *hear* the speaker speak. In the perceiver's experience, there is just the *da-da* thing. Phenomenal unity relates *already experiential* elements: what it's like to see the actor in the video say *da-da* and what it's like to hear him speak *da-da*.

Since phenomenal unity is a relation between the already-conscious, philosophers of phenomenal unity seem inclined toward holism. Among other things, if atomism were true, then one would expect disorders or impairments characterized by having a multitude of elements of experience not incorporated into any experiential whole, and there is no evidence that such cases exist (see Section 6 on the "disunity question").

The debate between atomism and holism would be more readily answered with an agreed-upon theory of consciousness at hand. There has been significant convergence around particular neurofunctional theories of *access* consciousness, perhaps especially the global neuronal workspace model of Dehaene and colleagues (e.g. Dehaene and Naccache 2001), developed out of Baars's "global workspace" model (Baars 1988). Some philosophers believe that access consciousness is the only kind of consciousness that exists (e.g. Dennett 2001), or at least that the machinery of access consciousness partially explains phenomenal consciousness (e.g. Carruthers 2017), though other philosophers maintain that phenomenal consciousness and conscious accessibility are distinct phenomena requiring distinct explanations (e.g. Block 2007, 1995). Another influential theory is Tononi's (2004) Integrated Information Theory (developed out of Edelman 2003), which Tononi presents as a theory of phenomenal consciousness, though because it casts consciousness as a kind of causal-informational integration, it rather looks like a theory of access consciousness.

Taking the theories of consciousness that are on the table, the natural approach is to see whether any one of them can also account for conscious unity. This seems possible for any of the major neurofunctional theories, including the two mentioned above. For example, if the entry of some content into the global workspace makes it an element of conscious experience, then perhaps two contents' simultaneous entry into the same global workspace makes their elements unified. Meanwhile, Tononi's (2004) theory of consciousness explicitly appeals to the integration of contents; whatever theory of phenomenal consciousness it offers is quintessentially holistic.

It's striking that these prominent theories of consciousness are so readily amended into theories of conscious unity. This is because they offer accounts of what happens to contents to make them contents of *experience*. They can then attribute experiential "togetherness" to the "togetherness" of experiential contents—to their being made conscious together.

Philosophers have occasionally expressed skepticism toward such content-centric explanations of consciousness. Consider Searle's (2000) contrast between *building block* and *unified field* accounts of consciousness. In the former, a subject's consciousness is constructed from elements understood to be conscious independently of their incorporation into a larger experiential whole. Searle suggests that this sort of account sails past consciousness itself, in pursuit of contingently conscious contents. According to a unified field account, the subject's consciousness itself is prior: specific contents become conscious only by modifying their subject's consciousness. The subject must already *be* conscious in order for their *phenomenal field* to be modified in any way—in order to experience anything at all.

One way of interpreting the unified field account is as taking phenomenal consciousness to be a kind of *creature consciousness* (Rosenthal 1993)—a condition of the entire subject. It cannot be the condition of wakeful alertness however (although this is the way Searle 2000 speaks), since, as Bayne (2007) points out, we are phenomenally conscious while dreaming. We might think of this creature consciousness as a state of readiness to enjoy phenomenal experiences.

It is not clear which questions about phenomenal unity would be resolved by adopting a unified field account. The account rules out the possibility of radical atomism, but does not entail that a subject can only have a single unified field at a time (a point conceded by Bayne 2007). For one thing, to know that what produces an X is a mechanism of type M isn't yet to know *how many* Xs each such mechanism can produce; nor is it to know how many mechanisms of that type there are within the brain.

The deeper issue is that we do think of conscious unity precisely in terms of relations between contents (again, see Bayne 2007). What after all is a *unified field*—the supposedly prior conscious "thing" that is only modified in this way and that by particular conscious contents? Searle essentially defines a unified field *as* an experiential whole. In that case, stating that two contents modify the same field does not explain their unity but is equivalent to stating that they are unified (as Prinz 2013 notes). If a phenomenal field is instead defined in a way that is neutral with respect to the relations between the contents modifying a field, then it is not clear that a subject with a single phenomenal field will necessarily have a unified consciousness. Why couldn't multiple contents modify their common field *so as to* generate a disunified consciousness?

6 The Disunity Question: Can Conscious Unity Fail?

Much of the philosophical literature on the unity of consciousness concerns candidate cases in which it fails. A subject's consciousness could, logically, fail to be unified in any of a number of ways, including the following three. First, a subject with a *radically atomistic* consciousness would have a multitude of elements of experience, none incorporated into any larger whole. In a subject with a *partially unified* consciousness, each element of experience would be unified with some but not all others. This would require that conscious unity not be a transitive relation between experiences or elements of experience. A subject with a *multiple consciousness* would instead possess multiple experiential wholes, not unified with each other, though the elements within each whole would be unified.

Candidate cases of this last sort have received the most philosophical attention. A subject with a multiple consciousness has multiple experiential wholes, rather than a single experiential whole incorporating everything the subject experiences. Most disruptions or impairments of consciousness do not have this characteristic, but instead rather reduce the number of elements encompassed by an experiential whole (simultagnosia, unilateral neglect) or else distort its contents (hallucinations, perhaps Capgras syndrome or apperceptive agnosia). Philosophers have, however, debated several kinds of cases that may be instances of multiple consciousness, most often *dissociative identity disorder* and the so-called *split-brain phenomenon*.

Dissociative identity disorder (DID)—formerly known as "multiple personality disorder"—is one of several dissociative conditions recognized by the Fifth Edition of the Diagnostic and Statistical Manual of Mental Disorders (DSM-5). All such conditions are characterized by impaired integration of consciousness, emotion, and memory, but DID attracts by far the most philosophical attention, because its disruptions and breakdowns are patterned in a unique way. A person with DID might, for instance, appear to suffer a kind of fluctuating amnesia about her gambling habit, but the fluctuations don't appear to be random: she might, say, *always* remember previous episodes of gambling, but *only* whenever she is already gambling or on her way to gamble; at other times she might always be amnesic for such episodes.

Many philosophers believe that this kind of personal and experiential memory is important to remaining the *same person* over time. Fluctuating amnesia might be viewed as compromising the kind of extended diachronic conscious unity that characterizes ordinary human experience. Certainly, one point of contention about dissociative identity disorder (Tye 2003; Braude 1991) concerns whether someone with DID has different, mutually interrupting perspectives or streams of consciousness at different times.

Another reason why many philosophers see DID as a candidate case of multiple consciousness, however, is that the condition is often characterized as one in which multiple persons somehow animate one body (Radden 1996; Humphreys and Dennett 1989). If we understand the condition in this way, it is not that the hypothetical DID subject mentioned above remembers previous episodes of gambling only during occurrent episodes; rather, her body is the body of at least *two* experiencing beings, one of whom gambles however often and naturally remembers having gambled before, and one who has never gambled and therefore has no memories of having done so. If understood in this way, DID appears to be a case of diachronic multiple consciousness within a human being—though not necessarily within an experiencing person (see Section 7 on "identity questions").

Although dissociative identity disorder is more striking than the split-brain phenomenon, DID has the disadvantage of continuing to be the object of fierce controversy among clinicians and cognitive psychologists (see, for instance, Lynn *et al.* 2014 vs. Dalenberg *et al.* 2014). In contrast, the basics of the split-brain phenomenon are well accepted. The split-brain phenomenon is a result of medical neurosurgery, colloquially known as "split-brain surgery," that cuts through the corpus callosum connecting the two cerebral hemispheres. Since the corpus callosum is a conduit for interhemispheric interaction and information exchange, the surgery naturally alters and impedes this interaction and exchange (though in no way prevents it). The split-brain phenomenon is generally discussed as a candidate case of synchronic conscious disunity: are all of a split-brain subject's experiential elements at every moment incorporated into a single experiential whole, or is each element instead incorporated into *one of two* experiential wholes—one associated with the right hemisphere and one with the left?

The question is difficult to resolve partly for straightforward empirical reasons: the cases are complicated, and while some kinds of experience appear to be cleanly interhemispherically dissociated in some subjects at some times, others do not. A more basic theoretical problem is that any evidence of conscious disunity takes the form of apparent lack of integration amongst experiential elements—yet lack of integration constitutes prima facie evidence that the elements were not in fact *conscious* in the first place.

In addition, having a *multiple* consciousness does not seem to be the sort of thing one could introspect. To think otherwise would be to suppose that a subject with a multiple consciousness could compare multiple elements of experience and judge that they were not elements of one whole. But then how would she have made the comparison? Any introspective act will itself be an element of some experiential whole or other, and can survey only the elements of that whole.

If introspection cannot reveal failures of phenomenal, access, or awareness unity, then even a neurotypical human being cannot know *via* introspection that her consciousness is perfectly unified, or even unitary. (See Marcel 1993 on the potential disunity of neurotypical consciousness over very short time scales; see also Dennett 1991.) Among other things, we could be subject to a version of the so-called *refrigerator light illusion*, if something about the very act of introspecting the elements of our experience itself unifies them (Prinz 2013). Perhaps consciousness is disunified whenever we aren't looking!

7 Identity Questions: Selves, Self-Consciousness, and Subjective Perspectives

From one perspective, it's hardly radical to suppose that neurotypical consciousness may not be wholly unified or even unitary. We already know that our minds are not wholly rationally unified, self-knowing, or self-controlled, and that introspection can mislead about all manner of things. Scientists and philosophers recognize that human reasoning and perception are subserved by a multitude of systems, perhaps even mind-like systems (Evans and Frankish 2009). So why not division or multiplicity within consciousness itself? If we feel that we have a better claim to our *phenomenal* unity, perhaps this is just pride speaking; perhaps it is just because our phenomenal properties are dimly felt to lie within the last realm that science has yet to conquer (Dennett 1988: 386).

On the other hand, phenomenal unity may be unique in the closeness of its connection to our first-personal ways of thinking about the identities of experiencing subjects. This connection is mediated by the concept of a subjective perspective, and it poses obstacles to understanding phenomenal disunity that we don't face when thinking about psychic disunity of other kinds.

Experiencing beings have perspectives: there are facts about the way the world *is*, and further facts about the way the world *appears to them*. Rocks do not have perspectives. There are no facts about the way the world appears to be to a rock; there is just the way the world is, including that portion of the world that is the rock.

Our first-personal way of understanding the conditions of being an experiencing subject is in terms of having a perspective. It is first-personal because the concept of a subjective perspective is itself one that we grasp only because each of us already has a perspective (Nagel 1974). We might conceptualize what it is for a being, X, to have a subjective perspective, in terms of there being something that would count as successfully imagining being X. For an agent Y at time t_y to successfully imagine being X at time t_x is for Y to (intentionally) experience at t_y all and only what X experiences at t_x. These success conditions are very stringent—presumably impossible to meet—but this is consistent with X's having a perspective. What is required for X to have a perspective is that an attempt to imagine being X has success conditions. In contrast, nothing would count as successfully imagining being a rock.

Now suppose that at some time t_t, some target being, T, does not have a single experiential whole, but rather has two of them, W1 and W2: the elements of W1 are all unified with each other, but not with those of W2, and analogously for the elements of W2. And suppose that agent Y wants to imagine at time t_y being T at t_t. Suppose that Y is fantastically good at such acts of empathic imagination: given enough third-personal information about a target, Y can undergo precisely those experiential elements undergone by Y's target. On the other hand, suppose that Y has perfect control over only the *contents*, and not the *structure*, of Y's own experience. So whatever the elements of Y's experience, these elements are always synchronically phenomenally unified, whether Y is trying to imagine being someone else or not.

Y then cannot successfully imagine being T-at-t_t by engaging in only a single act of empathic imagination. If Y chooses to undergo at t_y every experiential element of T at t_t, these elements would all be phenomenally unified, since this is the fixed structure of Y's consciousness. Y would thus experience everything T experienced at t_t, but not *only* what T experienced, since T does not experience unity between the elements of W1 and those of W2, and Y would. If Y instead chooses to undergo at t_y only every element of, say, W1 at t_t, and no element of W2, Y would thus experience *only* what T experienced at t_t, but not *everything* T experienced.

One could debate whether Y could succeed at occupying T-at-t_t via two successive imaginative acts, in which Y undergoes at t_y the elements of W1 and undergoes at t_{y+1} the elements of W2 (see Bayne 2010 and Schechter 2018). But it is at least clear that there is no other way for Y to succeed.

This is because T does not have *a* subjective perspective, singular, for Y to imaginatively take on. Rather T has *two* perspectives: W1 and W2. The success conditions for Y's imagining being T must be relativized to W1 and W2: there are success conditions for imagining being T-subject-to-W1 and T-subject-to-W2. But recall that these are conditions on imagining *being someone*. So now it begins to seem to us—and should to Y—that T is somehow *two* subjects of experience.

The connections between phenomenal unity and subjects of experience bear on the assumption that phenomenal unity is a transitive relation. Suppose that it weren't, and that two experiences not unified with each other, E1 and E2, were both unified with a third, E3. Because they are not unified with each other, E1 and E2 should be elements of distinct perspectives. We have seen that where there are distinct perspectives, P1 and P2, there is pressure to posit distinct subjects of them, S1 and S2. Perspectives presumably incorporate every element of experience that is unified with any element they do incorporate. Since E3 is unified with P1's E1 and with P2's E2, E3 should be an element of both P1 and P2. S1 and S2 should thus both be subjects of E3—yet if elements of experience get their identities partly from the subjects whose experiences they are, a single element of experience cannot belong to multiple subjects.

This contradiction creates pressure to insist that phenomenal unity must after all be transitive. Indeed the transitivity of (synchronic) phenomenal unity has something close to the status of an axiom in the literature. Lockwood's work (1989) defending the possibility of non-transitive phenomenal unity is perhaps both the major exception to this rule and the one that proves it, since in later writing he admitted to being unsure himself whether the idea of non-transitive phenomenal unity really makes sense (Lockwood 1994).

Because the identity conditions for subjects of experience and for subjective perspectives are so closely related, there will always be compelling reasons to deny that a human being has a disunified and especially a multiple consciousness (Bayne 2010). Again, recognizing multiple subjective perspectives within a single human being creates pressure to posit distinct subjects of each perspective. But subjects of experience are objects of moral concern, and it is not clear what it would mean for there to be multiple distinct objects of moral concern within a single living being. There will thus always be reason to deny that a single human being is more than a single subject of experience—whatsoever the empirical facts about consciousness may turn out to be.

References

Baars, B. (1988) *A Cognitive Theory of Consciousness*, Cambridge, UK: Cambridge University Press.

Bartels, A. and Zeki, S. (1998) "The theory of multistage integration in the visual brain," *Proceedings of the Royal Society of London B (Biological Sciences)* 265: 2327–2332.

Bayne, T. (2007) "Conscious states and conscious creatures: Explanation in the scientific study of consciousness," *Philosophical Perspectives* 21 (*Philosophy of Mind*): 1–22.

Bayne, T. (2010) *The Unity of Consciousness*, Oxford: Oxford University Press.

Bayne, T. and Chalmers, D. (2003) "What is the unity of consciousness?" In A. Cleeremans (ed.), *The Unity of Consciousness: Binding, Integration and Dissociation*, Oxford: Oxford University Press.

Block, N. (1978) "Troubles with functionalism," in C. Savage (ed.), *Perception and Cognition: Issues in the Foundations of Psychology*, Minneapolis, MN: University of Minnesota Press.

Block, N. (1995) "On a confusion about a function of consciousness," *Behavioral and Brain Sciences* 18: 227–287.

Block, N. (2007) "Consciousness, accessibility, and the mesh between psychology and neuroscience," *Behavioral and Brain Sciences* 30: 481–548.

Braude, S. (1991) *First Person Plural: Multiple Personality and the Philosophy of Mind*, New York: Routledge.

Bennett, D. and Hill, C. (2014) "A unity pluralist account of the unity of experience," in D. Bennett and C. Hill (eds.), *Sensory Integration and the Unity of Consciousness*, Cambridge, MA: MIT Press.

Brook, A. and Raymont, P. (2014) "The Unity of Consciousness," in E. Zalta (ed.), *The Stanford Encyclopedia of Philosophy*, URL = https://plato.stanford.edu/archives/win2014/entries/consciousness-unity/

Carruthers, P. (2017) "Block's overflow argument," *Pacific Philosophical Quarterly* 98: 65–70.

Chisholm, R. (1981) *The First Person: An Essay on Reference and Intentionality*, Minneapolis, MN: University of Minnesota Press.

Church, J. (1995) "Fallacies or analyses?" *Behavioral and Brain Sciences* 18: 251–252.

Dainton, B. (2000) *Stream of Consciousness*, London: Routledge.

Dainton, B. and Bayne, T. (2005) "Consciousness as a guide to personal persistence," *Australasian Journal of Philosophy* 83: 549–571.

Dalenberg, C., Brand, B., Lowenstein, R., Gleaves, D., Dorahy, M., Cardeña, E., et al. (2014) "Reality versus fantasy: Reply to Lynn et al. (2014)," *Psychological Bulletin* 140: 911–920.

Dehaene, S. and Naccache, L. (2001) "Towards a cognitive neuroscience of consciousness: Basic evidence and a workspace framework," *Cognition* 79: 1–37.

Dennett, D. (1988) "Quining qualia," in A. Marcel and E. Bisiach (eds.), *Consciousness in Contemporary Science*, Oxford: Clarendon Press.

Dennett, D. (1991) *Consciousness Explained*, New York: Little, Brown.

Dennett, D. (2001) "Are we explaining consciousness yet?," *Cognition* 79: 221–237.

Edelman, G. (2003) "Naturalizing consciousness: A theoretical framework," *PNAS* 100: 5520–5524.

Evans, J. and Frankish, K. (2009) "The duality of mind: A historical perspective," in J. Evans and K. Frankish (eds.), *In Two Minds: Dual Processes and Beyond*, Oxford: Oxford University Press.

Fodor, J. (1981) "The mind-body problem," *Scientific American* 244: 114–125.

James, W. (1890) *The Principles of Psychology*, New York: Dover.

Humphreys, N. and Dennett, D. (1989) "Speaking for our selves: An assessment of multiple personality disorder," *Raritan* 9: 68–98.
Hurley, S. (1998) *Consciousness in Action*, Cambridge, MA: Harvard University Press.
Levine, J. (1983) "Materialism and qualia: The explanatory gap," *Pacific Philosophical Quarterly* 64: 354–361.
Lockwood, M. (1989) *Mind, Brain and the Quantum*, Oxford: Blackwell Publishers.
Lockwood, M. (1994) "Issues of unity and objectivity," in C. Peacocke (ed.), *Objectivity, Simulation, and the Unity of Consciousness*, Oxford: Oxford University Press.
Logothetis, N. (1998) "Single units and conscious vision," *Philosophical Transactions of the Royal Society of London Series B* 353: 1801–1818.
Lynn, S. J., Lilienfeld, S. O., Merckelbach, H., Giesbrecht, T., McNally, R., Loftus, E., et al. (2014) "The trauma model of dissociation: Inconvenient truths and stubborn fictions. Comment on Dalenberg et al. (2012)," *Psychological Bulletin* 140: 896–910.
Marcel, A. (1993) "Slippage in the unity of consciousness," in G. Bock and J. Marsh (eds.), *Experimental and Theoretical Studies of Consciousness*, Chinchester, UK: John Wiley & Sons.
Marks, C. (1981) *Commissurotomy, Consciousness and Unity of Mind*, Cambridge, MA: MIT Press.
McGurk, H. and MacDonald, J. (1976) "Hearing lips and seeing voices," *Nature* 264: 746–748.
Nagel, T. (1971) "Brain bisection and the unity of consciousness," *Synthese* 22: 396–413.
Nagel, T. (1974) "What is it like to be a bat?," *Philosophical Review* 83: 435–450.
Prinz, J. (2013) "Attention, atomism, and the disunity of consciousness," *Philosophy and Phenomenological Research* 86: 215–222.
Radden, J. (1996) *Divided Minds and Successive Selves: Ethical Issues in Disorders of Identity and Personality*, Cambridge, MA: MIT Press.
Revonsuo, A. (1999) "Binding and the phenomenal unity of consciousness," *Consciousness and Cognition* 8: 173–185.
Rosenthal, D. (1986) "Two concepts of consciousness," *Philosophical Studies* 49: 329–359.
Rosenthal, D. (1993) "State consciousness and transitive consciousness," *Consciousness and Cognition* 2: 355–363.
Schechter, E. (2013) "The unity of consciousness: Subjects and objectivity," *Philosophical Studies* 165: 671–692.
Schechter, E. (2018) *Self-Consciousness and "Split" Brains: The Minds' I*, Oxford: Oxford University Press.
Searle, J. (2000) "Consciousness," *Annual Review of Neuroscience* 23: 557–578.
Shoemaker, S., (1996) "Unity of consciousness and consciousness of unity," in *The First-Person Perspective and Other Essays*, Cambridge: Cambridge University Press.
Tononi, G. (2004) "An information integration theory of consciousness," *BMC Neuroscience* 5: 42.
Treisman, A. (1998) "Feature binding, attention, and object perception," *Philosophical Transactions: Biological Sciences* 353: 1295–1306.
Treisman, A. and Gelade, G. (1980) "A feature integration theory of attention," *Cognitive Psychology* 12: 97–136.
Tye, M. (2003) *Consciousness and Persons: Unity and Identity*, Cambridge, MA: MIT Press.
Zihl, J., Von Cramon, D. and Mai, N. (1983) "Selective disturbance of movement vision after bilateral brain damage," *Brain* 106: 313–340.

Related Topics

Consciousness, Personal Identity, and Immortality
Multisensory Consciousness and Synesthesia
Consciousness and Psychopathology
Consciousness, Free Will, and Moral Responsibility
Representational Theories of Consciousness
The Global Workspace Theory
The Multiple Drafts Model
Consciousness and Attention
Consciousness, Time, and Memory

28

THE BIOLOGICAL EVOLUTION OF CONSCIOUSNESS

Corey J. Maley and Gualtiero Piccinini

1 Introduction

Human beings have consciousness, and some other organisms may have consciousness too. Organisms and their traits are products of biological evolution. This chapter is about how consciousness might have evolved, and why.

We begin in Section 2 by clarifying that our main topic is *phenomenal* consciousness—the ways our subjective experiences feel—rather than any functional or neural correlates thereof. In Section 3 we point out that there are several ways that a biological trait such as phenomenal consciousness might have arisen during evolution: it might have been selected for because it performs an adaptive function, it might have arisen as a byproduct of some other trait, or it might be the result of random genetic drift. If phenomenal consciousness performs a function, such a function must be identified; Section 4 considers what consciousness might be an adaptation for but adds that identifying the function of phenomenal consciousness is harder than it seems. Section 5 considers how phenomenal consciousness might have originated during evolution if it has no function. Section 6 wraps things up.

2 On "Consciousness"

Consciousness is a notoriously difficult topic. One reason is that the term is ambiguous: different philosophers and scientists use the term "consciousness" to pick out different mental phenomena. Another reason is that even when we disambiguate "consciousness," the referent of the term is often nebulous. The reason is twofold. First, whatever one means by "consciousness," it is a mental phenomenon, and mental phenomena are difficult to observe and conceptualize. Second, some phenomena that are referred to as "consciousness" are private in the sense that each subject experiences only her own consciousness and not the consciousness of anyone else. Because of this privacy, some theorists deny that consciousness is real, or at least that it's an appropriate subject of scientific investigation.

Here, by "consciousness" we mean *phenomenal consciousness* or *qualia*, as philosophers have sometimes labeled the phenomenon: the qualitative feel of being in, or having, particular mental states; the quality of "what-it's-like" to be in or have those states. Philosophers distinguish between *creature* consciousness and *state* consciousness. Creature consciousness is the property of an organism as a whole that differentiates between organisms that are awake or in REM sleep and organisms that are in non-REM sleep, in a coma, or otherwise unconscious. State consciousness is a property of some mental states such that there is a qualitative feel to them. Our main focus is on state consciousness, which can be illustrated by an example.

Consider a familiar event: biting into a ripe orange. The orange *tastes a certain way*. We might say, somewhat unhelpfully, that it tastes *like an orange*: that is, it tastes the way oranges taste, it tastes like other oranges, and so on. If a person has never tasted an orange before, then no amount of description or figurative language will fully enable that person to know *what it's like* to taste an orange: she simply has to experience that taste for herself. The qualitative feeling of tasting an orange is what we mean by phenomenal consciousness.

Before proceeding, it is also worth flagging what phenomenal consciousness is *not*. Phenomenal consciousness is not simply awareness or wakefulness: as mentioned above, this pertains to creature consciousness. By awareness, we mean the ability to detect and respond to stimuli—both internal and external stimuli. By wakefulness, we mean a global neural state of arousal that normally allows organisms to have awareness, although under special circumstances there can be limited awareness without wakefulness and wakefulness with limited or no awareness, as in unresponsive wakefulness syndrome (Laureys et al. 2010). Lack of both awareness and wakefulness occurs when one is asleep or otherwise made unconscious by, say, drugs or a concussion. Wakefulness and awareness—and more generally, creature consciousness—are probably neither necessary nor sufficient for phenomenal consciousness. They are not necessary because during REM sleep we are neither awake nor aware, yet we have phenomenal experiences. They might not be sufficient because very simple animals can be awake and aware, thus possessing creature consciousness, but it is far from clear that they are also capable of phenomenal consciousness.

In addition, phenomenal consciousness is not simply access consciousness (Block 1995). To have access consciousness is to have information available in the cognitive system for reasoning, speech, and intentional action control. In many circumstances, phenomenal consciousness and access consciousness go together: when we experience the taste of an orange, we can also reason about it, talk about it, and plan accordingly; conversely, when we have global access to information, we have a qualitative experience associated with that information. But there might be kinds of information that are accessible without being experienced, or experiences that are not accompanied by cognitive access. Whether phenomenal consciousness and access consciousness can truly be dissociated is controversial; what matters here is that the two notions are conceptually distinct, and that our main interest is phenomenal consciousness.

3 Evolutionary Explanations

Providing an evolutionary explanation of any biological trait faces two challenges. First, several types of evolutionary explanation are possible, and it's not always easy to determine which type applies to a given trait. Second, each kind of evolutionary explanation requires specific sorts of evidence, including evidence about the past, which is difficult to obtain.

Some traits are selected for, meaning that (i) their existence enhances the survival and fitness of organisms with the trait relative to those without that trait, (ii) the traits are genetically heritable, and (iii) organisms with the traits reproduce more than organisms without the trait *because* of the advantage conferred by the trait.[1] We will refer to the contribution to survival and fitness that a trait confers as its *function* (Maley and Piccinini, forthcoming). Traits that evolve via natural selection because of the function they perform—i.e., traits that are selected for—are known as *adaptations*.

For example, as Darwin discovered, having beaks of different shapes allows different finches to feed on different things. Sharp beaks allow some finches to pierce into cacti to get their seeds, whereas blunt beaks allow other finches to better eat seeds on the ground. Since beak shapes are heritable and have different functions, it is plausible that different beak shapes were selected for thanks to the functions they performed. Another example is coloration, which can have different functions in different organisms. Many species of geckos have remarkable coloration patterns that allow them to blend in with their environment, providing camouflage, which allows them to better avoid predators and capture prey. Saharan silver ants are a shiny silver color, but rather than functioning as camouflage, this color helps them reflect sunlight, keeping them relatively cool. By contrast, marine iguanas are nearly black, which helps them absorb heat from the sun. This is especially important for reptiles that spend considerable time swimming in the cold waters of the Pacific Ocean. Again, the colors of these different animals are heritable; they were selected because of the functions they performed. In short, providing an evolutionary explanation of traits that have been selected for requires a determination of their function within the ancestral environment as well as the traits' heritability.

One notable complication is that the same trait may either acquire new functions or lose some of the old functions, or both. On one hand, vestigial traits such as the human appendix used to have a function but no longer have any function at all. On the other hand, traits such as feathers may initially be selected for because they perform one function—thermal insulation, perhaps—and later acquire other functions, such as flying. Thus, developing an evolutionary explanation of a trait requires sensitivity to how the functions of a trait may change over time.

Until now, we have discussed traits that were selected for. Other traits may well be products of evolution, but not because their existence directly confers any adaptive advantage. Because they lack a function, these traits cannot be selected for. Yet they can be *byproducts* of other traits that themselves are adaptive. For example, the transparency of many insect wings might not itself be adaptive. Transparency is a byproduct of being made of thin films of chitin (unless a pigment is added during development), and chitin is the substance that arthropod exoskeletons are made of. Of course, the wings themselves do confer an adaptive advantage.

Byproducts can be either contingent or necessary, a distinction that enriches our understanding of how a trait that was *not* selected for came to be. Consider insect wings again. If the only way that insects could have had wings is for them to be made of thin films of chitin, then given that thin films of chitin are transparent, transparency is a *necessary* byproduct of insect wing evolution. Alternatively, if insect wings could have been made of some other substance that is not transparent, or if coloration can be added during development, then transparency is only a *contingent* byproduct of insect wing evolution.

An obvious question is how to distinguish what counts as necessary and what counts as contingent. How this distinction is made will be relative to different explanatory purposes. Relative to the actual evolutionary history of insects on Earth, it may be necessary that insect wings be transparent: for them to have been made of some other, non-transparent material would simply require a different evolutionary history. Relative to all *possible* flying organisms, insect wings need not be transparent: in fact, some insects add pigment to their wings during development, so their wings are not transparent. On the other hand, relative to the laws of physics, it is necessary that wings be solid at ordinary temperatures near the surface of the earth: there are no wings that have this trait accidentally.

Yet other traits may be simple evolutionary accidents. If a population possesses different variants of a trait, and none of the variants confer an evolutionary advantage, different variants may become distributed among the population simply due to random genetic drift. For example, blood type has no function in humans, but because the population of humans is large enough, the distribution of blood types is relatively fixed. In smaller populations, however, traits can end up dominating others merely because of chance, and become fixated.

Providing an evolutionary explanation of a psychological trait is especially challenging for two additional reasons. First, psychological traits leave few if any traces in the fossil record, so it is difficult to identify the psychological traits of ancestral organisms. Second, typical psychological traits are the result of a complex interplay between innate and environmental factors, making it difficult to identify variations in psychological traits that are genetically heritable and hence subject to evolutionary effects. Both of these challenges apply to phenomenal consciousness.

Finding a place for phenomenal consciousness within evolutionary history faces two additional challenges: there is little consensus on what the physical basis of consciousness is, and there is no consensus on how phenomenal consciousness relates to its physical basis. On the latter question, there are two relevant options. Option one is that phenomenal consciousness is produced by its physical basis but has no physical effects of its own. This view is known as *epiphenomenalism*; many metaphysical views about phenomenal consciousness are committed to it. Option two is the opposite of option one: phenomenal consciousness has physical effects of its own. If option two is correct, a further question is whether the physical effects of phenomenal consciousness confer adaptive advantages—i.e., they perform at least one function. If they do perform functions, a final question is whether phenomenal consciousness is indispensable to fulfilling its functions or its functions can also be fulfilled by non-conscious systems.

4 Consciousness as Adaptation

Many mental faculties and capacities have obvious functions. For example, perceptual systems have the function of acquiring usable information about the environment, and motor control systems are for initiating purposeful behavior. Mental faculties and capacities with clear functions are likely to have been selected for by evolution because of the advantages conferred on organisms by the functions they perform.[2] By contrast, it is especially difficult to see how phenomenal consciousness could have a function, and thus how it could have been selected for.

Nevertheless, various functions of phenomenal consciousness have been proposed, such as allowing organisms to construct an internal model of the world, perform certain inferences, learn in certain ways, perform voluntary actions, or represent themselves. If any of these adaptationist hypotheses is correct, it is likely that these functions are what phenomenal consciousness was selected for, and that would be the basis for the evolutionary explanation of consciousness. According to adaptationism about phenomenal consciousness, when phenomenal consciousness arose during evolutionary history, it conferred an adaptive advantage to organisms that had it; because of this adaptive advantage, phenomenal consciousness was selected for (Barron and Klein 2016; Feinberg and Mallatt 2013; Ginsburg and Jablonka 2010; Grinde 2013).

If phenomenal consciousness has a function, one follow-up question is whether phenomenal consciousness is nomologically necessary to perform that function. If so, then that function cannot be performed by any non-conscious system: the only way that evolution can generate organisms that perform that function is to select for phenomenally conscious organisms. Establishing that phenomenal consciousness is nomologically necessary to perform a function is hard: it would require determining the physical nature of phenomenal consciousness and establishing that such a nature is required to perform the function of consciousness. But there is no consensus on the physical nature of phenomenal consciousness—or on its function, for that matter. Therefore, the nomological necessity of phenomenal consciousness for its function is unlikely to be established any time soon.

This raises the possibility that, even if phenomenal consciousness has a function, that function can also be performed by non-conscious systems. If so, phenomenal consciousness may still have been selected for because of the function it performs, but this is just a contingent fact of evolutionary history. If evolutionary history had stumbled on a different, non-conscious way of performing the same function, we might have evolved to be zombies—entirely non-conscious creatures who behave indistinguishably from us. As if this were not unsettling enough, the possibility that the putative function of consciousness be performed by a non-conscious system raises another general problem for adaptationism about phenomenal consciousness.

Any candidate function of phenomenal consciousness might simply be a function of some other capacity or system that merely *correlates* with phenomenal consciousness. The putative functions of phenomenal consciousness that different theorists have proposed are undoubtedly important features of our psychology, and it may be that they are intimately related to phenomenal consciousness even if they are not themselves *functions* of phenomenal consciousness. In each case, it seems possible for these functions to be fulfilled in the complete absence of phenomenal consciousness. If that possibility obtains, these functions are performed by non-conscious correlates of phenomenal consciousness, and thus phenomenal consciousness *per se* plays no causal role in their performance. Perhaps because of this, some contemporary theories of consciousness have abandoned the strategy of positing a function of consciousness altogether. Nevertheless, it will be instructive to look in a bit more detail at some influential theories of consciousness that do have something positive to say about its function.

According to Global Workspace Theory and other theories like it, the function of consciousness is to make information available to the entire cognitive system, and consciousness simply *is* the availability of that information (Baars 1988). When information is globally available, the organism can flexibly act on the basis of that information, and the organism is phenomenally conscious. What it's like to be that organism is simply what it's like to have all of that information available to one's entire cognitive system.

A rival theory is Supramodular Interaction Theory (Morsella 2005), according to which the function of consciousness is to integrate different sources of information so as to select adaptive action. Sometimes information from different sources is compatible with multiple interpretations or actions. Consider the McGurk effect: the same visual information (a person moving their lips in a certain way) gives rise to different interpretations of what the person is saying, depending on the auditory information that co-occurs with the visual stimulus (whether a voice is heard saying the syllables "bah," "dah," or "gah"). This conflict is resolved automatically and unconsciously. Other conflicts cannot be resolved automatically and unconsciously, and on this theory, the function of consciousness is to allow organisms to resolve conflicts that involve instrumental action. For example, the pain a person feels while holding a hot plate informs the choice of whether to continue holding the plate or drop it. The choice is not automatic, and the person needs the conscious states of hunger and pain to weigh the cost of not eating versus the cost of potential tissue damage.

A third alternative is Flexible Response Mechanism Theory (Earl 2014). According to this theory, the function of consciousness is neither global availability nor integration of information. The function of consciousness is to provide organisms with input to their Flexible Response Mechanism—an executive mechanism by which responses to disparate stimuli can be made in a deliberate manner. For example, breaking a habit such as cigarette smoking requires inputs from a variety of sources, such as anticipating feeling good about quitting, memories of the bad health effects of smoking, or intending to keep a promise to a family member. Such inputs are given to the Flexible Response Mechanism, which can (in principle, if not always in practice) override the automatic, habit-driven behavior.

All of these theories have merits, and different aspects of them seem to accord well with our introspective experience. If they are correct, the functions they assign to phenomenal consciousness may be what consciousness was selected for. Each theory also faces the problem that many cognitive states and processes, which involve carrying and processing information, occur entirely unconsciously. It is not clear why the specific states or processes identified by these theories should require phenomenal consciousness.

On the other hand, simply because a function *could* be realized in a non-conscious system, that does not mean that the conscious system in question does not, in fact, have that function. For example, consider the Krebs cycle. In aerobic organisms, the Krebs cycle is a series of biochemical reactions that uses (among other things) oxygen to produce ATP, which is used by cells as a source of energy, and carbon dioxide, which is a waste product. It is also possible to produce energy without the Krebs cycle; for example, anaerobic organisms cannot use the Krebs cycle because it requires oxygen. It would be wrong to say that the Krebs cycle *lacks* the function to produce energy, just because some other mechanism also produces energy in a different way. In the case of consciousness, then, the mere fact that some other mechanism *could* perform whatever putative function one thinks consciousness has does not prove that consciousness *lacks* that function.

Still, there is an important point that follows from these considerations. While it is true that the Krebs cycle has the function of producing energy in aerobic organisms, it could have been some other mechanism. The fact that it *is* the Krebs cycle, and not some other mechanism, is a contingent fact of evolutionary history. Similarly, if consciousness has a function that some other mechanism could perform, then it might also be a contingent fact of evolutionary history that we have consciousness. Had evolutionary history been different, we might have had a mental life in which we are *not* conscious, yet our mental life would otherwise be identical.

From the point of view of evolution by natural selection, the possibility that we could have evolved without phenomenal consciousness—that is, that we could have evolved to be zombies—is not surprising: many traits are a matter of contingent historical happenings. But from the point of view of our mental life, this is quite surprising. If anything seems to be the core of who we are, of what it is to be human, and what makes life worth living, it is phenomenal consciousness. That consciousness is just another contingent trait is in tension with the idea that being conscious is essential to being human.

In summary, adaptationism about consciousness presupposes that consciousness has a function, but all of the extant theories that propose a function for consciousness face the same challenge: it seems possible that the proposed functions could also occur in a non-conscious system. If that is the case, the proposed functions might be functions of non-conscious systems that merely correlate with phenomenal consciousness, without being functions of phenomenal consciousness itself. If so, consciousness itself cannot be an adaptation. This leads us to theories that do not propose a function for consciousness. The theories that propose that consciousness *does* exist, yet has no function, are incompatible with adaptationism about consciousness. They are compatible with the view that consciousness is either a byproduct of other traits or an evolutionary accident.

5 Consciousness as a Byproduct or Evolutionary Accident

Phenomenal consciousness may have no function, and thus may have no adaptive value whatsoever. However, it may still be possible to explain consciousness using the resources of evolutionary biology; we simply have to look beyond natural selection. In particular, if phenomenal consciousness has no adaptive value, then it may be a byproduct of some other biological trait that itself *was* selected for—what Gould and Lewontin (1979) called a *spandrel*. Alternatively, consciousness may simply be an evolutionary accident. Either way, while consciousness itself has no function, it is intimately related to something else that does. We have referred to the view that consciousness is either a byproduct or an accident of evolution as the Byproduct or Accident View, i.e., BAV (Robinson, Maley, and Piccinini 2015). On this view, consciousness is not explained by appealing to its function. BAV takes consciousness to be similar to many other biological traits that are not explained via their direct adaptive value.

To begin with, let's consider the possibility that consciousness is a byproduct of some other trait. It may be that consciousness occurs when neural tissue of enough complexity is actively processing information in a certain spatiotemporal region. Consciousness itself does not *do* anything. Or, if it does do something, what it does has no adaptive value. In addition, consciousness is not identical to the neural activity that gives rise to it. Although neural activity does have many functions, it simply happens that consciousness is a byproduct of that kind of neural activity.

Consider a psychological process, such as human facial recognition, that (presumably) has been selected for. This capacity relies on neural structures that process complex information. These neural structures do many other things as well, such as generate small electrical fields (which is what EEGs are able to detect), produce small amounts of heat, and so on. Of the many things that these particular neural structures do, the ability to recognize faces was selected for, whereas producing electrical fields and heat were not. It just so happens that a certain amount of neural tissue functioning in a compact space generates electrical fields and heat. Consciousness may well be like that: it is simply a byproduct of neural systems functioning in a certain way.

If consciousness is a byproduct of some other trait, a further question is whether it is a contingent or necessary byproduct. It is a contingent byproduct if one of the following is the case: either there are organisms that function just as well as conscious ones but without consciousness, or there could have been such organisms had evolution proceeded differently. This is analogous to the transparency of insect wings. Again, many insect wings are transparent, but this is a contingent byproduct of what they are made of: add pigment during development and the wings cease to be transparent. By contrast, consciousness is a necessary byproduct if, as a matter of natural law, any physical process that performs certain cognitive functions gives rise to consciousness—even though the physical processes themselves are unconscious and consciousness has no functions of its own.

The other option is that consciousness is a frozen evolutionary accident. It may be that all cognitive functions can be performed entirely unconsciously, without giving rise to any conscious state or process. Yet, at some time during evolution, some harmless mutation arose that causes phenomenal consciousness to occur under certain circumstances—e.g., when certain cognitive functions are performed. If this mutation were lost, we would lose all phenomenal consciousness but keep on functioning exactly as before—just without consciousness. We'd become zombies. But this mutation became fixated through random genetic drift. Since it's harmless, it sticks around. This is analogous to the proportion of blood types within the human population. As far as we know, humans could do with only a single blood type, and no blood type appears to confer any advantage over any other. At the same time, no blood type appears to create any disadvantages either. Thus, the proportion of different blood types is an evolutionary accident.

There are theories of consciousness—the most prominent among them Information Integration Theory (IIT)—that attempt to explain consciousness without appealing to any supposed function of consciousness whatsoever. According to IIT, consciousness is an emergent property of certain systems that are capable of processing information in complex ways, irreducible to the activity of their mechanistic components (Tononi 2008; Koch et al. 2016). Instead of positing a function for consciousness, IIT begins with the supposition that phenomenal consciousness has a particular structure and proposes certain features and constraints on physical systems that could be the physical substrate of phenomenal consciousness. According to IIT, although consciousness itself does not (and perhaps, cannot) have a function, it may well be a necessary byproduct of any system that processes information in the highly integrated manner specified by the theory. It is the efficient processing of complex information that does the work, rather than its byproduct, phenomenal consciousness. Therefore, what was selected for by evolution was the efficient processing of complex information. Phenomenal consciousness came along for free.

Analogously, some empirical approaches to the evolution of consciousness do not presuppose a function of consciousness. They involve identifying the cognitive or neural systems that correlate with consciousness, identifying the species that possess those systems, and using that information to infer when consciousness might have arisen (Mashour and Alkire 2013).

6 Conclusion

Phenomenal consciousness might have evolved in one of three ways. If phenomenal consciousness performs a function—that is, if it has physical effects that confer an adaptive advantage—it was probably selected for. But it is difficult to establish that phenomenal consciousness has a function. In fact, many metaphysical and empirical theories of phenomenal consciousness posit that consciousness has no function at all.

If phenomenal consciousness has no function, it might be a byproduct of some other trait that does. For example, phenomenal consciousness might be a byproduct of a certain kind of information integration; when that kind of information integration was selected for, phenomenal consciousness came along for the ride. Another possibility is that phenomenal consciousness is a frozen evolutionary accident. Some adaptively neutral mutation might have caused phenomenal consciousness to arise; later, random genetic drift might have caused phenomenal consciousness to become fixated in the populations that had it.

There is no easy way to determine which of these explanations is correct. We need a better understanding of the metaphysics of consciousness as well as the related question of whether consciousness has a function. Philosophical work on the metaphysics of consciousness that informs, and is informed by, empirical work on the evolution of consciousness is a relatively nascent endeavor, with plenty of progress to be made. Although phenomenal consciousness and its evolution remain mysterious, understanding the place of consciousness in our evolutionary history may well lead to new insights into this central aspect of our mental life.[3]

Notes

1 Criterion (iii) may seem redundant, given that fitness—which is mentioned in criterion (i)—is often defined in terms of reproductive success. In principle, however, a trait could spread in a reproductively successful population due to factors *other* than the contribution of the trait to reproductive success. Although this line of reasoning has been questioned (Fodor and Piattelli-Palmarini 2010), most biologists find it unproblematic: which factors directly contribute to fitness increases is an empirical question (cf. Sober 2010).

2 In fact, etiological theories of function in philosophy attribute functions in organisms *only* to those traits and features that have been selected for.

3 This material is partially based upon work supported by the National Science Foundation under grant no. SES-1654982 to Gualtiero Piccinini.

References

Baars, B. (1988) *A Cognitive Theory of Consciousness*, Cambridge, UK: Cambridge University Press.

Barron, A.B., and Klein, C. (2016) "What insects can tell us about the origins of consciousness," *Proceedings of the National Academy of Sciences* 113: 4900–4908.

Block, N.J. (1995) "On a confusion about a function of consciousness," *Behavioral and Brain Sciences* 18: 227–287.

Earl, B. (2014) "The biological function of consciousness," *Frontiers in Psychology* 5. http://doi.org/10.3389/fpsyg.2014.00697

Feinberg, T., and Mallatt, J. (2013) "The evolutionary and genetic origins of consciousness in the Cambrian Period over 500 million years ago," *Frontiers in Psychology* 4. http://doi.org/10.3389/fpsyg.2013.00667

Fodor, J.A., and Piattelli-Palmarini, M. (2010) *What Darwin Got Wrong*, New York: Farrar, Straus and Giroux.

Ginbsurg, S., and Jablonka, E. (2010) "The evolution of associative learning: a factor in the cambrian explosion," *Journal of Theoretical Biology* 266: 11–20.

Gould, S.J., and Lewontin, R.C. (1979) "The spandrels of San Marco and the panglossian paradigm: a critique of the adaptationist programme," *Proceedings of the Royal Society B: Biological Sciences* 205: 581–598.

Grinde, B. (2013) "The evolutionary rationale for consciousness," *Biological Theory* 7: 227–236.

Koch, C., Massimini, M., Boly, M., and Tononi, G. (2016) "Neural correlates of consciousness: progress and problems," *Nature Reviews Neuroscience* 17: 307–321.

Laureys, S., Celesia, G.G., Cohadon, F., Lavrijsen, J., León-Carrión, J., Sannita, W.G., Schmutzhard, E., von Wild, K.R, Zeman, A., and Dolce, G., (2010) "Unresponsive wakefulness syndrome: a new name for the vegetative state or apallic syndrome," *BMC Medicine* 8: 292–295.

Maley, C. J., and Piccinini, G. (forthcoming) "A unified mechanistic account of teleological functions for psychology and neuroscience," In David Kaplan (ed.), *Explanation and Integration in Mind and Brain Science*, Oxford: Oxford University Press.

Mashour, G.A., and Alkire, M.T. (2013) "Evolution of consciousness: phylogeny, ontogeny, and emergence from general anesthesia," *Proceedings of the National Academy of Sciences* 110, supplement 2: 10357–10364.

Morsella, E. (2005) "The function of phenomenal states: Supramodular Interaction Theory," *Psychological Review* 112: 1000–1021.

Robinson, Z., Maley, C. J., and Piccinini, G. (2015) "Is consciousness a spandrel?" *Journal of the American Philosophical Association* 1: 365–383.

Sober, E. (2010) "Natural selection, causality, and laws: What Fodor and Piatelli-Palmarini got wrong," *Philosophy of Science* 77: 594–607.

Tononi, G. (2008) "Consciousness as integrated information: a provisional manifesto," *Biological Bulletin* 215: 216–242.

Related Topics

Materialism
Information Integration Theory
The Global Workspace Theory
Consciousness and Action
Animal Consciousness

29
ANIMAL CONSCIOUSNESS

Sean Allen-Hermanson

1 Introduction

Animal consciousness continues to draw the attention of philosophers, scientists and general audiences and is tethered to ongoing debates about fundamental questions of mind, knowledge, and morality. Phenomenal consciousness is very hard to define without reference to itself, and perhaps the best one can say is something along the lines of "states of mind with a qualitative feel" (Nagel 1974). Knowing which animals are sentient and knowing what it is like, that is, what kind of consciousness they possess, are respectively known as the Distribution and Phenomenological questions (Allen 1998) and essentially aspects of the problem of other minds extended to nonhumans (Allen and Trestman 1995/2016). Indeed, our ignorance about other species is arguably the quintessential formulation (Harnad 2016). Knowing anything about phenomenological feel is especially difficult, with the exception of when it is like nothing (Akins 1993) or for qualitative experiences regarding which we have first-hand acquaintance (Allen-Hermanson 2017), though see Thompson (1992), Thompson et al. (1992) and Matthen (1999) for reflections on alien perceptual qualities in nonhumans, especially colors. As much more has been written on the problem of distribution, this chapter will focus on providing an overview of the main philosophical responses to curiosity about which animals are conscious.

2 Basic Issues

Among the foundational matters not to be discussed here include whether consciousness is physical or non-physical, whether it is epiphenomenal, whether "punctate" minds are possible (made up of independent "atoms" of experience) or if clusters of conscious states must be bundled as a unified subject, whether it occurs on a gradient (like baldness, which comes in degrees) or is "binary" (like pregnancy, it's either there or it's not), whether there is an explanatory gap, whether consciousness is an irreducibly fundamental aspect of reality, its flow and relationship to time, and the nature of the relation between consciousness and intentionality, if any. These problems exacerbate the difficulty, perhaps intractably, of puzzlement about what it is and how it is distributed. Complicating matters further is the diversity of cognitive mechanisms, behaviors, and organisms to be considered, making it difficult to apply any single, all-encompassing treatment; "fish" is actually a vastly heterogeneous category, for instance (Allen 2013). But most basic is the matter of how the subject should be approached philosophically. Should we first sort out a "metaphysical" theory, which could then be applied to specific cases including various nonhuman species? Alternatively, perhaps we can table inquiry into what it is and proceed with our epistemological investigations (Allen and Bekoff 1997). Both types of approaches are taken up by scholars and researchers with examples of each to be canvassed next.

3 Epistemology First?

Though the mental lives of other people are not normally in serious doubt, we are often unsure if another thinks or feels as we do. There are also various uncertain cases, such as people in vegetative states, fetuses, and anencephalic infants. The status of animal consciousness is not just a philosophical problem either, and the difficulties concerning those incapable of speech are compounded by differences in anatomy and behavior. Overcoming the epistemic problems requires that we avoid both anthropomorphism (like Scylla, multifaceted and resilient) and excessive skepticism (like Charybdis, obliterating), though there are no generally accepted methods or principles for navigating these twin perils (see Fisher 1996 and Kennedy 1992 for contrast). Morgan's "Canon," for instance, was one reaction to overly generous anecdotes:

> [I]n no case may we interpret an action as the outcome of the exercise of a higher psychical faculty, if it can be interpreted as the outcome of the exercise of one which stands lower in the psychological scale.
>
> *(Morgan 1894: 53)*

The Canon was widely acknowledged as the scientific study of animal cognition began to develop in the beginning of the 20th century, particularly among behaviorists. In recent years it is mainly a historical curiosity, though there is a lively academic discussion about its proper interpretation, such as whether it is just a form of parsimony, or perhaps Ockham's razor (Burghardt 1985; Sober 1998; Allen-Hermanson, 2005; Sober 2005; Sober 2009; Fitzpatrick 2008).

Block (2002) argues that since the human case is the starting point for our investigations, it is unclear how we can have any grounds for attributing consciousness to beings physically and neurologically very different from ourselves, even if they are "common-sense functional isomorphs" acting as if they have experiences in virtue of internal states satisfying the causal roles making up our mental lives. There is, as Block puts it in earlier work, a "prima facie doubt" (1978). Indeed we cannot even form a conception of what those grounds could be in what he calls the "harder problem" of consciousness. Therein lies a tension between reconciling scientific understanding with the fact of subjective awareness: "these commitments do not fit together comfortably." If Block is correct then the hope that we may somehow merge first-person accounts and scientific approaches (Burghardt 1985) in order to "obtain testable hypotheses about private experience" (Burghardt 1997; Burghardt and Bekoff 2009) teeters on incoherence, despite an influential legacy (Morgan 1894; Griffin 1976, 1978, 2001). Block is trading on intuitions that may not be shared, though the stakes are high: his view suggests the study of animal consciousness cannot proceed with any confidence absent close scrutiny of neurological similarity and difference.

Leaving these preliminary matters aside, there are three main epistemic strategies available: Analogical (similarity) arguments, Inference to the best explanation, and Perceptualism.

4 Other Animal Minds: Historical Context

The use of comparative reasoning to ground our knowledge of other minds is often attributed to Mill (1889: 244) though its pedigree is much older, probably ancient.

Awareness of the problem of other minds may or may not be motivating Descartes' example about "hats and cloaks" in the second Meditation (1641a/1985)—though it is raised by Augustine, who is known to have influenced him, and where we also find one of the earliest expressions of the analogical solution (Matthews 1986). Although Matthews believes the problem "is not raised explicitly anywhere in Descartes" (1986: 144), we come close enough in his correspondence (see also *Discourse on the Method* Pt. V 1637a/1985). In a letter to the Marquess of Newcastle, Descartes writes "In fact, none of our external actions can show anyone who examines them that our body is not just a self-moving machine but contains a soul with thoughts..." with, of course, the exception of "words, or other signs" (1646/1991: 303). Descartes notoriously denies all aspects of mind to animals here and elsewhere.

Writing to More, Descartes argues we can have no absolute certainty on the matter of animal minds either way since "the human mind does not reach into their hearts" (1649/1991: 365). Despite this, many conform to the "preconceived opinion...accustomed from our earliest years...that dumb animals think" (ibid.). This popular view depends on "very obvious" analogical reasoning, in that "many of the organs of animals are not very different from ours...they have...sense-organs like ours, it seems likely that they have sensation like us...but there are other arguments...not so obvious...which strongly urge the opposite" (1649/1991: 265–6). Crucially, for Descartes "thought is included in our mode of sensation" (1649/1991: 365) by which is meant that phenomenal consciousness is part of rational judgment and a linguistic mode of representation (some, e.g., Cottingham 1978: 555, dispute this reading, though not convincingly).

Descartes' viewpoint is situated within his interactionist dualism, whereby human behavior cannot be explained unless we invoke both mechanism and mental substance (res cogitans). However, in the case of animal behavior the "mechanical and corporeal" (1649/1991: 365) is sufficient. Descartes thinks the everyday comparisons are revealed as superficial by way of three arguments: many animals move like machines (think of the haphazard flight of a butterfly); there are probably natural automatons ("art copies nature" 1649/1991: 366); and although animals are without language, "speech is the only certain sign of thought" (ibid.). In the *Discourse on the Method* Descartes emphasizes the creative aspect of language, noting an animal could never "produce different arrangements of words so as to give an appropriately meaningful answer to whatever is said in its presence" nor can they use what they know to respond flexibly in "all kinds of situations" as competent human beings can (1637a/1985: 140).

Although "*thinking* is to be identified...also with sensory awareness" (1644/1985: 195) Descartes' occasional reference to animal "sensations" (1649/1991: 365) such as "anger, fear, hunger, and so on" (1649/1991: 366; 1646/1991: 303) is sometimes taken to imply that he did in fact believe animals were phenomenally consciousness (Andrews 2015: 54). Yet it is hard to accept this was his considered view, since he often denied animals to have an incorporeal soul (1637b/1991: 62, 1641b/1991: 181, 1646/1991: 304, 1649/1991: 365; Cottingham 1978: 557). While it has been suggested that these remarks are better regarded as some kind of lapse (Cottingham 1978: 558), a more charitable reading, faithful to his texts, is that by "sensation" (sometimes "organic sensation" Seager 1999/2016: 4) Descartes only means certain movements of the body machine transmitted to the brain would have given rise to conscious episodes in the presence of res cogitans, so e.g., "hunger" in an animal is nothing more than internal muscle contractions and "brain commotions" (as mooted, though ultimately rejected by Cottingham 1978: 558). This would go some way towards making sense of his acquiescence in the dismemberment of living animals (such as dogs) for scientific and medical advancement. As he writes in a letter to Mersenne, "I do not explain the feeling of pain without reference to the soul" and so animals lack "pain in the strict sense" (1640/1991: 148).

Some have disputed that Descartes was a brute to the brutes in his actions. Dennett (1995: 692), via Leiber (1988), dismisses the "myth" Descartes was a "callous vivisector, completely indifferent to animal suffering..." To the contrary, Dennett portrays Descartes as "first victim" of a "lunatic fringe," continuing in the accusations of Mary Midgely and Peter Singer, against those innocently seeking to discover "how animals actually work!" Certainly, Leiber airs some second thoughts about the reliability of reports from Descartes' contemporaries. Yet concerning insensitivity to ill treatment he switches focus onto the fabulous, Disneyesque characterizations of animals' inner lives offered by Descartes' ideological opponents (Leiber 1988: 312ff.). But this is just irrelevant to the matter at hand, as is Leiber's rationalization of Descartes' vivisection of a rabbit (described in a letter to Plempius 1638/1991) as nothing other than "a most serious and painstaking pursuit of the truth" (Leiber 1988: 315). Could the tormented rabbit, chest opened, ribs removed, aorta pinched, also be said to have had a stake in the taking (and giving) of pains? In his defense, Descartes was probably no worse than many of us complicit in the infliction of suffering for a good cause, and unlike most beneficiaries of modern medicines, factory farming, and consumer products, his callousness is mitigated by the reasoning that animal behavior is dissimilar to movement under conscious direction.

This also demonstrates how the analogical strategy can backfire. In his letter to the Marquess of Newcastle, Descartes observes that animals can only "imitate or surpass us in those of our actions which are not guided by our thoughts, "such as when we "walk or eat without thinking" (1646/1991: 302). Certainly his skepticism has not been embraced, and indeed powerful arguments and evidence testify to quite sophisticated cognizing in many non-humans. Nevertheless, Descartes' dissatisfaction with the analogical solution to the problem of other animal minds finds many fellow travelers with philosophers and scientists continuing to debate the place of similarity-based reasoning, and his insight about adaptive response continues to be hugely influential.

5 Analogical Arguments

> [W]hen a living body is moved there is no way open to our eyes to see the mind... But we perceive something present in that mass such as is present in us to move our mass in a similar way; it is life and a soul.... Therefore we know the mind of anyone at all from our own.
>
> *(Augustine in Matthews 1986: 144)*

At least since Augustine, the analogical solution to the problem of other minds finds expression in the writings of such notable philosophers as Locke (1689/1975 bk. IV, ch. iii, par. 27), Hume (1739/1978 bk. I, pt. III, sec. xvi), Mill (1889), James (1912/1971), Broad (1925), Russell (1948) and Ayer (1956). Hume's (1739/1978) comparative reasoning about behavior led him to conclude that animals think, reason, and form associations between their sense impressions, though not with the same degree of sophistication as human beings. Others in their wake continue to employ analogical reasoning about animals (e.g., Singer 1975/1990, 1993; Perrett 1997).

"Behavior" is now understood widely to include physiological response (such as cardiac acceleration as an indicator of anticipation, or sensitivity to opioids for feelings of pain) as well as modulatory effects of cognitive states, such as emotions on learning and memory. Perceptual behavior includes reactions to ambiguous stimuli and awareness of threats. Metacognitive monitoring is investigated by way of perception (Smith et al. 1995), memory (Hampton 2001) and foraging (Call and Carpenter 2001), though interpretations positing high level awareness and control need to be carefully scrutinized against deflationary accounts (Hampton 2009).

The relative importance of the different bases for analogizing is a matter of debate, with some urging that "physiological data can play a qualitatively different and more definitive role" (Farah 2008) while others equally draw on "molar" behavior, in the sense of actions falling under everyday platitudes (e.g., Varner 2012, though not without other considerations mixed in). Besides behavior, two other important sources of human–animal continuity are neurocognitive mechanism and common evolutionary descent.

The structure of the argument from analogy for animal consciousness turns on the premise that conscious human beings are highly similar to most individuals of this-or-that species. Since this draws on features of large groups, this formulation avoids the "single case" problem for the analogical solution to the problem of other (human) minds, namely that reasoning on the basis of one, possibly unique case (i.e., my own), to a general conclusion about others makes for a weak induction (Malcolm 1962; Andrews 2008/2016). Then again, for the analogy about animals to get started we need to already know other human beings are conscious. If we don't need analogy to know that, why do we when it comes to animals?

A second problem area is that, unlike other inductions, e.g., the color of swans, the conclusion cannot be independently confirmed (Ryle 1949: 15; Pargetter 1984), though others dispute whether this matters. Hyslop and Jackson (1972) counter that since an induction that can be verified by other means can also be disconfirmed by other means, the fact that it can be checked adds nothing to the cogency of the inference (see also Plantinga 1967). Another limitation is that the analogical strategy runs the risk of chauvinism for sentients highly dissimilar to human beings (Graham, 1993/1998) raising the possibility of uncheckable type-2 errors (i.e., false negatives).

A third criticism concerns the difficulty in knowing which properties should factor into the comparison (Pargetter 1984). Returning to a theme from Descartes, animals resemble human beings only to a degree and it is hard to know when (and what) accumulated differences ought to defeat judgments about sentience (Allen 2004: 622). Instead of a comprehensive tally of all shared characteristics, only certain relevant properties—behavioral, physiological, neural or evolutionary—should be considered. Yet if we knew which ones counted there would be no problem of other minds! Even if the analogical solution does not depend on a full-blown theory of consciousness, knowledge of some crucial marks of structure and function do seem to be needed (Allen and Trestman 1995/2016). Others have argued for a hybrid account in which reasoning about competing hypothetical inferences is incorporated (Melnyk 1994).

6 Inference to the Best Explanation

Where the Argument by Analogy sought to extend what is given in introspection to other individuals, Inference to the Best Explanation (IBE) doesn't depend on self-observation. This is because posits in a successful empirical theory don't have to be directly observed: genes, the planet Neptune, electrons, and dinosaurs are all strongly evidenced, though only indirectly by way of their observable effects. Arguably "mentalism," or the positing of beliefs and desires (and perhaps states of consciousness) that are real internal causes of behavior can be the *best* explanation (Pargetter 1984).

Best explanation-style reasoning is often offered as a solution to epistemic doubts about animal minds (Bennett 1991; DeGrazia 1996; Allen and Bekoff 1997; Lurz 2009a: 7; Bermúdez 2003; Heyes 2008). Improvised paths are but one source of evidence for internal mental representations in animals, perhaps suggestive of "cognitive maps" (Tolman 1948; Gallistel 1990;

Gallistel and King 2009). Observations favor mentalism over skepticism (and behaviorism) when a test is passed that other hypotheses fail. The attempt to show animals make logical inferences by ruling out deflationary alternatives, such as the use of smell or other perceptual cuing (Call 2004) is another example (e.g., see Sober 2000, 2005, 2012, 2015 for a detailed examination of related considerations, including evolutionary propinquity and parsimony).

IBE has several virtues such as its anti-chauvinism, since a being need not be similar to me or even have a human form for the attribution of internal states with psychological roles to have the greatest, most unifying, explanatory power. It also doesn't depend on induction from a single case, and it accounts for the fuzzy persuasiveness of analogical reasoning in that my behavior and the behavior of beings similar to me are explained in a similar manner (i.e., in terms of beliefs and desires). Having said that, IBE isn't well understood as it is unclear what kinds of characteristics make an explanation best (Plantiga 1967; Lipton 1991), though simplicity, generality and coherence with the rest of our knowledge have been suggested (Harman 1965; Thagard 1978). As with other scientific theories, attributions of mind must be eligible for revision or perhaps complete overthrow (with respect to animals, behavioristic explanations typically threaten, e.g., Kennedy 1992; Wynne 2004). Another complication is that there may not be a clear demarcation between deflationary and non-deflationary hypotheses, such as associationist versus cognitivist (Penn and Povinelli 2007; see also Rescorla 2009; Allen and Bekoff 1997: 57–8; Buckner 2011; Mitchell et al. 2009).

Inferring the presence of mental states by their causal roles has also been applied to the problem of animal consciousness (Lurz 2002; Dretske 1995; Tye 1997, 2016), though this requires a grasp of its function (assuming it has one!). Unusual types of sensation (e.g., electroreception) may also lead to doubts: if we don't know what it is like, why assume it is like anything? The answer is that conscious feeling might still have objectively accessible causes and effects notwithstanding further facts about irreducibly qualitative character. As with mental state types in general, determining functions or causal roles might need to draw on everyday intuition, scientific investigations, or perhaps both. What relationship is there between propositional attitudes and attributions of phenomenal consciousness (Lurz 2009b)?

Perhaps consciousness best explains adjustment for perceptual error and can be evidenced by contrasting perception and belief (Allen and Bekoff 1997: 152). Although a fly can be "fooled" by the Müller-Lyer illusion (Geiger and Poggio 1975), there can be no mismatch between how things look versus what it takes to be true. Carruthers (2000) similarly proposes consciousness is a capacity for making an appearance–reality distinction. Various somewhat overlapping proposals about the cognitive role for consciousness include adaptive control (James 1912/1971; Block 1995; Dretske 1995), practical judgment (Kirk 1994), guidance by inner maps (Tye 1995), a central representation used to situate and move the body (Merker 2005, 2007), and higher-order awareness (Lycan 1987; Rosenthal 1986), perhaps in the form of self-report (Dennett 1991).

Cognitive interpretations of consciousness emphasize integrated use of sensory representations in control and movement, perhaps in the sense of being guided by reasons (Dretske 2006). When it comes to clarifying the functions of phenomenal consciousness on behalf of the epistemic strategy—whether it be in terms of higher order thinking, pain behavior, sensory integration, rational action, or what have you—trafficking in the metaphysical approach may be unavoidable. A lingering worry is that unlike beliefs and desires, awareness of states of consciousness in others is not strictly like a scientific inference. Qualitative features of mental representations may offer no added value to explanations of animal behavior (Carruthers 2005b: 203). It may be necessary to draw on one's direct acquaintance with phenomenal properties of experience, hence revisiting difficulties with the analogical solution to the other mind's problem (Melnyk 1994).

7 A Non-Inferential Solution?

In a third school of thought our knowledge of other minds is a matter of direct perception, much as when we make sensory judgments about everyday objects. Sometimes drawing on the phenomenology of Husserl (1913/1982) and Merleau-Ponty (1962), but also defended by disciples of Wittgenstein, behaviorists, and others (e.g., McDowell 1982), mentalistic attribution is not based on inference or theoretical judgment, so e.g., we don't infer pain from behavior, rather we just see it in the expression of wincing, moaning, etc. However, Perceptualism and its application to the problem of other animal minds (Cockburn 1994; Dupré 1996; Jamieson 1998; Cassam 2007) must meet several challenges.

How can we be sure inferences are not being made which are fast and unconscious? One reason for thinking this might be the case is because the direct perception of mental states would have trouble accounting for error. The Perceptualist can respond that these judgments depend on background knowledge helping to set default assumptions—such as that Martian marionettes (Peacocke 1998) and Blockheads (Block 1981) are highly atypical. But what background conditions ought to be assumed about fish, insects, and so on? How, that is, do we know what to frame as error given that we aren't sure which of our mentalistic attributions towards animals are correct? Relatedly, how do we adjudicate disagreement and balance skepticism against anthropomorphic bias? Indeed, why should there be so much variance in our views on animal minds (compare how there is little disagreement about whether there is a tree or a rock in the vicinity)? And what form does this perceptual access take? We can't literally see states of mind, though one could try to draw an analogy to our indirect awareness of the hidden surfaces of physical objects (Husserl 1913/1982; Smith 2010). In scene segmentation, for example, an object is perceived as a unity despite some of its parts being out of view: consider a housecat partially hidden by a blanket leaving only its head and tail in sight. For the Perceptualist the judgment about the cat at least seems automatic and immediate rather than the product of deliberate reasoning. On the other hand, this comparison is strained by the fact that mental states are not open to direct examination from other angles and perspectives. In light of the various difficulties, perhaps the role of inference in mental state attribution needs to be conceded after all, with some recommending a hybrid account (Roelofs 2017).

8 Metaphysics First?

Alternatively, perhaps we should work out a theory of consciousness before returning to the epistemic questions. One drawback with this approach is the heterogeneity of the philosophical menu, with widely divergent implications for animal subjectivity. In addition, it is unclear whether pre-theoretical (folk) judgments about animal minds ought to constrain our choice of theory. It is sometimes, for example, taken as a reductio of a theory of consciousness that it implies most animals are not phenomenally aware (e.g., Allen and Trestman 1995/2016; Gennaro 2009: 184; Tye 2016: 21). Or should one reject those folk judgments if they do not cohere with an account deemed plausible on other grounds (Lycan 1999)? As many leading theories can be loosely categorized as falling within the hardware/software distinction, this survey will examine ones that can be broadly characterized as either Neural-reductive or Functionalist.

9 Neuroreductive Approaches

Sidestepping some philosophical conundrums, Crick and Clark (1994) proposed researchers focus on the direct examination of neural mechanisms minimally sufficient for phenomenal awareness, as this can at least enlighten as regards the Neural Correlate of Consciousness (NCC). In outline, the strategy is simply to determine how brain processes differ between when subjects are aware as opposed to unaware of a stimulus. In application, extensive work has been done with brain imagining and psychophysics to investigate mechanisms and behavior, especially in vision, sleep, anesthesia, and neuropathology. The next step is also, in a sense, elementary though difficult to establish for empirical and conceptual reasons: one employs similarity-based reasoning to argue this-or-that species either has or does not have structures homologous to (that is, having a shared ancestry with) the human NCC. The chief dividend paid so far finds neocortex sufficient, especially thalmocortical regions working in conjunction with subcortical structures (Laureys et al. 2004; Laureys et al. 2009/2013). Mammalian hardware stands out as highly similar to the human case, though we cannot be sure what significance attaches to its absence in light of the possibility of humancentric chauvinism.

Though probably most neuroscientists believe consciousness depends on a functioning neocortex, some claim subcortical structures, such as the midbrain, suffice. The difference it makes is that consciousness would turn out to be much more widely distributed. Support for the midbrain view often draws on observations of children afflicted with congenital hydranencephaly (i.e., those born decorticate) who nevertheless evidenced strong signs of conscious awareness (Shewmon et al. 1999; Merker 2007, 2008; Aleman and Merker 2014). Some take this to mean consciousness could also be present in animals lacking a neocortex but having structures reminiscent of the midbrain (Merker 2007), such as many fish (Tye 2016: 84). Transferring these results to nonhumans (Barron and Klein 2016; Woodruff 2017) illustrates the dictum that "extrapolations require cautiousness" (Le Neindre et al. 2017) however.

Notably, the individuals in question weren't true hydranencephalics (Shewmon et al. 1999: 371). It appears that the brain was already investing in cortical resources and reassigned midbrain cell populations when problems arose. In mammals neurons are generated from generic progenitor (stem) cells (Gage 2000; Ming and Song 2005) and are known to pass through a critical period for plasticity (about 1–1.5 months, Ge et al. 2007) for adaptation to various subtypes (Molyneaux et al. 2007). This means early stage midbrain neurons could be reprogrammed to function differently. The identity of these cells and higher level structures could be verified by their "preferred" stimuli and other organizational features characteristic of auditory cortex (Schreiner et al. 2000; Kandel et al. 2013: 700)—in theory, that is, since the children in question died, no autopsies were performed, and the brain scans administered were imprecise. In short, cortical functioning isn't shown to be unimportant just because it has taken up residency at an unusual address.

No doubt, neocortex will continue to take a central place in debates over the identity of the NCC.

10 Representationalism

The leading "software" approaches are known as "Representationalist" theories reducing phenomenal consciousness to mental representations or intentional states of some sort. According to First-Order Representationalists (Kirk 1994; Dretske 1995; Tye 1995, 2000) conscious representations play a certain sort of cognitive role (especially being poised to make a difference to belief and action) emphasizing input integration and output flexibility typically along the lines of such views as Baars' Global Workspace (1988, 1997, 2005a) and Block's (1995) "access" consciousness. These ideas have been explicitly applied to the problem of sentience in nonhumans (Edelman and Seth 2009), such as birds and cephalopods (Edelman et al. 2005).

An important feature of the First Order Representationalist theory (FOR) is that it more or less straightforwardly implies many animals are conscious insofar as there is a sophisticated cognitive economy going beyond tropism and rudimentary learning. Whether this should be regarded as a feature or a bug is a matter of debate (Allen and Trestman 1995/2016). Tye (2000, 2016) finds that FOR most strongly suggests consciousness in mammals and birds, with a weaker though still reasonable case for teleost (bony) fish, reptiles, cephalopods (octopuses, squids and cuttlefish) and even one insect genus (*Apis,* i.e. honeybees)—though sharks, rays, most insects, and many "lower" vertebrates do not measure up.

Besides over-generosity, FOR has been criticized for having difficulty giving an account of conscious awareness of belief and desire (Lurz 2006), though the main reason some philosophers have sought an alternative formulation of Representationalism owes to the assumption that conscious mental states are simply ones the subject is aware of, hence requiring a higher-order awareness (Lycan 2001). Appeals to purported examples of unconscious perceptions, such as distracted driving (Armstrong 1968) and blindsight (Carruthers 1989; 1996), motivate similar objections, though it is not obvious to this author that there is nothing that it is like for the inattentive driver or that the requisite integration and flexibility demanded by FOR is present in those kinds of cases (Seager 1999/2016). A major worry on the horizon for First Order Representationalists ought to be robots and autonomous vehicles, which put pressure on the threshold for attributing attitude-like states of information processing. The F-16 drone that can "figure out" the safest path to a ground strike and respond to threats if interrupted (Lockie 2017) is not conscious, though as art continues to copy nature, sooner than later human contrivances guided by information states provided by "accredited receptor systems" (Dretske 2006) will exacerbate the metaphysical (and moral) dilemmas.

The other major version of Representationalism casts consciousness in terms of the mind's awareness of itself perhaps starting with Locke's proposal that "reflection" serves as inwardly oriented perception (1689/1975). It is uncertain whether Locke intended to equate reflection and consciousness (Thiel 2011), but as reflection consists in acts of inner observation, or "second-order" representing, his view has come to be associated with Higher Order Perception (HOP) or "inner sense" theories of consciousness as developed by Armstrong (1968, 1981), Lycan (1987, 1996) and others. On the assumption that perceptual higher-order awareness does not require any thoughts, the application of mentalistic concepts, or grasp of folk-psychology, the inner sense view is perhaps no less friendly to widespread consciousness in animals than First Order Representationalism. It remains unclear what this awareness consists in, however. In light of the transparency or "diaphanous" aspect of introspection (Moore 1903) some claim it has no distinctive phenomenology (Dretske 1995; Güzeldere 1995), and even if it does, which animals satisfy is not known (Lycan 1999).

HO theories are united in holding that what makes a mental state conscious is that it is taken as the representational object of a second (or higher) order mental state, though some philosophers argue these ought to be understood as thoughts rather than sensory perceptions (e.g., Rosenthal 1997 and Carruthers 2000 who differ over whether the HOT's need to be actual or merely dispositional states of the cognitive system). A close alternative view eschews the requirement of an additional state, replacing this with the idea that one and the same representation must be directed at some aspect of world while also being reflexively directed at itself (Kriegel 2009; Gennaro 2012).

A standard objection to the HOT theory is that most animals and even human infants don't possess the requisite concepts for tokening thoughts about mental representations (Dretske 1995; Kim 1996; Seager 2004; Bermúdez 2003, 2009; Proust 2009), with very few exceptions, possibly chimpanzees (Andrews 2012; Lurz 2009b, 2011; Tomasello and Call 2006; Call and

Tomasello 2008). However, even concerning chimps there is "very little consensus" (Fitzpatrick 2009: 258). Other, more dubious, candidates include "perspective taking" in corvids (such as ravens [Bugnyar and Heinrich 2005] and scrubjays [Dally et al. 2006]), "deceptive behavior" in squirrels (Steele et al. 2008), and "empathetic behavior" in rats (Bartal et al. 2011).

Some HOT theorists accept the denial of consciousness to nonhumans (Carruthers 1989, 2000, 2005a) or at least face the possibility with equanimity (Lycan, 1999). Others, such as Gennaro reject the claim that higher-order thoughts require robust first-personal concepts or language (1996, 2004a, 2009) with Lurz (1999) and Van Gulick (2006) similarly arguing that less sophisticated concepts (such as "looking" and "seeing") can be attributed to animals and suffice for tokening HOTs. This suggestion is also controversial (Carruthers 2000, 2005b) and discussion is ongoing (Gennaro 2004b, 2009; DeGrazia 2009).

11 A Critique of the Cambridge Declaration on Consciousness

In 2012 neuroscientists gathered to support "unequivocally" (Low et al. 2012: 1) a statement synthesizing main points of agreement, particularly that "humans are not unique in possessing the neurological substrates that generate consciousness," as these are possessed by "all mammals and birds, and many other creatures, including octopuses" (Low et al. 2012: 2). Little specific evidence is discussed in the supporting rationale though it mentions cortical activity in conjunction with subcortical regions in humans, adding that conscious states, such as emotions, "do not appear to be confined to cortical structures" (Low et al. 2012: 1). The preamble is unclear on whether subcortical regions must work in tandem with cortical activity, or, on their own, absent cortex, suffice for phenomenal awareness.

The Declaration itself vacillates between the highly plausible claim (with apologies to Descartes and certain HOT theorists) that consciousness is not unique to human beings with the more controversial idea that it can be extended far beyond mammals and birds to invertebrates, especially octopuses, and perhaps insects (the latter are only specified in the preamble). The Declaration begins: "The absence of a neocortex does not appear to preclude an organism from experiencing affective states. Convergent evidence indicates that non-human animals...have...conscious states..." (Low et al. 2012: 1). But since neocortex is a mammalian characteristic, the logical continuation ought to have been "non-mammals," not the broader category "non-human animals." Rhetorically, the understatement makes the Declaration seem less divisive than it is.

A second weakness concerns the awkward fit of octopuses and other invertebrates since it is doubtful they possess structures homologous to even mammalian subcortices, let alone neocortex—though birds are in a better position on both counts (see Karten 1997; Jarvis et al. 2005; Calabrese and Wolley 2015). The Declaration mentions attention, sleep and "decision making" (Low et al. 2012: 1) in insects and cephalopods, but this does not make a persuasive case for consciousness. One is left to wonder whether behavioral evidence (such as adaptive problem solving) is convincing on its own. After all, we knew other people were conscious long before we knew anything about brains, or is the case for consciousness in the octopus less strong than the Declaration would have us believe?

12 Animal Pain

Animal welfare is a major practical concern with legal frameworks and policy about conscious pain turning on outcomes of the scientific and philosophical debates. Answers are not straightforward, however, since the function of pain is not simply a matter of avoiding tissue damage.

The experience of pain in humans depends on sometimes competing subsystems (e.g., sensory versus affective-motivational, Langland-Hassan 2017: 251ff.) and can occur independently of nociceptive stimulation. Receptivity to opioids such as morphine can affect non-cortical areas of the nervous system including the brain stem (Rainville 2002). Protective responses such as "nocifensive" flexion withdrawal can occur without awareness (Allen 2004; Roy 2015). Strikingly, a dog's leg will scratch at the precise spot where an irritant has been applied despite a detached spinal cord (we know from paraplegics that there is no feeling below the point of lesion, Key 2016a; see also Allen 1998: 223). Similarly, the paw of a rat with a severed spine can even learn to distinguish noxious from other stimuli (Grau 2002). In these cases, deflationary hypotheses are favored.

Nevertheless, adaptive response to bodily damage such as guarding or favoring an injured limb, reduced activity, limping, seeking analgesics, or at least quiet spaces to lick one's wounds, and so on, do not seem best explained by automatic, unconscious mechanisms. Perhaps tellingly, protective and emotional responses are almost totally absent in arthropods (though see Elwood and Appel 2009 on hermit crabs), and likewise insects which don't appear to care about broken limbs and will continue to feed or mate even when cut in half (Eisemann et al. 1984). Another example is elasmobranch "fish," such as sharks, skates, and rays which are unperturbed by severe injuries (Rose 2002).

Notwithstanding a recycling of foundational issues, correlational evidence from brain imaging, effects of lesions, and direct stimulation suggest a central role for cortical structures, especially the anterior cingulate, somatosensory, and insular cortices in felt pain in human beings (Bushnell et al. 1999; Price 2000; Apkarian et al. 2005; Craig 2009; Key 2016a). Considerations of neuroanatomy, information processing, behavior and physiological changes, make a strong case for experienced pain in mammals (Shriver 2006; Le Neindre et al. 2017: 138) and birds (Gentle 1992; Wang et al. 2010; Prunier et al. 2013; Calbrese and Wolley 2015). For example, behavioral indicators of pain widely seen in farm animals include vocalizations, abnormal postures, rubbing, licking, and reductions in activity (Prunier et al. 2013). Some (Key 2015; 2016a; Walters 2016) employ a "structure-function argument" whereby empirically derived similarity in neural architecture forms the basis for inferences about shared cognitive functioning, including the experience of conscious pain. Necessary (though not sufficient) conditions for consciousness in humans are also found in nonhumans via lesion studies of mammals and birds and are suggestive of deficits in pain experience (LaGraize 2004; Allen 2005; Shriver 2016). Interpretation of the experimental results is a matter of ongoing dispute (e.g. Key 2016b; Shriver 2016) with the extent of plasticity (i.e., multiple realizability) of cortical functioning and neuroanatomy in need of further investigation.

13 Recent Developments

Several current debates about consciousness in nonhumans focus on various species of fish, cephalopods (squids, octopuses and cuttlefish), and insects with the open-access journal *Animal Sentience* emerging as a clearing-house for exchanges on these and other topics. For example, a target article by Key (2016a) about sentience in fish attracted over 40 responses, among them leading researchers and scholars.

Of vertebrates, fish and reptiles have long been points of controversy, with arguments noting the absence of neocortex, as argued by Rose (2002; see also Rose et al. 2014) setting the agenda for recent discussions. Others counter that teleost fish, such as trout, have nociceptors, respond favorably to painkillers (Sneddon 2003), suffer cognitive impairments such as attention deficits, and exhibit other abnormal behaviors when treated with noxious stimuli (Sneddon et al. 2003; 2011). Chandroo et al. (2004) and Braithwaite and Huntingford (2004) offered initial responses to Rose, while Allen (2013), Balcombe (2016), Brown (2016), Seth (2016) and Striedter (2016) continue the commentary. Woodruff (2017) offers a neuroanatomical approach with a focus on the pallial divisions of the fish telencephalon that is drawing a vigorous response from philosophers and scientists. Turning to lizards and reptiles, like fish, neocortex is not present though in some neuroanatomical respects they resemble birds (Lohman and Smeets 1991). Cabanac (1999) argues mainly from physiological and behavioral evidence that there are indications of felt pleasure in iguanas, though not toads or goldfish (Cabanac et al. 2009). Reptiles appear to be equipped with only a grab bag of encapsulated modules and Fixed Action Patterns inconsistent with representationalist and integrationist (e.g. Global Workspace) models. Snakes, such as boas and pythons, follow a tightly scripted hunting routine dispensing with any centralized representation of their prey. Even when coiled around a mouse the snake ignores proprioceptive feedback and is guided by smell and random probing in preparation for swallowing (Sjölander 1995; Gärdenfors 1995; Dennett 1995: 346).

As for invertebrates, the case is occasionally made that insects could be sentient, perhaps in virtue of neuroanatomical similarity to the mammalian midbrain (Barron and Klein 2016; Klein and Barron 2016). Physiological and behavioral evidence attesting to sensory integration, learning, and flexible response has also been adduced (Tye 2016). Sneddon et al. (2014) offer a mixture of considerations of varying plausibility concerning crabs. Others have noted that snails, earthworms, honeybees, and crustaceans release adrenal-like hormones when stressed (Elwood et al. 2009). That honeybees recognize faces even from unfamiliar perspectives might be suggestive of mental rotation (Dyer et al. 2005; Dyer and Vuong 2008; Knight 2010). Once again, however, nociception, learned responses to analgesics, and adaptive behavior such as cost-benefit tradeoffs and reduced activity can be accounted for by unconscious mechanisms. Some crabs detour appropriately (Vannini and Cannicci 1995; Cannicci et al. 2000) and spiny lobsters navigate home from novel locations (Boles and Lohmann 2003), perhaps using spatial maps. Yet these may only be automated domain-specific competencies. Even a very good GPS (or face-recognition system) is not conscious.

Finally, among cephalopods, the octopus is especially intriguing. These animals are highly intelligent, adept at learning (Mather 2001), and display a variety of complex cognizing (Godfrey-Smith 2013). The octopus is a notorious escape artist capable of unscrewing jars (from the inside!) and there is no end to the anecdotes about their idiosyncratic resourcefulness, such as the individual who ambushed trespassers with jets of water (Dews 1959). Despite its impressive reputation, caution is in order here as well. Consider that mating behavior suggestive to some of social intelligence (Godfrey-Smith 2013) is reminiscent of competitions between lizards, including a notorious type known as "sneakers" (see Cherfas 1977: 673 for a saltier sobriquet). The lizards employ "strategies" only in the nominal sense that adaptions for traits such as size, aggressiveness, and color are "competing" from the perspective of evolutionary game theory and frequency dependent fitness (Sinervo and Lively 1996). As such, neither consciousness nor (much) cognition is called for. Cephalopod neuroanatomy is also a far cry from mammalian architecture, with 600 million years separating us from a common ancestor (Godfrey-Smith 2013). Even in the grossest terms, the differences are striking. As over two-thirds of its neurons are located in the tentacles, the cephalopod nervous system is highly distributed compared to vertebrates. Unsurprisingly, the case for sentience rests on other considerations, and appeals to multisensory unification in cognitive information processing lead the way (Mather 2008).

14 Conclusion

This survey provides just a sample of the expanding literature on animal sentience (see also Baars 2005b; Merker 2007). A recent review sponsored by the European Food Safety Association runs the length of a short book and gives a sense of the enormity of the subject. Having turned up at least 2,000 extant works during their initial search, the authors found it practical to retain only a small fraction as reference materials (Le Neindre et al. 2017: 16). Besides its fecundity as a research topic, animal consciousness is a prime example of philosophy's relevance to a robust interdisciplinary conversation overlapping with matters of continual interest to the public and policy makers.

References

Akins, K. (1993) "What Is It Like to Be Boring and Myopic?" in M. Dahlbom (ed.) *Dennett and His Critics*, Oxford: Blackwell.

Aleman, B. and Merker, B. (2014) "Consciousness without Cortex: A Hydranencephaly Family Survey," *Acta Paediatrica* 103: 1057–1065.

Allen, C. (1998) "The Discovery of Animal Consciousness: An Optimistic Assessment," *Journal of Agricultural and Environmental Ethics* 10: 225–246.

Allen, C. (2004) "Animal Pain," *Noûs* 38: 617–643.

Allen, C. (2005) "Deciphering Animal Pain," in M. Aydede (ed.) *Pain: New Essays On the Nature of Pain and the Methodology of Its Study*, Cambridge, MA: MIT Press.

Allen, C. (2013) "Fish Cognition and Consciousness," *Journal of Agricultural and Environmental Ethics* 26: 25–39.

Allen, C. and Bekoff, M. (1997) *Species of Mind*, Cambridge, MA: MIT Press.

Allen, C. and Trestman, M. (1995/2016) "Animal Consciousness," *The Stanford Encyclopedia of Philosophy*, E. N. Zalta (ed.). https://plato.stanford.edu/archives/win2016/entries/consciousness-animal/. Accessed on April 15, 2017.

Allen-Hermanson, S. (2005) "Morgan's Canon Revisited," *Philosophy of Science* 72: 608–631.

Allen-Hermanson, S. (2017) "So *That's* What It's Like!" in K. Andrews and J. Beck (eds.) *Companion to the Philosophy of Animal Minds*, New York: Routledge.

Andrews, K. (2008/2016) "Animal Cognition," *The Stanford Encyclopedia of Philosophy*, E. N. Zalta (ed.). https://plato.stanford.edu/entries/cognition-animal/. Accessed on April 15, 2107.

Andrews, K. (2012) *Do Apes Read Minds? Toward a New Folk Psychology*, Cambridge, MA: MIT Press.

Andrews, K. (2015) *The Animal Mind: An Introduction to the Philosophy of Animal Cognition*, New York: Routledge.

Apkarian, A.V., Bushnell, M. C., Treede, R. D. and Zubieta, J. K. (2005) "Human Brain Mechanisms of Pain Perception and Regulation in Health and Disease," *European Journal of Pain* 9: 463–484.

Armstrong, D. M. (1968) *A Materialist Theory of the Mind*, London: Routledge and Kegan Paul.

Armstrong, D. M. (1981) *The Nature of Mind and Other Essays*, Ithaca, NY: Cornell University Press.

Ayer, A. J. (1956) *The Problem of Knowledge*, Harmondsworth: Penguin.

Baars, B. J. (1988) *A Cognitive Theory of Consciousness*, Cambridge, MA: Cambridge University Press.

Baars, B. J. (1997) "In the Theatre of Consciousness: Global Workspace Theory, A Rigorous Scientific Theory of Consciousness," *Journal of Consciousness Studies* 4(4): 292–309.

Baars, B. J. (2005a) "Global Workspace Theory of Consciousness: Toward a Cognitive Neuroscience of Human Experience," *Progress in Brain Research* 150: 45–53.

Baars, B. J. (2005b) "Subjective Experience is Probably Not Limited to Humans: The Evidence from Neurobiology and Behavior," *Consciousness and Cognition* 14: 7–21.

Balcombe, J. (2016) "In Praise of Fishes: Précis of *What a Fish Knows*, *Animal Sentience* 1(8): 1.

Barron, A. B. and Klein, C. (2016) "What Insects Can Tell Us about the Origins of Consciousness," *Proceedings of the National Academy of Sciences* 113: 4900–4908.

Bartal, I. B.A., Decety, J., Mason, P. (2011) "Empathy and Pro-Social Behavior in Rats," *Science* 334: 1427–1430.

Bennett, J. (1991) "How Is Cognitive Ethology Possible?" in C.A. Ristau (ed.) *Cognitive Ethology: The Minds of Other Animals*, Hillsdale: Lawrence Erlbaum.

Bermúdez, J. L. (2003) *Thinking Without Words*, Oxford: Oxford University Press.

Bermúdez, J. L. (2009) "Mindreading in the Animal Kingdom," in R. Lurz (ed.) *The Philosophy of Animal Minds*, New York: Cambridge University Press.

Block, N. (1978) "Troubles with Functionalism," in N. Block (ed.) *Readings in Philosophy of Psychology*, vol. 1, Cambridge, MA: Harvard University Press.

Block, N. (1981) "Psychologism and Behaviorism," *Philosophical Review* 90: 5–43.

Block, N. (1995) "On a Confusion about the Function of Consciousness," *Behavioral and Brain Sciences* 18: 227–247.
Block, N. (2002) "The Harder Problem of Consciousness," *The Journal of Philosophy* 99: 391–425.
Boles, L. C. and Lohmann, K. J. (2003) "True Navigation and Magnetic Maps in Spiny Lobsters," *Nature* 421: 60–63.
Braithwaite, V. A. and Huntingford, F. A. (2004) "Fish and Welfare: Do Fish Have the Capacity for Pain Perception and Suffering? *Animal Welfare* 13: 87–92.
Broad, C. D. (1925) *The Mind and Its Place in Nature*, London: Routledge and Kegan Paul.
Brown, C. (2016) "Comparative Evolutionary Approach to Pain Perception in Fishes," *Animal Sentience* 1(3): 5.
Buckner, C. (2011) "Two Approaches to the Distinction Between Cognition and Mere Association," *International Journal for Comparative Psychology* 24: 1–35.
Bugnyar, T. and Heinrich, B. (2005) "Ravens, Corvus Corax, Differentiate Between Knowledgeable and Ignorant Competitors," *Proceedings of the Royal Society B-Biological Sciences* 272: 1641–1646.
Burghardt, G. M. (1985) "Animal Awareness: Current Perceptions and Historical Perspective," *American Psychologist* 40: 905–919.
Burghardt, G. M. (1997) "Amending Tinbergen: A Fifth Aim for Ethology," in R. W. Mitchell and H. L. Miles (eds.) *Anthropomorphism, Anecdotes, and Animals*, Albany, NY: SUNY Press.
Burghardt, G. M. and Bekoff, M. (2009) "Animal Consciousness," in T. Bayne, A. Cleeremans and P. Wilken (eds.) *The Oxford Companion to Consciousness*, Oxford: Oxford University Press.
Bushnell, M. C., Duncan, G. H., Hofbauer, R. K., Ha, B., Chen, J. I. and Carrier, B. (1999) "Pain Perception: Is There a Role for Primary Somatosensory Cortex?" *Proceedings of the National Academy of Sciences* 96: 7705–7709.
Cabanac, M. (1999) "Emotion and Phylogeny," *Journal of Consciousness Studies* 6(6–7): 176–190.
Cabanac, M., Cabanac, A. J. and Parent, A. (2009) "The Emergence of Consciousness in Phylogeny," *Behavioural Brain Research* 198: 267–272.
Calabrese, A. and Woolley, S. M. N. (2015) "Coding Principles of the Canonical Cortical Microcircuit in the Avian Brain." *Proceedings of the National Academy of Sciences*, 112: 3517–3522.
Call, J. (2004) "Inferences about the Location of Food in the Great Apes (Pan paniscus, Pan troglodytes, Gorilla gorilla, and Pongo pygmaeus)," *Journal of Comparative Psychology* 118: 232–241.
Call, J. and Carpenter, M. (2001) "Do Apes and Children Know What They Have Seen?" *Animal Cognition* 3: 207–220.
Call, J. and Tomasello, M. (2008) "Does the Chimpanzee Have a Theory of Mind? 30 Years Later," *Trends in Cognitive Sciences* 12: 187–192.
Cannicci, S., Barelli, C. and Vannini, M. (2000) "Homing in the Swimming Crab Thalamita crenata: A Mechanism Based on Underwater Landmark Memory," *Animal Behavior* 60: 203–210.
Carruthers, P. (1989) "Brute Experience," *Journal of Philosophy* 86: 258–269.
Carruthers, P. (1996) *Language, Thought and Consciousness*, Cambridge, MA: Cambridge University Press.
Carruthers, P. (2000) *Phenomenal Consciousness: A Naturalistic Theory*, Cambridge, MA: Cambridge University Press.
Carruthers, P. (2005a) "Why the Question of Animal Consciousness Might Not Matter Very Much," *Philosophical Psychology* 18: 83–102.
Carruthers, P. (2005b) *Consciousness: Essays From a Higher-Order Perspective*, New York: Clarendon, Oxford University Press.
Cassam, Q. (2007) *The Possibility of Knowledge*, Oxford: Clarendon, Oxford University Press.
Chandroo, K., Duncan, I. and Moccia, R. (2004) "Can Fish Suffer? Perspectives on Sentience, Pain, Fear and Stress," *Applied Animal Behaviour Science* 86: 225–250.
Cherfas, J. (1977) "Games Animals Play," *New Scientist* 75: 672–673.
Cockburn, D. (1994) "Human Beings and Giant Squids," *Philosophy* 69: 135–150.
Cottingham, J. (1978) "'A Brute to the Brutes?' Descartes' Treatment of Animals," *Philosophy* 53: 551–559.
Craig, A. D. (2009) "How Do You Feel–Now? The Anterior Insula and Human Awareness," *Nature Reviews Neuroscience* 10: 59–70.
Crick, F. and Clark, J. (1994) "The Astonishing Hypothesis," *Journal of Consciousness Studies* 1(1): 10–6.
Dally, J. M., Emery, N. J. and Clayton, N. S. (2006) "Food-Caching Western Scrub-Jays Keep Track of Who Was Watching When," *Science* 312: 1662–1665.
DeGrazia, D. (1996) *Taking Animals Seriously: Mental Life and Moral Status*, Cambridge, MA: Cambridge University Press.

DeGrazia, D. (2009) "Self Awareness in Animals," in R. Lurz (ed.) *The Philosophy of Animal Minds*, Cambridge, MA: Cambridge University Press.
Dennett, D. C. (1991) *Consciousness Explained*, Boston: Little, Brown and Company.
Dennett, D. C. (1995) "Animal Consciousness: What Matters and Why," *Social Research* 62: 691–710.
Descartes, R. (1637a/1985) *Discourse on the Method*, in J. Cottingham, R. Stoothoff and D. Murdoch (trans.), *The Philosophical Writings of Descartes*, vol. 1, Cambridge, MA: Cambridge University Press.
Descartes, R. (1637b/1991) Letter to Plempius, 3 Oct. 1637 in J. Cottingham, R. Stoothoff and D. Murdoch (trans.), *The Philosophical Writings of Descartes*, vol. 3, Cambridge, MA: Cambridge University Press.
Descartes, R. (1638/1991) Letter to Plempius, 15 Feb. 1638 in J. Cottingham, R. Stoothoff and D. Murdoch (trans.), *The Philosophical Writings of Descartes*, vol. 3, Cambridge, MA: Cambridge University Press.
Descartes, R. (1640/1991) Letter to Mersenne, 11 June 1640 in J. Cottingham, R. Stoothoff and D. Murdoch (trans.), *The Philosophical Writings of Descartes*, vol. 3, Cambridge, MA: Cambridge University Press.
Descartes, R. (1641a/1985) *Meditations*, in J. Cottingham, R. Stoothoff and D. Murdoch (trans.), *The Philosophical Writings of Descartes*, vol. 2, Cambridge, MA: Cambridge University Press.
Descartes, R. (1641b/1985) Letter to Regius, May 1641 in J. Cottingham, R. Stoothoff and D. Murdoch (trans.), *The Philosophical Writings of Descartes*, vol. 3, Cambridge, MA: Cambridge University Press.
Descartes, R. (1644/1985) *Principles of Philosophy*, in J. Cottingham, R. Stoothoff and D. Murdoch (trans.), *The Philosophical Writings of Descartes*, vol.1, Cambridge, MA: Cambridge University Press.
Descartes, R. (1646/1991) Letter to the Marquess of Newcastle, 23 November 1646 in J. Cottingham, R. Stoothoff and D. Murdoch (trans.), *The Philosophical Writings of Descartes*, vol. 3, Cambridge, MA: Cambridge University Press.
Descartes, R. (1649/1991) Letter to More, 5 Feb. 1649 in J. Cottingham, R. Stoothoff and D. Murdoch (trans.), *The Philosophical Writings of Descartes*, vol. 3, Cambridge, MA: Cambridge University Press.
Dews, P. B. (1959) "Some Observations on an Operant in the Octopus," *Journal of Experimental Analysis and Behavior* 2: 57–63.
Dretske, F. (1995) *Naturalizing the Mind*, Cambridge, MA: MIT Press.
Dretske, F. (2006) "Perception Without Awareness," in T. Gendler and J. Hawthorne (eds.) *Perceptual Experience*, Oxford: Oxford University Press.
Dupré J. (1996) "The Mental Lives of Nonhuman Animals," in M. Bekoff and D. Jamieson (eds.) *Readings in Animal Cognition*, Cambridge, MA: MIT Press.
Dyer, A. G., Neumeyer, C. and Chittka, L. (2005) "Honeybee (*Apis mellifera*) Vision Can Discriminate Between and Recognise Images of Human Faces," *The Journal of Experimental Biology* 208: 4709–4714.
Dyer, A. G. and Vuong, Q. C. (2008) "Insect Brains Use Image Interpolation Mechanisms to Recognise Rotated Objects," *PLoS ONE*, 3: e4086.
Edelman, D. B., Baars, B. J. and Seth, A. K. (2005) "Identifying Hallmarks of Consciousness in Non-Mammalian Species," *Consciousness and Cognition* 14: 169–187.
Edelman, D. B. and Seth, A. K. (2009) "Animal Consciousness: A Synthetic Approach," *Trends in Neuroscience* 32: 476–484.
Eisemann, C. H., Jorgensen, W. K., Merritt, D. J., Rice, M. J., Cribb, B. W., Webb, P. D. and Zalucki, M. P. (1984) "Do Insects Feel Pain?—A Biological View," *Experientia*, 40: 164–167.
Elwood, R. W. and Appel, M. (2009) "Pain Experience in Hermit Crabs?" *Animal Behaviour* 77: 1243–1246.
Elwood, R. W., Barr, S. and Patterson, L. (2009) "Pain and Stress in Crustaceans," *Applied Animal Behaviour Science* 118: 128–136.
Farah, M. (2008) "Neuroethics and the Problem of Other Minds: Implications of Neuroscience for the Moral Status of Brain-Damaged Patients and Nonhuman Animals," *Neuroethics* 1: 9–18.
Fisher, J. A. (1996) "The Myth of Anthropomorphism," in C. Allen and D. Jamieson (eds.) *Readings in Animal Cognition*, Cambridge, MA: MIT Press.
Fitzpatrick, S. (2008) "Doing Away With Morgan's Canon," *Mind and Language* 23: 224–246.
Fitzpatrick, S. (2009) "The Primate Mindreading Controversy: A Case Study in Simplicity and Methodology in Animal Psychology," in R. Lurz (ed.) *The Philosophy of Animal Minds*, Cambridge, MA: Cambridge University Press.
Gage, F. H. (2000) "Mammalian Neural Stem Cells," *Science* 287: 1433–1438.
Gallistel, C. R. (1990) *The Organization of Learning*, Cambridge, MA: MIT Press.
Gallistel, R. and King, A. (2009) *Memory and the Computational Brain: Why Cognitive Science Will Transform Neuroscience*, Vol. 6. New York: John Wiley and Sons.
Gärdenfors, P. (1995) "Cued and Detached Representations in Animal Cognition," *Behavioural Processes*, 35: 263–273.

Ge, S., Yang, C. H., Hsu, K. S., Ming, G. L. and Song, H. (2007) "A Critical Period for Enhanced Synaptic Plasticity in Newly Generated Neurons of the Adult Brain," *Neuron* 54: 559–566.

Geiger, G. and Poggio, T. (1975) "The Muller-Lyer Figure and the Fly," *Science* 190: 479–480.

Gennaro, R. (1996) *Consciousness and Self-Consciousness: A Defense of the Higher-Order Thought Theory of Consciousness*, Amsterdam and Philadelphia: John Benjamins Publishers.

Gennaro, R. (2004a) "Higher-Order Thoughts, Animal Consciousness, and Misrepresentation: A Reply to Carruthers and Levine," in R. Gennaro (ed.) *Higher-Order Theories of Consciousness*, Amsterdam and Philadelphia: John Benjamins Publishers.

Gennaro, R. (2004b) (ed.) *Higher-Order Theories of Consciousness*, Amsterdam and Philadelphia: John Benjamins Publishers.

Gennaro, R. (2009) "Animals, Consciousness, and I-Thoughts," in R. Lurz (ed.) *The Philosophy of Animal Minds*, Cambridge, MA: Cambridge University Press.

Gennaro, R. (2012) *The Consciousness Paradox*, Cambridge, MA: MIT Press.

Gentle, M. J. (1992) "Pain in Birds," *Animal Welfare* 1: 235–247.

Godfrey-Smith, P. (2013) "Cephalopods and the Evolution of the Mind," *Pacific Conservation Biology* 19: 4–9.

Graham, G. (1993/1998) *Philosophy of Mind: An Introduction* (2nd ed.), Malden: Blackwell.

Grau, J. (2002) "Learning and Memory Without a Brain," in M. Bekoff, C. Allen and G. M. Burghardt (eds.) *The Cognitive Animal*, Cambridge, MA: MIT Press.

Griffin, D. R. (1976) *The Question of Animal Awareness: Evolutionary Continuity of Mental Experience*, New York: Rockefeller University Press.

Griffin, D. R. (1978) "Prospects for a Cognitive Ethology," *Behavioral and Brain Sciences* 1: 527–538.

Griffin, D. R. (2001) *Animal Minds: Beyond Cognition to Consciousness*, Chicago: University of Chicago Press.

Güzeldere, G. (1995) "Is Consciousness the Perception of What Passes in One's Own Mind?" in Thomas Metzinger (ed.) *Conscious Experience*, Paderborn: Ferdinand Schöningh.

Harman, G. (1965) "The Inference to the Best Explanation," *Philosophical Review* 74: 88–95.

Hampton, R. R. (2001) "Rhesus Monkeys Know When They Remember," *Proceedings of the National Academy of Sciences*," 98: 5359–5362.

Hampton, R. R. (2009) "Multiple Demonstrations of Metacognition in Nonhumans: Converging Evidence or Multiple Mechanisms?" *Comparative Cognition and Behavior Reviews* 4: 17–28.

Harnad, S. (2016) "Animal Sentience: The Other-Minds Problem," *Animal Sentience* 1(1): 1.

Heyes, C. (2008) "Reflections on Self-Recognition in Primates," *Animal Behaviour* 47: 909–919.

Hume, D. (1739/1978) *A Treatise of Human Nature* (eds.) L. A. Selby-Bigge and P. H. Nidditch. Oxford: Clarendon, Oxford University Press.

Husserl, E. (1982) *Ideas Pertaining to a Pure Phenomenology and to a Phenomenological Philosophy*, Bk. 1 (trans.) F. Kersten. Dordrecht, Holland: Kluwer.

Hyslop, A. and Jackson, F. (1972) "The Analogical Inference to Other Minds," *American Philosophical Quarterly* 9: 168–176.

James, W. (1912/1971) *Radical Empiricism and a Pluralistic Universe* (ed.) R. Bernstein. New York: E. P. Dutton.

Jamieson, D. (1998) "Science, Knowledge, and Animal Minds," *Proceedings of the Aristotelian Society* 98: 79–102.

Jarvis, E. D., Güntürkün, O., Bruce, L., Csillag, A., Karten, H., Kuenzel, W., Medina, L., Paxinos, G., Perkel, D. J., Shimizu, T. and Striedter, G. (2005) "Avian Brains and a New Understanding of Vertebrate Brain Evolution," *Nature Reviews Neuroscience* 6: 151–59.

Kandel, E. R., Schwartz, J. H. and Jessell, T. M. (2013) *Principles of Neural Science* (5th ed.), New York: McGraw Hill.

Karten, K. (1997) "Evolutionary Developmental Biology Meets the Brain: The Origins of Mammalian Cortex," *Proceedings of the National Academy of Sciences* 94: 2800–2804.

Kennedy, J. S. (1992) *The New Anthropomorphism*, New York: Cambridge University Press.

Key, B. (2015) "Fish Do Not Feel Pain and Its Implications for Understanding Phenomenal Consciousness," *Biology and Philosophy* 30: 149–165.

Key, B. (2016a) "Why Fish Do Not Feel Pain," *Animal Sentience* 3(1): 1.

Key, B. (2016b) "The Burden of Proof Lies with Proposer of Celestial Teapot Hypothesis," *Animal Sentience* 3(1): 44.

Kim, J. (1996) *Philosophy of Mind*, Boulder, CO: Westview Press.

Kirk, R. (1994) *Raw Feeling*, Oxford: Clarendon, Oxford University Press.

Klein, C. and Barron, A. B. (2016) "Insects Have the Capacity for Subjective Experience," *Animal Sentience* 2016 9(1): 1.

Knight, K. (2010) "Bees Recognize Faces Using Feature Configuration," *The Journal of Experimental Biology* 213: i.

Kriegel, U. (2009) *Subjective Consciousness*, Oxford: Oxford University Press.

LaGraize, S. C., LaBuda, C. J., Rutledge, M. A., Jackson, R. L. and Fuchs, P. N. (2004) "Differential Effect of Anterior Cingulate Cortex Lesion on Mechanical Hyperalgesia and Escape/Avoidance Behavior in an Animal Model of Neuropathic Pain," *Experimental Neurology* 188: 139–148.

Langland-Hassan, P. (2017) "Pain and Incorrigibility," in J. Corns (ed.) *The Routledge Handbook of Philosophy of Pain*, New York: Routledge.

Laureys, S., Owen, A. M. and Schiff, N. D. (2004) "Brain Function in Coma, Vegetative State, and Related Disorders," *Lancet Neurology* 3: 537–546.

Laureys, S., Gosseries, O. and Tononi, G. (2009/2013) "The Neurology of Consciousness: An Overview," *The Neurology of Consciousness* (2nd ed.), pp. 61–8; 407–449. San Diego, CA: Academic Press.

Leiber, K. (1988) "'Cartesian' Linguistics?" *Philosophia* 18: 309–346.

Le Neindre, P., Bernard, E., Boissy, A., Boivin, X., Calandreau, L., Delon, N., Deputte, B., Desmoulin-Canselier, S., Dunier, M., Faivre, N., Giurfa, M., Guichet J.-L., Lansade, L., Larrere, R., Mormede, P., Prunet, P., Schaal, B., Serviere, J. and Terlouw, C. (2017) "Animal Consciousness," *EFSA Supporting Publications*, 14(4), 169. doi:10.2903/sp.efsa.2017.EN-1196.

Lipton, P. (1991) *Inference to the Best Explanation*, London: Routledge.

Locke, J. (1689/1975). *An Essay Concerning Human Understanding* (ed.) P. H. Nidditch, Oxford: Clarendon, Oxford University Press.

Lockie, A. (2017) "The Air Force Just Demonstrated an Autonomous F-16 That Can Fly and Take Out a Target All By Itself." http://www.businessinsider.com/f-16-drone-have-raider-ii-loyal-wingman-f-35-lockheed-martin-2017-> April 11. Accessed on May 29, 2017.

Lohman, A. and Smeets, W. (1991) "The Dorsal Ventricular Ridge and Cortex of Reptiles in Historical and Phylogenetic Perspective" in Finlay B. L., Innocenti G. and Scheich H. (eds.) *The Neocortex* (vol. 200), Boston: Springer.

Low, P., Panksepp, J., Reiss, D., Edelman, D., Van Swinderen, B. and Koch, C. (2012) The Cambridge Declaration on Consciousness, *Francis Crick Memorial Conference:* University of Cambridge, MA.

Lurz, R. (1999) "Animal Consciousness," *Journal of Philosophical Research* 24: 149–168.

Lurz, R. (2002) "Reducing Consciousness by Making It HOT: A Review of Peter Carruthers' Phenomenal Consciousness," *Psyche* 8(5). http://psyche.cs.monash.edu.au/v8/psyche-8 05-lurz.html. Accessed on July 11, 2017.

Lurz, R. (2006) "Conscious Beliefs and Desires: A Same-Order Approach," in U. Kriegel and K. Williford (eds.) *Self-Representational Approaches to Consciousness*, Cambridge, MA: MIT Press.

Lurz, R. (2009a) *The Philosophy of Animal Minds*, New York: Cambridge University Press.

Lurz, R. (2009b) "Animal Minds," *Internet Encyclopedia of Philosophy*, accessed on 20 May 2017.

Lurz, R. (2011) *Mindreading Animals: The Debate Over What Animals Know About Other Minds*, Cambridge, MA: MIT Press.

Lycan, W. (1987) *Consciousness*, Cambridge, MA: MIT Press.

Lycan, W. (1996) *Consciousness and Experience*, Cambridge, MA: MIT Press.

Lycan, W. (1999) "A Response to Carruthers' 'Natural Theories of Consciousness,'" *Psyche* 5(11). http://psyche.cs.monash.edu.au/v5 /psyche-5-11-lycan.html. Accessed on July 11, 2017.

Lycan, W. (2001) "A Simple Argument for a Higher-Order Representation Theory of Consciousness," *Analysis* 61: 3–4.

Malcolm, N. (1962) "Knowledge of Other Minds," in V. C. Chappell (ed.) *The Philosophy of Mind*, Englewood Cliffs: Prentice-Hall.

Mather, J. (2001) "Animal Suffering: An Invertebrate Perspective," *Journal of Applied Animal Welfare Science.* 4: 151–156.

Mather, J. (2008) "Cephalopod Consciousness: Behavioural Evidence," *Consciousness and Cognition* 17: 37–48.

Matthen, M. (1999) "The Disunity of Color," *Philosophical Review* 108: 47–84.

Matthews, G. B. (1986) "Descartes and the Problem of Other Minds," in A. Rorty (ed.) *Essays on Descartes Meditations*, Berkeley, CA and Los Angeles: University of California Press.

McDowell, J. (1982) "Criteria, Defeasibility, and Knowledge," *Proceedings of the British Academy* 68: 455–479.

Melnyk, A. (1994) "Inference to the Best Explanation and Other Minds," *Australasian Journal of Philosophy* 72: 482–491.

Merker, B. (2005) "The Liabilities of Mobility: A Selection Pressure for the Transition to Consciousness in Animal Evolution," *Consciousness and Cognition* 14: 89–114.

Merker, B. (2007) "Consciousness Without a Cerebral Cortex: A Challenge for Neuroscience and Medicine," *Behavioral and Brain Sciences* 30: 63–134.
Merker, B. (2008) "Life Expectancy in Hydranencephaly," *Clinical Neurology and Neurosurgery*: 213–214.
Merleau-Ponty, M. (1962) *The Phenomenology of Perception*, C. Smith (trans.) London: Routledge and Kegan Paul.
Mill, J. S. (1889) *An Examination of Sir William Hamilton's Philosophy: And of the Principal Philosophical Questions Discussed in His Writings*, London: Longmans, Green, and Company.
Ming, G. and Song, H. (2005) "Adult Neurogenesis in the Mammalian Central Nervous System," *Annual Review of Neuroscience* 28: 223–250.
Mitchell, C. J., De Houwer, J. and Lovibond, P. F. (2009) "The Propositional Nature of Human Associative Learning," *Behavioral and Brain Sciences* 32: 183–198.
Molyneaux, B. J., Arlotta, P., Menezes, J. R. and Macklis, J. D. (2007) "Neuronal Subtype Specification in the Cerebral Cortex," *Nature Reviews Neuroscience* 8: 427–437.
Moore, G. E. (1903) "The Refutation of Idealism," *Mind* 12: 433–453.
Morgan, C. L. (1894) *An Introduction to Comparative Psychology*, New York: Scribner.
Nagel, T. (1974) "What Is It Like to Be a Bat?" *The Philosophical Review* 83: 435–450.
Pargetter, R. (1984) "The Scientific Inference to Other Minds," *Australasian Journal of Philosophy* 62: 158–163.
Peacocke, C. (1998) "Nonconceptual Content Defended," *Philosophy and Phenomenological Research* 58: 381–388.
Penn, D. and Povinelli, D. (2007) "Causal Cognition in Humans and Non-Human Animals: A Comparative, Critical Review," *Annual Review of Psychology* 58: 97–118.
Perrett, R. W. (1997) "The Analogical Argument for Animal Pain," *Journal of Applied Philosophy* 14: 49–58.
Plantinga, A. (1967) *God and Other Minds: A Study of the Rational Justification of Belief in God*, Ithaca, NY: Cornell University Press.
Price, D. (2000) "Psychological and Neural Mechanisms of the Affective Dimension of Pain," *Science* 288: 1769–1762.
Proust, J. (2009) "The Representational Basis of Brute Metacognition: A Proposal," in R. Lurz (ed.) *The Philosophy of Animal Minds*, Cambridge, MA: Cambridge University Press.
Prunier, A., Mounier, L., Le Neindre, P., Leterrier, C., Mormède, P., Paulmier, V., Prunet, P., Terlouw, C. and Guatteo, R. (2013) "Identifying and Monitoring Pain in Farm Animals: A Review," *Animal* 7: 998–1010.
Rainville, P. (2002) "Brain Mechanisms of Pain Affect and Pain Modulation," *Current Opinion in Neurobiology* 12: 195–204.
Rescorla, M. (2009) "Chrysippus' Dog as a Case Study in Non-linguistic Cognition," in R. Lurz (ed.) *The Philosophy of Animal Minds*, Cambridge, MA: Cambridge University Press.
Roelofs, L. (2017) "Seeing the Invisible: How to Perceive, Imagine and Infer the Minds of Others," *Erkenntnis*, doi:10.1007/s10670-017-9886-2.
Rose, J. D. (2002) "The Neurobehavioral Nature of Fishes and the Question of Awareness of Pain," *Reviews in Fisheries Sciences* 10: 1–38.
Rose, J. D., Arlinghaus, R., Cooke S. J., Diggles B. K., Sawynok, W., Stevens, E. D. and Wynne C. D. L. (2014) "Can Fish Really Feel Pain?" *Fish and Fisheries* 15: 97–133.
Rosenthal, D. (1986) "Two Concepts of Consciousness," *Philosophical Studies* 49: 329–359.
Rosenthal, D. M. (1997) "A Theory of Consciousness," in N. Block, O. Flanagan and G. Güzuldere (eds.) *The Nature of Consciousness: Philosophical Debates*, Cambridge, MA: MIT Press.
Rosenthal, D. M. (2004) "Varieties of Higher-Order Theory," in R. Gennaro (ed.) *Higher Order Theories of Consciousness*, Amsterdam and Philadelphia, PA: John Benjamins Publishers.
Roy, M. (2015) "Cerebral and Spinal Modulation of Pain by Emotions and Attention," in G. Pickering and S. Gibson (eds.) *Pain, Emotion and Cognition: A Complex Nexus*, Cham: Springer Publishing.
Russell, B. (1948) *Human Knowledge: Its Scope and Limits*, London: George Allan and Unwin.
Ryle, G. (1949) *The Concept of Mind*, Chicago: University of Chicago Press.
Schreiner, C. E., Read, H. L. and Sutter, M. L. (2000) "Modular Organization of Frequency Integration in Primary Auditory Cortex," *Annual Review of Neuroscience* 23: 501–529.
Seager, W. (1999/2016) *Theories of Consciousness: An Introduction and Assessment*, New York: Routledge.
Seager, W. (2004) "A Cold Look at HOT Theory," in R. Gennaro (ed.) *Higher-Order Theories of Consciousness*, Amsterdam and Philadelphia, PA: John Benjamins Publishers.
Seth, A. (2016) "Why Fish Pain Cannot and Should Not be Ruled Out," *Animal Sentience* 1(3): 14.

Shewmon, D. A., Holmes, G. L. and Byrne, P. A. (1999) "Consciousness in Congentially Decorticate Children: Developmental Vegetative State as Self-Fulfilling Prophecy," *Developmental Medicine and Child Neurology* 41: 364-374.

Shriver, A. (2006) "Minding Mammals," *Philosophical Psychology* 19: 433–442.

Shriver, A. (2016) "Cortex Necessary for Pain—but Not in Sense That Matters," *Animal Sentience* 1(3): 27.

Sinervo, B. and Lively, C. M. (1996) "The Rock-Paper-Scissors Game and the Evolution of Alternative Male Strategies," *Nature* 380: 240–243.

Singer, P. (1975/1990) *Animal Liberation* (2nd ed.), New York: Avon Books.

Singer, S. (1993) *Practical Ethics* (2nd ed.), Cambridge, MA: Cambridge University Press.

Sjölander, S. (1995) "Some Cognitive Breakthroughs in the Evolution of Cognition and Consciousness, and Their Impact on the Biology of Language," *Evolution and Cognition* 1: 3–11.

Smith, J. (2010) "Seeing Other People," *Philosophy and Phenomenological Research* 81: 731–748.

Smith, J. D., Schull, J., Strote, J., McGee, K., Egnor, R. and Erb, L. (1995) "The Uncertain Response in the Bottle-Nosed-Dolphin (Tursiops-truncatus), *Journal of Experimental Psychology-General* 124: 391–408.

Sneddon, L. U. (2003) "The Evidence for Pain in Fish: The Use of Morphine as an Analgesic," *Applied Animal Behavior Science* 83: 153–162.

Sneddon, L. U., Braithwaite, V. A. and Gentle, M. J. (2003) "Do Fish Have Nociceptors: Evidence for the Evolution of a Vertebrate Sensory System," *Proceedings of the Royal Society, B: Biological Sciences* 270: 1115–1121.

Sneddon, L. U. (2011) "Pain Perception in Fish: Evidence and Implications for the Use of Fish," *Journal of Consciousness Studies* 18(9–10): 209–229.

Sneddon, L. U., Elwood, R. W., Adamo, S. A. and Leach, M. C. (2014) "Defining and Assessing Animal Pain," *Animal Behaviour* 97: 201–212.

Sober, E. (1998) "Morgan's Canon," in C. Allen and D. Cummins (eds.) *The Evolution of Mind*, Oxford: Oxford University Press.

Sober, E. (2000) "Evolution and the Problem of Other Minds," *Journal of Philosophy* 97: 365–386.

Sober, E. (2005) "Comparative Psychology Meets Evolutionary Biology: Morgan's Canon and Cladistic Parsimony," in L. Datson and G. Mitman (eds.) *Thinking with Animals: New Perspectives on Anthropomorphism*, New York: Columbia University Press.

Sober, E. (2009) "Parsimony and Models of Animal Minds," in R. Lurz (ed.) *The Philosophy of Animal Minds*, Cambridge, MA: Cambridge University Press.

Sober, E. (2012) "Anthropomorphism, Parsimony, and Common Ancestry," *Mind and Language* 27: 229–238.

Sober, E. (2015) *Ockham's Razors: A User's Manual*, Cambridge, MA: Cambridge University Press.

Steele, M. A., Halkin, S. L., Smallwood, P. D., McKenna, T. J., Mitsopoulos, K. and Beam, M. (2008) "Cache Protection Strategies of a Scatter-Hoarding Rodent: Do Tree Squirrels Engage in Behavioural Deception?" *Animal Behaviour* 75: 705–714.

Striedter, G. (2016) "Lack of Neocortex Does Not Imply Fish Cannot Feel Pain," *Animal Sentience* 1(3): 15.

Thagard, P. (1978) "The Best Explanation: Criteria for Theory Choice," *Journal of Philosophy*, 75: 76–92.

Thiel, U. (2011) *The Early Modern Subject: Self-Consciousness and Personal Identity from Descartes to Hume*. Oxford: Oxford University Press.

Thompson, E. (1992) "Novel Colors," *Philosophical Studies* 68: 321–349.

Thompson E., Palacios, A. and Varela, F. J. (1992) "Ways of Coloring: Comparative Color Vision as a Case Study for Cognitive Science," *Behavioral and Brain Sciences* 15: 1–26.

Tolman, E. C. (1948) "Cognitive Maps in Rats and Men," *The Psychological Review* 55: 189–208.

Tomasello, M. and Call, J. (2006) "Do Chimpanzees Know What Others See—Or Only What They Are Looking At?" in S. Hurley and M. Nudds (eds.) *Rational Animals?* Oxford: Oxford University Press.

Tye, M. (1995) *Ten Problems of Consciousness*, Cambridge, MA: MIT Press.

Tye, M. (1997) "The Problem of Simple Minds: Is There Anything It Is Like to Be a HoneyBee?" *Philosophical Studies* 88: 289–317.

Tye, M. (2000) *Color, Consciousness, and Content*, Cambridge, MA: MIT Press.

Tye, M. (2016) *Tense Bees and Shell-Shocked Crabs: Are Animals Conscious?* Oxford: Oxford University Press.

Van Gulick, R. (2006) "Mirror, Mirror: Is That All?" in U. Kriegel and K. Williford (eds.) *Self-Representational Approaches to Consciousness*, Cambridge, MA: MIT Press.

Vannini, M. and Cannicci, S. (1995) "Homing Behaviour and Possible Cognitive Maps in Crustacean Decapods," *Journal of Experimental Marine Biology and Ecology* 193: 67–91.

Varner, G. E. (2012) *Personhood, Ethics, and Animal Cognition: Situating Animals in Hare's Two-Level Utilitarianism*, New York: Oxford University Press.

Walters, E. T. (2016) "Pain-Capable Neural Substrates May Be Widely Available in the Animal Kingdom," *Animal Sentience* 1(3): 37.
Wang, Y. A., Brzozowska-Prechtl, A. and Karten, H. J. (2010) "Laminar and Columnar Auditory Cortex in Avian Brain," *Proceedings of the National Academy of Sciences* 107: 12676–12681.
Woodruff, M. L. (2017) "Consciousness in Teleosts: There Is Something It Feels Like to Be a Fish," *Animal Sentience* 2(13): 1.
Wynne, D. L. (2004) "The Perils of Anthropomorphism," *Nature* 428: 606.

Related Topics

Materialism
Consciousness in Western Philosophy
Representational Theories of Consciousness
The Neural Correlates of Consciousness
The Biological Evolution of Consciousness

30
ROBOT CONSCIOUSNESS

Jonathan Waskan

We are rapidly approaching the time when we will have to take seriously the possibility that such technological contrivances as computers, automated personal assistants, and, most relevant to the present volume, *robots* have become conscious in the sense that things *seem* to them a certain way. For instance, we may soon find ourselves wondering: When my robot looks at a red tomato, does it see the distinctive quality that I see? Does it *look red* to it? And if my robot is damaged, does it *feel* pain? Does it *hurt*? From Asimov's *Robot* novels (which introduced the term 'robotics'), to *Star Trek* and *Terminator*, these issues have been treated at length in science fiction, but more often than not the robots involved, whether R. Daneel, Data, or the T-800, are either devoid of real experience or claim to only have something going on in them not unlike a genuine human experience—for instance, they might have something that *could be called* pain. But what we want to know is whether they really *do* (or could) feel pain, see red, or what have you.

Nor are such questions a matter of idle philosophical curiosity. Our beliefs about whether or not our robotic (or other) contrivances are conscious will impact how we ought to treat them. We may view them with less concern, or none at all, if we think they feel nothing: no pain, no distress at being bound, and no anguish at the prospect of deactivation. However sophisticated they become, we might also deny them anything in the way of rights.[1] And if they feel nothing, it could have practical ramifications for how *they* treat *us*. If robots are, so to speak, dead inside, they will experience no empathic understanding or concern, which, on a good day, constrains our treatment of one another. To keep robots from doing us harm, they may thus *have to be* bound by internal safeguards akin to Asimov's inviolable first law of robotics (do no harm to humans).

Consensus in science and philosophy is scarce at best, and with regard to the general question of whether (and if so, how) robots could become conscious in the above sense, we are as far from it as ever. Nevertheless, the many views on the matter cluster around one or another of several influential schools of thought, covered here. First, however, we should consider a few views on how, if at all, *biological* mechanisms produce consciousness.

1 The (Im-)Possibility of Conscious Machines from Galileo to McGinn

Galileo (1618/1957) was perhaps the first modern figure to assert, against the prevailing orthodoxy of his day, that some qualities of conscious perceptual experience, such as tastes, smells, warmth, and colors, are found only in us, as sentient beings, and not in external objects. The non-sentient world consists, he thought, merely of aggregates of tiny bits of matter with so-called *primary* (mechanical) properties like size, shape, location, motion, and contact. He outlined, quite presciently, how contact between these particles and our sense organs evoke *secondary* (conscious) properties in us such as the redness we see or the pain that we feel. But why should these secondary properties be produced in sentient beings? Galileo's direct intellectual descendants, Hobbes and Descartes, answered this question in vastly different ways.

Hobbes (1651/1958) adopted the Galilean view of nature and sensation, but he was more explicit in claiming that *even we sentient beings* consist only of bits of matter with primary properties. Secondary properties, he claimed, were somehow produced by these, specifically by the mechanisms of the brain, though he had no adequate theory as to why or how this occurs.

Descartes (1647/1988) agreed with Hobbes that the external world and human bodies consist only of tiny bits of matter, but to solve the mystery of how brains and consciousness are related he offered a mix of science and Christianity, a dualism of matter and soul. He claimed that the conscious mind exists outside of the natural realm and, like a drone pilot, controls and receives feedback from the human body via a tiny gland, a biological antenna of sorts, deep in the brain.

While Descartes is famous for his attempted knock-down deductive arguments in favor of mind/body dualism—in particular, his arguments assuming that the mind cannot be doubted and has no spatial extent—here his inductive arguments (1637/1988) hold far greater relevance. One involves a *viva voce* (live voice) test for distinguishing minded beings from mere mechanical entities such as humanoid contrivances (in essence, robots) and lesser animals like parrots and apes. He claimed that if one attempted to converse with these entities, one would quickly find that they have a finite verbal repertoire, whereas we humans can sensibly engage in *open-ended* conversations about *countless* topics. There is, he suggests, no possible way that a finite machine could exhibit this kind of infinite conversational ability. In humans, therefore, the best explanation for the mind behind the conversational ability is that it is a non-mechanical, spiritual substance. Descartes offered a parallel argument regarding our seemingly unique and boundless, so-called (remember this term) *universal* power to reason about countless different scenarios.

Decades later, Leibniz also despaired of a mechanical explanation for conscious experience. He wrote:

> Moreover, it must be confessed that perception and that which depends upon it are inexplicable on mechanical grounds… And supposing there were a machine, so constructed as to think, feel, and have perception, it might be conceived as increased in size, while keeping the same proportions, so that one might go into it as into a mill. That being so, we should, on examining its interior, find only parts which work one upon another, and never anything by which to explain a perception.
>
> *(1714/1968: 4)*

Put plainly, view my living brain through a microscope or any other instrument at any level of magnification and you will find nothing in there indicating an experience of blue wrist restraints or of the pain I feel as I try to writhe free. Indeed, many contemporary philosophers and neuroscientists think that Leibniz was basically right. Advances in microscopy and neuroimaging have made Leibniz' Mill a reality, and yet, it seems, we are no closer to cracking the mystery of how neural machinations produce those distinctive qualities of consciousness experience, *qualia* as they are now called.

A more recent argument that we *never will* solve the mystery of qualia has been offered by Jackson (1982). He claims that even if a completed neuroscience supplied all of the *physical* facts about the neural processes of perception, they would still be missing crucial facts such as how red looks to us. Thus, facts about qualia are non-physical facts, perhaps having to do with an ephemeral exhaust given off by neural processes (a view termed *epiphenomenalism*). Jackson maintains that our inability to understand qualia stems from the fact that we evolved to know only certain facts (which we term *physical*), the ones important to middle-sized creatures such ourselves. But these are only a subset of the facts. The truly surprising thing, he thinks, is how many facts about nature we *do* understand.

McGinn (1989) likewise argues that the truth about qualia is *cognitively closed* to limited beings such as ourselves. While he thinks they somehow result from a natural, biological process of evolution and development, an understanding of how this occurs may forever elude us. Just as there are truths about nature closed to sea slugs and monkeys, the truth about consciousness may be closed to us.

2 The Dawn of Robots

The discussion of *robot* consciousness became genuinely pressing with the advent of so-called artificial brains (electronic computers) mid-way through the 20th century. Turing (1937) laid the foundations for this endeavor about a decade earlier with his attempt to add precision to the fuzzy colloquial notion of *computation*. He defined 'computation' as the sort of formal symbol manipulation that can be automated by a hypothetical device (what we now call a *Turing machine*) that engages in simple operations such as writing or erasing a '1' from a memory tape or moving the tape left and right. Turing showed that such devices could be configured to perform any of the operations traditionally thought of as computations. One could also create a device taking two sets of inputs, say from distinct memory tapes, the data to be manipulated and the program for doing so. Thus, instead of a device built to perform one and only one type of calculation (e.g., addition), one could, through finite mechanical means, create a device with an effectively boundless capacity for formal reasoning, a *universal* Turing machine as it is now called.

Soon after, McCulloch and Pitts (1943) showed that structures with the rudimentary powers of neurons could be configured to carry out simple logical operations (e.g., reasoning with 'or' and 'and') and that a vast collection of these ersatz neurons could approximate a universal Turing machine. This work helped inspire von Neumann to show, quite concretely, that this kind of device could be created out of electronic components (e.g., vacuum tubes) (Boden 2006: 196). Thus was born the modern programmable computer, the basic architecture of which, the von Neumann architecture, is still in wide use today. This watershed event in human history very quickly led to the creation of computers that could generate novel proofs of mathematical theorems, converse about the manipulation of objects, and control robots as they navigated virtual and real environments.

Turing saw straightaway that this research would have as its ultimate goal the creation of artificial thinking machines. But could computing machines *really* think? And could they be conscious? Turing (1950) thought that if we tried to answer these questions directly, we would quickly get sidetracked into debating the meanings of our vaguely defined mental terms. He thus sought to do for these terms what he did for 'computation'—namely, define them (or redefine them) in a way that renders them useful and precise. For instance, instead of asking whether or not an electronic contrivance can think, we should ask whether it can exhibit the kinds of behaviors that normally lead us to attribute thinking to other humans. One good measure here is the ability to converse like a human. Specifically, Turing proposed that we should attribute thinking (and other mental states) to a device insofar as it performs well at the Imitation Game. Turing's variant of this parlor game involves two players, a human and a machine, in contact with a judge only via a teletype system—in essence, *texting*. The judge would ask questions with the goal of distinguishing human from contrivance. If the latter could fool the judge into thinking that *it* was the human a significant amount of the time, we would have about as much reason to attribute mental states to it as we have with regard to our fellow humans. This, you will recall, is the kind of test that Descartes thought no contrivance could pass on the grounds that *our* speaking and reasoning repertoire is unbounded. But computer scientists had already shown in some sense how to elicit a boundless repertoire from finite mechanisms.

With research advancing rapidly, it seemed that the creation of devices exhibiting *viva voce* behavior would not be far off. One important side note here is that by the 1960's, modelers worked less and less with the kind of basic machine code (e.g., 1's and 0's) discussed by Turing and more with high-level programming *languages* that had been implemented with that code. These artificial languages enabled researchers to focus on programming in sentence-like representations (e.g., 'on<ball, table>.') and rules (e.g., 'if on<x, y>, remove x') for making inferences from, and responding to, them. Soon, researchers needed to know little to nothing about the lower-level implementation details. All that mattered was the program. In fact, because a given program could be implemented by various machine architectures (e.g., Turing's or von Neumann's), and each of these could be realized by various physical substrates (e.g., McCulloch-Pitts neurons, vacuum tubes, transistors, gears and pistons, and, we shall see, *people*), many came to see implementation as only of ancillary concern, both in computers *and* in humans. The proper level for understanding and implementing conscious minds was the level of programs.

One early system that seemed promising with regard to *viva voce* behavior was Shank and Abelson's (1977) Script Applier Mechanism (SAM), which had a human-like ability to infer information only implicit in conversational context. For instance, told that Jonah dined at McDonald's, we might draw upon our generic knowledge of typical sequences of events (e.g., types of dining experience) to infer that Jonah retrieved his food from the counter, seated himself, and left no tip. SAM could infer these things as well, and even do so in different languages. Naturally there were those who felt that SAM *understood* language or was at least a precursor to the devices that would.

Enter Searle (1980) who, in perhaps the most widely cited philosophy essay of the 20th century, argued that this entire framework was wrongheaded. His simple thought experiment, termed the Chinese Room, is quite reminiscent of Leibniz' Mill. He imagines a next-generation SAM program that can achieve human-level performance on reading and question answering for a language (Chinese) that he, Searle, does not comprehend. But instead of imagining stepping into a mill-containing low-level brainware, he imagines stepping into a room containing SAM's high-level software, a vast set of syntactic expressions and inference rules, all in paper and book form. Searle's claim, in essence, was that by using the SAM program to process and manipulate Chinese expressions (i.e., the story input, the generic knowledge, and the subsequent questions about the story) he would *become* just another of the many possible implementations of the SAM program. Yet he would never, by virtue of this fact, comprehend a word of Chinese. He would only be *conscious of* following rules for manipulating expressions that were meaningless to him (squiggles). Now if *he*, in implementing SAM, does not thereby come to understand a word of Chinese, neither, he contended, would any other device (e.g., a so-called electronic brain) implementing SAM. Indeed, no matter what subsequent program is put forward, Searle could repeat the thought experiment with the same result. Running a program thus does not by itself produce an experience of understanding the meanings of the Chinese expressions; it does not produce what might be termed *comprehension qualia* (Waskan 2011). The argument, he thought, generalized to any other program designed to realize any mental state whatsoever, whether perceptual, volitional, cognitive, or emotional.

In one clean swing, Searle seemed to have demolished the foundations of this massive (and lucrative!) framework for explaining and duplicating conscious mental episodes. He claimed to have shown (a) that programs do not produce or explain conscious episodes (at best they simulate them much as computers simulate weather without *being* weather) and (b) that merely passing behavioral tests like the *viva voce* test never justifies attributing conscious mental events to a programmed contrivance.

3 Artificial Neural Networks

Searle's work, like Leibniz', highlights the fact that there is nothing about the entities and activities under consideration, in this case the application of rules to sentences, remotely indicative of the production of qualia. He parts ways with Leibniz, however, in claiming that low-level brain processes *are* important when it comes to qualia, that to replicate mental states one requires a system with the same (as yet unknown) *relevant* causal powers as real neurons.

Recall that McCulloch and Pitts (1943) did initiate a promising program of research into artificial neural networks (ANNs), simulated networks of simplified neurons. That program has since bloomed in a thousand ways. Today ANNs underlie the kinds of *deep learning* that many everyday contrivances such as smartphones and automated assistants utilize to support sophisticated linguistic interactions. Devices that can pass the *viva voce* test appear closer than ever, and it is entirely likely they will be realized through neural processing. But will *Siri* for instance, ever *really* understand? Or see? Or feel? Searle would offer a qualified "no," for the ANNs in question are mere simulations of neurons implemented by traditional programming. He could just as well carry out *those* programs and not be subject to any relevant qualia. However, neural computing is no longer limited to simulations run on von Neumann devices. IBM, for instance, has unveiled its SyNAPSE chip, containing 1 million artificial neurons and 256 million synaptic connections.[2] If there is something about the causal powers of real neurons that produces mental states, it is possible that SyNAPSE's neurons have it. To this somewhat vague possibility, Searle would not object.

4 The Limits of Searle's Critique

Searle, we saw, is not just criticizing the idea that programs can produce conscious mental events. He is also criticizing such behavioral tests for mindedness as *viva voce* tests. But let us imagine that a robot with a fairly sophisticated body and a full complement of sensors has been endowed with a program that makes it behave like a real human. Better still, imagine a robot constructed along the lines suggested by Turing (1950)—namely, one with perceptual systems and programmed with some amount of structure to facilitate language acquisition and reasoning, but that must otherwise learn on its own. Suppose this robot, call it Eimer, is protected from the corrupting intellectual influence of qualiaphiles like myself, but over time it nevertheless comes to say things like this:

> "How do I know that my red is not your blue?"
>
> "My experience of redness, that color I see, is produced in me somehow, but I don't see how it could possibly be explained by electronic states."

In light of its behavior, and given what we know of its etiology, we might have good reason to assert that Eimer does in fact have qualia. In other words, under suitable background conditions, perhaps a behavioral would persuade us to attribute qualia to a programmed contrivance.

Searle here would object that if he implemented Eimer's program himself he would find nothing resembling the production of qualia. However, as Clark (2004) points out, we can run the same type of thought experiment regarding the brain. In fact, Leibniz did just that. Blowing brains up to any size you like and walking around in them, one finds nothing resembling the production of qualia. But we know on independent grounds that brains do produce qualia (unless Cartesian dualism is true). So perhaps merely knowing how a machine works at some particular level of abstraction and not detecting qualia there does *not* justify denying that it is subject to qualia. Indeed, the real lesson may be that so long as we remain ignorant of how brains produce qualia, we cannot determine whether or not mechanical contrivances are capable of producing them. Put differently, if Eimer says it has qualia but can't fathom how, our response cannot be, "We looked inside your head and didn't see any, so there really are none." With as much justification, Eimer could say the same about us!

5 Stop Looking for Qualia

There are, on the other hand, those who would deny that Eimer has qualia on the simple grounds that they do not exist *anywhere*. Even after centuries of trying, there still seems to be no room for qualia in the mechanical world described by science. Particles (or waves or strings) in motion are all there is. The rest is aggregations. If it is painfully hard to fathom how bits of matter or energy could produce something as deeply mysterious as qualia, with properties that seem *non*-mechanical, perhaps it is because there are no qualia. Perhaps qualiaphiles secretly want humans to be special in some way and *not* mere machines. Or, to be more rigorous, perhaps our puzzlement is not due to being subject to mysterious qualia but to a kind of psychological illusion or cognitive defect, one arising from our distinctive way of processing information (Dennett 1991; Tye 1999; Fiala et al. 2011). And get this: If Eimer works like we do, we would expect it to be beset by the same psychopathology. So even if we produce robots like Eimer that spontaneously claim to have qualia, qualiaphobes claim that they are not to be believed. Neither are you.

6 Look Out for Qualia

Another response to the failure to find qualia in brains and robots would be to say that we are looking (mostly) in the wrong place, for qualia typically have one foot outside of the head. Some suggest that in order for any machine to have perceptual qualia it needs online, active control of a body existing in tightly coupled interaction with a real environment (see Noë 2006). This is not Galileo's now-mundane notion that qualia are typically *caused* by body and world, but the provocative thesis that they are constituted by cognition that is *embodied* and *embedded* (in the world). No wonder, then, that Leibniz and Searle found no relevant qualia. They were looking at just part of the relevant system.

Views of this sort seem on their face to be contradicted by fairly basic neurological considerations about the location of qualia in humans. Dreams and the phantom limb experiences had by amputees, both cited as evidence by Descartes (1647/1988), occur in the absence of any relevant external objects, and they are now thought to result from the same sort of neural activity that produces the corresponding veridical experiences. Or consider Penfield's many striking case studies of people who have vividly experienced bodily and worldly events, replete with sounds, smells and emotions, due to seizures or electrocortical stimulation. Patient M.M., for instance, reported the following: "'I had … a familiar memory, in an office somewhere. I could see the desks. I was there, and someone was calling to me, a man leaning on a desk with a pencil in his hand'" (Penfield 1958: 59). This suggests that (in humans) brain events *suffice* to generate qualia, that qualia are thus constituted by these events. If so, invoking body and world in the robot case would seem no more helpful or warranted.

One response to this worry is to opt for disjunctivism (Noë 2006), the view that there are two independent classes of qualia: the (aberrant) internal kind and the normal (partly) external kind. Still, one wonders: If there is a purely neural cause of qualia in the former case (with no relevant external objects), why add to the story—indeed, parsimony dictates otherwise—in the latter?[3] Surely the same neural events underpin qualia (e.g., experienced redness) in both cases.

Some maintain, however, that the extra-neural component of qualia is not the outside world in the usual sense, but the *contents* of our representations of it (Lycan 2001; Tye and Byrne 2006). Contents seem a good candidate for this role because they are widely considered to be (at least partly) external to the brain, part of the natural order, and at times present when their target objects are absent. On this way of thinking, if I experience a red tomato before me and there is none, there still is a (*ex hypothesi* external) mental state with content (e.g., *red tomato there*). Extrapolating to robots, if *their* internal states have appropriate (external) contents, they too can be expected to have qualia.

However, as a strategy for *explaining* qualia, content externalism appears as a kind of parlor trick as soon as we ask: Exactly what part of the external physical world produces, and thereby explains, the qualitative character of our experience of a red tomato? The experience is real. Now point to it! Lycan (2001) floats the idea that the relevant content is "a nonactual physical individual" or a set of possible worlds, but these intangible abstracta are poor candidates for the physical producers of, and thus nonstarters as explanations for, qualia. Perhaps recognizing this, Tye (1999) opts for the view that the seeming inability of contents to *physically explain* qualia is itself best *explained away* with a variant on the cognitive illusion strategy.

7 Look Up for Qualia

There is, admittedly, a certain pull to the idea that we are looking in the wrong place for qualia and that if we continue this course we will remain befuddled, confounded, and unable to replicate qualia in our robotic contrivances. But rather than looking out, perhaps we should be looking up, and not just past Leibniz' structural level, but also past Searle's programming level.

Consider all of the different ways of building a computer that implements MSWord. Starting at the bottom, there are various materials (e.g., from copper wire to nanotubes) that can be used to implement particular architectures (e.g., Turing, von Neumann, super-scalar)—that is, each of the latter is *multiply realizable*. Likewise, a program written in a particular language (e.g., C++, Lisp, Visual Basic) can be built from any of these lower-level architectures.

But have we now reached the pinnacle? Perhaps not. For instance, what if a relatively high-level C++ program realizes a Java virtual machine (a simulation of a type of machine), which runs a Java program, which implements the Minecraft virtual environment (a 3D sandbox game), which (through its proprietary array of blocks, redstone circuits, repeaters, and the like) implements a machine that approximates a universal Turing machine, which implements Visual Basic and, finally, MSWord? This is possible, at least in principle. There can thus be layers upon layers upon layers such that we may never be sure that we've reached the pinnacle, assuming there is one. And at each new layer we encounter entities and activities with properties that are (typically) unique to that level, and so we invent a new vocabulary to describe them. And not all of them involve sentences and rules.

Perhaps qualia too are high-level, multiply realizable properties. Indeed, the very possibility of *synthetic* robot consciousness may just require that they are. This proposal, in any case, has the great virtue of explaining why, when we view the brain in terms of low-level properties (neurons, neurotransmitters, action potentials, etc.) not only do we see 'nothing resembling a percept'—no qualia—but we also have trouble imagining how there *could be* any qualia inhabiting the system. Likewise, when we focus just on the hardware of a computer, it is hard to imagine how it could hold a virtual world with a landscape full of avatars.

But Searle would here object that *he* is not viewing the system at a low, Leibnizian level, but at a higher, programming level. But to be persuaded by his arguments about conscious mental events, we must accept, contrary to fact, that the program is the highest possible level of activity in a computational system. Indeed, if Searle were to *be* a certain C++ program, he could be entirely oblivious to the fact that his activities implement a certain type of virtual machine and, even more surely, have no capacity to appreciate how that machine runs a complex Java program that traffics in countless numerical data structures for tracking the 3D coordinates and properties of various coarse pixels, data structures that in turn realize a well-equipped avatar running around in a particular Minecraft seed. The virtual world Searle realizes might, purely from his low-level vantage, be for all intents and purposes *cognitively closed*. So perhaps where Searle's critique goes wrong is in assuming that the highest level of abstraction for understanding any given programmed device is the level of the program, the level of rules and sentences.

Consider, in addition, the many realistic computer models of physical systems that are being run across various supercomputers as we speak. Many are realized using commercially available physics modeling programs such as Lawrence Livermore Lab's LS-Dyna. At the highest programming level, they traffic in sentences that specify co-ordinates of basic building blocks (e.g., granular structures or polygon vertices) and rules constraining how the co-ordinates may change. In this way, they provide for a kind of virtual clay that can be used to create models of everything from tornados and hot springs to SUVs. The models are set in motion to see how things will play out, much as one uses physical modeling media (plastic, metal, etc.) to create and manipulate scale models. Indeed, these virtual models have many features widely taken to distinguish (roughly speaking) *imagistic* representations from their descriptive counterparts, including open-ended inferential powers (build anything you like and poke it any which way, and read off the results), a kind of *universal* character that is very much like what Descartes claimed no machine could possess (Waskan 2017).

Now what if, in the human case, the better part of how brains produce conscious experiences of our surroundings is to use the torrent of stimuli to our sense organs to form, through massive amounts of lower-level information processing, a high-level internal model of what is going on external to the brain. If that is the case, synthetic robot brains, even ones governed by programs at some non-terminal level, may be as capable of *this* as their mushy counterparts. But could endowing our robotic creations with such internal world models really make them conscious? Might it give them qualia?

As always, the answer is far from straightforward, but one finds reasons for optimism by returning to the beginning of our discussion, to Galileo. Recall his claim that some experienced properties (colors, tastes, etc.) inhere *only* in us, whereas others (shape, motion, etc.) inhere in external objects. But do the latter inhere *only* in those objects? Surely not. Cases like those supplied by Descartes and Penfield seem to show that we may have experiences of those properties (or *as* of them) even in their absence. The takeaway from Galileo may be then that the primary properties we experience often *also exist in a similar form* external to our minds whereas the secondary ones do not. As he suggests, color qualia may have more to do with what light *does* to us; they may reside *only* in us. But primary properties are both in us and (much of the time) out in the world. All of this is just to get to the point that there is also a mystery of how brains produce experiences of shape, motion, etc., a mystery of *primary* qualia alongside that of secondary qualia. In neither case, primary or secondary, is there anything about neural activity that betrays their instantiation. Indeed, just looking at the low level of neural circuitry, it is hard to see how brains even *could* produce primary qualia (see Waskan 2011).

But this is where computer simulations of physical systems are so instructive. They offer a framework for making sense of how electronic (or electrochemical) circuitry can give rise to distinct properties at distinct levels of activity, from logic gates, to numerical memory registers, to programs, to evolving coordinate specifications, to high-level models (or what have you) of primary properties. If our high-level internal models produce experiences of primary properties in us, then we should not look outward like Noë, Lycan and Tye, or downward like Leibniz and (as it turns out) Searle. We should be looking upward, past several levels of processing, to internal world models. Already inspired by computers, this view lends itself well to the creation of mechanical contrivances such as robots that have similar world models and, thereby, possess primary qualia.

But it is not obvious at first glance how this proposal could be extended to account for, let alone allow us to reproduce, secondary qualia. Still, it would be cosmically peculiar if primary qualia (e.g., the experience of a tomato's shape) turned out to be just the characteristic of some high-level internal model while secondary qualia (e.g., the experience of a tomato's redness) were best explained as, say, the physical properties of quantum events or, as Jackson (1982) hypothesized, as the non-physical exhaust of neural events. So if the above account of primary qualia proves out, we should expect other sorts of qualia to admit of a similar, high-level solution. And we should expect to make little progress in understanding qualia of any sort as long as we hold fast to our myopic lower-level wanderings. And again, such a high-level explanation for secondary qualia might be easily adaptable to the creation of robotic contrivances possessing secondary qualia like colors, smells, and love or hate.[4] Indeed, as already mentioned, the possibility of synthetic, robot consciousness may *require* a high-level account of qualia such as this.[5]

8 How Our Robot Overlords Might Overcome Cognitive Closure

As tempting as all of this seems to one of us, we must still allow for the possibility that the truth about qualia is far different from anything yet proposed. Indeed, it may be so strange as to remain effectively closed to feeble-minded creatures such as ourselves. Even so, there is another lesson to be gleaned from the advent of sophisticated computer models of the world, one that gives us renewed cause for optimism.

Consider that we began to produce finely detailed external models in the first place precisely because we *are* so feeble minded. Even if we understand, individually, how particular bits of matter can be configured relative to one another and the laws governing their behavior, it is often far beyond the powers of human imagination to envision in any detail how vast arrangements of low-level events give rise to higher-level phenomena. In isolation from artifacts and technology, plausibly our internal models of systems are always incomplete, piecemeal affairs (Norman 1983; Keil 2006; Hegarty 2004) and what passes for unaided mechanical *understanding* only extends only so far as we can find ways of compressing complicated information into useful approximations (Schwartz and Black 1996) or into simple and suggestive metaphors (Brown 2003).

However, *with* the aid of scale models and, more recently, computational models, we can watch complicated scenarios play out, interact with and manipulate them, and draw inferences in a way that opens windows onto events that would otherwise remain closed. Again, plausibly we produce these external models because our own offline (i.e., decoupled from external stimuli) internal modeling ability is so limited. At the same time, as external modeling complexity increases, the more limited becomes our ability to understand the models themselves. And so we must find ways of compressing *what the models do* into digestible form. ANNs are one case in point. They often display uncanny resemblances to human cognition (Bechtel and Abrahamsen 2002), but from the standpoint of *understanding* cognition their creation is typically only half the battle. The rest is figuring out how the models work, and to do so researchers invent techniques such as principal component analysis (the emphasis being on 'principal') that help us to tease out some of the major factors underlying a model's performance.[6] Thus, even *with* the aid of computer models, there can be many details that remain closed to us.

But taken in its extreme form, the cognitive closure thesis involves a dubious assumption—namely, that humans as we currently exist are the final word in terrestrial intelligence. Is it not likely that, as Kurzweil (2005) maintains, we are climbing a mountain, but getting stronger and faster as we go, toward the production of successors to ourselves, whether GMOs, cyborgs, or, straight to the present point, robotic contrivances. Surely some of these successors will utilize internal models that are not superficial and gappy like ours, but internal models with a degree of depth and detail far surpassing that of even our best present-day computer simulations. Whereas you and I might, in our mind's eye, see in a general, hazy way how floating plates and mantle currents can produce mountains and rift valleys, they might see in theirs how vast conglomerations of atoms do this, or how subatomic particles realize life, or how quantum events undergird general relativity, and yes, even how vast conglomerations of low-level processes, neural or electronic, conspire to realize qualia.[7]

Many phenomena that are closed to us should thus be wide open to our hyper-intelligent offspring, and qualia may be among them. If so, then we may find that the 'robot consciousness' question was on the wrong track from the get-go. Instead of asking whether synthetic contrivances can *have* qualia, we should have been asking when those contrivances will attain the level of sophistication needed to *explain* qualia, or at least our attestations thereof.

To take things a step further, we have just assumed that the problem of qualia is largely quantitative and can be solved with much brute force computation, plus, one imagines, significant theoretical insight. But what if the problem is not just quantitative but also qualitative? What if qualia can't be modeled under the constraints of a fundamentally Newtonian physical framework? Indeed, given their poor apparent fit with the mechanical view of things, they might be a different beast altogether. They might really be quantum in nature, or perhaps property dualism is correct, or panpsychism. Or maybe qualia reside in a realm of dark matter or energy. Indeed, it has become ever clearer that these are the dominant modes of existence in this universe, and some have suggested that the dark realm could be at least as variegated as that of 'normal' matter and energy. It might have its own dark chemistry, dark biology, or even dark cognition. Descartes could be more right than most give him credit for. Minds might be a kind of halo around certain machines in the way dark matter is an invisible halo around visible galaxies. Where extreme ignorance persists, anything seems possible.

But even in this case our robotic successors will have a real shot at understanding and confirming the truth. For though our own biologically evolved mental models have been shackled through inheritance and learned interactions with organism-level reality, the mathematical constructs we have devised, and which figure into the implementation of computer models, are significantly freer. We add new dimensions by adding variables, and we explore unthinkable non-Euclidean spaces by altering basic axioms. We even explore possible universes by tweaking fundamental constants and laws and modeling out their creation and evolution on computers.

In the same way, the worlds thought about by our synthetic descendants might range across possibilities far beyond anything our feeble brains are fit to envision. Add to this that we are witnessing the creation of computers that operate on the basis of 'spooky' quantum principles such as entanglement and superposition. When perfected, they will quite literally do things in hours that no von Neumann device could accomplish in a thousand years. In short, our synthetic progeny may find that there is far, *far* less closed to them than there is to us. This, in any case, is the picture that emerges when you sic Kurzweil on Jackson and McGinn.

The moral and practical considerations of such an eventuality are too numerous to mention. Elon Musk, founder of Tesla and SpaceX, puts the direst possibility succinctly: By creating such devices we are *summoning the demon* that will destroy us. No wonder he is trying to escape to Mars!

9 What If (for Us) It's *All* Synthetic?

Let me close with an *even more* exotic possibility. It could be that the answer to whether or not synthetic contrivances are capable of possessing qualia is an emphatic yes, for we *are* such contrivances. As Nick Bostrom (2003) has argued, if we extrapolate out, we find that (assuming we survive) our insatiable drive to create sophisticated computer models of the world will bring about a future brimming with massive, high-fidelity models of past events, events such as are now transpiring…or so we think. After all, says Bostrom, as a matter of simple probability, given the number of worlds we might be living in (i.e., the one real world plus the innumerable future virtual worlds) the chances are overwhelming that the future has already arrived and that we *are* conscious synthetic agents living out our paltry lives in a computer simulation. Indeed, for all we know, the mystery of qualia may have long since been solved by humanity's super-intelligent robotic descendants and all of this discussion is, well, just academic.

Notes

1 The *Star Trek: Next Generation* episode 'Sentient Being' provides an entertaining debate on this topic.
2 http://www.research.ibm.com/cognitive-computing/neurosynaptic-chips.shtml (last accessed 3/3/17).
3 For further discussion see Waskan (2011).
4 This approach does not require the controversial assumption that all conscious experiences (e.g., melancholy) have representational content. The central claim is that qualia are high-level happenings, representational or not.
5 Plausibly high-level representational (or other) processes must also be used, or *poised* for use, by other systems involved in attention, memory, reasoning, planning, motion guidance, and the like (Tye 2000; also see Rosenthal 1986).
6 See http://www.iep.utm.edu/connect/ (last accessed 3/6/17).
7 If it turns out that qualia are ineluctably biological, synthetic beings (who have none) might still take our testimony as evidence that qualia exist, or at least that we think they do, and go about trying to explain our attestations.

References

Bechtel, W., and Abrahamsen, A. (2002) *Connectionism and the Mind: An Introduction to Parallel Processing in Networks* (2nd ed.), Cambridge, MA: Basil Blackwell.
Boden, M. (2006) *Mind as Machine*, New York: Oxford University Press.
Bostrom, N. (2003) "Are You Living in a Computer Simulation?" *Philosophical Quarterly* 53: 243–255.
Brown, T. (2003) *Making Truth: Metaphor in Science*, Urbana, IL: University of Illinois Press.
Clark, A. (2004) *Mindware*, Oxford: Oxford University Press.
Dennett, D. (1991) *Consciousness Explained*, New York: Little, Brown, and Company.

Descartes, R. (1637/1988) *Discourse on the Method*, in J. Cottingham, R. Stoothoff, D. Murdoch (trans.) *The Philosophical Writings of Descartes*, Cambridge, UK: Cambridge University Press.
Descartes, R. (1647/1988) *Meditations on First Philosophy*, in J. Cottingham, R. Stoothoff, D. Murdoch (trans.) *The Philosophical Writings of Descartes*, Cambridge, UK: Cambridge University Press.
Fiala, B., Arico, A. and Nichols, S. (2011) "On the Psychological Origins of Dualism: Dual-process Cognition and the Explanatory Gap," in E. Slingerland and M. Collard (eds.) *Creating Consilience: Integrating Science and the Humanities*, Oxford: Oxford University Press.
Galilei, G. (1618/1957) "The Assayer," in S. Drake (trans.) *Discoveries and Opinions of Galileo*, New York: Anchor Books.
Hegarty, M. (2004) "Mechanical Reasoning by Mental Simulation," *Trends in Cognitive Sciences* 8: 280–285.
Hobbes, T. (1651/1958) *The Leviathan*, Amherst, NY: Prometheus.
Jackson, F. (1982) "Epiphenomenal Qualia," *The Philosophical Quarterly* 32: 127–136.
Keil, F. (2006) "Explanation and Understanding," *Annual Review of Psychology* 57: 227–254.
Kurzweil, R. (2005) *The Singularity Is Near: When Humans Transcend Biology*, New York: Penguin Books.
Leibniz, G. (1714/1898) *The Monadology*, in R. Latta (trans.) *The Monadology and Other Philosophical Writings*, Oxford: Clarendon Press.
Lycan, W. (2001) "The Case for Phenomenal Externalism," *Philosophical Perspectives* 15: 17–35.
McCulloch, W. and Pitts, W. (1943) "A Logical Calculus of the Ideas Immanent in Nervous Activity," *Bulletin of Mathematical Biophysics* 5: 115–133.
McGinn, C. (1989) "Can We Solve the Mind-Body Problem?" *Mind* 98: 349–366.
Noë, A. (2006) "Experience Without the Head," in T. Gendler and J. Hawthorne (eds.) *Perceptual Experience*, Oxford: Oxford University Press.
Norman, D. (1983) "Some Observations on Mental Models," in D. Gentner and A. Stevens (eds.) *Mental Models*, Hillsdale, NJ: Lawrence Erlbaum Associates.
Penfield, W. (1958) "Some Mechanisms of Consciousness Discovered During Electrical Stimulation of the Brain," *Proceedings of the National Academy of Sciences* 44: 51–56.
Rosenthal, D. (1986) "Two Concepts of Consciousness," *Philosophical Studies* 49: 329–359.
Schwartz, D. and Black, J. (1996) "Shuttling Between Depictive Models and Abstract Rules: Induction and Fall-back," *Cognitive Science* 20: 457–497.
Searle, J. (1980) "Minds, Brains, and Programs," *Behavioral and Brain Sciences* 3: 417–457.
Schank, R. and Abelson, R. (1977) *Scripts, Plans, Goals, and Understanding*, Hillsdale, NJ: Lawrence Erlbaum Associates.
Turing, A. (1937) "On Computable Numbers, With an Application to the Entscheidungsproblem," *Proceed ings of the London Mathematical Society* 2: 230–265.
Turing, A. (1950) "Computing Machinery and Intelligence," *Mind* 59: 433–460.
Tye, M. (1999) "Phenomenal Consciousness: The Explanatory Gap as a Cognitive Illusion," *Mind* 108: 705–725.
Tye, M. (2000) *Consciousness, Color, and Content*, Cambridge, MA: The MIT Press.
Tye, M., and Byrne, A. (2006) "Qualia Ain't In the Head," *Nous* 40: 241–255.
Waskan, J. (2011) "A Vehicular Theory of Corporeal Qualia," *Philosophical Studies* 152:103–125.
Waskan, J. (2017) "From Neural Circuitry to Mechanistic Model-based Reasoning," in L. Magnani and T. Bertolotti (eds.) *Springer Handbook of Model-Based Science*, Berlin: Springer-Verlag.

Related Topics

Consciousness in Western Philosophy
Materialism
Dualism
Representational Theories of Consciousness
Sensorimotor and Enactive Approaches to Consciousness
Quantum Theories of Consciousness
The Neural Correlates of Consciousness
The Biological Evolution of Consciousness
Consciousness and Dreams

31
CONSCIOUSNESS AND DREAMS
From Self-Simulation to the Simulation of a Social World

Jennifer M. Windt

1 Introduction

Sleep is phenomenologically rich, supporting diverse kinds of conscious experience as well as transient loss of consciousness. Sleep is also cognitively and behaviorally rich, with different sleep stages supporting different kinds of memory processing (Rasch and Born 2013; Stickgold and Walker 2013) as well as sleep behaviors ranging from subtle muscle twitches (Blumberg et al. 2013) to seemingly goal-directed behaviors, as in sleepwalking, sleep talking, and REM-sleep behavior disorder (Howell and Schenck 2015). This phenomenological, cognitive, and behavioral richness is flanked by a complex and cyclically organized sleep architecture, with sleep stages characterized by different levels of electroencephalogram (EEG) activity, regional patterns of brain activity, eye movements, and muscle tone (Pace-Schott 2009). Yet, how changes in conscious experience are associated with sleep stages and behavior continues to be poorly understood (Windt et al. 2016).

Progress in dream research was long hampered by lack of agreement about the target phenomenon. Different definitions—ranging from narrow definitions focused on certain types of narratively complex dreams to broad definitions classifying any kind of conscious experience in sleep as dreaming (Pagel et al. 2001)—were paralleled by disagreement about the sleep-stage correlates of dreaming. Early dream researchers assumed that dreaming can be identified with rapid eye movement (REM) sleep and that the contrast between REM and NREM (or non-REM) sleep marked the presence vs. absence of consciousness (Dement and Kleitman 1957). It is now, however, widely recognized that dreams occur in all stages of sleep (Nielsen 2000). There are also theoretical and empirical reasons for thinking that kinds of sleep experience exist that are distinct from dreaming (Windt et al. 2016).

This progress has been enabled by important conceptual and methodological advances. There is now increasing convergence on simulation views (Metzinger 2003, 2009; Nielsen 2010; Revonsuo 2006; Revonsuo et al. 2015; Windt 2010, 2015a; Windt et al. 2016; Thompson 2014, 2015), in which dreaming is defined by the experience of a self in a world. Methodologically, serial awakening paradigms (Noreika et al. 2009), in which participants are awakened multiple times throughout the night from different sleep stages and at short time intervals, coupled with high-density EEG recordings, are shedding light on the neural correlates of dreaming vs. non-dreaming (Siclari et al. 2013, 2017). Together with a fine-grained framework for describing types of dreamful and dreamless sleep experiences, this can enable a more precise mapping to neural and behavioral events during sleep.

In this chapter, I endorse a version of the simulation view that focuses on minimal forms of dreaming and argue that these coincide with minimal phenomenal selfhood, or the simplest form of experiencing oneself *as* a self. Yet the experience of being or having a self can take different forms in dreams, and I use examples from dream research to suggest how this minimal version of the simulation view can be scaled up to accommodate them. I then discuss this framework in light of current work on self-consciousness, suggesting that the analysis of self-experience in dreams can extend and enrich existing theories.

2 Probing the Phenomenology of Dreaming: Conceptual and Epistemological Considerations

Dreaming is notoriously heterogeneous, with different kinds of dreams having distinct phenomenological profiles. For example, lucid dreams, in which dreamers become aware that they are now dreaming, often additionally involve the ability to control the ongoing dream (Voss et al. 2013; Voss and Hobson 2015). There is also variation in dreams from different participant groups. While the vast majority of dreams involve visual imagery, congenitally blind subjects report spatial but nonvisual dreams (Kerr and Domhoff 2004). And there are developmental changes, with the degree of narrative organization and overall complexity gradually increasing from childhood into adolescence (Foulkes 2009).

This phenomenological diversity is flanked by different kinds of questions, often relating to diverging research interests and distinct disciplinary perspectives. In philosophy, the best-known discussion of dreaming is the epistemological problem of dream skepticism. Here, the question is how we can ever be certain that we are now awake rather than dreaming. In the *Meditations*, Descartes (1641/1901: I.4) tells us that we cannot: perceptually-based beliefs about the external world are potentially misleading. Descartes' assertion that he has *often* had the wake-like experience of sitting dressed by the fire even though he was in fact lying asleep in bed makes scenarios of dream deception psychologically gripping (Windt 2016). Yet what is needed to justify the theoretical possibility of dream deception is just that dreams can *potentially* mimic wake-like experience, not that they frequently or even typically do so (Windt 2016).

Emphasis on the *typical* features characterizing a *majority* of dreams plays an important role in scientific theories. For instance, Allan Hobson's (2009; Hobson et al. 2000) influential neuroscientific model characterizes dreaming through the predominance of visual imagery over other modalities, of negative over positive emotions, bizarreness, deficient reasoning and short- and long-term memory, and lack of metacognitive insight into the fact that one is now dreaming. The point here is not that strictly all dreams are captured by this definition, and lucid dreaming is a clear counterexample to the metacognitive deficit (Voss and Hobson 2015). Instead, the idea is that these stereotypical features can be mapped onto characteristic changes in neural activation patterns in sleep (Hobson et al. 2000).

This strategy can yield general insights into changes and continuities in experience across sleep-wake states, but cannot provide a strict definition in terms of necessary and sufficient conditions. For this type of project, it is more helpful to ask whether, underlying the variability of dreaming, there is something like a common phenomenal core that characterizes different types of dreams and dreams from different participant groups. To answer this question, it is more useful to focus on *minimal* forms of dreaming.

Elsewhere, I have argued that a plausible candidate for identifying the phenomenal core of dreaming is its immersive structure (Windt 2010, 2015a). Numerous studies have shown that dreams consistently involve the presence of a self (Strauch and Meier 1996; see also Occhionero et al. 2005; Speth et al. 2013).[1] Presence and the related concept of self-location can be given both a spatial reading (the experience of being located *here*, at a particular point in a larger spatial expanse) and a temporal reading (an experienced *now* plus the sense of duration). Dreams are organized around an internal first-person perspective, and the origin of the first-person perspective is what retrospectively, in dream reports, is described as the self.

Like other versions of the simulation view (Revonsuo et al. 2015), this focus on the immersive, *here-and-now* structure of dreaming highlights the similarity between dreams and presence in standard wake states, but also in virtual reality. However, the view I am proposing abstracts away from the features that are thought to standardly characterize both dreaming and waking self-experience. While most dreams involve visual imagery and strong emotions, they do not do so necessarily. Spatiotemporal self-location underwrites the subjective sense of presence as well as retrospective descriptions of having had a dream self even in the absence of modality-specific (visual or auditory) imagery (Windt 2010). This simplified account of presence brings us closer to a minimal definition that can also accommodate the variability of dreaming.

An important theoretical and methodological question concerns the relation between dream experience and dream reports. These can be verbal (written or oral) descriptions, but also drawings or answers to specific questions; together, they form the primary source of data about the phenomenology of dreaming. Skeptics have sometimes claimed that dream reports (Dennett 1976; see also Malcolm 1962) might equally well be the product of memories inserted into consciousness at the moment of awakening.[2] Even sincere reports describing the impression of having had certain experiences during sleep would then be systematically misleading. Such skepticism would undermine the use of dream reports to investigate the experience of dreaming. To make any substantial claims about the occurrence and phenomenal character of experience in sleep, dream reports must be transparent windows on what it is *actually* like to dream (Windt 2013, 2015a: chs. 3 and 4).

This does not mean that we should blindly trust dream reports, which would clearly be at odds with the elusive nature of dream recall. Laboratory research shows that spontaneous dream recall is a poor indicator of actual dream activity: most people only rarely remember their dreams and can recall no more than one dream per night. By contrast, dream recall in laboratory studies utilizing timed awakenings is much higher, reaching around 80% for REM sleep and about 40% for NREM sleep (Nielsen 2000). This suggests that we dream much more than we spontaneously remember, but also leads to the idea that at least under ideal conditions—for instance in the laboratory—dream reports are indeed trustworthy sources of evidence, for instance about the frequency of dreaming in different sleep stages.

Different factors can enhance or diminish the trustworthiness of dream reports. Most agree that to counteract dream amnesia—the fact that dreams are quickly forgotten unless special steps are taken to recall them—reducing the time lag between dream experience and dream reporting is important. Further factors may include the method of awakening, the way in which the dream is reported (oral vs. written), or motivational factors such as personal interest (Windt 2015a: ch. 4; Domhoff 2013). What counts as ideal reporting conditions is dependent on the specific research question, and further improving the collection and analysis of dream reports is an important goal for future research.

3 Dreaming the Self: Minimal Phenomenal Selfhood and Bodily Experience in Dreams

In the version of the simulation view I endorse, spatiotemporal self-location is the phenomenal core both of different kinds of dreaming and of the experience of being or having a self

(Windt 2010, 2015a). This is so because dreaming, as we have seen, is inextricably bound to the experience of presence, of being a self in a world. This core phenomenology of presence underlies richer forms of self-experience as well as different kinds of dreaming; but importantly, there are also instances in which self-experience can take the form of pure *here-and-now* experience. Dream reports may describe, for example, the feeling of identifying with an unextended point in space:

> I was inside a gigantic photocopying machine. I knew I was inside this machine, not as a physical human being but as an abstract entity, as a mind, so I couldn't see myself.
>
> *(Cicogna and Bosinelli 2001)*

This report is striking because it explicitly describes the feeling of lacking a body, or of being a *phenomenally disembodied self*. But bodily experience can also be lacking in dreams by way of an unnoticed absence. In such dreams, there is still the experience of spatiotemporal self-location, but the minimal sense of self this gives rise to is *phenomenally indeterminate* with respect to bodily experience: there is no *experienced* fact of the matter as to whether one has a body or not (Windt 2015a: ch. 6).

To make this point more vivid, try moving a playing card from the periphery of the visual field towards its center while fixating your eyes straight ahead. Notice how information about the card's suit, color, and value becomes available gradually. At least at the beginning, your visual experience is indeterminate with respect to color. As there is no reason to think that fixating your eyes straight ahead while attending towards the periphery causes the presence vs. absence of color, this simple experiment shows that indeterminacy with respect to color pervades a large part of the visual field (Dennett 1991)—it is just that in standard situations, we remain oblivious to this fact.

Phenomenal indeterminacy is probably more pervasive in dreams (but also, for instance, in memory and waking visual imagery) than in perception, which arguably inherits some of its detail from the external world. Phenomenal indeterminacy is also closely related to the epistemic notion of indeterminacy blindness (Windt 2015a: 329): phenomenal indeterminacy is pervasive, but inconspicuously so. Even in wakefulness, special attention (as in the playing card example) is required to notice phenomenal indeterminacy. In dreams, the attenuation of critical reflection and metacognitive insight might make phenomenal indeterminacy even harder to detect. A prediction would be that phenomenal indeterminacy should be more easily noticed and hence more frequently reported in lucid dreams, which are associated with metacognitive insight into the fact that one is now dreaming. While this question has not, to my knowledge, been studied systematically, it does seem that reports of phenomenal disembodiment, but also of visual imagery fading or taking on a washed-out quality often overlap with lucidity (LaBerge and DeGracia 2000). This leads to another important insight: an experienced absence (of the body, as in phenomenal disembodiment, or of color, as in the playing card example) is more sophisticated than an unnoticed absence in experience (cf. Dennett 1991: 359). The simplest way of lacking bodily experience is to remain oblivious to its absence.

Even if phenomenal indeterminacy, for instance for bodily experience, turns out to be more pervasive in dreams than in wakefulness, it would be false to say that bodily experiences are wholly lacking in most dreams. Movement sensations are frequent in dreams, second only to visual imagery. By contrast, thermal, tactile, pain, nausea, ticklish, and proprioceptive sensations are described in only 1–4% of laboratory reports (Hobson 1988; Schwartz 2000). Assuming these reports are transparent, this suggests that in a majority of dreams, bodily self-representation is schematic, associated mostly with movement, while sensations linked to detailed representations of individual body parts are rare. This is nicely illustrated by the dreams of specific participant groups.

First, dream reports of congenitally paraplegic subjects describe frequent whole-body movements such as walking, running, bike riding, or flying, and these descriptions are so similar to those from healthy participants that blind judges are unable to distinguish between them (Voss et al. 2011; Saurat et al. 2011). If we assume that whole-body movements in dreams exactly replicate their waking counterparts, it would be quite hard to explain how participants who have never had these experiences first-hand could nonetheless dream of them in a realistic and richly detailed way. Consider the example of a subject with congenital paraplegia for 40 years, who described a dream of learning to dance ballet, moving with a light step and wearing a tutu (Saurat et al. 2011: 1427). In wakefulness, such an experience would be associated with a sense of posture and limb position, tactile sensations, feelings of weight and balance, of effort and perhaps of losing one's breath. By contrast, in dreams, schematic movement representations, plus in some cases visual imagery of the body, may be enough. Phenomenal indeterminacy plus indeterminacy blindness would endow such experiences with a realistic gloss, while also relaxing the requirement for mimicking the exact phenomenological profile of their waking counterparts.

The second example is dreams of phantom limb patients. Following the loss of a limb, many people continue to have the vivid experience that the lost limb is still present. They also often continue to dream of having an intact body (Brugger 2008; Mulder et al. 2008). Both in waking and in dreaming, phantoms can vary in size, often shrinking over time. In dreams phantom limbs are mostly visually represented (Frank and Lorenzoni 1989), whereas the unpleasant bodily sensations (such as prickling, tingling, or pain) that characterize waking phantoms are typically missing (Vetrugno et al. 2009; Alessandria et al. 2011). In simplified terms, dream phantoms can be seen and moved but not felt, whereas waking phantoms are invisible and often paralyzed, as if frozen in an uncomfortable or even painful position. This nicely illustrates how the pattern of bodily experience in dreams departs from waking experience in systematic ways: the dream body is not just a whole-body analogue of waking phantom limbs.

Generalizing from these examples, we can conclude that dreams are *weakly phenomenally embodied states* (Windt 2015a: 338ff.). In a majority of dreams, visual and motor imagery predominate over tactile sensations, and body-part representations may occur in the absence of detailed whole-body representations. Also, where bodily experience in wakefulness is multimodal (de Vignemont 2014), dreams are often characterized by disturbances in multisensory integration, for example where a body part can be seen or moved but not felt. Just as bodily experiences vary across dreams, we should also expect phenomenal indeterminacy to be unevenly distributed: weak phenomenal embodiment and phenomenal indeterminacy complement each other. Figuratively speaking, we might describe body parts or bodily sensations in dreams as islands of determinacy occurring against a backdrop of phenomenal indeterminacy that, in turn, is clouded by indeterminacy blindness.

The next natural question to ask is how bodily experience in dreams relates to the physical body. Sleep is a state of reduced behavioral activity, and the processing of environmental stimuli and bodily sensations is attenuated. This is especially pronounced in REM sleep, which is associated with the most frequent, vivid, and narratively complex dreams (Hobson et al. 2000). Yet even in REM sleep, the processing of environmental and peripheral stimuli is not completely blocked. A familiar example is integrating the sound of an alarm clock into a dream. Incorporation rates are especially high for bodily stimulation. For example, a blood pressure cuff inflated on the leg leads to incorporation in 40–80% of dreams (Sauvageau et al. 1998; Nielsen et al. 1993). Vestibular stimulation (as when sleeping in a rotating chair or in a hammock) can lead to flying dreams or increased lucidity (Hoff and Plötzl 1937; Leslie and Ogilvie 1996), and thermal stimulation (Baldridge 1966; Baldridge et al. 1965) and sprays of water on the skin (Dement and Wolpert 1958) can prompt associated dream imagery.[3]

Sleep also involves a range of overt muscular activity, ranging from seemingly purposeful, goal-directed behaviors (as in sleepwalking or sleep talking) to subtler muscle twitching (Blumberg 2010; Blumberg and Plumeau 2016). These behaviors may have varying degrees of concordance and discordance with conscious experience in sleep. In so-called dream-enactment behaviors, there seems to be a particularly close correspondence between overt behavior and internally experienced dream movements (Nielsen et al. 2009). The common description of dreams as global offline states (Metzinger 2003, 2009; Revonsuo 2006; Hobson 2009) therefore seems oversimplified. Bodily self-experience in dreams typically does not arise completely independently of input from the sleeping body and muscular activity, but can be placed on a continuum with bodily illusions in wakefulness. Moreover, the unique pattern of bodily experience in dreams is best explained by appealing to the altered functional relationship between the physical body and the brain in sleep. Based on the available evidence, it seems plausible that dreams are both *weakly phenomenally* and *weakly functionally embodied states* (Windt 2015a: 382ff.).

The idea that illusory own-body perception plays an important role in dreams has a long history (for discussion of *Leibreiztheorie* or somatic source theory, see Wundt 1880; Schönhammer 2005). My proposal is not that dreams are caused by or strictly dependent on real-body stimulation, or that own-body perception characterizes all types of dreams and dream imagery. What I am suggesting is that investigating changes in the processing of external stimuli and motor behavior and how they are reflected on the level of dream experience is a fruitful explanatory and research strategy. Looking beyond the brain to real-body influences on dreams can yield a fine-grained framework for describing different kinds of (bodily) self-experience and inform a novel theory of dream imagery formation.

4 Dreaming Which Self? Self-Other Distinctions, Vicarious Dreams, and the Waking Self

Dreams are not just simulations of a self in a world, but of rich social realities. Dreams contain an average of 2–4 dream characters (Kahn et al. 2000), yet only one of these, at a given time, is experienced as the self. Self-other distinctions therefore play a central role in dreams: world simulation—both in a spatial and in a social sense—is necessarily grounded in the experience of a single self at its center. How then can we explain the experience of dream characters as distinct from the self? Rather than advancing a fully developed theory, I will just hint at some interesting parallels to self-experience in dreams.

Varying degrees of concordance between dream experience, bodily stimuli, and muscular activity in sleep may extend beyond the dream self to the experience of non-self dream characters. Of particular interest are cases where bodily sensations are projected to dream characters other than the self. Consider the following report from sleep onset:

> Someone in front of me is doubled over toward me, praying. Someone else reaches around from behind this person and quickly lifts him into an upright position. At the same time I feel my head nodding slightly forward and it awakens me.
>
> *(Nielsen 1992)*

Here, the forward movement of the head is not just externalized, but represented visually. Such externalization can also occur during full-fledged dreams in response to experimental body stimulation (Sauvageau et al. 1998; Nielsen et al. 1993). In a way, this is the flip side of weak phenomenal-functional embodiment: where self-experience in dreams is less strongly constrained by the physical body than in wakefulness, self-other distinctions are more porous, allowing real-body sensations to shape dream characters experienced as distinct from the self.

Self-other distinctions are also more fluid in dreams. An example is vicarious dreams, in which one dreams of being a different person than one is in wakefulness (Rosen and Sutton 2013).[4] Some of these dreams contain shifts in self-identification, with different dream characters successively being experienced as the self. An intriguing possibility is that these shifts again have a real-body basis. In the following report, again from sleep onset, it is almost as if a leg jerk, first represented visually and then in the form of motor imagery, were transporting the dreamer from the role of a passive observer to a self present in the dream.

> I am watching someone else walking. Then it is as if I was walking and I find myself about to start quickly up some stairs. I awaken with a very strong leg jerk.
>
> *(Nielsen 1992: 360)*

In keeping with the minimal version of the simulation view introduced earlier, a further and more basic factor underlying shifts in self-identification are shifts in experienced self-location:

> The person in the dream that so far had been me, now was suddenly my classmate J. Somehow I became (physically) detached from myself and I noticed that I was not me but him. This was accompanied by a funny feeling.
>
> *(Revonsuo 2005: 213ff.)*

It is tempting to attribute this funny feeling to a sudden perspectival shift in the phenomenal *here*.[5]

This brings us to another parallel between self- and other-experience in dreams. Just as the simplest form of self-experience in dreams requires only spatiotemporal self-location, there is also a purely spatial variant of experiencing non-self dream characters. Felt presence can occur in a number of conditions including heightened stress and emotional arousal (Nielsen 2007), but is particularly frequent in the vicinity of sleep. The experience that someone is present in the room can be associated with visual and auditory imagery—as in seeing shadows on the wall or hearing footsteps—but can also take an amodal, purely spatial form.

> It often happens that a hallucination is imperfectly developed: the person affected will feel a 'presence' in the room, definitely localized, facing in one particular way, real in the most emphatic sense of the word, often coming suddenly, and as suddenly gone; and yet neither seen, heard, touched, nor cognized in an of the usual 'sensible' ways.
>
> *(James 1902/2003: 51)*

Amodal, purely spatial variants of felt presence have been described as precursors to more complex experiences involving modality-specific imagery (Cheyne and Girard 2007a, 2007b; Nielsen 2007); yet even these minimal forms have a distinctly social flavor, with the presence being experienced not just as an undefined object, but as an agent having a definite spatial location as well as (frequently menacing) intentions towards the self. Felt presence therefore may involve a basic form of social imagery (Nielsen 2007). The convergence between purely spatiotemporal imagery and perceived intentions in felt presence complements the convergence between spatiotemporal self-location and the experience of selfhood in minimal forms of dreaming. Where in felt presence, spatiotemporal imagery underwrites the attribution of agency to another, in minimal dreams, spatiotemporal self-location underwrites the experience of selfhood. And because felt presence often occurs during sleep onset and precedes full-fledged dreaming, we might then say that the activation of *amodal agent models* (Windt 2015a: 570ff.) in felt presence is a prequel to *amodal self-models* in minimally immersive dreams. Likewise, shifts in self-identification towards a non-self dream character, either at sleep onset or within dreams, might minimally require shifts in spatiotemporal self-location.

As is the case for the dream self, this is not to say that dream characters typically take this minimal form. Just as more complex forms of self-experience are both possible and frequent in dreams, non-self characters are often embedded in a more complex dream narrative. They frequently represent people familiar to the dreamer and their identity is often recognized based on visual appearance and behavior (Kahn et al. 2000). Social interactions are even more frequent in dream reports than in randomly timed waking reports (McNamara et al. 2005) and are often described as subjectively realistic and emotionally engaging (Kahn et al. 2002; Revonsuo et al. 2015). Dreams of being chased, which are a common dream theme (Nielsen et al. 2003), are a good example. Dream characters are also treated as if they had a mind of their own, with the dream self often ascribing beliefs and desires to other dream characters (McNamara et al. 2007).

Yet, focusing on minimal kinds of self- and other experience in dreams can pave the way towards a parsimonious account that does not require dreams to exactly replicate waking experience. Tellingly, non-self dream characters are also often recognized by *just knowing* (Kahn et al. 2000, 2002), suggesting large-scale phenomenal indeterminacy may again complement more detailed, modality-specific representation of appearance or behavior. Both the dream self and non-self dream characters may be much flimsier, phenomenologically speaking, than their waking counterparts—and due to indeterminacy blindness, there may be a natural tendency to overlook this fact.

5 What Kind of Self? From Self-Simulation in Dreams to Theories of Self-Consciousness

Because dreams occur both frequently and spontaneously, their investigation can help identify core features of self-experience that are independent of behavioral state changes, such as the transition from wakefulness to sleep. Dreams therefore offer a contrast condition for standard and altered wake states.

Simulation views of dreaming have a natural affinity to virtual reality (VR) research. Here, a central question is under which conditions a virtual environment turns into an experienced reality. Again, a key concept is that of presence: the subjective experience of *being there*, in a world that is virtual but experienced as real. Dreaming has in fact been described as a natural experiment in presence and the gold standard to which VR design should aspire (Moller and Barbera 2006). Because participants in VR experiments maintain intellectual insight into the fact that what they are experiencing is not real, presence in VR is perhaps best compared to lucid dreaming.

In another way, however, presence in nonlucid dreams captures a key feature of standard waking experience. Antti Revonsuo's (2006) virtual reality metaphor of consciousness says that even standard waking experience is a kind of online hallucination, similar in its phenomenological features to dreaming and VR but additionally modulated by external sensory input. We don't directly experience mind-independent objects, but internally constructed world- and self-models (Metzinger 2003, 2009). Dreams and wakefulness are different in the degree to which external sensory stimuli modulate these internal models—yet in wakefulness, as in nonlucid dreams, we typically don't become aware of the simulational character of perception.

Do the factors that contribute to the experience of presence in VR play a similar role in dreaming? Or are these factors state-dependent, underlying the experience of presence only under the specific conditions of wakefulness? For presence in VR, three factors are thought to be particularly important: the quality (e.g., the resolution and overall realism) of computer-generated (mostly visual and auditory) inputs; the fluidity of sensorimotor interaction (i.e., the ability to move through the virtual environment and interact with virtual objects); and social interaction (Sanchez-Vives and Slater 2005; Slater 2009). Importantly, enabling participants to interact with avatars in VR increases the sense of presence even if their appearance is not realistic. Dreams might further relax these requirements on realistic, multimodal imagery and wake-like sensorimotor interaction. The importance of these factors for presence may be state-dependent, contingent on the close connection between bodily experience and motor activity that persists in VR but is attenuated during dreams. By contrast, social imagery and self-other distinctions appear to be closely associated with the experience of presence across sleep-wake transitions and may even be a precursor to fully immersive dreams.

The analysis of dreaming can also shed light on the minimal conditions that are both necessary and jointly sufficient for phenomenal selfhood. In the recent literature on self-consciousness, there is a strong emphasis on embodiment (de Vignemont 2016). Using the example of full-body illusions, Blanke and Metzinger (2009; but see Metzinger 2013) propose that embodiment is inextricably linked to even minimal forms of phenomenal selfhood. In these illusions, multisensory conflict is used to induce shifts in self-location and self-identification. In a standard setup, participants are stroked on their backs while seeing, through a head-mounted display, brush strokes being applied to the back of an avatar who appears to be standing in front of them (Lenggenhager et al. 2007). When the felt and seen strokes are applied synchronously, many participants report feeling localized towards the avatar, almost as if they were feeling the strokes where they are seeing them, on the avatar's back. This shift in self-location and self-identification towards the avatar is made possible through the continued feeling of body ownership and carefully administered stimulation of the physical body.

By contrast, in dreams, the relation between bodily experience and the physical body is loosened, and in minimal dreams, phenomenal selfhood does not require the experience of being an embodied self at all. The experience of owning a body is not, therefore, strictly necessary, and minimal phenomenal selfhood can attach to self-location in a purely spatiotemporal sense. Theories of consciousness and phenomenal selfhood focusing mostly on waking consciousness, including pathological wake states, VR, and full-body illusions, may suffer from wake-state bias, in which factors that are dependent on wakefulness are mistaken for general characteristics of conscious experience. To identify the simplest forms of self-experience, it may be necessary to look beyond wakefulness.

While minimal dreams suggest a thinner notion of minimal phenomenal selfhood than originally proposed by Blanke and Metzinger, this notion is still thicker than the concept of the minimal self often used in the phenomenological literature. There, the minimal or experiential self refers to a kind of first-personal givenness or mineness inherent in all conscious experience.[6] This contrasts with the narrative self, or the sense of being the same person over time (Gallagher 2000; Zahavi 2007, 2010b; Gallagher and Zahavi 2016). Unlike the narrative self, the minimal self cannot be lost or stand apart from the stream of experience, even in principle: an experience that lacked this minimal form of subjectivity would no longer *be* an experience.

Minimal phenomenal selfhood, as I use the term, does not indiscriminately refer to all experiences, but is specifically tied to spatiotemporal self-location. Spatiotemporal self-location characterizes experiences that have a particular perspectival structure: it helps organize experience around an internal, first-person perspective, and the origin of the fist-person perspective is experienced as the self. Unlike minimal subjectivity, this immersive, *here-and-now* structure does not characterize all kinds of phenomenal experience. In particular, the minimal version of the simulation view I defend gives a clear sense to types of experience that lack this structure. Examples are sleep thinking and isolated or static visuospatial or auditory imagery, such as visual imagery of faces arising seemingly out of nowhere but lacking, as is the case for images projected onto a screen, integration into a larger scene. While such experiences can still be perspectival, they are not organized around an *internal* first-person perspective and are not embedded in a larger hallucinatory context. Non-immersive imagery and sleep thinking, common in NREM sleep and at sleep onset, therefore do not fulfill the requirements for counting as even a minimal form of phenomenal selfhood—or of dreaming (Windt et al. 2016).

Another group of dreamless sleep experiences may lack not just the phenomenology of selfhood, but any specific imagery or conscious propositional thoughts (Thompson 2014, 2015; Windt 2015b). Experienced meditators who cultivate attention and meta-awareness sometimes report 'witnessing sleep'; in this state, conscious experience, alongside meta-awareness of the sleep state, is maintained even though any specific thought contents or imagery, including ones pertaining to the phenomenal self, are said to have disappeared (Thompson 2014, 2015). These subjective reports are accompanied by changes in EEG activity as compared to non-meditators (Mason et al. 1997; Ferrarelli et al. 2013; Dentico et al. 2016; Maruthai et al. 2016), including enhanced gamma-band activity—which is also associated with metacognitive insight in lucid dreams (Voss and Hobson 2013). Because such states have phenomenal character—there is something it is like to be in them—they count as minimally subjective. The sense in which they are subjective is, however, purely epistemological, referring to the first-personal mode of givenness of experience and a basic kind of ownership. Still, this does not require additionally experiencing oneself *as* a self. Such states are *phenomenologically* selfless in the sense that they lack the experience of being or having a self. While this may initially sound paradoxical, experiences that are minimally subjective in the epistemological sense can be selfless in the thicker phenomenological sense of lacking a positive representation of self. Distinguishing between such different readings of subjectivity and selfhood might involve looking to states in which they are lost, such as sleep.

The transition between states that are phenomenologically subjective, as is the case for minimal dreams, and those that are merely epistemologically subjective, as in dreamless sleep experience, can be fluid, involving a gradual dissolution of the self (or vice versa). The loss of a sense of self can be coupled with a sense of expansion—as if the phenomenal *here* were gradually expanding to the point at which any distinction from a larger environment is lost. These experiences are often described as having an indeterminate duration, suggesting that there is still some sense of the passage of time.[7] Indeed, it seems plausible that as long as some kind of phenomenal experience persists, there would still be an experienced *now* and at least a basic sense of duration. Such experiences would lack the spatial organization required for the experience of a phenomenal *here* and an internal first-person perspective, but would still have temporal dynamics. Minimal forms of phenomenologically selfless experience may therefore be associated with purely temporal experience (Windt 2015b).

Finally, just as there are two ways in which minimal phenomenal selfhood in dreams can lack bodily experience—phenomenal disembodiment, or the experience of lacking a body, and phenomenal indeterminacy with respect to the body—there may also be two ways in which minimal phenomenal experience can be phenomenologically selfless. One is the experience of lacking a self; another is phenomenal indeterminacy with respect not just to bodily experience, but even to the minimal kind of phenomenal selfhood associated with purely spatiotemporal self-location. There would then no longer be an *experienced* fact of the matter as to whether there was a self, even in a minimal sense. And again, the simplest way of lacking a self may be by way of an unnoticed absence. If this is on the right track, cases in which the experience of minimal phenomenal selfhood is lost entirely or becomes phenomenally indeterminate are perched in between dreamful states involving minimal phenomenal selfhood on the one hand and nonconscious sleep states on the other hand. Investigating these intermediate cases may help identify transitions between dreamful states and dreamless, phenomenologically selfless sleep experience, as well as transitions between dreamless sleep experience and nonconscious sleep states.

6 Conclusions

In this chapter, I defend a minimal version of the simulation view in which dreaming is defined by its immersive, *here-and-now* structure. Even in the simplest kinds of dreams, phenomenal selfhood involves spatiotemporal self-location. Because spatiotemporal self-location is inextricably bound to the perspectival structure of dreaming, it underwrites both the phenomenology of selfhood and the experience of a world, including the experience of a social reality. Coupled with phenomenal indeterminacy and indeterminacy blindness, this account of minimal dreams can be scaled up to offer a parsimonious explanation of richer kinds of dreams, including bodily experiences and self-other distinctions. In the last section, I identified several key points of contact between dream research and interdisciplinary research on consciousness and the self.

I want to end with a speculative observation: the ways in which dreams diverge from standard waking experience—including the characteristic flimsiness of self-experience and the fluidity of self-other distinctions—may in some cases function as a vehicle for the subjective significance and emotional impact of dreams on our waking lives. While only a small subset of dreams is subjectively meaningful—recall that most dreams are never even remembered in the first place—these are nonetheless the dreams that throughout history have fascinated theorists and laypersons alike. So I want to close by giving two examples in which the distinctive phenomenological profile of dreams enables them to reach beyond sleep to touch waking lives.[8]

The first example is from Aiha Zemp. Born without limbs, she experienced vivid phantoms in her lower arms and legs from early childhood. Neuropsychological research indicates that these reports of her phantoms were robust; for example, reports of moving her phantom hands correlated with bilateral activation in the premotor and parietal cortex (Brugger 2012). Aiha Zemp was also a skilled lucid dreamer and described how in her lucid dreams, she could perform different kinds of movements, including flying, dancing, jumping, kneeling, and using her hands (Windt 2015a: 344ff.). In these dreams, she also had tactile sensations in her hands, whereas this was not the case in her nonlucid dreams or in wakefulness. Towards the end of her life, when she was terminally ill, she used her lucid dreams in combination with meditation to experience the dissolution of self. She wrote,

> It [lucid dreaming] means a lot to me. It has really expanded my conscious awareness. In many of my lucid dreams I dissolve, everything dissolves, that is, these dreams are a way for me to practice dying. This makes me very happy.
>
> *(Unpublished interview with Aiha Zemp, conducted by Jennifer Windt and Bigna Lenggenhager; my translation)*

The second example is from John Hull. He gradually lost his eyesight in adulthood, eventually becoming enveloped in complete darkness. In later years, he also lost visual memories and the ability to intentionally conjure visual imagery. But in dreams, he was sometimes still able to see, and memories that had become lost in wakefulness—such as his wife's face—could resurface. In one dream, he had the experience of seeing his baby daughter for the first time:

> I had got out of bed. [...] This toddler came padding in and I could see her quite clearly in the dim light. [...] The first time I had been able to see her. I stared, full of wonder, taking in every detail of her face as she stood there, wreathed in smiles. 'So this is her, this is the smile they all talk about.' I had a wonderful sense of renewal of contact. [...] Then the dream faded.
>
> *(John Hull, quoted in Cole 1998: 30)*

To be sure, such dreams are rare, and they may also, as in the case of Aiha Zemp's lucid dreams, require attention to and interest in one's dreams. They may also occur against a backdrop of specific skills, such as long-term meditation practice. But these examples still show how sometimes, dreams can have a personal and emotional significance that reaches beyond sleep, enabling them to be continuous with our waking projects, interests, and concerns, and form part of our narrative self. To achieve this, dreams need not exactly replicate waking experience, including the phenomenology of (embodied) selfhood. Some dreams may owe their impact, in part, to the ways in which their phenomenological profile departs from waking experience. I think this is both a theoretically significant and a strangely beautiful point.

Notes

1 In speaking of the dream self, I am referring to the character one identifies with in the dream. The dream self should be distinguished from the dreamer, or the person lying asleep in bed. Speaking of the dream self also does not imply the existence of a substantive self or entity in any strong metaphysical sense. The dream self is just shorthand for the pattern of phenomenal experience that underwrites the sense of selfhood and its retrospective description in dream reports.

2 For discussion of how this relates to skepticism about introspection and first-person reports in consciousness research, see Schwitzgebel (2011).

3 Autosensory imagery, in which self-generated stimuli from muscle twitches, limb jerks, or snoring are integrated into dream experience, is frequent during sleep onset. While just a subset of sleep onset experiences qualifies as immersive and hence as dreamful (Windt 2015a: ch. 11), the investigation of so-called microdreams can help isolate core aspects of imagery formation, stimulus incorporation, and temporal dynamics that are crucial to full-fledged dreaming (Nielsen 2017).

4 An intriguing idea is that rare reports in which the dream self is described as *diverging* from the waking self require more complex kinds of self-representation: where a phenomenally indeterminate self would be described, simply, as *me*, in vicarious dreams a richer representation of self may be needed to ground the experience of being someone *other* than one's waking self.

5 Sudden, discontinuous jumps in dream narratives form a well-known subclass of dream bizarreness (Revonsuo and Salmivalli 1995): as in movies or novels, dream narratives can span spatially, but also temporally distant points. Such dreams can involve shifts in the experienced *here* and *now* without involving a shift in self-identification between different dream characters.

6 Zahavi (2010a) suggests that the minimal self is in fact closely bound up with both temporal experience and an embodied first-person perspective. This is closer to the concept of minimal phenomenal selfhood I propose, but does not account for potential dissociations between phenomenal selfhood and embodiment, as in minimal dreams.

7 The interlocked changes in self-experience, time, and space that characterize dreamless sleep experience are also reminiscent of certain psychedelic (Tagliazucchi et al. 2016) and deep meditative states (Berkovich-Ohana et al. 2013; Dor-Ziderman et al. 2013).

8 Appealing to the characteristic phenomenological profile of dreams is only part of the story. The themes and contents of dreams are often continuous with waking events, thoughts, and concerns (Schredl 2006; Domhoff 2013). There are also important phenomenological and neurophysiological similarities between dreams and spontaneous thought in wakefulness, or waking mind wandering (Fox et al. 2013). A full account will therefore also have to explain the narrative structure of dreams as well as the relationship between dream imagery and thoughts and beliefs, both in dreams and wakefulness (Windt 2015a: chs. 9 and 10).

References

Alessandria, M., Vetrugno, R., Cortelli, P., and Montagna, P. (2011) "Normal body scheme and absent phantom limb experience in amputees while dreaming," *Consciousness and Cognition* 20: 1831–1834.

Baldridge, B. J. (1966) "Physical concomitants of dreaming and the effect of stimulation on dreams," *Ohio State Medical Journal* 62: 1272–1275.

Baldridge, B. J., Whitman, R., and Kramer, M. (1965) "The concurrence of fine muscle activity and rapid eye movements during sleep," *Psychosomatic Medicine* 27: 19–26.

Berkovich-Ohana, A., Dor-Ziderman, Y., Glicksohn, J., and Goldstein, A. (2013) "Alterations in the sense of time, space, and body in the mindfulness-trained brain: a neurophenomenologically-guided MEG study," *Frontiers in Psychology* 4: 912.

Blanke, O., and Metzinger, T. (2009) "Full-body illusions and minimal phenomenal selfhood," *Trends in Cognitive Sciences* 13: 7–13.

Blumberg, M. S. (2010) "Beyond dreams: do sleep-related movements contribute to brain development?" *Frontiers in Neurology* 1: 140.

Blumberg, M. S., and Plumeau, A. M. (2016) "A new view of 'dream enactment' in REM sleep behavior disorder," *Sleep Medicine Reviews* 30: 34–42.

Blumberg, M. S., Marques, H. G., and Iida, F. (2013) "Twitching in sensorimotor development from sleeping rats to robots," *Current Biology* 23: R532–R537.

Brugger, P. (2008) "The phantom limb in dreams," *Consciousness and Cognition* 17: 1272–1278.

Brugger, P. (2012). "Phantom limb, phantom body, phantom self: a phenomenology of body hallucinations," in J. D. Blom and I. E. C. Sommer (eds.) *Hallucinations: Research and Practice*. New York: Springer.

Cheyne, J. A., and Girard, T. A. (2007a) "Paranoid delusions and threatening hallucinations: a prospective study of sleep paralysis experiences," *Consciousness and Cognition* 16: 959–974.

Cheyne, J. A., and Girard, T. A. (2007b) "The nature and varieties of felt presence experiences: A reply to Nielsen," *Consciousness and Cognition* 16: 984–991.

Cicogna, P., and Bosinelli, M. (2001) "Consciousness during dreams," *Consciousness and Cognition* 10: 26–41.

Cole, J. (1998) *About Face*, Cambridge, MA: MIT Press.

De Vignemont, F. (2014) "A multimodal conception of bodily awareness," *Mind* 123: 989–1020.

De Vignemont, F. (2016) "Bodily awareness," *The Stanford Encyclopedia of Philosophy* (Summer 2016 Edition), Edward N. Zalta (ed.), https://plato.stanford.edu/archives/sum2016/entries/bodily-awareness/.

Dement, W., and Kleitman, N. (1957) "Cyclic variations in EEG during sleep and their relation to eye movements, body motility, and dreaming," *Electroencephalography and Clinical Neurophysiology* 9: 673–690.

Dement, W., and Wolpert, E. A. (1958) "The relation of eye movements, body motility, and external stimuli to dream content," *Journal of Experimental Psychology* 55: 543–553.

Dennett, D. C. (1976) "Are dreams experiences?" *The Philosophical Review* 85: 151–171.

Dennett, D. C. (1991) *Consciousness Explained*, Boston: Little, Brown and Co.

Dentico, D., Ferrarelli, F., Riedner, B. A., Smith, R., Zennig, C., Lutz, A., Tononi, G., Davidson, R. J. (2016) "Short Meditation Trainings Enhance Non-REM Sleep Low-Frequency Oscillations," *PLoS ONE*, 11(2): e0148961.

Descartes, R. (1641/1901) *Descartes' Meditations* (John Veitch, Trans.). Retrieved from http://www.wright.edu/cola/descartes/mede.html.

Domhoff, G. W. (2013) *Finding Meaning in Dreams: A Quantitative Approach*, Springer Science and Business Media.

Dor-Ziderman, Y., Berkovich-Ohana, A., Glicksohn, J., and Goldstein, A. (2013) "Mindfulness-induced selflessness: a MEG neurophenomenological study," *Frontiers in Human Neuroscience* 7: 582.

Ferrarelli, F., Smith, R., Dentico, D., Riedner, B. A., Zennig, C., Benca, R. M., Lutz, A., Davidson, R. J., Tononi, G. (2013) "Experienced mindfulness meditators exhibit higher parietal-occipital EEG gamma activity during NREM sleep," *PLoS ONE* 8(8): e73417.

Foulkes, D. (2009). *Children's Dreaming and the Development of Consciousness*, Cambridge, MA: Harvard University Press.

Fox, K. C. R., Nijeboer, S., Solomonova, E., Domhoff, G. W., and Christoff, K. (2013) "Dreaming as mind wandering: evidence from functional neuroimaging and first-person content reports," *Frontiers in Human Neuroscience* 7: 412.

Frank, B., and Lorenzoni, E. (1989) "Experiences of phantom limb sensations in dreams," *Psychopathology* 22: 182–187.

Gallagher, S. (2000) "Philosophical conceptions of the self: Implications for cognitive science," *Trends in Cognitive Sciences* 4: 14–21.

Gallagher, S. and Zahavi, D. (2016) "Phenomenological approaches to self-consciousness," *The Stanford Encyclopedia of Philosophy* (Winter 2016 Edition), Edward N. Zalta (ed.), https://plato.stanford.edu/archives/win2016/entries/self-consciousness-phenomenological/.

Hobson, J. A. (1988) *The Dreaming Brain*, New York: Basic Books.

Hobson, J. A. (2009) "REM sleep and dreaming: towards a theory of protoconsciousness," *Nature Reviews Neuroscience* 10: 803–813.

Hobson, J. A., Pace-Schott, E. F., and Stickgold, R. (2000) "Dreaming and the brain: toward a cognitive neuroscience of conscious states," *Behavioral and Brain Sciences* 23: 793–842.

Hoff, H., and Pötzl, O. (1937) "Über die labyrinthären Beziehungen von Flugsensationen und Flugträumen," *European Neurology* 97: 193–211.

Howell, M. J., and Schenck, C. H. (2015) "REM sleep behavior disorder," in A. Videnovic and B. Högl (eds.) *Disorders of Sleep and Circadian Rhythms in Parkinson's Disease*, Wien: Springer.

James, W. (1902/2003) *The Varieties of Religious Experience: A Study in Human Nature*, With a new introduction by Peter J. Gomes. New York: Penguin.

Kahn, D., Pace-Schott, E., and Hobson, J. A. (2002) "Emotion and cognition: feeling and character identification in dreaming," *Consciousness and Cognition* 11: 34–50.

Kahn, D., Stickgold, R., Pace-Schott, E., and Hobson, J. (2000) "Dreaming and waking consciousness: a character recognition study," *Journal of Sleep Research* 9: 317–325.

Kerr, N. H., and Domhoff, G. W. (2004) "Do the blind literally 'see' in their dreams? A critique of a recent claim that they do," *Dreaming* 4: 230–233.

LaBerge, S., and DeGracia, D. J. (2000) "Varieties of lucid dreaming experience," in R. G. Kunzendorf and B. Wallace (eds.) *Individual Differences in Conscious Experience*, Amsterdam: John Benjamins.

Lenggenhager, B., Tadi, T., Metzinger, T., and Blanke, O. (2007) "Video ergo sum: manipulating bodily self-consciousness," *Science* 317: 1096–1099.

Leslie, K., and Ogilvie, R. (1996) "Vestibular dreams: the effect of rocking on dream mentation," *Dreaming* 6: 1–16.

McNamara, P., McLaren, D., Kowalczyk, S., Pace-Schott, E. F., Barrett, D., and McNamara, P. (2007) "'Theory of mind' in REM and NREM dreams," *The New Science of Dreaming, Content Recall, Personality Correlates* 2: 201–220.

McNamara, P., McLaren, D., Smith, D., Brown, A., and Stickgold, R. (2005) "A 'Jekyll and Hyde' within aggressive versus friendly interactions in REM and non-REM dreams," *Psychological Science* 16: 130–136.

Malcolm, N. (1962) *Dreaming*, London: Routledge and Kegan Paul.

Maruthai, N., Nagendra, R. P., Sasidharan, A., Srikumar, S., Datta, K., Uchida, S., and Kutty, B. M. (2016) "Senior Vipassana Meditation practitioners exhibit distinct REM sleep organization from that of novice meditators and healthy controls," *International Review of Psychiatry* 28: 279–287.

Mason, L. I., Alexander, C. N., Travis, F. T., Marsh, G., Orme-Johnson, D., Gackenbach, J., Walton, K. (1997) "Electrophysiological correlates of higher states of consciousness during sleep in long-term practitioners of the Transcendental Meditation program," *Sleep*: 102–110.

Metzinger, T. (2003) *Being No One: The Self-Model Theory of Subjectivity*, Cambridge, MA: MIT Press.

Metzinger, T. (2009) *The Ego Tunnel: The Science of the Soul and the Myth of the Self*, New York: Basic Books.

Metzinger, T. (2013) "Why are dreams interesting for philosophers? The example of minimal phenomenal selfhood, plus an agenda for future research," *Frontiers in Psychology* 4: 746.

Moller, H. J., and Barbera, J. (2006) "Media presence, consciousness and dreaming," in G. Riva, M. T. Anguera, B. K. Wiederhold, and F. Mantovani (eds.), *From Communication to Presence: Cognition, Emotions, and Culture. Towards the Ultimate Communicative Experience; Festschrift in Honor of Luigi Anolli*, Amsterdam: IOS Press.

Mulder, T., Hochstenbach, J., Dijkstra, P. U., and Geertzen, J. H. (2008) "Born to adapt, but not in your dreams," *Consciousness and Cognition* 17: 1266–1271.

Nielsen, T. (2007) "Felt presence: Paranoid delusion or hallucinatory social imagery?" *Consciousness and Cognition* 16: 975–983.

Nielsen, T. (2010) "Dream analysis and classification: The reality simulation perspective," in M. Kryger, T. Roth, and W. C. Dement (eds.), *Principles and Practice of Sleep Medicine* (5th ed.), New York: Elsevier.

Nielsen, T. (2017) "Microdream neurophenomenology," *Neuroscience of Consciousness* 3: 1–17.

Nielsen, T. (1992) "A self-observational study of spontaneous hypnagogic imagery using the upright napping procedure," *Imagination, Cognition and Personality* 11: 353–366.

Nielsen, T. (1993) "Changes in the kinesthetic content of dreams following somatosensory stimulation of leg muscles during REM sleep," *Dreaming* 3: 99–113.

Nielsen, T. (2000) "A review of mentation in REM and NREM sleep: "covert" REM sleep as a possible reconciliation of two opposing models," *Behavioral and Brain Sciences* 23: 851–866.

Nielsen, T., Zadra, A. L., Simard, V., Saucier, S., Stenstrom, P., Smith, C., and Kuiken, D. (2003) "The typical dreams of Canadian university students," *Dreaming* 13: 211–235.

Nielsen, T., Svob, C., and Kuiken, D. (2009) "Dream-enacting behaviors in a normal population," *Sleep* 32: 1629–1636.

Noreika, V., Valli, K., Lahtela, H., and Revonsuo, A. (2009) "Early-night serial awakenings as a new paradigm for studies on NREM dreaming," *International Journal of Psychophysiology* 74: 14–18.

Occhionero, M., Cicogna, P., Natale, V., Esposito, M. J., and Bosinelli, M. (2005) "Representation of self in SWS and REM dreams," *Sleep and Hypnosis* 7: 77–83.

Pace-Schott, E. F. (2009) "Sleep architecture," in R. Stickgold and M. P. Walker (eds.), *The Neuroscience of Sleep*, London: Elsevier Academic Press.

Pagel, J., Blagrove, M., Levin, R., Stickgold, B., and White, S. (2001) "Definitions of dream: a paradigm for comparing field descriptive specific studies of dream," *Dreaming* 11: 195–202.

Rasch, B., and Born, J. (2013) "About sleep's role in memory," *Physiological Reviews* 93: 681–766.

Revonsuo, A. (2005) "The self in dreams," in T. E. Feinberg and J. P. Keenan (eds.), *The Lost Self: Pathologies of the Brain and Identity*, Oxford: Oxford University Press.

Revonsuo, A. (2006) *Inner Presence: Consciousness as a Biological Phenomenon*, Cambridge, MA: MIT Press.

Revonsuo, A., and Salmivalli, C. (1995) "A content analysis of bizarre elements in dreams," *Dreaming* 5: 169–187.

Revonsuo, A., Tuominen, J., and Valli, K. (2015) "The avatars in the machine: dreaming as a simulation of social reality," in T. Metzinger and J. M. Windt (eds.), *Open MIND*: Frankfurt am Main: MIND Group.

Rosen, M., and Sutton, J. (2013) "Self-representation and perspectives in dreams," *Philosophy Compass* 8: 1041–1053.

Sanchez-Vives, M. V., and Slater, M. (2005) "From presence to consciousness through virtual reality," *Nature Reviews Neuroscience* 6: 332–339.

Saurat, M.-T., Agbakou, M., Attigui, P., Golmard, J.-L., and Arnulf, I. (2011) "Walking dreams in congenital and acquired paraplegia," *Consciousness and Cognition* 20: 1425–1432.

Sauvageau, A., Nielsen, T. A., and Montplaisir, J. (1998) "Effects of somatosensory stimulation on dream content in gymnasts and control participants: evidence of vestibulomotor adaptation in REM sleep," *Dreaming* 8: 125–134.

Schönhammer, R. (2005) "'Typical dreams': reflections of arousal," *Journal of Consciousness Studies* 12(4–5): 18–37.

Schredl, M. (2006) "Factors affecting the continuity between waking and dreaming: emotional intensity and emotional tone of the waking-life event," *Sleep and Hypnosis* 8: 1–5.

Schwartz, S. (2000) "A historical loop of one hundred years: similarities between 19th century and contemporary dream research," *Dreaming* 10: 55–66.

Schwitzgebel, E. (2011) *Perplexities of Consciousness*, Cambridge, MA: MIT Press.

Siclari, F., Baird, B., Perogamvros, L., Bernardi, G., LaRocque, J. J., Riedner, B., Boly, M., Postele, B. R., Tononi, G. (2017) "The neural correlates of dreaming," *Nature Neuroscience* : 1–10.

Siclari, F., LaRocque, J. J., Postle, B. R., and Tononi, G. (2013). "Assessing sleep consciousness within subjects using a serial awakening paradigm," *Frontiers in Psychology* 4: 542.

Slater, M. (2009) "Place illusion and plausibility can lead to realistic behaviour in immersive virtual environments," *Philosophical Transactions of the Royal Society B: Biological Sciences* 364: 3549–3557.

Speth, J., Frenzel, C., and Voss, U. (2013) "A differentiating empirical linguistic analysis of dreamer activity in reports of EEG-controlled REM-dreams and hypnagogic hallucinations," *Consciousness and Cognition* 22: 1013–1021.

Stickgold, R., and Walker, M. P. (2013) "Sleep-dependent memory triage: evolving generalization through selective processing," *Nature Neuroscience* 16: 139–145.

Strauch, I., and Meier, B. (1996) *In Search of Dreams: Results of Experimental Dream Research*, New York: State Univ. of New York Press.

Tagliazucchi, E., Roseman, L., Kaelen, M., Orban, C., Muthukumaraswamy, S. D., Murphy, K., Laufs, H., Leech, R., McGonigle, J., Crossley, N., Bullmore, E. (2016) "Increased global functional connectivity correlates with LSD-induced ego dissolution," *Current Biology* 26: 1043–1050.

Thompson, E. (2014) *Waking, Dreaming, Being: Self and Consciousness in Neuroscience, Meditation, and Philosophy*, New York: Columbia University Press.

Thompson, E. (2015) "Dreamless sleep, the embodied mind, and consciousness," in T. Metzinger and J. M. Windt (eds.), *Open MIND*, Frankfurt am Main: MIND Group.

Vetrugno, R., Arnulf, I., and Montagna, P. (2009) "Disappearance of 'phantom limb' and amputated arm usage during dreaming in REM sleep behaviour disorder," *Journal of Neurology, Neurosurgery, and Psychiatry* 79: 481–483.

Voss, U., and Hobson, A. (2015) "What is the state-of-the-art on lucid dreaming?" in T. Metzinger and J. M. Windt (eds.), *Open MIND*, Frankfurt am Main: MIND Group.

Voss, U., Schermelleh-Engel, K., Windt, J. M., Frenzel, C., and Hobson, A. (2013) "Measuring consciousness in dreams: the lucidity and consciousness in dreams scale," *Consciousness and Cognition* 22: 8–21.

Voss, U., Tuin, I., Schermelleh-Engel, K., and Hobson, A. (2011) "Waking and dreaming: related but structurally independent. Dream reports of congenitally paraplegic and deaf-mute persons," *Consciousness and Cognition* 20: 673–687.

Windt, J. M. (2010) "The immersive spatiotemporal hallucination model of dreaming," *Phenomenology and the Cognitive Sciences* 9: 295–316.

Windt, J. M. (2013) "Reporting dream experience: why (not) to be skeptical about dream reports," *Frontiers in Human Neuroscience* 7: 708.

Windt, J. M. (2015a) *Dreaming: A Conceptual Framework for Philosophy of Mind and Empirical Research*, Cambridge, MA: MIT Press.

Windt, J. M. (2015b) "Just in time—Dreamless sleep experience as pure subjective temporality," in T. Metzinger and J. M. Windt (eds.), *Open MIND*, Frankfurt am Main: MIND Group.

Windt, J. M., Nielsen, T., and Thompson, E. (2016) "Does consciousness disappear in dreamless sleep?" *Trends in Cognitive Sciences* 20: 871–882.

Windt, J. M. (2016) "Dreams and dreaming," *The Stanford Encyclopedia of Philosophy* (Winter 2016 Edition), Edward N. Zalta (ed.), https://plato.stanford.edu/archives/win2016/entries/dreams-dreaming/.

Wundt, W. M. (1880) *Grundzüge der physiologischen Psychologie (2nd ed.)*, Leipzig: Engelmann.

Zahavi, D. (2007) "Self and other: the limits of narrative understanding," *Royal Institute of Philosophy Supplement* 60: 179–202.

Zahavi, D. (2010a) "Minimal self and narrative self," in T. Fuchs, H. C. Sattel, and P. Henningsen (eds.) *The Embodied Self: Dimensions, Coherence, and Disorders*, Stuttgart: Schattauer Verlag.

Zahavi, D. (2010b) "Unity of consciousness and the problem of self," in S. Gallagher (ed.) *The Oxford Handbook of the Self*, Oxford: Oxford University Press.

Related Topics

Consciousness, Personal Identity, and Immortality
Biological Naturalism and Biological Realism
Sensorimotor and Enactive Approaches to Consciousness
The Unity of Consciousness
Consciousness and Psychopathology

32

MEDITATION AND CONSCIOUSNESS

Can We Experience Experience as Broken?

Jake H. Davis

1 Introduction

What can practices and theoretical analyses of meditation teach us about consciousness? And what can recent philosophical and psychological investigations of consciousness teach us about meditation?

The terms "meditation" and "consciousness," and related words in other languages, have each been used in many different ways. In order to begin to address the questions posed above effectively and in any depth, it is necessary at least initially to narrow the range of investigation. I will use "consciousness" to refer to "phenomenal consciousness" in the sense that that phrase has figured in debates in recent analytic philosophy (Block 1995; Chalmers 1995; Block 2007). In particular, Block (1995, 2007) has contrasted this notion of phenomenal consciousness, meaning what it is like for a conscious being to have some vivid experience (cf. Nagel 1974), with a different notion of consciousness that refers to the availability of information for recall or report—what he calls "access consciousness" or "cognitive access." With a similar kind of specificity, among the many practices that might be called mystical or meditative, my focus here will be on mindfulness practice in the context of the Theravāda Buddhist tradition of the meditation master Mahasi Sayadaw of Burma.

One might expect that a review of literature on meditation and consciousness would adopt a broader scope, and include a number of different meditative and mystical traditions. There are principled reasons as well as practical ones for my narrow focus here, however. First, while the recent surge of empirical studies of meditation has included a few with direct relevance to philosophical work on consciousness (see e.g., Slagter et al. 2007; van Vugt and Slagter 2016; Manuello et al. 2016), little of this work itself engages directly and substantively with debates in the contemporary literature on consciousness. One notable series of papers by Berkovich-Ohana and collaborators does develop a model of consciousness (Berkovich-Ohana and Glicksohn 2014), apply this to categorize types of meditation (Berkovich-Ohana and Glicksohn 2017), and examine the same type of meditative experience of the cessation of experience that I focus on here (Berkovich-Ohana 2017); yet even here it is not easy to see precisely how this empirical work would help us make progress on the questions posed by analytic philosophers of consciousness. And while there have been a handful of interesting examinations of isolated philosophical issues in the relation of consciousness and meditation (e.g., Dreyfus and Thompson 2007; Davis and Thompson 2013; Thompson 2014; Chadha 2015), there is nothing yet like a developed field in the modern academic literature with sustained debates to be surveyed. There are, of course, sustained debates within the separate theoretical literatures that accompanied meditative practices in various religious contexts. Drawing on these traditions, and putting them into conversation with each other and with contemporary philosophy can, I think, have great benefits for all sides.

This point, however, brings us to a deeper problem with reviewing the literature on meditation, mysticism, and consciousness. The concept of "meditation" does not refer straightforwardly enough to be of use in organizing a field of research, and the concept of "mysticism" is even worse; this is true in the empirical realm (Ospina et al. 2007) as in the philosophical.[1] In practice, the framing of certain categories of experiential states or traits as meditative, mystical, or contemplative often appeals implicitly to the sense of those doing the framing about which sorts of psychological development are ethically desirable. In this way, even with the concept of meditation, which is arguably more specific than that of mysticism, asking about its effects on and relations to consciousness is somewhat analogous to asking about the effects of exercise on health: it all depends on what you are doing. A different analogy, perhaps closer to home in the present context, is that it would do little good to survey debates in the field of consciousness studies where that was understood to include historical consciousness as a topic alongside phenomenal consciousness; there are interesting debates about each of these, but they are not debates about the same thing. The fact that a diverse array of practices such as thoughtful reflection on death, developing focal attention to the point of quieting thought entirely, or developing increased moment-to-moment awareness of all experience including thinking processes are all regarded as "mystical" or "meditative" does nothing to show that their empirical or conceptual relations to consciousness share even a family resemblance.

Employing "meditation" and "mysticism" as organizing concepts, then, has two problematic effects. First, it obscures aspects that *are* philosophically interesting about *specific* ways of being and training one's mind. Secondly, it marks certain ways of being and training one's mind (and not others) as exotic in ways that evidently serve as an implicit justification for their neglect by the mainstream of analytic philosophy. It is in order to counteract both of these trends that, instead of offering a general review, I aim here to offer a concrete demonstration of the philosophical benefit of surveying literature relevant to one specific philosophical proposal about consciousness arising from a specific meditative practice: that we can (and should) experience experience passing away. A more general review, I fear, would fail to make clear how any such proposal arising from meditative practice could really help contemporary philosophy of consciousness make progress on its central questions.

When we look from the perspective of recent analytic philosophy, it may seem to us that the concerns with the workings of consciousness as they are framed in the Buddhist philosophical context are quite specific and idiosyncratic to those historical conversations. What needs to be appreciated is that the concerns with the workings of consciousness as they are framed in recent analytic philosophy will appear equally specific and idiosyncratic to that tradition to someone not immersed in that context. For many Buddhist philosophers, for instance, questions about materialism have not seemed nearly as central or important to the philosophy of mind or of consciousness as they have for contemporary analytic philosophers. Instead, much of the discussion of consciousness in the Pāli texts and in Burmese Buddhist meditation traditions is embedded in and responsive to a framework whose central questions have to do with which states of mind we ought to cultivate and which we ought to train away, a framework that analytic philosophers would recognize as ethical rather than metaphysical. This state of affairs need not put an end to conversation between the two perspectives. Rather, the fact that there are deep philosophical differences between the respective background aims and assumptions of these two traditions is one of the most important reasons to cultivate such a conversation.

Regarding the metaphysics of mind and the ethical question of how to direct our minds, as in ethics more generally, really listening to and engaging respectfully with another, foreign, perspective can help us to see our own more clearly and to improve it (cf. Appiah 2010; Velleman 2015: 99). Such cosmopolitan conversations allow individuals immersed in each tradition to see more clearly where the blind spots of that tradition lie, to see how we might reframe and refine not only the answers we give to philosophical questions but also the questions we ask; and to refine not only the questions we ask, but also the habits of directing attention that give rise to certain sets of questions rather than others. This last point also brings to the fore the metaphilosophical significance of meditation, if we think of meditation as the intentional cultivation of habits of attention, and of habits of attention as giving urgency to certain sets of philosophical questions rather than others.

2 Experiencing Experience Arising and Passing

In order to contribute to the literature a concrete example of how rigorous and non-dismissive philosophical dialogue on meditation and consciousness might go, I have chosen to focus on a specific philosophical proposal. This proposal is drawn from a Burmese Theravāda Buddhist meditation tradition that is one of the most influential in the context of meditation in the modern West, both in its own right and through its impact on secular Mindfulness Based Interventions such as Mindfulness Based Stress Reduction (MBSR). In this Theravāda context it is the term *bhāvanā* that is usually rendered as "meditation." *Bhāvanā* refers to the intentional cultivation of specific types of mental states and character traits, in particular to the cultivation of wholesome qualities such as concentration of attention, goodwill of heart, and clear seeing of the characteristic nature of all experience. It is this last type of cultivation that will be of particular interest here.

Perhaps the most central aim of mindfulness practice, as it is characterized in Theravāda Buddhist mindfulness meditation practices of Burma, is to become vividly aware of the moment to moment changes in subjective experience. This is referred to as the development of insight understanding (*vipassana ñāna*). In beginning stages of insight understanding, one is primarily aware of the moment-to-moment change of the *contents* of phenomenal consciousness, sensations of heat changing to sensations of cool, experiences of hearing being followed by experiences of thinking (perhaps some thought triggered by the sound), and so on. To the degree meditators are paying attention, that is, cultivating mindfulness of these experiences, they are thus able to report on how experience changed; indeed this is how meditation teachers assess the development of students' ability to pay mindful attention. In the terms of the recent analytic debates indicated above, then, experiences of heat, cool, and so on are "access conscious"—they are available for recall, report, and so on (Block 1995)—but they are nonetheless only access conscious in virtue of being phenomenally felt. What is important and efficacious for the aim of mindfulness practice is not a recognition *that* one was previously feeling heat in the body, and is now feeling coolness, or movement, or whatever other sensation. That kind of knowledge would require recall of and comparison with past moments. Instead, one sustains awareness of the texture and phenomenal feel of a sensation such as heat as that sensation fades and another takes its place; one is phenomenally conscious of what it is like *as* that sensation changes to a different one rather than (just) thinking about that change.

More interesting philosophically is the further suggestion by meditation teachers and practitioners that as quietude of mind and discernment deepen over the course of dedicated cultivation of mindfulness, one develops a fine-grained experiential awareness not only of the inconstant, changing nature of the *contents* of phenomenal consciousness, but also the inconstant, changing nature of phenomenal consciousness *itself*, the vehicle of that phenomenal content. This awareness, it is claimed, manifests with all modalities of phenomenal consciousness: seeing, hearing, tasting, smelling, touching, and also experiences of thinking, wondering, remembering and so on.[2] One comes to consciously feel each of these experiences as oscillating and pulsating. As this awareness deepens, one comes to see this oscillating, pulsating, staccato-like nature of each instance of phenomenal consciousness on more and more fine-grained, subtle levels. At the deepest levels, it is claimed, it is possible to be experientially aware of discrete moments of phenomenal consciousness arising and completely passing away (Davis and Vago 2013).

In the course of an exciting recent exploration of Buddhist meditation and the cognitive neuroscience of consciousness, Lutz et al. (2007) mention a Tibetan Buddhist practice similar to the Burmese Mahasi method of attentiveness to conscious experience discussed above. In that Tibetan Buddhist practice of "Open Presence," as Lutz et al. describe it, one aims to attend not to the contents of consciousness but to the "invariant nature of consciousness" itself (2007: 514–515). While similar in this regard to mindfulness of consciousness in the Mahasi tradition as I have described it above, the Tibetan practice of "Open Presence" differs in important ways; most crucial for our purposes is that whereas the Mahasi tradition aims to see consciousness itself arising and also ceasing moment after moment, no such aim is evident in Lutz et al.'s description (2007) of "Open Presence" meditation. The proposal that consciousness is broken into discrete moments is mentioned, as a point of agreement among various traditions of Buddhist theoretical psychology (Abhidharma), in Dreyfus and Thompson's (2007) excellent recent survey of Indian Buddhist approaches to consciousness. As they also note, disagreements among Buddhist traditions on the exact time scale of these temporal units of consciousness suggest that these positions may owe more to theoretical development than to evidence from meditative experience. Nonetheless, the broad Buddhist position that consciousness can be experienced to be arising and passing on a momentary level is directly opposed with claims such as William James' that, "consciousness does not appear to itself chopped up in bits" (James 1981: 233, as quoted in Dreyfus and Thompson 2007: 95).[3]

In this and subsequent sections I will refer to two opposing positions on this issue as Unbrokenism and Brokenism. The former position claims that phenomenal consciousness is unbroken, at least for extended periods, such as while we are awake. Call this Metaphysical Unbrokenism. This has the implication that it is not possible to accurately experience phenomenal consciousness as broken into discrete momentary instances of consciousness. I will distinguish this latter claim as Epistemic Unbrokenism. An opposing set of views is held by certain Buddhist texts and teachers, to the effect that it *is* possible for a human being to accurately experience the arising and passing of phenomenal consciousness, that is, to be phenomenally conscious of phenomenal consciousness as oscillating, pulsating, having a staccato-like nature. Here too, we can separate two claims: first, the claim that phenomenal consciousness is broken in this way, Metaphysical Brokenism; and second, the claim that it is possible—through the attentional training of mindfulness meditation—to accurately perceive the brokenness of consciousness, Epistemic Brokenism. Interestingly, among these four positions, Epistemic Brokenism is the view most directly opposed to Metaphysical Unbrokenism, since the claim that it is possible to accurately experience consciousness as broken implies that consciousness is broken (but the converse does not hold), and the view that consciousness is unbroken implies that it is not possible to accurately experience consciousness as broken (but the converse of this also does not hold).

If true, Epistemic Brokenism offers one of the most promising avenues for experiential evidence from meditation to generate philosophically important questions and to challenge contemporary claims about consciousness. To take one example, Tye (2003: 108) says that "a stream of consciousness is just one temporally extended experience that represents a flow of things in the world. It has no shorter experiences as parts." And again, "with each experienced change in things and qualities, there is an experience of the change. But this does not necessitate that there be a new experience. The simplest hypothesis compatible with what is revealed by introspection is that, for each period of consciousness, there is only a single experience..." (2003: 97). But, if Epistemic Brokenism is true, then Tye's hypothesis—however parsimonious—is not compatible with what is revealed by introspection. Moreover, Tye puts this view together with a transparency thesis to the effect that we cannot introspect experience itself, we can only introspect the properties of the objects experience represents. According to Epistemic Brokenism, however, while one can experience (successive) moments of seeing as themselves oscillating, pulsating, arising and passing in a staccato, discontinuous manner, this experienced discontinuity of consciousness is not experienced as representing any feature of the object being seen.

Of course, there are ample reasons to doubt the reliability of introspective awareness in general. So if the majority of human beings (including Tye, for one) do not feel their experiences of seeing, hearing, and so on as having an oscillating, pulsating, staccato-like nature, when some (self-)selected group of people does claim to experience these commonplace phenomenal experiences in this remarkably different way, perhaps we are justified in eyeing Epistemic Brokenism with suspicion. Interestingly, there is a way in which the general unreliability of introspection may actually help the case for meditators' claims to introspective accuracy. Schwitzgebel's (2011) recent raft of critiques, to take a leading instance, are directed at the "naïve" introspection of "most" people. And he points out (2011: 118) that some Eastern meditative traditions combine an endorsement of this general skepticism about conclusions from untrained introspection with an optimism about properly attentive kinds of introspective awareness. Indeed the Theravāda Buddhist claims for Metaphysical and Epistemic Brokenism employ a tactic closely parallel to Schwitzgebel's arguments from error. He notes that "through more careful and thoughtful introspection, [subjects] seem to discover—I think they really do discover—that visual experience does not consist of a broad, stable field, flush with precise detail, hazy only at the borders. They discover that, instead, the center of clarity is tiny, shifting rapidly around a rather indistinct background. Most of my interlocutors confess to error in having originally thought otherwise" (Schwitzgebel 2011: 126). Similarly, by developing mindfulness, meditators take themselves to discover that their phenomenal experiences do actually have an oscillating, pulsating, staccato-like arising and passing away nature, and take themselves along with most everyone else to have been in error in originally perceiving these experiences as an unbroken flow of experience. That is, they go beyond Epistemic Brokenism as I have defined them above—the claim that it is possible to accurately experience the arising and passing of consciousness—to make the further claim that through such experience one *corrects* the naïve and erroneous view that phenomenal consciousness is continuous.

The early Buddhist discourses describe perversions (*vipallāsa*) of perception (*saññā*), thought (*citta*), and view (*diṭṭhi*) that reinforce one another, and offer mindfulness meditation as a means to counteract these perversions at the foundational level of perception (such as seeing consciousness as continuous) and to come to perceive rightly (such as seeing consciousness as discontinuous).[4] Evan Thompson and I have drawn on such texts and on the fast-growing body of empirical research to offer a two-part model of mindfulness, as involving on the one hand increases in generalized awareness, and on the other decreases in affective biases of attention (especially in Davis and Thompson 2015; see also Davis and Thompson 2013). To bolster claims for the accuracy of experiences of the discontinuity generated through mindfulness practice, one might appeal to results demonstrating that mindfulness practice improves subjects' ability to detect and report on rapidly presented visual stimuli (Slagter et al. 2007), predicts introspective accuracy (Fox et al. 2012), is correlated with more accurate first-person reports about emotional physiological response (Sze et al. 2010), is associated with decreased mind wandering (Brewer et al. 2011), and attenuates affective biases of attention and memory (Roberts-Wolfe et al. 2012; van Vugt et al. 2012).

Nonetheless, one alternative would be to suggest that rather than involving the correction of an error, the process of mindfulness meditation might instead serve to break up a stream of consciousness that was in fact continuous during an earlier period, and accurately perceived to be so. A different proposal from a generally anti-realist approach to consciousness would be to suggest that apart from the fact that things seem a certain way to me, there is no further thing "the seeming" whose continuity or discontinuity we could be correct or incorrect about (cf. Dennett 1991: 364). At its most charitable towards Brokenism, such a (broadly, anti-realist) view might allow that in the earlier period the ways things seemed to me (itself) seemed to me continuous, and grant that in the latter that the ways things seemed to me (itself) seemed to me discontinuous, but then insist that is all there is to say; there is no further question about whether conscious experience actually was continuous or discontinuous. Many, likely most, anti-realist theorists would likely go further and hold that to talk of the way things seem to me as *itself* seeming a certain way (continuous, or instead arising and passing, or whatever else) is to fall into a confusion.

Meditative experience of arising and passing in mindfulness practice might have implications for these debates about the metaphysics of consciousness. For instance, if I take myself to have experienced phenomenal consciousness as oscillating, pulsating, having a staccato-like arising and passing nature, then I will likely be motivated to find a way of talking that makes sense of this possibility. For that reason, one might be compelled to reject any (including anti-realist) accounts that would not make sense of such an experience. Possibly, one might be compelled to go further and endorse the ontological independence of phenomenal consciousness from the experience or introspection of it. Of course, these implications could be resisted, for instance through various strategies of explaining away subjects' sense that they are indeed phenomenally conscious of phenomenal consciousness arising and passing.

3 Non-self and Consciousness

It is often claimed that by mindfully investigating experience and finding no aspect of experience that lasts, meditators come to the realization that there is no lasting self. And this metaphysical conclusion, in turn, is often held to have ethical implications: if suffering is ownerless, then all of it ought to be avoided equally (see e.g., Goodman 2009; Siderits 2003). Yet a number of recent authors have appealed to philosophical considerations about meditation to reject this dominant interpretation.

Miri Albahari, for one, has offered a novel and creative interpretation in which the descriptions found in the early Buddhist Pāli suttas support a view on which the *contents* of consciousness that we identify with *are* impermanent, but the witness consciousness which directly experiences these changing contents is impersonal and ownerless, and also ever-present and unbroken. "When the Pāli sutras speak of consciousness as being impermanent, I take this to mean that the intentional content of consciousness—that to which consciousness is directed—is constantly changing," she writes (Albahari 2011: 95). The proposal that the notion of *viññāṇa* employed in the Pāli suttas amounts to an unbroken witness consciousness would, if correct, reveal an underappreciated convergence with Advaita-Vedanta claims about the understanding that emerges from meditative experience, as Albahari (2002) notes. Secondly, the considerations she raises would move us away from the standard reductionist, "bundle-theory" interpretation of non-self (*anattā*) in the Pāli Buddhist texts as a metaphysical denial of the self, and towards an understanding of the claims for *anattā* as a practical strategy for reducing identification with the contents of consciousness.

Albahari notes that on standard Theravāda Buddhist interpretations of the Pāli suttas, such as that of the meditation master Mahasi Sayadaw mentioned above, meditators must be experientially aware of the arising and passing nature of consciousness itself (Albahari 2011: 94–95). Indeed, it is by seeing even consciousness itself arising and passing that meditators are said to arrive at the conclusion that there is no self. Yet, interestingly, she charges this Brokenist proposal with incoherence. As she puts the point in an earlier manuscript (Albahari 2006: 45), if the discerning mind were impermanent, "such a mind, in order to directly experience (and hence know) its own impermanence, would have to be percipient of its own fleeting nature. That means it would have to be present while it directly discerns its own fleeting moments of absence (as well as presence). But then if present to its own absence, it cannot actually be *absent* during those moments; we arrive at a contradiction."

Moreover, Albahari contends that even if we were to allow numerically distinct moments of consciousness, there would be no phenomenological way to discern between the condition in which consciousness is unbroken and that in which it is broken.

> the observational component, which renders each moment of non-reflexive consciousness to be *conscious*, is qualitatively invariant, leaving no marker by which the contiguous numerical transition could be experientially discerned (it's not as if there will be a little jolt at each transition). The observational component to each conscious moment will thus seem, from the first person experiential perspective, to be unbroken—regardless of the underlying ontology.
>
> *(Albahari 2011: 97)*

If Albahari is right about this, Epistemic Brokenism fails even if we grant Metaphysical Brokenism.

Traditional Theravāda Buddhist proponents of Metaphysical and Epistemic Brokenism have some responses in store, however. First, they suggest in effect that—contra Albahari—there *is* "a little jolt" that will be experienced as advanced meditators become aware of the passing of one moment of consciousness, and the arising of another. This is precisely because, it is said, one directly experiences the cessation of (a momentary instance of) consciousness. This is experienced as a type of cessation that has been happening all along, in each moment, rather than something newly brought about by the meditative observation. And this is why, although consciousness is qualitatively invariant, one no longer regards it as an unbroken witnessing self, even an impersonal one. It is because we are not normally phenomenologically conscious of this cessation that we can correct our views by experiencing it through the training of mindful attention to more and more precise awareness.

Secondly, the classical commentarial manual *Visuddhimagga* (Path of Purification) suggests that after comprehending the impermanent (*anicca*), uneasy (*dukkha*), non-self (*anattā*) nature of physical phenomena (such as heat and cool in the body), the meditator comprehends that consciousness too (the one that had been contemplating the physical phenomena) as itself impermanent, uneasy, and non-self, *by means of a subsequent consciousness*.[5] Moreover, one can comprehend this second consciousness itself as impermanent, uneasy, and non-self by means of a third instance of consciousness, the third by means of a fourth, and so on.[6] So an initial response to Albahari is simply to question her use of the term 'it' to subsume multiple numerically distinct moments of consciousness into a single substance 'mind', thereby generating the paradox of the mind being aware of *its* own passing. Once we distinguish preceding and succeeding instances of consciousness, there is no incoherence in supposing that the latter can take as its object the absence of the preceding moment of consciousness.

It might be more problematic to insist that the subsequent consciousness comprehends not only the absence of the former, but also the process of its *passing*, or to say that the subsequent consciousness directly experiences the qualities of the preceding one—for instance experiences the preceding instance of consciousness *as* impermanent, uneasy, and non-self. To feel the 'jolt' of the cessation of one instance of consciousness or the arising of another would seem to require one instance of consciousness being aware of another, with both existing in some sense at the same time. This would be problematic if we were also to take on a strong form of another commitment that these traditional authors do subscribe to, namely the idea that there can only be one distinct instance of consciousness at a time. However, this particular notion of momentariness is not explicit in the Pāli suttas to which Albahari refers, and it is not clear philosophically that the account ought to be committed to such a principle. This commitment is present in the later commentaries on these early suttas, and Mahasi Sayadaw (2016: 364–365) does follow the commentarial commitment to this principle. He also follows the *Visuddhimagga* in suggesting that one apprehends a moment of consciousness by means of a subsequent one. He notes that the paradox generated by these two commitments (roughly the one Albahari raises) is also mentioned in the commentaries as a topic of debate. But Mahasi suggests that this paradox can be resolved by a further suggestion he finds in the commentaries to the effect that the experience of the immediately preceding moment of consciousness remains vivid enough to be the target of the present moment of consciousness. So there are multiple avenues open that each would resolve the worry Albahari raises. If we understand mindfulness practice as involving an impermanent instance of consciousness taking as its object another impermanent instance of consciousness—either a concurrent instance in the process of ceasing, or else (as the Mahasi tradition suggests) the still vivid experience of the moment of consciousness that has just ceased—then the contradiction Albahari points out does not arise.

For the reasons given above, I think that neither practical experience in mindfulness meditation nor textual evidence from Pāli suttas offer us reason to take phenomenal consciousness as unbroken—both suggest, on the contrary, that we ought to take phenomenal consciousness as discontinuous and thus impermanent.[7] For that reason the convergence Albahari sees between Buddhist and Advaita-Vedanta accounts is, I think, illusory. Nonetheless, these points should not distract us from what is right about Albahari's overall approach to the doctrine of non-self (*anattā*). Albahari and I share an aim of respecting and incorporating the epistemic value on direct experience over mere reasoning, a value found in various Buddhist practice traditions. Indeed, my overall strategy, like Albahari's, is to appeal to considerations from meditative practice and from the early Pāli suttas to show what is wrong with the kind of abstract, metaphysical approaches to the doctrine of non-self (*anattā*) arguably adopted later by Theravāda Buddhist commentators, and more explicitly by recent analytic philosophers. In particular, I have agreed with Albahari that in the context of the Pāli suttas, *anattā* is better understood as a practical strategy for not taking experience personally than as a reductionist metaphysical claim about persons. The gist of my argument (see Davis 2016) is that the *anattā* doctrine amounts to the claim that every aspect of experience can be seen to be impersonal and out of our control. Crucially, this is a perspective we can—and can only—take up from within our own subjective perspective.

In order to establish the further, metaphysical claim for reductionism or eliminitavism about persons, later Buddhist interpreters (e.g., Nāgasena in the *Milindapañha*) and recent analytical philosophers (e.g., Parfit 1984) take up a perspective on persons from the outside. It is only from that sort of a perspective that we can regard pleasure and pain, perceptions and consciousness as objects in the world that could make up persons. But this is not the perspective that is cultivated by mindfulness meditation, as it is described in the Pāli discourses or as it is taught by contemporary practitioners such as the Mahasi Sayadaw. Rather than abstracting away from one's individual perspective, I take the Pāli discourses and the Mahasi Sayadaw to be encouraging each of us instead to inhabit *more fully* our own subjective experience of the world (Davis 2016). My point is not that one *cannot* adopt a third person perspective on consciousness as such. Rather, I want to argue these traditions are suggesting that we *should not*, at least for the purposes of understanding non-self. Instead, we are to understand the *anattā* doctrine as a claim that by inhabiting our experience more fully we come to see each aspect of experience—including consciousness itself—as transitory, uneasy, and impersonal.

One might adopt the further premise that that is all there is to a person, metaphysically, and thereby conclude that ultimately there are no persons; as I understand the doctrine of *anattā* in the Pāli discourses, however, no stand is taken on either this further premise or this further conclusion. Rather, seeing each aspect of experience—including consciousness itself—as transitory, uneasy, and impersonal is all that there is to be done, ethically. The project of the Buddha in the Pāli discourses requires no more than this—and also no less.[8]

4 On the Very Idea of Experiencing Arising and Passing

Central to my discussions above has been the possibility that one can be phenomenally conscious of phenomenal consciousness arising and passing. I have noted that the Mahasi tradition of mindfulness practice takes its goal to be experiencing the momentary cessation of phenomenal consciousness, along with its contents. And I have detailed how the claims made in this context for the possibility of such experiences—what I have called Epistemic Brokenism—has a number of philosophically important implications for the metaphysics, ethics, and epistemology of consciousness.

However, as I have noted in passing, many anti-realist approaches to phenomenal consciousness emphasize that while there are ways things seem to us, the idea that there are thus *seemings* is a mistake. On such a view, it would seem to make little sense to speak of phenomenal consciousness as a thing of the sort that we could be phenomenally conscious of arising and passing. Perhaps the most interesting, rigorous, and sustained critique of this kind in the context of Buddhist meditation has been by Robert Sharf. In earlier work, Sharf (1995, 2000) charged that modern presentations that cast mindfulness as a type of bare attention leading to discrete, replicable, experiential realizations—he notes in particular the experience of cessation claimed in the Mahasi tradition—are problematic on a number of historical, sociological, and philosophical levels. Sharf claims that the emphasis given by Mahasi Sayadaw and others to rapid progress through meditative experiences, without study of Buddhist theory or deep concentrative practice, is a novel innovation not evidenced in premodern Asia. However, this point rests on an equivocation about historical periods. Even if such an emphasis on meditative experience over theoretical study is not attested to in the centuries immediately predating the modern meditation movement, this possibility is attested to in the early Buddhist texts within the Pāli Nikāyas and Chinese Āgamas not only in theoretical discussions of insight without deep concentration but also in stories of individuals attaining the goal rapidly and without theoretical study.[9] Secondly, Sharf notes sociological evidence that there are debates within and between traditions over whose experiences of cessation are the 'real' ones. This, I think, does present a more serious difficulty. Yet in Sharf's argument this point serves merely as circumstantial evidence for what is really a philosophical conclusion, that while different meditators may take themselves to be referring to the same discrete experience as each other when they make these claims, these claims may in fact be operating not referentially but instead performatively, in the service of legitimizing particular authority structures.

Sharf ties his critique of such claims for discrete shared meditative experiences—such as the experience of experience ceasing—to a more general philosophical critique, that the notion of bare attention as accessing conscious experience independent of conception and judgment requires the type of problematic picture of the mind that Dennett (1991) calls the "Cartesian theater" and Rorty (1979) calls "the mirror of nature." Noting these philosophical inspirations, Sharf gives a nod as well to Nietzsche, Heidegger, Wittgenstein, Sellars, and Derrida. In recent work Sharf also locates a line of critique within early Chinese Buddhist tradition that is closely aligned with his own. Describing the general position of the "subitists" in early Chinese Buddhism, who argued that enlightenment is sudden, Sharf (2014: 951–952) writes that these thinkers "reject any articulation of the path and any form of practice that takes the terms 'mind' and 'mindfulness' as referencing discrete and determinable states or objects or meditative experiences. For the Chan subitists, like the modern antifoundationalists, the image of the mind as mirror epitomizes a widespread but ultimately wrongheaded understanding of mind, cognition, and our relationship to the world." The metaphysical implication in Sharf's version of this critique is that while modern meditators might take their experiences—such as of experience arising and passing—to be presenting the way consciousness really is, independent of any socially conditioned theoretical framework, is a misconception.

Here again, I think the Mahasi tradition has resources to respond with. First, the evidence for the Mahasi tradition being committed to the mirror analogy, as Sharf conceives of it, is weak.[10] Nonetheless, it is plausible that the Mahasi tradition and the Theravāda more generally are committed to a philosophical distinction that Sharf would reject, between phenomenal consciousness itself and conceptualizations through which it is interpreted. I have elsewhere raised the possibility that the distinction between *viññāṇa* and *saññā* in these Buddhist contexts might map closely the distinction that analytic philosophers such as Block (1995, 2007) draw between phenomenal consciousness and cognitive access. Mahasi himself was obviously committed to some distinction between *viññāṇa* and *saññā*. However, it is less clear that his characterization of mindfulness was predicated on this. On the other hand it might be that in their characterization of bare awareness, modernist interpreters of the Mahasi tradition do assume some distinction along the lines of Block's phenomenal versus cognitive. If so, and if Block's opponents were to establish that this distinction is a mistake, that might count as well against such modernist presentations of mindfulness. However, the debate between realism and anti-realism about phenomenal consciousness is very much a live controversy, and if modernist interpretations of mindfulness cast their lot with Block, it is hardly clear that they have chosen the losing side. Secondly, although on an anti-realist view it would make little sense to speak of consciousness as a thing that can arise and pass away, this implication may be turned against the anti-realist. Thus one might suggest, in light of meditative experience, that since we evidently *can* be phenomenally conscious of phenomenal consciousness ceasing, this serves as evidence against anti-realist views of consciousness, if they cannot make sense of this possibility.

5 Conclusion

This chapter has aimed to demonstrate by example the value of bringing into conversation different traditions of investigating consciousness. I have focused especially on the philosophical interest of one claim found in the Mahasi tradition, among others, the proposal that it is possible for a human being to be phenomenally conscious of phenomenal consciousness as broken, arising and passing on a momentary level. Even in this area of focus, many, many questions remain. I do not hope to have demonstrated that the Mahasi tradition is correct in making this claim, much less that the questions raised in such an attempt are easy ones. On the contrary, my aim has been to show that the attempt to make sense of this aspect of meditative experience forces us to confront deep and difficult philosophical questions, and is capable of bringing fresh perspectives to bear on contemporary philosophical debates about the nature of consciousness. Indeed, I see the engagement between Buddhist meditative traditions and contemporary debates in academic philosophy, if it is done with mutual respect and with (what may amount to the same) a mutual willingness to question foundational assumptions, as capable of bringing immense benefits for both.

Notes

1 Indeed, perhaps the single most relevant, sustained debate in the existing literature is on the viability of mystical experience as an organizing concept; as detailed below, Sharf (2000) not only raises skeptical worries of this sort in regard to mysticism and meditation in general, he also expresses skeptical doubts as to whether there are discrete meditative experiences shared even among meditators in a single tradition.

2 I intend this characterization to be entirely neutral on the question of whether there is *distinctive* cognitive phenomenology.

3 In another recent article on "meditation and unity of consciousness" Chadha (2015) primarily discusses synchronic unity of consciousness, rather than diachronic unity as I do here, and (perhaps for this reason) draws mainly on the Yogācāra Buddhist tradition. For these reasons, the question of meditative experience of the cessation of consciousness, and the philosophical implications of this diachronic disunity, are not a focus of her discussion.

4 See e.g. *Vipallāsa Sutta*, Aṅguttara Nikāya II 52, in the Pali Text Society edition.

5 *Visuddimagga* XX 79.

6 *Visuddimagga* XX 80–81.

7 A note on exegetical approach: The Pali texts contain passages that leave room for multiple interpretations. Arguably, this leaves room for tying the nature of (some aspects of) *viññāṇa* closely to the nature of *nibbāna* along the general lines that Albahari suggests, and from which she moves to the conclusion that *viññāṇa*, like *nibbāna*, is unconditioned, and therefore not impermanent. Nonetheless, there is at least as good textual and philosophical reason to move in the opposite direction, from the premise that *viññāṇa* is conditioned and impermanent, to the conclusion that it cannot be equated with *nibbāna* in the ways Albahari suggests.

8 Indeed, the path of practice outlined in the Theravāda requires the cultivation of a certain kind of broken heartedness that arises through seeing Metaphysical Brokenness. In technical terms, this is the disenchantment (*nibbidhā*) that arises through being phenomenally conscious of the arising and passing of every aspect of experience—including phenomenal consciousness itself. By seeing each of these aspects of as arising and passing and out of our control, the tradition maintains, one abandons the implicit and misguided hope that any aspect of experience will give us lasting pleasure, and thereby finds a relief and freedom unavailable through pursuit of any kind of *content* of experience. The tradition thus takes a position in value theory that builds on and moves beyond Metaphysical Brokenism (the claim that all aspects of experience, including consciousness itself, are rapidly arising and passing) and further, beyond Epistemic Brokenism (the claim that it is possible to accurately experience this broken nature of consciousness and all other aspects of experience), to an ethical claim that we might call Heart Brokenism: the claim that we *ought* to experience the arising and passing away of consciousness itself because of the emotional release that that brings from the psychological causes of suffering.

9 See for instance the story of Bāhiya Daruciriyo (at Udāna 6ff., in the Pali Text Society edition), among others.

10 Sharf (2014: 951) notes versions of what he takes to be the mirror analogy as illustrating "the essential and unchanging nature of mind on the one hand and the transient, ephemeral, and ultimately unreal nature of what appears in the mind on the other. The reflections that appear on the surface of the mirror, whether beautiful or ugly, defiled or pure, leave the mirror's true nature unsullied." The Mahasi tradition takes the opposite position on both of these points, however. The mind is neither originally pure nor unchanging. As we have discussed in detail above, the Mahasi claim for Epistemic Brokenism, to the effect that we can see all aspects of mind, including consciousness itself, passing away moment after moment, is precisely to be free of the mistaken perception of consciousness as unbroken.

References

Albahari, M. (2002) "Against No-ātman Theories of Anatta," *Asian Philosophy* 12: 5–20.

Albahari, M. (2006) *Analytical Buddhism: The Two-Tiered Illusion of Self*, New York: Palgrave Macmillan.

Albahari, M. (2011) "Nirvana and Ownerless Consciousness" in M. Siderits, E. Thompson, and D. Zahavi (eds.) *Self, No Self? Perspectives from Analytical, Phenomenological and Indian Traditions*, Oxford: Oxford University Press.

Appiah, K. A. (2010) *Cosmopolitanism: Ethics in a World of Strangers* (issues of our time), New York: W. W. Norton and Company.

Berkovich-Ohana, A. (2017) "A Case Study of a Meditation-Induced Altered State: Increased Overall Gamma Synchronization," *Phenomenology and the Cognitive Sciences* 16: 91-106.

Berkovich-Ohana, A. and Glicksohn, J. (2014) "The Consciousness State Space (CSS)–A Unifying Model for Consciousness and Self," *Frontiers in Psychology* 5: 341.

Berkovich-Ohana, A. and Glicksohn, J. (2017) "Meditation, Absorption, Transcendent Experience, and Affect: Tying It All Together via the Consciousness State Space (CSS) Model," *Mindfulness* 8: 68–77.

Block, N. (1995) "On a Confusion about a Function of Consciousness," *Behavioral and Brain Sciences* 18: 227–247.

Block, N. (2007) "Consciousness, Accessibility, and the Mesh between Psychology and Neuroscience," *Behavioral and Brain Sciences* 30: 481–548.

Brewer, J. A., Worhunsky, P. D., Gray, J. R., Tang, Y.-Y., Weber, J., and Kober, H. (2011) "Meditation Experience Is Associated with Differences in Default Mode Network Activity and Connectivity," *Proceedings of the National Academy of Sciences U.S.A.* 108: 20254.

Chadha, M. (2015) "Meditation and Unity of Consciousness: A Perspective from Buddhist Epistemology," *Phenomenology and Cognitive Sciences* 14: 111–127.

Chalmers, D. J. (1995) "Facing Up to the Problem of Consciousness," *Journal of Consciousness Studies* 2: 200–219.

Davis, J. H. (2016) "The Scope for Wisdom: Early Buddhism on Reasons and Persons," in S. Ranganathan (ed.) *The Bloomsbury Research Handbook of Indian Ethics*, New York: Bloomsbury Academic.

Davis, J. H., and Thompson, E. (2013) "From the Five Aggregates to Phenomenal Consciousness: Towards Cross-Cultural Cognitive Science," in S. Emmanuel (ed.) *A Companion to Buddhist Philosophy*, Malden, MA: Wiley-Blackwell.

Davis, J. H., and Thompson, E. (2015) "Developing Attention and Decreasing Affective Bias," in K. Brown, J. Creswell, and R. Ryan (eds.) *Handbook of Mindfulness: Theory, Research, and Practice*, New York: Guilford Press.

Davis, J. H., and Vago, D. R. (2013) "Can Enlightenment Be Traced to Specific Neural Correlates, Cognition, or Behavior? No, and (a qualified) Yes," *Frontiers in Psychology* 4: 870.

Dennett, D. (1991) *Consciousness Explained*, Boston: Little, Brown and Company.

Dreyfus, G. and Thompson, E. (2007) "Philosophical Theories of Consciousness: Asian Perspectives," in P. Zelazo, M. Moscovitch, and E. Thompson (eds.) *The Cambridge Handbook of Consciousness*, Cambridge, UK: Cambridge University Press.

Fox, K., Zakarauskas, P., Dixon, M., Ellamil, M., Thompson, E., and Christoff, K. (2012) "Meditation Experience Predicts Introspective Accuracy," *PLoS ONE* 7: e45370.

Goodman, C. (2009) *Consequences of Compassion: An Interpretation and Defense of Buddhist Ethics*, New York: Oxford University Press.

Lutz, A., Dunne, J. D., and Davidson, R. J. (2007) "Meditation and the Neuroscience of Consciousness," in P. Zelazo, M. Moscovitch, and E. Thompson (eds.) *The Cambridge Handbook of Consciousness*, New York: Cambridge University Press.

Manuello, J., Vercelli, U., Nani, A., Costa, T., and Cauda, F. (2016) "Mindfulness Meditation and Consciousness: An Integrative Neuroscientific Perspective," *Consciousness and Cognition* 40: 67–78.

Nagel, T. (1974) "What Is It Like to Be a Bat?" *The Philosophical Review* 83: 435–450.

Ospina, M., Bond, K., Karkhaneh, M., Tjosvold, L., Vandermeer, B., Liang, Y., Bialy, L., Hooton, N., Buscemi, N., Dryden, D., Klassen, T. (2007) "Meditation Practices for Health: State of the Research," *Evidence Reports/Technology Assessments* 155: 1–263.

Parfit, D. (1984) *Reasons and Persons*, New York: Oxford University Press.

Roberts-Wolfe, D., Sacchet, M., Hastings, E., Roth, H., and Britton, W. (2012) "Mindfulness Training Alters Emotional Memory Recall Compared to Active Controls: Support for an Emotional Information Processing Model of Mindfulness," *Frontiers in Human Neuroscience* 6: 15.

Rorty, R. (1979) *Philosophy and the Mirror of Nature*, Princeton, NJ: Princeton University Press.
Sayadaw, M. (2016) *Manual of Insight*, Cambridge, MA: Wisdom Publications.
Schwitzgebel, E. (2011) *Perplexities of Consciousness*, Cambridge, MA: MIT Press.
Sharf, R. (1995) "Buddhist Modernism and the Rhetoric of Meditative Experience," *Numen* 42: 228–283.
Sharf, R. (2000) "The Rhetoric of Experience and the Study of Religion," *Journal of Consciousness Studies* 7: 267–287.
Sharf, R. (2014) "Mindfulness and Mindlessness in Early Chan," *Philosophy East and West* 64: 933–964.
Siderits, M. (2003) *Personal Identity and Buddhist Philosophy: Empty Persons*, Burlington, VT: Ashgate Publishing.
Slagter, H. A., Lutz, A., Greischar, L. L., Francis, A. D., Nieuwenhuis, S., Davis, J. M., and Davidson, R. J. (2007) "Mental Training Affects Distribution of Limited Brain Resources," *PLoS Biol* 5 (6): e138.
Sze, J. A., Gyurak, A., Yuan, J. W., and Levenson, R. W. (2010) "Coherence between Emotional Experience and Physiology: Does Body Awareness Training Have an Impact?" *Emotion* 10: 803–814.
Thompson, E. (2014) *Waking, Dreaming, Being: Self and Consciousness in Neuroscience, Meditation, and Philosophy*, New York: Columbia University Press.
Tye, M. (2003) *Consciousness and Persons: Unity and Identity*, Cambridge, MA: MIT Press.
Van Vugt, M. K, and Slagter, H. (2016) "Control over Experience? Magnitude of the Attentional Blink Depends on Meditative State," *Consciousness and Cognition* 40: 67–78.
Van Vugt, M. K., Hitchcock, P., Shahar, B., and Britton, W. (2012) "The Effects of Mindfulness-Based Cognitive Therapy on Affective Memory Recall Dynamics in Depression: A Mechanistic Model of Rumination," *Frontiers in Human Neuroscience* 6.
Velleman, D. (2015) *Foundations for Moral Relativism: Second Expanded Edition*, Cambridge, UK: Open Book Publishers.

Related Topics

Consciousness and the Mind-Body Problem in Indian Philosophy
The Neural Correlates of Consciousness
Consciousness, Time, and Memory
The Unity of Consciousness
Consciousness and Attention

33
CONSCIOUSNESS AND END OF LIFE ETHICAL ISSUES

Adina L. Roskies[1]

We can all appreciate the particular quality of the searing pain of touching a hot stove or the mouth-watering aroma of freshly-baked cookies. The capacity to experience these and other sensations, to react to them, and possibly to report on them is part of what it is to be conscious beings, beings for which it is like something to be. Against the background of normal states of consciousness, we can identify disorders of consciousness, perturbations of this awareness of self and environment that affect people as a consequence of traumatic and non-traumatic brain injury. A variety of global disorders of consciousness have been identified, including coma, persistent vegetative state (PVS), and minimally conscious state (MCS). Estimating the numbers of patients affected by these disorders is difficult, both because of difficulties in delineating and diagnosing them, and because of a lack of a formal reporting structure. However, a 2005 estimate of patients in PVS in the US ranged from 40–168 per million (Beaumont and Kenealy 2005), while another earlier estimate for PVS in the US was 14,000–35,000 (The Multi-Society Task Force on PVS, 1994). MCS prevalence was estimated to be between 45,000 and 250,000 (Fins et al. 2008). Regardless of exact numbers, it is clear that disorders of consciousness affect a great many people, few of whom are likely to regain normal consciousness. This chapter addresses the ethical issues raised by these cases.

1 Types of Disorders of Consciousness

Perhaps the most widely recognized disorder of consciousness is the coma, a state that occurs subsequent to brain injury. Comatose patients exhibit no evidence of wakefulness or arousal, no evidence of awareness, and no communication (Owen and Coleman 2008). They appear to be asleep, with eyes closed, but their brainwaves belie that interpretation, exhibiting no signs of normal sleep–wake patterns. Patients in comas sometimes transition to other states, occasionally regaining normal consciousness, but often transitioning to other recognized disorders of consciousness, such as PVS or MCS. A patient in PVS may open his or her eyes and appear to be awake, but nonetheless shows no evidence of awareness of self or environment, and is unable to communicate or to respond all but reflexively to stimuli. The vegetative state thus is a state of unconsciousness. PVS patients sometimes transition to a minimally conscious state. MCS differs from a vegetative state in that patients show some, albeit intermittent, signs of conscious mental activity. Patients may exhibit occasional visual tracking of stimuli, or they may respond to people around them with gestures or words (Kahane and Savulescu 2009: 6). Patients in minimally conscious states are more likely to regain normal consciousness, and the fact that they show some evidence of consciousness suggests to many that they should be treated differently than vegetative patients.

Because disorders of consciousness are typically diagnosed clinically on the basis of *lack* of certain types of behavior (Schnackers et al. 2009; Di Perri et al. 2014: 29), it is extremely important to distinguish PVS from another condition that manifests as virtually indistinguishable, but that does not involve a disorder of consciousness. Locked-in syndrome (LIS), a state of global paralysis, may be mistaken for PVS, for patients with LIS cannot respond behaviorally to stimuli except in minute and subtle ways. LIS results from systemic injury to voluntary motor neurons, either from damage to brainstem structures or by demyelination, as in amyotrophic lateral sclerosis (ALS) (Patterson and Grabois 1986: 760; Smith and Delargy 2005: 407). Patients who have locked-in syndrome experience sleep-wake cycles, as do PVS patients, but in contrast to PVS patients they are fully conscious and mentally competent. However, due to their motor dysfunction they are completely or almost completely paralyzed. Some, but not all, of them can voluntarily move only their eyes, and thus can communicate only with eye movements (Bauer et al. 1979; Owen and Coleman 2008: 236). Locked-in syndrome is thus not a disorder of consciousness at all, but rather a physical disorder that masquerades as a disorder of consciousness.

2 Theories of Consciousness

One of the difficulties facing bioethicists interested in addressing the relationship between consciousness and end of life issues lies in identifying the type of phenomenon that consciousness is. Different fields have different theories or frameworks for identifying consciousness, and they are often incommensurable. In addition, because the underlying theory of the phenomenon may have bearing on its moral significance, it may not be possible for theorists to remain neutral about committing to a particular theory. Although other chapters in this volume go into greater depth regarding theories of consciousness, a brief survey of some of the major theoretical approaches is necessary here as background to the ethical discussion.

Medical Distinctions

The medical community typically distinguishes between *wakefulness* and *awareness*. Wakefulness is produced by the activation and regulation of neural pathways in the brainstem, known as the *ascending reticular activating system* (Di Perri et al. 2014: 29). Mere wakefulness does not imply consciousness. Awareness, in contrast, is anatomically associated with regions in the frontoparietal cortex, and entails subjective first-person experience. In general, wakefulness precipitates awareness, but there are instances when the two can become dissociated. For example, in REM sleep one can be unawake, yet aware (one experiences one's dreams). In certain pathological states, such as the ones we discuss here, one can be awake, yet seem unaware (Di Perri et al. 2014: 28). In general, the medical term awareness maps onto what we refer to here as consciousness.

Philosophical and Scientific Distinctions

Philosophers and scientists have elaborated more fine-grained concepts of consciousness. No single theory is generally accepted, let alone completely explanatory, but some distinctions have gained widespread acceptance. Because it is possible that different types of consciousness should be accorded different levels of moral significance, careful fractionation of these concepts is an important precursor to a discussion of the ethical import of disorders of consciousness.

Perhaps the most influential taxonomy of consciousness distinguishes between "access consciousness" and "phenomenal consciousness" (Block 1995: 230–232). *Access consciousness* characterizes the availability of information in the brain that makes possible intelligent behaviors such as reasoning and executive function; it is an information-processing notion of consciousness. Contents of access consciousness are widely available and can be utilized in controlling actions or speech, and thus are "reportable." Because it is an informational construct with behavioral implications, access consciousness (or the contents thereof) is amenable to scientific study. Although the access/phenomenal distinction is a philosophical one, similar kinds of theoretical constructs to access consciousness are evident in the psychological literature (e.g., Baars 1988). Because access consciousness is the kind of consciousness that makes possible complex and goal-directed intelligent behaviors informational and measurable, Chalmers (1995: 201) has labeled the explanation of this form of consciousness as the "easy problem" of consciousness.

In contrast, *phenomenal consciousness*, or the "what-it-is-likeness" of subjective experience, gives rise to the "hard problem" (Chalmers 1995). Examples of phenomenal experiences include sensations, feelings, thoughts, emotions, or perceptions (Block 1995: 230). Nagel famously argued that because phenomenality is essentially subjective, we cannot know the phenomenal experience of an entity unlike ourselves (Nagel 1974). Various theories attempt to explain phenomenal consciousness, or at least to identify the nature of phenomenal content. *Intentionalism* or *representationalism* holds that the representational content of a subject's mental state determines the phenomenal character of the experience. *Phenomenalism,* on the other hand, rejects the idea that phenomenal character supervenes on representational content (Byrne 2001: 205), implying that there is something further to be said about the nature of phenomenal content.

Block contends that one can have access consciousness without phenomenal consciousness and phenomenal consciousness without access consciousness. Chalmers' thought experiment pointing to the conceivability of philosophical zombies, beings that are behaviorally indistinguishable from normal humans yet lack subjective states, is pointed to as an argument for the former type of dissociation (Block 1995: 233). However, we need not look to far-fetched conceivability arguments to grasp intuitively this dissociation. As artificial intelligence becomes more and more powerful, we can imagine machines that have sophisticated cognitive abilities, presumably fulfilling the informational demands of access consciousness, in the (supposed) absence of phenomenal experience. Block's other claim, that it is possible to have phenomenal consciousness in the absence of a sufficiently informationally complex organization to support access consciousness, is more contested. Tononi's (2008) computationally and neuroscientifically-inspired theory of consciousness postulates that (phenomenal) consciousness emerges from sufficient integrated informational complexity, and thus it implies that phenomenality and access consciousness, if these can be distinguished at all in this framework, are co-emergent.

Daniel Dennett's *Multiple Drafts Model* (MDM) of consciousness provides an alternative way of theorizing about consciousness (Dennett 1992). According to MDM, cognition involves concurrent processing of multiple streams of information that are subject to constant editing and re-editing. According to this view, there is no unitary locus or subject of consciousness, no dominant "central authority," "homunculus," or "Cartesian Theater" in which the contents of consciousness play out. In Dennett's view, phenomenality does not emerge as different from access consciousness: each is the result of different kinds of behavioral probes. Some have argued, however, that rather than explaining consciousness, Dennett explains it away (Roskies and Wood 1992).

The previously described theories are first-order theories of consciousness: consciousness depends on the obtaining of certain kinds of representational mental states. Higher order thought

(HOT) theories of consciousness maintain that phenomenal consciousness requires higher order mental states, or mental states that take first order mental states as their objects (Byrne 2001: 205; Rosenthal 2005). Thus, consciousness requires mental acknowledgement of an experience. HOT theories take *self-awareness or self-consciousness* to be a central aspect of consciousness (Carruthers 2000). How to unpack self-consciousness is itself a matter of debate. Self-consciousness can be unpacked as awareness of oneself as a self, termed *reflective* self-consciousness, or merely as awareness of oneself as a biological body in nature, termed *pre-reflective* self-consciousness. Pre-reflective self-consciousness emerges when an individual's biological body responds to stimuli in an external environment and performs sensorimotor actions, but in a self-identification-free manner (Legrand 2006: 92). Reflective self-consciousness, in which one contemplates one's own biological self-awareness, is related to HOT theories of consciousness. Finally, a variation of higher-order theories are *reflexive theories,* which hold that self-awareness exists directly *within* the conscious state, rather than within an associated meta-state that is merely *directed* at consciousness (Kriegel and Williford 2006).

3 New Methods of Assessing Consciousness

New noninvasive methods for imaging brain activity have the potential to revolutionize the diagnosis, care, and treatment of patients with disorders of consciousness. An enormous body of imaging studies in normal test subjects has identified signature patterns of brain activity associated with the performance of particular cognitive tasks. For example, motor imagery (imagining moving one's body) activates regions of premotor cortex, many of the same brain areas typically activated in tasks involving actual movement. Other brain areas are typically activated in navigation tasks, and yet others in other tasks, such as perceiving faces or body parts. In groundbreaking work, Adrian Owen and colleagues have developed a neuroimaging paradigm to assess the cognitive status of patients exhibiting no outward signs of consciousness. Owen and colleagues capitalized on known task-dependent regularities in brain activation to test for potential covert cognitive abilities in brain-damaged patients (Owen et al. 2006).

In the original study Owen and colleagues put a PVS patient in the fMRI scanner, and instructed her to perform different imaginative tasks, e.g., to imagine playing tennis or to imagine walking through her house (Owen et al. 2006). Surprisingly, this PVS patient, who had been unresponsive for five months, showed patterns of brain activation indistinguishable from the patterns found in a population of healthy controls doing the same imagery tasks. These results strongly suggest that this patient was aware of her surroundings, able to understand the experimenter's instruction, and, moreover, able to volitionally perform two complex mental tasks of significant duration. Owen et al. concluded that the patient had been misdiagnosed as being in PVS, and that rather than being unaware of her surroundings, she retained substantial cognitive abilities and consciousness. In further studies the group tested populations of normal and PVS patients. Unsurprisingly, the normal subjects showed reliable activation in canonical (and importantly, highly distinguishable) brain regions in these two tasks. While most of the brain damaged subjects lacked these canonical activations, a number of patients (17%) were able to follow the visualization instructions (Monti et al. 2010). In addition, by using these two visualization commands as proxies for answering "yes" or "no," researchers found that upwards of 15% of their tested patients believed to be in PVS were able to correctly answer questions about topics such as their family and their life history (Owen and Coleman 2008; Monti et al. 2010).

These findings suggest that a significant number of patients previously believed to be in PVS may actually occupy robust conscious states and may retain the ability to comprehend language, form intentions, and exercise executive control over their mental states. This realization carries with it consequential ethical implications. First, it means that a significant number of brain-damaged patients have been misdiagnosed as being in a persistent vegetative state when in fact they enjoy (or perhaps suffer) some significant level of consciousness. This raises directly the question of what the ethical implications are of having the capacity for various types of conscious states (see "The Moral Significance of Consciousness" section). Second, the ability to use these fMRI-discriminable cognitive tasks as proxies for "yes" and "no" answers to questions (and increasingly, with less costly and cumbersome neuroimaging techniques, see e.g., Cruse et al. 2011, 2012) opens up the possibility of using neurotechnologies not only to assess levels of consciousness, but also to communicate with such patients, potentially allowing them to play a role in determining their own futures. This possibility raises further ethical issues, which we explore below.

4 What Can We Conclude from These Studies about Consciousness?

The original Owen et al. studies strongly suggested that patients were conscious, but did they prove it? Earlier brain imaging studies had revealed stimulus-related increases in metabolism (de Jong et al. 1997) or increased brain activity in brain areas associated with semantic processing when spoken to (Dehaene et al. 1998), which had been argued to be indicative of consciousness. However, we have ample evidence from a variety of studies that significant automatic activation of neural processing occurs even in cases in which subjects are not conscious of a given stimulus. For example, subliminal (unconscious) primes can activate brain areas normally implicated in explicit or conscious processing of those stimuli (Meneguzzo et al. 2014). There is evidence for quite a lot of stimulus-specific neural activity in a variety of cognitive tasks, including some aspects of semantic processing, which can be independent of conscious state (Nigri et al. 2016). What arguably enabled Owen et al.'s original observations to circumvent this worry is that the brain activation he measured was not stimulus-locked in a straightforward way: his instructions had first to be comprehended, and then the patient had to sustain a cognitive task for a significant period of time (~30 seconds) in the absence of further stimulus. In contrast, automatic neural responses tend to be transient, lasting only a few seconds (Owen et al. 2006, 2013; Boly et al. 2007). In addition, the key responses observed in the Owen et al. studies were not in regions known to be automatically activated by semantic processing, but in those associated with the content of the visualization scenarios. Thus far, the best explanation for the observed activation patterns is that patients understood the instructions and deliberately and volitionally complied with them. Given our current understanding, such executive function implicates a significant degree of access consciousness, and is likely to implicate phenomenal consciousness as well.

More recent studies by Owen and colleagues (Sinai et al. 2017) that employ neuroimaging techniques to measure answers to yes/no questions by using cognitive tasks as proxies expand the scope of consciousness in these patients. Their ability to answer questions reliably and veridically suggests not only that they are capable of prolonged attention, but that they can retain old memories, can form new ones, and that they understand the norms of communication. These results are exciting and important, but they also raise important questions about how such neurotechnologies should be used and about the conditions under which severely brain-damaged patients ought to be able to make decisions regarding their lives and treatments (see e.g., Calabró et al. 2016).

Although one cannot definitively rule out all skeptical arguments that permit doubt about the levels of consciousness of these patients, these studies provide good evidence that there are a significant number of patients diagnosed as in PVS who remain conscious at least some of the time (Fernández-Espejo and Owen 2013; for more discussion about these arguments, see Peterson and Bayne, this volume). Suppose we take it as established that some PVS patients are indeed in some sense conscious. There are further, and perhaps more difficult, questions about the degree, nature of, and limits of their retained conscious and cognitive capacities. Monti et al. (2010) have established that some of these patients can correctly identify their names and current locations, as well as answer questions about their family, their history, etc. (see also Naci and Owen, 2013; Naci et al. 2017; Peterson et al. 2013). Their reliably correct answers indicate that they understand the questions asked. But what is the depth of their understanding, and the scope of it? How are we to assess understanding in cases in which we are unable to verify their answers?

This last question becomes important if the questions we have reason to ask them concern the nature of their phenomenal experience. Perhaps the most ethically pressing question we can ask these patients is whether they are in pain or are suffering, for we have the ability to rectify their suffering by modifying their treatment. However, the nature of phenomenal experience is subjective and in principle unverifiable. How can we assess whether a severely brain-damaged patient adequately understands these questions? What is the possibility that despite her ability to answer objective questions correctly she may not understand the meaning of subjective concepts such as "pain," and that her answers would thus not accurately portray whether she is experiencing pain? Must we be able to independently ensure that patients understand the meaning of terms for subjective states (such as pain, desire, hope, sadness, happiness) before we ask them questions about their experience? Must we be able to verify their answers to do so?

Is this a real problem? In theory, it is impossible to be certain about the subjective experience of fully conscious and healthy individuals, but in normal cases we have embodied cues and other behavioral information that can inform our understanding of their phenomenal states and the intensity of their emotions. For instance, if a child affirms that he is in excruciating pain, but he is sitting on the floor calmly sucking on his finger rather than writhing and screaming, we can infer that he probably does not fully understand our question. Because PVS patients do not exhibit overt behaviors, we do not have the possibility of this kind of behavioral corroboration.

Complicating the matter further, there is evidence that patients with brain damage who exhibit signs of consciousness fade in and out of consciousness, sometimes over short periods of time. Accordingly, we have no guarantee that when we use fMRI techniques for ascertaining consciousness, patients remain equally conscious from one scanning session to the next, or even from one question to the next. Because fMRI requires significant time to administer, it is possible that patient responses could range from reliable to unreliable, and if the subjective questions (or unverifiable questions) were to be administered during an unreliable phase, we may not be able to detect the change.

5 Epistemic Issues

Diagnosis of states of consciousness, whether by bedside examination, or by measurement of brain activity, is made on the basis of objective, physically manifest phenomena. But what it is to be conscious is to have a subjective perspective, for it to be like something to be that entity. We never have access to someone else's subjective experience, but instead infer it or assume it on the basis of behavior (and similarity to ourselves). As Peterson and Bayne discuss (this volume), assumptions must be made about what evidence is evidence of consciousness in order to infer the presence of consciousness from objective data.

Consciousness is usually assessed by observation of and communication with another agent. Indeed, most of the evidence we use to assess consciousness in humans comes from verbal reports of the subject. But verbal report is just a kind of behavior (though perhaps a privileged kind, since its content can be about subjective experience). The avenues that verbal report provide into subjectivity point to the importance of opening up avenues of communication with subjects who lack the ability to communicate through overt behavior. And although we may automatically assume that other humans are conscious, the assumption is defeasible. Once the consciousness of another is in doubt, it is a difficult question to determine exactly what evidence would be necessary or sufficient to warrant ascriptions of consciousness (think, for example, about the difficulty we would have in deciding and justifying our ascription of or denial of consciousness to an artificial intelligence that can match the complexity of human behavior, or to an octopus writhing when it is injured—is it in pain or just manifesting an outward behavior?). So one important question concerns whether neural data (in lieu of behavior) poses a different philosophical problem than does behavior itself for assessment of consciousness. To this we respond with the following suggestions: (1) Different behaviors provide different degrees of evidence for ascriptions of consciousness, and for different kinds, levels, or dimensions of consciousness. Which behaviors are diagnostic will depend upon one's theory of consciousness. For example, if intentionalism is right, then the contents of experience are fully specified by the mental representations giving rise to them. If we can identify those representations, whether verbally or via regular correlations with neural activity, we have evidence of consciousness. (2) If neural activations corresponding to a certain behavior were necessary and sufficient for the production of that behavior they should be accorded the same epistemic weight as the behavior in inductions about consciousness; and (3) since neural data rarely are so closely linked with behavior, their epistemic weight should be modulated by the degree to which the behavior can be reliably inferred by the neural signature in question. A Bayesian framework can thus be used to assess the value of the neural evidence.

The problem is perhaps slightly more difficult if the question is not just whether someone is conscious at all or capable of consciousness, but to what degree, in what respect, or on what dimensions they are conscious. In that case, a similar formula may be applied, but relative to the degree of evidence the behavior or neural data provides for ascribing a certain level or kind of consciousness. Because the measures we have are not perfectly correlated with the states or behaviors that we take to be evidence of awareness, we ought to look for agreement among different measures and types of evidence. This corroboration is what Peterson and Bayne term "consilience" (Peterson and Bayne, this volume; Gibson et al. 2014).

Even when we ascertain consciousness to a reasonable degree, and even if we feel sure that the patient understands the meaning of the questions we ask, we still face the problem of competence. It is hard enough to set the bar for competence for serious medical decisions (e.g., a choice for euthanasia, withholding treatment, etc.) with fully conscious patients without brain damage with whom we can communicate easily. A patient's answers obtained with fMRI to objective questions requiring reasoning or memory might be evidence of some level of rational competence, but perhaps the kind of competence we seek evidence of for consequential decisions needs to be broader and more holistic. The tools currently at our disposal may be too blunt to ascertain the kind of competence necessary for autonomy in life-or-death decisions.

A technique that enables us to ask only yes-or-no questions may not be sharp enough to give us the kind of confidence we need for consequential decisions, as they only scratch at the edge of phenomenal consciousness and deep rationality. In addition to yes-or-no answers, we may want to hear *reasons,* to understand *why* it is that a person answers as he does. Because of the limited periods of lucidity that these patients enjoy, they are unlikely to be able to use the kinds of labor-intensive communication techniques that people with locked-in syndrome do, such as eye-movement driven computer interfaces or brain-computer interfaces that enable them to transcribe full thoughts. As neural decoding with fMRI improves, it is possible that this problem can be ameliorated.

One way to increase confidence in results of asking questions by neuroimaging may be the following. While we cannot accurately discriminate, say, 26 different mental acts to correspond to different letters (nor could we reasonably expect patients to memorize such mappings), we may be able to broaden our options to 3: "Yes," "no," and "opt-out." It has been shown that use of an opt-out option in behavioral experiments with monkeys enables researchers to measure the animal's confidence in their answers (Fetsch et al. 2014; Kiani and Shadlen 2009; de Lafuente and Romo 2014). Appropriate use of an opt-out option when the patient is unsure or fails to comprehend the question may provide information about their confidence as well as their metacognitive abilities (and thus, at least on many views, a type of higher order consciousness).

Finally, a remaining important question involves what to make of PVS patients that do not present evidence of mental command-following with neuroimaging. Since phenomenal consciousness is subjective, there is no logical way to rule out its presence with objective data, whether behavioral or neural. Few would disagree that it would be tragic to mistake a conscious person who cannot overtly respond for someone who lacks the capacity for consciousness. It is this that leads some to place an extremely high bar on what is necessary to warrant a denial of consciousness, and to refuse to accept absence of evidence as evidence of absence. On the other hand, for pragmatic and social reasons it may be equally as important to determine that someone lacks the capacity for consciousness. What if all we can expect to acquire is evidence of absence?

Some researchers deny that evidence of absence is necessary. For example, Levy and Savulescu claim "we utterly reject the view that we need evidence for the absence of consciousness before we can justifiably conclude that consciousness is lacking. Sometimes absence of evidence is evidence of absence" (Levy and Savulescu 2009: 368). But when? We suggest that a Bayesian approach provides a principled answer to this question, and that it depends on what the hypothesis space is. If you think that the only possible evidence of consciousness is first-personal, then absence of evidence is not evidence of absence, because the relevant evidence is inaccessible. But if you accept that neural states can provide information about states of consciousness, you already reject that proposition. Absence of these is indeed evidence of absence. And depending on how many potential correlations with subjective states there are, and how reliable they are, continued absence then provides strong evidence of absence. Of course, the hypothesis space regarding the constructs of consciousness and their potential correlates is in flux: philosophers and neuroscientists are still in the process of developing and perhaps rethinking theories of consciousness, precluding a straightforward application of Bayesian principles.

6 The Moral Significance of Consciousness

The above concerned difficulties in establishing the presence and nature of consciousness. But there is a further question we must ask: What is the moral significance of consciousness? Many people find it intuitive that either (1) exhibiting clear signs of consciousness or (2) exhibiting some capacity for consciousness is a criterion for continuing care in cases of severely brain-damaged patients, and thus that the withdrawal of life-preserving treatment from patients satisfying either criterion would be morally prohibited. However, such a view rests upon an assumption about the moral significance of consciousness which may not be correct. Here we explore how different theories of consciousness may affect views of the moral significance of consciousness.

Kahane and Savulescu (2009) identify the Principle of the Moral Significance of Consciousness (SC) which states that the capacity for consciousness characterizes an important moral boundary that fundamentally separates conscious beings from their non-conscious counterparts. However, as many philosophers and scientists have realized, consciousness is not a single phenomenon, as there are arguably a variety of types or dimensions of consciousness. This raises the question whether all varieties of consciousness have the same moral importance. In order to address these questions, we must first ask what underlies the intuition that consciousness is morally significant? Rather than being a bedrock principle that requires no justification, perhaps there are underlying intuitions that explain the intuition underlying SC. In understanding these, we may also better come to understand how the presence of different types of consciousness might affect our moral duties to patients.

Pain and Suffering

Perhaps the most intuitive view is that the ability to feel pain is the criterion that lends an organism moral standing. Although the philosophical literature on pain is itself controversial (see e.g., Aydede 2009), let us accept the reasonably intuitive thesis that pain is bad, or has negative utility. If we further accept that pain is a paradigmatic phenomenal experience, and that phenomenal experience is the hallmark of phenomenal consciousness, then we can understand why consciousness delimits an important moral boundary: creatures that lack phenomenal consciousness will be unable to experience pain, and thus have no moral standing, whereas creatures that have the ability to feel pain require moral consideration. Levy and Savulescu explain it thus: "We are morally required to minimize the amount of pain suffered by any sentient being (to the extent to which this is compatible with our other moral obligations), where sentience is the ability to have phenomenally conscious states" (Levy and Savulescu 2009: 366).

Things may not, however, be even that simple. If it is not pain, but rather suffering (or the capacity to suffer) that is necessary for having moral standing, then the relation to consciousness becomes less clear. On this view, a suffering organism is morally significant because we have a moral duty to diminish suffering (this can be seen as definitionally the case or can be framed in terms of interests: "the sufferer has an interest in ameliorating its suffering") (for more on interests, see below). Experiencing suffering requires consciousness, but as we shall see, it may require something more than phenomenal consciousness. Let us call this "suffering-enabling consciousness." Therefore, suffering-enabling consciousness is morally significant insofar as it allows us to fulfill our moral duty to diminish suffering.

What is suffering-enabling consciousness? On the most basic view, suffering is just feeling pain, so the suffering view collapses to the simple pain view above. But what if suffering is something more than just feeling pain? It is reasonable to think that suffering involves something in addition to pain, or even perhaps something altogether different (for example, we speak of mental suffering in the absence of physical ([bodily] pain). However, there is indication from brain imaging studies that similar brain areas are involved in mental suffering and in the experience of physical pain (Eisenberger et al. 2003; Lamm et al. 2011). Just what else may be involved in suffering? Some have argued that suffering requires the conceptualization of oneself as an agent enduring through time, or an agent with life plans that could be realized or frustrated, or an ability to anticipate the future. These views of suffering may involve kinds of consciousness that go beyond pure phenomenal awareness. For example, if suffering requires self-conceptualization, then self-consciousness in addition to (or perhaps instead of) phenomenal consciousness may be required for morally significant consciousness. Or perhaps temporal awareness or high-level cognition and planning may be required in addition to phenomenal consciousness.

Peter Carruthers, in "Brute Experience" (1989), applies his preferred theory of consciousness in order to draw boundaries for morally significant consciousness in a way that is incompatible with those drawn by the simple pain view. His paper illustrates the interesting way in which substantive theories of consciousness can interact with ideas about the moral significance of consciousness in order to yield substantive views about the scope of our moral duties. Carruthers subscribes to the Higher Order Thought (HOT) view of consciousness: that we are only conscious of things that we have (or can have) HOTs about (Rosenthal 1986: 334). He also denies that animals have the capacity for HOT. In consequence, although he accepts that animals can feel pain, he thinks it is not *like* anything for animals to have these pain experiences: all their experiences are nonconscious. In addition, he holds that for a mental state to be an appropriate object of moral concern there must be *something it is like* for an organism to have it. Since he holds that animals only have non-conscious experiences, their lives and experiences are not appropriate objects of moral concern.

There are clearly many points at which the HOT argument could be contested. There is the viability of the HOT theory itself, the empirical claim about which animals have what kinds of capacities for HOTs (and particularly the great emphasis placed on linguistic competence in this debate), and the notion that only subjects with the kind of experiences enabled by HOTs are appropriate objects of moral consideration (see e.g., Gennaro 1993, 2012). While we think that Carruthers' argument is fundamentally flawed and may have potentially ethically pernicious consequences, it is easy to see how this argument could be applied to patients with extensive brain damage. While one might acknowledge that much of the neural machinery for registering pain remains in such patients, it is also the case that many patients with brain damage do not evidence the ability to think higher order thoughts (this doesn't obviously apply to the miscategorized PVS patients identified by Owen and colleagues). If this is the case, proponents of Carruthers' view would have to conclude that such patients likewise do not merit moral consideration.

Interests

Pain and suffering are very intuitive potential ways of grounding moral standing, but not the only ones. Kahane and Savulescu (2009) argue that what grounds moral standing are interests because "interests matter morally" (Kahane and Savulescu 2009: 11), and that interests come in many guises. The interest view subsumes the pain and suffering views, because not suffering (or not feeling pain) are only some of the interests an organism might have, and the reach of the interest view is much broader.

Kahane and Savulescu argue that merely identifying the presence or absence of consciousness is insufficient to settle the ethical issues surrounding end of life issues, because facts about our interests do not neatly line up with facts about consciousness. They argue against the assumption that consciousness is the basis for moral significance both by arguing that our interests don't necessarily have a phenomenal or experiential character and by arguing that the moral imperatives they ground do not necessarily point in the direction of extension of life. Indeed, they contend that the enjoyment of consciousness might actually give stronger moral reasons *not* to preserve a patient's life.

Kahane and Savulescu identify experiential or hedonic interests, desiderative interests, and objective interests. Experiential or hedonic interests refer to states of suffering or enjoyment, and thus are linked to experiences or phenomenal consciousness and perhaps to other cognitive abilities as well. Desiderative interests refer to the interests organisms have in satisfying their desires, and objective interests are interests that an organism may have in things that are objectively good for a life (for example, deep relationships, health etc.). Kahane and Savulescu argue that both desiderative and objective interests presuppose a variety of cognitive and motivational states and capacities that they term "sapience," and that sapience requires not phenomenal but access consciousness.

Kahane and Savulescu question whether phenomenal consciousness is sufficient for having interests, though they recognize that it may be necessary. They write,

> Indeed, it is doubtful that a mental life consisting only of a bare stream of consciousness — a sequence of random and hedonically neutral sensations — could be said to

> involve interests of any kind. A being cannot have desires, and thus desiderative interests, without a sufficient degree of cognitive capacity. Nor can one possess objective interests such as the interest in friendship or knowledge in the absence of such capacities (indeed many objective goods seem to require self-consciousness, not phenomenal consciousness). What about experiential interests? One cannot enjoy or suffer without being phenomenally conscious, but it is far from obvious that mere possession of phenomenal consciousness implies that one has the capacity to experience pain or pleasure. A being that lacked both cognitive capacities and the capacity to feel pleasure and pain might be a being without interests despite possessing phenomenal consciousness.
>
> *(Kahane and Savulescu 2009: 14)*

But even when the requisites for having interests are fulfilled, the moral imperatives do not always point in the way many people take for granted: the interests involved may actually not militate for preservation of life. For example, perhaps a person prior to brain injury valued her autonomy and desired not to live a life in which her survival was predicated upon being kept alive by a machine. Provided she had sufficient access consciousness and other capacities to preserve sapience and thus some desiderative and objective interests, satisfaction of her desiderative interests may require terminating life support. Or on the objective view, if we deem it objectively bad to continue to exist without the possibility of cultivating friendships, being creative, etc., or that burdening one's loved ones is objectively bad, then a person's objective interests may also not point in the direction of continued life. Kahane and Savulescu also consider the interesting possibility that a person could have interests that persist beyond the limits of their consciousness, just as the law acknowledges that persons can have legal interests that extend beyond their lifetimes. They argue that these interests often would tend not to favor life-preservation. On the interest view, then, the relationship between kinds of consciousness and our moral duties is complex, and may rest more upon access consciousness and other kinds of capacities than on phenomenal consciousness.

Personhood

The personhood view contests the standard assumption that evidence of consciousness provides a reason to grant a being full moral status. On this view what confers (full) moral status is personhood, and personhood requires self-consciousness. As Levy and Savulescu (2009) argue, only full moral status entails a right to life; merely being a moral patient (that is, having the capacity to possess mental states that matter morally) implies certain minimal duties, such as the duty to minimize aversive mental states via analgesics or other drugs, for example, but does not provide reason to value or preserve the life of the moral patient. Levy and Savulescu argue that beings that can experience aversive mental states are morally equivalent to animals, and they take for granted that we have no duties to preserve animal lives because of their intrinsic worth, unless they exhibit hallmarks of self-consciousness. Thus, while Levy and Savulescu disagree with Carruthers about the scope of suffering, allowing that phenomenal consciousness does matter morally and entails some minimal moral duties, they agree that higher-order or self-referential thoughts are necessary for full moral consideration. Such thoughts at a minimum require a type of sophisticated access consciousness. Thus they take the kind of evidence for consciousness found by Owen and colleagues in a subpopulation of PVS patients to be insufficient moral ground for life-preserving measures:

> But whether they are conscious or not, it can be argued that we have little reason to maintain them in existence (and perhaps even some reason to bring about the cessation

> of their lives), unless their mental states are at least as sophisticated as those exhibited by children, and, importantly, as connected across time. It is not merely consciousness that is required for what we shall call full moral status; it is self-consciousness, and we do not believe that we can (yet) attribute self-consciousness to any PVS patients.
>
> *(Levy and Savulescu 2009: 362)*

Finally, it bears mentioning that any theory we have about the moral significance of consciousness should be applicable across the board. If self-consciousness is what we deem necessary for moral consideration, it should apply to humans as well as nonhumans, if there are any that are self-conscious. But if we decide that phenomenal consciousness is what ultimately matters, then far more beings than humans should be accorded moral consideration in matters of life and death. Thus, the theories that govern the way we treat patients with varying kinds of disorders of consciousness will, if we are consistent, inform and guide our treatment of nonhuman animals.

7 Future Directions

Neuroimaging has opened up exciting possibilities for the diagnosis and care of patients with disorders of consciousness. Future work will involve improving the relevant technologies for cost-effective use at the bedside and developing more incisive diagnostic test paradigms. Theoretical work will involve refining and perhaps reconstructing theories of consciousness to reflect the novel scientific insights from continuing work on the neuroscience of consciousness, and developing more sophisticated views about the moral significance of consciousness and its relationship to what we value. Disorders of consciousness continue to provide a proving ground for neuroethics.

Note

1 This chapter benefitted immensely from the assistance of Guilliermo Gomez, Jett Oristaglio, Sushmita Sadhukha, Nikita Schiava, and Daniel Widawsky. The work was supported by a grant from the Templeton Foundation as part of the Philosophy and Science of Self-Control project led by Alfred Mele.

References

Aydede, M. (2009) "Pain," The Stanford Encyclopedia of Philosophy (Spring 2013 Edition), Edward N. Zalta (ed.). https://plato.stanford.edu/archives/spr2013/entries/pain/.

Baars, B. (1988) *A Cognitive Theory of Consciousness*, Cambridge, MA: Cambridge University Press.

Bauer, G., Gerstenbrand, F. and Rumpl, E. (1979) "Varieties of locked-in syndrome," *Journal of Neurology* 221: 77–91.

Beaumont, J. G. and Kenealy, P. M. (2005) "Incidence and prevalence of the vegetative and minimally conscious states," *Neuropsychological Rehabilitation* 15: 184–189.

Block, N. (1995) "On a confusion about a function of consciousness," *Brain and Behavioral Sciences* 18: 227–247.

Boly, M., Coleman, M. R., Davis, M. H., Hampshire, A., Bor, D., Moonen, G., Maquet, P. A., Pickard, J. D., Laureys, S., and Owen, A. M. (2007) "When thoughts become action: an fMRI paradigm to study volitional brain activity in non-communcaitve brain injured patients," *Neuroimage* 36: 979–992.

Byrne, A. (2001) "Intentionalism defended," *The Philosophical Review* 110: 199–240.

Calabró, R. S., Naro, A., De Luca, R., Russo, M., Caccamo, L., Manuli, A., Alagna, B., Aliquó, A., and Bramanti, P. (2016) "End-of-life decisions in chronic disorders of consciousness: sacrality and dignity as factors," *Neuroethics* 9: 85–102.

Carruthers, P. (1989) "Brute experience," *The Journal of Philosophy* 86: 258–269.

Carruthers, P. (2000) *Phenomenal Consciousness: A Naturalistic Theory*, Cambridge: Cambridge University Press.

Chalmers, D. (1995) "Facing up to the problem of consciousness," *Journal of Consciousness Studies* 2 (3): 200–219.

Cruse, D., Chennu, S., Chatelle, C., Bekinschtein, T., Fernández-Espejo, D., Pickard, J., Laureys, S. and Owen, A. (2011) "Bedside detection of awareness in the vegetative state: a cohort study," *The Lancet* 378 (9809): 2088–2094.

Cruse, D., Chennu, S., Fernández-Espejo, D., Payne, W., Young, G. and Owen, A. (2012) "Detecting awareness in the vegetative state: electroencephalographic evidence for attempted movements to command," *PLoS ONE* 7 (11), p.e49933.

Dehaene, S., Naccache, L., Le Clec, G., Koechlin, E., Mueller, M., Dahaene-Lambertz, G., van de Moortele, P., and Le Bihan, D. (1998) "Imaging unconscious semantic priming," *Nature* 395: 597–600.

Dennett, D. (1992) *Consciousness Explained*, London: Allen Lane, The Penguin Press.

Di Perri, C., Stender, J., Laureys, S., and Gosseries, O. (2014) "Functional neuroanatomy of disorders of consciousness," *Epilepsy and Behavior* 30: 28–32.

Eisenberger, N. I., Lieberman, M. D., and Williams, K. D. (2003) "Does rejection hurt? An fMRI study of social exclusion," *Science* 302: 290–292.

Fernández-Espejo, D. and Owen, A. M. (2013) "Detecting awareness after severe brain injury," *Nature Reviews Neuroscience* 14: 801–809.

Fetsch, C. R., Kiani, R., Newsome, W. T., and Shadlen, M. N. (2014) "Effects of cortical microstimulation on confidence in a perceptual decision," *Neuron* 84: 751–753.

Fins, J., Illes, J., Bernat, J., Hirsch, J., Laureys, S. and Murphy, E. (2008) "Neuroimaging and disorders of consciousness: envisioning an ethical research agenda," *The American Journal of Bioethics* 8: 3–12.

Gennaro, R. (1993) "Brute experience and the higher-order thought theory of consciousness," *Philosophical Papers* 22: 51–69.

Gennaro, R. (2012) *The Consciousness Paradox: Consciousness, Concepts, and Higher-Order Thoughts*, Cambridge, MA: MIT Press.

Gibson, R. M., Fernández-Espejo, D., Gonzalez-Lara, L. E., Kwan, B.Y., Lee, D. H., Owen, A. M., and Cruse, D. (2014) "Multiple tasks and neuroimaging modalities increase the likelikhood of detecting covert awareness in patients with disorders of consciousness," *Frontiers in Human Neuroscience* 8: 950.

de Jong, B. M., Willemsen, A. T. M., and Paans, A. M. J. (1997) "Regional cerebral blood flow changes related to affective speech presentation in persistent vegetative state," *Clinical Neurology and Neurosurgery* 99: 213–216.

Kahane, G. and Savulescu, J. (2009) "Brain damage and the moral significance of consciousness," *Journal of Medicine and Philosophy* 34: 6–26.

Kiani, R. and Shadlen, M. N. (2009) "Representation of confidence associated with a decision by neurons in the parietal cortex," *Science* 324: 759–764.

Kriegel, U. and Williford, K. (2006) *Self Representational Approaches to Consciousness*, Cambridge, MA: MIT Press.

de Lafuente, V. and Romo, R. (2014) "How confident do you feel?" *Neuron* 83: 797–804.

Lamm, C., Decety, J., Singer, T. (2011) "Meta-analytic evidence for common and distinct neural networks associated with directly experienced pain and empathy for pain," *NeuroImage* 54: S.2492–2502.

Legrand, D. (2006) "The bodily self: the sensori-motor roots of pre-reflective self-consciousness," *Phenomenology and the Cognitive Sciences* 5: 89–118.

Levy, N., and Savulescu, J. (2009) "Moral significance of phenomenal consciousness," *Progress in Brain Research* 177: 361–370.

Meneguzzo, P., Tsakiris, M., Schioth, H. B., Stein, D. J., and Brooks, S. J. (2014) "Subliminal versus supraliminal stimuli activate neural responses in anterior cingulate cortex, fusiform gyrus and insula: a meta-analysis of fMRI studies," *BMC Psychology* 2: 1–11.

Monti, M. M., Vanhaudenhuyse A., Coleman, M. R., Boly, M., Pickard, J. D., Tshibanda, L., Owen, A. M., and Laureys, S. (2010) "Willful modulation of brain activity in disorders of consciousness," *The New England Journal of Medicine* 362: 579–589.

Naci, L. and Owen, A. M. (2013) "Making every word count for nonresponsive patients," *JAMA Neurology* 70: 1235–1241.

Naci, L., Sinai, L., and Owen, A. M. (2017) "Detecting and interpreting conscious experiences in behaviorally nonresponsive patients," *Neuroimage* 145: 304–315.

Nagel, T. (1974) "What is it like to be a bat?" *Philosophical Review* 83: 435–450.

Nigri, A., Catricalà, E., Ferraro, S., Bruzzone, M.G., D'Incerti, L., Sattin, D., Sebastiano, D. R., Franceschetti, S., Marotta, G., Benti, R., Leonardi, M., Cappa, S. F.; CRC—Coma Research Centre members. (2016) "The neural correlates of lexical processing in disorders of consciousness," *Brain Imaging and Behavior* October 13: 1–12.

Owen, A. and Coleman, M. (2008) "Functional neuroimaging of the vegetative state," *Nature Reviews Neuroscience* 9: 235–243.

Owen, A. M. (2013) "Detecting consciousness: a unique role for neuroimaging," *Annual Review of Psychology* 64: 109–133.

Owen, A. M., Coleman, M. R., Boly, M., Davis, M. H., Laureys, S., and Pickard, J. D. (2006) "Detecting awareness in the vegetative state," *Science* 313: 1402.

Patterson, J. R. and Grabois, M. (1986) "Locked-in syndrome: a review of 139 cases," *Stroke* 17: 758–764.

Peterson, A., Naci, L., Weijer, C., Cruse, D., Fernández-Espejo, D., Graham, M., and Owen, A. M. (2013) "Assessing decision-making capacity in the behaviorally nonresponsive patient with residual covert awareness," *AJOB Neuroscience* 4: 3–14.

Rosenthal, D. M. (1986) "Two concepts of consciousness," *Philosophical Studies* 49: 329–359.

Rosenthal, D. M. (2005) *Consciousness and Mind*, New York: Oxford University Press.

Roskies, A. L. and C. C. Wood (1992) "A parliament of the mind," *The Sciences* May/June: 44–50.

Schnackers, C., Vanhaudenhuyse, A., Giacino, J., Ventura, M., Boly, M., Majerus, S., Moonen, G., and Laureys, S. (2009) "Diagnostic accuracy of the vegetative and minimally conscious state: clinical consensus versus standardized neurobehavioral assessment," *BMC Neurobiology* 9 (35): 1–5.

Sinai, L., Owen, A. M., Naci, L. (2017) "Mapping preserved real-world cognition in severely brain-injured patients," *Frontiers in Bioscience* 22: 815–823.

Smith, E. and Delargy, M. (2005) "Locked-in syndrome," *British Medical Journal* 330: 406–409.

The Multi-Society Task Force on PVS (1994) "Medical aspects of the persistent vegetative state," *New England Journal of Medicine* 330: 1499–1508.

Tononi, G. (2008) "Consciousness as integrated information: a provisional manifesto," *Biological Bulletin* 215: 216–242.

Related Topics

Materialism
Post Comatose Disorders of Consciousness
Representational Theories of Consciousness
The Neural Correlates of Consciousness

34
CONSCIOUSNESS AND EXPERIMENTAL PHILOSOPHY

Chad Gonnerman

Except in our most skeptical moods, it is hard for us philosophers to doubt the successes of science. Reflection on its achievements over the last 50 years alone is cause for optimism (e.g., the sequencing of the human genome, the development of a rich understanding of the architecture of human memory, and the discovery that the universe is rapidly expanding at an ever-increasing rate). It seems that science is quite good at achieving its epistemic ends—its goals with respect to the likes of knowledge, understanding, and explanation. But if we had to settle on one word that best represents the limits of science, at least as currently practiced, we could hardly do better than the one that unites this volume: consciousness.

But not all forms of consciousness are equally intractable. The one that has struck many as particularly resistant to our best theorizing is *phenomenal consciousness*. This is the form exhibited by mental states that have a subjective, qualitative, or experiential aspect to them. There is "something it is like" to be in them (Nagel 1974). At a minimum, these *phenomenal states* include perceptual experiences, bodily sensations, felt emotions, and felt moods (Tye 2016). So, think about the experience of sucking on a lemon, the sensation of stubbing your toe, and the elation of a hard-won A, and you'll have a pretty good idea of the type of mental state and form of consciousness that has seemed so puzzling to so many.

Just as we may note that some forms of consciousness seem more intractable than others, we may also note that some puzzles about phenomenal consciousness seem more puzzling than others. Certain "how" or "why" questions seem particularly perplexing. Here is how Chalmers (1995: 201) puts the mystery:

> The really hard problem is the problem of *experience*... Why is it that when our cognitive systems engage in visual and auditory information processing, we have visual or auditory experience: the quality of deep blue, the sensation of middle C? How can we explain why there is something it is like to entertain a mental image or to experience an emotion? It is widely agreed that experience arises from a physical basis, but we have no good explanation of why and how it so arises.

This quote captures what Chalmers calls "the hard problem of consciousness." It contrasts with what he deems to be the easier problems. These are problems primarily having to do with how the mind/brain accesses and deploys information. For instance, what underwrites our capacity to discriminate environmental stimuli? The hard problem is the problem of explaining how or why certain mental states feel the way they do—or why they feel any way at all.

When it comes to questions about what science can and cannot reveal about phenomenal consciousness, we are apt to think about the individual undergoing the states. Call him/her the *experiencer*. To illustrate, consider how Churchland responds to the claim that there is a hard problem of consciousness. According to her, we should simply "get on with the task of seeing how far we get when we address neurobiologically the problems of mental phenomena" (1995: 402). Or consider Chalmers' (2002) argument that there is a hard problem. According to him, neuroscience fails to provide any hints of how or why certain neural happenings are associated with particular experiences. Why do we have an experience of redness, not blueness, when we look at the typical fire engine? Indeed, why is there an experience at all? Why don't we simply register and process the color as a thermostat registers and processes ambient temperature? What is important for current purposes is that both Churchland and Chalmers are focused on the sciences of the experiencer. They just have different takes on how far the sciences will go.

Goldman and McGrath (2015) outline two ways of using cognitive science in epistemology. There are applications to the epistemic *subject* and applications to the epistemic *attributor*. We can draw a similar distinction here. We might then ask, "What could a science of consciousness *attribution* reveal? What could we learn from a science of the neural and cognitive systems that underwrite our capacities to think about phenomenal states?" Presumably, we could discover how these systems generate our attributions of phenomenal states to ourselves and to others. It would tell us, for instance, about the cognitive mechanisms and processes at play when we judge a person to be in pain. And, on some views of concepts, an account of this sort would also get at the contents, structure(s), and vehicle(s) of our phenomenal concepts. Results like these would be great. But could they also help in trying to understand phenomenal consciousness in general and the relevant phenomenal states in particular? Could they help dominant strands of philosophical research on consciousness? The view of many experimental philosophers is, yes. But, even if they are wrong, as we will see, the experimental philosophy of consciousness is fascinating for many reasons.

This chapter reviews research in this rapidly growing corner of the experimental philosophy of mind (for more on the general area, see Sytsma and Buckwalter 2016). Section 1 explores debates about how to characterize experimental philosophy. It ends by adopting a narrower conception that emphasizes cognitive scientific methods and the intuitions of ordinary people, or "the folk" (i.e., neurally typical adults with no extensive training in philosophy or the cognitive sciences). Section 2 gives an origins story. It discusses two papers that have significantly shaped the experimental philosophy of consciousness. Here, we will see one way in which experimental work can potentially inform dominant strands of philosophical research on consciousness, namely, by buttressing key lines of thought aimed at establishing a hard problem of consciousness. Section 3 gets into experimental research on sentences attributing phenomenal states to entire groups. The research landscape here is very unsettled. Some work suggests that people find these sentences to be acceptable at times, seeming to embrace the existence of group minds with phenomenal states. Other work suggests that people treat these sentences as attributions to individual group members. Finally, Section 4 discusses work looking into folk attributions of phenomenal states to individuals. The emphasis is on accounts of the psychological processes and mechanisms underlying these attributions.

1 A Framework

What is experimental philosophy? While the question has received much attention over recent years, experimental philosophers, evidently, have yet to converge on a single answer. There are some who give voice to a narrower conception of the area. The claim is that experimental philosophers use the methods of the cognitive sciences to study philosophical intuitions, especially of ordinary people (e.g., Nadelhoffer and Nahmias 2007; Knobe and Nichols 2008; Alexander 2012). There are also some who propose a broader conception. According to these, experimental philosophers use scientific methods, cognitive or not, to study whatever may help advance philosophical research, whether these are intuitions or not (e.g., Rose and Danks 2013; O'Neill and Machery 2014; Sytsma and Livengood 2016). Who is right?

There is some reason to think that the narrower conception is more often pursued. Silber and Knobe surveyed five years of publications in experimental philosophy (Knobe 2016). They culled all the studies and sorted them into two bins. The guiding question was: How did the paper use the study? Did it use the results to make a positive contribution to conceptual analysis? Or did it use the results to make a negative case against the intuitive practices of analytic philosophy? Reflecting on these findings, Knobe (2016) concludes that negative experimental philosophy is pretty rare. Instead, as the title of his paper goes, experimental philosophy is cognitive science. It is in the business of developing and testing hypotheses about the psychology responsible for our philosophical intuitions.

To illustrate the idea, consider Buckwalter and Phelan's (2014) first study. Participants read a short story about an agent who got in a car wreck while driving to his son's home in order to leave some photos in the treehouse. In one condition, the agent emerges from the wreck unscathed and walks the photos to the treehouse. In another, the agent dies in the wreck; undeterred, however, he floats the photos to the treehouse as a ghost. Buckwalter and Phelan found that attributions of felt emotions were indistinguishable across the two conditions. Participants were equally willing to attribute felt anger and felt happiness to the agent whether he was a human or a ghost. So, in view of Knobe's characterization of experimental philosophy, we might expect to see Buckwalter and Phelan give a cognitive-scientific spin to their results. And this is what they do. Buckwalter and Phelan claim that their results provide evidence against the *embodiment hypothesis*, the view "that unified *biological embodiment* is a major psychological factor that cues ordinary attributions of experiences, feelings, emotions, and so on, to other entities" (2014: 46; for a defense of the hypothesis or ones very close to it, see Knobe 2008; Knobe and Prinz 2008; Gray et al. 2011).[1] Thus Buckwalter and Phelan are in line with Knobe's proposal. They are engaged in what Arico (2010: 372) describes as the "new Folk Psychology project"—that of answering the question "how do we, as humans, go about attributing phenomenal states?"

Knobe's proposal does a decent job at capturing most work in experimental philosophy. Or at least it does so once we correct for one notable omission. It fails to make it perfectly obvious that experimental philosophy is, well, philosophy. After all, cognitive science is not philosophy. This is not to diminish the role that philosophy plays in cognitive science. And it is not to deny philosophy's long interest in minds. It is only to acknowledge a common thought: cognitive science is an interdisciplinary area of inquiry that goes beyond the contributing disciplines without amounting to a new discipline (Bechtel 1986). Thus, cognitive science is not philosophy, psychology, linguistics, etc. This claim, combined with Knobe's proposal as articulated, would imply that experimental philosophy is not philosophy. But this is false. The 'experimental' in 'experimental philosophy' is not like the 'fake' in 'fake diamonds.' It's more like the 'good' in 'good ideas.' It modifies the noun to identify a subset.

This view is not merely my own. Sytsma and Livengood surveyed philosophers for their views of experimental philosophy. They found nearly 90% agreement that addressing a philosophical problem or issue is needed for a paper to qualify as a work in experimental philosophy (2016: 22). So, we should revise Knobe's account. A good start would be to say that experimental philosophy is cognitive science that addresses something philosophical. Sometimes, the two simply overlap. This is especially true if Knobe (2007) is right in claiming that the giving of cognitive accounts represents a return to a traditional conception of philosophy practiced by the likes of Plato, Hume, and Nietzsche. And sometimes the author deploys her cognitive story for (other) philosophical ends.

But, even with this revision, we have yet to nail down an uncontroversial account of experimental philosophy. Many will regard our revised proposal as being too narrow. It fails, they'll insist, to capture all of the activity worthy of the label. This line of thought is especially likely to occur to experimental philosophers. They strongly tend to favor a broad conception of their area (Sytsma and Livengood 2016: 15–18). So, maybe, for the purposes of this chapter, we should adopt a broad conception of the experimental philosophy of consciousness. It would include the scientific study of intuitions about consciousness. It would also include the scientific study of anything else that might reasonably advance philosophical work on consciousness. Some may worry that the result is too broad; we are left an assemblage of work that is not deeply unified. But so what? Many areas of philosophy are loose assemblages, more or less. Still, loose assemblages can make for messy reviews.

The good news is that we can keep this review reasonably tidy, since most of the extant work on consciousness qualifies as experimental philosophy narrowly construed (Sytsma 2010a, 2014a). So, a good chunk of this work involves probing the intuitions of ordinary people to get at the underlying systems, processes, and mechanisms. And, while there are many types of phenomena to which we can apply the word 'consciousness' (Natsoulas 1978; Rosenthal 1986; Block 1995), most experimental philosophical work is on phenomenal consciousness in particular. So, for the purposes of this chapter, I am going to follow the herd, working with the following characterization:

> The experimental philosophy of consciousness involves (1) the methodical collection and analysis of empirical data pertaining to ordinary intuitions of, or related to, phenomenal consciousness or phenomenal states (2) using cognitive-scientific methods in order to contribute to research in (3) the cognitive sciences of these intuitions and (4) the philosophy of phenomenal consciousness, phenomenal states, and associated phenomena.

Of course, by working within this narrower framework, I am not denying that there is interesting work outside it. Reuter's (2011) corpus analysis of uses of pain terms, and his claim that people distinguish between pains and pain experiences, is an excellent example of the broader work. Here, I am simply trying to give the reader a decent sense of where the bulk of the literature is at the moment.

2 Origins

One work that has shaped the experimental philosophy of consciousness is Gray, Gray, and Wegner (2007). The paper reports the results of a study in which participants made pairwise comparisons of 13 characters for 18 mental capacities. They had to determine, for instance, "whether a girl of 5 is more or less likely to feel pain than a chimpanzee is" (2007: 619). According to Gray et al., the results indicate that people view minds as differing along two dimensions—Agency and Experience. It seems that from a folk perspective, minds can vary in their capacities for *agential* states (high-level cognitive states like those that figure in planning, communication, and thought) and *experiential* states (phenomenal states like feelings of hunger, pain, and embarrassment). Indeed, if Gray et al. are right, ordinary people recognize three types of minds: (1) those with high levels of Experience yet low levels of Agency (e.g., a fetus); (2) those that score high in Agency but low in Experience (e.g., God); and (3) those that register high in both (e.g., a man or a woman).

The claim that people distinguish between these three minds is philosophically interesting in many ways. To get at one, consider again the hard problem of consciousness. Chalmers suggests that we cannot respond to the problem by denying the phenomena. The reason is that "Experience is the central and manifest aspect of mental lives," and this gives experience a "status as an explanandum" (1995: 207). If true, we might expect people to have a (if not the) concept of phenomenal consciousness (Sytsma and Machery 2009, 2010). After all, we have a way of noticing things that are central and manifest in our lives. And such things often get channeled through our concept-formation processes in order to develop semi-stable bodies of information that can aid our thinking about these things down the road. So, the question arises, do people have a concept of phenomenal consciousness? The results of Gray et al. give us some reason to think that they do.

Their findings suggest that people take some beings to score high in Agency and low in Experience. God appears to be an example. This gives us some (as we'll see, imperfect) reason to think that the folk are willing to embrace the possibility of *purely intentional minds*.[2] Here, I am thinking about minds that only have intentional states. They never undergo phenomenal ones. Of course, phenomenal states are the stars of this chapter. They include your perceptual experiences, felt bodily states, and felt emotions. Intentional states are mental states that exhibit *intentionality*. They are about things. Examples include your beliefs, desires, and intentions. Notice that if the folk recognize the possibility of a purely intentional mind, then it seems that their attributions of phenomenal states to humans must be sensitive to cues (deemed to be) had by them but not (deemed to be) had by the purely intentional mind. The psychology of phenomenal state attribution must look different than the psychology of intentional state attribution. If so, this is a notable result. Some mindreading research—work on the systems responsible for our mental state attributions to ourselves and to others (for an introduction, see Nichols and Stich 2003)—deliberately base their accounts on findings limited to mindreading vis-à-vis intentional states (e.g., Apperly 2011: 4). Thus they may be only getting at part of the story, while masquerading as authors of the entire story. Moreover, we now have reason to think that ordinary people have a concept of phenomenal consciousness. Positing that they do would go a long way toward explaining why they distinguish between human minds and purely intentional minds. The claim would be that the distinction is driven by a concept that heavily weights cues had by the former and not the latter.

With that said, the underlying argument is not trivial. They rarely are when the goal is to establish a substantive claim about minds. One question we might ask is, 'To what extent do the results of Gray et al. show that people view some minds as scoring high in Agency and low in Experience?' There are reasons to think that the results are only moderately indicative at best. Notice that the generalization that people view some minds as scoring high in Agency and low in Experience stems from results suggesting that they tend to view God in particular as scoring in these ways. But the set of experiential states explicitly explored by Gray et al. was pretty narrow. It included the likes of hunger, fear, pain, rage, and desire. As Phelan, Arico, and Nichols (2013) note, these aren't exactly the types of states that seem appropriate for a being like God. How could God feel hungry, for example? It may be that had Gray et al. looked at a broader range of states, including love and maybe even anger, God would have come out looking rather differently (for a related worry, see Sytsma and Machery 2010). This line of thought raises the question of whether there is any additional evidence that people have a concept of phenomenal consciousness. Here, work by Knobe and Prinz might help.

Knobe and Prinz (2008) report the results of five studies. In one of the most discussed, they gave participants four sentences:

a Acme Corp. is feeling upset.
b Acme Corp. is upset about the court's recent ruling.
c Acme Corp. is feeling regret.
d Acme Corp. regrets its recent decision.

In (a) and (c), 'feeling' is used to attribute a mental state to a group agent. In (b) and (d), this word is not used, but a similar, if not identical, state is attributed to the group. Participants were asked to assess each sentence according to how weird or natural it sounded. They tended to say that (a) and (c) sounded weird while (b) and (d) sounded normal. Knobe and Prinz contend this shows that ordinary people have a concept of phenomenal consciousness. Their thought seems to go: that the results pattern as they do suggests that people are unwilling to attribute phenomenal states to group agents while perfectly willing to attribute intentional (or at least non-phenomenal) states to these agents. So, there is evidence that, from a folk perspective, group minds are purely intentional minds. We have thus landed on a discovery that calls for explanation. Knobe and Prinz maintain that the best explanation involves positing a folk concept. They suggest that when trying to determine whether a target is in phenomenal state P (say, a state of felt regret), people pull two concepts from long-term memory (LTM). The first carries information about the typical causes and effects of P—its *functional profile* (e.g., for regret, it might include increased rumination, feelings of self-blame, and a sense of loss for what could have been). The second concept pulled from LTM is one for phenomenal consciousness. Importantly, it is a concept that requires the target to have a spatially unified, physical body. Thus, according to Knobe and Prinz, the reason a sentence like 'Acme Corp. is feeling regret' rings odd to us is that, in building an interpretation of the sentence, we are forced to apply our concept of phenomenal consciousness to a target—here, a group agent—that we represent as not meeting the conditions of this concept.

But maybe this argument goes too fast. To underscore one worry, notice that the results reported by Knobe and Prinz are mainly about patterns of *linguistic* intuitions. Again, participants assessed sentences for how weird or natural they sounded. Thus, the extent to which folk concepts drive their results is unclear. It may be that the linguistic assessments stem, at least in part, from rather low-level syntactic processing, including the ease or difficulty that participants had in parsing the sentence. There is evidence that syntactic processing can be very sensitive. This includes attuning to *subcategory preference* information (i.e., the perceived likelihood that a given verb and parsing go hand-in-hand; see Garnsey et al. 1997). The possibility that Knobe and Prinz's data partly stem from syntactic processing is particularly relevant since the sentences they used were not *minimal pairs*. The 'feeling' sentences (a) and (c) above lacked "contextualizing" information had by the non-'feeling' sentences (b) and (d)—information that helps to specify what is being attributed. This detail matters. Arico (2010) reports that participants were more likely to say that a sentence ascribing to a group agent a mental state with contextualizing information sounds more natural than a similar sentence without this information, and this pattern emerged whether the 'feeling' locution was used or not (see also Sytsma and Machery 2009). It may be that what we are seeing here is, in part, an attunement to category frequency information of the sort explored in language sciences.

The early results don't quite establish a folk concept of phenomenal consciousness. Thus, as far as these results go, the status of the apparent upshot of a Chalmers-style argument for a hard problem of consciousness—namely, that there should be some such concept—remains unclear.

Still, the early results did help to launch a wide range of work in the experimental philosophy of consciousness. In the next two sections, I review work in two strands of this subsequent research—one on the folk psychology of group minds and the other on phenomenal state attributions in general. Of course, there are other strands of research. They include (1) work exploring potential connections between phenomenal state ascriptions and perceptions of moral patiency (for overview, see Theriault and Young 2014); (2) research probing ordinary conceptions of particular states commonly taken to be phenomenal such as pain (Sytsma 2010b; Reuter, Phillips, and Sytsma 2014; Sytsma and Reuter forthcoming); and (3) investigations looking into the role that consciousness plays in the folk concepts of free will and moral responsibility (e.g., Shepherd, 2012, 2015). Regrettably, space limitations prevent me from getting into these research strands.

3 Group Phenomenality

The first strand of research that the early results helped to launch is a contribution to the folk psychology of group minds (for review, see Huebner 2014: ch. 5). Research into this part of ordinary psychology is not new. Bloom and Veres (1999), for instance, found that people readily use intentional terms to describe the structured movements of groups as simple as three dots. What helps to set the philosophical work apart is its attention to intuitions of phenomenal states. Huebner, Bruno, and Sarkissian (2010), for example, report that both American and Chinese students found sentences ascribing phenomenal states to groups to sound less natural than sentences ascribing similar states to individuals; the difference, however, was less prominent with their Chinese participants. Huebner et al. go on to suggest that cultural factors may moderate a willingness to ascribe phenomenal states to groups. If so, we would have reason to be wary of Block's (1978) Chinese Nation thought experiment, which asks us to imagine the entire population of China duplicating a person's functional profile. We might worry that the intuitions it elicits are sensitive to cultural backgrounds in ways that are inappropriate for their philosophical deployment in arguments against functionalism (for discussion, see Nado 2014). Arico (2010) reports a similar pattern. He found that people were less willing to say that sentences ascribing phenomenal states to groups sounded natural than sentences ascribing these states to individuals. Yet, unlike in Huebner et al.'s results, we see clear a tendency in Arico's to treat phenomenal attributions to groups as natural sounding, at least when they include contextualizing information (e.g., 'McDonalds is feeling upset about the court's recent ruling' vs. 'McDonalds is feeling upset').

Results like these raise the question of how people interpret sentences ascribing phenomenal states to groups. Do they construe them literally? If not, what is the nonliteral reading? Arico et al. (2011) present the results of a pilot study that begins to examine these questions. After cutting participants who seemed to struggle with the literal-figurative distinction, they report that there was a decreased willingness to rate sentences attributing phenomenal states to groups as "literally true" compared to sentences attributing intentional states to groups. It appears that people are less willing to interpret phenomenal attributions to groups literally than intentional ones. Yet this doesn't quite answer our questions. A decrease in willingness is not a flat-out unwillingness. I may be less eager to eat a cookie after I just had one, but if history is any guide, I'm still game for another one. So, Arico et al.'s report is consistent with an overall, weakened tendency to interpret group phenomenal state ascriptions literally. We'll have to look elsewhere for more probative results.

Phelan et al. (2013) provide strong evidence on this score. For one study, they designed a pronoun replacement task. Participants saw a series of sentences with a dependent clause followed by an independent clause. Each clause had the same subject, ostensibly referring to a group agent. The task was to decide whether the subject in the independent clause is best replaced by a singular or plural pronoun. To illustrate, does 'it' or 'they' work best as a replacement for the underlined phrase: 'When MADD's Drunk Driving Prevention Act failed, MADD got extremely depressed?' Phelan et al. found that participants were more likely to pick 'it' for nonmental state ascriptions than for intentional and phenomenal state ascriptions. They take this as a sign that people tend to interpret group mental state ascriptions nonliterally, treating them *distributively*. That is, to a first approximation, their claim is that people usually walk away with a reading of such sentences in which the mental state is attributed to *each* group member, not the group itself.

Phelan et al.'s distributive view is certainly reasonable. But it also has some substantial empirical commitments. The traditional picture is that we settle on nonliteral readings only after our initial attempt at interpretation fails to deliver a literal reading that makes sense in the context (e.g., Clark and Lucy 1975). It is arguable that, in experimental contexts like those in Phelan et al.'s study, a literal, collective reading of group mental state ascriptions would make sense. Since their view is that people tend to walk away with a nonliteral reading, they are arguably committed to denying the traditional picture. But maybe this isn't a problem. There are many reasons to reject the traditional picture (e.g., Gibbs 1983). What may be more problematic is that Phelan et al. seem committed to predicting that we will see no differences in ordinary assessments of group mental state ascriptions and the corresponding distributive sentences. After all, if they are right, people tend to interpret the former as the latter. So, their respective assessments should match, modulo experimental noise. Do they? Not always, according to Jenkins et al. (2014). In response to their group-only vignettes, participants tended to agree that the group had the mental state at issue; however, they tended to disagree when asked whether any or each of the group members had the state (see also Waytz and Young 2012).

Stepping back and looking at the empirical record, I would wager that the psychology of how people interpret group mental state ascriptions is as complicated and variable as any other part of the cognitive science of language interpretation. Bringing this more general work to bear on group mental state ascriptions is likely to reveal similar principles. Yet there may be interesting differences as well. This is one issue that remains wide open in the experimental philosophy of consciousness.

4 Individual Phenomenality

The second strand of research that I want to highlight is work responding to Sytsma and Machery (2010). In this paper, the authors argue against Chalmers' approach to motivating the hard problem of consciousness. They contend that if the approach is correct, there should be a folk concept of phenomenal consciousness, but there is no such concept. To get to their negative conclusion, Sytsma and Machery asked participants to consider a simple robot, Jimmy. As part of a psychological experiment, Jimmy was put in a room that contained three boxes—red, blue, and green. In one condition, Jimmy correctly executed an order to put the red box in front of the door. In another, Jimmy received an electric shock as he grasped the box, after which he dropped it and backed away. While participants tended to affirm that Jimmy saw red, most denied that he felt pain. Sytsma and Machery take this as evidence that there is no folk concept of phenomenal consciousness. As articulated in Sytsma (2014a), the idea seems to be that if there is a folk concept, people should deploy it when trying to determine whether an entity is in a phenomenal state. And so if people tend to withhold attributions of pain to an entity that displays behaviors associated with this state, they should also tend to withhold attributions of seeing red even if the entity displays behaviors associated with this state. Yet this is not what we see. Again, the overall tendencies are to say that Jimmy saw red but did not feel pain (for similar results, see Sytsma and Machery 2012). Therefore, there is no folk concept. It seems that Chalmers' approach to establishing that there is a hard problem of consciousness needs some rethinking.

If ordinary people don't have a concept of phenomenal consciousness, then how do they go about attributing phenomenal states? The positive thesis of Sytsma and Machery (2010) is a response to this question. According to them, people attend to considerations of valence. If the mental state is associated with a valence (with being either pleasurable or unpleasurable), people will tend to attribute the state only if they hold that the target is capable of valenced states (with finding a mental state to be pleasurable or unpleasurable); otherwise, people will tend to attribute the state so long as the target shows behaviors associated with the state. Feelings of pain are associated with a valence. And so, Sytsma and Machery propose, the reason people tend to say that Jimmy did not feel pain is that they view simple robots as incapable of finding mental states to be pleasurable or unpleasurable. Matters are different with seeing red. While there may be situations in which we find experiences of seeing red to be pleasant or unpleasant (e.g., in a painting), we don't usually do so, at least not often enough for an association to form. And so, people tend to say that a simple robot like Jimmy sees red as long as it displays behaviors of the right sort.

Not everyone agrees with Sytsma and Machery. Buckwalter and Phelan (2013) respond to the positive thesis. They propose that ordinary attributions of mental states to a simple robot are sensitive to design details. Maybe when people consider the Jimmy vignettes, they ask themselves, "Has the robot been designed to carry out tasks associated with seeing red or feeling pain?" Buckwalter and Phelan present evidence that this question does matter. In other works (Phelan and Buckwalter 2012; Buckwalter and Phelan 2014), they argue that attributions of mental states are sensitive to functional details in a different sense as well. It appears that information about perceptual inputs, behavioral outputs, and other mental states are important to folk attributions, including phenomenal ones (cf. Huebner 2010). Thus, to the extent that Jimmy is thought to be incapable, say, of a wide range of behaviors associated with being in pain, we should see a tendency to withhold attributions of pain to the robot (for a similar suggestion, see Talbot 2012; cf. Sytsma and Machery 2012). And the results of Buckwalter and Phelan support this line of thought.

Another line of criticism comes from Fiala, Arico, and Nichols (2014). In other works, they defend a general picture of the psychology of ordinary mental state attributions (Arico et al. 2011; Fiala, Arico and Nichols 2012). It is a picture that portrays the underlying processes as responsive to rather low-level features. If the target has face-like qualities, displays interactive behavior, or moves in non-inertial ways, then people will be disposed to attribute a wide range of phenomenal and intentional states to the target. This is the Agency Model. According to Fiala et al., the model predicts that people will tend to attribute states of seeing red and feeling pain to Jimmy. But, again, they don't. Why is that? Fiala et al. suggest that there is another stream of processing at play. They characterize it as a "high road" pathway. It involves slow, deliberate, and introspectively accessible reasoning. And, according to them, it produces the widespread response that Jimmy does not feel pain. Their thought is that the high-road pathway overrides the deliverances of Agency-based processing. As they put it, "It is effectively a platitude in our culture that robots are incapable of pain or emotion" (2014: 37). But then why don't we see similar results with Jimmy seeing red? According to Fiala et al., what keeps participants from registering a denial here is the forced-choice nature of the response options used by Sytsma and Machery. Participants sense that Jimmy *detects* red, but the best that these options allow for is to affirm that Jimmy *sees* red. Fiala et al. report that when you give participants the option to say that Jimmy detected the color, they are unlikely to say that Jimmy saw it. But this result may itself be a product of the response options according to Sytsma (2014b). Once these issues are fixed, he finds that people are, once again, generally willing to say that Jimmy saw the color.

Still, Fiala et al. (2014) helps to bring out an important issue. A lot of work in the experimental philosophy of consciousness is driven by responses to vignettes involving simple robots. We might wonder about the extent to which these responses reflect processing of the sort discussed in this section. When participants read a story about a simple robot, they probably build a representation of the target that captures various details given in the story. But it probably includes other properties as well, such as ones deemed to be typical for simple robots. Consider Jimmy again. Bets are that you pictured him made of metal, not cardboard. And I'd wager that a CPU comes to mind before beer cans. Importantly, the properties automatically imputed to Jimmy could include an absence of mentality to some degree or other. If so, then responses to questions about his mental states may be affected by processes that extract information from participant-enriched representations of the target. To the extent that they are, cognitive accounts that depict the underlying processes more generally, as operating regardless of the type of entity in question, are apt to go awry. They'll depict participant responses as telling us, as it were, about how people apply psychological predicates—that they attune to details of valence, function, Agency, etc.—when in fact enriched representations of the subject are doing the heavy lifting. So, what we have here is a potential source of noise that has not received much attention by experimental philosophers of consciousness (for a related discussion of stereotypes in mindreading, see Epley 2014). Hopefully, it will receive more attention down the road.

While there is other work critical of the positive thesis of Sytsma and Machery, including Sytsma's own criticisms (2012), I want to briefly highlight some of the pushback given to their negative thesis. Is there a folk concept of phenomenal consciousness? Talbot (2012) notes that many approaches to this question focus on folk intuitions about the mental states of others, including Sytsma and Machery's approach. In his view, such approaches are ill suited for this end. Peressini (2014) adds to the exchange by arguing that there is a folk concept of phenomenal consciousness. But instead of probing intuitions elicited by short stories, he emphasizes more general intuitions. As interesting as these lines of thought are, in the space remaining, I want to raise one further possibility: perhaps the negative thesis is not all that surprising or troubling.

If we are looking for a concept whose content hews very closely to the philosopher's, maybe we shouldn't be too surprised to find that there is no folk concept of phenomenal consciousness. Again, the philosophical concept is of a mental state for which there is something it is like to be in the state. It covers the likes of biting into a lemon, stubbing a toe, and feeling regret. What it is like being in any of these states is different than being in any of the others. Still, they have something in common. They share the second-order property of there being something it's like to be in them. Should we expect ordinary people to have a concept that explicitly recognizes this commonality? If someone is remarkably introspective, sure, she might note it. But most people? In what arena of ordinary life is it important to draw a systematic distinction between mental states for which there is something it's like to be in them and for which there isn't? I would be surprised if there were a folk concept of this sort. And I am not so sure that it'd be much trouble for the hard problem of consciousness if there weren't. At first blush, it is a problem about phenomenal states, not their concept(s). If there is no folk concept, maybe phenomenal characters are not as central and manifest as some claim. But they don't seem to be entirely foreign to people either. When a doctor asks her patient whether her back pain is dull or sharp, few struggle to understand the question. And few are surprised to discover that pains can feel different ways. None of this is to say that experimental philosophical work on consciousness is uninteresting. Far from it! And it is not to say that this work cannot shed light on the hard problem. Fiala et al. (2012) help to show that it may by arguing that our sense that there is something hard about consciousness stems from a quirk in our psychology—materials of the sort that figure in physicalist explanations fail to trigger low-level mindreading processes that could otherwise intuitively corroborate the explanation, leaving us with a feeling that the explanation misses something. Instead, all that I am saying is that the connection between the experimental work and the hard problem is less direct than suggested by Sytsma and Machery.

5 Conclusion

We could skim only the surface of the experimental philosophy of consciousness in this chapter. Not only couldn't we give the works discussed herein their full due; there are many that we couldn't discuss. Among these are works more critical of experimental philosophy. The most forceful are works pushing the *expertise defense*—roughly put, the claim that folk intuitions about philosophical matters are less reliable than those of professional philosophers (for review, see Alexander 2016; see also Weinberg et al. 2010). The young experimental philosopher may be heartened to hear that often these criticisms turn on empirical claims that may be experimentally explored and supported—or not, as in some explorations of the expertise defense (e.g., Schwitzgebel and Cushman 2012). What this helps to suggest is that there is still a lot of exciting work to be done in experimental philosophy, including the experimental philosophy of consciousness.[3]

Notes

1 It is unclear to what extent these results falsify the hypothesis. Applying ideas that Cornwell, Barbey, and Simmons (2004) develop for other ends, ghosts are often depicted as affecting the world. We see this in Buckwalter and Phelan's story. This may trigger a sense that the ghost is embodied to some extent, in some way. Indeed, if our higher cognitive capacities are grounded in perceptual-motor simulations, as advocates of embodied cognition argue (Wilson and Foglia 2017), it may be psychologically impossible for us to fully and completely represent a person-like entity as disembodied.

2 This inference presupposes that the folk (implicitly) reject a version of the *phenomenal intentional theory* according to which all *possible* intentional states are "constituted by" phenomenal states. Also, when it comes to phenomenally tinged intentional states (e.g., an ordinary experience of seeing a bright red apple or maybe a thought associated with a feeling of understanding), the inference presupposes that the folk view these phenomenal characters as inessential to the intentional states, or at least they must be willing to say that a purely intentional being has intentional states even if its states are deeply unlike ours. For more on phenomenal intentional theory and cognitive phenomenology, see Bourget and Mendelovici (2017).

3 The author wishes to thank Morgan Dale, Jacob Robbins, and especially Rocco Gennaro for their helpful comments on an earlier version of this chapter.

References

Alexander, J. (2012) *Experimental Philosophy: An Introduction*, Cambridge: Polity Press.

Alexander, J. (2016) "Philosophical Expertise," in J. Sytsma and W. Buckwalter (eds.) *A Companion to Experimental Philosophy*, Malden, MA: Wiley.

Apperly, I. (2011) *Mindreaders: The Cognitive Basis of "Theory of Mind"*, New York: Psychology Press.

Arico, A. (2010) "Folk Psychology, Consciousness, and Context Effects," *Review of Philosophy and Psychology* 1: 371–393.

Arico, A., Fiala, B., Goldberg, R., and Nichols, S. (2011) "The Folk Psychology of Consciousness," *Mind and Language* 26: 327–352.

Bechtel, W. (1986) "The Nature of Scientific Integration," in W. Bechtel (ed.) *Integrating Scientific Disciplines*, Dordrecht: Martinus Nijhoff.

Block, N. (1978) "Troubles with Functionalism," *Minnesota Studies in the Philosophy of Science* 9: 261–325.

Block, N. (1995) "On a Confusion about the Function of Consciousness," *Behavioral and Brain Sciences* 18: 227–247.
Bloom, P., and Veres, C. (1999) "The Perceived Intentionality of Groups," *Cognition* 71: B1–B9.
Bourget, D., and A. Mendelovici (2017) "Phenomenal Intentionality," in E.N. Zalta (ed.), *The Stanford Encyclopedia of Philosophy* (Spring 2017 Edition). https://plato.stanford.edu/archives/spr2017/entries/phenomenal-intentionality.
Buckwalter, W., and Phelan, M. (2013) "Function and Feeling Machines: A Defense of the Philosophical Conception of Subjective Experience," *Philosophical Studies* 166: 349–361.
Buckwalter, W., and Phelan, M. (2014) "Phenomenal Consciousness Disembodied," in J. Sytsma (ed.) *Advances in Experimental Philosophy of Mind*, London: Bloomsbury.
Chalmers, D. (1995) "Facing Up to the Problem of Consciousness," *Journal of Consciousness Studies* 2: 200–219.
Chalmers, D. (2002) "The Puzzle of Consciousness Experience," *Scientific American* 12: 90–99.
Churchland, P. S. (1995) "The Hornswoggle Problem," *Journal of Consciousness Studies* 3: 402–408.
Clark, H. H., and Lucy, P. (1975) "Understanding What Is Meant from What Is Said: A Study in Conversationally Conveyed Requests," *Journal of Verbal Learning and Verbal Behavior* 14: 56–72.
Cornwell, B. R., Barbey, A. K., and Simmons, W. K. (2004) "The Embodied Bases of Supernatural Concepts," *Behavioral and Brain Sciences* 27: 735–736.
Epley, N. (2014) *Mindwise: How We Understand What Others Think, Believe, Feel, and Want*, New York: Alfred A. Knopf.
Fiala, B., Arico, A., and Nichols, S. (2012) "On the Psychological Origins of Dualism: Dual-Process Cognition and the Explanatory Gap," in E. Slingerland and M. Collard (eds.) *Creating Consilience: Issues and Case Studies in the Integration of the Sciences and Humanities*, Oxford: Oxford University Press.
Fiala, B., Arico, A., and Nichols, S. (2014) "You, Robot," in E. Machery and E. O'Neill (eds.) *Current Controversies in Experimental Philosophy*, New York: Routledge.
Garnsey, S. M., Pearlmutter, N. J., Myers, E., and Lotocky, M. A. (1997) "The Contributions of Verb Bias and Plausibility to the Comprehension of Temporarily Ambiguous Sentences," *Journal of Memory and Language* 37: 58–93.
Gibbs, R. W. (1983) "Do People Always Process Literal Meanings of Indirect Requests?" *Journal of Experimental Psychology: Learning, Memory, and Cognition* 9: 524–533.
Goldman, A. I., and McGrath, M. (2015) *Epistemology: A Contemporary Introduction*, New York: Oxford University Press.
Gray, H. M., Gray, K., and Wegner, D. (2007) "Dimensions of Mind Perception," *Science* 315: 619.
Gray, H. M., Knobe, J., Sheskin, M., Bloom, P., and Barrett, L. B. (2011) "More than a Body: Mind Perception and the Nature of Objectification," *Journal of Personality and Social Psychology* 101: 1207–1220.
Huebner, B. (2010) "Commonsense Concepts of Phenomenal Consciousness: Does Anyone Care about Functional Zombies?" *Phenomenology and the Cognitive Sciences* 9:133–155.
Huebner, B. (2014) *Macrocognition: A Theory of Distributed Minds and Collective Intentionality*, New York: Oxford University Press.
Huebner, B., Bruno, M., and Sarkissian, H. (2010) "What Does the Nation of China Think about Phenomenal States?" *Review of Philosophy and Psychology* 1: 225–243.
Jenkins, A. C., Dodell-Feder, D., Saxe, R., and Knobe, J. (2014) "The Neural Bases of Directed and Spontaneous Mental State Attributions to Group Agents," *PLoS ONE* 9(8): e105341. doi:10.1371/journal.pone.0105341
Knobe, J. (2007) "Experimental Philosophy and Philosophical Significance," *Philosophical Explorations* 10: 119–121.
Knobe, J. (2008) "Can a Robot, an Insect or God Be Aware?" *Scientific American Mind* 19: 68–71.
Knobe, J. (2016) "Experimental Philosophy Is Cognitive Science," in J. Sytsma and W. Buckwalter (eds.) *A Companion to Experimental Philosophy*, Malden, MA: Wiley.
Knobe, J., and Nichols, S. (2008) "An Experimental Philosophy Manifesto," in J. Knobe and S. Nichols (eds.) *Experimental Philosophy*, Oxford: Oxford University Press.
Knobe, J., and Prinz, J. (2008) "Intuitions about Consciousness: Experimental Studies," *Phenomenology and Cognitive Science* 7: 67–83.
Nadelhoffer, T., and Nahmias, E. (2007) "The Past and Future of Experimental Philosophy," *Philosophical Explorations* 12: 123–149.
Nado, J. (2014) "The Role of Intuition," in J. Sytsma (ed.) *Advances in Experimental Philosophy of Mind*, London: Bloomsbury.

Nagel, T. (1974) "What Is It Like to Be a Bat?" *Philosophical Review* 83: 435–450.
Natsoulas, T. (1978) "Consciousness," *American Psychologist* 33: 906–914.
Nichols, S., and Stich, S. (2003) *Mindreading: An Integrated Account of Pretense, Self-Awareness, and Understanding Other Minds*, Oxford: Oxford University Press.
O'Neill, E., and Machery, E. (2014) "Experimental Philosophy: What Is It Good For?" in E. Machery and E. O'Neill (eds.) *Current Controversies in Experimental Philosophy*, New York: Routledge.
Peressini, A. (2014) "Blurring Two Conceptions of Subjective Experience: Folk versus Philosophical Phenomenality," *Philosophical Psychology* 27: 862–889.
Phelan, M., Arico, A., and Nichols, S. (2013) "Thinking Things and Feeling Things: On an Alleged Discontinuity in Folk Metaphysics of Mind," *Phenomenology and the Cognitive Sciences* 12: 703–725.
Phelan, M., and Buckwalter, W. (2012) "Analytic Functionalism and Mental State Attributions," *Philosophical Topics* 40: 129–154.
Reuter, K. (2011) "Distinguishing the Appearance from the Reality of Pain," *Journal of Consciousness Studies* 18: 94–109.
Reuter, K., Phillips, D., and Sytsma, J. (2014) "Hallucinating Pain," in J. Sytsma (ed.) *Advances in Experimental Philosophy of Mind*, London: Bloomsbury.
Rose, D., and Danks, D. (2013) "In Defense of a Broad Conception of Experimental Philosophy," *Metaphilosophy* 44: 512–532.
Rosenthal, D.M. (1986) "Two Concepts of Consciousness," *Philosophical Studies* 49: 329–359.
Schwitzgebel, E., and Cushman, F. (2012) "Expertise in Moral Reasoning? Order Effects on Moral Judgment in Professional Philosophers and Non-Philosophers," *Mind and Language* 27: 135–153.
Shepherd, J. (2012) "Free Will and Consciousness: Experimental Studies," *Consciousness and Cognition* 21: 915–927.
Shepherd, J. (2015) "Consciousness, Free Will, and Moral Responsibility: Taking the Folk Seriously," *Philosophical Psychology* 28: 929–946.
Sytsma, J. (2010a) "Folk Psychology and Phenomenal Consciousness," *Philosophy Compass* 5: 700–711.
Sytsma, J. (2010b) "Dennett's Theory of the Folk Theory of Consciousness," *Journal of Consciousness Studies* 17: 107–130.
Sytsma, J. (2012) "Revisiting the Valence Account," *Philosophical Topics* 40: 179–198.
Sytsma, J. (2014a) "Attributions of Consciousness," *WIREs Cognitive Science* 5: 635–648.
Sytsma, J. (2014b) "The Robots of the Dawn of Experimental Philosophy of Mind," in E. Machery and E. O'Neill (eds.) *Current Controversies in Experimental Philosophy*, New York: Routledge.
Sytsma, J., and Buckwalter, W. (eds.) (2016) *A Companion to Experimental Philosophy*, Malden, MA: Wiley.
Sytsma, J., and Livengood, J. (2016) *The Theory and Practice of Experimental Philosophy*, Peterborough, Ontario: Broadview.
Sytsma, J., and Machery, E. (2009) "How to Study Folk Intuitions about Phenomenal Consciousness," *Philosophical Psychology* 22: 21–35.
Sytsma, J., and Machery, E. (2010) "Two Conceptions of Subjective Experience," *Philosophical Studies* 151: 299–327.
Sytsma, J., and Machery, E. (2012) "On the Relevance of Folk Intuitions: A Commentary on Talbot," *Consciousness and Cognition* 21: 654–660.
Sytsma, J., and Reuter, K. (forthcoming) "Experimental Philosophy of Pain," *Journal of Indian Council of Philosophical Research.*
Talbot, B. (2012) "The Irrelevance of Folk Intuitions to the 'Hard Problem' of Consciousness," *Consciousness and Cognition* 21: 644–650.
Theriault, J., and Young, L. (2014) "Taking an 'Intentional Stance' on Moral Psychology," in J. Sytsma (ed.) *Advances in Experimental Philosophy of Mind*, London: Bloomsbury.
Tye, M. (2016) "Qualia," in E.N. Zalta (ed.) *The Stanford Encyclopedia of Philosophy* (Winter 2016 Edition). https://plato.stanford.edu/archives/win2016/entries/qualia/.
Waytz, A., and Young, L. (2012) "The Group-Member Mind Trade-Off: Attributing Minds to Group versus Group Members," *Psychological Science* 23: 77–85.
Weinberg, J. M., Gonnerman, C. Buckner, C., and Alexander, J. (2010) "Are Philosophers Expert Intuiters?" *Philosophical Psychology* 23: 33–55.
Wilson, R. A., and Foglia, L. (2017) "Embodied Cognition," in E. N. Zalta (ed.), *The Stanford Encyclopedia of Philosophy* (Spring 2017 Edition). https://plato.stanford.edu/archives/spr2017/entries/embodied-cognition/.

Related Topics

Materialism
Dualism
Consciousness and Intentionality
Robot Consciousness
Consciousness, Free Will, and Moral Responsibility
Consciousness and Emotion

INDEX